U0298423

1980 年代中期，杨路先生（左）和张景中先生（右）在成都讨论学术问题

張景中 楊路 文集

CNS PUBLISHING & MEDIA 湖南教育出版社

图书在版编目(CIP)数据

张景中　杨路　文集/张景中,杨路著. —长沙:湖南教育出版社,2016.10

ISBN 978-7-5539-4393-0

Ⅰ.①张…　Ⅱ.①张…　②杨…　Ⅲ.①数学—文集　②计算机科学—文集　Ⅳ.①O1-53

中国版本图书馆 CIP 数据核字(2016)第 221680 号

张景中　杨路　文集

Zhang Jingzhong Yang Lu Wenji

书　　名:张景中　杨路　文集

作　　者:张景中　杨路

责任编辑:钟劲松

责任校对:崔俊辉　刘　源　尹雨璇

出版发行:湖南教育出版社(长沙市韶山北路 443 号)

网　　址:http://www.hneph.com

电子邮箱:hnjycbs@sina.com

客　　服:电话 0731—85486979

经　　销:湖南省新华书店

印　　刷:长沙超峰印刷有限公司

开　　本:787×1092　16 开

印　　张:37.75　插 1 页

字　　数:1 250 000

版　　次:2016 年 10 月第 1 版　2016 年 10 月第 1 次印刷

书　　号:ISBN 978-7-5539-4393-0

定　　价:95.00 元

序

1979 年我有幸考入了中国科学技术大学，成为数学系的一名学生。入学时便从系里的老师和高年级的同学那里得知张景中、杨路二位老师传奇的经历，也时常看见二位老师在校园中匆忙的身影。想不到的是，4 年后我和武河、李世海成了二位老师第一届的研究生。

1983 年的中国，百废待兴，科学的春天唤醒了知识分子沉睡多年的报国梦想。二位老师更是忘我地工作，科研成果不断涌现。那时他们关心的研究方向有距离几何、动力系统、数值分析、组合几何等诸多方面。我们三位同学选择的是动力系统方向，这也是我直到今日的研究方向。那时科大的研究条件十分有限，缺少文献和书籍，二位老师花费很多精力为我们上课，使得我们很快便进入到研究前沿。在科研选题方面，他们尊重学生自己的选择。我当时所选的问题是极小性的研究，虽然进展较慢，且最终的结果是我在莫斯科大学读副博士(PHD)时完成的，但这样的科研训练却使我终身受益。使我觉得遗憾的是，由于科研兴趣的不同，我与二位老师还没有合作任何文章。

1985 年，二位老师调到成都工作，之后又到广州大学等地工作。他们的研究方向不断扩展，很多是我完全不熟悉的领域，如数学机械化理论、算法和软件实现、符号计算、符号与数值混合计算、自动推理与机器证明、教育软件研究与应用、教育信息技术等等。他们在这些方向上都做出了出色的成绩，得到了学术界和社会的高度认可。二位老师在大学时便是好友，从事科研工作后更是进行了珠联璧合的合作。同时他们也培养了大量优秀的学生，现在这些学生分布在国内外的高校、科研院所、金融机构等诸多领域，成为这些单位的骨干，桃李满天下！

今年是二位老师 80 寿辰，湖南教育出版社决定出版二位老师的合作文集。这本文集收录了二位老师几乎所有合作的文献，共 58 篇。在此请容许我代表二位老师所有的学生对湖南教育出版社以及为文集的编辑付出辛劳的黄勇等师兄弟们表示衷心的感谢！同时，衷心祝愿二位老师健康长寿，永葆学术青春！

叶向东

2016 年 8 月于中国科学技术大学

目　录

关于林法收敛性的若干问题

张景中　杨　路

（中国科学技术大学）

多项式求根的劈因子法，肇端于林法[1]，在文献[2]中认为"当所求的二次因式对应于一对最小根时，林法是收敛的"；又说"两次林法迭代效果大致与一次劈因子法（指 Bairstow 法）效果相同". 本文将指出，文献[2]中的这两个论断都是不正确的.

设 $F(x)$ 是复系数无重根的 n 次多项式，n 是不小于 3 的自然数. $F(x)$ 的 n 个根 $x_1,\cdots,x_n$ 中，x_1、x_2 是模最小的两个根. 又令：

$$\omega(x)=(x-x_1)(x-x_2)=x^2+px+q, \tag{1}$$

则 $\omega(x)$ 是 $F(x)$ 的一个二次因式. 取 $\omega(x)$ 的一个足够好的近似二次式 $\omega_0(x)=x^2+p_0x+q_0$，令 $G_0(x)=x\omega_0(x)$，用 $G_0(x)$ 除 $F(x)$，得二次余式. 适当调整 $F(x)$ 之系数，使除法算式为

$$F(x)=Q(x)\cdot G_0(x)+\omega_1(x). \tag{2}$$

这里 $Q(x)$ 是 $(n-3)$ 次多项式而 $\omega_1(x)=x^2+p_1x+q_1$. 用 $\omega_1(x)$ 代替 $\omega_0(x)$，重复上述过程，得 $\omega_2(x),\cdots,\omega_m(x)$；在一定条件下，当 $m\to+\infty$，将有：

$$\omega_m(x)\to\omega(x),$$

这种求 $\omega(x)$ 的近似式的逐步逼近方法叫林法.

不妨再设 $F(x)$ 无 0 根. 下面来找出林法逼近过程中误差传递的规律.

在(2)中，令

$$Q(x)=c_0+c_1x+\cdots+c_{n-3}x^{n-3}, \tag{3}$$

并分别将 $x=x_i(i=1,2,\cdots,n)$ 代入(2)得

$$x_i^2+p_1x_i+q_1+G_0(x_i)[c_{n-3}x_i^{n-3}+\cdots+c_0]=0(i=1,2,\cdots,n). \tag{4}$$

把诸根 x_i 看成已知，把 $p_1,q_1,c_{n-3},\cdots,c_0$ 看成未知数，则(4)成为 n 个未知数的线性方程组. 它可以改写成矩阵形式：

$$\begin{pmatrix} x_1 & 1 & G_0(x_1)x_1^{n-3} & \cdots & G_0(x_1)x_1 & G_0(x_1) \\ x_2 & 1 & \vdots & & \vdots & \vdots \\ \vdots & \vdots & \vdots & & \vdots & \vdots \\ \vdots & \vdots & \vdots & & \vdots & \vdots \\ \vdots & \vdots & \vdots & & \vdots & \vdots \\ x_n & 1 & G_0(x_n)x_n^{n-3} & \cdots & G_0(x_n)x_n & G_0(x_n) \end{pmatrix}\begin{pmatrix} p_1 \\ q_1 \\ c_{n-3} \\ c_{n-2} \\ \vdots \\ c_0 \end{pmatrix}=\begin{pmatrix} -x_1^2 \\ -x_2^2 \\ \vdots \\ \vdots \\ \vdots \\ -x_n^2 \end{pmatrix} \tag{5}$$

本文刊于《中国科学技术大学学报》，第 10 卷第 4 期，1980 年.

设方程组(5)的行列式为Δ，把Δ的第i列换成“常数”列所得的行列式记为Δ_i. 下面来计算行列式Δ、Δ_1、Δ_2之值.

把Δ对前两列展开，利用范德蒙行列式的求值公式，可得

$$\Delta=\sum_{1\leqslant i<j\leqslant n}(-1)^{i+j}(x_j-x_i)\cdot\prod_{\substack{1\leqslant\alpha<\beta\leqslant n\\ \alpha\neq i,j\\ \beta\neq i,j}}(x_\alpha-x_\beta)\cdot\frac{\prod_{k=1}^{n}G_0(x_k)}{G_0(x_i)G_0(x_j)}$$

$$=\prod_{k=1}^{n}G_0(x_k)\sum_{1\leqslant i<j\leqslant n}\frac{(-1)^{i+j+1}(x_i-x_j)^2\prod_{1\leqslant\alpha<\beta\leqslant n}(x_\alpha-x_\beta)}{\left[(-1)^{i-1}\prod_{\substack{t\neq i\\1\leqslant t\leqslant n}}(x_i-x_t)\right](-1)^{j-1}\prod_{\substack{t\neq j\\1\leqslant t\leqslant n}}(x_j-x_t)}\cdot(G_0(x_i)G_0(x_j))^{-1}$$

$$=-\left[\prod_{k=1}^{n}G_0(x_k)\right]\prod_{1\leqslant\alpha<\beta\leqslant n}(x_\alpha-x_\beta)\cdot\sum_{1\leqslant i<j\leqslant n}\frac{a^2(x_i-x_j)^2}{F'(x_i)F'(x_j)G_0(x_i)G_0(x_j)}.\tag{6}$$

这里，a是$F(x)$的最高次项的系数.

类似地，有：

$$\begin{cases}\Delta_1=\left[\prod_{k=1}^{n}G_0(x_k)\prod_{1\leqslant\alpha<\beta\leqslant n}(x_\alpha-x_\beta)\sum_{1\leqslant i<j\leqslant n}\frac{a^2(x_i+x_j)(x_i-x_j)^2}{F'(x_i)F'(x_j)G_0(x_i)G_0(x_j)},\\ \Delta_2=-\left[\prod_{k=1}^{n}G_0(x_k)\prod_{1\leqslant\alpha<\beta\leqslant n}(x_\alpha-x_\beta)\sum_{1\leqslant i<j\leqslant n}\frac{a^2x_ix_j(x_i-x_j)^2}{F'(x_i)F'(x_j)G_0(x_i)G_0(x_j)},\end{cases}\tag{7}$$

为简便，令

$$f_{ij}=\frac{(x_i-x_j)^2}{F'(x_i)F'(x_j)},\tag{8}$$

则f_{ij}是仅与F有关的一些常数. 注意到：当$\omega_0(x)$和$\omega(x)$足够接近时，$|G_0(x_1)|$、$|G_0(x_2)|$远比$|G_0(x_i)|\ (i>2)$小，从而

$$\Delta\sim-\left[\prod_{k=1}^{n}G_0(x_k)\right]\left[\prod_{1\leqslant\alpha<\beta\leqslant n}(x_\alpha-x_\beta)\right]\cdot\frac{a^2(x_1-x_2)^2}{F'(x_1)F'(x_2)G_0(x_1)G_0(x_2)}$$

不为0，从而有

$$\begin{cases}p_1=-\dfrac{\sum_{1\leqslant i<j\leqslant n}f_{ij}\cdot(x_i+x_j)\cdot(G_0(x_i)G_0(x_j))^{-1}}{\sum_{1\leqslant i<j\leqslant n}f_{ij}\cdot(G_0(x_i)G_0(x_j))^{-1}},\\ q_1=\dfrac{\sum_{1\leqslant i<j\leqslant n}f_{ij}\cdot(x_ix_j)\cdot(G_0(x_i)G_0(x_j))^{-1}}{\sum_{1\leqslant i<j\leqslant n}f_{ij}\cdot(G_0(x_i)G_0(x_j))^{-1}}.\end{cases}\tag{9}$$

下面来估计误差p_1-p和q_1-q. 由(9)：

$$p_1-p=\frac{\sum_{1\leqslant i<j\leqslant n}(x_i+x_j)\cdot(G_0(x_i)G_0(x_j))^{-1}f_{ij}}{\sum_{1\leqslant i<j\leqslant n}(G_0(x_i)\cdot G_0(x_j))^{-1}f_{ij}}-(-x_1-x_2)$$

$$= \frac{\sum\limits_{1\leqslant i<j\leqslant n} [(x_1+x_2)-(x_i+x_j)]\cdot [G_0(x_i)\cdot G_0(x_j)]^{-1}\cdot f_{ij}}{\sum\limits_{1\leqslant i<j\leqslant n} [G_0(x_i)\cdot G_0(x_j)]^{-1}\cdot f_{ij}}$$

$$= \frac{A_1^* G_0(x_1)+A_2^* G_0(x_2)+Q_p^* G_0(x_1)G_0(x_2)}{f_{12}+B_1^* G_0(x_1)+B_2^* G_0(x_2)+Q^* G_0(x_1)G_0(x_2)}. \tag{10}$$

这里

$$\begin{cases} A_1^* = \sum\limits_{j=3}^{n}(x_1-x_j)\cdot [G_0(x_j)]^{-1}\cdot f_{2j}, \\ A_2^* = \sum\limits_{j=3}^{n}(x_2-x_j)\cdot [G_0(x_j)]^{-1}\cdot f_{1j}, \\ B_1^* = \sum\limits_{j=3}^{n}[G_0(x_j)]^{-1}\cdot f_{2j}, \\ B_2^* = \sum\limits_{j=3}^{n}[G_0(x_j)]^{-1}\cdot f_{1j}; \end{cases} \tag{11}$$

$$\begin{cases} Q^* = \sum\limits_{2\leqslant i<j\leqslant n}[G_0(x_i)G_0(x_j)]^{-1}\cdot f_{ij}, \\ Q_p^* = \sum\limits_{2\leqslant i<j\leqslant n}[(x_1+x_2)-(x_i+x_j)]\cdot [G_0(x_i)\cdot G_0(x_j)]^{-1}\cdot f_{ij}. \end{cases} \tag{12}$$

再令

$$Q_q^* = \sum_{2\leqslant i<j\leqslant n}(x_i x_j - x_1 x_2)[G_0(x_i)\cdot G_0(x_j)]^{-1}, \tag{13}$$

类似于(10),有

$$q_1 - q = \frac{-A_1^* G_0(x_1)x_2 - A_2^* G_0(x_2)x_1 + Q_q^* G_0(x_1)G_0(x_2)}{f_{12}+B_1^* G_0(x_1)+B_2^* G_0(x_2)+Q^* G_0(x_1)G_0(x_2)}. \tag{14}$$

由于$G(x)=x\omega_0(x)=x(x^2+p_0x+q_0)$,若令$\varepsilon_0=p_0-p$,$\delta_0=q_0-q$,由$x_1$,$x_2$是$\omega(x)=x^2+px+q$的根,故得:

$$\begin{aligned} G_0(x_1) &= x_1(\varepsilon_0 x_1+\delta_0), \\ G_0(x_2) &= x_2(\varepsilon_0 x_2+\delta_0). \end{aligned} \tag{15}$$

取$G(t)=t\omega(t)$,并令

$$\begin{cases} A_1 = \sum\limits_{j=3}^{n}(x_1-x_j)[G(x_j)]^{-1} f_{2j}, \\ A_2 = \sum\limits_{j=3}^{n}(x_2-x_j)[G(x_j)]^{-1} f_{1j}, \\ B_1 = \sum\limits_{j=3}^{n}[G(x_j)]^{-1} f_{2j}, \\ B_2 = \sum\limits_{j=3}^{n}[G(x_j)]^{-1} f_{1j}, \end{cases} \tag{16}$$

以及

$$\begin{cases} Q_p = \sum\limits_{2\leqslant i<j\leqslant n} [(x_1+x_2)-(x_i+x_j)]\cdot(G(x_i)\cdot G(x_i))^{-1}\cdot f_{ij}, \\ Q_q = \sum\limits_{2\leqslant i<j\leqslant n} (x_ix_j-x_1x_2)(G(x_i)\cdot G(x_j))^{-1}\cdot f_{ij}, \\ Q = \sum\limits_{2\leqslant i<j\leqslant n} (G(x_i)\cdot G(x_j))^{-1}\cdot f_{ij}. \end{cases} \tag{17}$$

取$\rho_0=\sqrt{\varepsilon_0^2+\delta_0^2}$,则当$\rho_0$足够小时,对给定的无重根的$F$,$A_i^*$、$B_i^*$、$Q_p^*$、$Q_q^*$、$Q^*$对应地是$A_i$、$B_i$、$Q_p$、$Q_q$、$Q$的近似值,其误差是$O(\rho_0)$型的量.从而由(10)、(11)及(12)得:

$$\begin{cases} p_1-p=\dfrac{1}{f_{12}}[A_1x_1(x_1\varepsilon_0+\delta_0)+A_2x_2(x_2\varepsilon_0+\delta_0)]+O(\rho_0^2), \\ q_1-q=-\dfrac{1}{f_{12}}[A_1x_1x_2(x_1\varepsilon_0+\delta_0)+A_2x_1x_2(x_2\varepsilon_0+\delta_0)]+O(\rho_0^2). \end{cases} \tag{18}$$

记$\varepsilon_1=p_1-p$,$\delta_1=q_1-q$,把(18)写成矩阵形式:

$$\begin{bmatrix}\varepsilon_1\\ \delta_1\end{bmatrix}=\frac{1}{f_{12}}\begin{bmatrix}A_1x_1^2+A_2x_2^2 & A_1x_1+A_2x_2\\ -(A_1x_1+A_2x_2)x_1x_2 & -(A_1+A_2)x_1x_2\end{bmatrix}\begin{bmatrix}\varepsilon_0\\ \delta_0\end{bmatrix}+\begin{bmatrix}O(\rho_0^2)\\ O(\rho_0^2)\end{bmatrix}$$

$$=(A_F)\begin{bmatrix}\varepsilon_0\\ \delta_0\end{bmatrix}+\begin{bmatrix}O(\rho_0^2)\\ O(\rho_0^2)\end{bmatrix},$$

于是,林法的收敛性与敛速,与矩阵(A_F)的特征值之大小有密切关系.显见:

1°　当(A_F)的两个特征值的模均小于1时,对足够好的初始因式,林法收敛.而且当这两个特征值均非0时,误差$\rho_m=\sqrt{\varepsilon_m^2+\delta_m^2}$按等比数列的量级缩小,所以不能认为林法两次相当于Bairstow法一次.

2°　当(A_F)的两个特征值模均大于1时,不论$\omega_0(x)(\neq\omega(x))$多次接近$\omega(x)$,林法均不会收敛.

下面将看到,即使$\omega(x)$对应于一对模最小的根,(A_F)的特征值模大于1的情形也会出现.

经整理,写出(A_F)的特征方程:

$$\lambda^2-\frac{1}{f_{12}}(x_1-x_2)(A_1x_1-A_2x_2)\lambda-\frac{1}{f_{12}^2}(x_1-x_2)^2x_1x_2A_1A_2=0,$$

它可分解为

$$\left(\lambda-\frac{(x_1-x_2)}{f_{12}}A_1x_1\right)\left(\lambda+\frac{(x_1-x_2)}{f_{12}}A_2x_2\right)=0.$$

故得两个特征值为

$$\lambda_1=\frac{x_1-x_2}{f_{12}}A_1x_1=\frac{x_1F'(x_1)}{(x_1-x_2)}\sum_{j=3}^{n}\frac{(x_2-x_j)}{x_jF'(x_j)},$$

$$\lambda_2=\frac{x_2-x_1}{f_{12}}A_2x_2=\frac{x_2F'(x_2)}{(x_2-x_1)}-\sum_{j=3}^{n}\frac{(x_1-x_j)}{x_jF'(x_j)}.$$

对$n=3$、4之情形,易算出:

$n=3$时,$\lambda_1=\dfrac{x_1}{x_3}$,$\lambda_2=\dfrac{x_2}{x_3}$;可见,对三次方程,当所求根是两个模最小的根时,林法确实

是收敛的.

$n=4$ 时,$\lambda_1=\frac{x_1}{x_3}+\frac{x_1}{x_4}-\frac{x_1^2}{x_2x_4}$,$\lambda_2=\frac{x_2}{x_3}+\frac{x_2}{x_4}-\frac{x_2^2}{x_3x_4}$,于是$[\lambda_1]$,$[\lambda_2]$可能大于1.

例如:方程

$$x^4+x^3-2.79x^2-1.37x+2.222=0$$

有一对最小共轭根,对应于因式$(x^2-2x+1.01)$,而另一因式$(x^2+3x+2.2)$.易算出,对应于一对最小根的林法特征值:

$$\lambda_1=\frac{-3.99-0.5\sqrt{-1}}{2.2},\lambda_2=\frac{-3.99+0.5\sqrt{-1}}{2.2},$$

$|\lambda_1|$、$|\lambda_2|$都大于1.无论初始因子取得和$\omega(x)=x^2-2x+1.01$多么接近,林法均不收敛.

但是,当(A_F)的特征值的模并非都大于1或都小于1的情形,林法的收敛性如何判断,仍是有待探讨的.

参考文献

[1] 赵访熊.求复数根的牛顿法.数学学报.1955,2(5):137-147.

[2] 清华大学、北京大学《计算方法》编写组.计算方法(上册).北京:科学出版社,1975:169-174,204-205.

单纯形构造定理的一个证明

杨　路　张景中
（中国科学技术大学）

我们知道，给定了长度为 a,b,c 的三个线段，当且仅当

$$a+b>c,b+c>a,c+a>b$$

同时成立时，可以用这三个线段为边，作一个三角形.

作为三角形在 n 维欧氏空间 E^n 中的推广，是 n 维单纯形 $A_0A_1\cdots A_n$，这里 $A_0,A_1,\cdots,A_n$ 是单纯形的顶点，它们不在同一个 $(n-1)$ 维超平面上，n 维单纯形有 C_{n+1}^2 条棱 $A_iA_j(i\neq j,i,j=0,1,2,\cdots,n)$. 令 $\overline{A_iA_j}=\rho_{ij}$ 表棱 A_iA_i 之长，并约定 $\rho_{ii}=0$. 那么，诸棱长 ρ_{ij} 之间应当有什么约束呢?“单纯形构造定理”回答了这个问题.

“单纯形构造定理”的第一个证明是由 Karl Menger 于 1928 年给出的[1]. 他的定理还包括了单形退化的情况，也就是说，他解决了点集在欧氏空间特等长嵌入的问题，给出了这种嵌入的充要条件.

本文对非退化单纯形构造定理给出一个较为有趣的证明. 证明中仅用到二次型、仿射变换及重心坐标的最基本的性质.

定理　预给棱长 $\rho_{ij}(i,j=0,1,2,\cdots,n;\rho_{ij}=\rho_{ji};\rho_{ii}=0)$ 的单纯形存在的充要条件是一组不等式

$$(-1)^k\Delta_k<0(k=0,1,2,\cdots,n) \tag{1}$$

成立，这里 Δ_k 是 $(k+2)$ 阶行列式

$$\Delta_k=\begin{vmatrix} 0 & 1 & 1 & 1 & \cdots & 1 \\ 0 & \rho_{00}^2 & \rho_{01}^2 & \rho_{02}^2 & \cdots & \rho_{0k}^2 \\ 0 & \rho_{10}^2 & \rho_{11}^2 & \rho_{12}^2 & \cdots & \rho_{1k}^2 \\ 0 & \rho_{20}^2 & \rho_{21}^2 & \rho_{22}^2 & \cdots & \rho_{2k}^2 \\ \vdots & \vdots & \vdots & \vdots & & \vdots \\ 0 & \rho_{k0}^2 & \rho_{k1}^2 & \rho_{k2}^2 & \cdots & \rho_{kk}^2 \end{vmatrix}(k=0,1,\cdots,n).$$

为了叙述方便，将定理的证明分成几个引理.

引理 1　令 $\omega_{ij}=\rho_{ij}^2-\rho_{i0}^2-\rho_{0i}^2(i,j=1,2,\cdots,n)$，则条件(1) 等价于 n 阶矩阵(ω_{ij}) 的严格负定性.

证　易知

本文刊于《数学的实践与认识》，第 1 期，第 1980 年.

$$\begin{vmatrix} \omega_{11} & \cdots & \omega_{1k} \\ \vdots & & \vdots \\ \omega_{k1} & \cdots & \omega_{kk} \end{vmatrix} = -\Delta_k (k = 1,2,\cdots,n). \tag{2}$$

这只要把行列式 Δ_k 的第 3 至第$(k+2)$ 行都减去第 2 行,再把第 3 至$(k+2)$ 列都减去第 2 列. 然后把所得之行列式按第一行展开,得一$(k+1)$ 阶行列式. 再把此$(k+1)$ 阶行列式按第一列展开,即得(2). 显见,条件(1) 等价于 n 阶矩阵(ω_{ij}) 的严格负定性. 引理 1 证毕.

引理 2 若(1) 成立,则在重心坐标(Möbius 坐标) 之下①,方程

$$\sum_{i=0}^{n} \sum_{j=0}^{n} p_{ij}^2 x_i x_j = 0 \tag{3}$$

所代表的二阶超曲面为一超椭球面,反之亦然.

证 只要证明"条件(1) 等价于下列事实:方程(3) 所表示的二阶超曲面上没有无穷远点" 就够了. 在重心坐标下,无穷远超平面方程为

$$x_0 + x_1 + \cdots + x_n = 0, \tag{4}$$

将(4) 与(3) 联立,消去 x_0,得方程

$$\sum_{i=1}^{n} \sum_{j=1}^{n} (p_{ij}^2 - p_{i0}^2 - p_{0j}^2) x_i x_j = \sum_{i=1}^{n} \sum_{j=1}^{n} \omega_{ij} x_i x_j = 0. \tag{5}$$

于是,超曲面(3) 上没有无穷远点,等价于方程(5) 除 $x_1 = x_2 = \cdots = x_n = 0$ 外无实解. 即二次型 $\sum\limits_{i=1}^{n} \sum\limits_{j=1}^{n} \omega_{ij} x_i x_j$ 是严格有定的,亦即矩阵(ω_{ij}) 是严格有定的. 经具体讨论可知其为严格负定的. 由引理 1,此与条件(1) 等价. 引理 2 证毕.

引理 3 在重心坐标下,若坐标单纯形 $A_0A_1\cdots A_n$ 各棱长 $\overline{A_iA_j} = p_{ij}$,$p_{ii} = 0$,则此单纯形之外接超球面方程为

$$\sum_{i=0}^{n} \sum_{j=0}^{n} p_{ij}^2 x_i x_j = 0. \tag{6}$$

证 首先,易知在二维情形时命题为真②,即当 $n = 2$ 时,坐标三角形 $A_0A_1A_2$ 外接圆之重心坐标方程为

$$tp_{01} x_0 x_1 + tp_{12}^2 x_1 x_2 + tp_{02}^2 x_0 x_2 = 0 (t \neq 0).$$

---

①重心坐标与仿射坐标之关系:设 A_i 的仿射坐标为$(a_{i1}, a_{i2}, \cdots, a_{in})(i = 0,1,2,\cdots,n)$,点 X 的仿射坐标为$(x_1, x_2, \cdots, x_n)$,点 X 的重心坐标为$(m_0, m_1, \cdots, m_n)$,则有

$$m_0 \begin{pmatrix} a_{01} \\ a_{02} \\ \vdots \\ a_{0n} \end{pmatrix} + m_1 \begin{pmatrix} a_{11} \\ a_{12} \\ \vdots \\ a_{1n} \end{pmatrix} + \cdots + m_n \begin{pmatrix} a_{n1} \\ a_{n2} \\ \vdots \\ a_{nn} \end{pmatrix} = M \begin{pmatrix} x_1 \\ x_2 \\ \vdots \\ x_n \end{pmatrix},$$

此处 $M = m_0 + m_1 + \cdots + m_n$,$M \neq 0$. 当 $M = 0$ 时,X 为无穷远点,没有仿射坐标,重心坐标为仿射齐次坐标,是射影坐标的一种.

②为了验证此命题当 $n = 2$ 的情形,可作 $\triangle A_0A_1A_2$ 的高 $A_2O \perp A_0A_1$,以 O 为笛卡尔坐标原点,A_2O,A_0A_1 为 x,y 二轴. 再把重心坐标方程化为笛卡尔坐标方程,马上可以看出所得方程之 x^2,y^2 项系数相同,而 xy 项系数为 0.

这只要把重心坐标换成笛卡尔坐标直接验证就知道了.

现在设坐标单纯形 $A_0A_1\cdots A_n$ 之外接超球面重心坐标方程为

$$\sum_{i=0}^{n}\sum_{j=0}^{n}a_{ij}x_ix_j=0,$$

往证必有 $a_{ij}=qp_{ij}^2$.

在 $A_0, A_1, \cdots, A_n$ 中任取三点 $A_{i_1}, A_{i_2}, A_{i_3}$,考虑此三点决定的平面截此超球面所得之截口,方程为

$$\begin{cases}\sum_{i=0}^{n}\sum_{j=0}^{n}a_{ij}x_ix_j=0,\\ x_k=0(k\neq i_1,i_2,i_3),\end{cases}$$

即

$$\begin{cases}a_{i_1i_2}x_{i_1}x_{i_2}+a_{i_1i_3}x_{i_1}x_{i_3}+a_{i_2i_3}x_{i_2}x_{i_3}=0,\\ x_k=0(k\neq i_1,i_2,i_3).\end{cases}$$

另一方面,此截口为 $\triangle A_{i_1}A_{i_2}A_{i_3}$ 之外接圆,而 $\triangle A_{i_1}A_{i_2}A_{i_3}$ 为坐标三角形,故此截口方程(由二维情形下的命题)为

$$\begin{cases}p_{i_1i_2}^2x_{i_1}x_{i_2}+p_{i_1i_3}^2x_{i_1}x_{i_3}+p_{i_2i_3}^2x_{i_2}x_{i_3}=0,\\ x_k=0(k\neq i_1,i_2,i_3).\end{cases}$$

比较系数,知有 $q_{i_1i_2i_3}\neq 0$,使

$$a_{i_1i_2}=q_{i_1i_2i_3}p_{i_1i_2}^2, a_{i_1i_3}=q_{i_1i_2i_3}p_{i_1i_3}^2, a_{i_1i_3}=q_{i_1i_2i_3}p_{i_2i_3}^2.$$

由 i_1, i_2, i_3 之任意性,可知 $q_{i_1i_2i_3}=q$ 与 i_1, i_2, i_3 无关. 于是得 $a_{ij}=qp_{ij}^2$. 引理 3 证毕.

现在来完成定理之证明.

条件之充分性. 任取一单纯形 $B_0B_1\cdots B_n$ 为坐标单纯形建立重心坐标系. 在此坐标系之下,由条件(1) 及引理 2,方程(3) 为一超椭球面方程. 设此超椭球面为 Ω;取一仿射变换 L,使 $L(\Omega)=\Omega'$ 为超球面. 令 $L(B_i)=C_i$,于是 Ω' 在坐标系 $C_0C_1\cdots C_n$ 下之重心坐标方程仍为(3). 又由于 Ω' 是坐标单纯形 $C_0C_1\cdots C_n$ 的外接超球面,故由引理 3,知 $p_{ij}^2=q\overline{C_iC_j}^2$,将单纯形 $C_0C_1\cdots C_n$ 按比例放大或缩小若干倍,即得各棱长为 p_{ij} 的所要的单纯形.

条件之必要性. 设有单纯形其各棱长为 p_{ij},作此单纯形之外接超球. 由引理 3,方程(3) 在此单纯形为标架时代表此超球. 再应用引理 2,即知(1) 成立. 定理证毕.

最后指出,进一步考虑会遇到两个问题:(1) 如果预给的诸棱长不满足条件(1),当然它不能实现为欧氏空间中的单纯形,但能否"等长" 地嵌入伪欧空间呢?(2) 如果预给的诸棱长是不带足标的,也即只给了 C_{n+2}^2 个正数而不限定组合方式,试求其能构成单纯形的充要条件.

前一个问题已得到肯定的回答[2]. 后一个问题难度似乎较大,至今尚未见诸文献.

参考文献

[1] Menger K. Untersuchungen über allgemeine Metri. Math. Ann. ,1928(100):75 - 163.
[2] 张景中,杨路. 有限点集在伪欧空间的等长嵌入. 数学学报,1981,24(4):481 - 487.

抽象距离空间的秩的概念

杨　路　张景中

（中国科学技术大学）

引　言

“距离几何”的基本思想应当说在19世纪末已经由比利时的Tilly[1] 明确提出了. Tilly引进的半度量(Semi-Metric)的概念后经Fréchet[2] 发展成为(在增加了三角形不等式之后)度量和度量空间(Metric Space)的重要概念

作为几何学的一个分支而言，距离几何(Distance Geometry)的真正基础是由K. Menger于1923年—1931年间的四篇论文中奠定的[3]. 在Menger工作的基础上，这一几何分支于30—40年代有一些发展，例如可参看L. M. Blumenthal的专著《*Theory and Applications of Distance Geometry*》[4]. Menger和Blumenthal所采用的行列式方法，肇源于A. Cayley[5] 发表在1841年剑桥数学杂志上的一个定理，这种方法在本文及作者的其它工作中得到了进一步的发展.

Menger和Blumenthal的某些工作主要是具有几何基础方面的意义. 但现在由于激光测距等直接测距技术的发展，物理学家已经对这方面的问题开始注意[6]. Einstein就表示希望有人能够对非正定的距离进行较深入的研究. 可惜目前从事这一几何分支工作的人很少，有些零星结果但无重大突破. [7]~[10]

本文属于非正定距离几何方面的基础性的工作，是到达比较深刻的结果的一个必不可少的阶梯. 文中引进了“秩”的概念并讨论了某些有限秩空间，其结论当然也适用于正定的几何.

§1　基础概念

1.1　定义　设$\mathscr{M}$为非空集，U为域；给了映射$g:\mathscr{M}\times\mathscr{M}\to U$，$g$满足$g(x,y)=g(y,x)\in U(x\in\mathscr{M},y\in\mathscr{M})$，则称$\mathscr{M}[U,g]$为$U$上的一个抽象距离空间；称$g(x,y)$为其中元素$x$、$y$的距离.

1.2　定义　设$\mathscr{S}=\{e_0,e_1,\cdots,e_m\}\subset\mathscr{M}[U,g]$，方阵$P=P[\mathscr{S}]=[g_{ij}]$($g_{ij}=g(e_i,e_j)$，$0\leqslant i,j\leqslant m$)叫做$\mathscr{S}$的距离阵；行列式$\|P\|$叫$\mathscr{S}$的判别式.

1.3　定义　设$e_0\in\mathscr{M}[U,g]$，如果

本文刊于《中国科学技术大学学报》，第10卷第4期，1980年.

1°　$\exists \mathscr{S}_N = \{e_0, e_1, \cdots, e_{N-1}\} \subset \mathscr{M}$，使 $\| P[\mathscr{S}_N] \| \neq 0$；

2°　对 $\mathscr{M}$ 中任意 $m(\geqslant N)$ 个元素 $e_1, \cdots, e_m$，$\mathscr{S}_{m+1} = \{e_0, e_1, \cdots, e_m\}$ 的判别式

$$\| P[\mathscr{S}_{m+1}] \| = 0;$$

则称元素 e_0 的秩为 N. 记作 $\mathrm{Rank} e_0 = N$.

1.4　定义　若抽象距离空间 $\mathscr{M}$ 中有一个元素 e_0，其秩为正整数，则称 $\mathscr{M}$ 为有限秩的. $\mathscr{M}$ 的元素中最高秩数称为 $\mathscr{M}$ 的上秩，最低的非 0 秩数称为 $\mathscr{M}$ 的下秩. 若 $\mathscr{M}$ 的上下秩相等，则称为齐秩的；而称此上(下) 秩为 $\mathscr{M}$ 的秩，记为 $\mathrm{Rank}\mathscr{M}$.

以下是一些有限秩抽象距离空间的例：

1.5 例　设 $\mathscr{M}$ 为 n 维欧氏空间 E^n 中所有 $n-1$ 维定向超平面之集. 两超平面 e_1、e_2 之夹角记为 $\widehat{e_1 e_2}$，定义 $g(e_1, e_2) = \cos \widehat{e_1 e_2}$；这样得到的抽象距离空间记为 R_L^n，其秩为 n.

1.6 例　设 $\mathscr{M}$ 是域 U 上的 n 维向量空间，取定一组基后，$\mathscr{M}$ 中的元素可用坐标表示.

$$x = (x_1, x_2, \cdots, x_n).$$

若定义抽象距离 $g(x, y) = \sum x_i y_i$，则得到一个秩为 n 的抽象距离空间.

1.7 例　取 $\mathscr{M}$ 为欧氏平面 E^2，$d(x,y)$ 记 x、y 两点间的欧氏距离，令 $g(x,y) = d^2(x, y)$，得到一个抽象距离空间 $E^2[d^2]$. 其中任意五点 e_0、e_1、e_2、e_3、e_4，恒有 $\| g_{ij} \| = 0$[5]. 可见 $E^2[d^2]$ 之上秩不超过 4. 易证 $E^2[d^2]$ 为齐秩. 更一般地，可定义 $E^n(d^2)$，其秩为 $n+2$.

1.8 例　在 $E^n(d^2)$ 中，增加一个假想元素 ϕ，并规定：若 $x \neq \phi$，则 $g(\phi, x) = g(x, \phi) = 1$，而 $g(\phi, \phi) = 0$；所得到的抽象距离空间记为 $\widetilde{E}^n(d^2)$. 由文献[4] 中结果可知，ϕ 之秩仍为 $n+2$.

1.9 例　在曲率为 $K(>0)$ 之 n 维球面型空间，设两点距离为 $\overline{e_i e_j}$，引入抽象距离 $g(e_i, e_j) = \cos(\sqrt{K}\,\overline{e_i e_j})$，则对任意 $m(\geqslant n+2)$ 个点 $e_1, e_2, \cdots, e_m$ 总有 $\| g_{ij} \| = 0$[5]；可见此空间之上秩不超过 $n+1$，易证它确是一个秩为 $n+1$ 的齐秩空间.

1.10 例　类似地，在曲率为 $K(<0)$ 之 n 维罗氏空间，设两点距离为 $\overline{e_i e_j}$，引入抽象距离：$g(e_i, e_i) = \mathrm{ch}(\sqrt{|K|}\,\overline{e_i e_j})$，则得到秩为 $n+1$ 的一个抽象距离空间.

以上诸例中所述事实，有些并不显然. 后面将在讨论其它问题时顺便给出证明. 这里先建立一个关于齐秩空间的一般引理，从而可以判断一系列有限秩空间的齐秩性. 为此，先要证明一个代数恒等式.

设 A 为 m 阶方阵；$1 \leqslant s \leqslant m$，从 A 中去掉第 i_1、i_2、$\cdots$、i_s 行和 j_1、j_2、$\cdots$、j_s 列所得之方阵记作 $A_{[i][j]}$；而去掉的这些行列相交处的元素所构成的 A 的子阵记为 $A\left[\begin{smallmatrix} i \\ j \end{smallmatrix}\right]$，这里：

$$[i] = (i_1, i_2, \cdots, i_s)(1 \leqslant i_1 < i_2 < \cdots < i_s \leqslant m).$$

$$[j] = (j_1, j_2, \cdots, j_s)(1 \leqslant j_1 < j_2 < \cdots < j_s \leqslant m).$$

则有：

1.11 引理　设 B 是由方阵 A 的各元素的余子式代替该元素所得之方阵，则

$$(*) \qquad \| B\left[\begin{smallmatrix} i \\ j \end{smallmatrix}\right] \| = \| A_{[i][j]} \| \cdot \| A \|^{s-1}$$

证　设使(∗) 成立的方阵 A 的全体构成集 $\mathscr{M}$，我们只要证明 $\mathscr{M}$ 是全体方阵之集. 易验证：若 A 为对角型，则 $A \in \mathscr{M}$. 然后逐条检验：

1°　若 $A \in \mathscr{M}$，把 A 的某行(列) 乘以 λ(λ 是 A 的基域中元素)，得 $\widetilde{A}$，则 $\widetilde{A} \in \mathscr{M}$.

2°　若 $A \in \mathscr{M}$，把 A 的某相邻两行(列) 互易后得 $\widetilde{A}$，则 $\widetilde{A} \in \mathscr{M}$.

3° 若 $A \in \mathscr{M}$,把 A 的某行(列) 加到另一行(列) 得 $\widetilde{A}$,则 $\widetilde{A} \in \mathscr{M}$.

即 $\mathscr{M}$ 对初等变换封闭,从而 $\mathscr{M}$ 是全体矩阵之集. 证毕.

由此引理立得:

1.12 推论 若 A^* 是 A 的伴随矩阵,则:

$$\| A^* {\scriptstyle\begin{bmatrix} i \\ j \end{bmatrix}} \| = (-1)^{i+j} \| A_{[i][j]} \| \cdot \| A \|^{s-1}.$$

$$(i = i_1 + i_2 + \cdots + i_s , j = j_1 + j_2 + \cdots + j_s)$$

特别地,取 $s = 2$,得到常用的.

1.13 推论 若 A 为 $m(> 2)$ 阶方阵,$i \neq j$,有:

$$\| A_{ii} \| \cdot \| A_{jj} \| - \| A_{ij} \| \cdot \| A_{ji} \| = \| A_{ij,ij} \| \cdot \| A \|.$$

由此可得关于秩的基本定理:

1.14 定理 设 $\mathscr{M} = \mathscr{M}[U,g]$ 为一抽象距离空间,$e_0 \in \mathscr{M}$,$e_1 \in \mathscr{M}$;若 e_0 之秩为 m,且 g_{00} 与 g_{01} 不同时为 0,则 e_1 之秩不超过 m.

证 任取 $\mathscr{M}$ 中元素列 e_2、e_3、…、e_{m+1},由定义,只要证明 $\| P(e_1 , e_2 , \cdots , e_{m+1}) \| = 0$ 即可.

令 $A = P(e_0 , e_1 , \cdots , e_{m+1})$,由 e_0 秩为 m,知 $\| A \| = 0$,且 $\| A_{m+1,m+1} \| = \| P(e_0 , e_1 , \cdots , e_m) \| = 0$. 在 1.13 推论中,取 $j = m + 1$,$i = 0$,由方阵对称性,得:

$$\| A_{0,m+1} \| = 0.$$

由于 e_1、e_2、…、e_{m+1} 顺序可换,可知对 $1 \leqslant j \leqslant m + 1$,

$$\| A_{0,j} \| = 0.$$

以下用反证法:设 $\| P(e_1 , e_2 , \cdots , e_{m+1}) \| \neq 0$,我们来推出 g_{00}、g_{01} 皆为 0. 考虑线性方程组:

$$(**)\quad \begin{bmatrix} g_{11} & g_{12} & \cdots & g_{1\ m+1} \\ g_{21} & g_{22} & \cdots & \vdots \\ \vdots & & & \vdots \\ g_{m-1\ 1} & \cdots & \cdots & g_{m+1\ m+1} \end{bmatrix} \begin{bmatrix} x_1 \\ x_2 \\ \vdots \\ x_{m+1} \end{bmatrix} = \begin{bmatrix} g_{1\ 0} \\ g_{2\ 0} \\ \vdots \\ g_{m+1\ 0} \end{bmatrix},$$

$(**)$ 的行列式为 $\| A_{00} \| = \| P(e_1 , e_2 , \cdots , e_{m+1}) \|$;当 $\| A_{00} \|$ 非 0 时,$(**)$ 有唯一的一组解

$$x_j = (-1)^{j+1} \| A_{0j} \| \cdot \| A_{00} \|^{-1} (j = 1,2,\cdots,m+1).$$

从而得 $x_j = 0$,代入$(**)$,得 $g_{j0} = 0$. 将 $\| A \|$ 对首列展开,得:

$$\| A \| = g_{00} \cdot \| A_{00} \|;$$

但已设 $\| A_{00} \| \neq 0$,由 $\| A \| = 0$ 推出 $g_{00} = 0$. 由反证法知 $\| A_{00} \| = 0$. 证毕.

由此定理立得:

1.15 推论 若抽象距离空间有一个秩为 m 的元素 e_0 满足 $g(e_0 , e_0) \neq 0$,则此空间上秩为 m.

1.16 推论 若抽象距离空间 $\mathscr{M}$ 中有一个秩为 m 的元素 e_0,且对任意 $e \in \mathscr{M}$,$e \neq e_0$,有 $g(e_0 , e) \neq 0$,则 $\mathscr{M}$ 为齐秩,秩为 m.

1.17 推论 若抽象距离空间 $\mathscr{M}$ 中有一个秩为 m 的元素 e_0 及一串元素 e_1、e_2、…、e_k、…,满足:

$$g(e_i , e_{i+1}) \neq 0 (i = 0,1,2,\cdots),$$

且对任意 $e \in \mathscr{M}$,总有 k 使 $g(e_k , e) \neq 0$,则空间 $\mathscr{M}$ 为齐秩,其秩为 m.

由以上推论易知前述各例均为齐秩的抽象距离空间.

§2　几个重要的有限秩空间

2.1 定义　设$\mathscr{M}$为由E^n中的点,$(n-1)$维定向超平面和一个假想元素ϕ所构成之集.令

$$g(e_1,e_2)=\begin{cases}-\frac{1}{2}\rho^2(e_1,e_2)(e_1,e_2\text{ 是点},\rho\text{ 表欧氏距离}),\\ \cos \widehat{e_1e_2}(e_1、e_2\text{ 是面},\widehat{e_1e_2}\text{ 表夹角}),\\ d(e_1,e_2)(e_1、e_2\text{ 是点和面},d\text{ 表带号距离}),\\ 1(e_1、e_2\text{ 中一个为点,另一个为 }\phi),\\ 0(e_1、e_2\text{ 中一个非点,另一个为 }\phi),\end{cases}$$

这样得到的抽象距离空间称为R_I^n.

2.2 定理　R_I^n为齐秩空间,其秩为$n+2$.

证　只要证明ϕ之秩为$n+2$即可.因为取ϕ和$(n+1)$个不共面的点,便得到1.17推论中所要的一串元素了.

先证ϕ的秩不小于$n+2$;任取点e_1和距离e_1为1的、两两正交的n个平面$e_2,\cdots,e_{n+1}$;则:

$$\| P(\phi,e_1,\cdots,e_{n+1})\| =\begin{vmatrix}0 & 1 & 0 & \cdots & 0\\ 1 & 0 & 1 & \cdots & 1\\ 0 & 1 & 1 & & \mathbf{0}\\ \vdots & \vdots & & \ddots & \\ 0 & 1 & \mathbf{0} & & 1\end{vmatrix}=-1,$$

可见ϕ之秩不小于$n+2$.

现往证ϕ之秩不大于$n+2$.取R_I^n中$N(\geqslant n+2)$个元素,要证的是$\| P(\phi,e_1,\cdots,e_N)\| =0$.不失一般性,设$e_1,\cdots,e_l$是面,$e_{l+1},\cdots,e_N$是点,$l<N$.取$e_N$为笛卡尔坐标原点,设$e_1,\cdots,e_l$的单位法向量为$\bar{a}_1,\cdots,\bar{a}_l$,由$e_N$垂引向$e_1,e_2,\cdots,e_l$的向量为$\bar{b}_1,\cdots,\bar{b}_l$;由$e_N$引向诸点之向量为$\bar{a}_{l+1},\cdots,\bar{a}_N$.则当$1\leqslant i\leqslant l$、$1\leqslant i\leqslant l$时:$g_{ij}=\bar{a}_i\cdot\bar{a}_j$;当$1\leqslant i\leqslant l$、$l<i\leqslant N$时:$g_{ij}=\bar{a}_i\cdot(\bar{a}_j-\bar{b}_i)$;当$l<i\leqslant N$、$l<j\leqslant N$时:$g_{ij}=-\frac{1}{2}(\bar{a}_i-\bar{a}_j)^2$.

于是:

$$\| P(\phi,e_1,\cdots,e_N)\| =\left|\begin{array}{c:c:c}0\ \cdots\ 0 & & 1\ \cdots\ 1\\ \vdots & \bar{a}_i\cdot\bar{a}_j & \bar{a}_i\cdot(\bar{a}_j-\bar{b}_i)\\ 0 & & \\ \hdashline 1 & & \\ \vdots & \bar{a}_j\cdot(\bar{a}_i-\bar{b}_j) & -\frac{(\bar{a}_i-\bar{a}_j)^2}{2},\\ 1 & \underbrace{\qquad}_{j\leqslant l} & \underbrace{\qquad}_{l<j\leqslant N}\end{array}\right|\begin{array}{l}\left.\right\}i\leqslant l\\ \\ \left.\right\}l<i\leqslant N,\end{array}$$

对 $k \leqslant l$,把第 0 行(列) 乘 $\bar{a}_k \cdot \bar{b}_k$ 加到第 k(行) 列上;对 $k > l$,把第 0 行(列) 乘 $\frac{1}{2}\bar{a}_k^2$ 加到第 k 行(列) 上,得

$$\| P(\phi, e_1, \cdots, e_N) \| = \begin{vmatrix} 0 & \delta_j \\ \delta_i & \bar{a}_i \bar{a}_j \end{vmatrix} (\delta_i \text{ 为 } 0 \text{ 或 } 1).$$

但 $\bar{a}_N = 0$,故末行列除 $\delta_N = 1$ 外均为 0. 故:

$$\| P(\phi, e_1, \cdots, e_N) \| = - \| \bar{a}_i \bar{a}_j \| (i, j = 1, 2, \cdots, N-1).$$

由 $N - 1 > n$,故 $\| \bar{a}_i \bar{a}_j \| = 0$. 证毕.

将 R_{I}^n 加以扩充,得

2.3 定义 在 R_I^n 上添加所有$(n-1)$ 维定向球面,以 O_i 记球 e_i 的心,r_i 记 e_i 的半径,点可看成半径为 0 的球,规定球与其它元素抽象距离为

$$g(e_i, e_j) = \begin{cases} 1 & (\text{若 } e_j = \phi), \\ -\frac{1}{2}[\overline{O_iO_j}^2 - (r_i - r_j)^2] & (\text{若 } e_j \text{ 为球}), \\ d(O_i, e_j) & (\text{若 } e_j \text{ 为超平面}), \end{cases}$$

所得的抽象距离空间,记为 R_{II}^n.

(注:球定向用 r 的正负表示. $r < 0$ 时球面单位法向量向外,外部为正侧. 定义中出现的 $[\overline{O_iO_j}^2 - (r_i - r_j)^2]$,其几何意义是两球公切线长之平方. 此意义下公切线长可能是纯虚数.)

显然,球退化为点时,(2.3) 与(2.1) 是一致的. 与(2.2) 类似,有

2.4 定理 R_{II}^n 中,$\mathrm{Rank}\phi = n + 3$,故 R_{II}^n 秩为 $n + 3$.

证 与(2.2) 同,只要证明 $\mathrm{Rank}\phi = n + 3$. 先证 $\mathrm{Rank}\phi \geqslant n + 3$:任取点 e_1,以 e_1 为心,半径为 -1 的球为 e_2,再取 n 个两两正交且与 e_2 相切的超平面 e_3、e_4、$\cdots$,e_{n+2},易算出

$$\| P(\phi, e_1, e_{n+2}) \| = \begin{vmatrix} 0 & 1 & 1 & 0 & \cdots & 0 \\ 1 & 0 & \frac{1}{2} & 1 & \cdots & 1 \\ 1 & \frac{1}{2} & 0 & 1 & \cdots & 1 \\ 0 & 1 & 1 & & & \\ \vdots & \vdots & \vdots & & E & \\ 0 & 1 & 1 & & & \end{vmatrix} = \begin{vmatrix} 0 & 1 & 1 \\ 1 & 0 & \frac{1}{2} \\ 1 & \frac{1}{2} & 0 \end{vmatrix} = 1,$$

可见 ϕ 之秩不小于 $n + 3$. 下面证 $\mathrm{Rank}\phi \leqslant n + 3$.

不失一般性,设 $e_1, e_2, \cdots, e_l$ 是超平面,$e_{l+1}, \cdots, e_N$ 为球(包括点),$l < N, N \geqslant n + 3$.

取 e_N 的心 O_N 为笛卡尔坐标之原点. 超平面 e_i 可以用由原点引向它的垂直向量 $a_i \bar{\theta}_i$ 表示;这里 $\bar{\theta}_i$ 是 e_i 的单位法向量,球 e_i 可以用由原点引向球心 O_i 的向量 $\bar{e}_i = \overrightarrow{O_NO_i}$ 和它的半径 r_i 表示. 考虑$(N-1)$ 阶行列式

$$\Delta=\begin{vmatrix}\bar{\theta}_1 & 0 & 0 \cdots\cdots 0\\ \vdots & \vdots & \vdots \quad\quad \vdots\\ \bar{\theta}_t & 0 & 0 \cdots\cdots 0\\ \bar{e}_{l+2} & \sqrt{-1}(r_{l+1}-r_N) & 0 \cdots\cdots 0\\ \vdots & \vdots & \vdots \quad\quad \vdots\\ \bar{e}_{N-1} & \sqrt{-1}(r_{N-1}-r_N) & 0 \cdots\cdots 0\end{vmatrix}=0,$$

把 Δ 与它的转置相乘

$$\Delta^2=\begin{vmatrix}\bar{\theta}_i\bar{\theta}_j & \bar{\theta}_i\cdot\bar{e}_j\\ \underbrace{\bar{\theta}_j\cdot\bar{e}_i}_{j\leqslant l} & \underbrace{\bar{e}_i\cdot\bar{e}_j-(r_i-r_N)(r_j-r_N)}_{l<j\leqslant N-1}\end{vmatrix}\begin{matrix}\}\ i\leqslant l\\ \\ \}\ l<i\leqslant N-1\end{matrix}=0.$$

对上述行列式镶边,使成 $(N+1)$ 阶而不变其值.

$$-\Delta^2=\Delta^2=\begin{vmatrix}0 & \cdots\cdots\ 0 & 0\ \cdots\cdots\cdots\cdots 0 & 1\\ \vdots & \bar{\theta}_i\bar{\theta}_j & \bar{\theta}_i\cdot\bar{e}_j & 0\\ 0 & & & \vdots\\ & & & 0\\ 1 & & & 0\\ \vdots & \bar{\theta}_j\bar{e}_i & \bar{e}_i\bar{e}_j-(r_i-r_N)(r_j-r_N) & \vdots\\ 1 & \underbrace{0\cdots 0}_{j\leqslant l} & \underbrace{0\cdots\cdots\cdots\cdots 0}_{l<j\leqslant N-1} & 0\end{vmatrix}\begin{matrix}\Big\}\ i\leqslant l\\ \\ \Big\}\ l<i\leqslant N-1\end{matrix}$$

对它再作不改变行列式值的变换:

1° 对 $i\leqslant l(j\leqslant l)$,把第 0 行(列)乘以 $-a_i(-a_j)$ 加到第 i 行(i 列).

2° 对 $i>l(j>l)$,把第 0 行(列)乘以 $\frac{1}{2}[(r_i-r_N)^2-\bar{e}_i^2]\left(或\frac{1}{2}[(r_j-r_N)^2-\bar{e}_j^2]\right)$

加到第 i 行(j 列),上行列式化为

$$\begin{vmatrix}0 & \cdots\cdots\ 0 & 1\cdots\cdots\cdots\cdots 1\\ \vdots & \bar{\theta}_i\bar{\theta}_j & \bar{\theta}_i(\bar{e}_i-a_i\bar{\theta}_i)\\ 0 & & \\ 1 & & \\ \vdots & \bar{\theta}_j\cdot(\bar{e}_i-a_j\bar{\theta}_j) & \frac{1}{2}[(r_i-r_j{}^2)]-(\bar{e}_i-\bar{e}_j)^2\\ 1 & \underbrace{\quad\quad}_{j\leqslant l} & \underbrace{\quad\quad\quad\quad}_{l<j\leqslant N}\end{vmatrix}\begin{matrix}\Big\}\ i\leqslant l\\ \\ \Big\}\ l<i\leqslant N\end{matrix}=\begin{vmatrix}0 & \delta_j\\ \delta_i & g_{ij}\end{vmatrix}=0$$

定理证毕.

上述定理给出了$(n-1)$维球几何中任意$(n+2)$个元素间的基本度量关系. $n=3$时,R^3_{II}可理解为 Minkowski 宇宙,从而对球几何的研究,与狭义相对论有密切联系. 华罗庚教授在文献[11]中着重指出:直到60年代,国内外的一些著作中并未注意到此点.

也可用别的方法引入球空间的抽象距离.

2.5 定义 设$\mathscr{M}$由E^n中的点、$n-1$维定向超平面和$n-1$维定向球面组成. 规定:

$$g(e_i,e_j)=\begin{cases}\cos\widehat{e_ie_j} & (e_i,e_j\text{ 为球或超平面}),\\ -\dfrac{1}{2}\rho^2(e_i,e_j) & (e_i,e_j\text{ 为点}),\\ d(e_i,e_j) & (e_i\text{ 为点},e_j\text{ 为超平面}),\\ \dfrac{1}{2r_i}(r_j^2-\overline{O_iO_j^2}) & (e_i\text{ 为球},e_j\text{ 为点}),\end{cases}$$

所成之抽象距离空间记为R^n_{III}.

顺便指出,这里定义的点到球的抽象距离,恰为 M. A. Bloch 所定义的"简化幂"的反号数.

2.6 定理 R^n_{III}为齐秩空间,秩为$n+2$.

证 任取一点e_1,以e_1为心,正数r为半径的球为e_2,过e_1再作n个两两正交的超平面:$e_3,e_4,\cdots,e_{n+2}$;由计算可知:

$$\|P(e_1,e_2,\cdots,_{n+2})\|=-\frac{r^2}{4}\neq 0,$$

从而知R^n_{III}的每个元素的秩均不小于$n+2$.

下面往证其元素之秩不大于$n+2$. 任取$N(>n+2)$个元素$e_1,e_2,\cdots,e_N$;不失一般性,设$e_1,\cdots,e_l$为超平面,$e_{l+1},e_{l+2},\cdots,e_k$为点,其余均为球$(0\leqslant l,k\leqslant N)$. 设其中球的半径顺次为$r_{k+1},\cdots,r_N$;球心为$O_{k+1},O_{k+2},\cdots,O_N$,考虑下列$(n+2)$阶行列式:

$$\Delta=\left|\begin{array}{cc|c|c|c}
0 & 1 & 0\cdots\cdots0 & 0\cdots\cdots0 & -\dfrac{r^2}{2}\\
1 & 0 & 0\cdots\cdots0 & 1\cdots\cdots1 & 1\cdots\cdots1\\
\begin{matrix}0\\ \vdots\\ 0\end{matrix} & \begin{matrix}0\\ \vdots\\ 0\end{matrix} & \cos\widehat{e_ie_j} & d(e_i,e_j) & d(e_i,O_j)\\
\hline
\begin{matrix}0\\ \vdots\\ 0\end{matrix} & \begin{matrix}1\\ \vdots\\ 1\end{matrix} & d(e_j,e_i) & -\dfrac{1}{2}\overline{e_ie_j^2} & -\dfrac{1}{2}\overline{e_iO_j^2}\\
\hline
-\dfrac{r_i^2}{2} & \begin{matrix}1\\ \vdots\\ 1\end{matrix} & d(e_j,O_i) & -\dfrac{1}{2}\overline{e_jO_i^2} & -\dfrac{1}{2}\overline{O_iO_j^2}\\
 & & \underbrace{\qquad}_{j\leqslant l} & \underbrace{\qquad}_{l<j\leqslant k} & \underbrace{\qquad}_{k<j\leqslant N}
\end{array}\right|
\begin{array}{l}
\\ \\ \}\,i\leqslant l\\ \}\,l<i\leqslant k\\ \}\,k<i\leqslant N\\ \\
\end{array}$$

注意到 Δ 去掉首行首列后,恰为 R_{I}^n 中元素列 $e_1,\cdots,e_l,e_{l+1},\cdots,e_k,O_{k+1},\cdots,O_N$ 的判别式,以 $\|A\|$ 记之. $\|A\|$ 的阶数 $N+1>n+3$. 由于 R_{I}^n 之秩为 $(n+2)$,仿(1.14)定理证明中的方法,并运用(1.13)可知 $\|A\|$ 的各元素的余子式除 A_{00} 外均为0. 因此,把 Δ 按首行首列展开时,各项皆为0,故 $\Delta=0$.

对 Δ 作不改变值的行列变换:设 Δ 的行(列)标号由 -1 算起,把第0行(列)乘以 $\frac{1}{2}r_i^2\left(\text{或}\frac{1}{2}r_j^2\right)$ 加到第 i 行(j 列),然后对首行首列展开,得

$$\Delta=-\begin{vmatrix}\cos\widehat{e_ie_j} & d(e_i,e_j) & d(e_i,O_j)\\ d(e_j,e_i) & -\frac{1}{2}\overline{e_ie_j^2} & \frac{1}{2}(r_j^2-\overline{e_iO_j^2})\\ d(e_j,O_i) & \frac{1}{2}(r_i^2-\overline{e_jO_i^2}) & \frac{1}{2}(r_i^2+r_j^2-\overline{O_iO_j^2})\\ j\leqslant l & \underbrace{\qquad}_{l<j\leqslant k} & \underbrace{\qquad}_{k<j\leqslant N}\end{vmatrix}\begin{matrix}\}i\leqslant l\\ \}l<i\leqslant k\\ \}k<i\leqslant N\\ \end{matrix}$$

$$=-(r_kr_{k+1}\cdots r_N)^2\|P(e_1,e_2,\cdots,e_N)\|=0.$$

定理证毕.

2.7 定义 设 $\mathscr{M}$ 是曲率为 K 的可定向 n 维常曲率空间中 $(n-1)$ 维定向球面及一个假想元素 ϕ 组成之集. 在 $\mathscr{M}$ 上引进抽象距离

$$g_{ij}=g(e_i,e_j)=\begin{cases}\cos\widehat{e_ie_j}\,(e_i,e_j\text{ 都是球}),\\ k_i\qquad(e_i\text{ 是球},e_i=\phi,k_i\text{ 是 }e_i\text{ 的带号曲率}),\\ -K\quad(e_i=e_j=\phi),\end{cases}$$

所得的抽象距离空间记为 $S^n(K)$.

2.8 定理 $S^n(K)$ 为齐秩,其秩不超过 $n+2$.

证 只要证明 $\mathrm{Rank}\phi\leqslant n+2$ 即可. 下面分三种情形讨论:

1° $K=0$,即欧氏情形:设 e_i 之半径为 r_i,e_i、e_j 连心线长为 d_{ij},则

$$\|P(\phi,e_1,\cdots,e_m)\|=\begin{vmatrix}0 & \frac{1}{r_j}\\ \frac{1}{r_i} & \frac{r_i^2+r_j^2-d_{ij}^2}{2r_ir_j}\end{vmatrix}$$

$$=\frac{1}{(r_1r_2\cdots r_m)^2}\begin{vmatrix}0 & 1\\ 1 & \frac{r_i^2+r_i^2-d_{ij}^2}{2}\end{vmatrix}=\frac{1}{(r_1\cdots r_m)^2}\begin{vmatrix}0 & 1\\ 1 & -\frac{d_{ij}^2}{2}\end{vmatrix}=0.$$

上面最后一步用到2.2定理中 e_i 全为点的情形.

2° $K<0$,即罗氏空间的情形:由罗氏平面上的余弦定理,

$$\|P(\phi,e_1,\cdots,e_m)\|=\begin{vmatrix}|K| & k_j\\ k_i & \frac{k_ik_j}{|K|}\left(1-\frac{\mathrm{ch}\sqrt{|K|}d_{ij}}{\mathrm{ch}\sqrt{|K|}r_i\,\mathrm{ch}\sqrt{|K|}r_j}\right)\end{vmatrix}$$

$$=\frac{(k_1k_2\cdots k_m)^2}{|K|^{m-1}}\begin{vmatrix}1 & 1\\ 1 & 1-\frac{\mathrm{ch}\sqrt{|K|}d_{ij}}{\mathrm{ch}\sqrt{|K|}r_i\,\mathrm{ch}\sqrt{|K|}r_j}\end{vmatrix}$$

$$= \frac{(k_1 k_2 \cdots k_m)^2}{|K|^{m-1}} \begin{vmatrix} 1 & 1 \\ 0 & \dfrac{-\operatorname{ch}\sqrt{|K|}d_{ij}}{\operatorname{ch}\sqrt{|K|}r_i \operatorname{ch}\sqrt{|K|}r_j} \end{vmatrix}$$

$$= \frac{(k_1 k_2 \cdots k_m)^2}{|K|^{m-1}} \cdot \frac{1}{(\operatorname{ch}\sqrt{|K|}r_i \cdots \operatorname{ch}\sqrt{|K|}r_m)^2} \| -\operatorname{ch}\sqrt{|K|}d_{ij} \|$$

$$= 0.$$

其中最后一步行列式为 0 的事实,见文献[4].

3° $K > 0$,即球面型空间情形:由球面上的余弦定理,

$$\| P(\phi, e_1, \cdots, e_m) \| = \begin{vmatrix} -K & k_j \\ k_i & \dfrac{\cos\sqrt{K}d_{ij} - \cos\sqrt{K}r_i \cos\sqrt{K}r_j}{\sin\sqrt{K}r_i \sin\sqrt{K}r_j} \end{vmatrix}$$

$$= \begin{vmatrix} -K & k_j \\ k_i & \dfrac{k_i k_j}{K}\left[\dfrac{\cos\sqrt{K}d_{ij}}{\cos\sqrt{K}r_i \cos\sqrt{K}r_j} - 1\right] \end{vmatrix}$$

$$= \frac{(k_1 k_2 \cdots k_m)^2}{K^{m-1}} \begin{vmatrix} -1 & 1 \\ 1 & \dfrac{\cos\sqrt{K}d_{ij}}{\cos\sqrt{K}r_i \cos\sqrt{K}r_j} - 1 \end{vmatrix}$$

$$= \frac{(k_1 k_2 \cdots k_m)^2}{K^{m-1}} \begin{vmatrix} -1 & 1 \\ 0 & \dfrac{\cos\sqrt{K}d_{ij}}{\cos\sqrt{K}\, r_i \cos\sqrt{K}r_j} \end{vmatrix}$$

$$= \frac{-1}{K^{n-1}} \cdot \frac{(k_1 k_2 \cdots k_m)^2}{(\cos\sqrt{K}r_1 \cdots \cos\sqrt{K}r_m)^2} \cdot \| \cos\sqrt{K}d_{ij} \| = 0.$$

最后一步 $\| \cos\sqrt{K}d_{ij} \| = 0$ 的事实见文献[4];定理证毕.

其实,可以证明 $S^n(K)$ 的秩为 $n+2$. 限于篇幅,此处从略.

§3 在欧氏空间的初步应用

在 2.2 定理中,取 $l = 0$ 的特例,可导出:

3.1 Cayley **定理**[5] E^n 中任意 m 个点 $e_1, e_2, \cdots, e_m$ 的 Cayley-Menger 行列式:

$$\| D(e_1, e_2, \cdots, e_m) \| = \begin{vmatrix} 0 & 1 \\ 1 & d_{ij}^2 \end{vmatrix} (i, j = 1, 2, \cdots, m; d_{ij} = | \overline{e_i e_j} |),$$

当 $m \geqslant n+2$ 时,其值为 0.

灵活运用 Cayley 定理,能解决一些从别的途径颇难入手的问题. 如文献[12]. 而 §2 的几个定理,比 3.1 更便于运用. 例如:设 E^n 中单形顶点为 $p_1, p_2, \cdots, p_{n+1}$, e 是由 $p_1, \cdots, p_n$ 决定的超平面,我们很容易求出 p_{n+1} 到 e 的高线 h.

令 $\mathscr{S}^* = \{\phi, p_1, p_2, \cdots, p_{n+1}, e\}$, $\mathscr{S} = \{\phi, p_1, \cdots, p_{n+1}\}$,又将 $\mathscr{S}$ 中去掉 p_j 后所得之元素列记

为 $\mathscr{S}_j$，由(2.2)：

$$\| P[\mathscr{S}^*] \| = \begin{vmatrix} 0 & 1 & \cdots & 1 & 0 \\ 1 & & & & \vdots \\ \vdots & & g_{ij} & & 0 \\ 1 & & & & h \\ 0 & \cdots & 0 & h & 1 \end{vmatrix} = 0(h = d(p_{n+1}, e)).$$

把此行列式对末行末列展开，即得：

$$h^2 = \frac{\| P[\mathscr{S}] \|}{\| P[\mathscr{S}_{n+1}] \|}.$$

一般地，若 h_i 为过顶点 p_i 的高，则：

3.2 $$h_i^2 = \frac{\| P[\mathscr{S}] \|}{\| P[\mathscr{S}_i] \|}.$$

由(3.2)出发用数学归纳法，易得单形体积公式：

3.3 $$V^2(p_1, p_2, \cdots, p_{n+1}) = -\frac{1}{(n!)^2} \| P(\phi, p_1, \cdots, p_{n+1}) \|.$$

此式在文献[13]中曾被用 Cayley-Menger 行列式给出：

3.4 $$V^2(p_1, p_2, \cdots, p_{n+1}) = \frac{(-1)^{n+1}}{2^n (n!)^2} \| D(p_1, p_2, \cdots, p_{n+1}) \|.$$

由(3.3)易知

3.5 推论 对 R_1^n 中任意 m 个点 $p_1, \cdots, p_m$，有

$$\| P(\phi, p_1, \cdots, p_m) \| \leqslant 0.$$

一般说来，若抽象距离空间的秩不超过 m，对其中的 $N(> m)$ 个元素 $e_1, \cdots, e_N$，有

$$\| g_{ij} \| = \| g(e_i, e_j) \| = 0,$$

若在这 $\frac{N(N-1)}{1}$ 个 $g_{i,j}$ 中仅有一个是未知数，便可利用上列等式把它解出来. 但这样得到的往往是关于这个未知数的二次方程，要对解得的两个根加以讨论，颇不方便. 下面提供一个此未知数所满足的一次方程 —— 单值公式.

3.6 引理 设 $\mathscr{M}$ 是秩不超过 $m+1$ 的抽象距离空间，$\mathscr{M}$ 是 $\mathscr{M}$ 的子空间(即 $\mathscr{M} \subset \mathscr{M}$，$\mathscr{M}$ 中的距离即 $\mathscr{M}$ 中的距离). 其秩不超过 m. 设 $\mathscr{S} = \{e_0, e_1, \cdots, e_{m+1}\} \subset \mathscr{M}$ 而 $\mathscr{S} \setminus e_j \subset \mathscr{M}$. $\| P_{ij}[\mathscr{S}] \|$ 记 $\| P[\mathscr{S}] \|$ 的代数余子式，则当 $i \neq j$ 时，

$$\| P_{ij}[\mathscr{S}] \| = 0.$$

证 由(1,13)

$$\| P_{ii}[\mathscr{S}] \| \cdot \| P_{jj}[\mathscr{S}] \| - \| P_{ij} \|^2 = \| P_{ii,jj} \| \cdot \| P \|,$$

由 $\mathscr{S} \subset \mathscr{M}$，$\| P \| = 0$；由 $\mathscr{S} \setminus e_i \subset \mathscr{M}$，$\| P_{ij} \| = 0$，于是立得 $\| P_{ij}[\mathscr{S}] \|^2 = 0$. 证毕.

下面，把方阵 A 的 i 行 j 列交叉处元素换为 0 后所得方阵记为 $A_{\langle i,j \rangle}$，又记

$$\mathscr{S}_{ij} = \mathscr{S} \setminus \{e_i, e_j\};$$

然后将 $\| P_{ij}[\mathscr{S}] \| = 0$ 对 g_{ij} 展开，得

$$g_{ij} \| P[\mathscr{S}_{ij}] \| + \| P_{\langle j,i \rangle i,j} \| [\mathscr{S}] = 0.$$

若 $\| P[\mathscr{S}_{ij}] \| \neq 0$,即得:

3.7［单值公式］（在 3.6 所设条件下）:

$$g_{ij}=g_{ji}=\frac{\| P_{\langle j,i\rangle i,j}[\mathscr{S}] \|}{\| P[\mathscr{S}_{ij}] \|}.$$

作为(3.7) 之应用,下面导出高维余弦定理.

设 R^n 中的点 $p_1,p_2,\cdots,p_{n+1}$ 是 R^n 中单形顶点,Ω_k 记$\{p_1,p_2,\cdots,p_{n+1}\}\backslash p_k$ 中诸点决定的超平面,并规定 Ω_k 的定向使 $d(\Omega_k,p_k)>0$,此时$\widehat{\Omega_i\Omega_j}$ 表单形之外角. 令 R_{I}^n 中假想元素 ϕ 为 $e_0,e_i=p_i\ (1\leqslant i\leqslant n+1),e_{n+2}=\Omega_i,e_{n+3}=\Omega_j$,并令 $\mathscr{S}^*=\{e_0,e_1,\cdots,e_{n+3}\}$,$\mathscr{S}=\{e_0,e_1,\cdots,e_{n+1}\}$;由(2.2) 定理,$R_{\mathrm{I}}^n$ 之秩为 $n+2$;在(3.7) 中取 $m=n+2,\mathscr{M}'=\mathscr{M}=R_{\mathrm{I}}^n$,得

$$\cos\widehat{\Omega_i\Omega_j}=g_{n+2,n+3}=-\frac{\| P_{\langle n+3,n+2\rangle n+2,n+3}[\mathscr{S}^*] \|}{\| P[\mathscr{S}^*_{n+2,n+3}] \|}.$$

而

$$\| P_{\langle n+3,n+2\rangle n+2,n+3}[\mathscr{S}^*] \|=\begin{vmatrix} & & & & & & 0\\ & & & & & & \vdots\\ & & & & & & 0\\ & & P[\mathscr{S}] & & & & g_{i,n+2}\\ & & & & & & 0\\ & & & & & & \vdots\\ 0 & \cdots & 0 & g_{n+3,j} & 0 & \cdots & 0\end{vmatrix}$$

$$=-g_{i,n+2}g_{j,n+3}\| P_{ij}[\mathscr{S}] \|,$$

$$\| P[\mathscr{S}^*_{n+2,n+3}] \|=\| P[\mathscr{S}] \|.$$

又 $g_{i,n+2}=d(Q_j,p_i)=d_i,g_{j,n+3}=d(Q_j,p_j)=d_j$,故得:

3.8
$$\cos\widehat{\Omega_i\Omega_j}=-\frac{\| Pij[\mathscr{S}] \|}{\| P[\mathscr{S}] \|}d_id_j.$$

按我们的定向,d_i,d_j 都是正的,由(3.2):

3.9
$$d_id_j=|h_ih_j|=-\frac{\| P[\mathscr{S}] \|}{(P[\mathscr{S}_i]P[\mathscr{S}_j])^{\frac{1}{2}}}.$$

这是因为:由(3.5),$\| P[\mathscr{S}] \|$、$\| P[\mathscr{S}_i] \|$、$\| P[\mathscr{S}_j] \|$ 均非正之故. 把(3.9) 代入(3.8) 得

3.10 余弦定理:

$$\cos\widehat{\Omega_i\Omega_j}=\frac{-\| P_{ij}[\mathscr{S}] \|}{\sqrt{\| P(\mathscr{S}_j) \|\cdot\| P[\mathscr{S}_j] \|}}=\frac{\| P_{ij}[\mathscr{S}] \|}{\sqrt{\| P[\mathscr{S}_i] \|}\sqrt{\| P[\mathscr{S}_j] \|}}.$$

应注意到,(3,10) 中的$\widehat{\Omega_i\Omega_j}$ 是单形外角,若以 θ_{ij} 表内角,则有

3.11
$$\cos\theta_{ij}=\frac{\| P_{ij}[\mathscr{S}] \|}{\sqrt{\| P[\mathscr{S}_i] \| P[\mathscr{S}_j]}}.$$

由此得:

3.12 广义勾股定理: 若 $\theta_{ij}=\frac{\pi}{2}$,则 $\| P_{ij}[\mathscr{S}] \|=0$.

3.13 广义布亚可夫斯基不等式：

$$P_{ij}^2[\mathscr{S}] \leqslant \| P[\mathscr{S}_i] \| \cdot P[\mathscr{S}_j] \|.$$

作为单值公式的又一应用，下面计算点到单形外接球的切线长.

设点列$\{e_1, e_2, \cdots, e_{n+2}\} \subset R_{\mathrm{I}}^n$，$\{e_1, e_2, \cdots, e_{n+2}\} \backslash e_i$ 中的$n+1$个点所决定的$n-1$维球记为Ω_i. 把R_{I}^n 看成R_{II}^n 的子空间，$\Omega_i \in R_{\mathrm{II}}^n$. 令$e_0 = \phi$，$e_{n+3} = \Omega_i$，取$\mathscr{S}^* = \{e_0, e_1, \cdots, e_{n+3}\}$，$\mathscr{S} = \mathscr{S}^* \backslash e_{n+3}$，$\mathscr{S}_i = \mathscr{S} \backslash e_i$. 由单值公式：

$$g(e_i, \Omega_i) = g_{i,n+3} = -\frac{P_{\langle i,n+3\rangle, n+3, i}[\mathscr{S}^*]}{\| P[\mathscr{S}_i] \|} = -\frac{\| P_{0i}[\mathscr{S}] \|}{P[\mathscr{S}_i]}.$$

又由(2.5)定义，$g(e_i, \Omega_i) = -\frac{1}{2}t_i^2$，$t_i$ 表e_i 到Ω_i 之切线长，因而得

3.14 切线公式：

$$t_i^2 = \frac{2\| P_{0j}[\mathscr{S}] \|}{\| P[\mathscr{S}_i] \|}.$$

从而马上得到：

3.15 推论 $\| P_{0i}[\mathscr{S}] \| > 0$、$< 0$或$= 0$，分别对应于$e_i$ 在球Ω_i 内、外或在球面上.

3.16 推论 以V_i 记$\{e_1, e_2, \cdots, e_{n+2}\} \backslash e_i$ 中各点为顶点的单形之体积，则e_i 到此单形外接球的幂的绝对值$|t_i^2|$与V_i 成反比.

证 由(1,13)：

$$\| P_{00}[\mathscr{S}] P_{ij}[\mathscr{S}] \| - \| P_{0i}[\mathscr{S}] \|^2 = 0,$$

$$\therefore \big| \| P_{0i}[\mathscr{S}] \| \big| = \sqrt{\| P_{00}[\mathscr{S}] \cdot P_{ij}[\mathscr{S}] \|}.$$

代入(3.14)，并用体积公式(3.3)，注意到按我们的记号：$P_{ij}[\mathscr{S}] = P[\mathscr{S}_i]$，故

$$|t_i^2| = 2\left| \frac{\| P_{00}[\mathscr{S}] \|}{\| P[\mathscr{S}_i] \|} \right|^{\frac{1}{2}} = \frac{2\big| \| P_{00}[\mathscr{S}] \| \big|^{\frac{1}{2}}}{n! V_i}$$

$$\therefore \quad |t_i^2| V_i = \frac{2}{n!} \big| \| P_{00}[\mathscr{S}] \|^{\frac{1}{2}}.$$

即$|t_1|^2 V_1 = |t_2|^2 V_2 = \cdots = |t_{n+2}|^2 V_{n+2}$. 证毕.

由于单值公式对一般的有限秩空间普遍有效，所以它在几何计算问题上有广泛意义. 欧氏及非欧几何中似不相同的问题，往往可以纳入单值公式的模式统一处理之. 其计算程序的编制则几乎相同.

§4 关于非欧几何的度量基础

下述的两个定理，可视为非欧几何的度量基础：

4.1 定理 设$\mathscr{M}$为曲率$K > 0$的n维球面型空间中所有点及$(n-1)$维定向超平面组成之集. 令

$$g(e_i, e_j) = \begin{cases} \cos(\sqrt{K}\, \overline{e_i e_j}) & (e_i, e_j \text{ 是点}, \overline{e_i e_j} \text{ 表距离}), \\ \cos \widehat{e_i e_j} & (e_i, e_j \text{ 是定向超平面}), \\ \sin \sqrt{K} h_{ij} & (e_i, e_j \text{ 一为点，一为面}, h_{ij} \text{ 表点到面的带号距离}), \end{cases}$$

则所得之抽象距离空间是齐秩的，秩为$n+1$.

4.2 定理 设$\mathscr{M}$为曲率$K<0$的n维罗氏空间中所有点及$(n-1)$维定向超平面组成之集,令

$$g(e_i,e_j)=\begin{cases}\mathrm{ch}(\sqrt{|K|}\cdot\overline{e_ie_j}) & (e_i,e_j\text{ 是点},\overline{e_ie_j}\text{ 表距离}),\\ \cos\widehat{e_ie_j} & (e_i,e_j\text{ 都是定向超平面}),\\ \sqrt{-1}\,\mathrm{sh}\sqrt{|K|}h_{ij} & (e_i,e_j\text{ 一为点,一为面},h_{ij}\text{ 表点到面的带号距离}),\end{cases}$$

则所得之抽象距离空间是齐秩的,秩为$n+1$.

从这两个定理出发,可以用比传统方法更为简捷的代数手段分别导出球面型几何和罗氏几何中的一系列和公式. 限于篇幅,上述两个定理的证明及其在非欧几何中的应用,将于另文中论述.

本文中提出的概念及方法,还有更多的应用. 例如:文献[14]、[15]、[16].

参考文献

[1]de Tilly J. Mémoires couronnés et autres mémoires publiesepar 1'Acade mie Royale de Belgique. 1892-1893(47),mémoire 5.

[2]Fréchet M. Rendiconti del Circolo Matematico di Paleermo. 1906(22):1-74.

[3]Menger K. Math. Ann. ,1928(100):75-163.

[4]Blumenthal L M. Theory and Applications of Distance Geometry. 2nd ed. New York,1970.

[5]Cayley A. Cambridge Math. J. ,1841(2):267-271.

[6]Misner C W. Gravitation. W. H. Freeman and Company,San Francisco,1973:306-309.

[7]Goldberg K. J. Res. Nat. Bur. Stand. (U. S.),77B(Math. Sci.),Nos 3 & 4(1972),145-152.

[8]Goldberg K J. Res. Nat. Bur. Stand. (U. S.),81 B(Match. Sci),Nos 1 & 2(1977),61-72.

[9]Valentine J E,et al. Trans. Amer. Math. Soc. ,1973(176):285-295.

[10]Valentine J E Pac. J Math. ,1970(34):817-825.

[11] 华罗庚. 从单位圆谈起. 北京:科学出版社,1977:103-105.

[12]Graham R L,et al. Amer. Math. Monthly,1974(81):21-25.

[13]Blumenthal L M,et al. Amer. Math. Monthly,1943(50):181-185.

[14]杨路,张景中. 数学学报,23(5):740-749.

[15]杨路,张景中. *Neuberg-Pedoe* 不等式的高维推广及应用. 数学学报,1981,24(3):401-408.

[16]杨路,张景中. 棱长为奇数之单形的一个注记,数学学刊,1982,3(3):343-347.

关于有限点集的一类几何不等式

杨路　张景中

（中国科学技术大学）

§1　引　言

设 $P_1,P_2,\cdots,P_N$ 是正 N 边形 S_N 的顶点，所有线段 P_iP_j 之长的平方和 $\sum r_{ij}^2$ 记为 N_1；所有 $\triangle P_iP_jP_k$ 的面积平方和 $\sum \Delta_{ijk}^2$ 记为 N_2. 对 $N=3,4,\cdots$ 进行计算表明，总有

$$\frac{N_1^2}{N_2}=16N;$$

若 S_N 是正多面体，则有

$$\frac{N_1^2}{N_2}=12N.$$

这些有趣的事实启发我们考虑：m 维空间 E^m 中的点集 $\mathfrak{S}_N=\{P_1,P_2,\cdots,P_N\}$ 的两个不变量 N_1,N_2 之间有什么关系？易证

$$\frac{N_1^2}{N_2}\geqslant\frac{8m}{m-1}N, \tag{1.1}$$

而等号，当 $\mathfrak{S}_N$ 具有某种对称性时成立.

更一般地，任取 $\mathfrak{S}$ 中的 $(k+1)$ 个点，以它们为顶点作一个 k 维单形，把所有这些 k 维单形的 k 维体积的平方和记作 $N_k(k=1,2,\cdots,m)$；可以证明诸如此类的不等式：

$$\frac{N_1^3}{N_3}\geqslant\frac{216m^2}{(m-1)(m-2)}N^2, \tag{1.2}$$

$$\frac{N_1N_{m-1}}{NN_m}\geqslant m^4, \tag{1.3}$$

$$N_{m-1}^2\geqslant 2\left(\frac{m}{m-1}\right)^3 N_mN_{m-2}; \tag{1.4}$$

在三维空间，有

$$N_1N_2\geqslant 81NN_3,\ N_2^2\geqslant\frac{27}{4}N_1N_3,$$

等等.

本文将给出这类不等式的更一般的形式，并找出等号成立的充要条件. 我们将看到：诸

本文刊于《数学学报》，第 23 卷第 5 期，1980 年.

不变量 N_k 与$\mathfrak{S}_N$ 的惯量椭球和密集椭球有关.

密集椭球与惯量椭球的概念,常见于统计学与力学中.为方便这里重新给出密集椭球的定义.

定义 1.1 设$\mathfrak{S}$是 E^m 中有限点集,H 是经过$\mathfrak{S}$的重心 O 的任一个$(m-1)$ 维超平面;把$\mathfrak{S}$中各点到 H 的距离平方和 I_H 叫做$\mathfrak{S}$关于 H 的转动惯量;过 O 引 H 的法线,在 H 两侧法线上截取等长线段 $P_HO=P'_HO=(\sqrt{I_H})^{-1}$,所有这些 P_H,P'_H 的轨迹是一个$(m-1)$ 维椭球面①,与这个椭球共主轴,而诸半轴为此椭球对应半轴之倒数的椭球,叫$\mathfrak{S}$的密集椭球.显然,二者中一个为球时另一个也是球.这时有

定义 1.2 若$\mathfrak{S}$的密集椭球是一个球,则称$\mathfrak{S}$是惯量等轴的.

本文主要结果是

定理 对 E^m 中点集$\mathfrak{S}_N=\{P_1,P_2,\cdots,P_N\}(N>m)$ 的诸不变量$\{N_k\}$,有不等式

$$\frac{N_k^l}{N_l^k}\geqslant\frac{[(m-l)!(l!)^3]^k}{[(m-k)!(k!)^3]^l}(m!N)^{l-k}\quad(1\leqslant k<l\leqslant m),\tag{1.5}$$

$$N_k^2\geqslant\left(\frac{k+1}{k}\right)^3\cdot\frac{m-k+1}{m-k}\cdot N_{k-1}N_{k+1}\quad(1\leqslant k\leqslant m,N_0=N).\tag{1.6}$$

其等号,当且仅当$\mathfrak{S}$惯量等轴时成立.

从(1.5),(1.6)取特例和作变换,易导出包括(1.1)~(1.4)在内的许多几何不等式,其等号都是当且仅当$\mathfrak{S}$为惯量等轴时成立.

取(1.5)的最简单的情况:$m=2$,$N=3$,得到三角形的三边平方和与面积Δ之间的关系

$$a^2+b^2+c^2\geqslant4\sqrt{3}\triangle.\tag{1.7}$$

不等式(1.7)早已被发现(Weitzenboeck,*Math. Zeit.*,1919(5),137-146),后来又有人提供了不同的证明(Finsier 等,*Math. Helv*,1937(10),316-326);近期又有人回顾了(1.7)并把它推广成为联系着两个三角形的面积与边长的一个不等式[1~2]

$$(-a_1^2+a_2^2+a_3^2)a_1'^2+(a_1^2-a_2^2+a_3^2)a_2'^2+(a_1^2+a_2^2-a_3^2)a_3'^2\geqslant16\triangle\triangle'.\tag{1.8}$$

但是,由(1.7)是不易看出它会自然推广为(1.1)那样的一类不等式的.因此,尽管(1.7)早经发现并几乎众所周知,长期以来,却未见有人提出过如(1.1)的猜想.

至于不等式(1.8),它的发现者 D. Pedoe 没有指出有何应用或怎样向高维情形推广,本文提供的方法解决这些问题也是有效的,将于另文中述及.

§2 两个引理

我们把 E^m 中的点和$(m-1)$ 维定向超平面都叫做 E^m 的基本元素;用 e_i 记基本元素.有限个基本元素之集$\mathfrak{S}_k=\{e_1,e_2,\cdots,e_k\}$ 叫 k 元基本图形.

用 $\rho(e_i,e_j)$ 表示两个点 e_i,e_j 的距离,$\widehat{e_ie_j}$ 表两超平面 e_i,e_j 之夹角.若 e_i,e_j 中一个为点,另一个为面,则以 $d(e_i,e_j)$ 记点到面的带号距离,并引入

① 特殊情况下,椭球的某几个半轴为 0,则它退化为更低维的椭球面.

定义 2.1 E^m 中两个基本元素 e_i,e_j 之间的抽象距离定义为

$$g_{ij}=g(e_i,e_j)=\begin{cases}-\dfrac{1}{2}\rho^2(e_i,e_j) & (\text{若 } e_i,e_j \text{ 都是点}),\\ \cos \widehat{e_ie_j} & (\text{若 } e_i,e_j \text{ 都是超平面}),\\ d(e_i,e_j) & (\text{若 } e_i,e_j \text{ 一为点,一为面}).\end{cases}$$

下述定理揭示出基本图形诸元间相关性:

引理 2.1 设$\mathfrak{S}_N=\{e_1,e_2,\cdots,e_N\}$是 E^m 中的基本图形,令 $\delta_i=1-g_{ii}$,并令

$$P[\mathfrak{S}_N]=P(e_1,e_2,\cdots,e_N)=\begin{vmatrix} 0 & \delta_1 & \cdots & \delta_N \\ \delta_1 & & & \\ \vdots & & g_{ij} & \\ \delta_N & & & \end{vmatrix}$$

则当 $N>m+1$ 时,有

$$P(e_1,e_2,\cdots,e_N)=0. \tag{2.1}$$

证 若$\mathfrak{S}_N$ 中没有点,(2.1) 显然.

不失一般性,设 $e_1,e_2,\cdots,e_l$ 是面,$e_{l+1},e_{l+2},\cdots,e_N$ 是点,$l<N$;取 e_N 为笛卡尔坐标原点. 设 $e_1,e_2,\cdots,e_l$ 的单位法向量为 $\bar{a}_1,\bar{a}_2,\cdots,\bar{a}_l$;由 e_N 引至 $e_{l+1},e_{l+2},\cdots,e_N$ 的向量为 $\bar{a}_{l+1},\bar{a}_{l+2},\cdots,\bar{a}_N$. 又设由 e_N 垂直引至 $e_1,e_2,\cdots,e_l$ 的向量为 $\bar{b}_1,\bar{b}_2,\cdots,\bar{b}_l$. 则有

当 $1\leqslant i\leqslant l,1\leqslant j\leqslant l$ 时,$g_{ij}=\bar{a}_i\cdot\bar{a}_j$;

当 $1\leqslant i\leqslant l,l<j\leqslant N$ 时,$g_{ij}=\bar{a}_i(\bar{a}_j-\bar{b}_i)$.

此时 $P[\mathfrak{S}_N]$ 可表为

$$P(e_1,e_2,\cdots,e_N)=\begin{vmatrix} 0 & 0\ \cdots\ 0 & 1\ \cdots\ 1 \\ \vdots & \bar{a}_i\cdot\bar{a}_j & \bar{a}_i(\bar{a}_j-\bar{b}_i) \\ 0 & & \\ \vdots & \bar{a}_j(\bar{a}_i-\bar{b}_j) & -\dfrac{(\bar{a}_i-\bar{a}_j)^2}{2} \\ 1 & & \end{vmatrix}\begin{matrix} \}i=1,2,\cdots,l \\ \\ \}i=l+1,l+2,\cdots,N\end{matrix}$$

$$\underbrace{\qquad}_{j=1,2,\cdots,l,}\quad\underbrace{\qquad}_{j=l+1,l+2,\cdots,N}$$

对上式作如下不改变行列式值的变换:对 $k\leqslant l$,把第 0 行(列) 乘 $\bar{a}_k\cdot\bar{b}_k$ 加到第 k 行(列) 上;对 $k>l$,把第 0 等行(列) 乘 $\dfrac{1}{2}\bar{a}_k^2$ 加到第 k 行(列) 上,得到

$$P(e_1,e_2,\cdots,e_N)=\begin{vmatrix} 0 & \delta_1 & \delta_2 & \cdots & \delta_N \\ \delta_1 & & & & \\ \delta_2 & & \bar{a}_i\cdot\bar{a}_j & & \\ \vdots & & & & \\ \delta_N & & & & \end{vmatrix}.$$

但 $\bar{a}_N=0$,故末行(列) 除 $\delta_N=1$ 外均为 0,故得

$$P(e_1,e_2,\cdots,e_N)=(-1)\,|\,(\bar{a}_i\cdot\bar{a}_j)\,|\,(i,j=1,2,\cdots,N-1).$$

设$\bar{a}_i=(a_{i1},a_{i2},\cdots,a_{im})$,由于$N>m+1$,以及在笛卡尔坐标系中的内积公式

$$\bar{a}_i\cdot\bar{a}_j=\sum_{k=1}^{m}a_{ik}\cdot a_{jk}$$

得到

$$P[\mathfrak{S}]=(-1)\begin{vmatrix}a_{11} & a_{12}\cdots & a_{1\ m} & 0\ \cdots\ 0\\ \vdots & & \vdots & \vdots\ \ \ \ \ \vdots\\ a_{N-1\ 1} & \cdots & a_{N-1\ m} & 0\ \cdots\ 0\end{vmatrix}\begin{vmatrix}a_{11} & a_{21} & \cdots & a_{N-1\ 1}\\ \vdots & & & \vdots\\ a_{1m} & a_{2m} & \cdots & a_{N-1\ m}\\ 0 & \cdots & \cdots & 0\\ \vdots & & & \vdots\\ 0 & \cdots & \cdots & 0\end{vmatrix}=0.$$

引理2.1证毕.

在引理2.1中取$l=0$的特例,可导出

Cayley **定理**[3]　E^m中任意N个点$P_1,P_2,\cdots,P_N$的Cayley-Menger行列式

$$D(P_1,P_2,\cdots,P_N)=\begin{vmatrix}0 & 1 & \cdots & 1\\ 1 & & & \\ \vdots & & \rho_{ij}^2 & \\ 1 & & & \end{vmatrix}\quad\begin{matrix}(i,j=1,2,\cdots,N)\\ (\rho_{ij}=|\overline{P_iP_j}|)\end{matrix}\tag{2.2}$$

当$N>m+1$时,其值为0.

灵活运用Cayley定理,能解决一些从别的途径很难入手的几何问题,如文献[4]中所述就是一例.而引理2.1显然比Cayley定理更便于应用.例如由(2.1)很容易导出单形高线公式:

设$\mathfrak{S}_{m+1}=\{P_1,P_2,\cdots,P_{m+1}\}$是$E^m$中单形的顶点之集,$e$是由$P_1,P_2,\cdots,P_m$决定的超平面,求由$P_{m+1}$到$e$的高线$h=d(P_{m+1},e)$.

考虑$(m+2)$元基本图形$\mathfrak{S}^*=\{P_1,P_2,\cdots,P_{m+1},e\}$;由(2.1)得到

$$P[\mathfrak{S}^*]=\begin{vmatrix}0 & 1 & \cdots & 1 & 0\\ 1 & & & & 1\\ \vdots & & g_{ij} & & \vdots\\ 1 & & & & h\\ 0 & 0 & \cdots & 0\ \ h & 1\end{vmatrix}=0,$$

把此式对末行末列展开,得

$$h^2=d^2(P_{m+1},e)=\frac{P(P_1,P_2,\cdots,P_{m+1})}{P(P_1,P_2,\cdots,P_m)}.\tag{2.3}$$

从(2.3)出发,用数学归纳法易得单形体积公式

$$V^2(P_1,P_2,\cdots,P_{m+1})=-\frac{1}{(m!)^2}P(P_1,P_2,\cdots,P_{m+1}).\tag{2.4}$$

这公式是早有的[5],曾用Cayley-Menger行列式表达:

$$V^2(P_1,P_2,\cdots,P_{m+1})=\frac{(-1)^{m+1}}{2^m(m!)^2}\cdot D(P_1,P_2,\cdots,P_{m+1}).\tag{2.5}$$

下面,再建立一个代数恒等式.

设A为n阶行列式,$1\leqslant s\leqslant n$,从A中去掉了$i_1,i_2,\cdots,i_s$和$j_1,j_2,\cdots,j_s$列所得之行列

式记为$A_{\substack{i_1 j_1\\ i_2 j_2\\ \vdots\ \vdots\\ i_s j_s}}$ 而这些行列相交处元素构成的行列式记为$A_{\substack{i_1 i_2 \cdots i_s\\ j_1 j_2 \cdots j_s}}$；于是有

引理 2.2[①]　设 B 是由行列式 A 的各元素的余子式构成的行列式.则

$$B_{\substack{i_1 i_2 \cdots i_s\\ j_1 j_2 \cdots j_s}} = A_{\substack{i_1 j_1\\ i_2 j_2\\ \vdots\ \vdots\\ i_s j_s}} \cdot A^{s-1}. \tag{2.6}$$

证　设使(2.6)成立的所有矩阵之集为 $\mathscr{M}$,我们应证明 $\mathscr{M}$ 是全体矩阵之集.

易验证,若(A)为对角型矩阵,$(A)\in\mathscr{M}$.

然后,逐条验证:

1°　若$(A)\in\mathscr{M}$,用任意λ乘(A)的某行(列)得到$(\widetilde{A})$,则$(\widetilde{A})\in\mathscr{M}$.

2°　若$(A)\in\mathscr{M}$,把(A)的某相邻两行(列)互换后得到$(\widetilde{A})$,则$(\widetilde{A})\in\mathscr{M}$.

3°　若$(A)\in\mathscr{M}$,把(A)的某行(列)加到另一行(列)后得到$(\widetilde{A})$,则$(\widetilde{A})\in\mathscr{M}$.

即 $\mathscr{M}$ 对初等变换封闭,从而 $\mathscr{M}$ 是全体矩阵之集.这就证明了引理 2.2.

从而立即可以得到

推论 2.1　若(A)是对称矩阵而(A^*)是(A)的伴随矩阵,即(A^*)的元素是(A)中对应元素的代数余子式,则

$$A^*_{\substack{i_1 i_2 \cdots i_s\\ j_1 j_2 \cdots j_s}} = (-1)^{i+j} A_{\substack{i_1 j_1\\ i_2 j_2\\ \vdots\ \vdots\\ i_s j_s}} \cdot A^{s-1} (i_1+i_2+\cdots+i_s=i, j_1+j_2+\cdots+j_s=j). \tag{2.7}$$

这个等式 §3 中将用到.

把引理 2.1 与引理 2.2 配合使用,解决一些几何问题是很方便的.例如,已知单形诸棱长计算它的各个二面角:

设 E^m 中某个单形顶点之集$\mathfrak{S}_{m+1}=\{P_1,P_2,\cdots,P_{m+1}\}$,每个点 P_k 所对的超平面记为e_k;$d(P_k,e_k)=d_k$,求 e_i,e_j 所成二面角的余弦 $\cos\theta_{ij}$.

考虑 $m+3$ 元基本图形$\mathfrak{S}^{i,j}_{m+1}=\{P_1,P_2,\cdots,P_{m+1},e_i,e_j\}$,对行列式 $P[\mathfrak{S}^{i,j}_{m+1}]$ 使用引理 2.2;由引理 2.1 知 $P[\mathfrak{S}^{i,j}_{m+1}]=0$,故

$$P^*[\mathfrak{S}^{i,j}_{m+1}]_{\substack{m+2,m+3\\ m+2,m+3}}=0,\text{(行列号码由 0 开始)}$$

把上式展开得到

$$P(P_1,P_2,\cdots,P_{m+1},e_i)P(P_1,P_2,\cdots,P_{m+1},e_j)-\begin{vmatrix} \boxed{\quad P[\mathfrak{S}_{m+1}]\quad} & \begin{matrix}0\\0\\\vdots\\d_i\\\vdots\end{matrix} \\ 0\ \ 0\ \cdots\ d_j\ \cdots & \cos\theta_{ij}\end{vmatrix}^2=0.$$

但 $P(P_1,P_2,\cdots,P_{m+1},e_i)=0$(引理 2.1)故得

① 这个命题在文献[6]中已有,但证法与此处不同.

$$\begin{vmatrix} 0 & 1 & \cdots\cdots & 1 & 0 \\ 1 & & & & 0 \\ \vdots & & g_{ij} & & \vdots \\ & & & & d_i \\ 1 & & & & \vdots \\ 0 & 0\cdots\cdots d_j & \cdots & & \cos\theta_{ij} \end{vmatrix} = 0. \tag{2.8}$$

将(2.8)对末行末列展开,得

$$\cos\theta_{ij} = d_i d_j \frac{A_{ij}}{A}, \tag{2.9}$$

这里$A = P[\mathfrak{S}_{m+1}]$,$A_{ij}$表$A$的对应于$g_{ij}$的代数余子式. 再应用高线公式(2.3)可得

$$\cos\theta_{ij} = \frac{A_{ij}}{\sqrt{A_{ii}} \cdot \sqrt{A_{jj}}} \tag{2.10}$$

等式(2.9)在§3中也将用到.

等式(2.10)中的θ_{ij},可以看作是单纯形的诸外角(将各超平面适当地定向);稍微改变一下形式就得到

高维余弦定理 设E^m中某个单形的诸内角为φ_{ij},单形的Cayley-Menger行列式是D,则成立着下列公式

$$\cos\varphi_{ij} = -\frac{D_{ij}}{\sqrt{D_{ii}} \cdot \sqrt{D_{jj}}} \quad (i,j = 1,2,\cdots,m+1). \tag{2.11}$$

高维余弦公式(2.11)在计算问题上有广泛的应用,也可以用来推导某些理论上的结果. 限于篇幅兹不赘述.

§3 密集椭球与$\{N_K\}$之关系

下述定理揭示出密集椭球与$\{N_k\}$之关系

预备定理 E^m中点集$\mathfrak{S}_N = \{P_1, P_2, \cdots, P_N\}(N > m)$的诸不变量$\{N_k\}$与$\mathfrak{S}_N$的密集椭球诸半轴的平方$a_1^2, a_2^2, \cdots, a_m^2$之间有关系.

$$N_k = \frac{N}{(k!)^2} \cdot \sigma_k(a_1^2, a_2^2, \cdots, a_m^2) \quad (k = 1,2,\cdots,m), \tag{3.1}$$

这里σ_k是k次初等对称函数. 或者说,$a_1^2, a_2^2, \cdots, a_m^2$是方程

$$\sum_{k=0}^{m} (-1)^k (k!)^2 N_k x^{m-k} = 0 \quad (N_0 = N) \tag{3.2}$$

的根.

证 把$\mathfrak{S}_N$看成E^{mN}中的点,则(3.1)两端都是定义于E^{mN}上的连续函数. 当$N=m+1$时,对应于E^m中非退化单形顶点集的$\mathfrak{S}_N$在E^{mN}中稠密;故只需证明(3.1)当$\mathfrak{S}_N$是非退化单形顶点之集时成立,即可断言它对$N = m+1$普遍成立.

而当$N > m+1$时,可以把$\mathfrak{S}_N$看成是E^{N-1}中的点集,由(3,1)在E^{N-1}中成立推知它在E^m中成立. 这是因为,诸不变量当$\mathfrak{S}_N$由E^{N-1}中退化到E^m中时,除了$(N-1-m)$个化为0外,其余均不变.

下面设 $N=m+1$，$\{P_1,P_2,\cdots,P_{m+1}\}$ 是 E^m 中非退化单形顶点之集，往证(3.1)。任取 E^m 中的一个 $(m-1)$ 维定向超平面 e，令 $h_i=d(P_i,e)$，由引理 2.1

$$P(P_1,P_2,\cdots,P_{m+1},e)=\begin{vmatrix} 0 & 1 & \cdots\cdots & 1 & 0 \\ 1 & & & & h_1 \\ \vdots & & g_{ij} & & \vdots \\ 1 & & & & h_{m+1} \\ 0 & h_1 & \cdots\cdots & h_{m+1} & 1 \end{vmatrix}=0. \tag{3.3}$$

令 $P[\mathfrak{S}_{m+1}]=A,h<A_{ij}$，$A_{ij}$ 记 A 的对应于 g_{ij} 的代数余子式，将(3.3)对末行末列展开，得

$$\sum_{i=1}^{m+1}\sum_{j=1}^{m+1}A_{ij}h_ih_j=A. \tag{3.4}$$

由于 $\mathfrak{S}_{m+1}$ 是非退化单形顶点之集，由(2.4)可知 $A\neq 0$，由(3,4)可写成

$$\sum_{i=1}^{m+1}\sum_{j=1}^{m+1}\frac{A_{ij}}{A}h_ih_j=1. \tag{3.5}$$

亦即，任一超平面 e 与 $P_1,P_2,\cdots,P_{m+1}$ 的距离 $h_1,h_2,\cdots,h_{m+1}$ 满足(3.5)。

反之，我们指出：对任一组满足(3.5)的实数 $\{h_k\}$，均有一超平面 e 使 $d(P_k,e)=h_k$。

令 $h_k^*=h_k-h_{m+1}$，我们先找到一个超平面 e^*，使 $d(P_k,e^*)=h_k^*$，然后，将 e^* 沿自己的法线向负侧作距离为 h_{m+1} 的平移，即得 e。

显然，$\{h_k^*\}$ 也满足约束(3.5)，这只要在(3.3)中把第0行(列)乘以 $-h_{m+1}$ 加到末行即易得出。

以 e_k 记 $\mathfrak{S}_{m+1}/P_k$ 的诸点所决定的超平面，$\bar{a}_k$ 记 e_k 的单位法向量，

$$\bar{a}_i\cdot\bar{a}_j=\cos\widehat{e_ie_j}=\cos\theta_{ij};$$

又记由 P_{m+1} 引向 P_k 之向量为 $\bar{P}_k$，则

$$d_k=d(e_k\cdot P_k)=\bar{a}_k\cdot\bar{P}_k\quad(k=1,2,\cdots,m).$$

而对 $i\neq j$，则有 $\bar{a}_i\cdot\bar{P}_j=0$。

取

$$\bar{a}=\sum_{k=1}^{m}\frac{h_k^*}{d_k}\cdot\bar{a}_k,$$

则有(用到(2.9))

$$|\bar{a}|^2=\bar{a}^2=\sum_{i=1}^{m}\sum_{j=1}^{m}\frac{\bar{a}_i\cdot\bar{a}_j}{d_i\cdot d_j}\cdot h_i^*h_j^*=\sum_{i=1}^{m}\sum_{j=1}^{m}\frac{A_{ij}}{A}\cdot h_i^*h_j^*=1.$$

故 $\bar{a}$ 为单位向量，取 $\bar{a}$ 为单位法向量过 P_{m+1} 作超平面 e^*，则

$$d(e^*,P_{m+1})=0=h_{m+1}^*,$$

$$d(e^*,P_k)=\bar{a}\cdot\bar{P}_k=\sum_{i=1}^{m}\frac{h_i^*}{d_i}\bar{a}_i\cdot\bar{P}_k=h_k^*.$$

把 e^* 沿方向 $\bar{a}$ 平移 $-h_{m+1}$，即得所求的 e。

在(3.3)左端，把第0行(列)乘 t 加到第 N 行(列)上，等式仍成立。可见当 $(h_1,\cdots,h_{m+1})$ 满足(3.5)时，$(h_1+t,h_2+t,\cdots,h_{m+1}+t)$ 亦满足(3.5)；因而，若把 $(h_1,h_2,\cdots,h_{m+1})$ 看成 E^N 中的点，则(3.5)表 E^N 中的以原点为中心的二阶柱面，柱面母线方向向量为

$$\bar{J}=(1,1,\cdots,1).$$

从而超平面(E^N 中的($N-1$) 维子空间)

$$H: h_1+h_2+\cdots+h_{m+1}=0$$

与柱面(3.5) 正交.

在 E^m 中考虑任一过$\mathfrak{S}_N$之重心的$m-1$维超平面e,它所对应的($h_1,h_2,\cdots,h_{m+1}$) 显然满足 $h_1+\cdots+h_{m+1}=0$,反之亦然. 因此,求$\mathfrak{S}_N$ 的密集椭球的诸轴平方 $a_1^2,\cdots,a_m^2$ 的问题,也正是约束(3.5) 及$\sum\limits_{i=1}^{m+1}h_i=1$下,求目标函数$\sum\limits_{i=1}^{m+1}h_i^2$ 的稳定值问题. 但由于(3.5) 是柱面,而$\sum\limits_{i=1}^{m+1}h_i=0$恰为过原点而与(3.5) 柱面正交的超平面,从而对目标函数$\sum\limits_{i=1}^{m+1}h_i^2$ 而言,条件$\sum\limits_{i=1}^{N}h_i=0$可以略去(因为不难算出:$\sum\limits_{i=1}^{N}h_i^2$ 沿方向$\bar{J}$ 的偏导数当且仅当$\sum\limits_{i=1}^{m+1}h_i=0$ 时才为 0).

因此,在约束(3.5) 下求$\sum\limits_{i=1}^{m+1}h_1^2$ 的稳定值的问题,等价于求$\mathfrak{S}_N$ 的密集椭球诸半轴的平方. 按照拉格朗日乘子法,这些稳定值应当是(3.5) 左端二次型的矩阵$\left[\dfrac{A_{ij}}{A}\right]$的非 0 特征值的倒数. 设 $B=[A_{ij}]$ 的非 0 特征值为 $\mu_1,\mu_2,\cdots$,则

$$a_k^2=\frac{A}{\mu_k}, \tag{3.6}$$

将 B 的特征方程 $|B-\mu E|=0$ 展开,得

$$\sum_{k=0}^{m+1}(-1)^{m+1-k}B_k\mu^{m+1-k}=0, \tag{3.7}$$

这里

$$B_0=1, B_k=\sum_{i_1<i_2<\cdots<i_k}\left|B_{\substack{i_1,i_2,\cdots,i_k\\ i_1,i_2,\cdots,i_k}}\right| \quad (k=1,2,\cdots,m+1). \tag{3.8}$$

若以$[A]$记$P[\mathfrak{S}_N]=A$所对应的矩阵,并约定$[A]$和$[A]^*$ 的行列足标由 0 开始,那么,$[A]$ 去掉第 0 行第 0 列后得 B. 由引理(2.2)

$$\left|B_{\substack{i_1,i_2,\cdots,i_k\\ i_1,i_2,\cdots,i_k}}\right|=\left|[A]^*_{\substack{i_1,i_2,\cdots,i_k\\ i_1,i_2,\cdots,i_k}}\right|=\left|[A]_{\substack{i_1,i_1\\ i_2,i_2\\ \vdots\ \vdots\\ i_k,i_k}}\right|A^{k-1},$$

对 $k=1,2,\cdots,m$ 应用体积公式(2.4),得

$$\begin{aligned}\sum_{i_1<i_2<\cdots<i_k}\left|B_{\substack{i_1,i_2,\cdots,i_k\\ i_1,i_2,\cdots,i_k}}\right|&=\sum_{i_1<i_2<\cdots<i_{N-k}}P(P_{j_1},P_{j_2},\cdots,P_{j_{N-k}})A^{k-1}\\ &=-((m-k)!)^2N_{m-k}A^{k-1}\quad (k=1,2,\cdots,m).\end{aligned} \tag{3.9}$$

在上式中 $N_0=m+1$,而且由引理 2.2 得到 $B_{m+1}=0$. 把这些结果代入(3.7) 得

$$\sum_{k=0}^{m}(-1)^kB_k\mu^{m+1-k}=\mu^{m+1}+\sum_{k=1}^{m}(-1)^{k+1}((m-k)!)^2N_{m-k}A^{k-1}\mu^{m+1-k}=0,$$

两端除以 A^m,得

$$A\cdot\left(\frac{\mu}{A}\right)^{m+1}+\sum_{k=0}^{m}(-1)^{k+1}((m-k)!)^2N_{m-k}\left(\frac{\mu}{A}\right)^{m+1-k}=0.$$

此方程有一个 0 根；约去 0 根，将它变为$\frac{A}{\mu}$的方程，并注意到

$$A=-(m!)^2N_m,$$

即得

$$\sum_{k=0}^{m}(-1)^k((m-k)!)^2N_{m-k}\left(\frac{A}{\mu}\right)^k=0.$$

即$\frac{A}{\mu}$满足(3.2). 也就是说(3.2) 的诸根是密集椭球诸半轴的平方，预备定理证毕.

§4 定理的证明

由 Maclaurin 定理[7]，对 m 个正实变元的 k 次和 l 次初等对称函数 σ_k 与 σ_l 有不等式

$$\left[\frac{k!(m-k)!}{m!}\sigma_k\right]^l\geqslant\left[\frac{l!(m-l)!}{m!}\sigma_l\right]^k(1\leqslant k<l\leqslant m). \tag{4.1}$$

由(3.1)，将

$$\sigma_j=\frac{(j!)^2N_j}{N}$$

代入(4.1) 整理即得(1.5).

类似地，根据 Newton 定理[7]，有

$$\left[\frac{k!(m-k)!}{m!}\sigma_k\right]^2\geqslant\left[\frac{(k-1)!(m-k+1)!}{m!}\sigma_{k-1}\right]\cdot\left[\frac{(k+1)!(m-k-1)!}{m!}\sigma_{k+1}\right]. \tag{4.2}$$

将(3.1) 中的 σ_j 代入(4.2) 即得(1.6). (4.1) 与(4.2) 等号成立的充要条件都是 m 个变数值相同. 在我们的情况下，即是惯量等轴.

最后，我们认为，引理 2.1 和预备定理比所导出的不等式更为深刻和重要，这些不等式的证明不过是引理 2.1 多种应用之一端.

而且，如果不追究等号成立的几何意义，仅仅证明(1.5) 和(1.6) 就有简单得多的方法，只要能证明(3.2) 所有的根都是正实数，再用(4.1) 和(4.2) 即可.

应用(2.4)，可将方程(3.2) 写成如下形式.

$$F(\lambda)=\begin{vmatrix}0 & 1 & 1 & \cdots & 1\\ 1 & \lambda & & & \\ 1 & & \lambda & \frac{\rho_{ij}^2}{2} & \\ \vdots & \frac{\rho_{ij}^2}{2} & & \ddots & \\ 1 & & & & \lambda\end{vmatrix}=0. \tag{4.3}$$

考虑

$$F_{\varepsilon}(\lambda)=\left|\begin{pmatrix}0 & 1 & 1 & \cdots & 1\\ 1 & & & & \\ 1 & & \frac{\rho_{ij}^2}{2} & & \\ \vdots & & & & \\ 1 & & & & \end{pmatrix}+\lambda\begin{pmatrix}\varepsilon & & & & \\ & 1 & \mathbf{0} & & \\ & & 1 & & \\ & \mathbf{0} & & \ddots & \\ & & & & 1\end{pmatrix}\right|=0. \qquad (4.4)$$

在行列式(4.4)中,前一矩阵是实对称的,后一矩阵是恒正型(令$\varepsilon>0$);根据二次型耦的一个熟知的定理得知(4.4)的根都是实的.令$\varepsilon\to 0$取极限,即知(4.3)也只有实根.又因其系数具有交错符号,故这些根都是正的.这就简捷地证明了(1.5),(1.6)等.

不过,由于缺少了等号成立的充分必要条件(几何意义),这个证明不能算是对几何不等式的一个完整的证明.

参考文献

[1] Pedoe D. Thinking Geometrically. Amer. Math. Monthly,1970(77):711-721.

[2]Carlitz L. An Inequality Involving the Area of two Triangles. Amer. Math. Monthly, 1971(78):772.

[3]Blumenthal L M. Theory and Applications of Distance Geometry. 2nd ed. New York,1970.

[4]Graham R L,et al. Are there $n+2$ Points in E^n with Odd Integeral Distance. Amer. Math. Monthly,1974(81):21-25.

[5]Blumenthal L M,Gillam B E. Distribution of Points in n-Space. Amer. Math. Monthly, 1943(50):181-185.

[6]Hodge W V,Pedoe D. Methods of Algebraic Geometry. Cambridge,1953(Ⅰ).

[7]Hardy G H,Littlewood J E,Pólya. Inequalities. 2nd ed. Cambridge,1952.

Neuberg-Pedoe 不等式的高维推广及应用

杨　路　张景中

（中国科学技术大学）

§0　引　　言

全文中我们用$\sum_A,\sum_B$表示n维欧氏空间E^n中的单形，其顶点分别为$a_1,a_2,\cdots,a_{n+1}$和$b_1,b_2,\cdots,b_{n+1}$，其棱长分别为$a_{ij}=|a_ia_j|$和$b_{ij}=|b_ib_j|$，其体积分别为$V(A),V(B)$.

令$\sum_A,\sum_B$的顶点集$\{a_i\},\{b_i\}$的 Cayley-Menger 阵分别为$n+2$阶方阵$\mathfrak{A},\mathfrak{B}$：

$$\mathfrak{A}=\begin{pmatrix}0 & 1 & \cdots & 1\\ 1 & & & \\ \vdots & & -\frac{1}{2}a_{ij}^2 & \\ 1 & & & \end{pmatrix},\quad \mathfrak{B}=\begin{pmatrix}0 & 1 & \cdots & 1\\ 1 & & & \\ \vdots & & -\frac{1}{2}b_{ij}^2 & \\ 1 & & & \end{pmatrix};$$

这个定义和一般文献略有出入——每个柱心元素多乘了一个因子$-\frac{1}{2}$.

记$\det\mathfrak{A}=A,\det\mathfrak{B}=B$，分别叫做点集$\{a_i\}$和$\{b_i\}$的 Gayley-Menger 行列式[1]. 同通常一致，用A_{ij},B_{ij}分别记对应于A,B的代数余子式，$i,j=0,1,\cdots,n+1$. 我们将证明下列一般形式的不等式：

定理 1　对E^n中的两个单形$\sum_A,\sum_B$，

$$\sum_{i=1}^{n+1}\sum_{j=1}^{n+1}a_{ij}^2B_{ij}\geqslant 2n(n!)^2V(A)^{\frac{2}{n}}V(B)^{2-\frac{2}{n}},\tag{0.1}$$

而且，当且仅当$\sum_A$相似于$\sum_B$且诸顶点a_i相似对应于b_i时，等式成立.

在定理 1 中取$n=2$，得到 Neuberg-Pedoe 不等式[2]：

$$a'^2(-a^2+b^2+c^2)+b'^2(a^2-b^2+c^2)+c'^2(a^2+b^2-c^2)\geqslant 16\triangle\triangle',\tag{$*$}$$

这里$\triangle,\triangle'$表两个三角形的面积，a,b,c和a',b',c'分别表两三角形的边；而等式成立的充分必要条件是这两个三角形相似.

不等式（∗）被称为“第一个有趣的关于两个三角形的不等式”. 1942 年 Pedoe[3] 在 Finsler 和 Hadwiger[4] 的一篇文章启发之下找到了（∗）的第一个证明. 此后数十年中 Pedoe 本人和别人又提供了许多新的证明，几何的或纯代数的[5~8]. Pedoe 的最近的一个证明发表

本文刊于《数学学报》，第 24 卷第 3 期，1981 年.

于 1976 年[7].

在国内,据作者所知,中国科技大学常庚哲在文献[9]中提供了一个使用复数的简单证明.

定理 1 是 Neuberg-Pedoe 不等式在高维空间的直接推广. 作为定理 1 的一个有趣的应用,我们将考虑下述问题:

我们知道,若单形 $\sum_A$ 的诸棱长 a_{ij} 各自不超过单形 $\sum_B$ 的对应的棱长 b_{ij} 时,一般而言,不能断言有 $V(A)\leqslant V(B)$. 那么,在 $a_{ij}\leqslant b_{ij}$ 之外再附加什么条件才能使 $V(A)\leqslant V(B)$ 一定成立呢?

当 $n=2$,即 $\sum_A$,$\sum_B$ 是三角形时,问题是平凡的:只要 $\sum_B$ 是非钝角三角形就够了. 在 $n\geqslant 3$ 的一般情况,可以引入非钝角单形的概念:如果某单形的每个(由单形的两个 $n-1$ 维棱所成的二面角)内角皆非钝角,则称之为非钝角的. 这样,应用定理 1 及某些属于距离几何的结果和方法,可得

定理 2 设 E^n 中两单形 $\sum_A$,$\sum_B$ 满足

1° $a_{ij}\leqslant b_{ij}(i,j=1,2,\cdots,n+1)$,

2° $\sum_B$ 是非钝角的,

则必有
$$V(A)\leqslant V(B).$$

在证明定理 2 的过程中,我们可以看到定理 1 也可写成另一种形式. 令 $S_i(B)$ 表 $\sum_B$ 的顶点 b_i 所对的 $n-1$ 维棱的 $n-1$ 维体积,$\theta_{ij}(B)$ 表 b_i 所对的$(n-1)$维棱与 b_j 所对的$(n-1)$维棱所成之内角,于是有

定理 1* 对 E^n 中的两个单形 $\sum_A$,$\sum_B$,

$$\sum_{i=1}^{n+1}\sum_{j=1}^{n+1}a_{ij}^2S_i(B)S_j(B)\cos\theta_{ij}(B)\geqslant 2n^3V(A)^{\frac{2}{n}}V(B)^{2-\frac{2}{n}}. \tag{0.1*}$$

(或 $\sum\limits_{i<j}a_{ij}^2S_i(B)S_j(B)\cos\theta_{ij}(B)\geqslant n^3V(A)^{\frac{2}{n}}V(B)^{2-\frac{2}{n}}$),而且等式当且仅当诸顶点 a_i 相似对应于 b_i 时成立.

(0.1*)与(0.1)的等价性将于 §2 中证明. 二者对比起来,(0.1*)具有鲜明的几何意义,而(0.1)则具有代数的简洁性.

附带提一下:有些文献中提到另外的"锐角单形"的概念. 与上述"非钝角单形"的概念有实质上的不同. 例如,Alexander 在文献[10]中引进的锐角单形概念同我们的概念互不蕴含. 在我们看来 Alexander 的概念与直观上锐角的概念颇有距离(这是可以用反例证明的),但他那样做是为了该文的特殊需要.

最后一节,讨论了定理 2 的非欧情况,并提出了一个猜想.

§1　定理 1 的证明

引入记号

$$\begin{cases} q_{ij} = \frac{1}{2}(a_{i,n+1}^2 + a_{j,n+1}^2 - a_{ij}^2), \\ r_{ij} = \frac{1}{2},(b_{i,n+1}^2 + b_{j,n+1}^2 - b_{i,j}^2) \end{cases} (i,j = 1,2,\cdots,n+1); \tag{1.1}$$

$$Q = (q_{ij}), R = (r_{ij}) \tag{1.2}$$

$$(Q,R \text{ 为 } n\times n \text{ 方阵}, i,j = 1,2,\cdots,n);$$

$$s_{ij}(\lambda) = q_{ij} + \lambda r_{ij}, \tag{1.3}$$

$$S(\lambda) = (s_{ij}(\lambda))$$

$$(S(\lambda) \text{ 为 } n\times n \text{ 方阵}, i,j = 1,2,\cdots,n);$$

$$f_{ij}(\lambda) = -\frac{1}{2}(a_{ij}^2 + \lambda b_{ij}^2), \tag{1.4}$$

$$F(\lambda) = \begin{pmatrix} 0 & 1 & \cdots & 1 \\ 1 & & & \\ \vdots & & f_{ij}(\lambda) & \\ 1 & & & \end{pmatrix}$$

$(F(\lambda)$ 为$(n+2)\times(n+2)$ 方阵, $i,j = 1,2,\cdots,n+1)$.

我们的方法是从考虑方程 $\det F(\lambda) = 0$ 的根入手，对 $\det F(\lambda)$ 作不改变值的行列变换：约定 $F(\lambda)$ 的行(列) 号是由 0 至 $n+1$，把第 0 行(列) 乘以 $-f_{i,n+1}(\lambda)(-f_{n+1,j}(\lambda))$ 后加到第 i 行(j 列)，即得

$$\det F(\lambda) = \begin{vmatrix} 0 & 1 & \cdots & 1 \\ 1 & & & \\ \vdots & & f_{ij}(\lambda) & \\ 1 & & & \end{vmatrix} = \begin{vmatrix} 0 & 1 & \cdots & 1 & 1 \\ 1 & & & & 0 \\ \vdots & & S_{ij}(\lambda) & & \vdots \\ 1 & & & & 0 \\ 1 & 0 & \cdots & 0 & 0 \end{vmatrix}$$

$$= -|s_{ij}(\lambda)| = -\det S(\lambda)$$

$$= -\det(Q+\lambda R).$$

于是可令

$$-\det F(\lambda) = \det(Q+\lambda R) = c_0\lambda^n + c_1\lambda^{n-1} + \cdots + c_n, \tag{1.5}$$

由于 Q,R 都是实的对称正定方阵，从而知诸系数 $c_0, c_1, \cdots, c_n$ 都是非负的，而且此方程的根都是非正的实根. 由 Maclaurin 不等式[11]

$$\frac{1}{n}\frac{c_1}{c_0} \geqslant \left(\frac{2}{n(n-1)}\cdot\frac{c_2}{c_0}\right)^{\frac{1}{2}} \geqslant \left(\frac{6}{n(n-1)(n-2)}\frac{c_3}{c_0}\right)^{\frac{1}{3}}$$

$$\geqslant \cdots \geqslant \left(\frac{c_n}{c_0}\right)^{\frac{1}{n}}, \tag{1.6}$$

这里只用其两端：

$$c_1 \geqslant n c_0^{1-\frac{1}{n}} c_n^{\frac{1}{n}}. \tag{1.7}$$

另一方面，将多项式(1.5) 按行列式展开得到

$$\begin{cases} c_0 = -\det \mathscr{B} = -B, \\ c_n = -\det \mathfrak{A} = -A, \\ c_1 = \sum_{i=1}^{n+1} \sum_{j=1}^{n+1} \frac{a_{ij}^2}{2} \cdot B_{ij}. \end{cases} \tag{1.8}$$

根据熟知的单纯形体积公式[12]

$$V(A)^2 = -\frac{1}{(n!)^2} A, V(B)^2 = -\frac{1}{(n!)^2} B, \tag{1.9}$$

把(1.8)，(1,9) 代入(1.7)，就得到

$$\sum_{i=1}^{n+1} \sum_{j=1}^{n+1} \frac{a_{ij}^2}{2} B_{ij} \geqslant n(n!)^2 V(A)^{\frac{2}{n}} V(B)^{2-\frac{2}{n}},$$

亦即

$$\sum_{i=1}^{n+1} \sum_{j=1}^{n+1} a_{ij}^2 B_{ij} \geqslant 2n(n!)^2 V(A)^{\frac{2}{n}} V(B)^{2-\frac{2}{n}}, \tag{0.1}$$

剩下的是证明等式成立的充要条件.

先证条件的充分性，假设$\sum_A$ 与$\sum_B$ 按顶点编号顺序相似，可令

$$a_{ij} = \mu_0 b_{ij} \quad (\mu_0 > 0, i, j = 1, 2, \cdots, n+1), \tag{1.10}$$

于是从(1.1) 和(1.2) 就有

$$q_{ij} = \mu_0 r_{ij}, Q = \mu_0 R, \tag{1.11}$$

$$\begin{aligned} \therefore -\det F(\lambda) &= \det(Q + \lambda R) = \det(\mu_0 R + \lambda R) = \det((\lambda + \mu_0) R) \\ &= (\lambda + \mu_0)^n \det R, \end{aligned}$$

可见，$-\mu_0$ 是 $\det F(\lambda) = 0$ 的 n 重根. 而诸根两两相等是 Maclaurin 不等式的等号成立的充要条件，故

$$\frac{1}{n} \cdot \frac{c_1}{c_0} = \left(\frac{c_n}{c_0}\right)^{\frac{1}{n}}, \tag{1.12}$$

从而(0.1) 中的等式成立，充分性证毕.

反之，若(0.1) 式的等号成立，即(1.12) 成立，那么，$\det(Q + \lambda R) = 0$ 有 n 重根. 但由于 Q, R 是对称阵，R 正定，故有合同变换 T，使

$$TRT^{\tau} = E, TQT^{\tau} = \begin{bmatrix} u_1 & & 0 \\ & \ddots & \\ 0 & & \mu_n \end{bmatrix},$$

于是

$$\det(Q + \lambda R) = \frac{1}{(\det T)^2} (\lambda + \mu_1) \cdots (\lambda + \mu_n).$$

由 $\det(Q + \lambda R) = 0$ 有 n 重根推知

故 $$\mu_1 = \mu_2 = \cdots = \mu_n = \mu,$$

$$TQT^{\tau} = \mu TRT^{\tau},$$

即 $Q = \mu R$. 再由(1.1)得

$$a_{ij} = \mu b_{ij}(i,j = 1,2,\cdots,n+1),$$

即 $\sum_A$ 与 $\sum_B$ 按顶点编号顺序相似，必要性证毕.

不难看出，把(1.6)中的 $c_2, c_3, \cdots, c_{n-1}$ 等计算出来之后，还可以从中得出许多几何不等式，其中每一个都可以看成 Neuberg-Pedoe 不等式的推广. 如果仅仅为了导出(0.1)，当然只用熟知的算术-几何平均不等式已经足够了.

§2 定理 2 的证明

令 $\theta_{ij}(B)$ 表单形 $\sum_B$ 的顶点 b_i 所对的 $(n-1)$ 维棱与 b_j 所对的 $(n-1)$ 维棱所成之内角，为证明定理 2 需要下述引理：

引理 在前面约定的记号之下

$$\cos\theta_{ij}(B) = \frac{B_{ij}}{\sqrt{B_{ii}B_{jj}}}(i,j = 1,2,\cdots,n+1). \tag{2.1}$$

这个引理(高维余弦定理)的证明作者已在文献[13]中给出. 由于篇幅较大，此地不再重复.

按前面所约定记号，$S_i(B)$ 和 $S_j(B)$ 分别表顶点 b_i, b_j 所对的 $(n-1)$ 维棱的 $(n-1)$ 维体积，由体积公式知

$$S_i(B)^2 = -\frac{1}{((n-1)!)^2}B_{ii}, S_j(B)^2 = -\frac{1}{((n-1)!)^2}B_{jj}, \tag{2.2}$$

把它代入(0.1)中就有

$$\sum_{i=1}^{n+1}\sum_{j=1}^{n+1} a_{ij}^2 S_i(B)S_j(B)\cos\theta_{ij}(B) \geqslant 2n^3 V(A)^{\frac{2}{n}} V(B)^{2-\frac{2}{n}}. \tag{0.1*}$$

这就是定理 1*. 现在往证定理 2.

考虑等式

$$\lambda^n B = \frac{1}{\lambda}\begin{vmatrix} 0 & \lambda & \cdots & \lambda \\ 1 & & & \\ \vdots & & -\frac{1}{2}\lambda b_{ij}^2 & \\ 1 & & & \end{vmatrix} = \begin{vmatrix} 0 & 1 & \cdots & 1 \\ 1 & & & \\ \vdots & & -\frac{1}{2}\lambda b_{ij}^2 & \\ 1 & & & \end{vmatrix},$$

两端对 λ 求微商再令 $\lambda = 1$，同时把右端由于分行微商所产生的行列式按求微商的那一行展开，即得

$$\sum_{i=1}^{n+1}\sum_{j=1}^{n+1}\frac{b_{ij}^2}{2}B_{ij} = -nB = n(n!)^2 V(B)^2. \tag{2.3}$$

由于定理 2 之假设条件 2°，$\sum_B$ 是非钝角的，即 $\cos\theta_{ij}(B) \geqslant 0$，从而

$$B_{ij} \geqslant 0(i,j = 1,2,\cdots,n+1), \tag{2.4}$$

再加上条件 1°，$a_{ij} \leqslant b_{ij}$，由(2.3)可推出

$$V^2(B)=\frac{1}{n(n!)^2}\sum_{i=1}^{n+1}\sum_{j=1}^{n+1}\frac{b_{ij}^2}{2}B_{ij}$$

$$\geqslant\frac{1}{n(n!)^2}\sum_{i=1}^{n+1}\sum_{j=1}^{n+1}\frac{a_{ij}^2}{2}B_{ij}\geqslant V(A)^{\frac{2}{n}}V(B)^{2-\frac{2}{n}},\qquad(2.5)$$

化简之即得

$$V(A)\leqslant V(B).$$

定理 2 证毕.

这个定理也可以写成下列形式而不必假定 $a_{ij}\leqslant b_{ij}$：

定理 2* 如果 $\sum_B$ 是非钝角的，则成立着不等式

$$\frac{V(A)}{V(B)}\leqslant\left(\max_{i,j}\frac{a_{ij}}{b_{ij}}\right)^n.\qquad(2.6)$$

显然此定理中的 $\sum_B$ 非钝角这一要求是不能取消的.

§3 关于非欧情况的讨论

本节考虑将定理 2 推广到非欧常曲率空间的可能性问题.

定理 2 在罗巴切夫斯基空间中并不成立. 即使在 2 维情形也存在反例. 事实上，有

定理 3.1 在罗巴切夫斯基平面上存在着两个这样的三角形 $\triangle_A$，$\triangle_B$：

1° $\triangle_A$ 的各边长不超过 $\triangle_B$ 的对应边长，

2° $\triangle_B$ 是非钝角的，

但是

$$\text{area }\triangle_A>\text{area }\triangle_B.$$

证明 令 $\triangle_A$ 是三个角都等于$\frac{\pi}{6}$的等边三角形，而 $\triangle_B$ 是等腰直角三角形，夹直角的两边与 $\triangle_A$ 的边等长.

由于罗氏平面上直角三角形仍然是斜边最长，因此有

1° $\triangle_A$ 的各边长不超过 $\triangle_B$ 的对应边长.

此外，既然 $\triangle_B$ 是直角三角形，它就不可能再有一个钝角，所以

2° $\triangle_B$ 是非钝角的.

最后来比较 $\triangle_A$ 与 $\triangle_B$ 的面积. 设这个罗氏平面曲率为 K. 由于 $\triangle_A$ 的角欠为$\frac{\pi}{2}$，故有

$$\text{area }\triangle_A=\frac{1}{|K|}\cdot\frac{\pi}{2}.$$

另一方面，$\triangle_B$ 的三内角和大于$\frac{\pi}{2}$，于是它的角欠小于$\frac{\pi}{2}$，故有

$$\text{area }\triangle_B<\frac{1}{|K|}\frac{\pi}{2},$$

从而定理 3.1 得证.

但是,定理 2 在球面上(具有正曲率的二维常曲率空间)却是正确的.亦即有

定理 3.2 在同一球面上的两个三角形 $\triangle_A$ 与 $\triangle_B$ 如果满足两个条件:

1° $\triangle_A$ 的各边长不超过 $\triangle_B$ 的对应边长,

2° $\triangle_B$ 是非钝角的,

则有不等式

$$\text{area}\,\triangle_A \leqslant \text{area}\,\triangle_B.$$

证 不妨设球半径为1,$\triangle_A$,$\triangle_B$ 的对应各边设为a_{ij},b_{ij}($i,j=1,2,3$),各边所对之内角设为 α_{ij},β_{ij}.又令

$$\mathfrak{A}^* = \begin{pmatrix} 1 & \cos a_{12} & \cos a_{13} \\ \cos a_{21} & 1 & \cos a_{23} \\ \cos a_{31} & \cos a_{32} & 1 \end{pmatrix}, \mathscr{B}^* = \begin{pmatrix} 1 & \cos b_{12} & \cos b_{13} \\ \cos b_{21} & 1 & \cos b_{23} \\ \cos b_{31} & \cos b_{32} & 1 \end{pmatrix}$$

$A^* = \det \mathfrak{A}^*$,$B^* = \det \mathscr{B}^*$.

由于 $\mathfrak{A}^*$,$\mathscr{B}^*$ 都是正定方阵,方程

$$\det(\mathfrak{A}^* + \lambda\mathscr{B}^*) = 0, \tag{3.1}$$

只能有负实根,将(3.1)展开得到

$$B^*\lambda^3 + \left(\sum_{i=1}^{3}\sum_{j=1}^{3} B_{ij}^* \cos a_{ij}\right)\lambda^2 + \left(\sum_{i=1}^{3}\sum_{j=1}^{3} A_{ij}^* \cos b_{ij}\right)\lambda + A^* = 0. \tag{3.2}$$

对方程(3.2)的根用算术-几何平均不等式得

$$\frac{1}{3B^*}\sum_{i=1}^{3}\sum_{j=1}^{3} B_{ij}^* \cos a_{ij} \geqslant \left(\frac{A^*}{B^*}\right)^{\frac{1}{3}}, \tag{3.3}$$

亦即

$$\sum_{i=1}^{3}\sum_{j=1}^{3} B_{ij}^* \cos a_{ij} \geqslant 3A^{*\frac{1}{3}}B^{*\frac{2}{3}}. \tag{3.4}$$

由球面上的余弦公式得到

$$B_{ij}^* = -\sqrt{B_{ii}^*}\sqrt{B_{jj}^*}\cos\beta_{ij}\,(i \neq j), \tag{3.5}$$

于是可将(3.4)改写成

$$\sum_{k=1}^{3} B_{kk}^* - \sum_{i\neq j}\sqrt{B_{ii}^*}\sqrt{B_{jj}^*}\cos a_{ij}\cos\beta_{ij} \geqslant 3A^{*\frac{1}{3}}B^{*\frac{2}{3}}. \tag{3.6}$$

由于条件 1°,$a_{ij} \leqslant b_{ij}$,则 $\cos a_{ij} \geqslant \cos b_{ij}$,又因条件 2°,$\triangle_B$ 是非钝角的,则 $\cos\beta_{ij} \geqslant 0$.于是在(3.6)中用 $\cos b_{ij}$ 代替 $\cos a_{ij}$ 时左端不会减少,即

$$\sum_{k=1}^{3} B_{kk}^* - \sum_{i\neq j}\sqrt{B_{ii}^*}\sqrt{B_{jj}^*}\cos b_{ij}\cos\beta_{ij} \geqslant 3A^{*\frac{1}{3}}B^{*\frac{2}{3}}.$$

亦即

$$\sum_{i=1}^{3}\sum_{j=1}^{3} B_{ij}^* \cos b_{ij} \geqslant 3A^{*\frac{1}{3}}B^{*\frac{2}{3}}. \tag{3.7}$$

另一方面,从直接展开得到

$$\sum_{i=1}^{3}\sum_{j=1}^{3} B_{ij}^* \cos b_{ij} = 3B^*, \tag{3.8}$$

比较(3.7)与(3.8)即得

$$A^* \leqslant B^*. \tag{3.9}$$

另一方面,因 $\cos\frac{a_{ij}}{2} \geqslant \cos\frac{b_{ij}}{2}(i,j=1,2,3)$,于是有

$$\frac{\sqrt{A^*}}{\cos\frac{a_{23}}{2}\cos\frac{a_{31}}{2}\cos\frac{a_{12}}{2}} \leqslant \frac{\sqrt{B^*}}{\cos\frac{b_{23}}{2}\cos\frac{b_{31}}{2}\cos\frac{b_{12}}{2}}, \tag{3.10}$$

由球面三角形面积公式

$$\begin{aligned} \sin\left(\frac{1}{2}\text{area } \triangle_A\right) &= \frac{\sqrt{A^*}}{4\cos\frac{a_{23}}{2}\cos\frac{a_{31}}{2}\cos\frac{a_{12}}{2}}, \\ \sin\left(\frac{1}{2}\text{area } \triangle_B\right) &= \frac{\sqrt{B^*}}{4\cos\frac{b_{23}}{2}\cos\frac{b_{31}}{2}\cos\frac{b_{12}}{2}}, \end{aligned} \tag{3.11}$$

最后得到

$$\sin\left(\frac{1}{2}\text{area } \triangle_A\right) \leqslant \sin\left(\frac{1}{2}\text{area } \triangle_B\right),$$

即 area $\triangle_A \leqslant$ area $\triangle_B$,定理 3.2 证毕.

由于在高维非欧常曲率空间中,单形的体积一般不能表为其诸棱长的初等函数,这里所用的方法不能直接推广于高维,作者提出了下列猜想,希望得到证实或否定:

猜想 对于 n 维球面型空间中的两个单形 $\sum_A^*$, $\sum_B^*$,如果满足条件

1° $a_{ij} \leqslant b_{ij}(i,j=1,2,\cdots,n+1)$,

2° $\sum_B^*$ 是非钝角的,

则有

$$V^*(A) \leqslant V^*(B).$$

参考文献

[1] Blumenthal L M. Theory and Applications of Distance Geometry. Oxford,1953:97－98.

[2]Bottma O,Klamkin M S. Joint Triangle Inequalities. Simon Stevin,1975(48):3－8.

[3]Finsler P,Hadwiger H. Einige Relationen im Dreieck. Comm. Math. Helv,1937(10):316－326.

[4]Pedoe D. An Inequality for two triangles. Proc. Camb. Phil. Soc. ,1942(38):397－398.

[5]Problem E 1562. Amer. Math. Monthly,1963(70):1012.

[6]Thinking Geometrically. Amer. Math. Monthly,1970(77):711－721.

[7]Inside-outside: the Neuberg-Pedoe Inequality. Univ. Begrad. Publ. Elektrotehn. Fak. Ser. Mat. Fiz. ,1976,No. 544－576,95－97.

[8]Carlitz L. An Inequality Involving the Area of Two Triangles. Amer. Math. Monthly, 1971(78):772.

[9] 常庚哲. Pedoe 定理的复数证明. 中学理科教学,1979(2).

[10] Alexander R. Two Notes on Metric Geometry. Proc. Amer. Math. Soc. ,1977,64(2):

317 - 320.
[11]Beckenbach E F, Bellman R. Inequalities. Springer-Verlag, 1961: 10 - 11.
[12]Menger K. New Foundation of Euclidean Geometry. Amer. J. Math., 1931(53): 721 - 745.
[13]杨路，张景中. 关于有限点集的一类几何不等式. 数学学报，1980，23(5)：740 - 749.

有限点集在伪欧空间的等长嵌入

张景中　　杨　　路

（中国科学技术大学）

§0　引　　言

有限点集等长嵌入于欧氏空间 E^n 的问题，首先由Menger[1] 于1928年解决；以后又有人给出各种不同的解法[2~5]. 而有限点集嵌入于伪欧空间的问题，则只见到 Schoenberg[2] 的一种解法. 问题的提法如下：

定义 0.1　指标为 k 的伪欧空间 $E_{m,k}$ 是 m 维的实向量空间；其每个向量的（固定的）某 $(m-k)$ 个分量为实数，其余 k 个为纯虚数或 0.

定义 0.2　一个集 $S=\{p_i\}$，如果有一个定义于 S 上的"距离"$d_{ij}=d(p_i,p_j)$，满足：

(i)　d_{ij}^2 是实数，

(ii)　$d_{ii}=0$，

(iii)　$d_{ij}=d_{ji}$，

则称$\{S,d\}$ 为一个伪欧点集.

定义 0.3　设$\{S,d\}$是一个伪欧点集，若有映射$\mathscr{K}:S\to E_{m,k}$，使 $p_x\in S$对应于$\bar{x}\in E_{m,k}$，这里 $\bar{x}=(x_1,x_2,\cdots,x_m)$，满足

$$d^2(p_x,p_y)=\sum_{i=1}^{m}(x_i-y_i)^2,$$

则称 $\mathscr{K}$ 为 S 到 $E_{m,k}$ 的一个等长嵌入.

所谓嵌入问题，是问：给定了一伪欧点集$\{S,d\}$和非负整数 m,k，是否存在 S 到 $E_{m,k}$ 的等长嵌入?若存在，如何具体给出这样的 $\mathscr{K}$?

本文给出了有限点集可嵌入 $E_{m,k}$ 的另一种（不同于文献[2]）充分必要条件及构造此嵌入 $\mathscr{K}$ 的具体步骤. 这些条件和步骤的特点是：它们紧密地联系着该点集的 Cayley-Menger 阵（简记为 CM 阵，定义见 §2），从而具有较明显的几何意义和几何应用. 这一特点尤其在本文最后一节可以反映出来.

§1　矩阵的次特征值和次特征向量

定义 1.1　设 A 是 $n+1$ 阶复元方阵，若有复数 λ 及非 0 向量 $\bar{u}$，$\bar{u}^\tau=(u_0,u_1,\cdots,u_n)$（$\tau$

本文刊于《数学学报》，第 24 卷第 4 期，1981 年.

表转置运算,下同)满足

$$A\bar{u} = \lambda \begin{bmatrix} 0 \\ u_1 \\ \vdots \\ u_n \end{bmatrix} \tag{1.1}$$

则称 λ 为 A 的次特征值,$\bar{u}^* = (u_1, \cdots, u_n)^\tau$ 为 A 的对应于 λ 的次特征向量.

次特征向量有一些与特征向量类似的性质.

引理 1.1　当且仅当 λ 是多项式

$$\| A(\lambda) \| = \| A - \lambda \widetilde{E} \| \left(\widetilde{E} = \begin{bmatrix} 0,0 \\ 0,E_n \end{bmatrix} \right) \tag{1.2}$$

的根时,λ 是 A 的次特征值. 这里 E_n 是 n 阶么阵.

证明略.

据此,今后对 A 的次特征值即理解为 $\| A(\lambda) \|$ 的根;当 λ 为其 s 重根时,认为是 A 的 s 个相重的次特征值.

引理 1.2　若 $A = A^\tau$,λ,μ 为 A 的次特征值,$\bar{u}^*$,$\bar{v}^*$ 分别是对应于 λ,μ 的次特征向量,若 $\lambda \neq \mu$ 则 $\bar{u}^* \cdot \bar{v}^* = 0$.

引理 1.3　若 $A = [a_{ij}]$ 为 $n+1$ 阶对称阵,而且 $a_{11} = 0$,但 A 的第一行不全为 0;$\check{\lambda}$ 为 $\| A(\lambda) \|$ 的 k 重根,则对应于 $\check{\lambda}$ 的次特征向量之全体构成 k 维线性空间.

这两个引理的证明与线性代数中关于特征向量的性质的证明类似. 故从略.

§2　用次特征向量解伪欧嵌入问题

设 $\{S;d\}$ 是 N 元素伪欧点集,令 $d_{ij} = d(p_i, p_j)(p_i, p_j \in S; i,j = 1, \cdots, N)$. 记 $g_{ij} = -\frac{1}{2} d_{ij}^2$,作方阵

$$P_S = \begin{bmatrix} 0,1 \\ 1,g_{ij} \end{bmatrix}_{(N+1)\times(N+1)}$$

称 P_S 为 S 的 CM 阵(即 Cayley-Menger 阵),由 S 为实对称阵,故 $\| P_S(\lambda) \|$ 只有实根.

由引理 1.2,1.3;从 P_S 的次特征向量中,可选出$(N-1)$个组成正交归范组$\{\bar{u}_i; i = 1,2, \cdots, N-1\}$. 设对应于 $\bar{u}_i$ 的次特征值为 λ_i,令

$$\bar{u}_i = (u_1^i, u_2^i, \cdots, u_N^i)(i = 1,2,\cdots,N-1).$$

取

$$\bar{x}_i = (\sqrt{\lambda_1} u_i^1, \sqrt{\lambda_2} u_i^2, \cdots, \sqrt{\lambda_{N-1}} u_i^{N-1} (i = 1, \cdots, N-1).$$

下面指出,映射 $\mathscr{K}: p_i \to \bar{x}_i$ 恰是 S 到 $E_{m,k}$ 的一个等长嵌入;这里 $m = N-1$,k 是 $\| P_S(\lambda) \|$ 的负根的个数.

事实上,由定义 1.1,对每个 $\bar{u}_i$ 有复数 u_0^i 使

$$P_S\begin{bmatrix}u_0^i\\ -\\ u_i\end{bmatrix}=\lambda_i\begin{bmatrix}0\\ -\\ u_i\end{bmatrix}$$

作 $N+1$ 阶方阵

$$U=\begin{bmatrix}0, & \cdots\cdots\ \cdots\cdots & ,0\\ 0, & \frac{1}{\sqrt{N}},\cdots, & \frac{1}{\sqrt{N}}\\ u_0^1, & \cdots\cdots\ \cdots\cdots & ,u_N^1\\ \vdots & & \vdots\\ u_0^{N-1} & \cdots\cdots\cdots\cdots & ,u_N^{N-1}\end{bmatrix}=\begin{bmatrix}0 & 0 & \cdots & 0\\ 0 & & & \\ u_0^1 & & U^* & \\ \vdots & & & \\ u_0^{N-1} & & & \end{bmatrix}$$

这里 U^* 显然是单位正交阵. 再令

$$A=U^{\tau}UP_SU^{\tau}U.$$

若先求 $B=UP_SU^{\tau}$,可得

(甲) $$A=U^{\tau}BU=\begin{bmatrix}* & * \cdots *\\ * & \\ \vdots & a_{ij}\\ * & \end{bmatrix},\quad\begin{pmatrix}a_{ij}=\bar{x}_i\cdot\bar{x}_j+\dfrac{g}{N},\\ \text{而 } g=\dfrac{1}{N}\sum\limits_{i,j}g_{ij}\end{pmatrix}$$

另一方面

(乙) $$A=(U^{\tau}U)P_S(U^{\tau}U)=\begin{bmatrix}* & * & \cdots & *\\ * & & & \\ \vdots & & g_{ij}+u_i+u_j & \\ * & & & \end{bmatrix}$$

这里,$u_j=\sum\limits_{i=1}^{N-1}u_0^iu_j^i$. 比较(甲)、(乙) 得

$$g_{ij}+u_i+u_j=\bar{x}_i\cdot\bar{x}_j+\frac{g}{N},$$

从而可得

$$(\bar{x}_i-\bar{x}_j)^2=\bar{x}_i\bar{x}_i+\bar{x}_j\bar{x}_j-2\bar{x}_i\bar{x}_j=d_{ij}^2;$$

这证明了 $\mathscr{K}:p_i\rightarrow\bar{x}_i$ 确是$\{S,d\}$ 到 $E_{m,k}$ 的等长嵌入.

注意到,对应于 $\|P_S(\lambda)\|$ 的 0 根的 $\bar{x}_i$ 中的分量为 0,可以抹去而降低 $E_{m,k}$ 之维数,得

定理 2.1 若 N 元素伪欧点集$\{S,d\}$ 的 CM 阵 P_S 的次特征值中有 m 个非 0,k 个负数,则对任意自然数 m',k',当 $k'\geqslant k$ 且 $m'-k'\geqslant m-k$ 时,$\{S,d\}$ 可等长嵌入于 $E_{m',k'}$ 中.

§3 嵌入条件之必要性

设 $\mathscr{K}:p_i\rightarrow\bar{x}_i$ 是 S 到 $E_{n,l}$ 的一个等长嵌入. 不妨设 $\bar{x}_i$ 的前$(n-l)$ 个分量为实,后 l 个为虚. 令 $\bar{x}_i^i=(x_1^i,\cdots,x_n^i)$. 再设原点在集$\{\bar{x}_i\}$ 之重心,因总可通过平移达到这一目的. 作 $N\times n$ 矩阵

$$X=(\bar{x}_1,\bar{x}_2,\cdots,\bar{x}_N),$$

再作 $n\times n$ 方阵

$$Q = XX^{\tau} = \begin{bmatrix} E_{n-1} & 0 \\ 0 & \sqrt{-1}E_l \end{bmatrix} Q^* \begin{bmatrix} E_{n-1} & 0 \\ 0 & \sqrt{-1}E_l \end{bmatrix},$$

则 Q^* 是正半定的，即 Q^* 之特征值全非负，于是 Q 的特征值中，正的不超过 $n-l$ 个，负的不超过 l 个.（s 重根作为 s 个）

下面指出：Q 的非 0 特征值恰与 P_S 之非 0 的次特征值一致，即 $n-l \geqslant m-k, l \geqslant k$.

考虑矩阵

$$\widetilde{X} = \begin{bmatrix} \sqrt{-N} & 1\cdots1 \\ 0 & \\ \vdots & \boxed{X} \\ 0 & \end{bmatrix},$$

由于原点是重心，故

$$\widetilde{X}\widetilde{X}^{\tau} = \begin{bmatrix} 0 & \cdots\cdots & 0 \\ \vdots & & \\ 0 & & \boxed{Q} \end{bmatrix}$$

这表明 $\widetilde{X}\widetilde{X}^{\tau}$ 的非 0 特征值与 Q 的一致. 但是 $\widetilde{X}\widetilde{X}^{\tau}$ 的非 0 特征值又与 $\widetilde{X}^{\tau}\widetilde{X}$ 的相一致，故只要证明 $\widetilde{X}^{\tau}\widetilde{X}$ 的非 0 特征值与 P_S 的相一致已足.

写出

$$\widetilde{X}^{\tau}\widetilde{X} = \begin{bmatrix} -N & \sqrt{-N}\cdots\sqrt{-N} \\ \sqrt{-N} & \\ \vdots & \boxed{\bar{x}_i \cdot \bar{x}_j + 1} \\ \sqrt{-N} & \end{bmatrix} = \widetilde{Q},$$

约定 $\widetilde{Q}$ 之行列标号为 0 至 N；考虑 $\widetilde{Q}$ 的特征多项式 $\widetilde{Q}(\lambda) = \|\widetilde{Q} - \lambda E\|$，把这个行列式

$$\widetilde{Q}(\lambda) = \begin{vmatrix} -(N+\lambda) & \sqrt{-N}\cdots\sqrt{-N} \\ \sqrt{-N} & \\ \vdots & \boxed{\bar{x}_i\bar{x}_j + 1 - \delta_{ij}\lambda} \\ \sqrt{-N} & \end{vmatrix} \left(\delta_{ij} = \begin{cases} 0, & (i \neq j) \\ 1, & (i = j) \end{cases}\right)$$

的 $(1-N)$ 行之和的 $1/\sqrt{-N}$ 倍加到第 0 行；再把所得行列式的第 0 行的 -1 倍加到其余各行；再把第 $(1-N)$ 列之和的 $-\dfrac{1}{N}$ 倍加到第 0 列；变换的结果为：

$$\widetilde{Q}(\lambda) = -\frac{\lambda^2}{N} \begin{vmatrix} 0 & 1 \quad \cdots\cdots \quad 1 \\ 1 & \\ \vdots & \boxed{\bar{x}_i \cdot \bar{x}_j - \delta_{ij}\lambda} \\ 1 & \end{vmatrix}$$

再把这个行列式第 0 行乘以 $-\dfrac{1}{2}\bar{x}_i^2$ 加到第 i 行，第 0 列乘以 $-\dfrac{1}{2}\bar{x}_j^2$ 加到第 j 列 $(i,j = 1, 2,\cdots,N)$，即得

$$\widetilde{Q}(\lambda)=-\frac{\lambda^2}{N}\parallel P_S-\lambda\widetilde{E}\parallel.$$

由此可见,$\widetilde{Q}$ 的非 0 特征值与 P_S 的非 0 的次特征值相一致. 从而把定理 2.1 改进为

定理 3.1 若 N 元素伪欧点集$\{S,d\}$ 的 CM 阵 P_S 的次特征值中有 m 个非 0,k 个为负(s 重根按 s 个计算),则$\{S,d\}$ 可等长嵌入于 $E_{n,l}$ 的充要条件是 $n-l\geqslant m-k$,且 $l\geqslant k$.

依据文献[4],若 S 可等长嵌入于空间 E 而不能等长嵌入于 E 的真子空间,则称 S 能不可约等长嵌入于 E. 由上述定理可得下述在形式上更为整齐之推论:

推论 1 若 N 元素伪欧点集$\{S,d\}$ 的 CM 阵 P_S 的次特征值中有 m 个非 0,k 个为负,则$\{S,d\}$ 能够不可约等长嵌入 $E_{n,l}$ 的充要条件是 $n=m,l=k$.

推论 2 N 元素伪欧点集$\{S,d\}$ 可嵌入于欧氏空间 E_n,其充要条件是它的 CM 阵的次特征值中没有负的,且正的不多于 n 个.

这个条件的不同的等价形式早已在文献[1] 中被得到.

§4 欧氏点集次特征值的几何意义

点集的等长嵌入如果可能,当然不会是唯一的. 今后称 §2 中借助于次特征向量而作的那种嵌入为"标准嵌入". 下面在欧氏空间来弄清楚这种嵌入的几何意义.

若$\{S,d\}$ 对应的 CM 阵的次特征值中 m 个非 0 者全为正,由定理 3.1 知$\{S,d\}$ 可嵌入于欧氏空间 E_m 而不能嵌入于 E_{m-1}. 此时称$\{S,d\}$ 为 m 维欧氏点集. 并常不加区别地亦称其嵌入象为$\{S\}$.

定义 4.1 N 元素欧氏点集$\{S\}$ 的一个"伴随坐标系"是指这样一个笛卡尔直角坐标系:其原点在该点集的重心,其坐标向量为该点集的 CM 阵的次特征向量的一个正交规范组.

这样,欧氏点集的标准嵌入,不过是找出点集中各点关于其伴随坐标系的诸坐标.

次之,定理从某一方面说明了点集的次特征向量、次特征值的几何意义:

定理 4.1 m 维欧氏点集$\{S,d\}$,$S=\{p_1,\cdots,p_N;N>m\}$,其 CM 阵的各个非 0 次特征值恰为 S 关于过其重心的超平面的转动惯量的稳定值.

(这里,与通常用语一致,把 S 中各点到某超平面距离的平方和称为点集 S 关于该超平面的转动惯量.)

证 设 S 的重心为 O,以 O 为原点任取笛卡尔坐标系,设各点坐标为

$$p_i=(x_1^i,x_2^i,\cdots,x_m^i)(i=1,\cdots,N),$$

由于重心在原点,故

$$\sum_{i=1}^{N}x_j^i=0(j=1,\cdots,m).$$

设 e 为任一过原点之超平面,e 的单位法向量记作 $\bar{x}^e=(x_1^e,x_2^e,\cdots,x_m^e)$,则 p_i 到 e 的带号距离为

$$h_i=h(p_i,e)=x_1^ix_1^e+\cdots+x_m^ix_m^e.$$

记 $\bar{h}=(h_1,h_2,\cdots,h_N)$,则 S 关于 e 之转动惯量为

$$I_e=h_1^2+h_2^2+\cdots+h_N^2=\bar{h}\bar{h}^{\tau}.$$

在 e 的过原点的法线上取一点 $\bar{x}$，使 $|\bar{x}| = \frac{1}{\sqrt{I_e}}$，考虑 $\bar{x}$ 的轨迹；由于 $|\bar{x}||\bar{x}| I_e = 1$，故 $|\bar{x}| \bar{x}^e = \bar{x}$，若令

$$X = [X_i^j]\begin{pmatrix} i = 1,\cdots,m \\ j = 1,\cdots,N \end{pmatrix}$$

记 m 行 n 列矩阵，得

$$\bar{x}XX^\tau\bar{x}^\tau = |\bar{x}| \bar{x}^e XX^\tau(\bar{x}^e)^\tau |\bar{x}| = |\bar{x}|^2 I_e = 1.$$

再令 $Q = XX^\tau$，得二次超曲面方程

$$\bar{x}Q\bar{x}^\tau = 1,$$

记此二次超曲面为 Γ. 按熟知的二次型理论，可知 Γ 的各轴的平方是 Q 的非 0 特征值的倒数. 换言之，由 Lagrange 乘子法，$|\bar{x}|^2 = \frac{1}{I_e}$ 当 $\bar{x}^e$ 变化时所取之诸稳定值，是 Q 的非 0 特征值的倒数. 即 I_e 的诸稳定值是 Q 的诸非 0 特征值. 由定理 3.1 的证明中我们知道 Q 的非 0 特征值恰与 S 的 CM 阵的非 0 次特征值一致. 证毕.

显见有

推论 4.1　有限元素欧氏点集关于其重心的惯量椭球是一个球的充要条件是：其 CM 阵的诸非 0 次特征值相等. 这里，球的维数即为非 0 的次特征值之个数.

推论 4.2　m 维欧氏点集 S 的 CM 阵的非 0 的次特征值 $\lambda_1 \geqslant \lambda_2 \geqslant \cdots \geqslant \lambda_m$ 和 S 的中心惯量椭球诸半轴 $b_1 \geqslant b_2 \geqslant \cdots \geqslant b_m$ 之间有下列关系：

$$b_i^2 = \frac{1}{(\lambda_1 + \cdots + \lambda_m) - \lambda_i}(i = 1,2,\cdots,m).$$

此二推论之证明略. 下面把一个欧氏点集叫做是"惯量等轴"的，如果它的中心惯量椭球是一个球，作为标准嵌入之应用，我们有

定理 4.2　E_m 中点集 $S = \{p_1,\cdots,p_N; N > m\}$ 为惯量等轴的充要条件是：S 是 E_m 的扩空间 E_{N-1} 中某个正单纯形顶点之集 $S^* = \{q_1,\cdots,q_N\}$ 在 E_m 上的正投影.

证　条件的充分性是显然的，下面证条件的必要性.

取 S 在 E_m 中的伴随坐标系，设 S 中各点在此系中之坐标为

$$p_i = (x_1^i, x_2^i, \cdots, x_m^i)(i = 1,2,\cdots,N)$$

这时，$\bar{v}_j = (x_j^1, x_j^2, \cdots, x_j^N)(j = 1,\cdots,m)$ 应为 S 的 CM 阵的次特征向量的一个正交无关组：当 $i \neq j$ 时 $\bar{v}_i \cdot \bar{v}_j = 0$，而 $\bar{v}_j^2 = \lambda_j$，λ_j 是非 0 的次特征值.

因为 S 是惯量等轴的，故 m 个非 0 的次特征值应当相同. 令

$$v_0 = \left(\sqrt{\frac{\lambda}{N}},\cdots,\sqrt{\frac{\lambda}{N}}\right),$$

这里，v_0 为 N 维向量且 $\lambda = \lambda_1 = \cdots = \lambda_m$. 于是得到一个正交组 $\bar{v}_0, \bar{v}_1, \cdots, \bar{v}_m$，组中向量之长度均为 $\sqrt{\lambda}$. 把此组扩充为 N 个向量的正交组 $\bar{v}_0, \bar{v}_1, \cdots, \bar{v}_{N-1}$，并设这 N 个向量长度相同. 考虑 E_{N-1} 中点集 $S^* = \{q_i, i = 1,2,\cdots,N\}$，这里

$$q_i = (x_1^i, x_2^i, \cdots, x_{N-1}^i),$$

显然，S^* 在 E_m 上的正投影就是 S.

为说明 S^* 是正单形顶点之集，我们来计算其中任意两点之距离，令

$$\bar{u}_i = \left(\sqrt{\frac{\lambda}{N}}, x_1^i, \cdots, x_{N-1}^i\right),$$

则诸 $\bar{u}_i$ 也构成正交组，当 $i \neq j$ 时有

$$|q_i - q_j|^2 = |\bar{u}_i - \bar{u}_j|^2 = \bar{u}_i^2 - 2\bar{u}_i\bar{u}_j + \bar{u}_j^2 = 2\lambda,$$

即 S^* 中不同的任两点之距离均为 $\sqrt{2\lambda}$，证毕.

根据惯量椭球的定义容易验证，任何维数的欧氏空间中的正多面体的顶点集都是惯量等轴的，因而它们都是更高维空间中正单纯形顶点集的正投影. 例如，E_3 中的正六面体、正八面体、正十二面体、正二十面体的顶点集，分别是 E_7, E_5, E_{19}, E_{11} 中正单形顶点集在 E_3 中的正投影；而 E_m 中正方体顶点之集，则是 E_{2^m-1} 中正单形顶点集在 E_m 中的正投影；等等.

应当指出，作者在文献[6]中也曾引进过“惯量等轴”的定义，它与本文所引进的定义是完全等价的. 作者在文献[6]中对有限点集$\mathfrak{S}$建立了一类几何不等式，这些不等式等号成立的充分必要条件是：“$\mathfrak{S}$是惯量等轴的.”因此，根据本节定理4.2，我们可以将文献[6]中的定理改述为如下形式：

定理 4.3 设$\mathfrak{S} = \{p_1, p_2, \cdots, p_N\}$是 E_m 中的有限点集，任取$\mathfrak{S}$的$(k+1)$个点为顶点作 k 维单形，将所有这样的 k 维单形的 k 维体积的平方和记为 $N_k (k = 1, 2, \cdots, m)$. 则有

$$\frac{N_k^l}{N_l^k} \geqslant \frac{[(m-l)!(l!)^3]^k}{[(m-k)!(k!)^3]^l}(m!N)^{1-k}(1 \leqslant k < l \leqslant m),$$

$$N_k^2 \geqslant \left(\frac{k+1}{k}\right)^3 \cdot \frac{m-k+1}{m-k} \cdot N_{k-1}N_{k+1}(1 \leqslant k \leqslant m, N_0 = N),$$

而等号成立的充分必要条件是：$\mathfrak{S}$是 E_m 的扩空间E_{N-1}中某个单纯形顶点之集$\mathfrak{S}^* = \{q_1, \cdots, q_N\}$在 E_m 上的正投影.

参考文献

[1] Menger K. Untersuchungen über allgemeine Metrik. Math. Ann. ,1928(100):75－163.

[2]Schoenberg I J. Remarks to Maurice Fréchet's article “Sur la définition…”. Ann. of Math. ,1935(36):724－732.

[3]Wilson W A. On the Imbedding of Metric Sets in Euclidean Space. Amer J. Math. , 1935(57):322－326.

[4]Blumenthal L M. Theory and Applications of Distance Geometry. 2nd ed. New York,1970.

[5] 杨路，张景中. 单纯形构造定理的一个证明. 数学的实践与认识，1980(1):43－45.

[6] 杨路，张景中. 关于有限点集的一类几何不等式. 数学学报，1980,23(5):740－749.

高维度量几何的两个不等式

杨　路　张景中

（中国科学技术大学）

本文建立了高维度量几何中与“度量加”概念有关的两个几何不等式——定理 2 和定理 3. 并为以上目的，事先引进了作为分析工具的定理 1.

在近期文献中发展了将矩阵不等式和二次型理论应用于度量几何的技巧，例如 R. Alexander 等[1~5] 的一系列工作. 几何的困难往往是维数的困难，高维情形难于凭借直观，更多依赖于抽象的工具，同时也促进了抽象理论的发展. 作者的定理 1 是较 Bergstrom 不等式更强的结果，它的应用范围并不限于文中所述的两个问题.

§1　Bergstrom 定理的改进

H. Bergstrom[6] 在一篇题为《*A Triangle Inequality for Matrices*》的文章中曾经建立下述命题：

定理　设 A 和 B 是两个正定矩阵，$|A|$ 和 $|B|$ 表示它们的行列式，$|A|_{ij}$，$|B|_{ij}$ 等表示对应的代数余子式，则有

$$\frac{|A+B|}{|A+B|_{ij}} \geqslant \frac{|A|}{|A|_{ij}} + \frac{|B|}{|B|_{ij}}. \tag{1}$$

这个定理的证明可参看文献[7]或[8]. 下面将要证明，Bergstrom 定理可进一步改进为如下的

定理 1　设 $A=(a_{ij})$ 和 $B=(b_{ij})$ 是 n 阶实对称矩阵，而且二次型 $\sum\sum a_{ij}x_i x_j$ 和 $\sum\sum b_{ij}x_i x_j$ 在约束条件

$$c_1x_1 + c_2x_2 + \cdots + c_nx_n = 0 \tag{2}$$

之下都是正定（或负定）的. 又分别用 $\overline{A}$，$\overline{B}$，$\overline{A+B}$ 表示如下的三个镶边矩阵：

$$\overline{A} = \begin{pmatrix} o & c_1\,c_2\cdots c_n \\ c_1 & \\ c_2 & A \\ \vdots & \\ c_n & \end{pmatrix}, \overline{B} = \begin{pmatrix} o & c_1\,c_2\cdots c_n \\ c_1 & \\ c_2 & B \\ \vdots & \\ c_n & \end{pmatrix}, \tag{3}$$

本文刊于《成都科技大学学报》，第 4 期，1981 年.

$$\overline{A+B}=\begin{pmatrix} o & c_1c_2\cdots c_n \\ \begin{matrix} c_1 \\ c_2 \\ \vdots \\ c_n \end{matrix} & \boxed{A+B} \end{pmatrix},$$

则成立下列不等式:

$$\frac{|A+B|}{|\overline{A+B}|}\leqslant\frac{|A|}{|\overline{A}|}+\frac{|B|}{|\overline{B}|}\left(\text{或}\frac{A+B}{|\overline{A+B}|}\geqslant\frac{|A|}{|\overline{A}|}+\frac{|B|}{|\overline{B}|}\right). \tag{4}$$

在定理 1 中取

$$C_k=\begin{cases}1 & (k=i),\\ 0 & (k\neq i)\end{cases}\quad (k=1,2,\cdots,n), \tag{5}$$

我们就立即得到 Bergstrom 定理. 这说明定理 1 确实是一个更强的命题.

为证明定理 1,我们先建立如下的

引理 1 设二次型$\sum\sum a_{ij}x_ix_j$在约束条件

$$c_1x_1+c_2x_2+\cdots+c_nx_n=0 \tag{6}$$

之下是正定(或负定)的,则该二次型在条件

$$c_1x_1+c_2x_2+\cdots+c_nx_n=1 \tag{7}$$

之下的最小(或最大)值,可表示为

$$\underset{\sum c_kx_k=1}{\text{Min}}\left(\sum\sum a_{ij}x_ix_j\right)=-\frac{|A|}{|\overline{A}|}. \tag{8}$$

$$\left(\text{或}\underset{\sum c_kx_k=1}{\text{Max}}\left(\sum\sum a_{ij}x_ix_j\right)=-\frac{|A|}{|\overline{A}|}\right).$$

证 对该二次型使用 Lagrange 乘子法,引进辅助函数

$$F=\sum\sum a_{ij}x_ix_j+2\lambda(c_1x_1+c_2x_2+\cdots+c_nx_n)$$

并对诸 x_k 求导,得方程组

$$\sum_{i=1}^{n}a_{ik}x_i+\lambda c_k=0(k=1,2,\cdots,n). \tag{9}$$

再加以约束条件(7),即可解出唯一的驻点:

$$x_1^*=-\frac{1}{|\overline{A}|}\cdot\begin{vmatrix} c_1 & a_{12} & \cdots & a_{1n}\\ c_2 & a_{22} & \cdots & a_{2n}\\ \vdots & \vdots & \cdots & \vdots\\ c_n & a_{n2} & \cdots & a_{nn}\end{vmatrix},x_2^*=-\frac{1}{|\overline{A}|}\cdot\begin{vmatrix} a_{11} & c_1 & \cdots & a_{1n}\\ a_{12} & c_2 & \cdots & a_{2n}\\ \vdots & \vdots & \cdots & \vdots\\ a_{n1} & c_n & \cdots & a_{nn}\end{vmatrix}, \tag{10}$$

$$\cdots,x_n^*=-\frac{1}{|\overline{A}|}\cdot\begin{vmatrix} a_{11} & a_{12} & \cdots & c_1\\ a_{12} & a_{22} & \cdots & c_2\\ \vdots & \vdots & \cdots & \vdots\\ a_{n1} & a_{n2} & \cdots & c_n\end{vmatrix}.$$

而二次型在这个驻点取值为

$$\sum\sum a_{ij}x_i^* x_j^* = -\frac{|A|}{|\overline{A}|}. \tag{11}$$

用配方法容易证明，当二次型在超平面 $\sum c_k x_k = 0$ 上正定(或负定) 时，它在条件 $\sum c_k x_k = 1$ 下必取到最小(或最大) 值. 那么，由于它只有唯一的驻点(10)，故这个最小(或最大) 值一定由(8) 式给出. 引理 1 证毕.

我们顺便指出，将引理 1 直接应用于几何学的一个例. 设 $P_1, P_2, \cdots, P_n$ 是 E^{n-1} 中一个单形的顶点，令 $\rho_{ij} = |P_i - P_j|$ $(i, j = 1, 2, \cdots, n)$ 表示单形各棱的长度，又以 $a_{ij} = \rho_{ij}^2$ 为元素构成 n 阶矩阵 A. 对于这样构成的 A，如果再取 $c_1 = c_2 = \cdots = c_n$ 并按(3) 的意义构作 $\overline{A}$，我们来看看引理 1 反映什么几何事实.

首先，二次型 $\sum\sum a_{ij}x_i x_j = \sum\sum \rho_{ij}^2 x_i x_j$ 在超平面 $x_1 + x_2 + \cdots + x_n = 0$ 上是负定的，这是 I. J. Schoenberg 的一个经典性的结果. 可参看文献[1] 或[9]. 于是我们可以援用引理 1 而得到

$$\underset{\sum x_k = 1}{\mathrm{Max}}\left(\sum\sum a_{ij}x_i x_j\right) = -\frac{|A|}{|\overline{A}|}. \tag{12}$$

其次，我们来看看这个等式右端有什么几何意义. 这里 $|\overline{A}|$ 恰好正是点集 $P_1, P_2, \cdots, P_n$ 的所谓"Gayley-Menger 行列式"(参看文献[10] 或[11]). 不难看出 $-|A|/|\overline{A}|$ 应等于单形 $P_1P_2\cdots P_n$ 的外接球半径平方的 2 倍，从而有：

推论 1　设单形 $P_1P_2\cdots P_n$ 的外接球半径为 R，则有

$$\underset{\sum x_k = 1}{\mathrm{Max}}\left(\sum_{i<j}\rho_{ij}^2 x_i x_j\right) = R^2. \tag{13}$$

证　由 A. Cayley 的一个定理(参看文献[11] 或[12])，单形的诸顶点 $P_1, P_2, \cdots, P_n$ 与外接球心之间的诸距离 ρ_{ij}^2, R 等满足下列方程：

$$\begin{vmatrix} 0 & 1 & 1 \cdots 1 & 1 \\ 1 & & & R^2 \\ 1 & & \rho_{ij}^2 & R^2 \\ \vdots & & & \vdots \\ 1 & & & R^2 \\ 1 & R^2 & R^2 \cdots R^2 & 0 \end{vmatrix} = 0. \tag{14}$$

将(14) 展开即得到

$$2R^2 = -\frac{|A|}{|\overline{A}|}. \tag{15}$$

将(15) 代入(12)，并考虑到 $\sum\sum a_{ij}x_i x_j = 2\sum_{i<j} a_{ij}x_i x_j$，就有

$$\underset{\sum x_k = 1}{\mathrm{Max}}\left(\sum_{i<j}\rho_{ij}^2 x_i x_j\right) = R^2.$$

推论 1 证毕.

就结论而言，推论 1 并不是新的，也存在着不依赖于引理 1 的证明. 然而引理 1 能够以这

种方式应用于几何题目却是可注意的.

下面我们来完成定理 1 的证明:

证 考虑二次型

$$\sum\sum(a_{ij}+b_{ij})x_ix_j=\sum\sum a_{ij}x_ix_j+\sum\sum b_{ij}x_ix_j. \tag{16}$$

由假设,等号右边两个二次型在超平面 $\sum c_kx_k=0$ 上是正(负)定的,因而左边的二次型也具有同样性质. 于是这三个二次型在条件 $\sum c_kx_k=1$ 之下分别取到极小(或极大)值(根据引理 1)

$$-\frac{|A+B|}{|\overline{A+B}|},-\frac{|A|}{|\overline{A}|},-\frac{|B|}{|\overline{B}|}.$$

另一方面,假设自变数的某一组值 $x_1^0,x_2^2,\cdots,x_n^0$ 在规定约束条件之下使 $\sum\sum(a_{ij}+b_{ij})x_ix_j$ 取到最小值,则有

$$\begin{aligned}-\frac{|A+B|}{|\overline{A+B}|}&=\sum\sum(a_{ij}+b_{ij})x_i^0x_j^0 \qquad (17)\\&=\sum\sum a_{ij}x_i^0x_j^0+\sum\sum b_{ij}x_i^0x_j^0\\&\geqslant-\frac{|A|}{|\overline{A}|}-\frac{|B|}{|\overline{B}|},\end{aligned}$$

即

$$\frac{|A+B|}{|\overline{A+B}|}\leqslant\frac{|A|}{|\overline{A}|}+\frac{|B|}{|\overline{B}|}.$$

对于负定的情形据同理可证:

$$\frac{|A+B|}{|\overline{A+B}|}\geqslant\frac{|A|}{|\overline{A}|}+\frac{|B|}{|\overline{B}|}.$$ 定理 1 证毕.

下节将述及如何将这个似属代数或分析的命题应用于几何不等式.

§2 两个几何不等式

R. Alexander[2] 在近期一篇文章中强调了两个点集“度量加”的概念. 这个概念可以严格定义如下:

定义 设 $\mathscr{A}=\{P_1,P_2,\cdots,P_n\}$ 和 $\mathscr{B}=\{Q_1,Q_2,\cdots,Q_n\}$ 是欧氏空间或 Hilbert 空间中的两个点集,则在 Hilbert 空间(甚至欧氏空间) 中必存在一个点集 $\{S_1,S_2,\cdots,S_n\}$,使得

$$|S_i-S_j|^2=|P_i-P_j|^2+|Q_i-Q_j|^2(i,j=1,2,\cdots,n). \tag{18}$$

这个存在性是 I. J. Schoenberg 早就证明了的. 我们将任何一个这样的点集叫做前两者的“度量和”,并记为 $\mathscr{A}^*+\mathscr{B}=\{S_1,S_2,\cdots,S_n\}$.

考虑 $\mathscr{A},\mathscr{B}$ 与度量和 $\mathscr{A}^*+\mathscr{B}$ 的某些不变量之间的联系是很自然的事,R. Alexander 提出[2],如果用 V,V',V'' 表示由 $\mathscr{A},\mathscr{B},\mathscr{A}^*+\mathscr{B}$ 所生成的$(n-1)$ 维单形的体积,他猜想有不等式:

$$V''^2\geqslant V^2+V'^2. \tag{19}$$

作者已经指出文献[11]Alexander 上述猜想不是真的,同时给出了正确的不等式:

$$V''^{\frac{2}{n-1}} \geqslant V^{\frac{2}{n-1}} + V'^{\frac{2}{n-1}} \tag{20}$$

代替单形的体积. 我们现在考虑$\mathscr{A}$,$\mathscr{B}$及其度量和$\mathscr{A}^* + \mathscr{B}$的外接球半径$R$,$R'$和$R''$,我们将证明

定理 2
$$R''^2 \leqslant R^2 + R'^2 \tag{21}$$

证　令 $\rho_{ij} = |P_i - P_j|$,$\rho'_{ij} = |Q_i - Q_j|$,$\rho''_{ij} = |S_i - S_j|$.
矩阵 $A = (\rho_{ij}^2)$,$B = (\rho'^2_{ij})$,$A + B = (\rho''^2_{ij}.)$ 根据(15) 我们有:

$$2R^2 = -\frac{|A|}{|\overline{A}|}, 2R'^2 = -\frac{|B|}{|\overline{B}|}, 2R''^2 = -\frac{|A+B|}{|\overline{A+B}|}, \tag{22}$$

其中镶边的各数 $c_1 = c_2 = \cdots = 1$.

另一方面,由于二次型$\sum\sum \rho_{ij}^2 x_i x_j$ 和$\sum\sum \rho'^2_{ij} x_i x_j$ 在超平面$\sum x_k = 0$ 上是负定的(同前可见文献[9]),由定理 1 有

$$\frac{|A+B|}{|\overline{A+B}|} \geqslant \frac{|A|}{|\overline{A}|} + \frac{B}{|\overline{B}|}.$$

再将(22) 代入,就得到

$$R''^2 \leqslant R^2 + R'^2.$$ 定理 2 证毕.

此外,在$\mathscr{A}$,$\mathscr{B}$或$\mathscr{A}^* + \mathscr{B}$所生成的单形中,经过顶点 P_i,Q_i 或 S_i 的高线长度分别以 h_i,h'_i或h''_i 来表示(设 $i = 0,1,\cdots,n$,这时各单形都是 n 维的). 则我们有:

定理 3
$$h''^2_i \geqslant h_i^2 + h'^2_i (i = 0,1,\cdots,n). \tag{23}$$

证　引进符号

$$\alpha_i = P_i - P_0, \beta_i = Q_i - Q_0, \sigma_i = S_i - S_0 \tag{24}$$

表示 n 维向量($i = 1,2,\cdots,n$).

用 $|A|$ 和 $|B|$ 表示($\alpha_1,\alpha_2,\cdots,\alpha_n$) 和($\beta_1,\beta_2,\cdots,\beta_n$) 的 Gram 行列式:

$$|A| = \begin{vmatrix} (\alpha_1,\alpha_1)(\alpha_1,\alpha_2)\cdots(\alpha_1,\alpha_n) \\ (\alpha_2,\alpha_1)(\alpha_2,\alpha_2)\cdots \\ \vdots \qquad \vdots \quad \cdots \quad \vdots \\ (\alpha_n,\alpha_1)(\alpha_n,\alpha_2)\cdots(\alpha_n,\alpha_n) \end{vmatrix}, \tag{25}$$

$$|B| = \begin{vmatrix} (\beta_1,\beta_1)(\beta_1,\beta_2)\cdots(\beta_1,\beta_n) \\ (\beta_2,\beta_1)(\beta_2,\beta_2)\cdots \\ \vdots \qquad \vdots \qquad \vdots \\ (\beta_n,\beta_1)(\beta_n,\beta_2)\cdots(\beta_n,\beta_n) \end{vmatrix}.$$

这时候,容易验证,$|A+B|$ 恰好是($\sigma_1,\sigma_2,\cdots,\sigma_n$) 的 Gram 行列式:

$$|A+B| = \begin{vmatrix} (\sigma_1,\sigma_1)(\sigma_1,\sigma_2)\cdots(\sigma_1,\sigma_n) \\ (\sigma_2,\sigma_1)(\sigma_2,\sigma_2)\cdots \\ \vdots \qquad \vdots \qquad \vdots \\ (\sigma_n,\sigma_1)(\sigma_n,\sigma_2)\cdots(\sigma_n,\sigma_n) \end{vmatrix}. \tag{26}$$

这是因为　　$(\alpha_i,\alpha_j) = \frac{1}{2}(|P_i - P_0|^2 + |P_j - P_0|^2 - |P_i - P_j|^2)$,

$$(\beta_i,\beta_j)=\frac{1}{2}(|Q_i-Q_0|^2+|Q_j-Q_0|^2-|Q_i-Q_j|^2),$$

以及 $$(\sigma_i,\sigma_j)=\frac{1}{2}(|S_i-S_0|^2+|S_j-S_0|^2-|S_i-S_j|^2).$$

现在我们将三个行列式镶边,镶边元素的取法仍按(5):

$$C_k=\begin{cases}1 & (k=i),\\ 0 & (k\neq i)\end{cases}\quad (k=1,2,\cdots,n),$$

于是立即得到

$$|\overline{A}|=-|A|_{ii},\ |\overline{B}|=-|B|_{ii},\ |\overline{A+B}|=-|A+B|_{ii} \tag{27}$$

(这里的 i 是一个固定的自然数,$i\leqslant n$).

然后来看一看(25) ~ (27) 中各行列式的几何意义. 若以 V,V',V'' 表示单形 $P_0P_1\cdots P_n$,$Q_0Q_1\cdots Q_n$ 和 $S_0S_1\cdots S_n$ 的体积,V_i,V'_i,V''_i 分别表示在各单形中与 P_i,Q_i,S_i 相对的表面(均为($n-1$) 维单形) 的体积,由 Gram 行列式的几何意义有

$$\begin{aligned}&|A|=(n!V)^2,\ |A|_{ii}=((n-1)!V_i)^2,\\ &|B|=(n!V')^2,\ |B|_{ii}=((n-1)!V'_i)^2,\\ &|A+B|=(n!V'')^2,\ |A+B|_{ii}=(n-1)!V''_i)^2\end{aligned} \tag{28}$$

应用定理 1,将(26) ~ (27) 代入(4) 式就得到

$$n^2\cdot\left(\frac{V''}{V''_i}\right)\geqslant n^2\cdot\left(\frac{V}{V_i}\right)^2+n^2\cdot\left(\frac{V'}{V'_i}\right)^2.$$

化简后有

$$h''^2_i\geqslant h_i^2+h'^2_i.$$ 定理 3 证毕.

显然,定理 3 的证明不一定必须用定理 1,而仅仅援用较弱的 Bergstrom 不等式就可以了.

定理2和定理3的最低维(2维)情形是由 A. Oppenheim首先提出和解决的,见文献[13]和[14]. 高维的讨论过去尚未见到.

最后,我们提及与"覆盖半径" 有关的一个问题. 一个点集的"覆盖半径",是指所有覆盖这个点集的各球的半径的最小值,覆盖半径的概念显然比外接球半径重要得多. 因为一般的点集往往不能生成非退化单形,这时外接球半径是无穷大;另一方面凡有界点集的覆盖半径必为确定的有限值.

下列命题是真的,但此地不加证明了.

定理 4 设 $\mathscr{A},\mathscr{B}$ 和 $\mathscr{A}^*+\mathscr{B}$ 表示两个点集及其度量和,令 $\mathscr{R},\mathscr{R}'$ 和 $\mathscr{R}''$ 依次表示此三个点集的覆盖半径,则总有

$$\mathscr{R}''^2\leqslant\mathscr{R}^2+\mathscr{R}'^2. \tag{29}$$

参考文献

[1] Alexander R,et al. Extremal Problems of Distance Geometry…. Trans. Amer. Math. Soc. ,1974(193):1-31.

[2]Alexander R. Metric Embedding Techniques Applied to Geometric Inequalies. Lecture

Notes in Math. ,no. 490. The Geometry of Metric and Linear Spaces,Springer-Verlag,1975:57 - 65.

[3]Alexander R. Two Notes on Metric Geometry. Proc. Amer. Math. ,1977,64(2):317 - 320.

[4]Alexander R. Generalized Sums of Distance. Pac. J. Math. ,1975(56):297 - 304.

[5]Alexander R. On the Sum of Distance between *n* Points on a Sphere. Acta. Math. Acad. Sci. Hunger. ,1972(23):443 - 448.

[6]Bergstrom H. A Triangle Inequality for Matrices,Den Elfte Skandinaviski Matematiker-kongress,Trondheim,1949,Oslo. Johan Grundt Tanums,1952.

[7]Beckenbach E F,Bellman R. Inqualities. Springer-Verlag,1961,Ch. Ⅱ,§17.

[8]Bellman R. An Inequality due to Bergstrom. Amer,Math. Monthly,1955(62):172 - 173.

[9] 杨路,张景中.单纯形构造定理的一个证明.数学的实践与认识,1980(1):43 - 45.

[10]Blumenthal L M. Theory and Applications of Distance Geometry. Oxford,1953.

[11] 杨路,张景中.关于有限点集的一类几何不等式.数学学报,1980,23(5):740 - 749.

[12] 杨路,张景中.抽象距离空间的秩的概念.中国科学技术大学学报,1980,10(4):52 - 62.

[13]Oppenheim A. Inequalities Involving Elements of Triangles,Quadrilaterals or Tetraheaedra. Univ. Beograd. Publ. Elektrotehn. Fak. Ser. Mat. F_{1z}. ,1974(461 - 497):257 -263.

[14]Oppenheim A. Advanced problem 5092. Amer. Math. Monthly,1963(70):444.

关于 Smale 马蹄及其 Ω 稳定性

张景中　　杨路

（中国科学技术大学）

由 Smale 所发现的马蹄型微分同胚是微分动力系的研究中的重要反例之一. 其意义及原始说明见文献[1]. 文献在[2]中用了整整一章的篇幅来证明二维马蹄的基本性质及其 Ω 稳定性，长达 30 余页. 这样一个具有基本意义的例子，其几何构造直观上相当简单，而分析的证明竟如此冗长，自然不能令人满意.

作者在文献[3]中曾给出一个简单的马蹄模型，但没有给出 Ω 稳定性的证明. 本文将对高维的一般情形，构造出多分支的马蹄，用较简单的方法，导出其基本性质和 Ω 稳定性.

§1　记号及定义

E^n 记 n 维欧氏空间，M 是 E^{m+1} 中的开集，$M \supset N = D^m \times [-(1+\delta), 1+\delta]$；这里 $\delta > 0$，D^m 是 E^m 中的圆盘而 $[a,b]$ 记 R^1 的闭区间. 又设

$$-1 < h_1 < h_2 < \cdots < h_{2k} < 1 (k \text{ 正整数}),$$
$$\sigma_j = 1 \text{ 或 } -1 (j = 1, 2, \cdots, k).$$

并记

$$S_j = D^m \times [h_{2j-1}, h_{2j}], S = \bigcup_j S_j,$$
$$(1 \leqslant j \leqslant k)$$
$$\partial^+ S_j = D^m \times \{h_{2j}\}, \partial^- S_j = D^m \times \{h_{2j-1}\}.$$

投射 $\pi: N \to R^1$ 为

$$\pi(x, y) = y (x \in D^m, y \in [-(1+\delta), 1+\delta]).$$

用以上记号，引入

定义 1　设 $0 < \varepsilon < 1$，微分同胚 $H: M \to M$ 叫做 ε- 马蹄，如果

(1) $H(S_j) \subset \operatorname{int} N$，且 $\sigma_j \pi [H(\partial^+ S_j)] > 1$，$\sigma_j \pi [H(\partial^- S_j)] < -1 (1 \leqslant j \leqslant k)$.

(2) 对 $x \in S$，有 $H(x) = L(x) + r(x)$，这里 $\| r(x) \| < \varepsilon$，$\| Dr(x) \| < \varepsilon$，而 $L(x) = (\lambda_j(x_1), \mu_j(x_2)) (x = (x_1, x_2)) \in S_j$，其中 $x_1 \in D^m$，$x_2 \in [-(1+\delta), 1+\delta]$. λ_j，μ_j 分别是 E^m、E^1 到自身的非退化线性映射，且有 $0 < p < q < 1$，使得 $P \| x_1 \| \leqslant \| \lambda_j(x_1) \| \leqslant q \| x_1 \|$，$q^{-1} \| x_2 \| \leqslant \| \mu_j(x_2) \| \leqslant P^{-1} \| x_2 \|$.

(3) $\exists P \in M/N$，M/N 表集 M，N 之差，P 有球形邻域 $B_i(P)$，使 H 在 $B_i(P)$ 上是以 P 为

本文刊于《数学进展》，第 10 卷第 2 期，1981 年.

不动点的压缩映象，满足

$$H(\overline{M/S}) \subset M/N,$$ ①

且有 l 使 $H^l(M/S) \subset B_{\frac{i}{2}}(P)$.

由定义 1 可知，若给定的某 H 是 ε- 马蹄，则当 $\varepsilon' > 0$ 足够小时，H 的 ε'—c^1 扰动是 ε_1 - 马蹄. 这里 $\varepsilon_1 = \varepsilon + \varepsilon'$.

定义 2 若 $T \subset N$，T 和任一个满足 $\pi(W) \supset [-1,1]$ 的 N 的连通子集 W 之交均非空，称 T 为隔集.

定义 3 若 $T \subset N$，T 包含连通子集 W 使 $\pi(W) \supset [-1,1]$，则称 T 为穿集.

定义 4

$$A_{i_0,i_1,\cdots,i_l} = \{x \mid H^{\alpha}(x) \in S_{i_\alpha}, 0 \leqslant \forall \alpha \leqslant l\},$$

$$A^{j_1,j_2,\cdots,j_l} = \{x \mid H^{-\beta}(x) \text{ 有定义}, \in S_{j_\beta}, 1 \leqslant \forall \beta \leqslant l\},$$

$$A^{j_1,j_2,\cdots,j_n}_{i_0,i_1,\cdots,i_l} = A^{j_1,j_2,\cdots,j_n} \cap A_{i_0,i_1,\cdots,i_l}.$$

（i_α, j_β 都是 1 到 k 的正整数）.

用文献[2]中所述几何构成方式，易知对某些 $\{\sigma_j\}$，定义 1 中的 H 存在. 这里不再赘述.

§2 几个引理

引理 1 ⅰ）$A_{i_0,i_1,\cdots,i_l}$，$A^{j_1,j_2,\cdots,j_l}$ 都是闭集.

ⅱ）$A_{i_0,i_1,\cdots,i_{l+1}} \subset A_{i_0,i_1,\cdots,i_l}$，$A^{j_1,j_2,\cdots,j_{l+1}} \subset A^{j_1,j_2,\cdots,j_l}$.

ⅲ）$H(A^{j_1,j_2,\cdots,j_n}_{i_0,i_1,\cdots,i_l}) = A^{i_0,j_1,\cdots,j_n}_{i_1,i_2,\cdots,i_l}$.

ⅳ）$A^{j_1,j_2,\cdots,j_n}_{i_0,i_2,\cdots,i_l} = H^{-1}(A^{j_0,j_1,\cdots,j_n}_{i_1,i_2,\cdots,i_l})$.

证 由 H 是同胚及 S_i 为闭集，得 ⅰ）；由定义 4，ⅱ）显然. 下面证 ⅲ）.

若 $x \in A^{j_1,j_2,\cdots,j_n}_{i_0,i_1,\cdots,i_l}$，则 $\alpha = 1,2,\cdots,l$，$\beta = 1,2,\cdots,n$，有

$$H^{\alpha-1}[H(x)] \in S_{i_\alpha} = S_{i'_{\alpha-1}} \text{（约定 } i'_{\alpha-1} = i_\alpha, j'_{\beta+1} = j_\beta, j'_1 = i_0\text{）},$$

$$H^{-1}[H(x)] = x \in S_{i_0} = S_{j'_1},$$

$$H^{-(\beta+1)}[H(x)] = H^{-\beta}(x) \in S_{j_\beta} = S_{i'_{\beta+1}},$$

故由定义有

$$H(x) \in A^{j'_1,j'_2,\cdots,j'_{n+1}}_{i'_0,i'_1,\cdots,i'_{l-1}} = A^{i_0,j_1,\cdots,j_n}_{i_1,i_2,\cdots,i_l}.$$

反之，若 $y \in A^{i_0,j_1,\cdots,j_n}_{i_1,i_2,\cdots,i_l}$，由定义知 $H^{-\beta}(y)(1 \leqslant \beta \leqslant n+1)$ 有定义，令 $x = H^{-1}(y)$，则 $x \in A^{j_1,j_2,\cdots,j_n}_{i_0,i_1,\cdots,i_l}$ 且 $H(x) = y$. 于是 ⅲ）获证.

在 ⅲ）的两端同用 H^{-1} 作用之，得 ⅳ），于是引理 1 证毕.

引理 2 ⅰ）$A^{j_1,j_2,\cdots,j_{n+1}}$ 为穿集.

ⅱ）$A_{i_0,i_1,\cdots,i_l}$ 为隔集.

证 先证 ⅰ）：我们对 n 行数学归纳. 当 $n = 0$ 时，有 $A^{j_1} = H(A_{j_1}) = H(S_{j_1})$，由 S_i 之

---

① $\overline{M/S}$ 表 M/S 的闭包.

定义及定义1(1)，知$H(S_{j_1})$为穿集. 设对$n-1$，ⅰ)已真，往证它对n亦真. 由引理1

$$A^{j_1,j_2,\cdots,j_{n+1}}=H(A^{j_2,j_3,\cdots,j_{n+1}}_{j_1})=H(A^{j_2,\cdots,j_{n+1}}\cap A_{j_1})$$
$$=H(A^{j_2,\cdots,j_{n+1}}\cap S_{j_1}),$$

由归纳假设，$A^{j_2,\cdots,j_{n+1}}$为穿集，故它有连通集T，使$\pi(T)\supset[-1,1]$，从而有$T_1\subset T\cap S_i$，使T_1满足$\pi(T_1)\supset[h_{2i-1},h_{2i}]$. 由$H$之定义1(1)，知$\pi[H(T_1)]\supset[-1,1]$，取$i=j_1$，即得

$$H(T_1)\subset H(T\cap S_{j_1})\subset H(A^{j_2,\cdots,j_{n+1}}\cap A_{j_1})=A^{j_1,\cdots,j_{n+1}}.$$

由数学归纳法，知ⅰ)为真.

再用归纳法证ⅱ)：当$l=0$时，$A_{i_0}=S_{i_0}$显然为隔集. 设ⅱ)对$l-1$为真，下面往证其对l亦真. 由引理1

$$A_{i_0,i_1,\cdots,i_l}=H^{-1}(A^{i_0}_{i_1,\cdots,i_l})=H^{-1}(A^{i_0}\cap A_{i_1,\cdots,i_l})$$
$$=H^{-1}[H(S_{i_0})\cap A_{i_1,\cdots,i_l}].$$

由归纳假设，$A_{i_1,i_2,\cdots,i_l}$为隔集. 若$T\subset N$是满足$\pi(T)\supset[-1,1]$的连通集，则T有连通子集$T_1\subset T\cap S_{i_0}$，使$\pi(T_1)\supset[h_{2i_0-1},h_{2i_0}]$，从而$\pi[H(T_1)]\supset[-1,1]$. 由于

$$H(T_1)\subset H(T\cap S_{i_0})\subset H(S_{i_0})\subset N,$$

由隔集之定义：$H(T_1)\cap A_{i_1,\cdots,i_l}\neq\varnothing$，从而

$$H(T\cap S_{i_0})\cap A_{i_1,\cdots,i_l}\neq\varnothing.$$

即

$$H(T)\cap H(S_{i_0})\cap A_{i_1,\cdots,i_l}\neq\varnothing.$$

取H^{-1}作用之，得

$$T\cap H^{-1}[H(S_{i_0})\cap A_{i,\cdots,i_l}]=T\cap A_{i_0,i_1,\cdots,i_l}\neq\varnothing.$$

即$A_{i_0,i_1,\cdots,i_l}$为隔集，由数学归纳法，ⅱ)为真. 引理2证毕.

引理3 给定了满足定义1的ε-马蹄H，如果$\varepsilon<\dfrac{p^2}{4q}$，则存在正数$\theta<1$，使得当$x,y$同属于$A^{j_1,j_2,\cdots,j_n}_{i_0,i_2,\cdots,i_n}$时，有$\|x-y\|<2\theta^n$.

证 设$u=(u_1,u_2)$，$v=(v_1,v_2)$是同一个S_i中的两个点，由定义1(2)得

$$\|H(u)-H(v)\|=\|L(u)+r(u)-L(v)-r(v)\|$$
$$\geqslant\|L(u)-L(v)\|-\|r(u)-r(v)\|$$
$$\geqslant\frac{1}{q}\|u_2-v_2\|-\varepsilon\|u-v\|;$$

另一方面，若$H^{-1}(u)$，$H^{-1}(v)$也同在某一S_j中，则由$H^{-1}(x)=L^{-1}(x)-L^{-1}[r[H^{-1}(x)]]$可得

$$\|H^{-1}(u)-H^{-1}(v)\|=\|L^{-1}(u)-L^{-1}(v)+L^{-1}[r[H^{-1}(v)]]-L^{-1}[r[H^{-1}u]]\|$$
$$\geqslant\|L^{-1}(u)-L^{-1}(v)\|-\frac{\varepsilon}{p}\|H^{-1}(v)-H^{-1}(u)\|$$
$$\geqslant\frac{1}{q}\|u_1-v_1\|-\frac{\varepsilon}{p}\cdot\frac{1}{p-\varepsilon}\|u-v\|^{①}.$$

---

① 由Lipschitz反函数定理，若L为线性，则：$\mathrm{Lip}((L+\varnothing)^{-1})\leqslant\frac{1}{|L^{-1}|^{-1}-\mathrm{Lip}\,\phi}$.

两式相加得

$$\| H(u)-H(v) \| + \| H^{-1}(u)-H^{-1}(v) \| \geqslant \frac{1}{q}(\| u_1-v_1 \| + \| u_2-v_2 \|)-$$

$$\left(1+\frac{1}{p(p-\varepsilon)}\right)\varepsilon \| u-v \|$$

$$\geqslant \left[\frac{1}{q}-\left(1+\frac{1}{p(p-\varepsilon)}\right)\right]\varepsilon \| u-v \| = f(p,q,\varepsilon) \| u-v \| .$$

若 $f(p,q,\varepsilon)>2$,则不等式

$$\| u-v \| \leqslant \frac{2}{f(p,q,\varepsilon)} \| H(u)-H(v) \| < \| H(u)-H(v) \| ,$$

$$\| u-v \| \leqslant \frac{2}{f(p,q,\varepsilon)} \| H^{-1}(u)-H^{-1}(v) \| < \| H^{-1}(u)-H^{-1}(v) \|$$

中至少有一个成立.

注意到 u,v 是同在某个 S_i 中且使 $H^{-1}(u)$,$H^{-1}(v)$ 也同在某个 S_j 中任两点,故上述不等式中的 u,v 可分别换成 $H^l(x)$,$H^l(y)(-n<l<n)$,这是因为 x,y 同属于 $A_{i_0,i_1,\cdots,j_n}^{j_1,j_2,\cdots,i_n}$ 之故.

令 $x_l=H^l(x)$,$y_l=H^l(y)$,若

$$\| x_0-y_0 \| \leqslant \frac{2}{f(p,q,\varepsilon)} \| x_1-y_1 \| ,$$

则对一切 $0\leqslant l<n$,都有

$$\| x_l-y_l \| \leqslant \frac{2}{f(p,q,\varepsilon)} \| x_{l+1}-y_{l+1} \| ,$$

从而 $\| x-y \| = \| x_0-y_0 \| \leqslant \left(\frac{2}{f(p,q,\varepsilon)}\right)^n \| x_n-y_n \| \leqslant 2\left(\frac{2}{f(p,q,\varepsilon)}\right)^n$,反之,若

$$\| x_0-y_0 \| \leqslant \frac{2}{f(p,q,\varepsilon)} \| x_{-1}-y_{-1} \| \text{,则也可推出 } \| x-y \| \leqslant 2\left(\frac{2}{f(p,q,\varepsilon)}\right)^n,$$

以上假定了 $f(p,q,\varepsilon)>2$. 这一点何时成立?易知当 $\varepsilon<\dfrac{p^2}{4q}$ 即可. 取 $\theta=2f(p,q,\varepsilon)^{-1}$,即完成了引理 3 之证明.

实际上,对 ε 的限制还可放宽为

$$\varepsilon<\frac{p(p-\varepsilon)}{q[1+p(p-\varepsilon)]}.$$

§3 马蹄的基本性质及其 Ω 稳定性

设给定了 ε- 马蹄 H,我们来考虑 H 的非游荡集 $\Omega(H)$. 由定义 1(3) 可知,对 $x\in M$,一旦 $H^n(x)$ 不属于 S,就再也回不到 S 中而必须趋于渊点 P. 因此,$\Omega(H)$ 仅包含那些对一切整数 n 都有 $H^n(x)\in S$ 的那些点,令

$$A=\{x \mid H^n(x)\in S, \forall n\},$$

$$A_{i_0,i_1,i_2,\cdots}^{j_1,j_2,\cdots}=\left\{x \mid H^{\alpha}(x)\in S_{i_\alpha}, H^{-\beta}(x)\in S_{i_\beta}; \begin{matrix}\alpha=0,1,2,\cdots \\ \beta=1,2,\cdots\end{matrix}\right\}.$$

则显然有

$$A=\bigcup_{\substack{\forall(i_\alpha)\\ \forall(j_\beta)}}A_{i_0,i_1,i_2,\cdots}^{j_1,j_2,\cdots}$$

$$A_{i_0,i_1,i_2,\cdots}^{j_1,j_2,\cdots}=\bigcap_{n=1}^{\infty}A_{i_0,i_1,\cdots,i_n}^{j_1,j_2,\cdots,j_n}.$$

由引理1,2可知$A_{i_0,i_1,\cdots,i_n}^{j_1,j_2,\cdots,j_n}$是非空闭集,而且

$$A_{i_0,i_1,i_2,\cdots,i_n}^{j_1,j_2,\cdots,j_n}\supset A_{i_0,i_1,\cdots,i_{n+1}}^{j_1,j_2,\cdots,j_{n+1}}.$$

故$A_{i_0,i_1,i_2,\cdots}^{j_1,j_2,\cdots}$也是非空闭集.由引理3可知,给定了无穷序列$\{i_\alpha\}$、$\{j_\beta\}$之后,对应的$A_{i_0,i_1,i_2,\cdots}^{j_1,j_2,\cdots}$中只有一个点.反之,对任意$x\in A$,$x=A_{i_0,i_1,i_2,\cdots}^{j_1,j_2,\cdots}$对应的两个无穷序列$\{i_\alpha\}$、$\{j_\beta\}$也是唯一确定的(以下即把$A_{i_0,i_1,i_2,\cdots}^{j_1,j_2,\cdots}$和它的唯一元素视为等同).

下面证明

1° 若$x=A_{\dot{i}_0,i_1,\cdots,\dot{i}_n}^{i_n,i_{n-1},\cdots,i_0}$,则$x$是$H$的周期点.这里"$\dot{i}_0,i_1,\cdots,\dot{i}_n$"上方的两个圆点"."表示循环节.

2° H的周期点在A中稠密,故$\Omega(H)=A$.

3° $\Omega(H)$是无处稠密的完全集.更进一步,它是完全不连通的.

事实上,1°马上由引理1得出,为证明2°,我们指出,对任意的

$$x=A_{i_0,i_1,i_2,\cdots}^{j_1,j_2,\cdots}$$

取$x'=A_{i_0,i_1,\cdots,i_n,j_n,j_{n-1},\cdots,j_1}^{j_1,j_2,\cdots,j_n,i_n,i_{n-1},\cdots,i_0}$,则$x'$为$H$之周期点且$\|x-x'\|<2\theta^n$,可见2°为真.

为证明3°,引入映射

$F:A\to E^2$,

$(F(x)=(0.i_0i_1i_2\cdots,0.j_1j_2\cdots),x=A_{i_0,i_1,i_2,\cdots}^{j_1,j_2,\cdots}$,

这里$0.t_1t_2\cdots$表示E^2中点的坐标的$(k+2)$进无穷小数.由于i_α,j_β中不取$0,k+1$,故F的像是E^2中的Cantor完全集.记$F(A)=C$,我们指出,F是A与C之间的同胚.事实上,双方单值性是刚才已经证明了的,由引理3,F^{-1}是连续的,剩下的证明F的连续性.事实上,由于H是同胚,而且诸S_j是不相交的闭集,故对给定的n只要$\|x-x'\|$充分小时,$H^n(x)$,$H^n(x')$不可能分属于不同的S_{j_n},$S_{j'_n}$.这证明,当A的两点x,x'充分接近时,$F(x)$、$F(x')$的前面充分多位的小数数字必须相同,从而$F(x)$连续.即F是同胚,从而A是完全集.

至于A的完全不连通性,可这样来证,若x,y都是A的点,而T是包含x,y的A的连通子集.若$x\neq y$,则有n使$H^n(x)$,$H^n(y)$属于不同的S_i,S_j.由于S_i,S_j是不相交的闭集,$H^n(T)$中必有N/S的点,此与$T\subset A$矛盾.

最后来证Ω稳定性.由定义可知,H的$\varepsilon'-c^1$扰动H^*,当ε'足够小时也是ε_1-马蹄.$\varepsilon_1=\varepsilon+\varepsilon'$.取$\varepsilon'$足够小使$\varepsilon+\varepsilon'<\frac{p^2}{4q}$,则引理1,2,3对$H^*$也成立.对应地,记$\Omega(H^*)=A^*$.类似于对$H$所做的,定义一个$A^*$和$C=F(A)$之间的同胚$F^*$,

$F^*:A^*\leftrightarrow C$,

$F^*(x)=(0.i_0i_1i_2\cdots,0.j_1j_2\cdots)(x=A^{*\,j_1,j_2,\cdots}_{\ \,i_0,i_1,i_2,\cdots})$.

再引进一个C的自身的同胚

$G: C \leftrightarrow C$,

$G(0.i_0 i_1 i_2 \cdots, 0.j_1 j_2 \cdots) = (0.i_1 i_2 \cdots, 0.i_0 j_1 j_2 \cdots)$.

显然有

$$F[H(x)] = G[F(x)], (x \in A);$$
$$F^*[H^*(x)] = G[F^*(x)], (x \in A^*).$$

从而,若取

$$\varnothing = F^{-1} \circ F^*,$$

则,对 $x \in A^*$,有

$$\varnothing \circ H^* = F^{-1} \circ F^* \circ H^* = F^{-1} \circ G \circ F^* = H \circ F^{-1} \circ F^* = H \circ \varnothing.$$

即 H 与 H^* 在 Ω 集上共轭. Ω 稳定性证毕.

参考文献

[1] Smale S. Diffeomorphisms with many periodic points, in Differential and Combinatorial Topology (S. S. Cairns, ed.), Princeton University Press, 1965: 63 - 80.

[2] Nitecki Z, Differentiable Dynamics. The M. I. T. Press. Cambridge, Massachusetts, and London, England, 1971: 118 - 158.

[3] 张景中,杨路. Smale 马蹄的一个简单模型. 科学通报, 1981(12).

编辑部注 作者 1981 年 9 月 10 日来信,对本文作如下修改.

(1) 引理 3 的叙述改为:

引理 3 给了满足定义 1 的 ε- 马蹄 H,当 ε 足够小时,存在正数 $\theta < 1$,使得当 x, y 同属于 $A_{i_0, i_1, \cdots, i_n}^{j_1, j_2, \cdots, j_n}$ 时,有 $|x - y| < C\theta^n$. 此处 $C > 0$ 是仅与 H 有关的常数.

(2) 在引理3的证明中,最后有"以上假定了 $f(p,q,\varepsilon) > 2$,这一点何时成立?",这句话的前后均需修改,改为:…… 从而 $\|x - y\| = \|x_0 - y_0\| \leqslant \left(\frac{2}{f(p,q,\varepsilon)}\right)^n \|x_n - y_n\| \leqslant C\left(\frac{2}{f(p,q,\varepsilon)}\right)^n$,反之,若 $\|x_0 - y_0\| \leqslant \frac{2}{f(p,q,\varepsilon)} \|x_{-1} - y_{-1}\|$,则可推出: $\|x - y\| \leqslant \left(\frac{2}{f(p,q,\varepsilon)}\right)^n \|x_{-n} - y_{-n}\| \leqslant C\left(\frac{2}{f(p,q,\varepsilon)}\right)^n$.

以上假定了 $f(p,q,\varepsilon) > 2$. 这一点何时成立?易知当 $q < \frac{1}{2}$ 且 ε 足够小即可. 若 $q \geqslant \frac{1}{2}$,由于 $q < 1$,故可取自然数 k 使 $q^k < \frac{1}{2}$,然后在以上的论证中 $\widetilde{H}$ 代替 H^k,可得:

$$|x_0 - y_0| \leqslant C\left(\frac{2}{\tilde{f}(p,q,\varepsilon)}\right)^{\frac{n}{k}},$$

这里 $\tilde{f}(p,q,\varepsilon) > 2$,再令 $\theta = \left(\frac{2}{\tilde{f}(p,q,\varepsilon)}\right)^{\frac{1}{k}}$ 即可.

(以下即接原稿 §3)

关于质点组的一类几何不等式

张景中　杨　路
（中国科学技术大学）

作者在文献[1]中用代数方法获得了一类几何不等式. 本文通过新的途径导出了较文献[1]更为广泛的结果. 所凭藉的工具仍然是代数的.

首先约定记号. 设 $\mathscr{S}=\{A_i(m_i), i=1,2,\cdots,N\}$ 是 E^n 中的质点组, $m_i \geqslant 0$ 是点 A_i 所赋有的质量. 取 $\mathscr{S}$ 的质心 O 为坐标原点. 设 H 是过 O 的任一个 $(n-1)$ 维定向超平面, $\bar{e}_H$ 是 H 的单位法向量. 又令 $\bar{a}_i=\overrightarrow{OA_i}$ 表示点 A_i 的坐标向量. 按常例将

$$I_H=m_1(\bar{a}_1\cdot\bar{e}_H)^2+m_2(\bar{a}_2\cdot\bar{e}_H)^2+\cdots+m_N(\bar{a}_N\cdot\bar{e}_H)^2$$

叫做 $\mathscr{S}$ 关于 H 的转动惯量. 令

$$\frac{\bar{e}_H}{\sqrt{I_H}}=\bar{x}=(x_1,x_2,\cdots,x_n),$$

则

$$\bar{e}_H=\frac{\bar{x}}{\|\bar{x}\|}=\frac{1}{\|\bar{x}\|}(x_1,x_2,\cdots,x_n).$$

设
$$\bar{a}_i=(a_{i1},a_{i2},\cdots,a_{in})\quad i=1,2,\cdots,N,$$

$$Q=\begin{pmatrix}\sqrt{m_1}\ a_{11} & \sqrt{m_2}\ a_{21} & \cdots & \sqrt{m_N}\ a_{N1}\\ \sqrt{m_1}\ a_{12} & \sqrt{m_2}\ a_{22} & \cdots & \sqrt{m_N}\ a_{N2}\\ \vdots & \vdots & \vdots & \\ \sqrt{m_1}\ a_{1n} & \sqrt{m_2}\ a_{2n} & \cdots & \sqrt{m_N}\ a_{Nn}\end{pmatrix},$$

则由 I_H 之定义有

$$\bar{e}_H QQ^T \bar{e}_H{}^T=I_H=\frac{1}{\|\bar{x}\|^2},$$

这里 $\bar{e}_H^T, Q^T$ 表示 $\bar{e}_H, Q$ 之转置. 从而

$$\|\bar{x}\|\bar{e}_H QQ^T(\|\bar{x}\|\bar{e}_H)^T=1.$$

即
$$\bar{x}\,QQ^T\,\bar{x}^T=1.$$

由此可见, 当 H 取遍所有过 O 之 $(n-1)$ 维超平面时, 点 $\bar{x}$ 之轨迹为一个二阶曲面. 不妨假定 $\{A_i\}$ 不在同一超平面上, 此时 QQ^τ 是正定阵而该二阶超曲面为一椭球, 记之为 $\mathscr{B}$. 如果 $\mathscr{U}$ 是另外一个椭球, 它与 $\mathscr{B}$ 有相同的主轴而且它的各半轴是 $\mathscr{B}$ 的对应半轴的倒数, 我们就将 $\mathscr{U}$ 叫做 $\mathscr{S}$ 的"密集椭球".

本文刊于《中国科学技术大学学报》, 第 11 卷第 2 期, 1981 年.

任取$\mathscr{S}$中的$k+1$个点$A_{i_0},A_{i_1},\cdots,A_{i_k}$，将其所支撑的单形的$k$维体积记为

$$V_{i_0 i_1}\cdots,$$

令

$$M_k=\sum_{i_0<i_1<\cdots<i_k}\sum\cdots\sum m_{i_0}m_{i_1}\cdots m_{i_k}V_{i_0 i_1\cdots i_k}\ (1\leqslant k\leqslant n),$$

$$M_0=m_1+m_2+\cdots+m_N.$$

则下述命题显然是文献[1]中定理的推广.

定理 1　对E^n中质点组$\mathscr{S}=\{A_i(m_i),i=1,2,\cdots,N\}(N>n)$的诸不变量$\{M_k\}$有不等式

$$\frac{M_k^l}{M_l^k}\geqslant\frac{[(n-l)!(l!)^3]^k}{[(n-k)!(k!)^3]^l}(n!M_0)^{l-k}\quad(1\leqslant k<l\leqslant n),\tag{1}$$

$$M_k^2\geqslant\left(\frac{k+1}{k}\right)^3\cdot\frac{n-k+1}{n-k}\cdot M_{k-1}M_{k+1}\quad(1\leqslant k\leqslant n),\tag{2}$$

其等号当且仅当$\mathscr{S}$的密集椭球为球时成立.

这里顺便指出，一个质点组的密集椭球为球的充分必要条件是：它关于质心的惯量椭球是一个球.因此定理1中不等式成立的充分必要条件可改述为："$\mathscr{S}$关于其质心的惯量椭球是一个球."

证　令$P=Q^TQ$，则P的非零特征值应与QQ^T一致，共有n个，都是正实数.现考虑P的特征多项式：

$$\begin{aligned}P(\lambda)&=\|Q^TQ-\lambda E\|\\&=\|\sqrt{m_i}\sqrt{m_j}\,\bar{a}_i\bar{a}_j-\delta_{ij}\lambda\|\left(\delta_{ij}=\begin{cases}0&(i\neq j),\\1&(i=j)\end{cases}\right)\\&=\frac{1}{M_0}\begin{vmatrix}M_0&0\quad 0&\cdots&0\quad 0\\\sqrt{m_1}&&&\\\sqrt{m_2}&&&\\\vdots&&\sqrt{m_i}\sqrt{m_j}\,\bar{a}_i\bar{a}_j-\delta_{ij}\lambda&\\\sqrt{m_N}&&&\end{vmatrix}.\end{aligned}$$

约定行列式的行列号由0算起.将第i行乘以$-\sqrt{m_i}$加到第1行，援用$\sum m_i\bar{a}_i=0$(质心性质)就得到：

$$P(\lambda)=\frac{\lambda}{M_0}\begin{vmatrix}0&\sqrt{m_1}\quad\sqrt{m_2}&\cdots&\sqrt{m_N}\\\sqrt{m_1}&&&\\\sqrt{m_2}&&&\\\vdots&&\sqrt{m_i}\sqrt{m_j}\,\bar{a}_i\bar{a}_j-\delta_{ij}\lambda&\\\sqrt{m_N}&&&\end{vmatrix}.$$

再将第0行(列)乘以$-\frac{1}{2}\sqrt{m_i}\,\bar{a}_i^2$(或$-\frac{1}{2}\sqrt{m_j}\,\bar{a}_j^2$)加到第$i$行(第$j$列)，得到

$$P(\lambda)=\frac{\lambda}{M_0}\begin{vmatrix}0 & \sqrt{m_1} & \sqrt{m_2} & \cdots & \sqrt{m_N}\\ \sqrt{m_1} & & & & \\ \sqrt{m_2} & & -\frac{\sqrt{m_i}\sqrt{m_j}}{2}(\bar{a_i}-\bar{a_j})^2-\delta_{ij}\lambda & & \\ \vdots & & & & \\ \sqrt{m_N} & & & & \end{vmatrix}.$$

若令 ρ_{ij} 表示线段 A_iA_j 之长度,则有

$$P(\lambda)=\frac{\lambda}{M_0}\begin{vmatrix}0 & \sqrt{m_1} & \sqrt{m_2} & \cdots & \sqrt{m_N}\\ \sqrt{m_1} & & & & \\ \sqrt{m_2} & & -\frac{\sqrt{m_i}\sqrt{m_j}}{2}\rho_{ij}^2-\delta_{ij}\lambda & & \\ \vdots & & & & \\ \sqrt{m_N} & & & & \end{vmatrix}.$$

利用单形体积公式[2]

$$V^2(A_1,A_2,\cdots,A_{k+1})=\frac{(-1)^{k+1}}{2^k(k!)^2}\begin{vmatrix}0 & 1 & \cdots & 1\\ 1 & & & \\ \vdots & & \rho_{ij}^2 & \\ 1 & & & \end{vmatrix}$$

可知

$$\begin{vmatrix}0 & \sqrt{m_{i_0}} & \sqrt{m_{i_1}} & \cdots & \sqrt{m_{i_k}}\\ \sqrt{m_{i_0}} & & & & \\ \sqrt{m_{i_1}} & & -\frac{1}{2}\sqrt{m_{i_\alpha}}\sqrt{m_{i_\beta}}\cdot\rho_{i_\alpha i_\beta}^2 & & \\ \vdots & & & & \\ \sqrt{m_{i_k}} & & & & \end{vmatrix}=-m_{i_0}m_{i_1}\cdots m_{i_k}V_{i_0i_1\cdots i_k}^2(k!)^2.$$

现将方程 $P(\lambda)=0$ 展开.由于它只有 n 个非零根,故展开后形为

$$\Big(\sum_{k=0}^{n}(-1)^k(k!)^2M_k\lambda^{n-k}\Big)\cdot\lambda^{N-n}=0.$$

于是 $P(\lambda)=0$ 的 n 个非零根 $\lambda_1,\lambda_2,\cdots,\lambda_n$ 满足方程

$$\sum_{k=0}^{n}(-1)^k(k!)^2M_k\lambda^{n-k}=0.$$

从而得到 $\lambda_1,\lambda_2,\cdots,\lambda_n$ 的各阶初等对称多项式 σ_k 的表达式

$$\sigma_k(\lambda_1,\lambda_2,\cdots,\lambda_n)=(k!)^2\frac{M_k}{M_0}.$$

由 Maclaurin 定理[3]

$$\left[\frac{k!(n-k)!}{n!}\sigma_k\right]^l\geqslant\left[\frac{l!(n-l)!}{n!}\sigma_l\right]^k(l>k),$$

得到所欲证之(1).由 Newton 定理[3]

$$\left[\frac{k!(n-k)!}{n!}\sigma_k\right]^2\geqslant\left[\frac{(k-1)!(n-k+1)!}{n!}\sigma_{k-1}\right]\left[\frac{(k+1)!(n-k-1)!}{n!}\sigma_{k+1}\right],$$

得到所欲证之(2). 等号成立的充要条件是:$\lambda_1=\lambda_2=\cdots=\lambda_n$,即$\mathscr{S}$的密集椭球是一个球. 定理 1 证毕.

容易看出,如果$\mathscr{S}$不是有限质点组而是某个具有有限质量的区域,设质量分布函数为

$$m(x)(x\in\mathscr{S}),$$

即可定义

$$M_k=\frac{1}{k!}\iint\cdots\int m(x_0)m(x_1)\cdots m(x_k)V^2(x_0,x_1,\cdots,x_k)dx_0dx_1\cdots dx_k,$$

$$M_0=\int m(x)dx.$$

通过极限过程可以证明,定理 1 中的 M_k,M_0 理解为这里的积分值时,两个不等式仍成立.

下面举例说明定理 1 的应用.

1970 年 D. Veljian 提出了如下猜测:在 E^n 中一个单形的体积 V 和它的诸棱长

$$\rho_{ij}(i,j=1,2,\cdots,n+1)$$

之间有不等式:

$$n!V\leqslant\left(\frac{n+1}{2^n}\right)^{\frac{1}{2}}\prod_{1\leqslant i<j\leqslant n+1}\rho_{ij}{}^{\frac{2}{n+1}}.$$

且当该单形为正则时等号成立. 这个猜想在 1974 年被 Korchmáros 所证实.[4]

这导致我们考虑类似的一个问题:设 n 维单形 $A_1A_2\cdots A_{n+1}$ 中每个顶点 A_i 所对的"侧面"(它是$n-1$维单形)的$(n-1)$维体积是 v_i,那么在原单形的体积V和这些侧面体积 v_i 之间是否成立着类似于 Veljian 猜想的不等式呢?答案是肯定的. 事实上,援用定理 1 我们容易得到:

系 1 在 n 维单形的体积 V 及其诸侧面体积 v_i 之间有不等式

$$V\leqslant\sqrt{n+1}\left[\frac{(n-1)!^3}{n^{3n-2}}\right]^{\frac{1}{2(n-1)}}\cdot\left(\prod_{i=1}^{n+1}v_i\right)^{\frac{n}{n^2-1}},\tag{3}$$

且当该单形为正则时等号成立.

证 在不等式(1) 中令 $l=n,k=n-1$,并令

$m_1=v_1^2,m_2=v_2^2,\cdots,m_{n+1}=v_{n+1}^2$.

就有

$$M_0=\sum_{i=1}^{n+1}v_i^2,$$

$$M_{n-1}=(n+1)\prod_{i=1}^{n+1}v_i^2,$$

$$M_n=\left(\prod_{i=1}^{n+1}v_i^2\right)\cdot V^2.$$

统统代入(1) 式中,经过移项整理后有

$$\frac{(n+1)^n\cdot(n-1)!^2}{n^{3n-2}}\prod_{i=1}^{n+1}v_i^2\geqslant V^{2(n-1)}\cdot\sum_{i=1}^{n+1}v_i^2.$$

由算术-几何平均不等式

$$\sum_{i=1}^{n+1}v_i^3\geqslant(n+1)\prod_{i=1}^{n+1}v_i{}^{\frac{2}{n+1}}$$

故有

$$\frac{(n+1)^{n-1}\cdot(n-1)!^2}{n^{3n-2}}\prod_{i=1}^{n+1}v_i^{\frac{2n}{n+1}}\geqslant V^{2(n-1)}.$$

从而得到(3)式.今取棱长为1的n维和$(n-1)$维正则单形.其体积公式是熟知的,将对应的V和v_i的值代入(3),即知等号对正则单形确实成立.系1证毕.

系2 令S_n是E^n中所有n维单形组成之集,一个单形K的体积及其各侧面体积分别用$V(K)$和$v_i(K)$来表示,则有

$$\lim_{n\to\infty}\inf_{K\in S_n}\frac{1}{V(K)}\prod_{i=1}^{n+1}v_i(K)^{\frac{n}{n^2-1}}=e. \tag{4}$$

证 由(3)式立即得到

$$\inf_{K\in S_n}\frac{1}{V(K)}\prod_{i=1}^{n+1}v_i(K)^{\frac{n}{n^2-1}}=\frac{1}{\sqrt{n+1}}\left[\frac{n^{3n-2}}{(n-1)!^2}\right]^{\frac{1}{2(n-1)}}.$$

运用Stirling公式显然有

$$\lim_{n\to\infty}\frac{1}{\sqrt{n+1}}\left[\frac{n^{3n-2}}{(n-1)!^2}\right]^{-\frac{1}{2(n-1)}}=e.$$

于是系2证毕.

从(3)还可以导出若干别的有趣的不等式.首先将其中诸v_i的几何平均代之以算术平均:

$$V\leqslant\sqrt{n+1}\left[\frac{(n-1)!^2}{n^{3n-2}}\right]^{\frac{1}{2(n-1)}}\left(\frac{1}{n+1}\sum_{i=1}^{n+1}v_i\right)^{\frac{n}{n-1}},$$

令此单形内切球半径为r.将关系

$$\frac{1}{n}\sum_{i=1}^{n+1}v_i=\frac{1}{r}V$$

代入前一式,移项整理后得到:

系3 在n维单形的内切球半径r和体积V之间有不等式

$$r\leqslant\left[\frac{n!^2}{n^n(n+1)^{n+1}}\right]^{\frac{1}{2n}}\cdot V^{\frac{1}{n}}. \tag{5}$$

且当该单形为正则时等号成立.

由(5)式取极限可以得到

$$\lim_{n\to\infty}\sup_{K\in S_n}r(K)/V(K)^{\frac{1}{n}}=\frac{1}{e}. \tag{6}$$

从而又有:

系4 对于一切具有单位体积的单形而言,当空间的维数趋于无穷时,其内切球半径的上限是$1/e$.

我们还可以考虑单形体积和它各条高线的长度h_i之间的关系.由于

$$v_i=\frac{1}{h_i}nV,i=1,2,\cdots,n+1,$$

将此式代入(3)中,经移项整理后得到:

系5 在n维单形的体积V及其各条高线长度h_i之间有不等式

$$V \geqslant \frac{1}{n!}\left[\frac{n^n}{(n+1)^{n-1}}\right]^{\frac{1}{2}} \cdot \prod_{i=1}^{n+1} h_i^{\frac{n}{n+1}}, \tag{7}$$

且当单形为正则时等号成立.

由以上的例子可以看到,系 1 是一个颇为有用的命题.事实上从它也可以直接推出前面提到过的 Veljian-Korchmáros[1] 不等式:

系 6　E^n 中一个单形的体积 V 和它的诸棱长 ρ_{ij} 之间有关系

$$n!V \leqslant \left(\frac{n+1}{2^n}\right)^{\frac{1}{2}} \prod_{1\leqslant i<j\leqslant n+1} \rho_{ij}^{\frac{2}{n+1}}, \tag{8}$$

且当该单形为正则时等号成立.

证　显然命题对 E^1 是成立的.现在假定它对 E^{n-1} 已成立,往证它对 E^n 为真.

沿用系 1 中的符号,将 n 维单形的各个$(n-1)$维界面的体积记为 $v_r(r=1,2,\cdots,n+1)$.既然(8) 对 E^{n-1} 成立,那就有

$$(n-1)!v_r \leqslant \left(\frac{n}{2^{n-1}}\right)^{\frac{1}{2}} \prod_{\substack{1\leqslant i<j<n+1 \\ i\neq r, j\neq r}} \rho_{ij}^{\frac{2}{n}} (r=1,2,\cdots,n+1).$$

乘起来得到

$$(n-1)!^{n+1} \prod_{r=1}^{n+1} v_r \leqslant \left(\frac{n}{2^{n-1}}\right)^{\frac{n+1}{2}} \prod_{1\leqslant i<j\leqslant n+1} \rho_{ij}^{\frac{2(n-1)}{n}}.$$

另一方面由系 1 及(3) 有

$$V \leqslant \sqrt{n+1}\left[\frac{(n-1)!^2}{n^{3n-2}}\right]^{\frac{1}{2(n-1)}} \left(\prod_{r=1}^{n+1} v_r\right)^{\frac{n}{n^2-1}}.$$

将前一式代入后一式得到

$$V \leqslant \frac{1}{n!}\left(\frac{n+1}{2^n}\right)^{\frac{1}{2}} \prod_{1\leqslant i<j\leqslant n+1} \rho_{ij}^{\frac{2}{n+1}},$$

此即(8) 式.当单形为正则时等号显然成立.系 6 证毕.

提出如下的问题是自然的:当定理 1 中的质点组 $\mathscr{S}$的部分质点带有负质量时,我们能够说些什么?有关的不等式是否仍然成立?事实上我们可以建立下述的

定理 2　设 $\mathscr{S}=\{A_i(m_i), i=1,2,\cdots,N\}$ 是 E^n 中的质点组$(N>n)$,各点 A_i 所赋有的质量 m_i 是可正可负的实数.令 $M_0=m_1+m_2+\cdots+m_N\neq 0$,$M_k(1\leqslant k\leqslant n)$ 的意义如前所述.则有

$$M_k^2 \geqslant \left(\frac{k+1}{k}\right)^3 \cdot \frac{n-k+1}{n-k} \cdot M_{k-1}M_{k+1}. \tag{2}$$

而等号成立的充分必要条件仍然是:$\mathscr{S}$关于其质心的惯量椭球是一个球.

证　我们基本上沿用定理 1 的证明,只需注意哪些地方需要修改.首先遇到的问题是:由于质量是可正可负的,矩阵 Ω 中出现一些纯虚数,但是 QQ^T 仍是实对称矩阵,所以它的特征值必然是实的.虽然 $P=Q^TQ$ 可能是一个含有纯虚数的矩阵,但它的非零特征值应与 QQ^T 的一致,因而 P 的特征值也都是实的.于是前面这部分论证,完全可以照搬.只是到了最后一步,由于方程 $P(\lambda)=0$ 的 n 个非零根仅仅是实的而不一定是正的,所以不能援用 Maclaurin 定理而只能用 Newton 定理,从而得到(2).

这里不准备花很多篇幅来说明定理 2 的种种应用了，只举一个简单的例子. 若令 $n=2$，$N=3$，三角形 $A_1A_2A_3$ 的三条对应边是 a,b,c，其面积为 $\triangle$. 我们将证明：对于任意三个实数 λ,μ,v 成立着不等式

$$(\lambda a^2+\mu b^2+\nu c^2)^2\geqslant 16(\mu v+\nu\lambda+\lambda\mu)\triangle^3. \tag{9}$$

证 不失一般性，不妨假定 $\lambda\mu\nu>0$.（否则考虑 $-\lambda,-\mu,-\nu$）从方程组

$$m_2m_3=\lambda, m_3m_1=\mu, m_1m_2=\nu$$

中可以解出一组实数 m_1,m_2,m_3. 另一方面，在(2)式中取 $k=1, n=2$，就有

$$M_1^2\geqslant 16M_0M_2.$$

将解得之 m_1,m_2,m_3 代入此式即得(9)式.

试将此例应用于特殊情况. 设另有一个三角形边长为 a_1,b_1,c_1，面积为 $\triangle_1$. 在(9)中令

$$\lambda=-a_1^2+b_1^2+c_1^2, \mu=a_1^2-b_1^2+c_1^2, \nu=a_1^2+b_1^2-c_1^2.$$

便又得到众所熟知的 Pedoe[5] 不等式：

$$a^2(-a_1^2+b_1^2+c_1^2)+b^2(a_1^2-b_1^2+c_1^2)+c^2(a_1^2+b_1^2-c_1^2)\geqslant 16\triangle\triangle_1.$$

参考文献

[1] 杨路，张景中. 数学学报，1980，23(5)：740－749.

[2] Blumenthal L M. Theory and Applications of Distance Geometry. Oxford，1953：98－99.

[3] Hardy G U 等. 不等式. 科学出版社，1965：53－57.

[4] Korchmáro C. Atti Accad. Nav. Lincei Rend. Cl. Sci. Fis. Mat. Nature.，1974，8(56)：876－879.

[5] Pedoe D. Proc. Camb. Phil. Soc.，1942(38)：397－398.

A Criterion of Existence and Uniqueness of Real Iterative Groups with One Parameter

Zhang Jing-zhong and Yang Lu

(*Department of Mathematics*)

Definition 1. If $y = G(x,t)$ is a continuous function on both x and $t(a < x < b, 0 \leqslant t < \infty)$ such that

1) $G(x,0) = x$,

2) $b > G(x,t) > x > a$, (for $t > 0$),

3) $G(x_1,t) > G(x_2,t)$, (for $x_1 > x_2$),

4) $G(G(x,t_1),t_2) = G(x,t_1+t_2)$,

then we say that $G(x,t)$ is a regular iterative family on (a,b) with parameter t.

Definition 2. Suppose $G(x,t)$ is a regular iterative family on (a,b) and $F(x)$ a function on (a,b); if the inequality $F(x) \geqslant x$ holds on a neighborhood of the point $x = b$ and

$$\lim_{x \to b} T(x, F(x)) = \alpha > 0,$$

(here and later let $t = T(x,y)$ denote the function satisfying the equation $y = G(x,t)$) then we say that $F(x)$ is comparable in family G and denote this situation by $F \sim (G)$. And the number $\alpha = \alpha_F$ is called the index of F in $\{G\}$.

Let $F^0(x) = x, F^{n+1}(x) = F(F^n(x))$ as usual. We have:

Theorem. Suppose $G(x,t)$ is a regular iterative family; $F(x)$ is continuous and increasing on (a,b), $F(x) > x$ and $F \sim (G)$; then the following five statements are equivalent:

(i) There exists an infinite sequence of natural numbers $m_1 < m_2 < \cdots < m_k < \cdots$ such that for every natural number k there is some continuous function $f_k(x)$ on (a,b), $f_k(x) \sim \{G\}$, satisfying $f_k^{m_k}(x) = F(x)$ and $\lim\limits_{k \to +\infty} f_k(x) = x$.

(ii) The sequence of function

$$\Phi_n(x,y) = T(F^n(x), F^n(y)),$$

converges to a continuous and increasing function $\Phi(x,y)$ on the domain

---

本文刊于《中国科学技术大学学报》,第 11 卷第 4 期,1981 年.

$$D:\{(x,y)\mid a<x<b, x\leqslant y<b\}.$$

(iii) For every function $g(x)$ on (a,b), if $g\sim(G)$, then the sequence of function

$$H_n(x)=F^{-n}(g(F^n(x)))$$

converges to a function $H(x)$, such that $H\sim(G)$, $\alpha_H=\alpha_g$, and

$$\Phi(x,H(x))=\alpha_g.$$

(iv) For every real number $c\in[0,+\infty)$, there is one and only one function $g(x,c)$ which satisfies:

1) $g(x,c)$ is monotonous for $x\in(a,b)$,

2) $g(x,c)\sim\{G\}$ and $\alpha_{g(x,c)}=c$,

3) $g(F(x),c)=F(g(x,c))$.

(v) For every natural number m there is one and only one function $f(x)\sim\{G\}$ such that $f^m(x)=F(x)$ and this function $f(x)$ is continuous on (a,b).

A Geometrical Proof of an Algebraic Theorem

Yang Lu and Zhang Jing-zhong

(Department of Mathematics)

There are a series of powerful inequalities about the positive definite matrices[1], but we seldom find corresponding results for general real matrices. The purpose of this note is, by means of geometrical method, to prove a theorem concerning the upper bounds of modulus of a class of non-negative real matrices. And we don't know still how to find an algebraical proof.

Theorem 1. Let $A=(a_{ij})$ be an $n\times n$ matrix$(i,j=1,2,\cdots,n)$, such that

i) $a_{ij}=a_{ji}\geqslant 0$,

ii) $a_{ii}=0$,

iii) $\operatorname{sgn} D_k=(-1)^{k-1},(k=2,3,\cdots,n)$,

where

$$D_k=\begin{vmatrix} a_{11} & a_{12} & \cdots & a_{1k} \\ a_{21} & a_{22} & \cdots & a_{2k} \\ \vdots & \vdots & & \vdots \\ a_{k1} & a_{k2} & \cdots & a_{kk} \end{vmatrix}.$$

Then

$$|\det A| \leqslant (n-1)\prod_{1\leqslant i<j\leqslant n} a_{ij}^{\frac{2}{n-1}} \tag{1}$$

And the equality holds if and only if there exist n positive numbers $c_r>0\,(r=1,2,\cdots,n)$ such that

$$a_{ij}=c_i c_j,\ (1\leqslant i<j\leqslant n). \tag{2}$$

Proof. First, we take an identity transformation of D_k:

$$D_k=(a_{12}a_{13}\cdots a_{1k})^2=\begin{vmatrix} 0 & 1 & 1 & \cdots & 1 \\ 1 & & & & \\ 1 & & \frac{a_{ij}}{a_{1i}a_{1j}} & & \\ \vdots & & & & \\ 1 & & & & \end{vmatrix}. \tag{3}$$

Let

本文刊于《中国科学技术大学学报》,第 11 卷第 4 期,1981 年.

$$p_{ij}=\left(\frac{a_{ij}}{a_{1i}a_{1j}}\right)^{\frac{1}{2}}(i,j=2,3,\cdots,n), \tag{4}$$

then

$$D_k=(a_{12}a_{13}\cdots a_{1k})^2=\begin{vmatrix} 0 & 1 & 1 & \cdots & 1 \\ 1 & & & & \\ 1 & & p_{ij}^2 & & \\ \vdots & & & & \\ 1 & & & & \end{vmatrix}. \tag{5}$$

From the Simplex Construction Theorem[2] and the above condition iii) we conclude that there exists an $(n-2)$-dimensional simplex $A_2A_3\cdots A_n$ such that

$$p_{ij}=|A_iA_j|\,(i,j=2,3,\cdots,n) \tag{6}$$

Let V denote the volume of this simplex. By the Korchmáros Inequality[3] we have

$$V\leqslant\frac{1}{(n-2)!}\left(\frac{n-1}{2^{n-2}}\right)^{\frac{1}{2}}\prod_{2\leqslant i<j\leqslant n}p_{ij}^{\frac{2}{n-1}} \tag{7}$$

and hence is obtained:

$$\begin{aligned}|D_n| &= (a_{12}a_{13}\cdots a_{1n})^2 2^{n-1}(n-2)!^2V^2 \\ &\leqslant (n-1)(a_{12}a_{12}\cdots a_{1n})^2\prod_{2\leqslant i<j\leqslant n}p_{ij}^{\frac{4}{n-1}}.\end{aligned} \tag{8}$$

Here the volume formula[4] of the simplex is used. From (4) and (8) we obtain

$$|\det A|=|D_n|\leqslant(n-1)\prod_{2\leqslant i<j\leqslant n}a_{ij}^{\frac{2}{n-1}}.$$

This is exactly (1). Since the equality in Korchmáros Inequality holds if and only if the simplex is regular, the equality in (1) also under the same condition. i. e. while all p_{ij} $(2\leqslant i<j\leqslant n)$ are equal to one another.

If (2) holds, we have

$$p_{ij}=\left(\frac{a_{ij}}{a_{1i}a_{1j}}\right)^{\frac{1}{2}}=\frac{1}{c_1}(2\leqslant i<j\leqslant n). \tag{9}$$

Hence the condition (2) is sufficient for making the equality hold in (1). Conversely, if all p_{ij} are equal to one another, letting.

$$\frac{a_{ij}}{a_{1i}a_{1j}}=p_{ij}^2=p^2(2\leqslant i<j\leqslant n), \tag{10}$$

and

$$c_1=\frac{1}{p},C_r=pa_{1r}(r=2,3,\cdots,n), \tag{11}$$

then, it is obvious:

$$a_{ij}=c_ic_j(1\leqslant i<j\leqslant n).$$

Hence the condition (2) is necessary for making the equality hold in (1). Theorem 1 is proved.

Corollary. Let D_{ij} denote the edge-lengths of a simplex in E^n, V and R denote its volume and circumradius. Then

$$R \leqslant \left(\frac{n}{2^{n+1}}\right)^{\frac{1}{2}} \cdot \frac{1}{n!V} \prod_{1 \leqslant i < j \leqslant n+1} \rho_{ij}^{\frac{2}{n}} \tag{12}$$

And the equality holds if and only if there exist $n+1$ positive numbers $\mu_r > 0(r=1,2,\cdots,n+1)$ such that

$$\rho_{ij} = \mu_i \mu_j \quad (1 \leqslant i < j \leqslant n+1). \tag{13}$$

Reference

[1]Beckenbach E F, Bellman R. Inequalities. Springer-Verlag, 1961.

[2]Yang Lu, Zhang Jingzhong. A Proof of the Construction Theorem of Simplex. Math. in Practice & Cognition(Chinese), 1980(1): 43 - 45.

[3]Korchmáros G. Atti Accad. Naz. Lincei Rend. Cl. Sci. Fis. Mat. Natur., (8) 56 (1974), No. 6, 876 - 879.

[4]Yang Lu, Zhang Jingzhong. A Class of Geometric Inequalities on Finite Points, Acta Math. Sinica(Chinese), 1980, 23(5): 740 - 749.

A Simplified Model of Smale's Horseshoe

Zhang Jing-zhong(张景中)and Yang Lu(杨　路)

(*University of Science and Technology of China*)

In the study of differentiable dynamics, "Smale's horseshoe" is a well-known and important example[1]. This example presents a diffeomorphism which has an infinite number of periodic points, and the non-wandering set is a "Cantor set".

But the presentation of this example in [1] is rather complicated. Here we shall present a simpler model of the "horseshoe". The proof of the Ω-stability is omitted, since it can be derived from the general result of [1].

First, some rectangle domains Q, Q_1, Q_2, P_1, P_2 in R^2 are defined as follows:

$$Q=\left\{(x,y)\left|\begin{array}{l}x\in(0,1)\\ y\in(0,1)\end{array}\right.\right\},$$

$$Q_1=\left\{(x,y)\left|\begin{array}{l}x\in[0,1]\\ y\in[0.1,0.2]\end{array}\right.\right\},\qquad Q_2=\left\{(x,y)\left|\begin{array}{l}x\in[0,1]\\ y\in[0.8,0.9]\end{array}\right.\right\},$$

$$P_1=\left\{(x,y)\left|\begin{array}{l}x\in[0.1,0.2]\\ y\in[0,1]\end{array}\right.\right\},\qquad P_2=\left\{(x,y)\left|\begin{array}{l}x\in[0.8,0.9]\\ y\in[0,1]\end{array}\right.\right\}.$$

Suppose that diffeomorphism f defined on the set $M\supset\bar{Q}$, (the closure of Q), $f: M\to R^2$, $f(M)\subset M$ satisfies

1) If $z\in M/Q_1\cup Q_2$, then $f^n(z)\overline{\in}Q$. ($\forall n>0$, n is an integer, and $f^n=f\circ f^{n-1}$ as usual.)

2) If $z=(x,y)\in Q_1\cup Q_2$, then

$$f(z)=\begin{cases}\left(\dfrac{1+x}{10},10y-1\right), & (z\in Q_1)\\ \left(\dfrac{9-x}{10},9-10y\right), & (z\in Q_2)\end{cases}\tag{1}$$

then the following statements are true.

1. If $z=(x,y)\in Q$ such that x, y can be represented by the denary periodic decimals as

$$\left.\begin{array}{l}x=0.\dot{a}_{m-1}a_{m-2}\cdots a_1\dot{a}_m,\\ y=0.\dot{a}_1a_2\cdots\dot{a}_m.\end{array}\right\}\tag{2}$$

(Here $m\geqslant 1$, $a_m=1$ and $a_i=1$ or 8 for other i.) Then $f^m(z)=z$, in other words, z are periodic points.

本文刊于《科学通报》,第27卷第2期,1982年.

2. Let $\Omega(f)$ denote the non-wandering set of f as usual, then

$$Q \cap \Omega(f) = \left\{ (x, y) \middle| x = \sum_{k=1}^{\infty} \frac{x_k}{10^k}, y = \sum_{k=1}^{\infty} \frac{y_k}{10^k}; \right.$$

$$x_k = 1 \text{ or } 8 \text{ and } y_k = 1 \text{ or } 8 \text{ for all } k\}. \tag{3}$$

It is obvious that the diffeomorphism f satisfying 1), 2) can be formed easily. And statements 1, 2 express that f has the essential properties of the "horseshoe". These statements will be proved separately in the following.

For $z=(x,y)\in Q$, let the denary decimals of x, y be

$$\begin{cases} x=0.\, x_1 x_2 \cdots x_n \cdots, \\ y=0.\, y_1 y_2 \cdots y_n \cdots. \end{cases} \tag{4}$$

By (1), we have

$$f(z)=(f_x, f_y)=\begin{cases} (0.\, y_1 x_1 \cdots x_n \cdots, 0.\, y_2 \cdots y_n \cdots), (z\in Q_1) \\ (0.\, y_1 \bar{x}_1 \cdots \bar{x}_n \cdots, 0.\, \bar{y}_2 \cdots \bar{y}_n \cdots). (z\in Q_2) \end{cases} \tag{5}$$

(Here $\bar{x}_i=9-x_i$ and $\bar{y}_i=9-y_i$.)

In order to prove statement 1, we shall point that if z has the form (2) then $f^m(z)=z$. In fact, for any $z=(x,y)\in Q_1\cup Q_2$, if $y_1=y_2=\cdots=y_l=1$, then by (5) we have

$$\begin{cases} f_x^l=0.\, y_l y_{l-1} \cdots y_1 x_1 x_2 \cdots x_n \cdots, \\ f_y^l=0.\, y_{l+1} y_{l+2} \cdots y_n \cdots. \end{cases} \quad (y_1=y_2=\cdots=y_l=1) \tag{6}$$

(Here and below $f^l(z)=(f_x^l, f_y^l)$.) If $y_1=y_2=\cdots=y_l=8$ and $y_{l+1}=1$, then

$$\begin{cases} f_x^{l+1}=0.\, y_l y_{l-1} \cdots y_1 1 x_1 x_2 \cdots x_n \cdots, \\ f_y^{l+1}=0.\, y_{l+2} y_{l+3} \cdots y_n \cdots. \end{cases} \quad (y_1=\cdots=y_l=8 \text{ and } y_{l+1}=1) \tag{7}$$

From (6) and (7), it is sure that if $y_{l+1}=1$ and $y_i=1$ or 8 for $1\leqslant i\leqslant l$, then

$$\begin{cases} f_x^{l+1}=0.\, y_l y_{l-1} \cdots y_1 y_{l+1} x_1 \cdots x_n \cdots, \\ f_y^{l+1}=0.\, y_{l+2} \cdots y_n \cdots. \end{cases} \quad (y_{l+1}=1, y_i=1 \text{ or } 8 \text{ for } i\leqslant l) \tag{8}$$

This fact shows that $f^m(z)=z$ if z has the form (2). Thus the proof of statement 1 is complete.

In order to prove statement 2, we have to point out first that if the denary decimals of x, y consist of only digits 1 and 8, then $z=(x,y)\in Q$ are non-wandering. In fact, set

$$\tilde{z}_m=(\tilde{x}, \tilde{y})=(0, \dot{x}_1 \cdots x_m y_m y_{m-1} \cdots y_1 \dot{1}, 0.\, \dot{y}_1 \cdots y_m x_m x_{m-1} \cdots x_1 \dot{1})$$

then $f^{2m+1}(\tilde{z}_m)=\tilde{z}_m$ by (8), i. e. $\tilde{z}_m$ are periodic points. But for any $\varepsilon>0$ we have $|\tilde{z}_m-z|<\varepsilon$ if $m>\lg\left(1+\frac{2}{\varepsilon}\right)$, which means that z is non-wandering.

Consequently, we have to prove that if the denary decimals of x, y consist not only of digits 1 and 8, then $z=(x,y)\in Q$ must be wandering. There exist two cases.

If $\{y_k\}$ consist not only of 1 and 8, let k^* denote the first number such that $(y_k-1)(y_k-8)\neq 0$. By (5), $f_y^{k^*-1}=0.\, y_{k^*} \cdots$ or $0.\, \bar{y}_{k^*} \cdots\cdots$, therefore $f^{k^*-1}(z)\,\overline{\in}\, Q_1\cup Q_2$, Hence z are wandering by 1).

If $\{y_k\}$ consist of only 1 and 8 but $\{x_k\}$ do not, then we can find the neighborhood of z small enough, say $N(z)$, such that there exists some number l, for any $\tilde{z}=(\tilde{x},\tilde{y})\in N(z)$, $\{\tilde{y}_1,\cdots,\tilde{y}_{2l}\}$ consists of only 1 and 8 but $\{\tilde{x}_1,\cdots,\tilde{x}_l\}$ do not. We assert that $f^n(\tilde{z})\overline{\in}N(z)$ if $n\geqslant 2l$. In fact, if $\{\tilde{y}_1,\cdots,\tilde{y}_n\}$ consist of not only 1 and 8, by 1) $f^n(\tilde{z})$ will be outside of $\tilde{Q}$. Otherwise, let $f^n(\tilde{z})=(0.\omega_1\omega_2\cdots, *)$, then $\{\omega_1,\cdots,\omega_l\}$ will consist of only 1 and 8, hence $f^n(\tilde{z})\overline{\in}N(z)$. Therefore z are wandering.

We are looking forward to a simpler and direct proof of the Ω-stability of this model to be found some day.

Reference

[1]Nitecki Z. Differential Dynamics. the M. I. T. Press, Cambridge, Mass., 1971.

On a Conjecture of R. Alexander's

Yang Lu(杨　路)and Zhang Jing-zhong(张景中)

(University of Science and Technology of China)

Ⅰ. Introduction

In a recent paper of R. Alexander's[1] a conjecture is proposed as follows: Suppose two n-simplices have vertices $p_1, p_2, \cdots, p_{n+1}$ and $p'_1, p'_2, \cdots, p'_{n+1}$ respectively, a new simplex p''_1, $p''_2, \cdots, p''_{n+1}$ is made such that

$$|p''_i - p''_j|^2 = \frac{1}{2}(|p_i - p_j|^2 + |p'_i - p'_j|^2), i, j = 1, 2, \cdots, n+1, \tag{1}$$

then follows the inequality

$$V''^2 \geqslant \frac{1}{2}(V^2 + V'^2), \tag{2}$$

where V, V', V'' are the respective volumes.

This operation which makes the third simplex as above is called "metric addition". It is shown in [1] how effectively this method for "metric addition" of simplices is applied to certain geometric inequalities. For this reason the relations between the mutual volumes of these simplices are considered.

The purpose of this note is first to show that the conjecture is false, and secondly, to present a correct proposition.

We put

$$|p'_i - p'_j| = 3|p_i - p_j|, \quad (i, j = 1, 2, \cdots, n+1). \tag{3}$$

In this case, three simplices are similar to each other. Hence

$$V' = 3^n V, \tag{4}$$

and

$$\frac{1}{2}(V^2 + V'^2) = \frac{1}{2}(3^{2n} + 1)V^2. \tag{5}$$

On the other hand. from (1) and (3) we have

$$|p''_i - p''_j| = 5^{\frac{1}{2}}|p_i - p_j|, \quad (i, j = 1, 2, \cdots, n+1), \tag{6}$$

so that

本文刊于《科学通报》,第 27 卷第 7 期,1982 年.

$$V''^2=5^nV^2. \tag{7}$$

When $n\geqslant 2$, it is obvious $5^n<\frac{1}{2}3^{2n}$. Therefore

$$V''^2<\frac{1}{2}(V^2+V'^2), \tag{8}$$

which shows that Alexander's conjecture is false.

However, the following weaker proposition holds.

Theorem 1 *In n-dimensional Euclidean space we have*

$$V''^{\frac{2}{n}}\geqslant\frac{1}{2}(V^{\frac{2}{n}}+V'^{\frac{2}{n}}), \tag{9}$$

and the equality holds if and only if these simplices are similar to each other.

Ⅱ. THE PROOF OF THEOREM 1

Theorem 1 can be written into the following form.

Theorem 1*. *Let Ω,Ω',Ω'' be simplices in E^n, and a_{ij},b_{ij},c_{ij} $(i,j=1,2,\cdots,n+1)$, $|\Omega|$, $|\Omega'|$, $|\Omega''|$ denote the edge-lengths and volumes of these simplices respectively. If*

$$c_{ij}^2=a_{ij}^2+b_{ij}^2 \qquad (i,j=1,2,\cdots,n+1), \tag{10}$$

then

$$|\Omega''|^{\frac{2}{n}}\geqslant|\Omega|^{\frac{2}{n}}+|\Omega'|^{\frac{2}{n}}. \tag{11}$$

And the equality holds if and only if these simplices are similar to each other.

Proof. Notations:

$$\left.\begin{aligned} q_{ij}&=\frac{1}{2}(a_{i,n+1}^2+a_{j,n+1}^2-a_{ij}^2),\\ r_{ij}&=\frac{1}{2}(b_{i,n+1}^2+b_{j,n+1}^2-b_{ij}^2),\end{aligned}\right\}(i,j=1,2,\cdots,n) \tag{12}$$

$$Q=(q_{ij}),R=(r_{ij}),(Q,R \text{ are } n\times n \text{ matrices}) \tag{13}$$

$$s_{ij}(\lambda)=q_{ij}+\lambda r_{ij}, \qquad (i,j=1,2,\cdots,n) \tag{14}$$

$$s(\lambda)=(s_{ij}(\lambda)),(s(\lambda) \text{ is } n\times n \text{ matrix,})$$

$$f_{ij}(\lambda)=-\frac{1}{2}(a_{ij}^2+\lambda b_{ij}^2), \qquad (i,j=1,2,\cdots,n+1) \tag{15}$$

$$F(\lambda)=\begin{pmatrix}0 & 1 & \cdots & 1\\ 1 & & & \\ \vdots & & f_{ij}(\lambda) & \\ 1 & & & \end{pmatrix},(F(\lambda) \text{ is } (n+2)\times(n+2) \text{ matrix.})$$

First, we consider the roots of equation $\det F(\lambda)=0$. Multilply its first row(column) by $-f_{i,n+1}(\lambda)(-f_{n+1,j}(\lambda))$, then add it to the i-th row (j-th column), we get

$$\det F(\lambda)=\begin{vmatrix} 0 & 1 & 1 & \cdots & 1 \\ 1 & & & & \\ 1 & & f_{ij}(\lambda) & & \\ \vdots & & & & \\ 1 & & & & \end{vmatrix}=\begin{vmatrix} 0 & 1\cdots 1 & 1 \\ 1 & & 0 \\ \vdots & s_{ij}(\lambda) & \vdots \\ 1 & & 0 \\ 1 & 0\cdots 0 & 0 \end{vmatrix}$$

$$=-\det S(\lambda)=\det(Q+\lambda R).$$

Put

$$-\det F(\lambda)=\det(Q+\lambda R)=C_0\lambda^n+C_1\lambda^{n-1}+\cdots+C_n, \tag{16}$$

where both Q and R are real symmetric positive definite matrices. Hence all coefficients C_0, $C_1, \cdots, C_n$ are non-negative, the roots of Eq. (16) are real and non-positive. By the Maclaurin[2] inequality we obtain:

$$\frac{C_1}{C_0}\Big/\binom{n}{1}\geqslant\left[\frac{C_2}{C_0}\Big/\binom{n}{2}\right]^{\frac{1}{2}}\geqslant\left[\frac{C_3}{C_0}\Big/\binom{n}{3}\right]^{\frac{1}{3}}\geqslant\cdots\geqslant\left[\frac{C_n}{C_0}\right]^{\frac{1}{n}}. \tag{17}$$

And it follows

$$C_k\geqslant\binom{n}{k}C_0^{1-\frac{k}{n}}C_n^{\frac{k}{n}}\ (k=0,1,\cdots,n). \tag{18}$$

On the other hand, by direct evaluation we have

$$C_0=\det R, C_n=\det Q, \tag{19}$$

where Q and R are the Gramians of Ω and Ω' respectively, hence

$$C_0=n!^2|\Omega'|^2, C_n=n!^2|\Omega|^2. \tag{20}$$

In (16) we put $\lambda=1$, so that

$$C_0+C_1+\cdots+C_n=-\det F(1)$$

$$=-\begin{vmatrix} 0 & 1 & 1 & \cdots & 1 \\ 1 & & & & \\ 1 & & -\frac{1}{2}c_{ij}^2 & & \\ \vdots & & & & \\ 1 & & & & \end{vmatrix}$$

$$=n!^2|\Omega''|^2. \tag{21}$$

In the last step of (21), the volume formula of simplex (cf. [3]or [4]is used.)

Next, summing up the inequalities of (18) for k, we have

$$C_0+C_1+\cdots+C_n\geqslant\sum_{k=0}^{n}\binom{n}{k}C_0^{1-\frac{k}{n}}C_n^{\frac{k}{n}}$$

$$=(C_0^{\frac{1}{n}}+C_n^{\frac{1}{n}})^n. \tag{22}$$

And substituting (21) and (20) into (22), we obtain (11)

$$|\Omega''|^{\frac{2}{n}}\geqslant|\Omega|^{\frac{2}{n}}+|\Omega'|^{\frac{2}{n}}.$$

Finally, we consider the sufficient and necessary condition, for the equality holds in (11). It is obvious that the equality holds if Ω and Ω' are similar. Conversely, if in (11) the equality holds, the equality in (18) holds also. By the Maclaurin inequality, $\det(Q+\lambda R)$

should have n-multiple root $-\mu_0$. Hence rank $(Q-\mu_0 R)=0$, i. e. $Q=\mu_0 R$ and

$$q_{ij}=\mu_0 r_{ij} \quad (i,j=1,2,\cdots,n). \tag{23}$$

then, by (12) we get

$$a_{ij}=\mu_0 b_{ij} \quad (i,j=1,2,\cdots,n). \tag{24}$$

therefore, Ω and Ω' are similar (so it Ω''), and the proof of Theorem 1^* is complete.

Further, suppose $m+1$ n-simplices have vertices $p_1^{(k)}, p_2^{(k)}, \cdots, p_{n+1}^{(k)}$ $(k=0,1,\cdots,m)$ respectively, such that

$$|p_i^{(m)}-p_j^{(m)}|^2=\frac{1}{m}\sum_{k=0}^{m-1}|p_i^{(k)}-p_j^{(k)}|^2,(i,j=1,\cdots,n+1) \tag{25}$$

and $V, V', \cdots, V^{(m)}$ are respective volumes. Obviously, by Theorem 1^* it is quite easy to show that

Corollary 1.

$$V^{(m)\frac{2}{n}}\geqslant\frac{1}{m}(V^{\frac{2}{n}}+V'^{\frac{2}{n}}+\cdots+V^{(m-1)\frac{2}{n}}). \tag{26}$$

And the equality holds if and only if these simplices are similar to each other.

Corollary 2.

$$V^{(m)}\geqslant(VV'\cdots V^{(m-1)})^{\frac{1}{m}}. \tag{27}$$

Ⅲ. FURTHER PROBLEMS

Problem 1. Suppose two simplices have vertices $p_1, p_2, \cdots, p_{n+1}$ and $p_1', p_2', \cdots, p_{n+1}'$ such that

$$|p_i''-p_j''|=(|p_i-p_j||p_i'-p_j'|)^{\frac{1}{2}}.$$

And V, V', V'' are the respective volumes. then shall we have the following inequality

$$V''\geqslant(VV')^{\frac{1}{2}}?$$

And how can we say that if

$$|p_i''-p_j''|^k=\frac{1}{2}(|p_i-p_j|^k+|p_i'-p_j'|^k)?$$

Problem 2. Suppose $|p_i'-p_j'|=|p_i-p_j|^{\frac{1}{2}}$, can we obtain

$$V'^2\geqslant\frac{1}{n!}\left(\frac{n+1}{2^n}\right)V?$$

And how can we say that if $|p_i'-p_j'|=|p_i-p_j|^{\frac{1}{k}}$?

Reference

[1]Alexander R. the Geometry of Metric and Linear Space. Springer-Verlag, 1975: 57 - 65.
[2]Beckenbach E F, Bellman R. Inequalities. Springer-Verlag, 1961: 10 - 11.
[3]Blumenthal L M. Theory and Applications of Distance Geometry. Oxford, 1953: 97 - 98.
[4]杨路,张景中. 关于有限点集的一类几何不等式. 数学学报,1980,23(5):740 - 749.

单变元实迭代半群的存在唯一准则

张景中 杨 路

（中国科学技术大学）

§1 引 言

给了一个定义于集 M 而取值于 M 的函数

$$y=F(x)(x\in M,y\in M),$$

可以定义它的 n 次迭代

$$F^{[0]}(x)=x,F^{[n+1]}(x)=F[F^{[n]}(x)](n=0,1,2,\cdots).$$

如果 $F(x)$ 有反函数，则记为 $F^{[-1]}(x)$，并记

$$F^{[-n]}(x)=F^{[-1]}[F^{[-n+1]}(x)](n=1,2,\cdots),$$

这时，把 n 叫做 $F^{[n]}(x)$ 关于 $F(x)$ 的迭代指数. 显然，迭代指数与数的乘幂指数有类似的运算规律：

$$\begin{cases}F^{[n]}(F^{[m]}(x)]=F^{[n+m]}(x),\\(F^{[m]})^{[n]}(x)=F^{[mn]}(x),\end{cases}\tag{1.1}$$

等等.

由于迭代运算在数学中的重要性，很早就有人考虑过迭代的表达问题. 对于实函数 $F(x)$，1871 年（见文献[1]中）提出：

1° 如何找出 $F^{[n]}(x)$ 关于 n 的分析表达式？

2° 函数方程

$$f^{[n]}(x)=F(x)(n\text{ 正整数},f\text{ 未知})\tag{1.2}$$

是否有解？或何时有解？

3° 能不能把迭代指数 n 推广到任意实数？

文献[1]中也提出了解决这些问题的想法：对给定的 $F(x)$，考虑所谓 Schröder 函数方程

$$\varphi[F(x)]=a\varphi(x);\tag{1.3}$$

如果(1.3)有解 $\varphi(x)$ 且 $\varphi^{[-1]}(x)$ 存在，则有：

$$F^{[n]}(x)=\varphi^{[-1]}[a^n\varphi(x)].\tag{1.4}$$

这就找到了 $F^{[n]}(x)$ 的分析表达式；如果 $a>0$，取 n 为任意实数，就得到了迭代指数推广的途径，而且当令 $f(x)=\varphi^{[-1]}(a^{\frac{1}{n}}\varphi(x))$ 时，$f(x)$ 显然满足方程(1.2).

本文刊于《北京大学学报（自然科学版）》，第 6 期，1982 年.

令 $F^{[t]}(x)=\varphi^{[-1]}[a^t\varphi(x)],t\in(-\infty,+\infty)$，
这样由 $F(x)$ 出发而得到的函数族 $\{F^{[t]}(x)\}_t$，在复合运算之下成为一个单参数的 Abel 群.
一般说来，若单参数函数族 $\{F(x,t)\}_t$ 满足：

1° $F[F(x,t_1),t_2]=F(x,t_1+t_2)(t_1,t_2\in(-\infty,+\infty))$，

2° $F(x,0)=x,F(x,1)=F(x)$；

则称之为由 F 生成的迭代群，若限定参数 $t\geqslant 0$，则称为迭代半群.

一个基本的问题是：对于什么样的 $F(x)$，存在由 F 生成的迭代群或迭代半群？这样的迭代群或迭代半群是否唯一？或者，对族中的函数加上哪些限制才能使之成为唯一的？

这方面早期的工作有文献[2]～[5]，近期的则有文献[6]～[16]. 早期工作中，偏重于研究 $F(x)$ 为解析或多次可微的情况；近来，则更多地关心严格单调的连续函数. 通常，都假定 $F(x)$ 定义于一个不动点的邻域或半邻域，并要求 F 在不动点可微且微商不为 0 或 ± 1. 尽管对 F 有了这些限制，仍未得到由 F 生成的迭代半群存在且唯一的充要条件. 特别是，对参数族中的函数 $F^{[t]}(x)$ 究竟应当加上什么限制，至今也没有统一的看法.

另一方面，近年来崛起的微分动力体系的研究中，进一步讨论了流形上的变换 F 所生成的迭代群的性质[13]. 其中经常用到的 Hartman 线性化定理[13]～[14]，可以看成是 Schröder 方程(1.3)在高维情况下连续解的存在定理. 但是，前述基本问题，即使在一维的似乎很简单的情形，却仍然没得到满意的回答.

本文将指出：若对 $F(x)$ 及由它生成的 $F(x,t)$ 中的函数在不动点邻域的渐近性态加以分类的描述，就可以对上述的存在唯一问题作出看来是恰当的提法；并获得一大批函数的迭代群存在唯一的充要条件. 其中还包括了不动点处微商为 0、1 的情况，从而使这一古典课题有所进展.

§2　正规迭代族

(2.1)　**定义 1**　若二元连续函数 $y=G(x,t)(a<x<b)(0\leqslant t<\infty)$ 满足：

1° $G(x,0)=x$；

2° $a<x<G(x,t)<b$(当 $t>0$)；

3° 当 $x_1<x_2$，有 $G(x_1,t)<G(x_2,t)$；

4° $G[G(x,t_1),t_2]=G(x,t_1+t_2)$；

则称 $G(x,t)$ 为 (a,b) 上的以 t 为参数的正规迭代族，简称正规族.

例　1) $G_0(x,t)=x+t(x\in(-\infty,+\infty))$，

2) $G_1(x,t)=a^tx(0<a<1,x\in(-\infty,0))$，

3) $G_2(x,t)=x^{2^t}(1<x<+\infty)$，

4) $G_4(x,t)=\dfrac{x}{\sqrt{1+tx^2}}(-1<x<0)$.

(2.2)　**引理 1**　若 $y=G(x,t)$ 为 (a,b) 上的正规族，则存在唯一的二元函数 $t=T(x,y)(a<x<b,x\leqslant y<b)$，满足 $y=G[x,T(x,y)]$，且 $T(x,y)$ 连续，关于 y 递增，关于 x 递减.

证明　对任意的$t\geqslant 0$，$\Delta t>0$，由(2.1)中的条件4°及2°，得

$$G(x,t+\Delta t)=G[G(x,t),\Delta t]>G(x,t),$$

可见$G(x,t)$关于t递增，又因$G(x,0)=x$，故当x固定时，$G(x,t)$取到的最小值为x. 下面往证$G(x,t)$可取到$[x,b)$中间的一切值. 由连续性，只要证明

$$\lim_{t\to+\infty}G(x,t)=b$$

即可. 事实上，若令$X_n=G(x,n)$，则$X_{n+1}>X_n$，若$X_n\to X^*<b$，由

$$G(x_n,1)=G[G(x,n),1]=G(x,n+1)=x_{n+1},$$

取极限得$C(x^*,1)=x^*$，此与条件2°矛盾. 故必有$x^*=b$(这里，也并没有排除$b=+\infty$的情形).

因此，对任一$y\in[x,b)$，有唯一的$t=T(x,y)$使$y=G[x,T(x,y)]$，由G连续知T连续，剩下的是证明$T(x,y)$关于y递增、关于x递减：

1)若有$x\leqslant y_1<y_2<b$，使$t_1=T(x,y_1)\geqslant T(x,y_2)=t_2$，则：

$y_1=G(x,t_1)=G[G(x,t_2),t_1-t_2]\geqslant G(x,t_2)=y_2$，此与$y_1<y_2$矛盾. 从而$T(x,y)$关于$y$递增.

2)若有$a<x_1<x_2<y$，使$t_1=T(x_1,y)\leqslant T(x_2,y)=t_2$，则

$$G(x_1,t_1)\leqslant G(x_1,t_2),$$

即

$$y=G(x_1,t_1)\leqslant G(x_1,t_2)<G(x_2,t_2)=y,$$

所推出的矛盾证明$T(x,y)$关于x递减. 引理1证毕.

(2.3)　**定义2**　若$G(x,t)$为(a,b)上的正规族；$F(x)$为(a,b)上的函数，$a<F(x)<b$，$T(x,y)$如引理2.1所述. 如果在$x=b$的邻域有$F(x)\geqslant x$，且

$$\lim_{x\to b}T[x,F(x)]=\alpha_F>0$$

存在，则称F关于族G可度，记之以$F\sim\{G\}$，并把$\alpha=\alpha_F$叫做F关于G的指标.

(2.4)　**引理2**　若$F\sim\{G\}$，则对任给的$0<\varepsilon\leqslant\alpha_F$，必可找到$a^*\in(a,b)$，使当$x\in(a^*,b)$时有：

$$G(x,\alpha_F-\varepsilon)<F(x)<G(x,\alpha_F+\varepsilon).$$

证明　由α_F之定义，可知有a^*，使当x在(a^*,b)中时有：

$$\alpha_F-\varepsilon<T[x,F(x)]<\alpha_F+\varepsilon,$$

又由$G(x,t)$关于t递增，知：

$$G(x,\alpha_F-\varepsilon)<G[x,T(x,F(x))]<G(x,\alpha_F+\varepsilon),$$

由$T(x,y)$之定义，得$G[x,T(x,F(x))]=F(x)$，证毕.

(2.5)　**引理3**　若$F_1\sim\{G\}$，$F_2\sim\{G\}$，则必有$F_1[F_2]\sim\{G\}$，且：

$$\alpha_{F_1[F_2]}=\alpha_{F_1}+\alpha_{F_2}.$$

证明　对任给的足够小的$\varepsilon>0$，由引理2，可找到a^*，使当$x\in(a^*,b)$时有：

$$G(x,\alpha_{F_i}-\varepsilon)\leqslant F_i(x)\leqslant G(x,\alpha_{F_i}+\varepsilon)\ (i=1,2).$$

由$G(x,t)\geqslant x$，知$F_i(x)\in(a^*,b)$；由G之单调性得：

$$F_1[F_2(x)]\leqslant G[F_2(x),\alpha_{F_1}+\varepsilon]\leqslant G[G(x,\alpha_{F_2}+\varepsilon),\alpha_{F_1}+\varepsilon]=G[x,\alpha_{F_1}+\alpha_{F_2}+2\varepsilon].$$

上式两端代入 $T[x,y]$，得：

$$T[x,F_1[F_2(x)]]\leqslant \alpha_{F_1}+\alpha_{F_2}+2\varepsilon,$$

类似地得：

$$T[x,F_1[F_2(x)]]\geqslant \alpha_{F_1}+\alpha_{F_2}-2\varepsilon,$$

此二式说明：

$$\lim_{x\to b}T[x,F_1[F_2(x)]]=\alpha_{F_1}+\alpha_{F_2},$$

即 $\alpha_{F_1[F_2]}$存在且等于 $\alpha_{F_1}+\alpha_{F_2}$，证毕.

(2.6) **引理 4** 若 $f\sim\{G\}$，$g\sim\{G\}$，$\alpha_f<m\alpha_g$，这里 m 为自然数，则有 $a^*\in(a,b)$，使当 $x\in(a^*,b)$时有 $f(x)<g^{[m]}(x)$.

证明 令 $g^{[m]}(x)=h(x)$，由引理 3，$h\sim\{G\}$且 $\alpha_h=m\alpha_g$；由引理 2，有 a^*使当 $x\in(a^*,b)$时：

$$f(x)<G(x,\alpha_f+\varepsilon),$$

$$g^{[m]}(x)>G(x,\alpha_h-\varepsilon),$$

如果取 $\varepsilon=\dfrac{\alpha_h-\alpha_f}{3}$，则由 G 关于 t 递增及不等式 $\alpha_f+\varepsilon<\alpha_h-\varepsilon$ 即得：

$$f(x)<G(x,\alpha_f+\varepsilon)<G(x,\alpha_h-\varepsilon)<g^{[m]}(x),$$

即所欲证.

(2.7) **引理 5** 若 $g\sim\{G\}$，$f\sim\{G\}$，f 连续且 $f(x)>x$，则对 $x_0\in(a,b)$，有

$$\lim_{n\to+\infty}T[f^{[n]}(x_0),g[f^{[n]}(x_0)]]=\alpha_g.$$

证明 显然只需指出 $\lim\limits_{n\to+\infty}f^{[n]}(x_0)=b$，由于在$(a,b)$上有 $x<f(x)<b$ 及 $f(x)$连续，这是显然的事实.

引理 5 中的条件“f 连续”，显然也可代之“$f(x)\geqslant h(x)>x$，而 $h(x)$连续”.

§3 基本定理

应用前面的定义及引理，可得：

(3.1) **定理 1** 设 $G(x,t)$为(a,b)上的正规迭代族，$F(x)$在(a,b)上连续、递增，$F(x)>x$ 而且 $F(x)\sim\{G\}$，则下列五条件等价：

(Ⅰ) 存在自然数列 $m_1<m_2<\cdots<m_k<\cdots$，使对任一自然数 k，有连续函数 $f_k(x)$定义于(a,b)，$f_k(x)\sim\{G\}$，满足 $f_k^{[m_k]}(x)=F(x)$，且：

$$\lim_{k\to+\infty}f_k(x)=x.$$

(Ⅱ) 二元函数列

$$\Phi_n(x,y)=T[F^{[n]}(x),F^{[n]}(y)]\quad(n=1,2,\cdots)$$

在区域 D：$\begin{cases}a<x<b\\ x\leqslant y<b\end{cases}$上收敛到连续、严格单调的函数 $\Phi(x,y)$.

(Ⅲ) 对任给的(a,b)上的函数 $g(x)$，若 $g\sim\{G\}$，则函数列

$$H_n(x)=F^{[-n]}[g[F^{[n]}(x)]](n=1,2,\cdots)$$

当 $n\to+\infty$ 时有极限 $H(x)\sim\{G\}$，且 $\alpha_H=\alpha_g$. 而且，这个 $H(x)$ 恰是由(Ⅱ)中的 $\Phi(x,y)=\alpha_H$ 所确定的隐函数，即

$$\Phi(x,H(x))=\alpha_H.$$

(Ⅳ)　对任给的实数 $c\in(0,+\infty)$，有唯一的函数 $g(x,c)\quad(x\in(a,b))$，满足：

1) $g(x,c)$ 在 $x\in(a,b)$ 单调；

2) $g(x,c)\sim\{G\}$，且 $\alpha_{g(x,c)}=c$；

3) $g[F(x),c]=F[g(x,c)]$.

(Ⅴ)　对任意自然数 m，有唯一的 $f(x)\sim\{G\}$，满足

$$f^{[m]}(x)=F(x),$$

且 $f(x)$ 在 (a,b) 连续.

证明　分五步进行：

(Ⅰ)⇨(Ⅱ)：

1°　证明 $\Phi_n(x,y)$ 在区域 D 收敛.

任取 $(x_0,y_0)\in D$，若 $x_0=y_0$，序列显然收敛，故不妨设 $y_0>x_0$，取充分大的 m_k，考虑点列：

$$x_0,x_1=f_k(x_0),\cdots,x_l=f_k(x_{l-1}),\cdots$$

由于 $f_k^{[m_k]}(x)=F(x)$，从 $F(x)$ 在 (a,b) 内无不动点知 $f_k(x)$ 在 (a,b) 内无不动点. 又若 $f_k(x')=f_k(x'')$ 则有 $F(x')=F(x'')$，从而由 $F(x)$ 递增及 f_k 连续推得 f_k 递增及 $f_k(x)>x$，故有：

$$\lim_{l\to+\infty}x_l=b,$$

于是有 l^* 使

$$x_l^*=f_k^{[l^*]}(x_0)\leqslant y_0\leqslant f_k^{[l^*+1]}(x_0)=x_{l^*+1},$$

从而

$$\begin{aligned}\Phi_n(x_0,y_0)&=T[F^{[n]}(x_0),F^{[n]}(y_0)]\leqslant T[F^{[n]}(x_0),F^{[n]}(f_k^{[l^*+1]}(x_0))]\\&=T[F^{[n]}(x_0),f_k^{[l^*+1]}(F^{[n]}(x_0))].\end{aligned}$$

同理：

$$\Phi_n(x_0,y_0)\geqslant T[F^{[n]}(x_0),f_k^{[l^*]}(F^{[n]}(x_0))],$$

令 $n\to+\infty$，由 $F^{[n]}(x_0)\to b$ 及 $f_k\sim\{G\}$ 得：

$$\overline{\lim_{n\to+\infty}}\Phi_n(x_0,y_0)\leqslant\alpha_{f_k}^{[l^*+1]}=(l^*+1)\alpha_{f_k},$$

$$\underline{\lim_{n\to+\infty}}\Phi_n(x_0,y_0)\geqslant\alpha_{f_k}^{[l^*]}=l^*\alpha_{f_k},$$

即 $\Phi_n(x_0,y_0)$ 上下极限之差不超过 α_{f_k}. 由引理 3，及 $f_k^{[m_k]}(x)=F(x)$，知 $\alpha_{f_k}=\dfrac{\alpha_F}{m_k}$；但 m_k 可任意大，故 $\Phi_n(x_0,y_0)$ 的上下极限相等，即 $\Phi_n(x,y)$ 收敛. 设其极限函数为 $\Phi(x,y)$.

2°　证明 $\Phi(x,y)$ 连续.

由 $\Phi_n(x,y)$ 单调知 $\Phi(x,y)$ 单调，故只要证明：对任给的 y_0，$\Phi(x,y_0)$ 是 x 的连续函数；且对任给的 x_0，$\Phi(x_0,y)$ 也是 y 的连续函数已足.

因为 $\Phi(x,y_0)$ 单调，故它的间断点是第一类的. 又因 $\Phi(x,y_0)$ 关于 x 非增，且 $\Phi(y_0,y_0)$

$=0$,故若 $\Phi(x,y_0)$ 不连续,则必有 $t_1>0$ 及 $\delta>0$,使

$$(t_1,t_1+\delta)\subset(0,\Phi(a,y_0)),$$

但 $\Phi(x,y_0)$ 取不到 $(t_1,t_1+\delta)$ 内的值. 这里:

$$\Phi(a,y_0)=\lim_{x\to a}\Phi(x,y_0).$$

在 (a,y_0) 内取足够靠近于 a 的点 x_1,使:

$$\Phi(x_1,y_0)\geqslant t_1+\delta,$$

作序列 $f_k^{[l]}(x_1)$,$l=1,2,\cdots$,由于

$$\lim_{l\to+\infty}f_k^{[l]}(x_1)=b,$$

故可找到 s,使 $f_k^{[s]}(x_1)\leqslant y_0\leqslant f_k^{[s+1]}(x_1)$,由 Φ 的单调性得:

$$(3.2)\quad\begin{cases}\Phi[f_k^{[l]}(x_1),y_0]\leqslant\Phi[f_k^{[l]}(x_1),f_k^{[s+1]}(x_1)],\\ =\lim\limits_{x\to b}T[x,f_k^{[s+1-l]}(x)]=\dfrac{s+1-l}{m_k}\alpha_F,\\ \Phi[f_k^{[l]}(x_1),y_0]\geqslant\Phi[f_k^{[l]}(x_1),f_k^{[s]}(x_1)]=\dfrac{s-l}{m_k}\alpha_F,\end{cases}$$

因而有:

$$0\leqslant\Phi[f_k^{[l]}(x_1),y_0]-\Phi[f_k^{[l+1]}(x_1),y_0]\leqslant\frac{\alpha_F}{m_k}.$$

当 $l=0$ 时,有 $\Phi[x_1,y_0]\geqslant t_1+\delta$,而当 $l=s$ 时又有

$$\frac{\alpha_F}{m_k}\Phi[f_k^{[s]}(x_1),y_0]\geqslant 0,$$

故当 k 充分大,以致使 $\dfrac{2\alpha_F}{m_k}<\delta$ 时,数列

$$\Phi[F_k^{[l]}(x_1),y_0](l=0,1,\cdots,s)$$

中必取到 $(t_1,t_1+\delta)$ 内之值. 所推出的矛盾证明了 $\Phi(x,y_0)$ 对 x 的连续性.

至于 $\Phi(x_0,y)$ 对 y 的连续性,更易证明. 仍用反证法:若不然,必有 $\delta>0$,$t_2>0$,使得 $\Phi(x_0,y)$ 取不到 $(t_2,t_2+\delta)$ 内的值,但若令

$$y_l=f_k^{[l]}(x_0),$$

则有

$$\Phi[x_0,f_k^{[l]}(x_0)]=\lim_{n\to\infty}T\Big(F^{[n]}(x_0),f_k^{[l]}[F^{[n]}(x_0)]\Big)=\frac{l\alpha_F}{m_k},$$

由于 $\dfrac{\alpha_F}{m_k}$ 可以任意小而 $l=0,1,2,\cdots$,故 $\Phi[x_0,f_k^{[l]}(x_0)]$ 一定可以取到 $(t_2,t_2+\delta)$ 内的值,从而 $\Phi[x_0,y]$ 关于 y 也连续.

3° 证明 $\Phi(x,y)$ 在区域 D 严格单调,亦即当 $x_0\leqslant y_1<y_2<b$ 时有 $\Phi(x_0,y_1)<\Phi(x_0,y_2)$,当 $a<x_1<x_2\leqslant y_0$ 时有 $\Phi(x_1,y_0)>\Phi(x_2,y_0)$.

首先,由条件(Ⅰ)中之假设知 $\lim\limits_{k\to\infty}f_k(x)=x$;由于 $f_k(x)$ 连续、单调及极限函数之连续性,对 (a,b) 内任两点 $a'<b'$,$f_k(x)$ 在 $[a',b']$ 上一致收敛于 x;故有 k_0,使当 $k\geqslant k_0$ 时,对一切 $x\in[x_0,y_2]$,有:

$$|f_k(x)-x|<\frac{y_2-y_1}{3}.$$

考虑序列 $x_l=f_k^{[l]}(x_0)$，对 $x_l\in[x_0,y_2]$有

$$|x_{l+1}-x_l|=|f_k(x_l)-x_l|\subset\frac{y_2-y_1}{3}.$$

由于当 $l\to+\infty$时 $x_l\to b$，故有 l^*，使：

$$y_1\leqslant x_{l^*}<x_{l^*+1}\leqslant y_2;$$

于是

$$\Phi(x_0,y_1)\leqslant\Phi(x_0,x_{l^*})=\lim_{n\to\infty}T[F^{[n]}(x_0),f_k^{[l^*]}(F^{[n]}(x_0))]=\frac{l^*}{m_k}\alpha_F,$$

$$\Phi(x_0,y_2)\geqslant\Phi(x_0,x_{l^*+1})=\frac{l^*+1}{m_k}\alpha_F,$$

$\therefore\quad \Phi(x_0,y_2)>\Phi(x_0,y_1).$

类似地，为了证明当 $x_1<x_2$ 时 $\Phi(x_1,y_0)>\Phi(x_2,y_0)$，可考虑序列 $f_k^{[l]}(x_1)(l=1,2,\cdots)$. 和刚才一样，取 k 足够大时，有 $f_k^{[l]}(x_1)<x_2$，又设

$$f_k^{[s]}(x_1)\leqslant y_0\leqslant f_k^{[s+1]}(x_1),$$

由前述不等式(3.2)得：

$$\Phi[f_k^{[2]}(x_1),y_0]\leqslant\frac{s+1-2}{m_k}\alpha_F=\frac{s-1}{m_k}\alpha_F,$$

$$\Phi[x_1,y_0]\geqslant\frac{s-0}{m_k}\alpha_F=\frac{s}{m_k}\alpha_F;$$

$$\therefore\Phi(x_1,y_0)-\Phi(x_2,y_0)\geqslant\Phi(x_1,y_0)-\Phi[f^{[2]}(x_1),y_0]\geqslant\frac{\alpha_F}{m_k}>0.$$

(Ⅰ)⇨(Ⅱ)证毕.

(Ⅱ)⇨(Ⅲ)之证明：

首先指出，当 n 充分大时，$H_n(x)=F^{[-n]}[g(F^{[n]}(x))]$在$(a,b)$内确有定义，这是因为 $g(x)\sim\{G\}$，故由引理 2，存在 $a^*\in(a,b)$，使当 $x\in(a^*,b)$时，成立着 $g(x)>G(x,\alpha_g-\varepsilon)>x$ 之故.

下面往证 $H_n(x)$收敛.

在(a,b)内任取 x_0，作序列$\{H_n(X_0)\}$，由于当 n 足够大时 $g[F^{[n]}(x_0)]>F^{[n]}(x_0)$，故得：

$$H_n(x_0)\geqslant F^{[-n]}[F^{[n]}(x_0)]=x_0.$$

但由于 $g\sim\{G\}$，故有自然数 k 使 $\alpha_g\leqslant k\alpha_F$，由引理 4，对足够大的 n 有

$$g[F^{[n]}(x_0)]\leqslant F^{[n+k]}(x_0),$$

故 $H_n(x_0)\leqslant F^{[k]}(x_0)$，即$\{H_n(x_0)\}$有界. 设 y_0 是序列$\{H_n(x_0)\}$的一个极限点，我们来证明它只有这一个极限点.

为此，先证明对(Ⅱ)中给出的 $\Phi(x,y)$，有：$\Phi(x_0,y_0)=\alpha_g$.

对任给的 $\varepsilon>0$，取 $\delta>0$，使 $\Delta=[y_0-\delta,y_0+\delta]\subseteq(x_0,b)$，并使当 $y\in\Delta$ 时：

$$|\Phi(x_0,y_0)-\Phi(x_0,y)|<\varepsilon;$$

取 N_1 充分大，使当 $m\geqslant N_1$ 时有：

$$|\Phi_m(x_0, y_0 \pm \delta) - \Phi(x_0, y_0 \pm \delta)| < \varepsilon,$$

从而：

$$|\Phi_m(x_0, y_0 \pm \delta) - \Phi(x_0, y_0)| < 2\varepsilon.$$

由于 $\Phi_m(x_0, y)$关于 y 在 Δ 上单调，故：

(3.3) $$|\Phi_m(x_0, y) - \Phi(x_0, y_0)| < 2\varepsilon \qquad (y \in \Delta, m \geqslant N_1).$$

由 $g \sim \{G\}$知$\lim\limits_{n\to\infty} T[F^{[n]}(x_0), g[F^{[n]}(x_0)]] = \alpha_g$，取 N_2 足够大，使当 $l \geqslant N_2$ 时：

(3.4) $$|T[F^{[l]}(x_0), g[F^{[l]}(x_0)]] - \alpha_g| < \varepsilon.$$

由于 y_0 是$\{H_n(x_0)\}$的极限点，故可找到比 N_1, N_2 都大的 N，使

$$|H_N(x_0) - y_0| < \delta,$$

于是由(3.3)：

$$|\Phi_N[x_0, H_N(x_0)] - \Phi[x_0, y_0]| < 2\varepsilon,$$

又由(3.4)：

$$\begin{aligned}|\Phi_N[x_0, H_N(x_0)] - \alpha_g| &= |T[F^{[N]}(x_0), F^{[N]}[H_N(x_0)] - \alpha_g| \\ &= |T[F^{[N]}(x_0), g[F^{[N]}(x_0)]] - \alpha_g| < \varepsilon,\end{aligned}$$

从而得：

$$|\Phi(x_0, y_0) - \alpha_g| < 3\varepsilon,$$

但 ε 可以任意小，故 $\Phi(x_0, y_0) = \alpha_g$. 由于 $\Phi(x_0, y)$关于 y 严格单调，故满足 $\Phi(x_0, y_0) = \alpha_g$ 的 y_0 唯一地被 x_0 决定；亦即$\{H_n(x_0)\}$的极限点 $H(x_0)$是唯一的. 所以

$$\therefore \lim_{n\to\infty} H_n(x) = H(x); \Phi[x, H(x)] = \alpha_g.$$

顺便知道：$H(x)$连续，仅与 α_g 有关，而且是递增的.

下面证明 $H(x) \sim \{G\}$，且 $\alpha_H = \alpha_g$，为此要用到如下之命题.

(3.5)**命题**　若函数列$\{f_n(x, y)\}$在有界闭区域 D^2 上收敛于连续的 $f(x, y)$，且对任一点(x_0, y_0)有 $\delta > 0$，使当 $\delta \geqslant \delta_1 > 0$ 时，在区域

$$\Delta = \begin{Bmatrix} x_0 - \delta_1 \leqslant x \leqslant x_0 + \delta_1 \\ y_0 - \delta_1 \leqslant y \leqslant y_0 + \delta_1 \end{Bmatrix}$$

上有两点$(x_{0,1}, y_{0,1})$，$(x_{0,2}, y_{0,2})$，对一切 n 和一切$(x, y) \in \Delta$，都有

$$f_n(x_{1,0}, y_{1,0}) \leqslant f_n(x, y) \leqslant f_n(x_{2,0}, y_{2,0}),$$

则 $f_n(x, y)$在 D^2 上一致收敛于 $f(x, y)$.

应用有限覆盖定理即可证明(3.5)，兹从略.

由 $\Phi_n(x, y)$的严格单调性及 $\Phi(x, y)$之连续性，应用命题(3.5)，即知 $\Phi_n(x, y)$在

$$D_{a,b} = \begin{Bmatrix} a < x < b \\ x \leqslant y < b \end{Bmatrix}$$

的任一闭子域上一致收敛，特别是在闭集

$$S_{x_0} = \{a < x_0 \leqslant x \leqslant F(x_0), y = H(x)\}$$

上一致收敛，即 $T[F^{[n]}(x), F^{[n]}(H(x))]$在$[x_0, F(x_0)]$上一致收敛，由 F 之连续性：

$$F[H(x)]=F[\lim_{n\to\infty}H_n(x)]=\lim_{n\to\infty}F[H_n(x)]$$
$$=\lim_{n\to\infty}H_{n-1}[F(x)]=H[F(x)].$$

故在$[x_0,F(x_0)]$上一致地有

$$\lim_{n\to\infty}T[F^{[n]}(x),H[F^{[n]}(x)]]=\Phi[x,H(x)]=\alpha_g.$$

这说明由隐函数方程$\Phi(x,y)=\alpha$所确定的函数$y=\widetilde{H}(x)$即是我们由$H_n(x)$取极限得到的$H(x)$. 由于$F^{[N]}(x)$是1-1地连续地把$[x_0,b)$变到$[F^{[N]}(x_0),b)$，而又有$\lim_{n\to\infty}F^{[n]}(x_0)=b$,故：

$$\lim_{x\to b}T[x,H(x)]=\lim_{n\to\infty}T[F^{[n]}(x),H(F^{[n]}(x))]=\alpha_g,$$

由定义2,$H\sim\{G\},\alpha_H=\alpha_g$.（Ⅱ）⇨（Ⅲ）证毕.

（Ⅲ）⇨（Ⅳ）之证明：

若（Ⅲ）成立，对于给定的c，令：

$$g(x,c)=\lim_{n\to+\infty}F^{[n]}[G[F^{[n]}(x),c]],$$

由于$G(x,c)\sim\{G\},\alpha_{G(x,c)}=c$，故$g(x,c)$存在，$g(x,c)\sim\{G\},\alpha_{g(x,c)}=c$.

由F之连续性：

$$F[g(x,c)]=g[F(x),c].$$

（这一点，在（Ⅱ）⇨（Ⅲ）的证明中已提到过）另外，由$F(x)$及$G(x,c)$之单调性知道$g(x,c)$关于x单调，剩下的是：证明这样的函数是唯一的. 若不然，设有$g_1(x),g_2(x)$都满足：

$$\begin{cases}g_1(x)\sim\{G\},\alpha_{g_1}=\alpha_{g_2}=c\geqslant 0,\\ g_1[F(x)]=F[g_1(x)],\end{cases}$$

往证必有$g_1(x)\equiv g_2(x)$. 若不然，设在某点$x_0\in(a,b)$，有$g_1(x_0)\neq g_2(x_0)$，我们构造一个函数$\psi(x)$，使之满足

$$\psi(x)\sim\{G\},\alpha_\psi=c,$$

而$\lim_{n\to+\infty}F^{[-n]}[\psi(F^{[n]}(x))]$却不存在，从而推出与条件（Ⅲ）的矛盾. 令

$$\psi(x)=\begin{cases}g_1(x),x\in\{F^{[2m]}(x_0)\}(m=0,1,\cdots),\\ g_2(x),\text{其它 } x\in(a,b).\end{cases}$$

显然$\psi\sim\{G\},\alpha_\psi=c$，但当$m\to+\infty$时：

$$\lim_{m\to+\infty}F^{[-2m]}[\psi[F^{[2m]}(x_0)]]=\lim_{m\to+\infty}F^{[-2m]}[g_1[F^{[2m]}(x_0)]]=g_1(x_0),$$

$$\lim_{m\to+\infty}F^{[-2m-1]}[\psi[F^{[2m]}(x_0)]]=\lim_{m\to+\infty}F^{[-2m-1]}[g_2[F^{[2m+1]}(x_0)]]=g_2(x_0),$$

从而推出了矛盾.（Ⅲ）⇨（Ⅳ）证毕.

（Ⅳ）⇨（Ⅴ）之证明：

由（Ⅳ），对任意正整数m，有$f(x)$满足：$f(x)$单调，$f\sim\{G\},\alpha_f=\frac{1}{m}\alpha_F$，及

$$f[F(x)]=F[f(x)],$$

而且这样的f是唯一的.

令$F_1(x)=f^{[m]}(x)$，则由引理3,$F_1\sim\{G\},\alpha_{F_1}=\alpha_F$，又显然有：

$$F_1[F(x)]=F[F_1(x)],$$
$$F[F(x)]=F[F(x)],$$

于是由(Ⅳ)中所设之唯一性知 $F_1(x)=F(x)$.

另一方面,若有 $f_1\sim\{G\}$ 也满足 $f_1^{[m]}(x)=F(x)$,则必有 $f_1[F(x)]=F[f_1(x)]$ 及 $\alpha_{f_1}=\frac{1}{m}\alpha_F$,由(Ⅳ)知亦必有 $f_1(x)=f(x)$.

最后证明 $f(x)$ 连续:由于 $f^{[m]}(x)=F(x)$,从 $F(x)$ 严格单调及 $f(x)$ 单调,可推知 $f(x)$ 是严格单调的.若 $f(x)$ 有不连续点,则必有 x_1,x_2 满足 $a<x_1<x_2<b$,使 $f(x)$ 取不到 $f(x_1)$ 和 $f(x_2)$ 之间的某些值 y,从而 $F(x)$ 取不到 $F(x_1)$ 和 $F(x_2)$ 之间的某些值 $f^{[m-1]}(y)$,此与 $F(x)$ 的连续性矛盾.(Ⅳ)⇨(Ⅴ)证毕.

(Ⅴ)⇨(Ⅰ)之证明:

以 $F^{\left[\frac{1}{m}\right]}(x)$ 记(Ⅴ)中给出的,满足方程

$$f^{[m]}(x)=F(x),f\sim\{G\}$$

的函数 $f(x)$,且记 $f_k(x)=F^{\left[\frac{1}{2^k}\right]}(x)$,要证明的只是:

$$\lim_{k\to+\infty}F^{\left[\frac{1}{2^k}\right]}(x)=\lim_{k\to+\infty}f_k(x)=x.$$

由条件(Ⅴ)中的唯一性要求,知

$$f_{k+1}^{[2]}(x)=f_k(x), \tag{3.6}$$

另外,由 $F(x)>x$ 及 $f_k(x)$ 单调性知:

$$f_k(x)>f_{k+1}(x)>x, \tag{3.7}$$

故 $\lim\limits_{k\to+\infty}f_k(x)=\varphi(x)\geqslant x$ 存在,且有

$$x\leqslant\varphi(x)<f_k(x). \tag{3.8}$$

我们的目的是证明 $\varphi(x)\equiv x$.用反证法,设对某点 $x_0\in(a,b)$ 有 $\varphi(x_0)>x_0$,我们将构造一个连续函数 $f_*(x)$,使它满足:

$$f_*[f_*(x)]=F(x),f_*\sim\{G\},f_*(x)\not\equiv f_1(x), \tag{3.9}$$

从而破坏(Ⅴ)中所设的唯一性以推出矛盾.

取 $d=\varphi(x_0)-x_0>0$,在 (a,b) 内取某个 y_0 使:

$$F^{\left[\frac{1}{2}\right]}(y_0)>x_0\geqslant y_0; \tag{3.10}$$

作序列:

$$\cdots<y_{-2}<y_{-1}<y_0<y_1<y_2<\cdots<y_k<\cdots,$$

使

$$F^{\left[\frac{1}{2}\right]}(y_k)=y_{k+1}(k=\cdots,-2,-1,0,1,2,\cdots), \tag{3.11}$$

则此序列的右端是无穷的,左端当 $\lim\limits_{x\to a}F^{\left[\frac{1}{2}\right]}(x)=a^*>a$ 时有穷而当 $a^*=a$ 时无穷.在左端有穷时,设序列中左端第一个点是 $y_{k^*}=a^*(k^*\leqslant 0)$,这当然是能够办到的.

取 $\varepsilon>0$ 充分小,使

$$\varepsilon<\min\{d,1,F^{\left[\frac{1}{2}\right]}(y_0)-x_0\}, \tag{3.12}$$

在$[y_0,y_1]$上定义：

$$\psi_1(x)=\begin{cases}x-(x-x_0)(x-x_0-\varepsilon) & x\in(x_0,x_0+\varepsilon),\\ x & x\in(y_0,y_1]/(x_0,x_0+\varepsilon),\end{cases}\tag{3.13}$$

容易验证：$\psi_1(x)$在$[y_0,y_1]$上连续.由(3.12)，$\psi_1(x)$递增，在$(x_0,x_0+\varepsilon)$上满足

$$x\leqslant\psi_1(x)\leqslant x_0+\varepsilon,$$

而在其它点有$\psi_1(x)\equiv x$；下面在(a,b)上定义：

$$\psi(x)=F^{\left[\frac{k}{2}\right]}\left[\psi_1\left[F^{\left[-\frac{k}{2}\right]}(x)\right]\right]\qquad(x\in(y_k,y_{k+1}]),\tag{3.14}$$

显然，$\psi(x)$在(a,b)上连续，递增，且满足

$$\psi\left[F^{\left[\frac{1}{2}\right]}(x)\right]=F^{\left[\frac{1}{2}\right]}[\psi(x)]\tag{3.15}$$

再取：

$$f_*(x)=\begin{cases}\psi\left[F^{\left[\frac{1}{2}\right]}(x)\right] & x\in(y_{2k+1},y_{2k+2}],\\ \psi^{[-1]}\left[F^{\left[\frac{1}{2}\right]}(x)\right] & x\in(y_{2k},y_{2k+1}),\end{cases}$$

此处$2k,2k+1$取遍序列$\{y_l\}$的下标l，由于在y_k处有：

$$\psi(y_k)=\psi^{[-1]}(y_k)=y_k,$$

故可推出f_*严格单调、连续.此外，由于(3.15)并注意到下述复合函数变量取值范围的合理性，便有：

$$\begin{aligned}f_*[f_*(x)]&=\psi\left[F^{\left[\frac{1}{2}\right]}\left[\psi^{[-1]}\left[F^{\left[\frac{1}{2}\right]}(x)\right]\right]\right]=F^{\left[\frac{1}{2}\right]}\left[\psi\left[\psi^{[-1]}\left(F^{\left[\frac{1}{2}\right]}(x)\right)\right]\right]\\&=F^{\left[\frac{1}{2}\right]}\left[F^{\left[\frac{1}{2}\right]}(x)\right]=F(x),\end{aligned}\tag{3.16}$$

由于$\psi(x)\not\equiv x$，故由$f_*(x)$之定义$f_*(x)\not\equiv F^{\left[\frac{1}{2}\right]}(x)$，剩下的是$f_*(x)\sim\{G\}$，即证明

$$\lim_{x\to b}T[x,f_*(x)]\tag{3.17}$$

存在，而(3.17)存在等价于下列极限式

$$\lim_{n\to+\infty}T\left[F^{\left[\frac{n}{2}\right]}(x),f_*\left[F^{\left[\frac{n}{2}\right]}(x)\right]\right]\tag{3.18}$$

在$x\in(y_0,y_1)$关于x一致收敛，即等价于

$$\lim_{m\to+\infty}T\left[F^{[m]}(x),f_*\left[F^{[m]}(x)\right]\right],\tag{3.19}$$

$$\lim_{m\to+\infty}T\left[F^{[m]}\left[F^{\left[\frac{1}{2}\right]}(x),f_*[F^{[m]}\left[F^{\left[\frac{1}{2}\right]}(x)\right]\right]\right]\tag{3.20}$$

两个极限过程对$x\in(y_0,y_1]$都一致收敛.若$x\in(y_0,y_1]$而不属于$(x_0,x_0+\varepsilon)$，由定义$\psi(x)=x$，故：

$$\begin{aligned}F_*[F^{[m]}(x)]&=F^{[m]}[f_*(x)]=F^{[m]}\left[\psi^{[-1]}(F^{\left[\frac{1}{2}\right]}(x))\right]=F^{[m]}\left[F^{\left[\frac{1}{2}\right]}\left[\psi^{[-1]}(x)\right]\right]\\&=F^{[m]}\left[F^{\left[\frac{1}{2}\right]}(x)\right]=F^{\left[\frac{1}{2}\right]}[F^{[m]}(x)].\end{aligned}$$

同理：

$$f_*\left[F^{[m]}\left[F^{\left[\frac{1}{2}\right]}(x)\right]\right]=F^{\left[\frac{1}{2}\right]}\left[F^{[m]}\left[F^{\left[\frac{1}{2}\right]}(x)\right]\right].$$

于是,由 $F^{\left[\frac{1}{2}\right]}\sim\{G\}$知(3.19)、(3.20)在$[y_0,y_1]$上除$(x_0,x_0+\varepsilon)$内的点以外一致收敛于$\frac{1}{2}\alpha_F$.

下面证明,(3.19)和(3.20)在$(x_0,x_0+\varepsilon)$上也一致收敛于$\frac{1}{2}\alpha_F$. 事实上,由(3.13)以及不等式(3.12)和(3.8),可知在$(x_0,x_0+\varepsilon)$上对一切 k 有:

$$x_0\leqslant x\leqslant\psi(x)\leqslant x_0+\varepsilon\leqslant x_0+d=\varphi(x_0)\leqslant f_k(x_0),$$

从而在$(x_0,x_0+\varepsilon)$上有:

$$\begin{aligned}f_*\left[F^{[m]}\left[F^{\left[\frac{1}{2}\right]}(x)\right]\right]&=F^{[m]}\left[f_*\left[F^{\left[\frac{1}{2}\right]}(x)\right]\right]=F^{[m]}[\psi[F(x)]]\\&=F^{[m+1]}[\psi(x)]\leqslant F^{[m+1]}[f_k(x_0)].\end{aligned}$$

注意到 $T(x,y)$关于变元 x,y 分别是递减和递增,并由 f_*之定义,得:

$$\begin{aligned}T\left[F^{\left[m+\frac{1}{2}\right]}(x),F^{[m+1]}(x)\right]&\leqslant T\left[F^{\left[m+\frac{1}{2}\right]}(x),f_*\left[F^{\left[m+\frac{1}{2}\right]}(x)\right]\right]\\&\leqslant T\left[F^{\left[m+\frac{1}{2}\right]}(x_0),F^{[m+1]}[f_k(x_0)]\right],\end{aligned}$$

两端令 $m\to+\infty$得:

$$\begin{aligned}\frac{\alpha_F}{2}&\leqslant\varliminf_{m\to+\infty}T\left[F^{\left[m+\frac{1}{2}\right]}(x),f_*\left[F^{\left[m+\frac{1}{2}\right]}(x)\right]\right]\\&\leqslant\varlimsup_{m\to+\infty}T\left[F^{\left[m+\frac{1}{2}\right]}(x),f_*\left[F^{\left[m+\frac{1}{2}\right]}(x)\right]\right]\leqslant\frac{\alpha_F}{2}+\frac{\alpha_F}{2^k}.\end{aligned}\tag{3.21}$$

另一方面,在$(x_0,x_0+\varepsilon)$内有:

$$x_0\leqslant\psi^{[-1]}(x)\leqslant x\leqslant x_0+\varepsilon\leqslant x_0+d=\varphi(x_0)\leqslant f_k(x_0),$$

结合着:

$$f_*\left[F^{\left[m+\frac{1}{2}\right]}(x)\right]=F^{[m]}\left[f_*\left[F^{\left[\frac{1}{2}\right]}(x)\right]\right]=F^{[m+1]}[\psi(x)]$$

及

$$\begin{aligned}f_*[F^{[m]}(x)]&=F^{[m]}[f_*(x)]=F^{[m]}\left[\psi^{[-1]}\left[F^{\left[\frac{1}{2}\right]}(x)\right]\right]\\&=F^{\left[m+\frac{1}{2}\right]}[\psi^{[-1]}(x)],\end{aligned}$$

类似地得到:

$$\begin{aligned}T[F^{[m]}(f_k(x_0)),F^{[m]}(x_0)]&\leqslant T[F^{[m]}(x),f_*[F^{[m]}(x)]]\\&\leqslant T\left[F^{[m]}(x),F^{\left[m+\frac{1}{2}\right]}(x)\right],\end{aligned}$$

两端取极限得到:

$$\begin{aligned}\frac{\alpha_F}{2}-\frac{\alpha_F}{2^k}&\leqslant\varliminf_{m\to+\infty}T[F^{[m]}(x),f_*[F^{[m]}(x)]]\\&\leqslant\varlimsup_{m\to+\infty}T[F^{[m]}(x),f_*[F^{[m]}(x)]]\leqslant\frac{1}{2}\alpha_F.\end{aligned}\tag{3.22}$$

综合(3.21)及(3.22)即得(3.18)(3.17),亦即 $f_*\sim\{G\}$.(Ⅴ)⇨(Ⅰ)得证.

基本定理毕.

§4　单参迭代群的存在唯一条件

在前面的定义 2 中我们要求 $F(x)\geqslant x$，为了去掉这个限制，我们引入：

(4.1)　**定义 3**　若 $F(x)$ 在 (a,b) 上连续递增，而且 $F^{[-1]}(x)$ 关于 (a,b) 上的正规族 G 为可度，则称 $F(x)$ 关于 G 为逆可度，记 $\alpha_F=-\alpha_{F^{[-1]}}$；称负数 α_F 为 $F(x)$ 关于 G 的指标. 若函数 $f(x)$，$f^{[-1]}(x)$ 中有一个关于 G 可度，即称 f 关于 G 广义可度.

(4.2)　**定义 4**　若 $F(x,t)$ 为由 $F(x)$ 生成的单参迭代群，若对任一个 $t_0\neq 0$，$F(x,t_0)$ 关于 G 是广义可度的，则称 $F(x,t)$ 为由 F 生成的 G 可度迭代群.

由 §3 的基本定理易得：

(4.3)　**定理 2**　设 G 是 (a,b) 上的正规族，$F(x)$ 在 (a,b) 上连续、递增、没有不动点，F 把 (a,b) 变为 (a,b) 且 F 关于 G 可度. 则由 F 生成的关于 x 和参数 t 连续的 G 可度单参迭代群存在而且唯一的充要条件是：二元函数列

$$\Phi_n(x,y)=T_G[F^{[n]}(x),F^{[n]}(y)]\,(n=1,2,\cdots)$$

在区域 $D:\begin{cases}a<x<b\\ x\leqslant y<b\end{cases}$ 上收敛到连续、严格单调的函数 $\Phi(x,y)$.

当然，这个条件可用定理 1 中的五条件中的任一个另外的代替. 后面将指出，这里采用的(Ⅱ)，乃是比较易于检验的一个条件. 此外，若去掉“F 把 (a,b) 变为 (a,b)”这个假设，只要把定理中的“迭代群”改为“迭代半群”，命题仍真. 这一点，从下面的证明中即可看出.

证明　先证条件的必要性，设由 F 生成的这个 G 可度迭代群为 $F[x,t]$，由于 F 没有不动点而且关于 G 可度，故必有 $F(x)>x$，令 $f_k(x)=F\left(x,\dfrac{1}{k}\right)$，由群的性质知：

$$f_k^{[k]}(x)=F(x),$$

从而由 $f_k(x)$ 连续推得 $f_k(x)$ 递增且 $f_k(x)>x$. 由于 $F(x,t)$ 可度，$f_k(x)$ 是广义可度的，由 $f_k(x)>x$ 知 $f_k(x)$ 不是逆可度，故 $f_k(x)\sim\{G\}$. 由 $F(x,t)$ 关于参数的连续性得：

$$\lim_{k\to+\infty}f_k(x)=\lim_{k\to+\infty}F\left(x,\frac{1}{k}\right)=F(x,0)=x,$$

于是定理 1 中的条件(Ⅰ)得到满足. 由等价性，可知所述条件是必要的.

现往证这个条件也是充分的. 如果所述条件成立，由定理 1 之条件(Ⅳ)，对每个 $t>0$，取 $c=t\alpha_F$，则有唯一的 G 可度的函数 $g(x,c)$，满足(Ⅳ)中的 1)、2)、3)，令：

$$F(x,t)=g(x,t\alpha_F)\,(t>0),$$

我们来证明 $F(x,t)$ 关于参数 t 有群的性质：

$$F[F(x,t_1),t_2]=F(x,t_1+t_2).$$

事实上，按我们的定义：

$$F[F(x,t_1),t_2]=g[g(x,t_1\alpha_F),t_2\alpha_F],$$
$$F[x,t_1+t_2]=g(x,(t_1+t_2)\alpha_F),$$

由引理 3 得：

$$\alpha_{F[F(x,t_1),t_2]}=\alpha_{F[x,t_1+t_2]}=(t_1+t_2)\alpha_F,$$

同时，$F(F(x,t_1),t_2)$和$F(x,t_1+t_2)$都满足定理1中条件(Ⅳ)下的1)和3)，由(Ⅳ)中所要之唯一性知两者恒等.

又由(Ⅲ)，可知

$$F(x,t)=F^{[-n]}[g[F^{[n]}(x),t\alpha_F]]=\lim_{n\to+\infty}F^{[-n]}[g(F^{[n]}(x),t\alpha_F)]$$

恰是由$\Phi(x,y)=t\alpha_F$所确定的隐函数，于是由$\Phi(x,y)$的连续与严格单调可得$F(x,t)$关于x,t的连续性，顺便得到：$F(x,t)$关于x,t都是递增的. 由定理1的(Ⅰ)可知

$$\lim_{t\to0}F(x,t)=\lim_{n\to+\infty}F\left(x,\frac{1}{n}\right)=x,$$

于是令：$F(x,0)=x$；对任一个$t_0<0$，令：

$$F(x,t_0)=F^{[-1]}(x,-t_0),$$

易验证这样得到的是关于x,t连续的G可度迭代群. 实际上，只要指出每个$F(x,t)(t>0)$都把(a,b)变为(a,b)就可以了. 当$t=\frac{1}{n}$，n是自然数时，由

$$F^{[n]}\left(x,\frac{1}{n}\right)=F(x)$$

这是显然的，从而$t=\frac{m}{n}$时也有$F(x,t)$把(a,b)变为(a,b)，由连续性，对一切$t>0$都对. 从而$F^{[-1]}(x,t)=F(x,-t)$在(a,b)上有定义.

至于这样的单参迭代群的唯一性，由定理1的(Ⅴ)及连续性即可推出. 定理2证毕.

若$F(x)$不是可度而是逆可度，则有：

(4.4) **定理3** 设G是(a,b)上的正规族，$F(x)$在(a,b)上连续、递增、没有不动点，F把(a,b)变为(a,b)，且$F^{[-1]}$关于G可度，则由F生成的关于x和t连续的G可度单参迭代群存在而且唯一的充要条件是：二元函数列

$$\Phi_n(x,y)=T_G[F^{[-n]}(x),F^{[-n]}(y)](n=1,2,\cdots)$$

在区域$D:\begin{cases}a<x<b\\x\leqslant y<b\end{cases}$上收敛到连续、严格单调的函数$\Phi(x,y)$.

事实上，只要把$F^{[-1]}$生成的G可度迭代群中的参数t改换一下正负号，便得到由F生成的G可度迭代群. 因而定理3可由定理2直接导出. 兹从略.

(4.5) **定义5** 设$F(x)$在(a,b)上连续而G是(a,b)上的正规族. 令$\widetilde{F}(x)=(b+a)-F(b+a-x)$，若$\widetilde{F}(x)\sim\{G\}$，则称$F$关于$G$下可度. 若$F^{[-1]}$存在关于$G$下可度，则称$F$为关于$G$逆下可度. 记$\beta_F=\alpha_{\widetilde{F}}$，称$\beta_F$为$F$关于$G$的下指标. 若$F,F^{[-1]}$二者之中有一个为关于$G$下可度，即称$F$为关于$G$广义下可度的. 若对每个$t_0\neq0$，由$F$生成的迭代群中的函数$F(x,t_0)$都是关于$G$广义下可度的，则称$F(x,t)$是由$F$生成的$G$下可度的单参迭代群.

注意到，若$\widetilde{F}(x)=(b+a)-F(b+a-x)$生成关于$G$可度的单参迭代群或迭代半群$\widetilde{F}(x,t)$，则令

$$F(x,t)=a+b-\widetilde{F}(a+b-x,t),$$

便得到由F生成的关于G下可度的单参迭代群$F(x,t)$，反之亦然. 从这种对应关系可得：

(4.6) **定理4** 若$F(x)$在(a,b)上连续、递增，$F(x)$把(a,b)变为(a,b)，且没有不动点，G

是(a,b)上的正规族而F关于G下可度，则由F生成的关于x和t连续的G下可度单参迭代群存在而且唯一的充要条件是二元函数列：

$$\Phi_n(x,y)=T_G[\widetilde{F}^{[n]}(x),\widetilde{F}^{[n]}(y)] \qquad (n=1,2,\cdots)$$

在区域$D:\begin{cases}a<x<b\\x\leqslant y<b\end{cases}$上收敛到连续、严格单调的函数$\Phi(x,y)$. 此处$\widetilde{F}(x)=a+b-F(a+b-x)$.

如果把条件"$F(x)$把(a,b)变为(a,b)"去掉，相应地把"迭代群"改为"迭代半群"，命题仍真.

定理4的证明从略. 类似地，若把F关于G下可度改为"逆下可度"，则相应地把$\widetilde{F}^{[n]}$改为$\widetilde{F}^{[-n]}$，命题仍真，兹不赘述.

如果F在(a,b)上连续、递增而且没有不动点，当$F(x)>x$时，$F(x)$可能关于G可度而且逆下可度；(当$F(x)<x$时$F(x)$可能关于G逆可度而且下可度)这时，我们可以得到由F生成的两个连续的单参迭代群：一个关于G可度，另一个关于G下可度. 而且，这两个单参迭代群一般说来是不同的. 这方面，本文不再作更多的讨论了.

由于递减函数偶次迭代后成为递增，故若$F(x)$递减，我们不能期望它生成连续的单参迭代半群. 但如果$F(x)$在(a,b)的某个子区间上递增而且生成单参迭代半群，那么，是有可能把它开拓为(a,b)上的单参迭代半群的. 这就是：

(4.7)　**定理5**　若$F(x)$在(a,b)上连续而且在(a,b)的子区间(a^*,b^*)上递增，$F[a,b]\subseteq F(a^*,b^*)\in(a^*,b^*)$，以$F^*(x)$记$F(x)$在$(a^*,b^*)$上的限制函数，若$F^*$在$(a^*,b^*)$上生成连续的单参迭代半群$F^*(x,t)(t\geqslant 0)$，则$F^*(x,t)$可以唯一地拓广为$F$在$(a,b)$上的连续的单参迭代半群.

证明　由于群的性质，有

$$F^{[n]}(x)=F^{*[n]}(x)=F^*(F^*(x,t),n-t)(x\in(a^*,b^*),n\text{为自然数},n>t),$$

于是由F^*严格单调可知$F^*(x,t)$严格单调. 又由连续性

$$\lim_{t\to 0^+}F^*(x,t)=x,$$

可见对充分小的$t>0$，$F^*(x,t)$关于x递增. 但对任何t_0都有

$$F^*(x,t_0)=F^{*[n]}\left(x,\frac{t_0}{n}\right).$$

可见$F^*(x,t)$对任一固定的t关于x递增. 此外，群的性质要求$F^*(x,t_1)$和$F^*(x,t_2)$可以在(a^*,b^*)上复合，因而$F^*(x,t)$取值于(a^*,b^*). 现在令：

$$F(x,t)=F^{*[-1]}(F^*(F(x),t)), \tag{4.8}$$

由于对任意$x\in(a,b)$，总有$F^*(a^*)\leqslant F(x)\leqslant F^*(b^*)$，由$F^*(x,t)$递增得：

$$F^*(F^*(a^*),t)\leqslant F^*(F(x),t)\leqslant F^*(F^*(b^*),t),$$

由群的性质，又有：

$$F^*(F^*(a^*),t)=F^*(F^*(a^*,t))\geqslant F^*(a^*),$$

$$F^*(F^*(b^*),t)=F^*(F^*(b^*,t))\leqslant F^*(b^*).$$

这说明$F^*(F(x),t)$在$F^{*[-1]}(x)$的定义域内，因而(4.8)右端的复合是合理的. 显然，(4.8)

确定的函数 $F(x,t)$ 关于 x,t 连续，而且：

$$\begin{aligned}F[F(x,t_1),t_2]&=F^{*[-1]}[F^*[F[F^{*[-1]}[F^*[F(x),t_1]]],t_2]]\\&=F^{*[-1]}[F^*[F^*[F(x),t_1]t_2]]=F^{*[-1]}[F^*[F(x),t_1+t_2]]\\&=F(x,t_1+t_2).\end{aligned}$$

即 $F[x,t_1+t_2]$ 确有群的性质，易验证 $F(x,1)=F(x)$ 和 $F(x,0)=x$. 此外，由可交换性应当有：

$$F^*(F(x),t)=F(F(x,t))=F^*[F(x,t)],$$

从而必须有：

$$F(x,t)=F^{*[-1]}[F^*[F(x),t]].$$

即定义(4.8)是唯一可能的. 定理 5 证毕.

§5　不动点处导数模小于 1 之情形

在有关迭代方程(1.2)和 Schröder 方程(1.3)的研究中，$F(x)$ 具有不动点 $F(x_0)=x_0$ 且 $F'(x_0)$ 存在，$0<F'(0)<1$ 的情形受到了最多的注意. 我们把 §3 的结果应用于这种情况. 不失一般性，我们和这方面的许多作者一样，设 $F(x)$ 仅在定义区间的端点有一个不动点.

由定理 1 取特殊情况可得：

(5.1)　**推论 1**　若 $F(x)$ 在 $[-1,0]$ 上连续，严格单调、有唯一的不动点 $x=0$ 且有 $0<F'(0)<1$，则下列诸条件等价：

(Ⅰ)有自然数列 $m_1<m_2<\cdots<m_k<\cdots$ 使方程

$$f_k^{[m_k]}(x)=F(x)$$

对每个 k 至少有一个在 $[-1,0]$ 连续而且在 $x=0$ 可微的解 $f_k(x)$，且这一列 $f_k(x)$ 满足

$$\lim_{k\to+\infty}f_k(x)=x.$$

(Ⅱ)二元函数列：

$$\Phi_n^*(x,y)=\frac{F^{[n]}(y)}{F^{[n]}(x)}$$

在区域 $D:\begin{cases}-1<x<0\\x\leqslant y<0\end{cases}$ 上收敛于连续、严格单调的 $\Phi^*(x,y)$.

(Ⅲ)对任给的 $[-1,0]$ 上的函数 $g(x)$，只要 $g(0)=0,0<g'(0)<1,x\leqslant g(x)\leqslant 0$，则有：

$$\lim_{n\to+\infty}F^{[-n]}[g[F^{[n]}(x)]]=H(x)$$

在 $[-1,0]$ 上存在，且 $H'(0)=g'(0)$，而且 $H(x)$ 恰为由 $\Phi^*(x,y)=g'(0)$ 所确定的隐函数.

(Ⅳ)对任给的实数 c，$0<c<1$，在 $[-1,0]$ 上有唯一的单调函数 $g(x)=g(x,c)$，$g'(0)$ 存在等于 c，且 $g[F(x)]=F[g(x)]$.

(Ⅴ)对任意正整数 m，方程

$$f^{[m]}(x)=F(x)$$

有唯一的在 $x=0$ 可微的解. 且此解在 $[-1,0]$ 上连续.

证明 在基本引理中，取(a,b)为$(-1,0)$，取正规族为：

$$y=G(x,t)=c^t x \quad (c=F'(0)),$$

则

$$T_G=T(x,y)=\frac{1}{\ln c}\ln\frac{y}{x},$$

于是，条件"$f(x)\sim\{G\}$"等价于"$0<f'(0)<1$"，事实上，按定义 2，$f\sim\{G\}$相当于下列极限存在且为正：

$$\lim_{x\to 0^-}T(x,f(x))=\lim_{x\to 0^-}\frac{1}{\ln c}\ln\frac{f(x)}{x}=\frac{\ln f'(0)}{\ln c}>0,$$

这当然等价于 $0<f'(0)<1$.

于是，推论中一切关于 $0<f'(0)<1$ 的要求等价于定理 1 中 $f\sim\{G\}$的要求. 而条件(Ⅱ)中，序列

$$\Phi_n^*(x,y)=\frac{F^{[n]}(y)}{F^{[n]}(x)}$$

与基本定理中之序列 Φ_n 之关系为：

$$\Phi_n(x,y)=\frac{1}{\ln c}\ln\Phi_n^*(x,y),$$

由此可见，推论(5.1)不过是定理 1 之特例.

如果 $F(x)$的不动点是定义区间的左端，我们可以像前面引入"下可度"概念那样来讨论. 例如，若 $F(x)$定义于$[0,1]$，$0<F'(0)<1$，$F(0)=0$，且在$[0,1]$上没有其它不动点，我们取：

$$F^*(x)=-F(-x),$$

则 $F^*(x)$满足推论 1 中的种种要求，这时可把推论 1 逐字逐句地改成：

(5.2) **推论 2** 若 $F(x)$定义于$[0,1]$上，$F(x)$连续、严格单调，有唯一不动点 $x=0$ 且有 $0<F'(0)<1$，则下列诸条件等价：

(Ⅰ)有自然数列 $m_1<m_2<\cdots<m_k<\cdots$，使方程

$$f_k^{[m_k]}(x)=F(x)$$

对每个 k 至少有一个在$[0,1]$上连续，在 $x=0$ 处可微的解 $f_k(x)$，且这一列 $f_k(x)$满足：

$$\lim_{k\to+\infty}f_k(x)=x.$$

(Ⅱ)二元函数列

$$\Phi_n^*(x,y)=\frac{F^{[n]}(y)}{F^{[n]}(x)}\quad(n=1,2,\cdots)$$

在区域 $D:\begin{cases}0<x<1\\0<y\leqslant x\end{cases}$上收敛于连续、严格单调的函数 $\Phi^*(x,y)$.

(Ⅲ)对任给的$[0,1]$上的函数 $g(x)$，只要 $g(0)=0$，$0<g'(0)<1$，$0\leqslant g(x)\leqslant x$，则有：

$$\lim_{k\to+\infty}F^{[-n]}[g[F^{[n]}(x)]]=H(x)$$

在$[0,1]$上存在，$H'(0)=g'(0)$，且 $H(x)$恰是由方程 $\Phi^*(x,y)=g'(0)$确定的隐函数.

(Ⅳ)对任给的实数 c，$0<c<1$，在$[0,1]$上有唯一的单调函数 $g(x)=g(x,c)$满足：

$$g'(0)=c,g[F(x)]=F[g(x)].$$

（Ⅴ）对任意正整数 m，方程

$$f^{[m]}(x)=F(x)\quad (x\in[0,1])$$

有唯一的在 $x=0$ 可微的解，且此解在$[0,1]$上连续.

在以上 5 个条件中，只有条件（Ⅱ）是便于检验的. 因此，条件（Ⅱ）的满足与否可以当成 $F(x)$能否生成关于 x,t 连续，而且在 t 固定时关于 x 在 $x=0$ 可微的迭代群 $F(x,t)$的存在唯一性的判别准则，像§4 中所证明的那样. 作为更具体的、易于检验的充分条件，有：

1° 若 $F(x)$在 $x=0$ 处有二级微商，则推论 1 及推论 2 中条件（Ⅱ）成立.（事实上，在文献[5]中早已指出）这时对应的 Schröder 方程

$$\varphi[F(x)]=F'(0)\varphi(x)$$

有在 $x=0$ 邻域可微而且微商为正的解 $\varphi(x)$，显然有

$$\frac{\varphi(y)}{\varphi(x)}=\Phi^*(x,y).$$

2° 另一个充分条件是："在 $x=0$ 的某个邻域$\dfrac{F(x)}{x}$单调"，这时，推论 1 及推论 2 中的（Ⅱ）成立.

我们设 $F(x)$是定义于$[0,1]$上的，满足推论 2 中前提的函数，（对$[-1,0]$上的，满足推论 1 中条件的函数，证明类似）令：

$$F(x)=cx[1+\gamma(x)]\quad (c=F'(0))$$

则"$\dfrac{F(x)}{x}$单调"等价于"$\gamma(x)$单调趋于 0". 取 $0<y_0\leqslant x_0\leqslant 1$，设 $F^{[N]}(x_0)<y_0$，则

$$\Phi_n(x_0,y_0)\leqslant\Phi_n(x_0,x_0)=1,$$

$$\Phi_n(x_0,y_0)>\Phi_n[x_0,F^{[N]}(x_0)].$$

又因

$$\lim_{n\to+\infty}\Phi_n[x_0,F^{[N]}(x_0)]=c^N>0,$$

故对充分大的 n，$\Phi_n(x_0,y_0)\geqslant\varepsilon>0$，同时：

$$\Phi_{n+1}(x_0,y_0)=\Phi_n(x_0,y_0)\left[1+\frac{\gamma[F^{[n]}(y_0)-\gamma[F^{[n]}(x_0)]}{1+\gamma[F^{[n]}(x_0)]}\right]\leqslant\Phi_n(x_0,y_0),$$

这是由于 $\gamma[F^{[n]}(y_0)]\leqslant\gamma[F^{[n]}(x_0)]$之故. 于是函数列 $\Phi_n(x_0,y_0)$当 $n\to+\infty$时有非 0 的极限. 且当 $y_0<x_0$ 时有 $\Phi(x_0,y_0)<1$，这是因为：

$$\Phi_{n+1}(x_0,y_0)\leqslant\Phi_1(x_0,y_0)=\frac{F(y_0)}{F(x_0)}<1$$

之故. 但又因：

$$\Phi(x,y)=\lim_{n\to+\infty}\frac{F^{[n]}(y)}{F^{[n]}(x)},$$

故：

$$\frac{\Phi(x,y_1)}{\Phi(x,y_2)}=\Phi(y_2,y_1),$$

当 $y_1<y_2$ 时，$\Phi(y_2,y_1)<1$，从而得 $\Phi(x,y_1)<\Phi(x,y_2)$，即知 $\Phi(x,y)$是严格单调的.

为了保证 $F(x)$ 生成在不动点可微的单参迭代群是唯一的，文献[10]中曾要求 $F'(x)$ 在 $[0,1]$ 上递增，这时，显然有 $F(x)/x$ 递增. 可见，这里的结果蕴含了文献[10]中的对应结果.

对应于§4诸定理，显然可以建立关于不动点处微商为正，且不为1的函数的一系列命题，这些工作是完全机械的，这里不再叙述了.

§6 不动点导数为1的情况

在定理1中，取正规迭代族为：

$$G(x,t)=\frac{x}{\sqrt[k]{1+tax^k}}\left(\begin{matrix}k\text{ 为自然数},(-1)^k a>0,\\ x\in(-1,0)\end{matrix}\right) \tag{6.1}$$

我们来看，若 $F\sim\{G\}$，F 应当是哪一类函数？

由(6.1)解出 $t=T(x,y)$：

$$T(x,y)=\frac{1}{a}\left[\frac{1}{y^k}-\frac{1}{x^k}\right]. \tag{6.2}$$

若 $F\sim\{G\}$，由定义

$$\lim_{x\to 0^-}\frac{1}{a}\left[\frac{1}{[F(x)]^k}-\frac{1}{x^k}\right]=\alpha_F, \tag{6.3}$$

这个式子可以改写成

$$\frac{1}{[F(x)]^k}-\frac{1}{x^k}=a\,\alpha_F+o(1)\quad(x\to 0^-),$$

即

$$\frac{x^k}{[F(x)]^k}-1=a\,\alpha_F x^k+o(x^k)\quad(x\to 0^-), \tag{6.4}$$

由此可得

$$\lim_{x\to 0^-}\left(\frac{F(x)}{x}\right)^k=1.$$

按 $F\sim\{G\}$ 之定义，应有 $-1\leqslant F(x)\leqslant 0$，故 $F(x)$ 与 x 同号，从而得：

$$F'(0)=\lim_{x\to 0^-}\frac{F(x)}{x}=1.$$

另外，由(6.4)又得：

$$\frac{x}{F(x)}=(1+a\,\alpha_F x^k+o(x^k))^{\frac{1}{k}},$$

从而得：

$$F(x)=x-\frac{a\,\alpha_F}{k}x^{k+1}+o(x^{k+1}). \tag{6.5}$$

这就是说：若 $F\sim\{G\}$，则 F 在 $x=0^-$ 处有形如(6.5)之展式. 反之，若(6.5)成立，也易验证(6.3)成立，即 $F\sim\{G\}$.

当 $F(x)$ 在 $x=0$ 处有 $(k+1)$ 阶导数时，(6.5)的意义是：$F'(0)=1$，$F''(0)=\cdots=F^{(k)}(0)=0$，而：

$$F^{(k+1)}(0)=\frac{(k+1)!}{k}\cdot a\,\alpha_F\neq 0.$$

但仅由(6.5)并不能推出 $F^{(k+1)}(0)$存在;仅能推出 $F(x)$在 $x=0$ 处的$(k+1)$阶广义导数存在且不为 0,而当 $1<l<k+1$ 时,l 阶广义导数为 0;$F'(0)=1$.

在作了这些说明之后,马上可由基本定理得到:

(6.6) **推论 3** 设 $F(x)$在$[-1,0]$上连续、递增,$F(x)\geqslant x$ 且等号仅当 $x=0$ 时成立,$F'(0)=1$ 且在 $x=0^-$ 邻域有渐近式:

$$(*)\quad F(x)=x-cx^{k+1}+o(x^{k+1})(k\text{ 为自然数},(-1)^k c>0),$$

则以下五条件等价:

(Ⅰ) 有自然数列 $m_1<m_2<\cdots<m_k<\cdots$,在使对任一自然数 k,有连续函数 $f_k(x)$定义于$[-1,0]$,$-1\leqslant f_k(x)\leqslant 0$,$f_k'(0)=1$,$f_k$ 在 $x=0^-$ 处有直到$(k+1)$阶的广义导数.满足:

$$f_k^{[m_k]}(x)=F(x)\text{ 及 }\lim_{k\to+\infty}f_k(x)=x.$$

(Ⅱ) 二元函数列

$$\Phi_n(x,y)=[F^{[n]}(y)]^{-k}-[F^{[n]}(x)]^{-k}$$

在区域 $D:\begin{cases}-1<x<0\\ x\leqslant y<0\end{cases}$收敛到连续、严格单调的函数 $\Phi(x,y)$.

(Ⅲ) 对任给的$(-1,0)$上的函数 $g(x)$,若 $g(x)$取值于$(-1,0)$,在 $x=0^-$ 处有展式:

$$g(x)=x+\alpha x^{k+1}+o(x^{k+1}),$$

则函数列 $H_n(x)=F^{[-n]}[g[F^{[n]}(x)]]$当 $n\to+\infty$时有极限 $H(x)$,且 $H(x)$亦有展式

$$H(x)=x+\alpha x^{k+1}+o(x^{k+1}),$$

而且,$H(x)$恰为方程 $\Phi(x,y)=-k\alpha$ 所确定的隐函数.

(Ⅳ) 对任给的 λ,$(-1)^k\lambda>0$,有唯一的函数 $g(x,\lambda)$存在,满足条件

1) $g(x,\lambda)$在$(-1,0)$上单调,

2) $g(x,\lambda)=x-\lambda x^{k+1}+o(x^{k+1})$,

3) $g[F(x),\lambda]=F[g(x,\lambda)]$.

(Ⅴ) 对任意自然数 m,有唯一的在 $x=0^-$ 处有$(k+1)$阶广义导数的 $f(x)$,满足:

$$f^{[m]}(x)=F(x),\quad x\in[-1,0],$$

且 $f(x)$连续.

以上推论的证明中,仅须注意的是:由 f 在 $x=0^-$ 处$(k+1)$阶广义导数之存在及条件

$$f^{[m]}(x)=F(x),$$

马上可推出 $f(x)$有形如(6.5)之渐近展式.剩下的事,几乎是照搬定理 1.

完全类似地,可以讨论$[0,1]$上的,以 0 为不动点,$F'(0)=1$,$0\leqslant F(x)<x$ 的函数 $F(x)$而得出相应的结论.因为如令 $F^*(x)=-F(-x)$,$F^*(x)$就是$[-1,0]$上的函数了.

§7 关于其它类型的正规族

1° 若取正规族为无穷区间上的:

$$(7.1)\qquad G(x,t)=x^{k^t}\quad(k>1,1<x<\infty),$$

则当 $F\sim\{G\}$时必有 $F(x)\sim x^a$.这是因为,由(7.1)解出

$$T(x,y)=\ln\frac{\ln y}{\ln x},$$

若 $F\sim\{G\}$,则

$$\lim_{x\to+\infty}\ln\frac{\ln F(x)}{\ln x}=\alpha_F>0,$$

即

(7.2) $$F(x)=X^{e^{\alpha}F+o(1)}\quad(x\to+\infty).$$

也就是说,当 $x\to+\infty$ 时,$F(x)$的无穷大的阶为 e^{α_F}. 对这一类函数——包括了首项系数为 1 的多项式——也可以建立类似于推论 1、2、3 的定理.

2° 对 1°中的正规族作变换,令:

(7.3) $$G^*(x,t)=A^{-1}G(Ax,t)=A^{K^t-1}x^{K^t}\quad(A>0),$$

又得到更广泛的一类正规族$\{G^*\}$,若 $F\sim\{G\}$,则 $F(x)$具有更一般的形式

(7.4) $$F(x)=Bx^{\lambda+o(1)}\quad(\lambda>1).$$

于是对形如(7.4)的函数类,也可建立对应于基本定理之结果.

3° 对(7.3)所示的正规族再作变换:

(7.5) $$G^{**}(x,t)=G^*(x^{-1},t)^{-1}=A^{1-K^t}x^{K^t}\quad(0<x<1),$$

形式上没有变,但定义区间却变为(0,1)了. 再取 $\bar{G}(x,t)=-A^{1-K^t}(-x)^{K^t}$,可得到(−1,0)上的正规族 $\bar{G}$,若定义于(−1,0)上的 $F\sim\{\bar{G}\}$,则 $F(x)$在 $x=0$ 处导数为 0,从而对这一类的 $F(x)$可以建立对应的结果.

4° 若取正规迭代族为

(7.6) $$G(x,t)=x+t\quad x\in(0,+\infty),$$

那么,对应地

(7.7) $$T(x,y)=y-x$$

从而,$F\sim\{G\}$意即 $\lim\limits_{x\to+\infty}(F(x)-x)=\alpha_F$,即

$$F(x)=x+\alpha_F+o(1)\quad(x\to+\infty).$$

对这一类函数,也可以建立对应于基本定理之结果.

参考文献

[1]Schröder E. Ueber Iterirte Funktionen. Math. Ann. ,1871(3):296 - 322.

[2]Koenigs G. Recherches sur les intergrales de certaines équations fonctionelles. Ann. de I′ Ecole Norm Sup. 1884(3):3 - 41.

[3]Hadamard J. Two works on iteration and related questions. Bull. Amer. Math. ,1944(50):67 - 75.

[4]Bödewadt, U. T. : Zur Iteration realler Funktionen. Zeitsc. Math. , 1944, 49(3): 497 - 516.

[5]Kneser H. Reelle analytisehe Lösungen der Gleichung $\varphi[\varphi(x)]=e^x$ Verwandter Funktionalgleichungen. J. reine angew Math. ,1950(187):56 - 57.

[6]Zdun M C. Some remarks on iteration semigroups. Univ. slaski w Katowieaeh Prace

Nawkowe. № 158 Prace Mat. ,1977(7):65－69.
[7]Roznowski M. Approximate Solutions of the functional equation $\varphi^k(x)=f(x)$. Uniw slaski w Katowicach Prace Mat. ,1973(3):53－73.
[8]Kuczma M, Smajdor A. Regular fractional iteration. Bull. Acad. Polon. Sci. Ser. Sci. Math. Astronom. Phys. ,1971(19):203－207.
[9]Kuezma M. Fractional iteration of differentiabie functions. Ann. Polon. Math. 22(1969/70)217－227.
[10]Reznick B A. A uniqueness criterion for fractional iteration. Ann. Polon. Math. ,1974(30):219－224.
[11]Kuczma M and Smajdor A. Fractional iteration in the class of convex functions. Bull. Acad. Polon. Sci. Sér. Sci. Math. Astronom. Phys. ,1968(16):717－720.
[12]Smajdor A. On convex iteration groups. Bull. Acad. Polon. Sei. Sér. Sei. Math. Astronom. Phys. ,1967(15):325－328.
[13]Z Nitecki. Differential dynamics. The. M. I. T. Press. Cambridge. mass. 1971.
[14]P Hartman. Ordinary differential equations. wiley,New York,1964,MR 30. № 1270.
[15]Lam Ping-Fun. J. of Differentiable Equations. 1978(30):31－40.
[16]张筑生. 圆周上自同胚的嵌入流与变换群的作用. 数学学报,1981(24):953－957.

求多项式根的 2^n 阶劈因子法

张景中　杨　路
（中国科学技术大学）

2^n-Order Cut-factor Method to Find the Roots of Polynomial

Zhang Jing-zhong　Yang Lu
(University of Science and Technology of China)

Abstract

In this paper, a 2^n-order process $\{w_k(z)\}$ is given to cut a factor of degree 2 from polynomial $F(z)$ as follows:

$$\left\{\begin{array}{l}(\text{i})\text{Let } \xi_0(z)=w_0(z)=(z-z_1^{(0)})(z-z_2^{(0)}).\\ (\text{ii})\text{If } \xi_j(z)=(z-z_1^{(j)})(z-z_2^{(j)}),j=0,1,\cdots,i-1 \text{ has given suppose}\\ \quad F(z)\equiv R_{i-1}(z)=\alpha_0 z^{2i+1}+\alpha_1 z^{2i}+\cdots+\alpha_{2i+1}\,(\text{mod } w_0\prod_{j=0}^{i-1}\xi_j),\\ \quad R_{i-1}(z)\equiv P_1(z)=\beta_0 z^3+\beta_1 z^2+\beta_2 z+\beta_3\,(\text{mod } \xi_{i-1}^2),\\ \quad \xi_{i-1}^2(z)\equiv\alpha(z-z_1^{(i)})(z-z_2^{(i)})\,(\text{mod } P_1),\\ \quad \text{Then let } \xi_i(z)=(z-z_1^{(i)})(z-z_2^{(i)}).\\ (\text{iii})\text{Let } w_1(z)=\xi_n(z). \text{ Replace } w_0(z) \text{ by } w_1(z),\\ \quad \text{and circulate again.}\end{array}\right. \tag{1}$$

It is pointed out that (1) is more efficient than the Bairstow[1] and 3-order methods[2].

求多项式根的劈二次因子法，除熟知的所谓 Bairstow 法[1]外，近来又提出了辗转相除法[3]及三阶方法[2]．就计算效能而论，辗转相除法稍优于 Bairstow 法，而三阶方法，当多项式的次数 $m\geqslant 34$ 时优于 Bairstow 法，$m\geqslant 97$ 时优于辗转相除法．但是，即使 $m\to+\infty$，三阶方

本文刊于《计算数学》，第 4 期，1982 年．

法的计算效能仅比其他两种方法高约 6%，也就是说，三种方法的计算效能是差不多的.

本文提出的 2^n 阶劈因子法，计算效能比以上三种方法高 30%～40%；所用的思想，是[3]，[4]中想法的综合与发展.

注. 按文献[2]，Bairstow 法的计算效能如下：

$$E_{ff_2}=\frac{602}{4m+4}=\frac{150.5}{m+1},$$

三阶方法的计算效能如下：

$$E_{ff_3}=\frac{954}{6m+18}=\frac{159}{m+3},$$

辗转相除法的计算效能如下：

$$E^*_{ff_2}=\frac{2000\cdot\log 2}{4m-3}\doteq\frac{602}{4m-3}.$$

§1 方 法

设 $F(z)$ 是实系数或复系数 m 次多项式，$m\geqslant 3$，欲求 $F(z)$ 的二次因子：

$$w^*(z)=(z-z_1^*)(z-z_2^*).$$

先取一个近似二次因子 $w_0(z)$，然后逐步加以修正，方法如下.

$$\begin{cases}\text{(i) 令 } \xi_0(z)=w_0(z)=(z-z_1^{(0)})(z-z_2^{(0)}).\\ \text{(ii) 若已确定 } i \text{ 个二次多项式：}\\ \quad \xi_j(z)=(z-z_1^{(j)})(z-z_2^{(j)}),\ j=0,1,\cdots,i-1,\\ \quad \text{我们用 } w_0\prod_{j=0}^{i-1}\xi_i \text{ 除 } F\text{，除法算式为}\\ \quad F=(w_0\prod_{j=0}^{i-1}\xi_j)S_{i-1}(z)+R_{i-1}(z).\\ \quad \text{将 } R_{i-1}(z) \text{ 与 } \xi_{i-1}^2(z) \text{ 辗转相除，所得之二次}\\ \quad \text{余式记为 } \alpha\xi_i(z)=\alpha(z-z_1^{(i)})(z-z_2^{(i)}).\\ \text{(iii) 令 } w_1(z)=\xi_n(z)\text{，用 } w_1(z) \text{ 代替 } w_0(z)\text{，继续进行.}\end{cases}\tag{1}$$

显然，当行 $n=1$ 时，得到的正是辗转相除法. 如文献[3]所述，这是 2 阶方法. 后面将证明，一般说来，(1)给出的是一个 2^n 阶迭代程序.

为保证定义(1)的合理性，应要求 $R_{i-1}(z)$ 与 $\xi_{i-1}^2(z)$ 辗转相除时产生二次余式. 关于此点，下面将证明：只要 $w_0(z)$ 和 $w^*(z)$ 足够接近，而且 $F'(z_1^*)\cdot F'(z_2^*)\neq 0$，这个要求必可满足.

§2 有关的记号与引理

为简便，以下略去某些足标，记 $R=R_{i-1}$，$S=S_{i-1}$，并以 z_1，z_2 记 $\xi_{i-1}(z)=0$ 的两个根，$\tilde{z}_1$，$\tilde{z}_2$ 记 $\xi_i(z)=0$ 的两个根，这里 $i\geqslant 1$ 是某个固定的整数. 由(1)，有

$$F(z)=\left[w_0(z)\prod_{j=0}^{i-1}\xi_j(z)\right]S(z)+R(z). \tag{2}$$

显然，$R(z)$是以$\{z_1^{(0)},z_2^{(0)},z_1^{(j)},z_2^{(j)},j=0,1,\cdots,i-1\}$为基点的 F 的插补多项式，从而

$$S(z)=F(z_1^{(0)},z_2^{(0)},z_1^{(0)},z_2^{(0)},z_1^{(1)},z_2^{(1)},\cdots,z_1^{(i-1)},z_2^{(i-1)},z), \tag{3}$$

这里 $F(x,y,z,\cdots)$是均差记号.

按定义(1)，$R(z)$与 $\xi_{i-1}^2(z)$辗转相除求二次余式之过程，也可这样进行：令

$$\begin{cases}R(z)=(\xi_{i-1}(z))^2Q(z)+P_1(z),\\ zR(z)=(\xi_{i-1}(z))^2[zQ(z)+q]+P_2(z).\end{cases} \tag{4}$$

设 a_1,a_2 分别为 P_1,P_2 的三次项系数，令

$$\begin{cases}a_2P_1(z)-a_1P_2(z)=\alpha z^2+\beta z+\gamma,\\ (a_2-a_1z)R(z)=(\xi_{i-1}(z))^2[(a_2-a_1z)Q(z)-qa_1]+(\alpha z^2+\beta z+\gamma),\end{cases} \tag{5}$$

则只要 $\alpha\neq0$，必有 $\alpha z^2+\beta z+\gamma=\alpha\xi_i(z)$，亦即当 $\alpha\neq0$ 时，可保证迭代过程(1)能继续进行.

如上所述，用 z_1,z_2 简记 $z_1^{(i-1)},z_2^{(i-1)}$. 以 z_1,z_2 为基点，按厄米特公式将 $R(z)$展开：

$$\begin{aligned}R(z)=&b_0+b_i(z-z_1)+b_2(z-z_1)(z-z_2)+\\ &b_3(z-z_1)^2(z-z_2)+R(z,z_1,z_1,z_2,z_2)\xi_{i-1}^2(z),\end{aligned} \tag{6}$$

这里 $R(x_1,x_2,\cdots)$是均差记号，而且

$$\begin{aligned}&b_0=R_1,b_1=R_{1,2},b_2=R_{1,1,2},b_3=R_{1,1,2,2},\\ &R_i=R(z_i),R_{i,j}=R(z_i,z_j),R_{i,j,k}=R(z_i,z_j,z_k),\\ &R_{i,j,k,l}=R(z_i,z_j,z_k,z_l),R_{i,j,k,l^*}=R(z_i,z_j,z_k,z_l^*),\cdots.\end{aligned}$$

将(6)与(4)比较，得

$$\begin{aligned}&Q(z)=R(z,z_1,z_1,z_2,z_2),\\ &a_1=b_3=R_{1,1,2,2},\\ &a_2=b_2+b_3z_2=R_{1,1,2}+z_2R_{1,1,2,2},\\ &q=b_3=R_{1,1,2,2},\\ &\alpha=a_2[b_2-b_3(z_2+2z_1)]-a_1[b_1-b_2(z_1+2z_2)+z_2(b_2-b_3(z_2+2z_1))]\\ &\quad=b_2^2-b_1b_3+b_2b_3(z_2-z_1)=R_{1,1,2}^2+R_{1,1,2,2}(R_{1,1,2}(z_2-z_1)-R_{1,2})\\ &\quad=R_{1,1,2}^2-R_{1,1,2,2}\cdot R_{1,1}.\end{aligned} \tag{7}$$

假设 $F'(z_1^*)F'(z_2^*)(z_1^*-z_2^*)\neq0$，可以找到 z_1^* 和 z_2^* 的半径为 r 的圆形邻域 $B_r(z_1^*)$和 $B_r(z_2^*)$，使当 $z\in B_r(z_1^*)\cup B_r(z_2^*)$时有

$$|F'(z)|\geqslant m>0. \tag{8}$$

再设 F 在 $B_r(z_1^*)\cup B_r(z_2^*)$上的各阶微商及各阶均差之上界为 $M>0$，又令$|z_1^*-z_2^*|=d>0$. 这时有

引理 1　对任给的 $\varepsilon>0$，可以找到 $\delta>0$，使当

$$|z_t^{(j)}-z_t^*|<\delta,j=0,1,\cdots,i-1;t=1,2 \tag{9}$$

时，有

$$|\alpha-(F_{1,1,2}^2-F_{1,1}\cdot F_{1,1,2,2})|<\varepsilon,$$

这里 $F_{1,1}=F(z_1,z_1)$，$F_{1,1,2}=F(z_1,z_1,z_2)$，$F_{1,1,2,2}=F(z_1,z_1,z_2,z_2)$等记 F 的均差.

证明 由于 $\alpha=R_{1,1,2}^2-R_{1,1,2,2}\cdot R_{1,1}$，故只需证明，对任给的 $\varepsilon>0$，存在 $\delta>0$，当(9)式成立时有

$$|R_{1,1,2}-F_{1,1,2}|<\varepsilon,\ |R_{1,1}-F_{1,1}|<\varepsilon,\quad |R_{1,1,2,2}-F_{1,1,2,2}|<\varepsilon \tag{10}$$

就足够了. 我们分别对 $R_{1,1}$，$R_{1,1,2}$，$R_{1,1,2,2}$加以讨论.

由(2)及 $z_1=z_1^{(i-1)}$ 得

$$\begin{aligned}
|R_{1,1}-F_{1,1}| &= \left|\frac{d}{dz}\left(w_0(z)\prod_{j=0}^{i-1}\xi_j(z)S(z)\right)\Big|_{z=z_1}\right| \\
&= \left|\left(w_0(z_1)\prod_{j=0}^{i-2}\xi_j(z_1)\right)S(z_1)\right|\cdot|z_1-z_2| \\
&\leqslant |(z_1-z_1^{(0)})(z_1-z_2^{(0)})|\cdot\left|\prod_{j=0}^{i-2}(z_1-z_1^{(j)})(z_1-z_2^{(j)})\right|\cdot \\
&\quad |S(z_1)|\cdot|z_1-z_2| \\
&\leqslant M\cdot(2\delta)^i\cdot(d+2\delta)^i\to 0,\text{当 }\delta\to 0;
\end{aligned}$$

$$\begin{aligned}
|R_{1,1,2}-F_{1,1,2}| = |(R-F)_{1,1,2}| &= \left|\frac{1}{z_1-z_2}((R-F)_{1,1}-(R-F)_{1,2})\right| \\
&= \left|\frac{1}{z_1-z_2}(R-F)_{1,1}\right| \leqslant \frac{M(2\delta)^i(d+2\delta)^i}{|d-2\delta|}\to 0,\text{当 }\delta\to 0;
\end{aligned}$$

$$\begin{aligned}
|R_{1,1,2,2}-F_{1,1,2,2}| &= \left|\frac{1}{z_1-z_2}((R-F)_{1,2,1}-(R-F)_{1,2,2})\right| \\
&= \left|\frac{1}{(z_1-z_2)^2}((R-F)_{1,1}-(R-F)_{2,2})\right| \\
&\leqslant \frac{2M(2\delta)^i(d+2\delta)^i}{(d-2\delta)^2}\to 0,\text{当 }\delta\to 0.
\end{aligned}$$

于是引理 1 证毕. 由此立得

推论 1 任给 $\varepsilon>0$，必存在 $\delta>0$，使当(9)式成立时，有

$$\left|\alpha-\frac{F'(z_1^*)\cdot F'(z_2^*)}{(z_1^*-z_2^*)(z_2^*-z_1^*)}\right|<\varepsilon. \tag{11}$$

这由

$$\begin{aligned}
\lim_{\substack{z_1\to z_1^*\\ z_2\to z_2^*}}(F_{1,1,2}^2-F_{1,1}\cdot F_{1,1,2,2}) &= F^2(z_1^*,z_1^*,z_2^*)-F(z_1^*,z_1^*)F(z_1^*,z_1^*,z_2^*,z_2^*) \\
&= -\frac{F'(z_1^*)F'(z_2^*)}{(z_1^*-z_2^*)^2}
\end{aligned}$$

立即可以推出.

推论 2 若 $F'(z_1^*)F'(z_2^*)\neq 0$，则存在 $\delta>0$，使当(9)式成立时有 $\alpha\neq 0$，即 $R(z)$与 $\xi_{i-1}^2(z)$辗转相除过程中必然产生二次余式.

下面的引理，更具体地给出了误差传递关系.

引理 2 对任给的 $q\in(0,1)$，必有 $\delta>0$，使当

$$|z_t^{(0)}-z_t^*|<\delta, t=1,2 \tag{12}$$

时,可以适当规定 $\tilde{z}_1,\tilde{z}_2$ 的足标,使有

$$|\tilde{z}_t-z_t^*|<q^i\delta. \tag{13}$$

证明　对 i 作归纳. $i=0$ 时显然. 设命题对 $1,2,\cdots,i-1$ 为真,往证它对 i 亦真.

在(5)的第二式中,令 $z=z_1^*,z_2^*$,得

$$\begin{cases}(a_2-a_1z_1^*)R(z_1^*)=(\xi_{i-1}(z_1^*))^2[(a_2-a_1z_1^*)Q(z_1^*)-qa_1]+\\ \qquad\qquad \alpha(z_1^*-\tilde{z}_1)(z_1^*-\tilde{z}_2),\\ (a_2-a_1z_2^*)R(z_2^*)=(\xi_{i-1}(z_2^*))^2[(a_2-a_1z_2^*)Q(z_2^*)-qa_1]+\\ \qquad\qquad \alpha(z_2^*-\tilde{z}_1)(z_2^*-\tilde{z}_2),\end{cases} \tag{14}$$

由(2)及 $F(z_t^*)=0(t=1,2)$,得

$$\begin{aligned}|R(z_t^*)|&=\left|w_0(z_t^*)\left(\prod_{j=0}^{i-1}\xi_j(z_t^*)\right)S(z_t^*)\right|\\&\leqslant Mq^{1+2+\cdots+(i-1)}\delta^{i+1}(d+\delta)^{i+1}.\end{aligned} \tag{15}$$

另一方面,由引理 1 之证明知,可取 $\delta>0$ 足够小,使

$$\begin{cases}|a_2-a_1z_t^*|\leqslant L,\\ |(a_2-a_1z_t^*)Q(z_1^*)-qa_1|\leqslant L;\end{cases} \tag{16}$$

$$|\alpha|\geqslant\frac{1}{2}\left(\frac{m}{d}\right)^2, \tag{17}$$

这里 L 是仅与 d,m,M 有关的常数. 故由(13)得

$$\begin{aligned}|z_t^*-\tilde{z}_1|\cdot|z_t^*-\tilde{z}_2|&\leqslant 2L\left(\frac{d}{m}\right)^2(Mq^{1+2+\cdots+(i-1)}\delta^{i+1}(d+\delta)^{i+1}+q^{2(i-1)\delta^2})\\&<H(M(d+\delta)^{i+1}\delta^i+\delta)(q^i\delta), t=1,2,\end{aligned} \tag{18}$$

这里 H 是归并而得的一个常数.

显然,对于给定的 H,M,d,可以取 $\delta>1$ 足够小,使得

$$H(M(d+\delta)^{i+1}\delta^i+\delta)<\sqrt{\delta}<1$$

对一切 i 成立. 如果再有 $\delta\leqslant\frac{1}{4}d^2$,则有

$$\begin{cases}|z_1^*-\tilde{z}_1|\cdot|z_1^*-\tilde{z}_2|\leqslant\sqrt{\delta}\cdot q^i\delta,\\ |z_2^*-\tilde{z}_1|\cdot|z_2^*-\tilde{z}_2|\leqslant\sqrt{\delta}\cdot q^i\delta.\end{cases} \tag{19}$$

不妨设 $|z_1^*-\tilde{z}_1|\leqslant|z_1^*-\tilde{z}_2|$,于是有

$$|z_1^*-\tilde{z}_1|\leqslant\sqrt{\delta}.$$

故由第二式知,$|z_2^*-\tilde{z}_1|\geqslant d-\sqrt{\delta}>\frac{d}{2}$,从而

$$|z_2^*-\tilde{z}_2|\leqslant\frac{2\sqrt{\delta}}{d}q^i\delta<q^i\delta.$$

同时

$$|z_1^*-\tilde{z}_2|\geqslant d-q^i\delta\geqslant\sqrt{\delta}.$$

同理得

$$|z_1^* - \tilde{z}_1| < q^i \delta.$$

引理 2 证毕.

推论 3 对任意 i,可适当规定 $\xi_i(z)$ 的两个根 $\tilde{z}_1, \tilde{z}_2$ 的足标,使得

$$\lim_{\substack{z_1^{(0)} \to z_1^* \\ z_2^{(0)} \to z_2^*}} (|\tilde{z}_1 - z_1^*| + |\tilde{z}_2 - z_2^*|) = 0.$$

§3 主要结果

我们把本文的主要结果写成定理形式.

定理 设 $F(z)$ 是给定的实或复系数多项式,$w^*(z) = (z - z_1^*)(z - z_2^*)$ 是 $F(z)$ 的二次因子,$\xi_j(z), z_1^{(j)}, z_2^{(j)}$ $(j = 0, 1, \cdots, n)$ 由程序(1)给出,则

1) $\xi_0(z)$ 的两根与 z_1^*, z_2^* 充分接近时,程序(1)有意义.

2) 存在常数 $C_t^{(n)}$ $(t = 1, 2; n = 1, 2, \cdots)$,使得能够适当选取若 $\tilde{z}_1 = z_1^{(n)}, \tilde{z}_2 = z_2^{(n)}$ 之下标,有

$$\lim_{\substack{z_1^{(0)} \to z_1^* \\ z_2^{(0)} \to z_2^*}} \frac{z_1^* - \tilde{z}_1}{(z_1^{(0)} - z_1^*)^{2^n}} = C_1^{(n)},$$

$$\lim_{\substack{z_1^{(0)} \to z_1^* \\ z_2^{(0)} \to z_2^*}} \frac{z_2^* - \tilde{z}_2}{(z_2^{(0)} - z_2^*)^{2^n}} = C_2^{(n)}.$$

证明 由引理 1,2 及其推论可知,1)为真.

为证结论 2),可对 n 作数学归纳. $n=1$ 时,由文献[3]知命题为真. 设命题对 $1, 2, \cdots, n-1$ 已真,往证它对 n 亦真.

在(14)中取 $i = n$,由引理 2 知,可适当选择下标,使 $z_1^{(0)} \to z_1^*, z_2^{(0)} \to z_2^*$ 时有 $\tilde{z}_1 \to z_1^*$, $\tilde{z}_2 \to z_2^*$. 从(14)解出 $(z_1^* - \tilde{z}_1)$,考虑下列极限过程:

$$\lim_{\substack{z_1^{(0)} \to z_1^* \\ z_2^{(0)} \to z_2^*}} \frac{(z_1^* - \tilde{z}_1)}{(z_1^* - z_1^{(0)})^{2^n}} = \lim_{\substack{z_1^{(0)} \to z_1^* \\ z_2^{(0)} \to z_2^*}} \frac{1}{\alpha(z_1^* - \tilde{z}_2)(z_1^* - z_1^{(0)})^{2^n}} [(a_2 - a_1 z_1^*) R(z_1^*) - (\xi_{n-1}(z_1^*))^2 ((a_2 - a_1 z_1^*) Q(z_1^*) - q a_1)]$$

$$= \lim_{\substack{z_1^{(0)} \to z_1^* \\ z_2^{(0)} \to z_2^*}} \frac{1}{\alpha(z_1^* - z_2^*)} \cdot \lim_{\substack{z_1^{(0)} \to z_1^* \\ z_2^{(0)} \to z_2^*}} \left[(a_2 - a_1 z_1^*) S(z_1^*) \frac{w_0(z_1^*) \prod_{j=0}^{n-1} \xi_j(z_1^*)}{(z_1^* - z_1^{(0)})^{2^n}} - \frac{(\xi_{n-1}(z_1^*))^2}{(z_1^* - z_1^{(0)})^{2^n}} ((a_2 - a_1 z_1^*) Q(z_1^*) - q a_1) \right].$$

但我们有(当 $z_1^{(0)} \to z_1^*, z_2^{(0)} \to z_2^*$):

$$\alpha \to -\frac{F'(z_1^*) F'(z_2^*)}{(z_1^* - z_2^*)^2},$$

$$(a_2-a_1z_1^*)\to F(z_1^*,z_1^*,z_2^*)+z_2^*F(z_1^*,z_1^*,z_2^*,z_2^*)-z_1^*F(z_1^*,z_1^*,z_2^*,z_2^*)$$
$$=F(z_1^*,z_1^*,z_2^*)+F(z_1^*,z_2^*,z_2^*)-F(z_1^*,z_2^*,z_1^*)$$
$$=F(z_1^*,z_2^*,z_2^*)=\frac{F'(z_2^*)}{(z_2^*-z_1^*)}.$$

由(3)

$$S(z_1^*)\to F(\underbrace{z_1^*,z_2^*,z_1^*,z_2^*,\cdots,z_1^*,z_2^*}_{n\text{对}},z_1^*),$$

将右式记为

$$F(z_1^{*(n+1)},z_2^{*(n)}),$$

由(7)及(2)有

$$Q(z_1^*)\to\lim_{\substack{z_1^{(0)}\to z_1^*\\ z_2^{(0)}\to z_2^*}}R(z_1^*,z_1^*,z_1^*,z_2^*,z_2^*)$$
$$=F(z_1^*,z_1^*,z_1^*,z_2^*,z_2^*)=F(z_1^{*(3)},z^{*(2)}).$$

又由(7)及引理1,有

$$qa_1\to(F(z_1^*,z_2^*,z_1^*,z_2^*))^2.$$

同时,由归纳假设,

$$\lim_{\substack{z_1^{(0)}\to z_1^*\\ z_2^{(0)}\to z_2^*}}\frac{w_0(z_1^*)\prod\limits_{j=0}^{n-1}\xi_j(z_1^*)}{(z_1^*-z_1^{(0)})^{2^n}}=\lim_{\substack{z_1^{(0)}\to z_1^*\\ z_2^{(0)}\to z_2^*}}\frac{(z_1^*-z_1^{(0)})(z_1^*-z_2^{(0)})}{(z_1^*-z_1^{(0)})}\times\prod_{j=0}^{n-1}\frac{(z_1^*-z_1^{(j)})(z_1^*-z_2^{(j)})}{(z_1^*-z_1^{(0)})^{2^j}}$$
$$=(z_1^*-z_2^*)^{n+1}C_1^{(1)}C_1^{(2)}\cdots C_1^{(n-1)},$$

$$\lim_{\substack{z_1^{(0)}\to z_1^*\\ z_2^{(0)}\to z_2^*}}=\frac{(\xi_{n-1}(z_1^*))^2}{(z_1^*-z_1^{(0)})^{2^n}}=\left[\lim_{\substack{z_1^{(0)}\to z_1^*\\ z_2^{(0)}\to z_2^*}}\frac{(z_1^*-z_1^{(n-1)})(z_2^*-z_2^{(n-1)})}{(z_1^*-z_1^{(0)})^{2^{n-1}}}\right]^2=[(z_1^*-z_2^*)C_1^{(n-1)}]^2.$$

故得

$$\lim_{\substack{z_1^{(0)}\to z_1^*\\ z_2^{(0)}\to z_2^*}}\frac{(z_1^*-\tilde{z}_1)}{(z_1^*-z_1^{(0)})^{2^n}}=\frac{(z_2^*-z_1^*)}{F'(z_1^*)F'(z_2^*)}\Big[\frac{F'(z_2^*)}{(z_2^*-z_1^*)}F(z_1^{*(n+1)},z_2^{*(n)})(z_1^*-z_2^*)^{n+1}\times$$
$$C_1^{(1)}C_1^{(2)}\cdots C_1^{(n-1)}-(z_1^*-z_2^*)^2(C_1^{(n-1)})^2\times$$
$$\left(\frac{F'(z_2^*)}{(z_2^*-z_1^*)}F(z_1^{*(3)},z_2^{*(2)})-(F(z_1^{*(2)},z_2^{*(2)}))^2\right)\Big]$$
$$=\frac{(z_1^*-z_2^*)^2}{F'(z_1^*)}\Big[F(z_1^{*(n+1)},z_2^{*(n)})(z_1^*-z_2^*)^{n-1}C_1^{(1)}C_1^{(2)}\cdots C_1^{(n-1)}$$
$$-(C_1^{(n-1)})^2\left(F(z_1^{*(3)},z_2^{*(2)})-\frac{(z_2^*-z_1^*)}{F'(z_2^*)}(F(z_1^{*(2)},z_2^{*(2)})^2\right)\Big]$$
$$=\frac{1}{F'(z_1^*)F'(z_2^*)}\Big[F(z_1^{*(n+1)},z_2^{*(n)})F'(z_2^*)(z_1^*-z_2^*)^{n+1}C_1^{(1)}\cdots C_1^{(n-1)}-$$
$$(C_1^{(n-1)})^2\left(F''(z_1^*)F'(z_2^*)-\frac{1}{z_1^*-z_2^*}(F'(z_1^*)^2+4F'(z_1^*)F'(z_2^*)+3F'(z_2^*)^2)\right)\Big]$$
$$=C_1^{(n)}.$$

同理有

$$\lim_{\substack{z_1^{(0)}\to z_1^*\\ z_2^{(0)}\to z_2^*}}(z_1^*)\frac{(z_2^*-\bar{z}_2)}{(z_2^*-z_2^{(0)})^{2^n}}=\frac{1}{F'(z_1^*)F'(z_2^*)}\Big[F(z_1^{*(n)},z_2^{*(n+1)})F'(z_1^*)(z_2^*-z_1^*)^{n+1}\times$$

$$C_2^{(1)}\cdots C_2^{(n-1)}-(C_2^{(n-1)})^2\Big[F''(z_2^*)F'(z_1^*)-$$

$$\frac{1}{(z_2^*-z_1^*)}\left[F'(z_2^*)^2+4F'(z_1^*)F'(z_2^*)+3F'(z_1^*)^2\right]\Big]\Big]$$

$$=C_2^{(n)}.$$

定理证毕.顺便指出,若 $z_1^*=z_2^*$ 恰为二重根,类似的估计可知定理的结论仍真.由于这种情形不重要,就不再详加证明了.

§4　计算效能之比较

现在来估计一下,对于(1)给出的 2^n 阶迭代程序,每迭代一次至多用多少次(乘、加)运算.

设 F 是 m 次多项式,为了从 $w_0=\xi_0$ 求取 ξ_1,需进行的运算次数与文献[3]中之辗转相除法相同,共$(4m-3)$次.

我们从已有的除法算式:

$$F=w_0^2S_0+R_0=w_0\xi_0S_0+R_0$$

出发,求取 S_1 和 R_1.方法是:用 ξ_1 除 S_0,得

$$S_0=\xi_1S_1+(\lambda_1z+\mu_1).$$

从而

$$F=w_0\xi_0\xi_1S_1+w_0\xi_0(\lambda_1z+\mu_1)+R_0=w_0\xi_0\xi_1S_1+R_1.$$

这样,为求取 5 次多项式 R_1 所用的运算为:

1)　用 ξ_1 除 S_0,因 ξ_1 为二次,S_0 为$(m-4)$次,故需 $2(m-6)$次运算.

2)　把 $R_1=w_0\xi_0(\lambda_1z+\mu_1)+R_0$ 整理成降幂形式,所需之运算次数约为 5 次乘加.

然后,用 ξ_1^2 与 R_1 辗转相除求取 ξ_2,共需运算次数为 $4\times5-3=17$(次).这样,由 ξ_1 到 ξ_2,共需运算 $2(m-6)+5+17=(2m+10)$(次).

一般地,已求得 $\xi_0,\xi_1,\cdots,\xi_{i-1}$时,我们已有等式:

$$F=w_0\xi_0\xi_1\cdots\xi_{i-2}S_{i-2}+R_{i-2},$$

这里 S_{i-2}是$(m-2i)$次多项式,而 R_{i-2}是$(2i-1)$次多项式.我们用 ξ_{i-1}除 S_{i-2},得

$$S_{i-2}=\xi_{i-1}S_{i-1}+(\lambda_{i-1}z+\mu_{i-1}).$$

从而

$$\begin{aligned}F&=w_0\xi_0\xi_1\cdots\xi_{i-1}S_{i-1}+w_0\xi_0\xi_1\cdots\xi_{i-2}(\lambda_{i-1}z+\mu_{i-1})+R_{i-2}\\&=w_0\xi_0\xi_1\cdots\xi_{i-1}S_{i-1}+R_{i-1}.\end{aligned}$$

于是,为求取$(2i+1)$次多项式 R_{i-1}所用的运算为

1)　用 ξ_{i-1}除 S_{i-2},因 ξ_{i-1}为 2 次,S_{i-2}为$(m-2i)$次,故此除法需 $2(m-2i-2)$次运算.

2)　把 $R_{i-1}=w_0\xi_0\xi_1\cdots\xi_{i-2}(\lambda_{i-1}z+\mu_{i-1})$ 整理成降幂形式，所需之运算次数约为 $(2i+1)$ 次乘加.

然后，用 ξ_{i-1}^2 与 R_{i-1} 辗转相除，共需运算次数为 $4(2i+1)-3=8i+1$(次). 因此，由 ξ_{i-1} 到 ξ_i 共需运算次数为

$$2(m-2i-2)+2i+1+8i+1=2m+6i-2.$$

因此，由 $w_0=\xi_1$ 到 $w_1=\xi_n$ 共需之运算次数为

$$\begin{aligned}&(4m-3)+(2m+10)+(2m+16)+\cdots+(2m+6n-2)\\&=2m(n+1)+10(n-1)+3(n-1)(n-2)\\&=2mn+2m+(n-1)(3n+4),\end{aligned}$$

故程序(1)对于 m 次多项式之计算效能为

$$\begin{aligned}E_{ff}(m,n)&=2000\times\frac{\log 2^n}{2mn+2m+(n-1)(3n+4)}\\&=\frac{2000\times\log 2}{2m+2\,\frac{m}{n}+(3n+4)\left(1-\frac{1}{n}\right)}\\&=\frac{2000\times\log 2}{(2m+1)+\frac{2(m-2)}{n}+3n}\doteq\frac{602}{(2m+1)+\frac{2(m-2)}{n}+3n}.\end{aligned}$$

取 $n=2$，得到一个 4 阶迭代程序，其计算效能为

$$E_{ff}(m,2)\doteq\frac{602}{3m+5}\doteq\frac{201}{m+1.7}.$$

对任意的 m，它都比 Bairstow 法的 $E_{ff_2}=\frac{602}{4m+4}$ 和三阶法的 $E_{ff_3}=\frac{954}{6m+18}\doteq\frac{159}{m+3}$ 要大，而且当 m 趋于无穷时，它分别比两法高出 33%和 27%.

取 $n=3,4$，分别得到 8 阶，16 阶迭代程序，其计算效能分别为

$$E_{ff}(m,3)\doteq\frac{1806}{8m+2b}\doteq\frac{226}{m+3.3},$$

$$E_{ff}(m,4)\doteq\frac{2408}{10m+48}\doteq\frac{241}{m+5}.$$

它们都比三阶法的 E_{ff_3} 为大. 当 $m>5$ 时也比 Bairstow 法大. 当 m 足够大时，$E_{ff}(m,3)$ 分别比 Bairstow 法和三阶方法高 51%和 42%，而 $E_{ff}(m,4)$ 则分别比两法高 61%和 52%.

最后讨论一下，当 m 给定之后，取多少阶的方法最合算？由

$$E_{ff}(m,n)\doteq\frac{602}{(2m+1)+\frac{2(m-2)}{n}+3n},$$

当 m 固定时，总有

$$\begin{aligned}&(2m+1)+\frac{2(m-2)}{n}+3n\geqslant(2m+1)+2\sqrt{\frac{2(m-2)}{n}\cdot 3n}\\&=(2m+1)+2\sqrt{6(m-2)},\end{aligned}$$

即当 $n=\sqrt{\frac{2}{3}(m-2)}$ 时，$E_{ff}(m,n)$ 最大. 由于 n 是整数，故可取

$$n=\left[\sqrt{\frac{2}{3}(m-2)}\right]+1.$$

这时，其计算效能可列表如下：

m	5	8	14	20	26	56	98
n	2	2	3	3	4	6	8
$E_{ff}(m,n)$	30	21	13	10	8	4	2.4
E_{ff_2}	25	17	11	7	6	2.6	1.5
E_{ff_3}	20	15	9	7	6	2.7	1.6

显见，当 $m\to+\infty$ 时，取 $n_m=\left[\sqrt{\frac{2}{3}(m-2)}\right]+1$，$E_{ff}(m,n_m)$ 大致是 E_{ff_2} 和 E_{ff_3} 的两倍.

参考文献

[1]清华大学、北京大学《计算方法》编写组. 计算方法（上册）. 北京：科学出版社，1975：169－174.

[2]叶贻才. 求多项式根的三阶劈因子法. 计算数学，1980，2(3)：269－272.

[3]井中. 求多项式根的新劈因子法. 计算数学，1979：1(1)：31－34.

[4]张景中. 求解超越方程的 $n+1$ 信息的 2^n 阶迭代. 计算数学，1980，2(4)：350－355.

棱长为奇数之单形的一个注记

杨　路　张景中

（中国科学技术大学）

与本文所述同一类型的问题，首先是由美国知名的离散数学家 Graham 加以考虑的.

为了方便，我们将 n 维欧氏空间中$(n+1)$个点构成的图形叫做单形，它可以是退化的.

1974 年，在 R. L. Graham 等[1]三人联合署名的一篇文章中，解决了如下的问题：

在 E^n 中是否存在这样一个单形，使得

ⅰ)此单形的各棱长都是奇数；

ⅱ)此单形的各顶点到某一个点的距离都是奇数？（叙述上似与文献[1]略异，其实是等价的.）

Graham 等人证明了，只有当 $n\equiv-2 \pmod{16}$时这样的单形确实存在，否则无解.

我们这篇短文将要解决的问题是：

在 E^n 中是否存在这样一个单形，使得

(ⅰ)此单形的各棱长都是奇数；

(ⅱ)此单形的各顶点到某一超平面的距离都是整数？

我们将要证明：只有当 $m=\pm1 \pmod{8}$时这样的单形确实存在，否则这种单形不存在.

Graham 等人的工作，依整于 A. Cayley[2]. 的一个著名的定理. 但是 Cayley 定理对于现在这个问题是不够用的. 我们的证明主要依赖于如下的结果：

引理　设 E^n 中某$(n+1)$个点 $A_1, A_2, \cdots, A_{n+1}$ 两两间的距离 $A_iA_j=d_{ij}$，各顶点 A_i 到某一个超平面Π的带号距离为 $h_i(i,j=1,2,\cdots,n+1)$，则必然有

$$\begin{vmatrix} 0 & 1 & 1 & 1 & \cdots & 1 & 0 \\ 1 & 0 & -\frac{d_{12}^2}{2} & -\frac{d_{13}^2}{2} & \cdots & -\frac{d_{1,n+1}^2}{2} & h_1 \\ 1 & -\frac{d_{21}^2}{2} & 0 & -\frac{d_{23}^2}{2} & \cdots & -\frac{d_{2,n+1}^2}{2} & h_2 \\ 1 & -\frac{d_{31}^2}{2} & -\frac{d_{32}^2}{2} & 0 & & \vdots & \vdots \\ \vdots & \vdots & \vdots & & \ddots & \vdots & \vdots \\ 1 & -\frac{d_{n+1,1}^2}{2} & -\frac{d_{n+1,2}^2}{2} & \cdots & & 0 & h_{n+1} \\ 0 & h_1 & h_2 & \cdots & & h_{n+1} & 1 \end{vmatrix}=0. \tag{1}$$

本文刊于《数学年刊(中文版)》，第 3 卷第 3 期，1982 年.

这个引理是[3]中的引理 2.1 的一个特殊情形. 此地只需引用而不再重复其证明. 经过极简单初等变换,(1)式可变形为

$$\begin{vmatrix} 0 & 1 & 1 & 1 & \cdots & 1 & 0 \\ 1 & 0 & d_{12}^2 & d_{13}^2 & \cdots & d_{1,n+1}^2 & h_1 \\ 1 & d_{21}^2 & 0 & d_{23}^2 & \cdots & d_{2,n+1}^2 & h_2 \\ 1 & d_{31}^2 & d_{32}^2 & 0 & & \vdots & \vdots \\ \vdots & \vdots & \cdots & & \ddots & \vdots & \vdots \\ 1 & d_{n+1,1}^2 & d_{n+1,2}^2 & \cdots & & 0 & h_{n+1} \\ 0 & h_1 & h_2 & \cdots & & h_{n+1} & -\frac{1}{2} \end{vmatrix}=0. \tag{2}$$

我们令 D 表示下述所谓 Cayley-Menger 行列式[2]

$$D=\begin{vmatrix} 0 & 1 & 1 & 1 & \cdots & 1 \\ 1 & 0 & d_{12}^2 & d_{13}^2 & \cdots & d_{1,n+1}^2 \\ 1 & d_{21}^2 & 0 & d_{23}^2 & \cdots & d_{2,n+1}^2 \\ 1 & d_{31}^2 & d_{32}^2 & 0 & & \vdots \\ \vdots & \vdots & \vdots & & \ddots & \vdots \\ 1 & d_{n+1,1}^2 & d_{n+1,2}^2 & \cdots & & 0 \end{vmatrix},$$

这是一个 $n+2$ 阶行列式,我们可以约定它的行和列的编号都是从 0 到 $n+1$,然后用 D_{ij} 表示 D 的各个代数余子式($i,j=0,1,2,\cdots,n+1$). 在这些约定下可将(2)式展开而得到

$$\frac{1}{2}D+\sum_{i=1}^{n+1}\sum_{j=1}^{n+1}D_{ij}h_ih_j=0, \tag{3}$$

考虑到 D 是对称行列式,故(3)还可改写为

$$\frac{1}{2}D+\sum_{i=1}^{n+1}D_{ii}h_i^2+2\sum_{1\leqslant i<j\leqslant n+1}D_{ij}h_ih_j=0. \tag{4}$$

下面进入主题:

定理 当且仅当 $n\equiv\pm1 \pmod 8$时,在 E^n 中存在着一个单形使得

(ⅰ) 此单形的各棱长都是奇数;

(ⅱ) 此单形的各顶点到某一超平面的距离都是整数.

证 假定满足条件(ⅰ),(ⅱ)的单形 $A_1A_2\cdots A_{n+1}$ 和超平面 Π 确已存在. 完全沿用引理中的记号,从(3)式可知 D 必为偶数.

另一方面,我们引进记号 $\Delta(m)$ 表示这样一个 m 阶行列式:它的主对角线上的元素皆为 0 而其余元素皆为 1. 容易算出

$$\Delta(m)=(-1)^{m-1}(m-1). \tag{5}$$

又令 $\Delta_{ij}(m)$ 表示 $\Delta(m)$ 的各个代数余子式($i,j=0,1,\cdots,m-1$),也不难算出

$$\Delta_{ij}(m)=\begin{cases}(-1)^{m-2}(m-2) & (i=j), \\ (-1)^{m-1} & (i\neq j).\end{cases} \tag{6}$$

因 $d_{ij}(i\neq j)$都是奇数,故有 $d_{ij}^2=1 \pmod 8$. 于是行列式 D 的每个元素与 $\Delta(n+2)$ 的对应元素都是关于 mod 8 同余的,从而有

$$D\equiv\Delta(n+2) \pmod 8. \tag{7}$$

同理也有

$$D_{ij}\equiv\Delta_{ij}(n+2)\pmod{8}.\tag{8}$$

参照(5)～(7)可知

$$D\equiv(-1)^{n+1}(n+1)\pmod{8}.\tag{9}$$

$$D_{ij}\equiv\begin{cases}(-1)^{n}\cdot n & (i=j)\\(-1)^{n+1} & (i\neq j)\end{cases}\pmod{8}.\tag{10}$$

前面已说明 D 必是偶数，故由(9)，n 必是奇数，所以我们只需考虑 $n\equiv\pm3$ 和$n\equiv\pm1\pmod{8}$这四个同余类.

令 l 表示 $h_1,h_2,\cdots,h_{n+1}$ 中奇数的个数.

假定 $n\equiv3\pmod{8}$，这时 $D\equiv4\pmod{8}$，从而

$$\frac{D}{2}\equiv2\pmod{4}.\tag{11}$$

那么(4)中的 $\sum D_{ii}h_i^2$ 必须是偶数，注意到 D_{ii} 是奇数（由(10)式），可知 l 必为偶数，不外乎两种可能，$l=4k$ 或 $l=4k+2$.

显然，所有形如 $h_ih_j\,(1\leqslant i<j\leqslant n+1)$ 的各数中奇数的个数应为 $C_l^2=\frac{1}{2}l(l-1)$. 如果 $l=4k$，则 $C_l^2=2k(4k-1)$，即 h_ih_j 各数（从而 $D_{ij}h_ih_j$ 各项，因这时 D_{ij} 是奇数）中奇数的个数为偶数. 于是有

$$2\sum\sum_{1\leqslant i<j\leqslant n+1}D_{ij}h_ih_j\equiv0\pmod{4}.\tag{12}$$

另一方面，对于每个奇的 h_i，由 $h_i\equiv1\pmod{8}$ 和 $D_{ii}\equiv-n$ 知 $D_{ii}h_i^2\equiv-n\pmod{8}$. 考虑到这样的项共有 $l=4k$ 个，另外注意到当 h_i 为偶数时 $D_{ii}h_i^2$ 是 4 的倍数，故有

$$\sum_{i=1}^{n+1}D_{ii}h_i^2\equiv0\pmod{4}.\tag{13}$$

将(11)～(13)三式相加即与(4)式矛盾.

如果 $l=4k+2$，则 $C_l^2=(2k+1)(4k+1)$，从而 $D_{ij}h_ih_j\,(1\leqslant i<j\leqslant n+1)$ 各项中奇数的个数为奇数，故有

$$2\sum\sum_{1\leqslant i<j\leqslant n+1}D_{ij}h_ih_j\equiv2\pmod{4}.\tag{14}$$

另一方面，在形如 $D_{ij}h_1^2$ 的各项中 $l=4k+2$ 个满足 $D_{ii}h_i^2\equiv-n\pmod{8}$ 而另一些是 4 的倍数，故有

$$\sum_{i=1}^{n+1}D_{ii}h_i^2\equiv2\pmod{4}.\tag{15}$$

将(11)，(14)，(15)相加即与(4)矛盾，故 $n\not\equiv3\pmod{4}$.

假定 $n=-3\pmod{8}$，这时 $D=-2\pmod{8}$，从而

$$\frac{D}{2}\equiv-1\pmod{4}.\tag{16}$$

那么(4)中的 $\sum D_{ii}h_i^2$ 必须是奇数. 注意到 D_{ii} 都是奇数，故 l 必为奇数. 不外乎两种可能，$l=4k+1$或 $l=4k-1$.

如果 $l=4k+1$，则 $C_l^2=2k(4k+1)$，从而 $D_{ij}h_ih_j\,(1\leqslant i<j\leqslant n+1)$ 各项中奇数的个数为偶数，这时(12)成立. 另一方面形如 $D_{ii}h_i^2$ 的各项中有 $l=(4k+1)$ 个满足 $D_{ii}h_i^2\equiv-n\pmod{8}$ 而另一些是 4 的倍数，故有

$$\sum_{i=1}^{n+1} D_{ii} h_i^2 \equiv -1 \pmod 4 . \tag{17}$$

将(12),(16),(17)相加即与(4)矛盾.

如果 $l=4k-1$,则 $C_l^2=(4k-1)(2k-1)$,从而 $D_{ij}h_ih_j(1\leqslant i<j\leqslant n+1)$各项中奇数的个数为奇数,这时(14)式成立. 另一方面在形如 $D_{ii}h_i^2$ 的各项中有 $l=(4k-1)$个满足 $D_{ii}h_i^2\equiv -n \pmod 8$而另一些是 4 的倍数,故有

$$\sum_{i=1}^{n+1} D_{ii} h_i^2 \equiv 1 \pmod 4 . \tag{18}$$

将(14),(16),(18)相加则与(4)矛盾.

综上所述可知 $n\not\equiv \pm 3 \pmod 8$.

下面将要指出,当 $n\equiv \pm 1 \pmod 8$时,满足条件(ⅰ)(ⅱ)的单形确实存在.

如果 $n=8s-1$,s 是一个正整数,我们将构造出满足(ⅰ)(ⅱ)的 n 维单形:在 E^n 中任取一超平面Π,在Π上作一个$(n-1)$维正则单形 $A_1A_2\cdots A_n$,令 $A_iA_j=(8s-1)((i\neq j;i,j=1,2,\cdots,n)$. 众所周知,任何一个 m 维正则单形的顶点到其中心的距离R 与单形的棱长 a 之间有关系(可参看文献[1])

$$R=\sqrt{\frac{m}{2(m+1)}}\cdot a,$$

这叫正则单形的"顶心距"公式,据此容易算出单形 $A_1A_2\cdots A_n$ 各顶点到其中心 O 的距离

$$OA_i=\sqrt{(8s-1)(4s-1)} \quad (i=1,2,\cdots,n). \tag{19}$$

过 O 作直线垂直于超平面Π,在此直线上取一点 A_{n+1},使得

$$OA_{n+1}=2s. \tag{20}$$

于是显然有

$$A_{n+1}A_i=\sqrt{OA_i^2+OA_{n+1}^2}=6s-1 \quad (i=1,2,\cdots,n). \tag{21}$$

这样得到的单形 $A_1A_2\cdots A_{n+1}$的棱长均为奇数($8s-1$ 或 $6s-1$),而其各顶点到超平面Π的距离均为整数(0 或 $2s$),故合乎所示.

如果 $n=16s+1$,我们在 E^n 中取一超平面Π并在Π上作$(n-1)$维正则单形 $A_1A_2\cdots A_n$,令 $A_iA_j=16s+1(i\neq j;i,j=1,2,\cdots,n)$. 设单形中心为 O,由"顶心距"公式可算出

$$OA_i=\sqrt{8s(16s+1)} \quad (i=1,2,\cdots,n). \tag{22}$$

过 O 作直线垂直于Π,在此直线上取点 A_{n+1}使得

$$OA_{n+1}=32s^2+2s-1. \tag{23}$$

于是有

$$A_{n+1}A_i=\sqrt{OA_i^2+OA_{n+1}^2}=32s^2+2s+1 \quad (i=1,2,\cdots,n). \tag{24}$$

这样得到的单形 $A_1A_2\cdots A_{n+1}$合乎所求.

最后,如果 $n=16s+9$,要作出符合条件的单形不像以上两种情况那么容易,步骤如下:

先在 E^{n-3}中作一个中心在原点的正则单形 $B_1B_2\cdots B_{n-2}$,使 $B_iB_j=48s+21(1\leqslant i<j\leqslant n-2)$. 令

$$B_i=(b_{i1},b_{i2},\cdots,b_{i,n-3}),$$

则其顶心距

$$OB_i=\sqrt{9(8s+3)(16s+7)}=(b_{i1}^2+b_{i2}^2+\cdots+b_{i,n-3}^2)^{\frac{1}{2}} \quad (i=1,2,\cdots,n-2). \tag{25}$$

然后按下列方式给出 E^n 中单形 $A_1A_2\cdots A_{n+1}$ 诸顶点之坐标

$$\begin{aligned}&A_i=(b_{i1},b_{i2},\cdots,b_{i,n-3},b,b,b)\quad(1\leqslant i\leqslant n-2),\\&A_{n-1}=(0,0,\cdots,0,a,-a,0),\\&A_n=(0,0,\cdots,0,0,a,-a),\\&A_{n+1}=(0,0,\cdots,0,-a,0,a).\end{aligned}\tag{26}$$

此处取

$$b=\frac{1}{\sqrt{3}}(38s+17),a=\sqrt{\frac{3}{2}}(16s+7).$$

容易验证下列事实：

1° 当 $1\leqslant i<j\leqslant n-2$ 时，$A_iA_j=B_iB_j=48s+21$；

2° 当 $n-1\leqslant i<j\leqslant n+1$ 时，$A_iA_j=\sqrt{6}a=48s+21$；

3° 当 $1\leqslant i\leqslant n-2,n-1\leqslant j\leqslant n+1$ 时

$$\begin{aligned}A_iA_j&=(b_{i1}^2+b_{i2}^2+\cdots+b_{i,n-3}^2+(a+b)^2+(a-b)^2+b^2)^{\frac{1}{2}}\\&=(OB_i^2+2a^2+3b^2)^{\frac{1}{2}}\\&=(9(8s+3)(16s+7)+3(16s+7)^2+(38s+17)^2)^{\frac{1}{2}}\\&=58s+25.\end{aligned}\tag{27}$$

故此单形 $A_1A_2\cdots A_{n+1}$ 诸棱长均为奇数. 取超平面 Π

$$x_{n-2}+x_{n-1}+x_n=3b,\tag{28}$$

显然 $A_1,A_2,\cdots,A_{n-2}$ 均在 Π 上，而 A_{n-1},A_n,A_{n+1} 到 Π 之距离为

$$\frac{1}{\sqrt{3}}3b=\sqrt{3}b=38s+17$$

是整数. 于是单形 $A_1A_2\cdots A_{n+1}$ 符合要求.

证毕.

参考文献

[1]Graham R L, et al. Amer. Math. Monthly, 1974(81): 21 - 25.

[2]Blumenthal L M. Theory and Applications of Distance Geometry. Oxford, 1953. or 2nd ed. New York, 1970.

[3]杨路，张景中. 关于有限点集的一类几何不等式. 数学学报，1980，23(5)：740 - 749.

Spanning-Radius of a Compact Set of Hyperbolic Space

Yang Lu(杨 路) and Zhang Jing-zhong(张景中)

(*University of Science and Technology of China*)

ABSTRACT

Let $\mathfrak{S}$ be a compact set of hyperbolic n-dimensional space $\boldsymbol{H}^n$ with constant curvature $K<0$. The n-dimensional ball is denoted by B. We put

$$\rho(\mathfrak{S})=\inf_{B\supset\mathfrak{S}}\{\text{radius of } B\},$$

and call it"the spanning-radius of $\mathfrak{S}$". On the other hand, let m be a prescribed family of non-negative Borel measures of total mass supported by $\mathfrak{S}$; and let $g(x,y)$ be the distance of x from $y(x,y\in\boldsymbol{H}^n)$. We consider the family of integrals having the form

$$I(\mathfrak{S},\mu)=\int_{\mathfrak{S}}\int_{\mathfrak{S}}\mathrm{ch}(\sqrt{-K}g(x,y))d\mu(x)d\mu(y),\mu\in m.$$

The following equality is proved:

$$(\mathrm{ch}\sqrt{-K}\rho(\mathfrak{S}))^2=\max_{\mu\in m}I(\mathfrak{S},\mu).$$

And some applications of this equality are given, e. g. a point set of diameter d of $\boldsymbol{H}^n$ with curvature $K<0$ can be covered by a ball of diameter

$$\frac{2}{\sqrt{-K}}\mathrm{sh}^{-1}\left(\sqrt{\frac{2n}{n+1}}\mathrm{sh}\sqrt{-\mathrm{K}}\frac{d}{2}\right).$$

Ⅰ. INTRODUCTION

Let $\boldsymbol{M}$ be a metric space and $\mathfrak{S}$ be a bounded subset of $\boldsymbol{M}$. If $\mathfrak{S}\subset\mathrm{B}\subset\boldsymbol{M}$, where B is a closed metric ball in $\boldsymbol{M}$, we say"B covers $\mathfrak{S}$". The infimum of the radii of all the closed balls covering $\mathfrak{S}$

$$\rho=\inf_{\mathrm{B}\supset\mathfrak{S}}\{\text{radius of } B\}$$

本文刊于《中国科学(A辑)》,第26卷第2期,1983年.

is called the spanning-radius of $\mathfrak{S}$. And $\delta=2\rho$ is called the spanning-diameter.

On the spanning-diameters of the subsets of Euclidean space $\boldsymbol{E}^n$, there were many results previously. such as the Jung Theorem[12]:

"In $\boldsymbol{E}^n$, any point set of diameter not exceeding 1 can be covered by a closed ball of diameter not exceeding

$$\sqrt{\frac{2n}{n+1}}."$$

It was posed in [1] whether the Jung Theorem could be generalized to other metric spaces. In reality, almost half century after the publishing of Jung Theorem, L. A. Santaló[2] already generalized this result to the spherical space. In the present paper, we obtain the corresponding result on hyperbolic space as a simple consequence of the main theorem:

"In a hyperbolic n-dimensional space with curvature $K(<0)$, a point set of diameter d can be covered by a ball of diameter not exceeding

$$\frac{2}{\sqrt{-K}}\mathrm{sh}^{-1}\left(\sqrt{\frac{2n}{n+1}}\mathrm{sh}\sqrt{-K}\,\frac{d}{2}\right)."$$

In this paper, we apply the method of energy integral introduced by G. Björck[3], which has been developed in a series of the recent works[4-6] by R. Alexander, and is applied to some extremal problems of distance geometry.

In Theorem 6 of [3], Björck showed a connection between the spanning-radius of an Euclidean point set and the maximum of the energy integral as a functional. To do analogously, this paper has established two foundational theorems in See. Ⅱ, and applies them to some covering problems on hyperbolic space.

Ⅱ. Foundational Theorems

Theorem 1. *Let $\boldsymbol{H}^n$ be the hyperbolic space with constant curvature $K<0$, and $\mathfrak{S}=\{p_1,p_2,\cdots,p_l\}$ be a set which consists of l points in $\boldsymbol{H}^n$. By g_{ij} we denote the distance between p_i and p_j in $\boldsymbol{H}^n$. Then, the quadratic form*

$$I(x)=\sum_{1\leqslant i,j\leqslant l}(\mathrm{ch}\sqrt{-K}g_{ij})x_ix_j \tag{1.1}$$

under the conditions

$$x_i\geqslant 0(i=1,2,\cdots,l),\ \sum_{i=1}^{l}x_i=1 \tag{1.2}$$

takes $(\mathrm{ch}\sqrt{-K}\,\rho)^2$ *as the absolute maximum, where ρ is the spanning-radius of $\mathfrak{S}$.*

Theorem 2. *Let $\boldsymbol{H}^n$ be the hyperbolic space with curvature $K(<0)$, and $\mathfrak{S}$ be a compact subset of $\boldsymbol{H}^n$. By $g(x,y)$ we denote the distance between x and y in $\boldsymbol{H}^n$, and μ's denote the non-negative Borel measures on $\mathfrak{S}$. Then, the energy integral*

$$I(\mu) = \iint_{\mathfrak{S}\mathfrak{S}} (\operatorname{ch} \sqrt{-K} g(x,y)) d\mu(x) d\mu(y) \tag{1.3}$$

under the restriction

$$\int_{\mathfrak{S}} d\mu(x) = 1 \tag{1.4}$$

takes $I(\mu^*) = (\operatorname{ch} \sqrt{-K}\rho)^2$ *as the absolute maximum, where* ρ *is the spanning-radius of* $\mathfrak{S}$.

Clearly, Theorem 1 is a special case of Theorem 2, and an important one.

Ⅲ. SEVERAL LEMMAS

$\boldsymbol{H}^n$ denotes the hyperbolic space with curvature $K(<0)$, and $\mathfrak{S}$ a compact subset of, $\boldsymbol{H}^n$. If $\mathfrak{S} = \{p_1, p_2, \cdots, p_l\}$, then let g_{ij} be the distance between p_i and p_j. Generally, for $x, y \in \boldsymbol{H}^n$, let $g(x,y)$ be the distance between x and y and put

$$c_{ij} = \operatorname{ch} \sqrt{-K}\, g_{ij},$$

$$c(x,y) = \operatorname{ch} \sqrt{-K}\, g(x,y).$$

Then we have the following lemmas:

Lemma 1. *Let* $\mathfrak{S}$ *be a point set which consists of the vertices of a simplex (non-degenerate) in* $\boldsymbol{H}^n$. *Then the quadratic form*

$$I(x) = \sum_{1 \leqslant i,j \leqslant n+1} c_{ij} x_i x_j \tag{2.1}$$

under the restriction

$$S: x_i \geqslant 0 \quad (i = 1, 2, \cdots, n+1), \sum_{i=1}^{n+1} x_i = 1,$$

takes $(\operatorname{ch} \sqrt{-K} R_m)^2$ *as the absolute maximum, where* R_m *is the circumradius of the* m-*dimensional simplex spanned by some subset* $\mathfrak{S}^*$ *of* $\mathfrak{S}$. *And simplex* $\mathfrak{S}^*$ *contains its circumcenter.*

Proof. It is obvious that there exists the absolute maximum. Let x^* be the point at which (2.1) takes its absolute maximum. Without loss of generality, suppose the first $m+1$ components of x^* are positive and the rest equal to zero, i. e.

$$x^* = (x_1^*, x_2^*, \cdots, x_{m+1}^*, x_{m+2}^*, \cdots, x_{n+1}^*),$$

$$x_1^* > 0, x_2^* > 0, \cdots, x_{m+1}^* > 0; x_{m+2}^* = \cdots = x_{n+1}^* = 0.$$

We consider the following quadratic form

$$I^*(\tilde{x}) = \sum_{1 \leqslant i,j \leqslant m+1} c_{ij} x_i x_j, (\tilde{x} \in \boldsymbol{R}^{m+1}). \tag{2.2}$$

Under the restriction

$$\tilde{S}: x_i \geqslant 0 (i = 1, 2, \cdots, m+1), \sum_{i=1}^{m+1} x_i = 1,$$

it has to take the absolute maximum at $\tilde{x}^*=(x_1^*,x_2^*,\cdots,x_{m+1}^*)$. But $\tilde{x}^*$ is, obviously, an interior point of the region of $\boldsymbol{R}^{m+1}$

$$\{\tilde{x}\,|\,x_1\geqslant 0,x_2\geqslant 0,\cdots,x_{m+1}\geqslant 0\},$$

then $I^*(\tilde{x})$ takes the absolute maximum at x^*.

$$\max_{\tilde{S}} I^*(\tilde{x})=\max_{S} I(x)=I^*(\tilde{x}^*)=I(x^*).$$

Using the Lagrange multiplier method, we construct an associated function

$$F(x)=I^*(\tilde{x})+2\lambda(x_1+x_2+\cdots+x_{m+1});$$

we derive from x_j and obtain the system satisfied by the critical points

$$\sum_{i=1}^{m+1} c_{ij}x_i+\lambda=0,(j=1,2,\cdots,m+1). \tag{2.3}$$

Let

$$D=\begin{vmatrix} c_{11}\,c_{12}\cdots c_{1m+1} \\ c_{21}\,c_{22}\cdots c_{2m+1} \\ \vdots\quad\vdots\qquad\vdots \\ c_{m+11}\cdots c_{m+1m+1}\end{vmatrix},\overline{D}=\begin{vmatrix} 0 & 1\;1\cdots 1 \\ 1 & \\ 1 & D \\ \vdots & \\ 1 & \end{vmatrix}, \tag{2.3}$$

and $\overline{D}_{ij}$ denotes the cofactor of c_{ij} in $\overline{D}$, $(i,j=0,1,\cdots,m+1)$. Solving(2.3), we have

$$x_i^*=\lambda\overline{D}_{0i}/D,(i=1,2,\cdots,m+1), \tag{2.4}$$

where D can be a denominator because of the following classical result:

"*Imbedding Theorem of Hyperbolic space.*[7] Given $\binom{n+1}{2}$ positive numbers c_{ij} $(1\leqslant i<j\leqslant n+1)$, there exist $n+1$ points $p_1,p_2,\cdots,p_{n+1}$ in $\boldsymbol{H}^m$ $(m<n)$ such that $c_{ij}=\operatorname{ch}\sqrt{-K}g_{ij}$ if and only if

$$\begin{cases}(-1)^{k+1}\det(c_{ij})_{i,j=1}^k\geqslant 0,(k=1,2,\cdots,m+1),\\(-1)^{k+1}\det(c_{ij})_{i,j=1}^k=0,(k=m+2,\cdots,n+1).\end{cases}$$

And if $p_1,p_2,\cdots,p_k$ are the vertices of a non-degenerate simplex,

$$(-1)^{k+1}\det\,(c_{ij})_{i,j=1}^k>0."$$

Since $p_1,p_2,\cdots,p_{m+1}$ are the vertices of non-degenerate simplex, then $D\neq 0$.

From the restriction $\tilde{S}$, we obtain

$$1=x_1^*+x_2^*+\cdots+x_{m+1}^*=\frac{\lambda\overline{D}}{D};\lambda=\frac{D}{\overline{D}}.$$

Multiplying Eqs. (2.3) by x's respectively, and calculating the total of them, we have

$$\max_{\tilde{S}} I^*(\tilde{x})=I^*(\tilde{x}^*)=\sum_{i,j=1}^{m+1} C_{ij}x_ix_j=-\frac{D}{\overline{D}}. \tag{2.5}$$

Also, $c_{ij}>1$ while $i\neq j$, we know $I^*(\tilde{x})^*>1$, hence

$$I^*(x^*)=-\frac{D}{\overline{D}}>1. \tag{2.6}$$

One may let

$$(\operatorname{ch}\sqrt{-K}R_m)^2=I^*(\tilde{x}^*)=-\frac{D}{\overline{D}}, \tag{2.7}$$

i. e. $$R_m=\frac{1}{\sqrt{-K}}\mathrm{ch}^{-1}\sqrt{-\frac{D}{\overline{D}}}, \text{and put } c_{00}=1,$$

$$c_{0i}=c_{i0}=\mathrm{ch}\sqrt{-K}\,R_m=\sqrt{-\frac{D}{\overline{D}}},(i=1,2,\cdots,m+1).$$

Then it follows

$$\begin{vmatrix} c_{00}\,c_{01}\cdots c_{0\,m+1} & \\ c_{10} & \\ \vdots & D \\ c_{m+1\,0} & \end{vmatrix}=c_{00}D+c_{01}^2\overline{D}=0. \tag{2.9}$$

Because $p_1,p_2,\cdots p_{m+1}$ belong to $\boldsymbol{H}^m$, we have

$$(-1)^{k+1}\det(c_{ij})_{i,j=1}^{k}\geqslant 0(k=1,2,\cdots,m+1),$$

by which together with (2.9) the conditions of Imbedding Theorem are satisfied. Hence there exists a point $p_0\in\boldsymbol{H}^m$ such that $g_{0i}=R_m(i=1,2,\cdots,m+1)$, i. e. p_0 is the circumcenter of $\mathfrak{S}^*$, and R_m is its circumradius.

In the following, we are going to prove that simplex $\mathfrak{S}^*$ contains its circumcenter as an interior point. According to a theorem of L. M. Blumenthal et al[8],

"Let $p_0,p_1,\cdots,p_{m+1}$ be $m+2$ points in $\boldsymbol{H}^m$, then p_0 and p_i are on the same or the opposite sides of the hyperplane which is determined by the rest of these points if and only if the cofactor of the element c_{0i} in determinant (2.9) has the sign of $(-1)^{m+1}$ or $(-1)^m$, respectively."

Moreover, noticing that by (2.8) the cofactor of c_{0i} in determinant(2.9) has the same sign as $\overline{D}_{0i}$ by (2.4) the equality $D=\lambda\overline{D}$ holds, we obtain

$$\overline{D}_{0i}=x_i^*\overline{D}.$$

Since $x_i^*>0$ and $\overline{D}$ has the opposite sign to D, then D has the sign of $(-1)^m$ from "Imbedding Theorem", so $\overline{D}_{0i}$ has the sign of $(-1)^{m+1}$. Thus p_0 and p_i are on the same side if the hyperplane determined by the rest of the points of $\mathfrak{S}^*$, i. e. the point p_0 is in the interior of the simplex spanned by $\mathfrak{S}^*$, Lemma 1 is proved.

Lemma 2. *Let $\mathfrak{S}'=\{p_1,p_2,\cdots,p_{l+1}\}$ be a finite subset of $\mathfrak{S}$, and suppose the l-dimensional simplex spanned by $\mathfrak{S}'$ contains its circumcenter as an interior point. Then there exists a point*

$$x'=(x_1',x_2',\cdots,x_{l+1}')\in\boldsymbol{R}^{l+1}$$

satisfying

$$x_i'>0\quad(i=1,2,\cdots,l+1),\sum_{i=1}^{l+1}x_i=1$$

such that

$$\sum_{i,j=1}^{l+1}c_{ij}x_i'x_j'=(\mathrm{ch}\sqrt{-K}R_l)^2,$$

where R_l is the circumradius of $\mathfrak{S}'$.

Proof. Put

$$D'=\begin{vmatrix} c_{11}c_{12}\cdots c_{1l+1} \\ c_{21}c_{22}\cdots c_{2l+1} \\ \vdots \quad \vdots \\ c_{l+11}\cdots c_{l+1l+1} \end{vmatrix}, \overline{D}'=\begin{vmatrix} 0 & 1 & 1 & \cdots & 1 \\ 1 & & & & \\ 1 & & D' & & \\ \vdots & & & & \\ 1 & & & & \end{vmatrix}.$$

By Imbedding Theorem, since $\mathfrak{S}'$ is a simplex, then $(-1)^l D'>0$. Because $\mathfrak{S}'$ contains its circumcenter as an interior point, quoting Blumenthal's theorem which has been cited in Lemma 1, the cofactor $\overline{D}'_{0i}$ in $\overline{D}'$ satisfies

$$(-1)^{l+1}\overline{D}'_{0i}>0. \tag{2.10}$$

Let $c'_{00}=1, c'_{i0}=c'_{0i}=\text{ch}\sqrt{-K}R_l$. Since the circumcenter p'_0 of $\mathfrak{S}'$ together with $p_1, p_2, \cdots, p_{l+1}$ belongs to the same space $\boldsymbol{H}^l$, it follows from Imbedding Theorem that

$$\begin{vmatrix} c'_{00}\cdots c'_{0l+1} \\ \vdots \quad D' \\ c_{l+10} \end{vmatrix}=D'+(\text{ch}\sqrt{-K}R_l)^2\overline{D}'=0. \tag{2.11}$$

From $(-1)^l D'>0$, we have $(-1)^{l+1}\overline{D}'>0$.

Put

$$x'_i=\frac{\overline{D}'_{0i}}{\overline{D}}, \tag{2.12}$$

then $x'_i>0$ and $x'_1+x'_2+\cdots+x'_{i+1}=1$. On the other hand, $(x'_1, x'_2, \cdots, x'_{l+1})$ is clearly a solution of the system

$$\sum_{i=1}^{l+1} c_{ij}x'_i=-\frac{D'}{\overline{D}'}, (j=1,2,\cdots,l+1). \tag{2.13}.$$

Multiplying Eqs. (2.13) by $x'_j(j=1,2,\cdots,l+1)$ respectively, calculating the total of them, and using (2.11), we obtain

$$\sum_{i=1}^{l+1}\sum_{j=1}^{l+1} c_{ij}x'_i x'_j=-\frac{D'}{\overline{D}'}=(\text{ch}\sqrt{-K}R_l)^2.$$

Lemma 2 is proved.

By Lemmas 1 and 2, it follows immediately

Lemma 3. *Let $\mathfrak{S}=\{p_1, p_2, \cdots, p_{n+1}\}$ be the point set which spans an n-simplex in $\boldsymbol{H}^n$, and R be the circumradius of a subsimplex of $\mathfrak{S}$. If this subsimplex contains its circumcenter, then*

$$(\text{ch}\sqrt{-K}R)^2\leqslant\max_s I(x)=I(x^*). \tag{2.14}$$

In other words, the R_m defined in Lemma 1 is the largest of the circumradii of the subsimplices of $\mathfrak{S}'$, which contain their circumcenter.

Lemma 4. *Let R and ρ be the circumradius and spanning-radius of the n-simplex spanned by $\mathfrak{S}=\{p_1, p_2, \cdots, p_{n+1}\}$, respectively. Then $R=\rho$ if this simplex contains its circumcenter; otherwise $R>\rho$.*

Proof. First, let the opposite sides of the vertices of a hyperbolie triangle $\triangle ABC$ be a,

b, c, respectively, then we have the cosine theorem[9]

$$\operatorname{ch}\sqrt{-K}\,c=\operatorname{ch}\sqrt{-K}\,a\ \operatorname{ch}\sqrt{-K}\,b-\operatorname{sh}\sqrt{-K}\,a\ \operatorname{sh}\sqrt{-K}\,b\ \cos C,$$

and so the opposite side of a right angle (or an obtuse angle) is longer than each of the other sides.

Let p_0 be the circumcenter of the simplex $\mathfrak{S}$. If p_0 is in the exterior of $\mathfrak{S}$, one may take a hyperplane π which separates p_0 from $\mathfrak{S}$. By p_0' we denote vertical projection of p_0 onto π. Then $\angle p_0 p_0' p_i$ is an obtuse angle in $\triangle p_0 p_0' p_i (1 \leqslant i \leqslant n+1)$, hence

$$g(p_0, p_i) > g(p_0', p_0), (i=1,2,\cdots,n+1).$$

Thus it can be seen that $R > \rho$.

If p_0 is contained in the simplex spanned by $\mathfrak{S}$, we proceed to prove that the radius of any sphere $B \supset \mathfrak{S}$ is not less than R. In fact, let p_0' be the center of any sphere $B \supset \mathfrak{S}$. If $p_0' = p_0$, the conclusion holds obviously. If $p_0' \neq p_0$, through p_0 we draw a hyperplane π perpendicular to the segment $p_0 p_0'$. Because p_0 is in the interior of simplex $\mathfrak{S}$, both sides of π contain some vertices of $\mathfrak{S}$. Let p_k, p_0' locate on the opposite sides of π, then $\angle p_0' p_0 p_k$ is an obtuse angle in $\triangle p_0' p_0 p_k$. By reason of that $g(p_0', p_k) > g(p_0, p_k)$, i. e. the radius of this sphere covering $\mathfrak{S}$ is larger than R. Thus it can be seen that $\rho \geqslant R$. On the other hand, it is impossible by the definition of spanning-radius that $\rho > R$, hence $\rho = R$. Lemma 4 is proved.

Lemma 5. *The spanning-radius ρ of the simplex spanned by $\mathfrak{S} = \{p_1, p_2, \cdots, p_{n+1}\}$ must be equal to the circumradius of a simplex which contains its circumcenter.*

Proof. Let p_0 be the center of the minimum sphere covering $\mathfrak{S}$, and $\mathfrak{S}^* = \{p_1, p_2, \cdots p_m\} (1 \leqslant m \leqslant n+1)$ be the subset of $\mathfrak{S}$ such that $g(p_0, p_i) = \rho$. We claim that there is no hyperplane π such that p_0 and $\mathfrak{S}^*$ locate on the opposite sides of π. In fact, suppose there exist such a π, and let p_0' be the vertical projection of p_0 onto π. We take a point q_0 in segment $p_0 p_0'$ such that the distance of q_0 from p_0 is small enough and the distance of q_0 from any point of $\mathfrak{S} \backslash \mathfrak{S}^*$ is less than ρ. Draw a hyperplane π' through q_0 so that $\pi' \perp p_0 q_0$ then p_0 and $\mathfrak{S}$ are on the opposite sides of π'. Using the method analogous to the proof of Lemma 4, we see that the distance of q_0 from any point of $\mathfrak{S}^*$ is less than ρ, so the spanning-radius of $\mathfrak{S}$ is less than ρ. This contradiction shows that there is no hyperplane separating p_0 from $\mathfrak{S}^*$. Hence p_0 belongs to the subspace determined by $\mathfrak{S}^*$ and is in the interior of the subsimplex spanned by $\mathfrak{S}^*$. And p_0 is the circumcenter of this subsimplex because the distances of p_0 from all points of $\mathfrak{S}^*$ equal one another. Lemma 5 is proved.

Ⅳ. Proof of Foundational Theorems

Proof of Theorem 1. First, we consider the case that the points of $\mathfrak{S}$ are the vertices of a simplex. Since the spanning-radius of a point set is irrelevant to the dimensional of the space spanned by this point set, there is no harm to assume the number of the points of $\mathfrak{S}$ to be $l=n+1$. By Lemma 5, the spanning-radius of $\mathfrak{S}$ is equal to the circumradius of a subsimplex of $\mathfrak{S}$'s which contains its circumcenter; and by Lemma 4, the circumradius of any subsimplex which contains its circumcenter is equal to its spanning-radius, thus being not larger than ρ. Hence ρ is the largest of the circumradii of all the subsimplices, each of which contains its circumcenter. By Lemma 3, ρ is equal to R_m determined in Lemma 1. This shows Theorem 1 to be held in case the points of $\mathfrak{S}$ are the vertices of a simplex.

In the following the general case is considered. Put $\mathfrak{S}=\{p_1, p_2, \cdots, p_l\}$ in $\boldsymbol{H}^{l-1}$; both the spanning-radius $\rho(\mathfrak{S})$ of $\mathfrak{S}$ and the absolute maximum $\boldsymbol{M}(\mathfrak{S})$ of (1.1) under condition (1.2) can be regarded as the functions defined on $(\boldsymbol{H}^{l-1})^l$. Moreover, both the two functions are continuous. We have that

$$\boldsymbol{M}(\mathfrak{S})=(\operatorname{ch}\sqrt{-K}\rho(\mathfrak{S}))^2, \tag{3.1}$$

if $\mathfrak{S}$ consists of the vertices of an $l-1$ dimensional simplex, i. e. (3.1) holds on a dense subset of $(\boldsymbol{H}^{l-1})^l$. And from the continuity, (3.1) holds on $(\boldsymbol{H}^{l-1})^l$ everywhere. Specially, it holds on the subspace $(\boldsymbol{H}^n)^l$ of $(\boldsymbol{H}^{l-1})^l$. Theorem 1 is proved.

Proof of Theorem 2. For a positive number δ being small enough, we partition $\mathfrak{S}$ into a finite number of the Borel measurable subsets $\mathscr{B}_1, \mathscr{B}_2, \cdots, \mathscr{B}_l$ whose diameters are less than δ. Taking any point p_i from $\mathscr{B}_i$, we put

$$\mathfrak{S}_\delta=\{p_1, p_2, \cdots, p_l\},$$

and consider the point

$$x_\mu=(x_{\mu 1}, x_{\mu 2}, \cdots, x_{\mu l})$$

of $\boldsymbol{R}^l$, where μ is a non-negative Borel measure satisfying (1.4) and

$$x_{\mu i}=\mu(\mathscr{B}_i).$$

When δ is small enough, $|\rho(\mathfrak{S})-\rho(\mathfrak{S}_\delta)|$ can be less than any prescribed positive number. On the other hand,

$$\left|\sum_{1\leqslant i,j\leqslant 1}(\operatorname{ch}\sqrt{-K}g_{ij})x_{\mu i}x_{\mu j}-\iint_{\mathfrak{S}\mathfrak{S}}(\operatorname{ch}\sqrt{-K}g(x,y)d\mu(x)d\mu(y)\right|$$

can be less than any prescribed positive number for the μ satisfying (1.4). Hence, the difference between the absolute maximums $I(\mu^*)$ of (1.3) under (1.4) and $I_\delta(x^*)$ of (1.1) corresponding to $\mathfrak{S}_\delta$ under (1.2) can be less than any prescribed positive number when δ is small enough. However, it is shown by Theorem 1 that

$$I_\delta(x^*)=(\operatorname{ch}\sqrt{-K}\,\rho(\mathfrak{S}_\delta))^2,$$

therefore the difference of $I(\mu^*)$ and $(\operatorname{ch}\sqrt{-K}\rho(\mathfrak{S}))^2$ can be arbitrarily small. Theorem 2 is proved.

V. Some Consequences

Let $\mathfrak{S}$ still be a compact subset of the hyperbolic, n-dimensional space $\boldsymbol{H}^n$ with curvature $K(<0)$, and suppose its image $\varphi(\mathfrak{S})$ under the map $\varphi:\mathfrak{S}\to \boldsymbol{H}^n$ is compact as well. $\delta=2\rho$ and $\delta_\varphi=2\rho_\varphi$ denote the spanning-diameters of $\mathfrak{S}$ and $\varphi(\mathfrak{S})$ respectively. The following theorem follows easily from Theorems 1 and 2.

Theorem 3.

$$\operatorname{ch}\sqrt{-K}\,\delta-\operatorname{ch}\sqrt{-K}\,\delta_\varphi\leqslant\frac{2n}{n+1}\sup_{x,y\in\mathfrak{S}}\left[\operatorname{ch}\sqrt{-K}\,g(x,y)-\operatorname{ch}\sqrt{-K}\,g(\varphi(x),\varphi(y))\right].$$

Proof. First, one can choose a finite subset $\mathfrak{S}'=\{p_1,p_2,\cdots,p_m\}(m\leqslant n+1)$ of $\mathfrak{S}$ such that $\mathfrak{S}'$ and $\mathfrak{S}$ have the same spanning-radius. In fact, choose a sequence of finite subsets $\{\mathfrak{S}_t\}(t=1,2,\cdots)$ such that the spanning-radius $\rho(t)$ of $\mathfrak{S}_t$ satisfies $|\rho(t)-\rho|<1/t$. According to Lemma 5, $\rho(t)$ is equal to the circumradius of the simplex which contains its circumcenter and is spanned by some subset of $\mathfrak{S}_t$. Denote this subset by $\mathfrak{S}_t^*$, then the number of elements of $\mathfrak{S}_t^*$ is not larger than $n+1$ and the spanning-radius of $\mathfrak{S}_t^*$ is $\rho(t)$. One can find, from the compactness of $\mathfrak{S}$, a subsequence $\{\mathfrak{S}_{t_j}^*\}$ such that the points of $\mathfrak{S}_{t_j}^*$ converges to the points of some finite subset $\mathfrak{S}'$ of $\mathfrak{S}$. The spanning-radius of $\mathfrak{S}'$ is ρ because $\rho(t_j)\to\rho$.

Again put $g_{ij}=g(p_i,p_j)$, and $f_{ij}=g[\varphi(p_i),\varphi(p_j)]$, where $i,j=1,2,\cdots,m(m\leqslant n+1)$ and $p_i,p_j\in\mathfrak{S}'$. Then

$$(\operatorname{ch}\sqrt{-K}\,\rho)^2=\max_{\substack{\Sigma x_i=1\\ x_i\geqslant 0}}\sum_{i,j=1}^{m}(\operatorname{ch}\sqrt{-K}\,g_{ij})x_ix_j,$$

$$(\operatorname{ch}\sqrt{-K}\,\rho_\varphi)^2=\max_{\substack{\Sigma x_i=1\\ x_i\geqslant 0}}\sum_{i,j=1}^{m}(\operatorname{ch}\sqrt{-K}\,f_{ij})x_ix_j,$$

hence

$$\begin{aligned}
&\operatorname{ch}\sqrt{-K}\,\delta-\operatorname{ch}\sqrt{-K}\,\delta_\varphi=2((\operatorname{ch}\sqrt{-K}\,\rho)^2-(\operatorname{ch}\sqrt{-K}\,\rho_\varphi)^2)\\
&\leqslant 2\left(\max_{\substack{\Sigma x_i=1\\ x_i\geqslant 0}}\sum_{i,j=1}^{m}(\operatorname{ch}\sqrt{-K}\,g_{ij})x_ix_j-\max_{\substack{\Sigma x_i=1\\ x_i\geqslant 0}}\sum_{i,j=1}^{m}(\operatorname{ch}\sqrt{-K}\,f_{ij})x_ix_j\right)\\
&\leqslant 2\max_{\substack{\Sigma x_i=1\\ x_i\geqslant 0}}\left(\sum_{i,j=1}^{m}(\operatorname{ch}\sqrt{-K}\,g_{ij}-\operatorname{ch}\sqrt{-K}\,f_{ij})x_ix_j\right)\\
&\leqslant 4\sum_{1\leqslant i<j\leqslant m}x_i^*x_j^*\cdot\max_{1\leqslant i<j\leqslant m}(\operatorname{ch}\sqrt{-K}\,g_{ij}-\operatorname{ch}\sqrt{-K}\,f_{ij}),
\end{aligned}$$

where $(x_1^*,x_2^*,\cdots,x_m^*)$ is a system of real numbers satisfying $\sum_{i=1}^{m}x_i^*=1$ and $x_i^*\geqslant 0$. By the Maclaurin Inequality[10], we have

$$\sum_{1\leqslant i<j\leqslant m}x_i^*x_j^*\leqslant\frac{m-1}{2m}\left(\sum_{i=1}^{m}x_i^*\right)^2\leqslant\frac{n}{2(n+1)}.$$

On the other hand, it is obvious that

$$\max_{1\leqslant i<j\leqslant m}(\operatorname{ch}\sqrt{-K}\,g_{ij}-\operatorname{ch}\sqrt{-K}\,f_{ij})$$

$$\leqslant \sup_{x,y\in \mathfrak{S}}[\operatorname{ch}\sqrt{-K}\, g(x,y)-\operatorname{ch}\sqrt{-K}\, g[\varphi(x),\varphi(y)]].$$

Hence we obtain

$$\operatorname{ch}\sqrt{-K}\,\delta-\operatorname{ch}\sqrt{-K}\,\delta_{\varphi}$$
$$\leqslant \frac{2n}{n+1}\sup_{x,y\in \mathfrak{S}}[\operatorname{ch}\sqrt{-K}\, g(x,y)-\operatorname{ch}\sqrt{-K}\, g[\varphi(x),\varphi(y)]].$$

Theorem 3 is proved.

It can be seen, immediately, from Theorem 3 that the spanning-radius of the image $\varphi(\mathfrak{S})$ is not less than the spanning-radius of $\mathfrak{S}$, if φ is an expanding map. Actually, it can be seen directly from Theorems 1 and 2 that the spanning-radius is non-decreasing under an expanding map and non-increasing under a contracting map. These facts is well known in the Euclidean space[11].

Futhermore, if $\varphi(\mathfrak{S})$ contains only one point, by Theorem 3 we obtain

$$\operatorname{ch}\sqrt{-K}\,\delta-1\leqslant\frac{2n}{n+1}\sup_{x,y\in \mathfrak{S}}[\operatorname{ch}\sqrt{-K}\, g(x,y)-1],$$

i. e.

$$\operatorname{sh}\sqrt{-K}\,\rho\leqslant\sqrt{\frac{2n}{n+1}}\sup_{x,y\in \mathfrak{S}}\operatorname{sh}\sqrt{-K}\,\frac{g(x,y)}{2},$$

hence it follows

Theorem 4. *Let d be the diameter of $\mathfrak{S}$, then the spanning-diameter δ of $\mathfrak{S}$ satisfies the following inequality*

$$\operatorname{sh}\sqrt{-K}\frac{\delta}{2}\leqslant\sqrt{\frac{2n}{n+1}}\operatorname{sh}\sqrt{-K}\frac{d}{2}.$$

Theorem 4 is the result corresponding to the famous Jung Theorem[12], but on hyperbolic space. In Theorem 4, clearly, $\mathfrak{S}$ can be noncompact because one may consider the closure $\hat{\mathfrak{S}}$ of $\mathfrak{S}$ and they have the same diameter and spanning-diameter.

Theorem 4 may be restated as follows:

In a hyperbolic n-dimensional space with curvature $K(<0)$, a point set of diameter d can be covered by a ball of diameter

$$\frac{2}{\sqrt{-K}}\operatorname{sh}^{-1}\left(\sqrt{\frac{2n}{n+1}}\operatorname{sh}\sqrt{-K}\,\frac{d}{2}\right).$$

It is seen easily that the equal-sign in the inequality of Theorem 4 can be held, such as when $\mathfrak{S}$ consists of the vertices of a regular simplex, the equal-sign is held.

Theorem 5. *Suppose the three subsets*

$$\mathfrak{S}_1=\{p_1,p_2,\cdots,p_m\},\mathfrak{S}_2=\{q_1,q_2,\cdots,q_m\}$$

and

$$\mathfrak{S}=\{r_1,r_2,\cdots,r_m\}$$

of the hyperbolic space with curvature $K(<0)$ satisfy

$$g(r_i,r_j)\leqslant\frac{1}{2}[g(p_i,p_j)+g(q_i,q_j)],$$

then the spanning-diameters δ_1, δ_2 *and* δ_3 *of* $\mathfrak{S}_1, \mathfrak{S}_2$ *and* $\mathfrak{S}_3$ *are related as follows*:

$$\operatorname{ch}\sqrt{-K}\,\delta_3 \leqslant \frac{1}{2}(\operatorname{ch}\sqrt{-K}\,\delta_1 + \operatorname{ch}\sqrt{-K}\,\delta_2).$$

Proof. Since the function ch x is convex and increasing when $x>0$, one obtain

$$\sum_{i,j=1}^{m}[\operatorname{ch}\sqrt{-K}g(r_i, r_j)]x_i x_j$$
$$\leqslant \sum_{i,j=1}^{m}\left[\operatorname{ch}\sqrt{-K}\,\frac{1}{2}(g(p_i, p_j)+g(q_i, q_j))\right]x_i x_j$$
$$\leqslant \frac{1}{2}\sum_{i,j=1}^{m}[\operatorname{ch}\sqrt{-K}\,g(p_i, p_j)]x_i x_j + \frac{1}{2}\sum_{i,j=1}^{m}[\operatorname{ch}\sqrt{-K}\,g(q_i, q_j)]x_i x_j.$$

Take the absolute maximums on the two sides of this inequality, then it follows by Theorem 1 that

$$\operatorname{ch}^2\sqrt{-K}\,\frac{\delta_3}{2} \leqslant \frac{1}{2}\left(\operatorname{ch}^2\sqrt{-K}\,\frac{\delta_1}{2} + \operatorname{ch}^2\sqrt{-K}\,\frac{\delta_2}{2}\right).$$

And using the half angle formula of hyperbolic cosine, we have Theorem 5.

References

[1]Danzer L, et al. Helly's theorem and its relatives. Proc. of Symposia in Pure Math., Amer. Math. Soc., 7, "Convexity", 1968: 101 - 180.

[2]Santaló L A. Convex regions on the spherical surface, Ann. of Math., 1946(47): 448 - 459.

[3]Björek G. Distribution of positive mass which maximize a certain generalized energy integral. Ark. Math., 1956(3): 255 - 269.

[4]Alexander R, et al. Extremal problems of distance geometry related to energy integrals. Trans. Amer. Math. Soec., 1974(195): 1 - 31.

[5]Alexander R. Generalized sums of distance. Pac. J. Math., 1975(56): 297 - 304.

[6]Two notes on metric geometry. Proc. Amer. Math. Soe., 1977(64): 317 - 321.

[7]Blumeathal L M. Theory and Applications of Distance Geometry. 2nd ed. New York, 1970.

[8]Blumenthal L M, et al. Distribution of points in n-space. Amer. Math. Monthly, 1943(50): 181 - 185.

[9] Martin G E. The Foundations of Geometry and the Non-Euclidean Plane. Intext Educationsl Publishers, 1975: 432 - 433.

[10]Beckenbach E F, et al. Inequalities. Springer-Verlag, 1961: 10 - 11.

[11]Valentine FA. Convex Sets. McGraw-Hill, 1964: 172 - 173.

[12] Hadwiger H, et al. Combinatorial Geometry in the Plane. Holt, Rinehart and Winston, 1964: 46.

度量方程应用于 Sallee 猜想

杨　路　张景中

（中国科学技术大学）

§0　引　　言

在 n 维欧氏空间 E^n 中，一个有界凸体 K 的宽度是这样定义的：对于每个单位向量 $\boldsymbol{u}$，将 K 的一对与 $\boldsymbol{u}$ 垂直的支撑超平面之间的距离记作 $\tau(K,\boldsymbol{u})$. 令

$$\omega(K)=\min_{u}\tau(K,\boldsymbol{u}), \tag{0.1}$$

我们将 $\omega(K)$ 叫做 K 的宽度.

Sallee 在 1974 年提出猜想说[1]，“内接于球的所有单形中，正则单形具有最大宽度”；他当时并指出，甚至对三维情形也未能证明. 随后，这个猜想被 R. Alexander[2] 所证实.

本文获得了比 Sallee-Alexander 定理更强的结果：“一切维数相同体积相等的单形中，正则单形具有最大宽度”. 从这个命题可以直接导出文献[2]中的结果，反之则不能.

更详细地讲，令 $V(\triangle_n)$ 记 n 维单形 $\triangle_n$ 的体积，我们证明了

$$\omega(\triangle_n)\leqslant C_nV(\triangle_n)^{\frac{1}{n}}. \tag{0.2}$$

这里 C_n 是仅与维数 n 有关的绝对常数，其中等式当且仅当 $\triangle_n$ 为正则单形时成立.

我们的方法与文献[2]完全不同，所用主要工具是作者曾在文献[3]中给出的两个命题.

§1　基本图形的度量方程

把 E^n 中的每个点或者每个定向超平面都叫做基本元素，由 k 个基本元素构成的集叫做 E^n 中的一个 k 元基本图形.

设 $\mathfrak{S}$ 是一个基本图形，$\mathfrak{S}=\{e_1,e_2,\cdots,e_N\}$，这些 $e_i(1\leqslant i\leqslant N)$ 是基本元素——点或定向超平面. 在 $\mathfrak{S}$ 上定义一个二元实值函数 $g:\mathfrak{S}\times\mathfrak{S}\to R$，使得

$$g(e_i,e_j)=\begin{cases}-\dfrac{1}{2}\rho^2(e_i,e_j) & (\text{当 } e_i,e_j \text{ 都是点})；\\ \cos\widehat{e_ie_j} & (\text{当 } e_i,e_j \text{ 都是超平面})；\\ d(e_i,e_j) & (\text{当 } e_i,e_j \text{ 中一为点，一为面}).\end{cases}$$

这里，$\rho(x,y)$ 表示两点 x,y 间的距离，$\widehat{xy}$ 表示两定向超平面 x,y 之间的夹角，而 $d(x,y)$ 则

本文刊于《数学学报》，第 26 卷第 4 期，1983 年.

表示点 x(或 y)到定向超平面 y(或 x)的带号距离.

简单地记 $g_{ij}=g(e_i,e_j)$,我们有下述命题(文献[3]中的引理 2.1):

定理 1 设$\mathfrak{S}=\{e_1,e_2,\cdots,e_N\}$是 E^n 中的基本图形,令 $\delta_i=1-g_{ij}(1\leqslant i\leqslant N)$,并令$[\mathfrak{S}]$表下列$(N+1)$阶方阵:

$$[\mathfrak{S}]=\begin{bmatrix}0 & \delta_1 & \delta_2\cdots\delta_N \\ \delta_1 & & \\ \delta_2 & & g_{ij} \\ \vdots & & \\ \delta_N & & \end{bmatrix} \tag{1.1}$$

记 $P[\mathfrak{S}]=\det[\mathfrak{S}]$,则当 $N>n+1$ 时,有

$$P[\mathfrak{S}]=0. \tag{1.2}$$

定理 1 的证明见文献[3].

形如(1.2)的方程,叫做基本图形的度量方程,这类方程在度量几何中扮演着极其重要的角色,有关的更一般的讨论可参看文献[4].

§2 几个引理

现在我们回到单形的宽度上来.

沿用§0 的记号,用 $\tau(\triangle_n,\boldsymbol{u})$表示$\triangle_n$ 在方向 $\boldsymbol{u}$ 的宽度,我们有

引理 1 设 n 维单形$\triangle_n$ 的顶点之集为 S,$S=\{p_1,p_2,\cdots,p_{n+1}\}$,则对 S 的每个非空的、不同于 S 的子集 A,必存在一个定向超平面 H,使得 $S\backslash A\subset H$,而且,A 中的各点到 H 有相等的带号距离,若以 $\boldsymbol{v}$ 记 H 的法量,则这个带号距离的绝对值,就是 $\tau(\triangle_n,\boldsymbol{v})$.

证 不妨设 $A=\{p_2,p_2,\cdots,p_k\}(1\leqslant k\leqslant n)$,考虑向量组 $\mathscr{B}_A=\{p_2-p_1,p_3-p_1,\cdots,p_k-p_1;p_{k+2}-p_{k+1},p_{k+3}-p_{k+1},\cdots,p_{n+1}-p_{k+1}\}$,$\mathscr{B}_A$ 由 $n-1$ 个线性无关的向量组成,它生成 E^n 的一个 $n-1$ 维子空间 E_A,E_A 的正交补子空间是一维的,亦即,存在一个单位向量 $\boldsymbol{v}$ 与 $\mathscr{B}_A$ 中所有向量正交,而且这样的 $\boldsymbol{v}$ 不计正负号时是唯一的,作超平面

$\pi_1:\boldsymbol{v}\cdot(x-p_1)=0$,

$\pi_2:\boldsymbol{v}\cdot(x-p_{k+1})=0$,

则 $p_1,\cdots,p_k$ 在 π_1 上,$p_{k+1},\cdots,p_{n+1}$ 在 π_2 上,而且 $\pi_1/\!/\pi_2$,这就得到了所要的结论.

以下令 $I=\{1,2,\cdots,n+1\}$是开始的$(n+1)$个正整数的集合. I 的一切 m 元子集所组成的集合记为ϑ_m(用$|\sigma|$表示 σ 的元素的个数):

$$\vartheta_m=\{\sigma|\sigma\subset I,|\sigma|=m\}. \tag{2.1}$$

这样,单形$\triangle_n$ 的顶点集 $S=\{p_1,p_2,\cdots,p_{n+1}\}$的每个子集 S_σ 可以和 I 的一个子集 σ 对应:

$$S_\sigma=\{p_\alpha|\alpha\in\sigma,\sigma\subset I\}. \tag{2.2}$$

由引理 1 可知,对 S_σ,当 $1\leqslant|\sigma|\leqslant n$ 时,存在一个定向超平面 H_σ,$S\backslash S_\sigma\subset H_\sigma$,使得 S_σ 中的一切点到 H_σ 的带号距离相等,这个带号距离仅与 H_σ 有关,故可记作 $d(H_\sigma)$. 若以 $\boldsymbol{v}_\sigma$ 记 H_σ 的单位法向量,则当$\triangle_n$ 取定时,$\tau(\triangle_n,\boldsymbol{v}_\sigma)$仅与 σ 有关,故可记作

$$\tau_\sigma=\tau(\triangle_n,\boldsymbol{v}_\sigma)=|d(H_\sigma)|. \tag{2.3}$$

下面，我们给出 τ_σ 的计算公式.

引理 2　若$\triangle_n$ 是 n 维单形，$\triangle_n$ 之顶点集 $S=\{p_1,p_2,\cdots,p_{n+1}\}$，则对 S 的任一非空真子集 S_σ，有等式

$$\tau_\sigma^{-2}=-\frac{1}{P[S]}\begin{vmatrix} [S] & \begin{matrix}0\\ \beta_1\\ \vdots\\ \beta_{n+1}\end{matrix} \\ 0\quad \beta_1\cdots\beta_{n+1} & 0\end{vmatrix} \tag{2.4}$$

这里 $[S]$，$P[S]$ 见(1.1)及(1.2)，τ_σ 见(2.3)，而

$$\beta_i=\begin{cases}1 & (i\in\sigma),\\ 0 & (i\notin\sigma).\end{cases}$$

证　考虑 $n+2$ 元基本图形 $\mathfrak{S}_\sigma=\{p_1,\cdots,p_{n+1},H_\sigma\}$. 如果令 $\alpha_i=d(p_i,H_\sigma)$，则有

$$\alpha_i=\begin{cases}d(H_\sigma) & (i\in\sigma),\\ 0 & (i\notin\sigma).\end{cases}$$

由定理 1，对 $\mathfrak{S}_\sigma$ 有

$$P[\mathfrak{S}_\sigma]=\begin{vmatrix} [S] & \begin{matrix}0\\ \alpha_1\\ \vdots\\ \alpha_{n+1}\end{matrix} \\ 0\quad \alpha_1\cdots\alpha_{n+1} & 1\end{vmatrix}=0. \tag{2.5}$$

把行列式的末行末列除以 $d(H_\sigma)$，并注意到有 $\tau_\sigma=|d(H_\sigma)|$，由(2.5)立得(2.4).

引理 3　对 n 维单形$\triangle_n$，有

$$\omega(\triangle_n)=\min_{\substack{\sigma\subset I\\ \varnothing\neq\sigma\neq I}}\{\tau_\sigma\}. \tag{2.6}$$

这里 $\omega(\triangle_n)$ 表$\triangle_n$ 之宽度，$I=\{1,2,\cdots,n+1\}$，而 τ_σ 如(2.3)所述.

证　设 $\boldsymbol{u}$ 是 E^n 中任一个单位向量，π 是以 $\boldsymbol{u}$ 为法向量的$\triangle_n$ 的支撑平面. 不妨设$\triangle_n$ 的顶点都在 π 的正侧或在 π 上. 令 $d_i=d(p_i,\pi)$. 这里 $p_1,\cdots,p_{n+1}$ 仍记$\triangle_n$ 的顶点，而 $d(p_i,\pi)$ 表 p_i 到 π 的带号距离，则显然有

$$\tau(\triangle_n,\boldsymbol{u})=\max_{1\leqslant i\leqslant n+1}\{d_i\}. \tag{2.7}$$

对 $(n+2)$ 元基本图形 $\mathfrak{S}_\pi=\{p_1,p_2,\cdots,p_{n+1},\pi\}$ 应用定理 1，得

$$P[\mathfrak{S}_\pi]=\begin{vmatrix} [S] & \begin{matrix}0\\ d_1\\ \vdots\\ d_{n+1}\end{matrix} \\ 0\quad d_1\cdots d_{n+1} & 1\end{vmatrix}=0. \tag{2.8}$$

这里 $S=\{p_1,p_2,\cdots,p_{n+1}\}$，简记 $\tau(\triangle_n,\boldsymbol{u})=\tau_{\boldsymbol{u}}$，用 $\tau_{\boldsymbol{u}}$ 除(2.8)中行列式的末行末列，得

$$\begin{vmatrix} & & & 0 \\ & [S] & & t_1 \\ & & & \vdots \\ & & & t_{n+1} \\ 0 & t_1 & \cdots t_{n+1} & \tau_u^{-2} \end{vmatrix}=0. \tag{2.9}$$

由(2.7)以及诸 d_i 非负,可知 $0\leqslant t_i\leqslant 1$,而且诸 t_i 中至少有一个为 0,至少有一个为 1.

从(2.9)中解出(P_{ij} 表 P 的对应于 g_{ij} 的代数余子式):

$$\tau_u^{-2}=-\frac{1}{P[S]}\begin{vmatrix} & & & 0 \\ & [S] & & t_1 \\ & & & \vdots \\ & & & t_{n+1} \\ 0 & t_1 & \cdots t_{n+1} & 0 \end{vmatrix}$$

$$=-\frac{1}{P[S]}\sum_{i=1}^{n+1}\sum_{j=1}^{n+1}(-P_{ij}[S])t_i t_j. \tag{2.10}$$

我们指出,若某个 $t_j\neq 0$ 或 1,总可以把它换成 0 或 1 而使(2.10)的右端变得更大或不变. 事实上,考察(2.10)的和号中与 t_i 有关的那些项之和:

$$\left(-2\sum_{i\neq j}P_{ij}[S]t_i t_j\right)-P_{jj}[S]t_j^2=t_j(-P_{jj}[S]t_j+C),$$

当$(-P_{jj}[S]t_j+C)\leqslant 0$ 时,我们把 t_j 换成 0;反之,当$(-P_{jj}[S]t_j+C)>0$ 时,我们把 t_j 换成 1. 由文献[3]中的公式(2.4)可知,$P[S]$和 $P_{jj}[S]$都是负的,故这样的代换下(2.10)右端不减. 经过有限次代换后,诸 t_j 都变成 0 或 1,于是由引理 2 即得

$$\tau_u^{-2}\leqslant\tau_\sigma^{-2}.$$

这里 σ 是 $I=\{1,2,\cdots,n+1\}$的某个非空真子集. 引理 3 得证.

以下对 $m=1,2,\cdots,n$,记

$$D_m=\sum_{\sigma\in\vartheta_m}\tau_\sigma^{-2}. \tag{2.11}$$

这里 ϑ_m 由(2.1)给出. 我们来计算 D_m 之值.

引理 4 设 $V(F_i)$表 n 维单形$\triangle_n$ 各侧面 F_i 的$(n-1)$维体积,$V(\triangle_n)$表$\triangle_n$ 的体积,$\binom{n}{k}$按通常表示组合数,则有

$$D_m=\binom{n-1}{m-1}\frac{\sum_{i=1}^{n+1}V^2(F_i)}{n^2V^2(\triangle_n)}. \tag{2.12}$$

证 按引理 2 的(2.4),我们有

$$\tau_\sigma^{-2}=(P[S])^{-1}\sum_{i\in\sigma}\sum_{j\in\sigma}P_{ij}[S], \tag{2.13}$$

这里 $P_{ij}[S]$仍记 $P[S]$的对应于元素 g_{ij} 的代数余子式$(i,j=1,2,\cdots,n+1)$. 于是有

$$D_m = \sum_{\sigma\in\vartheta_m}\tau_\sigma^{-2} = (P[S])^{-1}\sum_{\sigma\in\vartheta_m}\sum_{i\in\sigma}\sum_{j\in\sigma}P_{ij}[S]$$
$$= P_{ij}[S]^{-1}\left(\binom{n}{m-1}\sum_{i=1}^{n+1}P_{ii}[S]+\binom{n-1}{m-2}\sum_{\substack{i=1\\(i\neq j)}}^{n+1}\sum_{j=1}^{n+1}P_{ij}[S]\right)$$
$$= (P[S])^{-1}\left(\binom{n-1}{m-1}\sum_{i=1}^{n+1}P_{ii}[S]+\binom{n-1}{m-2}\sum_{i=1}^{n+1}\sum_{j=1}^{n+1}P_{ij}[S]\right).$$

但是

$$\sum_{i=1}^{n+1}\sum_{j=1}^{n+1}P_{ij}[S] = -\begin{vmatrix} 0 & 1 & \cdots & 1 & 0 \\ 1 & & & & 1 \\ \vdots & & g_{ij} & & \vdots \\ 1 & & & & 1 \\ 0 & 1 & \cdots & 1 & 0 \end{vmatrix} = 0,$$

故

$$D_m = \binom{n-1}{m-1}(P[S])^{-1}\sum_{i=1}^{n+1}P_{ii}[S]. \tag{2.14}$$

另一方面,由单形体积公式(文献[3]中)§2,式(2.4))

$$P[S] = -(n!)^2V^2(\triangle_n),$$
$$P_{ii}[S] = -(n-1)!^2V^2(F_i),$$

代入(2.14),即得所欲证之(2.12).引理 4 证毕.

引理 5 按引理 4 中记号,有

$$\sum_{i=1}^{n+1}V^2(F_i) \geqslant n^3\left(\frac{n+1}{n!^2}\right)^{\frac{1}{n}}V(\triangle_n)^{2-\frac{2}{n}}. \tag{2.15}$$

其中等号当且仅当$\triangle_n$ 为正则单形时成立.

引理 5 不过是文献[3]中定理的一个特例,在文献[3]的(1.5)中取

$$m=l=k+1=N-1=n,$$

即得这里的不等式(2.15).

§3 联系着宽度与体积的不等式

本文的主要目的是导出下列的

定理 2 若 $\omega(\triangle_n)$和 $V(\triangle_n)$分别记 n 维单形的宽度和体积,则有

$$\omega(\triangle_n)\leqslant C_nV(\triangle_n)^{\frac{1}{n}}. \tag{0.2}$$

其中

$$C_n = \frac{(n!)^{\frac{1}{n}}(n+1)^{\frac{n-1}{2n}}}{\left[\frac{n+1}{2}\right]^{\frac{1}{2}}\left(n+1-\left[\frac{n+1}{2}\right]\right)^{\frac{1}{2}}}, \tag{3.1}$$

而且(0.2)中的等号当且仅当$\triangle_n$ 是正则单形时取到.

证 对一切$\sigma\in\vartheta_m$ 计算 τ_σ^{-2} 的算术平均$\underset{|\sigma|=m}{\mathrm{A.M.}}(\tau_\sigma^{-2})$,由引理 4 得

$$\underset{|\sigma|=m}{\mathrm{A.M.}}(\tau_\sigma^{-2})=\binom{n+1}{m}^{-1}D_m=-\frac{\binom{n-1}{m-1}}{n^2\binom{n+1}{m}}\cdot\frac{\sum_{i=1}^{n+1}V^2(F_i)}{V^2(\triangle_n)}$$

$$=\frac{m(n+1-m)}{n^3(n+1)}\cdot\frac{\sum_{i=1}^{n+1}V^2(F_i)}{V^2(\triangle_n)},\tag{3.2}$$

显然,当 $m=\left[\frac{n+1}{2}\right]$时,(3.2)的右端取到最大值,即

$$\max_{1\leqslant m\leqslant n}\{\underset{|\sigma|=m}{\mathrm{A.M.}}(\tau_\sigma^{-2})\}=\underset{|\sigma|=\left[\frac{n+1}{2}\right]}{\mathrm{A.M.}}(\tau_\sigma^{-2}).\tag{3.3}$$

另一方面,由引理 3 知

$$\omega(\triangle_n)=\min_{\substack{\sigma\subset I\\ \varnothing\neq\sigma\neq I}}\{\tau_\sigma\}=\min_{0<|\sigma|\leqslant n}\{\tau_\sigma\},$$

故

$$\omega^{-2}(\triangle_n)=\max_{0<|\sigma|\leqslant n}\{\tau_\sigma\}\geqslant\underset{|\sigma|=\left[\frac{n+1}{2}\right]}{\mathrm{A.M.}}(\tau_\sigma^{-2}).\tag{3.4}$$

由(3.4)与(3.2)得

$$\omega^{-2}(\triangle_n)\geqslant\frac{\left[\frac{n+1}{2}\right]\left(n+1-\left[\frac{n+1}{2}\right]\right)}{n^3(n+1)}\cdot\frac{\sum_{i=1}^{n+1}V^2(F_i)}{V^2(\triangle_n)}.\tag{3.5}$$

再应用引理 5,把(2.15)代入(3.5),化简之后即可得到(0.2),并可确定 C_n 之取值如(3.1).在推导过程中,$\triangle_n$ 为正则单形是每一步骤中的"≥"号取等号的充分条件,而且至少在使用(2.15)时是必要条件,故当且仅当$\triangle_n$ 为正则单形时,(0.2)中的等式成立. 定理 2 证毕.

这也就证实了我们在引言中的断言:一切维数相同、体积相等的单形中,以正则单形具有最大宽度.

现在让我们顺便导出 Sallee-Alexander 的定理:

推论 内接于球的所有单形中,正则单形具有最大的宽度.

证 只要指出,内接于一个球(不妨设其为单位球)的一切单形中,以正则单形体积最大就够了. 而这个事实是熟知的[5].

参考文献

[1]Guy R K. Problems, Lecture Notes in Math. 490,"The Geometry of Metric and Linear Spaces". Springer-Verlag, 1975:233 - 244.

[2]Alexander R. The width and diameter of a simplex. Geometriae Dedicata, 1977, 6(1):87 - 94.

[3]杨路,张景中. 关于有限点集的一类几何不等式. 数学学报,1980,23(5):740 - 749.

[4]杨路,张景中. 抽象距离空间的秩的概念. 中国科学技术大学学报,1980,10(4):52 - 65.

[5]Tanner R M. Some content maximizing properties of the regular simplex. Pac. J. Math., 1974(52):611 - 616.

非欧双曲几何的若干度量问题 I 等角嵌入和度量方程

杨 路 张景中
（中国科学技术大学）

1 引 言

度量嵌入问题一向是距离几何所关注的问题. 近期文献[1~5]已将度量嵌入方法有效地应用于积分几何，从而说明它不仅仅具有数学基础方面的意义. 关于单形在欧氏空间的等角嵌入问题，作者在文献[6]中给出了充分必要条件和一个有趣的应用. 本文解决双曲型空间的相应问题. 文中用到的双曲几何的基本度量工具定理 2，在文献[7]中已有明确叙述. 本文将给出证明.

“等角嵌入”相对于“等长嵌入”来说是讨论得较少的一个课题，特别对双曲型空间更是如此，通常只限于 2 维即双曲平面的某些讨论. 例如近期 A. F. Beardon 的文章[8]中给出了预给内角的凸多角形在双曲平面上可以实现的充分必要条件.

我们的问题的提法是：预给了一组实数，其中每个数都介于 0 和 π 之间. 能否在双曲型空间 $\boldsymbol{H}^n$ 中找到一个单形，使得单形的诸内角按指定的次序分别等于预给的那些实数？

令 Ω 表示 $\boldsymbol{H}^n$ 中的某个 n 维单形. $\Omega_1, \Omega_2, \cdots, \Omega_{n+1}$ 表示 Ω 的侧面，它们是 $n-1$ 维子单形. 每一对侧面 Ω_i 与 Ω_j 作为两超平面组成两个互补的二面角，其中包含 Ω 的那个二面角叫做 Ω 的一个内角，记为 $\Omega_i\hat{\Omega}_j$. 显然一个 n 维单形应当有 $\binom{n+1}{2}$ 个内角.

设 Θ 是一个由 $\binom{n+1}{2}$ 个实数组成的数组：

$$\Theta=\{\theta_{ij}, 0<\theta_{ij}<\pi, 1\leqslant i<j\leqslant n+1\},$$

对应于 Θ 我们可以构作一个 $(n+1)$ 阶方阵 $A[\Theta]$ 如下，当 $j>i$ 时我们置 $\theta_{ji}=\theta_{ij}$，而令

本文刊于《中国科学技术大学学报(数学专辑)》，1983 年.

$$A[\Theta]=\begin{pmatrix} 1 & -\cos\theta_{12} & -\cos\theta_{13} & \cdots & -\cos\theta_{1,n+1} \\ -\cos\theta_{21} & 1 & -\cos\theta_{23} & \cdots & -\cos\theta_{2,n+1} \\ -\cos\theta_{31} & -\cos\theta_{32} & 1 & \cdots & -\cos\theta_{3,n+1} \\ \vdots & \vdots & & \ddots & \vdots \\ -\cos\theta_{n+1,1} & -\cos\theta_{n+1,2} & \cdots & \cdots & 1 \end{pmatrix}. \tag{1}$$

在这些约定下，我们建立下述的

定理 1 预先给定了数组 Θ，则在 $\boldsymbol{H}^n$ 中存在一个单形 Ω 使得

$$\widehat{\Omega_i\Omega_j}=\theta_{ij} \tag{2}$$

对一切 $1\leqslant i<j\leqslant n+1$ 成立的充分必要条件是：

（ⅰ）$A[\Theta]$ 的行列式取负值.

（ⅱ）$A[\Theta]$ 的其它各阶主子式均取正值.

（ⅲ）$A[\Theta]$ 的 n 阶代数余子式均取正值.

为证明此命题需要一系列引理. 我们将在下节叙述这些引理，然后完成定理 1 的证明.

2 几个引理

首先需要的引理纯粹是代数的. 设 D 是一个实对称行列式而 D_{ij} 表示 D 的各个代数余子式，又用 $D_{\substack{ii\\jj}}$ 表示从 D 中去掉 i、j 两行和 i、j 两列后所剩下子行列式①，则有

引理 1 当 $i\neq j$ 时成立着

$$D_{ii}\cdot D_{jj}-D_{ij}^2=D\cdot D_{\substack{ii\\jj}}. \tag{3}$$

这个引理在距离几何中有广泛的应用，它是一个更普遍的命题的一个特例，其证明也很简单，可参看文献[7]或[9].

引理 2 如果 $D=0$，又对某个 i 有 $D_{ii}=0$，则对一切 j 有

$$D_{ij}=0.$$

其证明从引理 1 立即得到. 引理 2 在距离几何中用以建立所谓“单值公式”（可参见文献[7]的 3.6～3.7）.

下面回到几何上来. 仍设 $\boldsymbol{H}^n$ 表具有常数曲率 $-\frac{1}{r^2}$ 的 n 维双曲典型空间. 又设 $\rho_{ij}>0$ 是一组给定的实数，$1\leqslant i<j\leqslant n+1$. 对这个数构作一个 $(n+1)$ 阶方阵 Λ 如下：约定 $\rho_{ii}=0$（$i=1,2,\cdots,n+1.$），当 $j>i$ 时置 $\rho_{ji}=\rho_{ij}$. 然后令

① 我们约定，空行列式（即所有元素被删去的行列式）的值为 1.

$$\Lambda=\left(\operatorname{ch}\frac{\rho_{ij}}{r}\right)=\begin{pmatrix} 1 & \operatorname{ch}\frac{\rho_{12}}{r} & \cdots & \operatorname{ch}\frac{\rho_{1,n+1}}{r} \\ \operatorname{ch}\frac{\rho_{21}}{r} & 1 & \cdots & \operatorname{ch}\frac{\rho_{2,n+1}}{r} \\ \vdots & \vdots & \ddots & \vdots \\ \operatorname{ch}\frac{\rho_{n+1,1}}{r} & \operatorname{ch}\frac{\rho_{n+1,2}}{r} & & 1 \end{pmatrix}. \tag{4}$$

下述命题是距离几何的经典结果之一:

引理 3 对预给的实数组 $\rho_{ij}>0,1\leqslant i<j\leqslant n+1$,在 $\boldsymbol{H}^n$ 中存在一个以 $p_1,p_2,\cdots,p_{n+1}$ 为顶点之单形使得

$$p_ip_j=\rho_{ij}(\text{这里 } p_ip_j \text{ 表 } p_i \text{ 到 } p_j \text{ 的距离}) \tag{5}$$

对一切 $1\leqslant i<j\leqslant n+1$ 成立的充分必要条件是,方阵 Λ 的 k 阶顺序主子式(记为以 Λ_k)有符号 $(-1)^{k-1}$. 也即

$$\operatorname{sign}\Lambda_k=(-1)^{k-1}\quad(k=1,2,\cdots,n+1). \tag{6}$$

这叫做双曲型空间单形构造定理,是与欧氏空间单形构造定理相对应的结果,其证明可参看 Blumenthal 的专著[10].

设单形 Ω 的顶点为 $p_1,p_2,\cdots,p_{n+1}$. 如果 ρ_{ij} 的确表示 Ω 的棱长 $p_ip_j(1\leqslant i<j\leqslant n+1)$,则我们将(4)所规定的矩阵叫做是 Ω 的"度量矩阵".

引理 4 设 Ω 是 $\boldsymbol{H}^n$ 中的单形,其度量矩阵为 Λ,则 Ω 的诸内角 $\Omega_i\widehat{\ }\Omega_j$ 可由下式给出:

$$\cos\Omega_i\widehat{\ }\Omega_j=(-1)^n\frac{\Lambda_{ij}}{\sqrt{|\Lambda_{ii}|}\sqrt{|\Lambda_{jj}|}}(1\leqslant i<j\leqslant n+1), \tag{7}$$

这里 $\Lambda_{ij},\Lambda_{ii},\Lambda_{jj}$ 等表示 Λ 中对应的代数余子式. (7)给出了根据棱长计算单形内角的"余弦公式",这一引理的证明放到文章的最后一部分.

3 定理 1 的证明

充分性的证明.

对于给定的数组

$$\Theta=\{\theta_{ij};0<\theta_{ij}<\pi,1\leqslant i<j\leqslant n+1\},$$

按(1)所述构作方阵 $A[\Theta]$,其行列式记为 A. 又将 A 的诸代数余子式记为 $A_{ij},i,j=1,2,\cdots,n+1$. 并令

$$\lambda_{ij}=\frac{A_{ij}}{\sqrt{A_{ii}}\sqrt{A_{jj}}},i,j=1,2,\cdots,n+1. \tag{8}$$

由条件(ⅲ),$A_{ij}>0$,故 $\lambda_{ij}>0$,

此处由引理 1 当 $i\neq j$ 时有

$$A_{ii}\cdot A_{jj}-A_{ij}^2=A\cdot A_{\substack{ii\\jj}}. \tag{9}$$

再由条件(ⅰ),$A<0$,由条件(ⅱ),$A_{\substack{ii\\jj}}>0$,故有

$$A_{ii}\cdot A_{jj}<A_{ij}^2\,(i\neq j),$$

从而得到

$$\lambda_{ij}>1(i\neq j). \tag{10}$$

又显然

$$\lambda_{ii}=1(i=1,2,\cdots,n+1).$$

这时可以令

$$\rho_{ij}=r\cdot \mathrm{ch}^{-1}\lambda_{ij}\,(i,j=1,2,\cdots,n+1). \tag{11}$$

即

$$\lambda_{ij}=\mathrm{ch}\,\frac{\rho_{ij}}{r} \tag{12}$$

然后以诸 λ_{ij} 为元素构作一个$(n+1)$阶方阵 Λ:

$$\Lambda=(\lambda_{ij})=\left(\mathrm{ch}\,\frac{\rho_{ij}}{r}\right),$$

这正是公式(4)所定义的,依赖于数组 ρ_{ij} 的那个矩阵.

以下我们打算采取两个步骤.

第一步,证明由公式(11)给出的数组 ρ_{ij} 作为棱长确实可以构成一个单形 Ω.

第二步,证明我们所构成的单形 Ω 的诸内角恰好按原定的次序分别等于预给数组 Θ 中的各数.

首先来完成第一步.考虑两个 $n+1$ 阶方阵 Λ 与 $A[\Theta]$之间的关系,令 $A^*[\Theta]$表示 $A[\Theta]$的伴随矩阵,又置

$$P=\begin{pmatrix} A_{11}^{-\frac{1}{2}} & & & \mathbf{0} \\ & A_{22}^{-\frac{1}{2}} & & \\ & & \ddots & \\ \mathbf{0} & & & A_{n+1,n+1}^{-\frac{1}{2}} \end{pmatrix}, \tag{13}$$

此为对角型矩阵,当然有 $P=P^{\tau}$(τ 表转置).这时由(8)可知

$$\Lambda=(\lambda_{ij})=P\cdot A^*[\Theta]\cdot P^{\tau}. \tag{14}$$

前面已经约定用 A 表示 $A[\Theta]$的行列式之值.于是用实数$\dfrac{1}{A}$乘(14)的两端就有

$$\frac{1}{A}\Lambda=P\cdot A^{-1}[\Theta]\cdot P^{\tau}. \tag{15}$$

据二次型的惯性律,矩阵$\dfrac{1}{A}\Lambda$ 与 $A^{-1}[\Theta]$的正特征值的个数相同.但由条件(ⅰ),$A<0$,于是我们可以断言:Λ 的正特征值的个数恰好等于 $A[\Theta]$的负特征值的个数.

为了确定 $A[\Theta]$的特征值的符号,我们需要以下这个众所熟知的事实[11].

引理 5 设某个实对称方阵的各阶顺序主子式均不为 0,令 $D_1,D_2,\cdots,D_k,\cdots$表示各阶顺序主子式.则该方阵的特征值的符号数与数组

$$\left\{\frac{D_1}{1},\frac{D_2}{D_1},\frac{D_3}{D_2},\cdots,\frac{D_k}{D_{k-1}},\cdots\right\} \tag{16}$$

的符号数完全一致.

据此,由条件(ⅱ)和(ⅰ)可以确定 $A[\Theta]$有 n 个正特征值和一个负特征值. 从而 Λ 有一个正特征值和 n 个负特征值. 又由 A_k 非零推知以 Λ_k 非零. 于是

$$\left\{\frac{\Lambda_1}{1},\frac{\Lambda_2}{\Lambda_1},\frac{\Lambda_3}{\Lambda_2},\cdots,\frac{\Lambda_{n+1}}{\Lambda_n}\right\} \tag{17}$$

中有 1 个正数和 n 个负数.

由 Λ 之定义知 $\Lambda_1=1$,即(17)中的第一个数是正的. 于是其余各数都是负的. 从而有

$$\operatorname{sign}\Lambda_k=(-1)^{k-1}(k=1,2,\cdots,n+1).$$

此即(6)式. 据引理 3 可知在 $\boldsymbol{H}^n$ 中确实存在一个以诸 ρ_{ij} 为棱长的单形 Ω.

其次来完成第二步. 由引理 4 可以计算 Ω 的诸内角之余弦:

$$\cos\Omega_i\widehat{\ }\Omega_j=(-1)^n\frac{\Lambda_{ij}}{\sqrt{|\Lambda_{ii}|}\sqrt{|\Lambda_{jj}|}}.$$

为此须先计算以 Λ_{ij}、Λ_{ii}、Λ_{jj} 之值. 由(14)

$$\Lambda=P\cdot A^*[\Theta]\cdot P,$$

(注意 P 是对角型矩阵)两端取代数余子式有

$$\Lambda_{ij}=P_{ii}\cdot A_{ij}^*\cdot P_{jj}, \tag{18}$$

这时 P_{ii},P_{ij},A_{ij}^* 都是代数余子式而非矩阵,由于伴随矩阵与逆矩阵二者代数余子式之间有如下关系

$$A_{ij}^*=A^n\cdot(A^{-1})_{ij}, \tag{19}$$

此处$(A^{-1})_{ij}$表示 $A[\Theta]$的逆矩阵的代数余子式,据(18)就有

$$\Lambda_{ij}=P_{ii}\cdot P_{jj}\cdot A^n\cdot(A^{-1})_{ij}, \tag{20}$$

又显然有

$$(A^{-1})_{ij}=\begin{cases}\dfrac{1}{A}\cdot(-\cos\theta_{ij}) & (\text{当 } i\neq j),\\ \dfrac{1}{A} & (\text{当 } i=j).\end{cases} \tag{21}$$

于是从(20)得到

$$\begin{cases}\Lambda_{ij}=-P_{ii}\cdot P_{jj}\cdot A^{n-1}\cdot\cos\theta_{ij}(\text{当 } i\neq j),\\ \Lambda_{ii}=P_{ii}^2\cdot A^{n-1},\Lambda_{jj}=P_{jj}^2\cdot A^{n-1}.\end{cases} \tag{22}$$

然后运用余弦公式(7)而有

$$\cos\Omega_i\widehat{\ }\Omega_j=(-1)^{n+1}\cdot\left(\frac{A}{|A|}\right)^{n-1}\cos\theta_{ij}=\cos\theta_{ij}. \tag{23}$$

于是$\Omega_i\widehat{\ }\Omega_j=\theta_{ij}(1\leqslant i<j\leqslant n+1)$,即 Ω 正是我们所要求的单形. 条件充分性证毕.

必要性的证明:

设 Ω 是 $\boldsymbol{H}^n$ 中的单形,$\Omega_i\widehat{\ }\Omega_j=\theta_{ij}(1\leqslant i<j\leqslant n+1)$. 我们往证矩阵 $A[\Theta]$满足条件(ⅰ),(ⅱ),(ⅲ). 作一个$(n+1)$阶对角型矩阵 Q:

$$Q=\begin{pmatrix} \Lambda_{11}^{-\frac{1}{2}} & & & \mathbf{0} \\ & \Lambda_{22}^{-\frac{1}{2}} & & \\ & & \ddots & \\ \mathbf{0} & & & \Lambda_{n+1,n+1}^{-\frac{1}{2}} \end{pmatrix}, \tag{24}$$

这里诸 Λ_{ii} 仍表 Ω 的度量矩阵 Λ 的主代数余子式. 对 $A[\Theta]$中的元素 $\cos\theta_{ij}$ 运用余弦公式(7)就得到(注意到 $\operatorname{sign}\Lambda_{ij}=(-1)^{n-1}$)

$$\begin{aligned} A[\Theta] &= (-1)^{n+1}Q\cdot\Lambda^{*}\cdot Q^{\tau} \\ &= (-1)^{n+1}(\det\Lambda)\cdot Q\cdot\Lambda^{-1}\cdot Q^{\tau} \\ &= -|\det\Lambda|\cdot Q\cdot\Lambda^{-1}\cdot Q^{\tau}, \end{aligned} \tag{25}$$

此处 Λ^{*},Λ^{-1}表示 Λ 的伴随阵及逆阵. 已知 Λ^{-1}有 n 个负特征值和 1 个正特征值,故由(25)可知 $A[\Theta]$有 n 个正特征值和 1 个负特征值. 而且由 Λ_k 非零可推知(仍根据(25))$A[\Theta]$的各阶顺序主子式 A_k 非零. 于是数组

$$\left\{\frac{A_1}{1},\frac{A_2}{A_1},\frac{A_3}{A_2},\cdots,\frac{A_{n+1}}{A_n}\right\} \tag{26}$$

中有 n 个正数和一个负数.

将(25)两端取行列式有

$$A=-|\det\Lambda|^{n}\cdot(\det Q)^{2}<0. \tag{27}$$

将(25)两端取 n 阶主子式有

$$\begin{aligned} A_{ii} &= (-1)^{n}|\det\Lambda|^{n}\cdot Q_{ii}^{2}\cdot\Lambda_{ii}^{-1} \\ &= (-1)^{n}|\det\Lambda|^{n}\cdot Q_{ii}^{2}\cdot\det\Lambda>0. \end{aligned} \tag{28}$$

即 $A_{n+1}=A<0$,$A_n=A_{n+1,n+1}>0$. 故数组(26)中最后一数是负的,而前面 n 个数都是正的. 由递推可知

$$A_1>0,A_2>0,\cdots,A_n>0, \tag{29}$$

于是从(27),(29)得知条件(ⅰ),(ⅱ)成立.

其次对(25)两端取代数余子式,当 $i\neq j$ 时有

$$\begin{aligned} A_{ij} &= (-1)^{n}|\det\Lambda|^{n}\cdot Q_{ii}\cdot\Lambda_{ij}^{-1}\cdot Q_{jj} \\ &= (-1)^{n}|\det\Lambda|^{n}\cdot Q_{ii}Q_{jj}\cdot\det\Lambda\cdot\operatorname{ch}\frac{\rho_{ij}}{r}>0. \end{aligned} \tag{30}$$

即(ⅲ)成立. 至此定理 1 证毕.

在定理 1 的证明中没有直接用到引理 2. 但下面将看到,为了证明引理 4,前者是需要的.

4　双曲型空间的度量方程

这一节的主要目的是为双曲度量几何提供一个有效率的工具,而作为应用例题顺便给出引理 4 的一个证明.

我们从双曲型空间的射影模型开始,n 维实向量空间 $\boldsymbol{R}^n$ 中取所有那些满足条件

$$x_1^2+x_2^2+\cdots+x_n^2<1$$

的点作成一个子集 B. 在 B 上定义一个距离，使其中任意两点 x 与 y 之间的距离(记作 xy)由下式规定之：

$$\operatorname{ch}\frac{xy}{r}=\frac{1-x_1y_1-x_2y_2-\cdots-x_ny_n}{\sqrt{1-x_1^2-x_2^2-\cdots-x_n^2}\sqrt{1-y_1^2-y_2^2-\cdots-y_n^2}}. \tag{31}$$

对 B 赋予这个距离之后得到一个度量空间，记为 $\boldsymbol{H}^n$，通常叫做 n 维双曲型空间，并将 $-\frac{1}{r^2}$ 称为空间的曲率.

$\boldsymbol{H}^n$ 中的超平面即指 $\boldsymbol{R}^n$ 中的超平面与 B 的交. 设 $\boldsymbol{R}^n$ 中超平面 u 的方程为

$$u_0+u_1x_1+u_2x_2+\cdots+u_nx_n=0,$$

显然只有当

$$u_1^2+u_2^2+\cdots+u_n^2>u_0^2$$

成立时 u 与 B 的交才是非空的. 因此 $\boldsymbol{H}^n$ 中每个超平面 u 必须满足这个不等式. $\boldsymbol{H}^n$ 中两个相交的超平面 u,v 所成角度按如下计算：

$$\cos\widehat{uv}=\frac{-u_0v_0+u_1v_1+u_2v_2+\cdots+u_nv_n}{\sqrt{-u_0^2+u_1^2+u_2^2+\cdots+u_n^2}\sqrt{-v_0^2+v_1^2+v_2^2+\cdots+v_n^2}}. \tag{32}$$

而点 x 到超平面 u 的距离由下式给出：

$$\operatorname{sh}\frac{xu}{r}=\frac{|u_0+u_1x_1+u_2x_2+\cdots+u_nx_n|}{\sqrt{-u_0^2+u_1^2+u_2^2+\cdots+u_n^2}\sqrt{1-x_1^2-x_2^2-\cdots-x_n^2}}. \tag{33}$$

注意以上所考虑的超平面是没有定向的. 这时如果用向量 $\boldsymbol{u}(\boldsymbol{u}_0,\boldsymbol{u}_1,\cdots,\boldsymbol{u}_n)^\tau$ 来表示超平面的坐标，则这种坐标是齐次的.

现在众所周知的那样引进定向超平面的概念. 我们认为 $n+1$ 维非零向量 $\boldsymbol{u}$ 的一切正数倍 $\rho_{\boldsymbol{u}}$ 均代表同一个定向超平面，而其一切负数倍 $-\rho_{\boldsymbol{u}}$ 则代表另一个定向超平面，虽然二者所描划的点的轨迹是合同的. 点到定向超平面 u 的距离规定为

$$\operatorname{sh}\frac{xu}{r}=\frac{u_0+u_1x_1+u_2x_2+\cdots+u_nx_n}{\sqrt{-u_0^2+u_1^2+u_2^2+\cdots+u_n^2}\sqrt{1-x_1^2-x_2^2-\cdots-x_n^2}}. \tag{34}$$

由于等式右端可正可负，所以点到定向超平面的距离是带符号的. 每个定向超平面 u 将 $\boldsymbol{H}^n$ 划分为两个半空间，凡使 $x\boldsymbol{u}>0$ 的点 x 组成 u 的"正侧"，凡使 $y\boldsymbol{u}<0$ 的点 y 组成 u 的"负侧".

比较起来，(34)去掉了(33)中的绝对值符号，从而消除了许多度量问题的不确定性，而这种不确定性往往使事情难办得多. 椭圆型空间由于不能定向，著名的"合同阶"的问题[12]迄今未能解决. 虽然该课题在半个世纪以前距离几何发展初期就受到密切关注，而且欧氏和双曲型空间的对应问题早就容易地解决了.

下面进入本节的主题.

定义 1　$\boldsymbol{H}^n$ 中的一个点或一个定向超平面叫做一个"基本元素". 可以笼统地用 e_1，e_2，…表示 $\boldsymbol{H}^n$ 的基本元素，而不必区别它们是点或是定向超平面.

由 $\boldsymbol{H}^n$ 中某些基本元素组成的有限序列

$$\mathscr{S}=(e_1,e_2,\cdots,e_m) \tag{35}$$

叫做 $\boldsymbol{H}^n$ 的一个"基本形"，该序列的项数 m 也叫做基本形 $\mathscr{S}$ 的项数.

由$\mathscr{S}$中所出现的基本元素组成之集叫做$\mathscr{S}$的“生成集”.

定义 2　x,y,u,v 是 $\boldsymbol{H}^n$ 中的基本元素，令

$$\left.\begin{array}{l}
1)\quad g(x,y)=\operatorname{ch}\dfrac{xy}{r}，当 x,y 都是点；\\
2)\quad g(x,u)=g(u,x)=\sqrt{-1}\operatorname{sh}\dfrac{xu}{r}，当 x 是点而 u 是定向超平面；\\
3)\quad 当 u=(u_0,u_1,\cdots,u_n)^\tau,v=(v_0,v_1,\cdots,v_n)^\tau 都是定向超平面时，
\end{array}\right\}\tag{36}$$

$$g(u,v)=\frac{-u_0v_0+u_1v_1+\cdots+u_nv_n}{\sqrt{-u_0^2+u_1^2+\cdots+u_n^2}\cdot\sqrt{-v_0^2+v_1^2+\cdots+v_n^2}}.$$

由(32)可知，当 u,v 相交时，$g(u,v)=\cos\widehat{uv}$，当 u,v 不相交时，$g(u,v)$的几何意义暂不论及.

对 $\boldsymbol{H}^n$ 的基本元素偶 e,e'，我们将 $g(e,e')$叫做 e 到 e'的“g-距离”.

定义 3　设 $\mathscr{S}=(e_1,e_2,\cdots,e_m)$是一个 m 项基本形. 则记 $g_{ij}=g(e_i,e_j)(i,j=1,2,\cdots,m)$，而将下列的 m 阶方阵

$$M[\mathscr{S}]=(g_{ij})=\begin{pmatrix} g_{11} & g_{12}\cdots g_{1m}\\ g_{21} & g_{22}\cdots g_{2m}\\ \vdots & \vdots\\ g_{m1} & g_{m2}\cdots g_{mm}\end{pmatrix}\tag{37}$$

叫做基本形 $\mathscr{S}$的“度量矩阵”，也可直接记为

$$M(e_1,e_2,\cdots,e_m).$$

在这些定义之下我们将证明

定理 2　$\boldsymbol{H}^n$ 中任何一个基本形的度量矩阵的秩不超过 $n+1$.

证　设 $\mathscr{S}=(e_1,e_2,\cdots,e_m)$是一个基本形. 当 $m\leqslant n+1$ 时结论不需证明，故设 $m\geqslant n+2$. 我们不妨假定$(e_1,e_2,\cdots,e_m)$中所有的点都排在每个定向超平面之前，这样做并不影响结论. 即我们可以设

$$\mathscr{S}=(x,y,z,\cdots,u,v,w,\cdots),\tag{38}$$

我们引进一些记号：

$$\left.\begin{array}{l}\|x\|=\sqrt{1-x_1^2-x_2^2-\cdots-x_n^2}，当 x 是点，\\ \|u\|=\sqrt{-u_0^2+u_1^2+u_2^2+\cdots+u_n^2}，当 u 是超平面.\end{array}\right\}\tag{39}$$

令 T 表下列 m 阶对角矩阵：

$$T=\begin{Bmatrix}\|x\| &&&&&&\\ & \|y\| &&&& \text{0} &\\ && \|z\| &&&&\\ & \text{0} && \ddots\ \|u\| &&&\\ &&&& \|v\| &&\\ &&&&& \|\omega\| &\\ &&&&&& \ddots\end{Bmatrix}.\tag{40}$$

又 U 表示下列 m 阶方阵：

$$U=\begin{pmatrix} 1 & \sqrt{-1}x_1 & \sqrt{-1}x_2 & \cdots & \sqrt{-1}x_n & 0\cdots 0 \\ 1 & \sqrt{-1}y_1 & \sqrt{-1}y_2 & \cdots & \sqrt{-1}y_n & 0\cdots 0 \\ 1 & \sqrt{-1}z_1 & \sqrt{-1}z_2 & \cdots & \sqrt{-1}z_n & 0\cdots 0 \\ \vdots & \vdots & \vdots & & \vdots & \vdots \quad \vdots \\ \sqrt{-1}u_0 & u_1 & u_2 & \cdots & u_n & 0\cdots 0 \\ \sqrt{-1}v_0 & v_1 & v_2 & \cdots & v_n & 0\cdots 0 \\ \sqrt{-1}\omega_0 & \omega_1 & \omega_2 & \cdots & \omega_n & 0\cdots 0 \\ \vdots & \vdots & \vdots & & \vdots & \vdots \quad \vdots \end{pmatrix}. \tag{41}$$

作矩阵积

$$T^{-1}UU^{\tau}T^{-1},$$

并将公式(31),(32),(34)代入这个乘积可知

$$M(x,y,z,\cdots,u,v,\omega)=T^{-1}UU^{\tau}T^{-1}. \tag{42}$$

显然 U 的秩不超过 $n+1$,故 $M[\mathscr{S}]$的秩不超过 $n+1$. 定理 2 证毕.

系 1　当基本形 $\mathscr{S}$的项数 $m\geqslant n+2$ 时必有

$$\det M[\mathscr{S}]=0. \tag{43}$$

关系(43)叫做这类基本形的“度量方程”.

现在考虑 $\boldsymbol{H}^n$ 中的一个单形 Ω,其顶点是 $p_1,p_2,\cdots,p_{n+1}$. 令 p_i 所对之侧面为 Ω_i. 对 Ω_i 所在之超平面进行定向使得 p_i 在它的正侧,该定向超平面记为 f_i. 考虑$(n+2)$项基本图形

$$\mathscr{S}=(p_1,p_2,\cdots,p_{n+1},f_i) \tag{44}$$

的度量矩阵. 根据(36),(37)和(4)可知 $M[\mathscr{S}]$恰好是矩阵 Λ 镶上一个边:(令 $\mu=\sqrt{-1}\ \mathrm{sh}\dfrac{p_if_i}{r}$)

$$M[\mathscr{S}]=\begin{pmatrix} & & & 0 \\ & & & \vdots \\ & & & 0 \\ & \Lambda & & \mu \\ & & & 0 \\ & & & \vdots \\ & & & 0 \\ 0\cdots 0 & \mu & 0\cdots 0 & 1 \end{pmatrix} \begin{matrix} \\ \\ \\ \leftarrow\cdots\text{第 } i \text{ 行} \\ \\ \\ \\ \\ \end{matrix} \tag{45}$$

（μ 所在列为第 i 列）

按度量方程(43)展开就有

$$\det\Lambda-\mu^2\Lambda_{ii}=0. \tag{46}$$

我们将 $h_i=p_if_i$ 叫做单形 Ω 的一个“高”,则由(46)可以算出

$$\mathrm{sh}^2\frac{h_i}{r}=\mathrm{sh}^2\frac{p_if_i}{r}=-\mu^2=-\frac{\det\Lambda}{\Lambda_{ii}}. \tag{47}$$

于是我们有

系 2　单形 Ω 的每个侧面 Ω 上的高 h_i 由下式给出

$$h_i = r \cdot \mathrm{sh}^{-1}\sqrt{-\frac{\det \Lambda}{\Lambda_{ii}}}, i=1,2,\cdots,n+1. \tag{48}$$

我们进而考虑两个这样的定向超平面 f_i 和 f_j. 对两个基本图形

$$\begin{aligned} \mathscr{S}^* &= (p_1, p_2, \cdots, p_{n+1}, f_i, f_j) \\ \text{和} \qquad \mathscr{S} &= (p_1, p_2, \cdots, p_{n+1}, f_i) \end{aligned} \tag{49}$$

应用度量方程. 若设

$$D = \det M[\mathscr{S}^*], \tag{50}$$

就有

$$D=0, D_{n+3,n+3}=0. \tag{51}$$

于是由引理 2 得到

$$D_{n+2,n+3}=0. \tag{52}$$

仍根据(36),(37)和(4)写出 $D_{n+2,n+3}$ 的表达式就有

$$\begin{array}{c} \left|\begin{array}{cccccccc} & & & & & & & 0 \\ & & & & & & & \vdots \\ & & & & & & & 0 \\ & & & \Lambda & & & & \sqrt{-1}\,\mathrm{sh}\dfrac{h_j}{r} \\ & & & & & & & 0 \\ & & & & & & & \vdots \\ & & & & & & & 0 \\ 0 & \cdots & 0 & \sqrt{-1}\,\mathrm{sh}\dfrac{h_i}{r} & 0 & \cdots & 0 & \cos\widehat{f_i f_j} \end{array}\right| \begin{array}{l} \\ \\ \\ \leftarrow\cdots \text{第 } j \text{ 行} \\ =0. \\ \\ \\ \\ \end{array} \\ \uparrow \\ \vdots \\ \text{第 } i \text{ 列} \end{array} \tag{53}$$

展开得

$$\cos\widehat{f_i f_j} = -\frac{\Lambda_{ij}}{\det \Lambda} \cdot \mathrm{sh}\frac{h_i}{r}\mathrm{sh}\frac{h_j}{r}. \tag{54}$$

又由(47)可知

$$\mathrm{sh}\frac{h_i}{r} = \sqrt{\frac{|\det \Lambda|}{|\Lambda_{ii}|}}, \mathrm{sh}\frac{h_j}{r} = \sqrt{\frac{|\det \Lambda|}{|\Lambda_{jj}|}}, \tag{55}$$

将此代入(54)即得(注意 sign $\det \Lambda = (-1)^n$)

$$\cos\widehat{f_i f_j} = (-1)^{n+1}\frac{\Lambda_{ij}}{\sqrt{|\Lambda_{ii}|}\sqrt{|\Lambda_{jj}|}}. \tag{56}$$

于是得到

系 3　设 f_i, f_j 是单形 Ω 的两侧面所在的两个定向超平面,其定向使得 Ω 在它们二者的正侧,则 $\widehat{f_i f_j}$ 由(59)给出.

将(56)与(7)比较可知,为了证明引理 4,只需验明 $\widehat{f_i f_j}$ 是单形 Ω 的外角就行了. 让我们

回到本节的开始,如果我们对 $\boldsymbol{R}^n$ 的那个子集 B 以及与有交的超平面赋与通常的欧氏度量,而且按欧氏度量将超平面进行定向,容易看出它同按双曲度量进行的定向是完全一致的. 这是因为一个点 x 是否属于定向超平面 u 的正侧,无论在欧氏的或双曲的情形都只决定于 $u_0+u_1x_1+u_2x_2+\cdots+u_nx_n$ 这个数是否大于零. 另一方面,单形的某个角是否内角,这是 $\boldsymbol{R}^n$ 中的仿射性质,与引进何种度量无关. 于是我们只需考虑在 $\boldsymbol{E}^n$ 中单形的两个侧面 f_i,f_j,如果其定向使得单形在它们二者的正侧,这两个定向超平面组成的角 $\widehat{f_if_j}$ 是否一定是外角? 答案是肯定的而且显然,只要作出 f_i,f_j 的法向量即可.

至此引理 4 证明完毕. 经过类似的推导我们也可以建立由单形诸内角计算棱长的公式.

$$\operatorname{ch}\frac{p_ip_j}{r}=\frac{A_{ij}}{\sqrt{A_{ii}}\sqrt{A_{jj}}}(i,j=1,2,\cdots,n+1).$$

这里 A_{ij},A_{ii},A_{jj} 所指与上节(8)中相同.

参考文献

[1]杨路,张景中. 双曲型空间紧致集的覆盖半径. 中国科学 A 辑,1982(8):683-692.

[2]Alexander R. Lecture Notes in Math. ,Springer-Verlag,1975(490):57-65.

[3]Proc. Amer. Math Soc. ,1977(64):317-321.

[4]Pac. J. of Math. ,1979(85):1-9.

[5]Alexander R,et al. Trans. Amer. Math. Soc. ,1974(193):1-31.

[6]杨路,张景中. 预给二面角的单形嵌入 $\boldsymbol{E}^n$ 的充分必要条件. 数学学报,1983(26):250-256.

[7]杨路,张景中. 抽象距离空间的秩的概念. 中国科学技术大学学报,1980,10(4):52-65.

[8]Beardon A F. Hyperbolic Polygons and Fuchsian Groups. J. London Math. soc. ,1979,2(20):247-254.

[9]杨路,张景中. 关于有限点集的一类几何不等式. 数学学报,1980(23):740-749.

[10]Blumenthal L M. Theory and Applications of Distance Geometry. Oxford,1953.

[11]Bellmen R. Introduction to Matrix Analysis. McGraw-Hill,1970,§5.3.

[12]Seidel J J. Metric Problems in Elliptic Geometry. Lecture Notes in Math. ,No. 490,Springer. Verlag,1975:32-43.

关于凸体的一个不等式的简单证明

杨　路　张景中

（中国科学技术大学）

本文涉及 Grünbaum 一个已被解决的问题.

设 K 是 R^n 中的紧致凸体，p 是 K 的一个内点. 过 p 任作 K 的一条弦 $C(p)$，并以 $C'(p)$ 记平行于 $C(p)$ 的任意一条弦. $|C(p)|$ 和 $|C'(p)|$ 分别表 $C(p)$ 与 $C'(p)$ 的长. 令

$$F^p(K)=\min\left\{\frac{|C(p)|}{|C'(p)|}: p\in C(p),\quad C(p)/\!/C'(p)\right\},$$

则 $F^p(K)$ 在某种意义上反映了凸体 K 关于其内点 p“中心对称”的程度. 而关于凸体对称性的量度，是受到许多作者关心的有趣的问题. 例如，可参看 B. Grünbaum[1] 的长篇综述.

在 $F^p(K)$ 中取 p 为 K 的重心 g，就得到 K 关于其重心的对称程度的指标之一的 $F^g(K)$. Grünbaum[1] 指出，对 $K\subset R^2$，易证 $F^g(K)$ 之下确界是 2/3. 他猜想在 R^3 中 $F^g(K)$ 的下确界是 1/2，但未能证明. 同时，他提出了在 R^n 中确定 $F^g(K)$ 的下确界的一般问题.

洪加威[2] 完全解决了这一问题，获得了下述 Grünbaum -洪加威不等式：

定理 1

$$\inf_{K\subset R^n} F^g(K)=\frac{2}{n+1}. \tag{1}$$

此外，洪加威[2] 还得到了更为一般的结果.

本文将给定理 1 一个简单而初等的证明. 从我们的证明中，不仅回答了文献[1]中的这个问题，还导出其他一些不等式. 当然，本文不能包含文献[2]的全部结果，但后面将要看到，我们的定理 3 也不被文献[2]所包含.

定理 1 的证明　设 p 为凸体 K 之一内点. 过 p 的任一弦 $C(p)$ 端点为 a、b，而平行于 $C(p)$ 的任一弦 $C'(p)$ 端点为 a'、b'. 作过 p 的另两根弦 $a'c$ 和 $b'd$，设 $a'd$，$b'c$ 分别交 ab 于 e，f，于是显然有（见图 1）：

$$pe : a'b'=dp : db',$$
$$pf : a'b'=cp : ca',$$

故

$$ab=pa+pb\geqslant pe+pf=\left(\frac{dp}{db'}+\frac{cp}{ca'}\right)a'b',$$

$$\frac{ab}{a'b'}\geqslant\frac{dp}{db'}+\frac{cp}{ca'}. \tag{2}$$

以上是整个证明的第一步骤.

本文刊于《数学学报》，第 26 卷第 1 期，1983 年.

第二个步骤是证明下列事实：当 $p=g$ 是 K 的重心时，(2)式右端的两上比值都不小于 $\frac{1}{n+1}$. 我们只对其中一个验证这一断言即可.

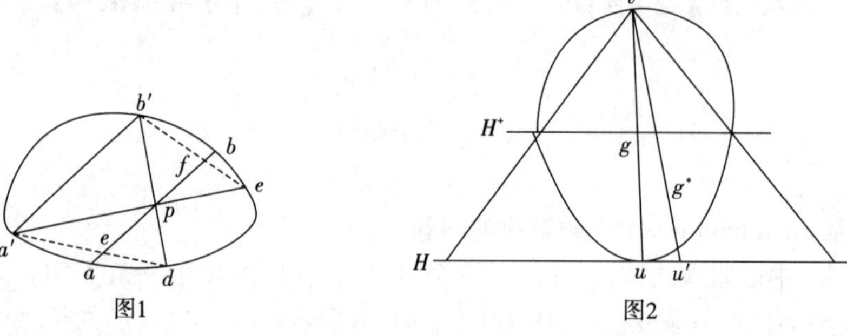

图1　　　图2

如图 2，g 为凸体 K 的重心，过 g 任取 K 的一条弦 uv，我们往证

$$\frac{ug}{gv}\geqslant\frac{1}{n}\left(\text{亦即 }\frac{ug}{uv}\geqslant\frac{1}{n+1}\right).\tag{3}$$

过 u 作 K 的一个支撑超平面 H，再过 g 作超平面 $H^*/\!/H$. H^* 分 K 为两部分：设含 v 的部分为 K_v，含 u 的部分为 K_u. 再以 $\mathscr{D}^*$ 记截面 $H^*\cap K$，以 v 为投影中心将 $\mathscr{D}^*$ 投影于 H 上得 $\mathscr{D}$，以 v 为顶点，$\mathscr{D}$ 为底作锥体 K^0. K^0 被 H^* 分为两部分：包含 v 的部分记为 K_v^0，包含 u 的部分记为 K_u^0，显然有 $K_v^0\subset K_v$，$K_u\subset K_u^0$. 设 g^* 为 K^0 之重心，设直线 vg^* 交 H 于 u'.

以 H^* 为第一坐标超平面，v 点所在的一侧作为正侧在 R^n 中建立笛卡尔坐标系. 由重心之定义及 $g\in H^*$ 可知，g 的第一坐标为 0，即

$$\int_K x_1 dV=0.\tag{4}$$

这里及下面 x_1 表 K 中点的第一坐标. 设 g^* 的第一坐标为 x^*，则有

$$x^*=\frac{1}{V_{K^0}}\int_{K^0}x_1dV=\frac{1}{V_{K^0}}\left[\int_K x_1dV-\int_{K_v\setminus K_v^0}x_1dV+\int_{K_u^0\setminus K_u}x_1dV\right]\leqslant 0,\tag{5}$$

这是因为括号内的三个积分第一个为 0，第二个被积函数非负，而第三个被积函数非正之故. 可见 x^* 在超平面 H^* 上或和 u(及 u')在 H^* 同侧，又由 g^* 是 n 维锥体 K^0 之重心可知，

$$\frac{1}{n}=\frac{u'g^*}{g^*v}\leqslant\frac{ug}{gv},\tag{6}$$

即得不等式(3). 联合(3)与(2)得，当 $p=g$ 时，

$$\frac{ab}{a'b'}\geqslant\frac{2}{n+1},\tag{7}$$

从而知

$$\inf_{K\subset R^n}F^g(K)\geqslant\frac{2}{n+1}.\tag{8}$$

另一方面，若 K 为 n 维单形$\triangle^n$，取$\triangle^n$ 的任一条棱为 $C'(p)$，过$\triangle^n$ 重心且与 $C'(p)$平行之弦为 $C(p)$，简单计算可知$\frac{|C(p)|}{|C'(p)|}=\frac{2}{n+1}$，故有

$$F^g(\triangle^n)\leqslant\frac{2}{n+1}.\tag{9}$$

联合(9)与(8),就得到了定理1所要的等式(1).

在证明过程中所获得的不等式(2),实质上把 $F^p(K)$ 和所谓凸体对称性的 Minkowski 量度联系起来. 所谓凸体 K 的 Minkowski 量度 $M^p(K)$ 是这样定义的:过 K 的内点 p 任作弦 $C(p)$,设 p 分 $C(p)$ 为两段 $C_1(p)$,$C_2(p)$,令

$$M^p(K)=\min\{|C_1(p)|/|C_2(p)|\},$$

则由(2)显然可得到与维数无关的不等式:

定理2 设 p 是 R^n 中紧凸体 K 的内点,则

$$F^p(K)\geqslant\frac{2}{1+\frac{1}{M^p(K)}}. \tag{10}$$

定理2的证明虽然简单,但很有用. 兹举一例:把所有包含凸体 K(或被 K 包含)的椭球中体积最小(或最大)者的中心叫做 K 的外心(或内心);分别以 c 和 i 记 K 的外心和内心,关于其 Minkowski 量度的下列结果是熟知的:

[John 定理][3~4]

$$\inf_{K\subset R^n}M^c(K)=\frac{1}{n};\ \inf_{K\subset R^n}M^i(K)=\frac{1}{n}.$$

把 John 定理和定理2结合起来易得

定理3

$$\inf_{K\subset R^n}F^c(K)=\inf_{K\subset R^n}F^i(K)=\frac{2}{n+1}.$$

这个结果已经不属于 Grünbaum 问题的范围.

此外,定理2提示我们去考虑凸体的各种对称性指标相互之间的关系(在文献[1]中这类指标列举了9种之多). 要想全部确定这些关系看来不是简单的事,有待于进一步研究.

参考文献

[1]Grünbaum B. Measures of symmetry for convex sets. Proc. of Symposia in Pure Math., 1963(3):233-270.

[2]洪加威. 一个几何的不等式. 数学学报,1974,17(3):205-213.

[3]John F. Extremum problems with inequalities as subsidiary conditions, Courant Anniv. Volume, 1948:187-204.

[4]Leichtweiss K. Ueber die affine Exzentrlzität konvexer Körper, Arch. Math., 1959(10):187-199.

论逐段单调连续函数的迭代根

张景中　杨　路

（中国科学技术大学）

设 E 是一个集合，f 和 g 是将 E 映射到自身的函数 $f\circ g$ 表示 f 和 g 的复合函数.

$$(f\circ g)(x)=f(g(x)),x\in E.$$

f 的迭代函数 f^n 的定义是

$$f^0(x)=x,f^{n+1}=f\circ f^n,n=0,1,2,\cdots.$$

如果对一个整数 $r\geqslant 2$ 和一切 $x\in E$ 成立着

$$f^r=g,$$

我们就说 f 是 g 的一个 r 阶的迭代根.

关于迭代根的研究至少可以上溯到 Abel[1]，甚至更早的 Babbage[2]. 多年以来这问题一直引起许多作者的注意. 1950 年 R. Isaacs[3] 在一篇精辟的论文中完成了一个奠基性的工作，给出了将抽象集 E 映射到自身的函数的迭代根存在的一个充分必要条件. 这个结果最近又有所发展[4].

复变函数迭代根的研究，首先要提到 Koenigs[5] 的经典工作，但其结果限于解析函数的不动点邻域. 为获得全局性的结果，必须作解析开拓. 例如 Kneser[6] 关于整函数 e^x 的二次迭代根的一个结果. 最近 Rice[7~10] 作了某些新的贡献，其中包括他在 1977 年的博士论文[8].

实变函数方面的相应工作，首先要提到 Böedewadt[11] 和 Fort[12]，这是属于单实变元的. 多实变函数方面结果较少. Sternberg[13~15] 关于可微变换与线性变换"共轭"的研究，实际上就是求解 Schröder 方程[16]，因而与迭代根问题密切相关.

近年来波兰的学派在函数方程和迭代根方面的研究是卓有成效的，特别在单实变的情形有 Kuczma[17] 们的专著和其他论文可参考[18~22]. 但单实变的迭代根的研究，一般限于单调连续函数，非单调情形只讨论了个别特例. 即如 x^2-2 这样一个简单函数是否有 2 次迭代根的问题，也长期不为人知[22].

以下设函数 $F(x)$ 在 $[a,b]$ 上连续，且当 $x\in[a,b]$ 时 $F(x)\in[a,b]$. 如果 F 在 $[a,b]$ 上只有有限个极值点，且在两个极值点之间严格单调，则称之为逐段严格单调连续函数. 所有这样的 F 之集记为 $S_{[a,b]}$.

如上所述，考虑 F 的迭代根问题，即考虑函数方程

$$f^n=F(F\in S_{[a,b]})\tag{1}$$

的解 f 的问题. 自然，当 F 连续时，我们易于提出(1)是否有连续解的问题.

本文刊于《数学学报》，第 26 卷第 4 期，1983 年.

这是一个整体性问题. 因为如果把$[a,b]$分段考虑,则(1)本身在某些段上失去了意义. 所以,我们不能简单地把关于单调连续的F对应的(1)的结果推广到逐段单调.

本文将给出(1)对所有自然数$n>1$有连续解的充要条件. 且对不满足此条件的F给出了当且仅当n为奇数时(1)有连续解的充要条件. 同时证明:对其余的$F\in S_{[a,b]}$,(1)至多对有限个n有连续解.

§1 逐段严格单调的连续函数之迭代

定义 1 若$F(x)$定义于$[a,b]$上,$x_0\in[a,b]$. 如果存在$\varepsilon>0$,使$F(x)$在$[x_0-\varepsilon,x_0+\varepsilon]$上严格单调,则称$x_0$是$F(x)$的一个"单调点". 否则就称$x_0$是$F(x)$的一个非单调点.

由定义 1,易知下列诸显然之事实:

引理 1 若$F(x)$在$[a,b]$上连续,则

$F(x)\in S[a,b]$的充分必要条件是:$F(x)$在$[a,b]$内只有有限个非单调点.

其证明略去. 我们看到,对于逐段严格单调的连续函数,$[a,b]$内的极值点是它仅有的非单调点. 为了考虑复合函数的非单调点,我们将下列显然事实也作为引理列出:

引理 2 设$F(x)$于$[a,b]$上连续,$x_0\in[a,b]$,则x_0是$F(x)$的非单调点的充分必要条件是:对于任给的$\varepsilon>0$,总有$x_1,x_2\in[a,b]$满足$x_1\neq x_2$,$|x_1-x_0|<\varepsilon$,$|x_2-x_0|<\varepsilon$,且$F(x_1)=F(x_2)$.

由引理 2 可推得

引理 3 设$F_1(x)$,$F_2(x)$皆在$[a,b]$上连续并取值于$[a,b]$上. 令$F(x)=F_1(F_2(x))$,则对于任一点$x_0\in(a,b)$,x_0是$F(x)$的单调点的充分必要条件是,x_0是$F_2(x)$的单调点而且$F_2(x_0)$是$F_1(x)$的单调点.

引理 3 也可改写为:

引理 3′ 在引理 3 的条件下,以M,M_2,M_1分别记F,F_2,F_1的非单调点之集,并令

$$M_3=\{x\mid F_2(x)\in M_1\},$$

则$M=M_2\cup M_3$.

由引理 3′可知M的势不小于M_2的势. 故若M为有穷集,则M_2也为有穷集. 结合引理 1 便得:

引理 4 设F_1,F_2在$[a,b]$上连续,且取值于$[a,b]$. 则由$F_1(F_2)\in S[a,b]$推知$F_2\in S[a,b]$. 特别地,对于$[a,b]$上的连续函数$f(x)$,$a\leqslant f(x)\leqslant b$,由$f^{[m]}\in S[a,b]$可推知,$f(x)\in S[a,b]$.

引理 4 说明当$F(x)$在$[a,b]$上逐段严格单调且连续时,方程(1)的连续解也必然是逐段严格单调的,反过来,我们也有

引理 5 若$F_1\in S[a,b]$,$F_2\in S[a,b]$,则

$$F_1(F_2)\in S[a,b].$$

作为特例有

$$f\in S[a,b]\Rightarrow f^m\in S[a,b].$$

§2 迭代时极值点个数变化情况

现在我们引进记号 $N(F)$. 此处 $F\in S[a,b]$, $N(F)$ 表 $F(x)$ 在 $[a,b]$ 内的极值点个数. 当 $F(x)$ 在 $[a,b]$ 上严格单调时 $N(F)=0$. 根据引理 5, $F^m\in S[a,b]$; 再由引理 3′就有

$$N(F^0)\leqslant N(F^1)\leqslant N(F^2)\leqslant\cdots\leqslant N(F^m)\leqslant\cdots.$$

如果对于任意的非负整数 m, 恒有

$$N(F^m)<N(F^{m+1}),$$

则我们称 $F(x)$ 为"升级"的. 若 $F(x)$ 非升级的, 则至少有一个 m 使 $N(F^m)=N(F^{m+1})$, 以 $H(F)$ 记这样的数 m 中的最小者. 对于升级的 $F(x)$ 我们约定 $H(F)=\infty$. 为了弄清 $H(F)$ 的性质, 引入

引理 6 若 $F\in S[a,b]$, $A=\min\limits_{a\leqslant x\leqslant b}F(x)$, $B=\max\limits_{a\leqslant x\leqslant b}F(x)$, 则对任意的 $c\in(A,B)$ 必有 $x_0\in(a,b)$, 使得 x_0 是 F 的单调点, 而且 $f(x_0)=c$ (以下 A,B 的意义沿用).

事属显然(图象上看更直观), 证明从略.

引理 7 设 $F_1\in S[a,b]$, $F_2\in S[a,b]$, 而且 $a\leqslant\min F_2=A<B=\max F_2\leqslant b$, 则 $N(F_1(F_2))=N(F_2)$ 的充分必要条件是 $F_1(x)$ 在 $[A,B]$ 上严格单调.

证 充分性: $F_1(x)$ 在 $[A,B]$ 上单调说明, 如果 $F_2(x_0)$ 是 $F_1(x)$ 的极值点, 则必有 $F_2(x_0)=A$ 或 $F_2(x_0)=B$, 即 x_0 是 $F_2(x)$ 的极值点. 沿用引理 3′的记号, 就有 $M_3\subseteq M_2$, 即 $M=M_2$. 因而 $N(F_1(F_2))=N(F_2)$.

必要性: 若 $F_1(x)$ 在 $[A,B]$ 上非严格单调, 则它在 $[A,B]$ 内至少有一个极值点 $x=c$. 根据引理 6 存在 $x_0\in(a,b)$, x_0 是 F_2 的单调点, 而且 $F_2(x_0)=c$. 再由引理 3′, x_0 是 $F_1(F_2)$ 的极值点, 即 $M_3\not\subseteq M_2$, 因而, $M=M_2\cup M_3\supset M_2$. 此即 $N(F_1(F_2))>N(F_2)$, 与假设矛盾, 故 $F_1(x)$ 在 $[A,B]$ 上严格单调. 引理 7 证毕.

引理 8 设 $F\in S[a,b]$, 且对 $x\in[a,b]$ 有 $F(x)\in[a,b]$, $H(F)=m<\infty$, 则对任意自然数 l 常有 $N(F^m)=N(F^{m+l})$.

证 由 $H(F)$ 之定义, 知

$$N(F^m)=N(F^{m+l})=N(F\circ F^m).$$

以 A_k, B_k 记 $F^k(x)$ 在 $[a,b]$ 的下、上确界. 由引理 7 可知 $F(x)$ 在 $[A_m,B_m]$ 上严格单调. 但显然有

$$A_m\leqslant A_{m+l-1}<B_{m+l-1}\leqslant B_m\ (l\geqslant 1),$$

故 $F(x)$ 也在 $[A_{m+l-1},B_{m+l-1}]$ 上严格单调. 由引理 7 得

$$N(F^{m+l-1})=N(F^{m+l})\quad(l=1,2,3,\cdots),$$

即所欲证.

由此可得一个关系式:

引理 9 以 $\{x\}$ 记 x 的分数部分, 并令 $[x]=x-\{x\}$, 则在引理 8 条件下必有等式

$$H(F^k)=\left[\frac{m}{k}\right]+\operatorname{sgn}\left\{\frac{m}{k}\right\}.$$

证 令 $m_k=\left[\dfrac{m}{k}\right]+\operatorname{sgn}\left\{\dfrac{m}{k}\right\}$，$F_1=F^k$，只需往证

$$N(F_1^{m_k})=N(F_1^{m_k+1}),$$
$$N(F_1^{m_k})>N(F_1^{m_k-1}),$$

显然 $k(m_k-1)<m\leqslant km_k<k(m_k+1)$.

而 $$F_1^{m_k}=F^{km_k},F^{m_k-1}=F^{k(m_k-1)},F_1^{m_k+1}=F^{k(m_k+1)}.$$

故由 $H(F)=m$，结合引理 8，即得引理 9.

现在可以获得关于方程(1)的一个初步结论：

定理 1 设 $F(x)$在$[a,b]$上连续，逐段严格单调，$x\in[a,b]$时 $F(x)\in[a,b]$. 则当 $H(F)>1$，$n>N(F)$时，函数方程(1)没有连续解.

证 若(1)有连续解 $f(x)$，由引理 4，$f\in S[a,b]$. 由假设 $H(F)>1$，知 $N(F(F))>N(F)$，即 $N(f^{2n})>N(f^n)$，由引理 8 即知 $H(f)>n$. 由 $H(f)$之定义知

$$0=N(f^0)<N(f)<N(f^2)<\cdots<N(f^n).$$

故 $N(f^n)=N(F)\geqslant n$，与假设矛盾. 证毕.

由定理 1 可知，对于单峰函数(即只有一个极值点的连续函数)$F(x)$，当 $H(F)>1$ 时，方程(1)没有连续解.

§3 $H(F)\leqslant 1$ 时(1)有连续解的必要条件

在上节中证明了，只在 $H(F)\leqslant 1$ 的情况下(1)才可能对 n 的无穷个数值都有连续解，现在就进一步探讨满足 $H(F)\leqslant 1$ 的函数 $F(x)$的性质. 由定义，$H(F)\leqslant 1$ 与等式 $N(F)=N(F(F))$等价. 由引理 7，$F(x)$在$[A,B]$上严格单调，此处 A,B 仍记 $F(x)$在$[a,b]$上的下、上确界. 于是可引入：

定义 2 令 $F(x)$在$[a,b]$的下、上确界为 A,B，$a\leqslant A<B\leqslant b$，$F(x)\in S[a,b]$，而 $H(F)\leqslant 1$. 如果

ⅰ) $[\bar{a},\bar{b}]\subseteq S[a,b]$，$[A,B]\subseteq S[\bar{a},\bar{b}]$；

ⅱ) $\bar{a}$ 和 $\bar{b}$ 都是 $F(x)$的极值点；

ⅲ) $(\bar{a},\bar{b})$中没有 $F(x)$的极值点.

则将$[\bar{a},\bar{b}]$叫做 $F(x)$的“特征区间”.

本节的主要目的是建立

定理 2 设 $F\in S[a,b]$，而 $H(F)\leqslant 1$，$f(x)$是方程(1)的连续解. 则若将 $F(x)$，$f(x)$都看成仅定义于 $F(x)$的特征区间$[\bar{a},\bar{b}]$的函数时，$f(x)$仍为(1)的解(换句话说，$f(x)$可以在$[\bar{a},\bar{b}]$上进行迭代即当 $x\in[\bar{a},\bar{b}]$时有 $f(x)\in[\bar{a},\bar{b}]$).

在证明此定理之前，先重申一下众所熟知的“周期点”的概念.

定义 3 设 $g(x)$定义于$[a,b]$上，而 $a\leqslant g(x)\leqslant b$，$k$ 是正整数. 若 $x_0\in[a,b]$，$g^k(x_0)=x_0$，则称 x_0 是 $g(x)$的一个“周期点”. 如果 k 是使 $g^k(x_0)=x_0$ 成立的最小的正整数，则称 x_0 是 $g(x)$的一个 k 阶周期点. 特别地，一阶周期点就是 $g(x)$的不动点.

定理 2 的证明 仍以 A_l,B_l 记 $F^l(x)$ 在 $[a,b]$ 上的下、上确界. 显然当 l 增加时,A_l 不减而 B_l 不增. 从而若 x_0 是 $F(x)$ 的一个周期点,则对一切正整数 l 都有 $x_0\in[A_l,B_l]$. 于是 $F(x)$ 所有的周期点都在其特征区间 $[\bar{a},\bar{b}]$ 上. 由定义可知 $f(x)$ 的周期点都是 $F(x)$ 的周期点,故 $f(x)$ 的周期点也都在 $[\bar{a},\bar{b}]$ 上. 又由 $F(x)$ 在 $[\bar{a},\bar{b}]$ 上严格单调可知 $f(x)$ 在 $[\bar{a},\bar{b}]$ 上严格单调. 下面分两种情况来证明当 $x\in[\bar{a},\bar{b}]$ 时有 $f(x)\in[\bar{a},\bar{b}]$.

1° 若 $f(x)$ 在 $[\bar{a},\bar{b}]$ 上严格递增. 当 $x\in[\bar{a},\bar{b}]$ 时,若 $f(x)$ 取值超出了 $[\bar{a},\bar{b}]$,即 $f(\bar{a})<\bar{a}$ 或 $f(\bar{b})>\bar{b}$,但 $f(x)$ 取值于 $[a,b]$,故 $f(a)\geqslant a, f(b)\leqslant b$. 若 $f(\bar{a})<\bar{a}$,由介值定理必存在 $x_0\in[a,\bar{a})$,使得 $f(x_0)=x_0$,即在 $[\bar{a},\bar{b}]$ 外存在着 f 的周期点,矛盾. 同理不能有 $f(\bar{b})>\bar{b}$.

2° 若 $f(x)$ 在 $[\bar{a},\bar{b}]$ 上严格递减,则 $F(x)$ 必在 $[f(\bar{b}),f(\bar{a})]$ 上严格单调. 否则,如果 F 在这个区间内有极值点,由引理 3,$F(f(x))=f^{n+1}(x)$ 必在 $[\bar{a},\bar{b}]$ 内有极值点. 从而 $F(F(x))=f^n(f^n(x))$ 在 $[\bar{a},\bar{b}]$ 内有极值点,于是 $N(F(F))>N(F)$,这与 $H(F)\leqslant 1$ 矛盾.

其次因 F 在 $\bar{a},\bar{b}$ 取到极值,故 $\bar{a},\bar{b}$ 均不能是 $[f(\bar{b}),f(\bar{a})]$ 的内点. 那么,只要能证明 $[\bar{a},\bar{b}]$ 与 $[f(\bar{b}),f(\bar{a})]$ 有公共内点,则有 $[f(\bar{b}),f(\bar{a})]\subseteq[\bar{a},\bar{b}]$. 用反证法:若当 x 遍历 $[\bar{a},\bar{b}]$ 时 $f(x)$ 取不到 $[\bar{a},\bar{b}]$ 内之值,则当 x 遍历 $[a,b]$ 时 $f(F(x))$ 取不到 $[\bar{a},\bar{b}]$ 内之值,即 $F(f(x))=f(F(x))$ 取不到 $[\bar{a},\bar{b}]$ 内之值. 另一方面,$F(f(x))$ 必取值于 $[\bar{a},\bar{b}]$. 故必有 $F(f(x))\equiv\bar{a}$ 或 $F(f(x))\equiv\bar{b}$,与其逐段严格单调性矛盾. 定理 2 证毕.

定理 2 告诉我们,当 $H(F)\leqslant 1$ 时,(1)有连续解的必要条件是

$$f^n(x)=F(x)\quad(x\in[\bar{a},\bar{b}])\tag{1'}$$

有连续解. 但此条件非充分的.

定理 3 若 $F\in S[a,b]$,$H(F)\leqslant 1$,F 的特征区间为 $[\bar{a},\bar{b}]$. 则对某正整数 $n>N(F)+1$,方程(1)有连续解的必要条件是:若 $F(x)$ 在 $[a,b]$ 上某一点取到值 $\bar{a}$(或 $\bar{b}$),则 $F(x)$ 必在 $[\bar{a},\bar{b}]$ 上取到此值.

证 首先对 $n>N(F)$ 必有 $H(f)<n$,此处 f 是(1)的连续解. 若不然,将有 $N(f^n)\geqslant n>N(F)$. 由 $H(f)<n$ 得 $N(f^{n-1})=N(F)$. 由引理 8

$$N(f^{n-1})=N(f^{n-1}\circ f^{n-1}),$$

故 $H(f^{n-1})\leqslant 1$. 这说明 f^{n-1} 也以 $[\bar{a},\bar{b}]$ 为特征区间;这是因为 $[A,B]\in[\min f^{n-1},\max f^{n-1}]$,而 f^{n-1} 与 $f^n=F$ 有共同的极值点之故. 于是当 x 走遍 $[a,b]$ 时 f^{n-1} 取值于 $[\bar{a},\bar{b}]$. 欲使 F 在 $[a,b]$ 上取到 $\bar{a}$ 或 $\bar{b}$,必须 f 在 $[\bar{a},\bar{b}]$ 上取到此值. 下面分两种情形:

1° 若 f 在 $[\bar{a},\bar{b}]$ 递增,f 在 $[\bar{a},\bar{b}]$ 上取到 $\bar{a}(\bar{b})$,必有 $f(\bar{a})=\bar{a}(f(\bar{b})=\bar{b})$,从而 $F(\bar{a})=\bar{a}(F(\bar{b})=\bar{b})$.

2° 若 f 在 $[\bar{a},\bar{b}]$ 递减,由 $n>N(F)+1$,可以推知 $H(f)<n-1$,从而 $H(f^{n-2})\leqslant 1$. 于是对一切 $x\in[a,b]$,有 $f^{n-2}(x)\in[\bar{a},\bar{b}]$. 想要 F 在 $[a,b]$ 上取到 $\bar{a}(\bar{b})$,必须 f^2 在 $[\bar{a},\bar{b}]$ 上取到此值. 由 f 的递减性质,及 $f(\bar{a})\leqslant\bar{b}, f(\bar{b})\geqslant\bar{a}$,想要 f^2 在 $[\bar{a},\bar{b}]$ 取到 $\bar{a}$ 或 $\bar{b}$ 中之一,必须同时有

$$f(\bar{a})=\bar{b},\ f(\bar{b})=\bar{a}.$$

从而 f 将 $[\bar{a},\bar{b}]$ 变为自身,故 F 将 $[\bar{a},\bar{b}]$ 变为自身,即 F 在 $[\bar{a},\bar{b}]$ 上取到 $\bar{a},\bar{b}$. 这结论比要证的还强. 定理 3 证毕.

下面将指出:这里给出的必要条件合起来就成为充分条件 3.

§4 解的开拓及在特征区间上有解之必要条件

定理 4 若 $F\in S_{[a,b]}$，$H(F)\leqslant 1$，$[\bar{a},\bar{b}]$是 F 的特征区间. 仍以 A、B 和 $\bar{A}$、$\bar{B}$ 分别记 F 在 $[a,b]$上和在$[\bar{a},\bar{b}]$上的下确界和上确界. 若有定义于$[\bar{a},\bar{b}]$上的连续函数 f^*，$f^*(x)\in[\bar{a},\bar{b}]$，且当 x 在$[A,B]$上变化时 $f^*(x)$在$[\bar{A},\bar{B}]$上变化，并满足

$$f^{*n}=F,x\in[\bar{a},\bar{b}]. \tag{2}$$

则 f^* 可连续地开拓为$[a,b]$上的函数 f，f 满足(1)

证 令 F^* 是 F 在$[\bar{a},\bar{b}]$上的限制，则 F^* 的反函数在$[\bar{A},\bar{B}]$上唯一确定且连续. 记之以 F^{*-1}. 令

$$f=F^{*-1}\circ f^*\circ F \quad (x\in[a,b]).$$

因为当 $x\in[a,b]$时，$F(x)\in[A,B]\subseteq[\bar{a},\bar{b}]$，而 $y=F(x)$在$[A,B]$上变化时，$f^*(y)\in[\bar{A},\bar{B}]$，从而$(F^{*-1}\circ f^*)(y)$有定义；故 f 在$[a,b]$上确定而且连续，且

$$f^n(x)=(F^{*-1}\circ f^{*n}\circ F)(x)=(F^{*-1}\circ F^*\circ F)(x)=F(x).$$

定理 4 证毕.

由定理 4，我们只要在$[\bar{a},\bar{b}]$上求出(2)的连续解 f^*，并满足：当 $x\in[A,B]$时 $f^*(x)\in[\bar{A},\bar{B}]$，就可以经开拓而得到(1)在$[a,b]$上的连续解. 而 F 在$[\bar{a},\bar{b}]$上是严格单调的，故问题化归为对严格单调的连续函数 F 求(1)的连续解. 下面给出此时(1)有连续解的必要条件.

定理 5 若 F 在$[a,b]$上递增、连续，且有 $F(x)\in[a,b]$，则当 n 为奇数时，(1)的连续解必为递增的. 当 n 为偶数时，(1)有递减的连续解的必要条件是：有 $x_0\in(a,b)$使 $F(x_0)=x_0$，且使 F 在$[a,x_0)$的不动点之集 E_1 与 F 在$(x_0,b]$内的不动点之集 E_2 之间可以建立反序的一一对应

$$D:E_1\to E_2,D(e)=e'(e\in E_1,e'\in E_2)$$
$$(\text{当 } e_1<e_2 \text{ 时 } D(e_1)>D(e_2)).$$

且使得：若 $F(x)-x$ 在(e_1,e_2)上为正(负)，则它在$(D(e_2),D(e_1))$上为负(正). 同时有 $F(a)-a$ 与 $F(b)-b$ 反号或都为 0.

证 由于递减函数的奇次迭代仍为递减，故若(1)有递减解，则 n 为偶数. 要证的是定理的其余部分. 令 $g=f^2$，则 g 递增，且满足

$$g^{\frac{n}{2}}=F,$$

故 g 与 F 有相同的不动点集且 $g(x)-x$ 与 $F(x)-x$ 同号，故对 $n=2$ 证明所要之结论已足.

由于 f 递减、连续且 $f(x)\in[a,b]$，故 f 在$[a,b]$上有唯一不动点 $x_0\in[a,b]$，显然 $g(x_0)=x_0$. 分别以 E_1，E_2 记 g 在$[a,x_0)$，$(x_0,b]$上的不动点之集. 若 $e\in E_1$，则 $e<x_0$，令 $e'=f(e)=D(e)$，则 $D(e)=f(e)>f(x_0)=x_0$，且 $g(D(e))=g[f(e)]=f[g(e)]=f(e)=e'$，从而 $e'\in E_2$. 这就建立了 E_1 与 E_2 之间的反序的一一对应. 若 $g(x)-x$ 在(e_1,e_2)为正(负)，任取 $\bar{x}\in(e_1,e_2)$，则 $f(\bar{x})=\bar{x}'\in(e_2',e_1')$，由 $g(\bar{x})=f(\bar{x}')>\bar{x}(g(\bar{x})<\bar{x})$，可知 $f(g(\bar{x}))<f(\bar{x})$，即 $g(\bar{x}')<\bar{x}'(g(\bar{x}')>\bar{x}')$. 即 $g(x)-x$ 在(e_1,e_2)与(e_2',e_1')上反号.

最后，若 $f(a)=b$，且 $f(b)=a$，则 $a\in E_1$，$b\in E_2$. 若 $f(a)<b$，则 $f^2(a)=g(a)>f(b)\geqslant$

a,从而

$$g(b)=f[f(b)]\leqslant f(a)<b.$$

定理 5 证毕.

定理 6　若 F 在 $[a,b]$ 上连续、递减,且有 $F(x)\in[a,b]$,则当 n 为偶数时(1)无连续解. n 为奇数时(1)有连续解之必要条件是 $F(a)=b$ 与 $F(b)=a$ 同时成立或同时不成立. 且此时(1)之连续解必为递减.

定理 6 至为显然,证明略去.

§5　F 递增时,(1)的递增连续解

我们需要一个早已为人所知的事实:

引理 10　(W. Chayoth[12])若 F,G 都是 $[a,b]$ 上的连续递增函数. $F(a)=G(a)=a$, $F(b)=G(b)=b$,且对 $x\in(a,b)$, $(G(x)-x)(F(x)-x)>0$,则有 $[a,b]$ 上的连续递增函数 φ,满足 $\varphi(a)=a$, $\varphi(b)=b$ 和初始条件

$$\varphi(x)=\varphi^*(x)(x\in[F(x_0),x_0]\text{或}[x_0,F(x_0)])\tag{3}$$

和函数方程

$$\varphi\circ F=G\circ\varphi(x\in[a,b]).\tag{4}$$

(这里 x_0 是 (a,b) 的任一点, φ^* 是满足 $(\varphi^*\circ F)(x_0)=(G\circ\varphi^*)(x_0)$ 的任一个给定的 $[F(x_0),x_0]$(或 $[x_0,F(x_0)]$)上的连续递增函数,且 $a<\varphi^*(x_0)<b$.)

由于文献[12]中没有明确指出 φ 的递增性和它满足初值的可能性,故我们仍简略地给出下列之证明.

证　不妨先设在 (a,b) 上有 $G(x)<x$ 和 $F(x)<x$. 令 $y_0=\varphi^*(x_0)$,作点列 $\{x_n\}$, $\{y_n\}$:

$$x_n=F^n(x_0),y_n=G^n(y_0)(n=0,\pm1,\pm2,\cdots),$$

则 $x_n>x_{n+1}$, $y_n>y_{n+1}$,且当 $n\to+\infty$ 时 $x_n\to a$, $y_n\to a$,而当 $n\to-\infty$ 时, $x_n\to b$, $y_n\to b$.

令 $\varphi_0(x)=\varphi^*(x)(x\in[x_1,x_0])$,再取

$$\varphi_1(x)=G[\varphi_0[F^{-1}(x)]](x\in[x_2,x_1]),$$

$$\varphi_{-1}(x)=G^{-1}[\varphi_0[F(x)]](x\in[x_0,x_{-1}]),$$

显然, φ_{-1}, φ_0, φ_1 连续递增,且

$$\varphi_1(x_1)=y_1,\varphi_{-1}(x_0)=y_0,$$

一般地,若在 $[x_{n+1},x_n]$ 上和 $[x_{-n+1},x_{-n}]$ 上(n 是自然数)分别给定了 φ_n 和 φ_{-n},它们连续递增,且

$$\varphi_n(x_{n+1})=G[\varphi_n(x_n)],\varphi_n(x_n)=y_n,$$

$$\varphi_{-n}(x_{-n})=G^{-1}[\varphi_{-n}(x_{-n+1})],\varphi_{-n}(x_{-n+1})=y_{-n+1},$$

则可以归纳地定义

$$\varphi_{n+1}(x)=G[\varphi_n(F^{-1}(x))],x\in[x_{n+2},x_{n+1}],$$

$$\varphi_{-(n+1)}(x)=G^{-1}[\varphi_{-n}(F(x))],x\in[x-n,x_{-(n+1)})].$$

然后令

$$\varphi(x)=\begin{cases}x & (x=a\text{ 或 }b),\\ \varphi_n(x) & (x\in[x_{n+1}x_n]),\end{cases}$$

易知这个定义是合理的，且 φ 连续，递增，满足 $\varphi[F(x)]=G[\varphi(x)]$.

如果在 (a,b) 上有 $G(x)>x$ 和 $F(x)>x$，可令

$$G_1(x)=(b+a)-G(b+a-x),$$
$$F_1(x)=(b+a)-F(b+a-x),$$

则 $G_1(x)<x$，$F_1(x)<x$ 在 (a,b) 上成立. 由已证之事实，有 φ^* 使 $\varphi^*\circ F_1=G_1\circ\varphi^*$，令

$$\varphi(x)=a+b-\varphi^*[a+b-x],$$

则 $\varphi(x)$ 满足 $\varphi\circ F=G\circ\varphi$，等等. 引理 10 证毕.

借助于引理 10，我们可以给出 Hardy[23]-Bödewadt[11]定理的一个新证明. 这里，仍对解的初始条件给了限制.

定理 7 若 F 在 $[a,b]$ 上连续、递增，且 $F(x)\in[a,b]$，则对任给的自然数 $n\geqslant2$ 和满足不等式 $a<A<B<b$ 的 A,B，有 $[a,b]$ 上的连续递增函数 f，满足(1)：

$$f^n=F,\quad x\in[a,b]$$

和不等式

$$F(a)\leqslant f(A)<f(B)\leqslant F(b).\tag{5}$$

证 不妨设 F 在 (a,b) 内没有不动点. 因为若 F 在 (a,b) 内有不动点，我们可以分别在每个以不动点为端点，而内部没有不动点的区间上考虑(1)的连续递增解，再把它们拼凑成(1)在 $[a,b]$ 上的解.

先考虑 a、b 都是 F 的不动点的情况. 令 $G=F^n$，由引理 10，有 $[a,b]$ 上的连续递增函数 φ，满足 $\varphi(a)=a$，$\varphi(b)=b$，$\varphi\circ F=G\circ\varphi$，亦即

$$F=\varphi^{-1}\circ G\circ\varphi,$$

现在令 $f=\varphi^{-1}\circ F\circ\varphi$，则 $f^n=\varphi^{-1}\circ F\circ\varphi=\varphi^{-1}\circ G\circ\varphi=F$，显然 f 是(1)的连续递增解. 由 $F(a)=a$ 和 $F(b)=b$ 可知不等式(5)自然成立.

若 a,b 不是或不全是 F 的不动点. 为确定起见，设 $F(a)=a$，而 $F(b)<b$. 任取 $c>b$，令

$$F_1(x)=\begin{cases}F(x),x\in[a,b),\\ \dfrac{F(b)}{b-c}(x-c)+\dfrac{c}{c-b}(x-b), & x\in[b,c],\end{cases}$$

显然 F_1 在 $[a,c]$ 上连续，递增，以 a,c 为不动点，而在 (a,c) 内则有 $F_1(x)<x$. 令 $G=F_1^n$，应用引理 10，有在 $[a,c]$ 上连续递增的 φ，$\varphi(a)=a$，$\varphi(c)=c$，使 $\varphi\circ F_1=G\circ\varphi$. 按我们已用过的方法，可取 $f_1=\varphi^{-1}\circ F_1\circ\varphi$，则 $f_1^n=\varphi^{-1}\circ G\circ\varphi=F_1$. 但是，我们还要求有不等式(5)，对这种情况，就是要证明对某个指定的 $B\in(a,b)$，要求 $f_1(B)\leqslant F(b)$. 为此，我们对 φ 的初始条件加以规定.

为了 $f_1(B)\leqslant F(b)$，应当有

$$(\varphi^{-1}\circ F_1\circ\varphi)(B)\leqslant F(b),$$

即

$$(F_1\circ\varphi)(B)\leqslant(\varphi\circ F)(b)=(G\circ\varphi)(b)=(F_1^n\circ\varphi)(b).$$

为此，令 $\varphi^*(b)=b$，$(\varphi^*\circ F)(b)=F^n(b)$，而在 $[F(b),b]$ 内，任意插入连续递增的 $\varphi^*(x)$，并

使之在 B 满足(注意:若 $B\leqslant F(b)$,则不等式(5)自然得到满足,故此处不妨设 $B\in(F(b),b)$):

$$\varphi^*(B)=F^{n-1}(b)-\varepsilon\left(0\leqslant\varepsilon\leqslant\frac{F^{n-1}(b)-F^n(b)}{2}\right),$$

则对于满足初值:

$$\varphi(x)=\varphi^*(x),x\in[F(b),b]$$

的 φ,$f_1=\varphi^{-1}\circ F_1\circ\varphi$ 满足 $f_1(B)\leqslant F(b)$.

这时 f_1 在 $[a,c]$ 上满足 $a\leqslant f_1(x)\leqslant x$,特别地,$f_1(b)<b$,故当 $x\in[a,b]$ 时,$f_1(x)\in[a,b]$. 对于 $F(a)>a$ 的情形,不等式(5)中的 $F(a)\leqslant f(A)$ 可类似地做到. 故从略. 定理 7 证毕.

结合定理 7 与定理 4,可得

定理 8　若 $F\in S_{[a,b]}$,$H(F)\leqslant 1$,F 在其特征区间上递增,且当 F 在特征区间端点 $\bar{a}(\bar{b})$ 处不取值 $\bar{a}(\bar{b})$ 时,F 在 $[a,b]$ 上也取不到值 $\bar{a}(\bar{b})$,则对任意的自然数 $n\geqslant 2$,方程(1)有连续解. 如果 $n>N(F)+1$,则此条件也是必要的.

§6　F 递增时(1)的递减连续解

在 §4 定理 5 中,我们给出了 F 递增时(1)有递减连续解的必要条件. 现在来证明这个条件也是充分的.

定理 9　若 F 在 $[a,b]$ 上连续递增,$F(x)\in[a,b]$,F 在 (a,b) 内有不动点 x_0,且 $F(a)=a$ 与 $F(b)=b$ 同时成立或不成立. 以 E_1,E_2 记 F 在 $[a,x_0]$ 和在 $[x_0,b]$ 上的不动点之集. 存在 E_1,E_2 之间的反序的一一对应

$$D:E_1\to E_2,D(e)=e'(e\in E_1,e'\in E_2)$$
$$(\text{当 } e_1<e_2 \text{ 时},e_2'<e_1'),$$

且当 $F(x)-x$ 在 (e_1,e_2) 上为正(负)时,它在 $(e_2'<e_1')$ 上为负(正). 则对任意自然数 $n\geqslant 1$,存在在 $[a,b]$ 上连续的递减函数 f,满足 $f^{2n}=F$. 且当 $F(a)>a$,$F(b)<b$ 时,可对预先给定的满足 $a<A<B<b$ 的 A,B,使有

$$F(a)\leqslant f(B)<f(A)\leqslant F(b).\tag{5'}$$

为证明定理 9,我们引入

引理 11　在定理 9 的条件下,存在 $[a,b]$ 上的连续递减函数 $\varphi(x)$,满足 $\varphi(x_0)=x_0$,$\varphi(a)=b$,$\varphi(b)=a$,且

$$\varphi[\varphi(x)]=x,\varphi\circ F=F\circ\varphi.$$

证　先证 $F(a)=a$,$F(b)=b$ 同时成立之情形. 令 F_1,F_2 分别是 F 在 $[a,x_0]$,$[x_0,b]$ 上的限制,则 E_1,E_2 分别是 F_1,F_2 的不动点之集. 令 $\varphi_1(x)=x_0+b-x(x\in[a,b])$,则置

$$F_3=\varphi\circ F_2\circ\varphi_1,x\in[x_0,b]$$

之后,φ_1 将 E_2 映为 F_3 的不动点之集 E_3. 且若

$$e_1,e_2\in E_1,e_1<e_2,$$

由假设 $e_1',e_2'\in E_2$,$e_2'<e_1'$,于是 $\varphi_1(e_1')$,$\varphi_1(e_2')\in E_3$,$\varphi_1(e_1')<\varphi_1(e_2')$. 从而 φ_1 是 E_2、E_3 之间

的反序的一一对应. 这说明 E_1,E_3 之间存在保序的一一对应. 我们指出:此时若 $F_1(x)-x$ 在 (e_1,e_2) 上为正(负),则 $F_3(x)-x$ 在 $(\varphi_1(e_1'),\varphi_1(e_2'))$ 上亦为正(负). 为此只要证明 $F_2(x)-x$ 之符号在 (e_2',e_1') 内与 $F_3(x)-x$ 在 $(\varphi_1(e_1'),\varphi_1(e_2'))$ 上相反. 事实上

$$\begin{aligned}F_2(x)-x&=(x_0+b-x)-(x_0+b)+F_2(x)\\&=\varphi_1(x)-\varphi_1(F_2(x))\\&=\varphi_1(x)-\varphi_1(F_2(\varphi_1(\varphi_1(x))))\\&=\varphi_1(x)=F_3\circ\varphi_1(x)=-[F_3\circ\varphi_1(x)-\varphi_1(x)].\end{aligned}$$

由于 E_1,E_2 是闭集,故它们之间的保序的一一对应可开拓为 $[a,x_0]$ 到 $[x_0,b]$ 的连续递增的函数 $g(x)$. 令

$$F_4(x)=g^{-1}\circ F_3\circ g(x),x\in[a,x_0],$$

则 F_4 是 $[a,x_0]$ 上的连续递增函数,且

$$\operatorname{sgn}(F_4(x)-x)=\operatorname{sgn}(F_1(x)-x).$$

事实上,若对某个 x,$F_1(x)-x>0(<0)$,由前面已指出的,$(F_1(x)-x)$ 与 $(F_3(x)-x)$ 在 E_1 与 E_3 的余集的对应的构成区间上同号,故

$$F_3(g(x))-g(x)>0(<0),$$

由 g 之递增性,得

$$g^{-1}\circ F_3\circ g(x)>g^{-1}\circ g(x)(g^{-1}\circ F_3\circ g(x)<g^{-1}\circ g(x)),$$

即 $F_4(x)-x>0(<0)$.

对 F_1,F_4 应用引理 10,有 $[a,x_0]$ 上的连续递增函数中,φ_2 满足 $\varphi_2(a)=a,\varphi_2(x_0)=x_0$,使

$$\varphi_2\circ F_1=F_4\circ\varphi_2,$$

从而

$$\varphi_2\circ F_1=g^{-1}\circ\varphi_1\circ F_2\circ\varphi_1\circ g\circ\varphi_2.$$

若令 $\varphi^*=\varphi\circ g\circ\varphi_2$,则 $F_1=\varphi^{*-1}\circ F_2\circ\varphi^*$. 令

$$\varphi=\begin{cases}\varphi^*,x\in[a,x_0),\\\varphi^{*-1},x\in[x_0,b],\end{cases}$$

则 φ 在 $[a,b]$ 上连续、递减,满足

$$\varphi[\varphi(x)]=x,\varphi\circ F=F\circ\varphi.$$

对于 $F(a)>a$、$F(b)<b$ 的情形,设 a^*,b^* 是 F 的最小和最大的不动点,我们只要再给出 φ 在 $[a,a^*)$ 和 $(b^*,b]$ 上的定义即可. 令 F_a,F_b 分别为 F 在 $[a,a^*]$ 和 $[b^*,b]$ 上的限制,并设 g^* 是把 $[a,a^*]$ 映为 $[b^*,b]$ 的任一个连续递减函数. 令 $\widetilde{F}_b=g^{*-1}\circ F_b\circ g^*$,则 $\widetilde{F}_b$ 是 $[a,a^*]$ 上的连续递增函数,以 a^* 为不动点,而且

$$\widetilde{F}_b(x)>x,x\in[a,a^*).$$

任取 $c<a$,把 F_a 和 $\widetilde{F}_b$ 分别连续地开拓为 $[c,a^*]$ 上的递增函数 $\overline{F}_a$ 和 $\overline{F}_b$,使

$$\overline{F}_a(c)=\overline{F}_b(c)=c,$$

$$\overline{F}(x)-x>0,\overline{F}_b(x)-x>0,x\in[c,a^*],$$

由引理 10,有在 $[c,a^*]$ 上连续递增的 $\overline{\varphi}^*$,满足 $\overline{\varphi}^*(c)=c,\overline{\varphi}^*(a^*)=a^*,\overline{\varphi}^*\circ\overline{F}_a=\overline{F}_b\circ\overline{\varphi}^*$ 及初

始条件 $\bar{\varphi}^*(a)=a$.

令 φ^* 为 $\bar{\varphi}^*$ 在 $[a,a^*]$ 上的限制,则在 $[a,a^*]$ 上有

$$F_a=\varphi^{*-1}\circ\widetilde{F}_b\circ\varphi^*=\varphi^{*-1}\circ g^{*-1}\circ F_b\circ g^*\circ\varphi^*,$$

然后,只要把 φ 在 $[a,a^*]$ 上定义为 $g^*\circ\varphi^*$,而在 $[b^*,b]$ 上定义为 $\varphi^{*-1}\circ g^{*-1}$ 已足.引理 11 证毕.

定理 9 之证明　由定理 7,有在 $[a,x_0]$ 上的连续递增的 f_a,满足 $f_a^{2n}=F$,且

$$f_a(x_0)=x_0.$$

当 $F(a)=a$ 时 $f_a(a)=a$,当 $F(a)>a$ 时,$f_a(x)$ 可以使之满足条件 $f_a(\alpha)\geqslant f(a)$,这里 α 是任意取定的 $(a,F(a))$ 内的一点.

由引理 11,有在 $[a,b]$ 上连续递减的 φ,使 $\varphi(x_0)=x_0$、$\varphi(a)=b$、$\varphi(b)=a$,$\varphi\circ\varphi(x)=x$,并且有 $\varphi\circ F=F\circ\varphi$,令

$$f=\begin{cases}\varphi\circ f_a, & x\in[a,x_0),\\ f_a\circ\varphi, & x\in[x_0,b].\end{cases}$$

由于 $\varphi\circ f_a(x_0)=x_0=f_a\circ\varphi(x_0)$,故 f 在 $[a,b]$ 上连续递减,且 $\varphi\circ f=f\circ\varphi$;对 $f\in[a,x_0]$,有

$$f^{2n}=(f_a\circ\varphi\circ\varphi\circ f_a)^n=f_a^{2n}=F.$$

对 $x\in[x_0,b]$,则有

$$f^{2n}=(\varphi\circ f_a\circ f_a\circ\varphi)^n=\varphi_0 f_a^{2n}\circ\varphi=\varphi\circ F\circ\varphi=F\circ\varphi\circ\varphi=F.$$

剩下要证的是可以选择初始条件,使(5′)满足.不妨设 $a<A<F(a)$,$b>B>F(b)$,因若 $A\geqslant F(a)$,$B\leqslant F(b)$,(5′)是显然的,这时要证的是

$$f(B)=f_a\circ\varphi(B)\geqslant F(a),$$
$$f_a(A)=\varphi\circ f(A)\geqslant\varphi\circ F(b)=F(a),$$

取 α 为 $\varphi(B)$,A 中之小者,则当 $f_a(\alpha)\geqslant F(a)$ 时,以上两不等式均被满足.定理 9 证毕.

§7　F 递减时,(1)的连续递减解

现在往证定理 6 中(1)有连续递减解的必要条件同时也是充分的.

定理 10　若 F 在 $[a,b]$ 上连续递减,$a\leqslant F(x)\leqslant b$,$F(a)=b$ 与 $F(b)=a$ 同时成立或同时不成立,则对任意的 $n\geqslant 1$,有在 $[a,b]$ 上的连续递减函数 f,满足

$$f^{2n+1}=F,$$

且当 $F(b)>a$,$F(a)<b$ 时,对任意给定的 A,B,$a<A<B<b$,可以使 f 满足

$$F(b)\leqslant f(B)<f(A)\leqslant F(a).\tag{5″}$$

证　先考虑 $F(a)=b$,$F(b)=a$ 的情况.令 $G=F\circ F$,则 G 在 $[a,b]$ 上连续递增,以 a,b 为不动点.设 F 在 $[a,b]$ 内的唯一不动点为 x_0,由定理 7,在 $[a,x_0]$ 上有连续递增的 g 满足

$$g^{2(2n+1)}(x)=G(x),$$

令

$$g_1=\begin{cases}g, & x\in[a,x_0),\\ F^{-1}\circ g\circ F, & x\in[x_0,b],\end{cases}$$

则 g_1 在$[a,b]$上连续递增，满足

$$g_1^{2(2n+1)}=G, g_1\circ F=F\circ g_1,$$

令 $g_1^{2n+1}=F_1$，$\varphi=F_1^{-1}\circ F$，则 φ 是$[a,b]$上的连续递减函数，满足

$$\varphi^2(x)=F_1^{-1}\circ F\circ F_1^{-1}\circ F(x)=F_1^{-2}\circ F^2(x)=G^{-1}\circ G(x)=x,$$

且 $\varphi\circ F=F\circ\varphi$. 令 $f=\varphi\circ g_1$，则 f 在$[a,b]$上连续递减. 我们断言：

$$f^{2n+1}=F.$$

事实上，若令 $f^{2n+1}=\Phi$，则 Φ 满足

$$\Phi^2=F^2, \Phi\circ F=F\circ\Phi,$$

因而 $\Phi^{-1}\circ F\circ\Phi^{-1}\circ F=\Phi^{-2}\circ F^2=F^0$，故由于 $\Phi^{-1}\circ F$ 是连续递增而得

$$\Phi^{-1}\circ F(x)=x,$$

即 $F=\Phi$.

下面讨论 $F(b)>a$ 且 $F(a)<b$ 的情形. 设 a^*，b^* 是 $G=F^2$ 的最小与最大不动点. 由前一部分之证明可知，我们只给出 f 在$[a,a^*]$，$[b^*,b]$上之定义即可. 这是因为有 $F(a^*)=b^*$，$F(b^*)=a^*$ 之故. 我们用直接插值法来确定$[a,a^*]$上的一个函数 F_a，使 $F_a^2=G$，方法如下：

令 $a_m=G^m(a)(m=0,1,2,\cdots)$，则 $a=a_0<a_1<a_2<a_3<\cdots$，且 $a_n\to a^*$. 在$[a,F(b)]$上任取一个连续递增的函数 F_0，并使 F_0 满足

$$F_0(a)=F(b), F_0(F(b))=G(a),$$

及不等式

$$x<F_0(x)<G(x).$$

再令 $c_0=a_0$，$c_{2k}=a_k$，$c_1=F_0(c_0)$，$c_{2k+1}=G^k(c_1)(k=1,2,\cdots)$，则 $a=c_0<c_1<c_2<\cdots$，且 $c_k\to a^*$.

已知 F_0 将$[c_0,c_1]$映为$[c_1,c_2]$，令

$$F_1(x)=G\circ F_0^{-1}(x), x\in[c_1,c_2],$$

则 F_1 把$[c_1,c_2]$映为$[c_2,c_3]$. 引入归纳定义：若已有连续递增的 F_{k-1}，把$[c_{k-1},c_k]$映为$[c_k,c_{k+1}]$，定义

$$F_k=G\circ F_{k-1}^{-1}, x\in[c_k,c_{k+1}],$$

则 F_k 在$[c_k,c_{k+1}]$连续递增，把$[c_k,c_{k+1}]$映为$[c_{k+1},c_{k+2}]$. 令

$$F_a(x)=\begin{cases}a^*, x=a^*,\\ F_{k-1}(x), x\in[c_{k-1},c_k],\end{cases}$$

则 F_0 在$[a,a^*]$上连续递增，满足

$$F_a(a^*)=a^*, F_a(a)=F(b), F_a^2=G.$$

再在$[b^*,b]$上定义

$$F_b=F\circ F_a^{-1}\circ F, x\in[b^*,b].$$

易验证 $F_b(b)=F(a)$，$F_b(b^*)=b^*$，且 $F_b^2=G$. 仍令 g 记$[a^*,x_0]$上满足

$$g^{2(2n+1)}=G$$

的连续递增函数. 令

$$F^*=\begin{cases}F_a,x\in[a,a^*],\\F_b,x\in[b^*,b],\\g^{2n+1},x\in(a^*,x_0],\\F^{-1}\circ g^{2n+1}\circ F,x\in(x_0,b^*).\end{cases}$$

则 F^* 在$[a,b]$上连续递增，满足

$$F^*(a)=F(b),F^*(b)=F(a),F^{*2}=F^2=G.$$
$$F^*\circ F=F\circ F^*.$$

在$[a,x_0]$上求出一个满足 $f_1^{2n+1}=F^*$ 的连续递增的 f_1，再令

$$f=\begin{cases}F^{-1}\circ f_1^{2n+2}, & x\in[a,x_0],\\f_1^{-2n}\circ F, & x\in(x_0,b],\end{cases}$$

此处定义之合理性由不等式

$$f_1^{2n}(a)<f_1^{2n+1}(a)=F(b)<f_1^{2n+2}(a)$$

来保证，不难验证，f 在$[a,b]$上连续递减，满足 $f^{2n+1}=F$.

最后，为使不等式

$$f(A)\leqslant F(a),f(B)\geqslant F(b) \tag{5''}$$

得到满足，应有而且只需有

$$f_1^{2n+2}(A)\geqslant G(a);f_1^{-2n}\circ F(B)\geqslant F(b),$$

前者即 $f_1\circ F^*(A)\geqslant G(a)$，后者即

$$f_1\circ F(B)\geqslant F_a\circ F(b)=F_a\circ F_a(a)=G(a),$$

取 α 为 $F^*(A)$，$F(B)$中之小者，由于 $F^*(A)$，$F(B)$均大于 $F^*(a)$，故可选取初值，使

$$f_1(\alpha)\geqslant G(a),$$

此时两不等式均得到满足. 定理 10 证毕.

于是得(由解的连续开拓定理——定理 4)

定理 11　若 $F\in S_{[a,b]}$，$H(F)\leqslant 1$，$a\leqslant F(x)\leqslant b$，$F$ 在其特征区间上递减. 则当 n 为奇数时，(1)有连续解的充分条件是：或者 $F(\bar{a})=\bar{b}$，$F(\bar{b})=\bar{a}$ 同时成立，或者在$[a,b]$上恒有 $\bar{a}<F(x)<\bar{b}$.

§8　总　　结

把全文的结果综述如下：

1°　若 F 是$[a,b]$上的逐段严格单调的连续函数，$a\leqslant F(x)\leqslant b$，则 F^n 也是$[a,b]$上的逐段严格单调函数. 且(1)的连续解也必是逐段严格单调的.

2°　以 $N(F)$ 记 F 在$[a,b]$内的极值点个数，则当 $N(F)\neq N(F^2)$ 时，对 $n>N(F)$，(1)无连续解.

3°　以 A，B 记 F 在$[a,b]$上的下、上确界，则 $N(F)=N(F^2)$ 的充要条件是：有$[a,b]$中的两点 $\bar{a}$，$\bar{b}$，使 $\bar{a}\leqslant A<B\leqslant\bar{b}$，$F$ 在$[\bar{a},\bar{b}]$上严格单调而以 $\bar{a}$，$\bar{b}$ 为极值点. 此时若 $n>N(F)+1$，F 在$[a,b]$上某点取到 $\bar{a}$(或 $\bar{b}$)，而在$[\bar{a},\bar{b}]$取不到该值，则(1)无连续解.

$4°$ 若 F 在 $[\bar{a},\bar{b}]$ 递增，则(1)对所有的 n 有连续解的充要条件是：$A=\bar{a}$ 蕴含 $f(\bar{a})=\bar{a}$，且 $B=\bar{b}$ 蕴含 $f(\bar{b})=\bar{b}$.

$5°$ 若 F 在 $[\bar{a},\bar{b}]$ 递减，则(1)当 n 为偶数时无连续解. 当 n 为奇数时，如果 $F(\bar{a})=\bar{b}$，$F(\bar{b})=\bar{a}$ 同时成立；或 $F(\bar{a})<\bar{b}$，$F(\bar{b})>\bar{a}$ 同时成立，且 $\bar{a}<A<B<\bar{b}$，则(1)有连续解.

至于当 $N(F)<N(F^2)$ 而 $n\leqslant N(F)$ 的情形，(1)什么时候有连续解，尚是一个待解决的问题. 还有，当 $N(F)=N(F^2)$ 而 $n\leqslant N(F)+1$ 时，若 F 在 $[a,b]$ 取到值 $\bar{a},\bar{b}$ 而在 $[\bar{a},\bar{b}]$ 上取不到此值，(1)的连续解存在的条件仍属未知.

作为结束，我们不加证明地举一个例子：

若 F 是 m 次多项式，$a=-\infty$，$b=+\infty$，则当 m 为奇数时，如果(1)对某个 $n>N(F)$ 有连续解，则 F' 在 $(-\infty,+\infty)$ 上不变号；当 m 为偶数时，(1)对所有 n 有解的充要条件是 $F'\circ F\geqslant 0$；否则，对 $n>N(F)$，(1)无连续解.

特别地，若 F 是二次多项式 $F(x)=ax^2+bx+c(a\neq 0)$，则(1)有连续解的充要条件是

$$b^2-4ac\leqslant 2b.$$

这些断言，只要和前文略一对照，便可证实.

与本文论题有关的，区间上的离散流嵌入连续流的问题，近期文献可参阅[24]，[25].

参考文献

[1]Abel N H. Oeuvres Complètes, t. II. Christiana, 1881: 36 - 139.

[2]Dubbey J M. The Mathematical Work of Oharles Babbage. Cambridge Univ. Press, 1978.

[3]Isaacs R. Iterates of fractional order. Canad. J. Math., 1950(2): 409 - 416.

[4]Riggert G. n-te iterative Wurzeln von beliebigen Abbildungen (abstract). Aequations Math., 1976(14): 208.

[5]Koenigs G. Recherches sur les intergrales de certailnes equations fonetionelles. Ann. de l'Ecole Norm. Sup., 1884(3): 3 - 41.

[6]Kneser H. Realle analytische Lösungen der Gleichung $\varphi(\varphi(x))=e^x$ Verwandter Funktional gleichungen. J. Reine Angew Math., 1950(187): 51 - 57.

[7]Mice R E. An upper bound for the order of an iterative root of function. Notices Amer. Math. Soc. 23(Oct. 1976)A—609. Abstract 738 - 810.

[8]Rice R E. Fractional Iterates, Ph. D. Dissertation, University Massachusetts. Amherst, Mass. 1977.

[9]Iterative square roots of Čebyšev Polynomials, Stochastica, 3. 2(1979), 1 - 14.

[10]Rice R E et al. When is $f(f(z))=ax^2+bz+c$? Amer. Math. Monthly, 1980(87): 252 - 263.

[11]Boedewadt U T. Zur Iteration realler Funktionen. Zeitsc. Math. 1944, 49(3): 497 - 516.

[12]Fort M, K Jr. The embedding of homeomorphisms in flows. Proceedings of Amer. Math. Soc., 1955(6): 960 - 967.

[13]Sternberg S. Local contractions and a theorem of Poincare. Amer. J. Math., 1957(79):

809 - 824.

[14]The Structure of Local Homeomorphisms of Euclidean n-space Ⅱ. Amer. J. Math. , 1958(80):623 - 631.

[15]The Structure of Local Homeomorphisms III. Amer. J. Math. ,1959(81):578 - 603.

[16]Schröder E. Ueber Iterate Funktionen. Math,Ann. ,1871(3):296 - 322.

[17]Kuczma M. Functional Equations in a Single Variable. Monografic Matematyczne,Tom 46,Warszwa,1968.

[18]Fractional iteration of differentiable functions. Ann. Polon. Math. ,22(1969/70),217 - 227.

[19] Kuczma M, Smajdor A. Fractional iteration in the class of convex functions. Bull. Acad. Polon. Sci. Sér. Sei. Math. Astronom. Phys. ,1968(16):717 - 720.

[20]Regular fractional iteration. ibid,1971(19):203 - 207.

[21]Reznick B A. A uniqueness criterion for fractional iteration. Ann Polon. Math.

[22]Rosenbaum J. An Iterated Function. Amer. Math. Monthly,1980(87):303.

[23]Hardy G H. Orders of Infinity. Cambridge,1924:31.

[24]Lain P F. Embedding homeomorphisms in differential flows. Colloq. Math. ,1976(35):275 - 287.

[25]Embedding a differentiable homeomorphism in a flow subject to a regularity condition on the derivatives of the positive transition homeomorphisms. J. of Differential Equations,1978(30):31 - 40.

预给二面角的单形嵌入 E^n 的充分必要条件

杨 路 张景中

（中国科学技术大学）

§1 引 言

将预给棱长的单形嵌入于 n 维欧氏空间 E^n 的问题是距离几何的经典问题之一. 这一问题首先由 Fréchet 提出而被 K. Menger[1] 圆满解决. Menger 所找到的构成 n 维单形的充分必要条件是，预给的 $\binom{n+1}{2}$ 条棱长满足一组具有行列式形式的不等式. 作者最近在文献[2]中对于 Menger 的嵌入定理曾给出一个新的证明. 各种互相等价的嵌入条件的介绍可以参看文献[3]；这本专著的主要篇幅都是用于讨论各式各样的嵌入问题. 这一课题直至近期仍然是有兴趣的[4~6]. 本文解决的是一个完全新的类型的嵌入问题. 其提法是：预给了 $\binom{n+1}{2}$ 个二面角，这些角度须且仅必须满足什么样的条件，它们才能实现为欧氏空间中某个 n 维单形的诸内角？这种"预给内角"的嵌入条件形式上和实质上都不同于上列熟知的那些"预给棱长"的嵌入条件.

我们将在 §3，定理 1 中给出的，预给二面角的单形嵌入 E^n 的充分必要条件（ⅰ）～（ⅲ），实际上包括了一个等式和若干个不等式. 根据这一组等式和不等式来对任何预给内角的单形的可嵌入性进行判别完全是切实可行的. 定理 2 就是一个卓有成效的应用例题. 特别要指出的是，定理 1 的条件具有十分简洁的代数形式，这往往可能对我们解决高维空间中较复杂的问题提供较大的方便.

§2 某些引理

引理 1 在 E^n 中取定的重心坐标系中，坐标单形 Ω 的内切球与其各侧面 f_i 的切点 $t_i(i=1,2,\cdots,n+1)$ 的齐次重心坐标为 $(t_{i1},t_{i2},\cdots,t_{in+1})$，其中

$$T_{i1}=\begin{cases}0 & (i=j),\\ |f_i||f_j|\cos^2\dfrac{\theta_{ij}}{2} & (i\neq j),\end{cases}\tag{1}$$

这里 $|f_\alpha|$ 表侧面 f_α 的 $(n-1)$ 维体积，$\theta_{ij}=\widehat{f_if_j}$ 是侧面 f_i，f_j 所夹之内角.（因为是齐次重心

本文刊于《数学学报》，第 26 卷第 2 期，1983 年.

坐标①,故诸分量之和不必为 1,可以差一个比例因子)

证 设内切球半径为 ρ,球心为 O,过 O,t_i,t_j 作二维平面 π_{ij},则 $\pi_{ij}\perp f_i$,$\pi_{ij}\perp f_j$. 设 π_{ij} 与 $f_i\cap f_j$ 交于点 s,又令 s 到 t_i 的距离为 l,则自 t_i 所作的三角形 st_it_j 的高 h_{ij} 也就是 t_i 到 f_j 的距离,当 $i\neq j$ 时,显然有

$$h_{ij}=l\sin\theta_{ij}=\rho\operatorname{ctg}\frac{\theta_{ij}}{2}\cdot\sin\theta_{ij}=2\rho\cos^2\frac{\theta_{ij}}{2}.$$

也就是

$$h_{ij}=\begin{cases}0 & (i=j),\\ 2\rho\cos^2\dfrac{\theta_{ij}}{2} & (i\neq j).\end{cases}\tag{2}$$

再由重心坐标的定义及其齐次性可知(1)成立. 引理 1 证毕.

引理 2 单形的内切球心必然在以各切点为顶点的单形之内部.

证 由引理 1,我们已经知道若以原单形为坐标单形建立重心坐标系,则切点 t_i 之坐标向量为

$$(t_i)=(t_{i1},t_{i2},\cdots,t_{in+1})=|f_i|\left(|f_1|\cos^2\frac{\theta_{1i}}{2},|f_2|\cos^2\frac{\theta_{2i}}{2},\cdots,|f_{n+1}|\cos^2\frac{\theta_{n+1,i}}{2}\right).$$

但由于显然的几何事实:

$$|f_i|=\sum_{\substack{j=1\\ j\neq i}}^{n+1}|f_j|\cos\theta_{ij},$$

故得

$$t_{i1}+t_{i2}+\cdots+t_{i,n+1}=\frac{1}{2}|f_i|\sum_{i=1}^{n+1}|f_j|>0.\tag{3}$$

注意到 $t_{ij}=t_{ji}$

$$\frac{(t_1)+(t_2)+\cdots+(t_{n+1})}{\dfrac{1}{2}\sum\limits_{j=1}^{n+1}|f_j|}=(|f_1|,|f_2|,\cdots,|f_{n+1}|).\tag{4}$$

而(4)的右端正好是内切球心的重心坐标向量. 由(3)可知内切球心的坐标向量是诸 t_i 的坐标向量的正系数线性组合,即球心在$\{t_i\}$为顶点的单形之内部. 引理 2 证毕.

引理 3[7] 设 $\boldsymbol{E}^n$ 中单位球球面上有$(n+1)$个点 $t_1,t_2,\cdots,t_{n+1}$. 令 t_i 与 t_j 之间的球面距离为 φ_{ij},置

$$A=(\cos\varphi_{ij})_{i,j=1}^{n+1}=\begin{bmatrix}1 & & \\ & 1 & \cos\varphi_{ij} \\ & & \ddots & \\ \cos\varphi_{ij} & & & 1\end{bmatrix},\tag{5}$$

而且 A_{ij} 表 A 的相应的代数余子式,则当且仅当有

$$A_{ij}>0\tag{6}$$

①全文中凡提到重心坐标之处,均系指齐次重心坐标.

时，t_i, t_j 两点被其它各点生成的径面分隔于两个半球上.

引理 4 给了 $\binom{n+1}{2}$ 个实数 $0<\varphi_{ij}<\pi(1\leqslant i<j\leqslant n+1)$，则在球面型空间 $S_{n-1,1}$[①] 中存在 $(n+1)$ 个点 $t_1, t_2, \cdots, t_{n+1}$ 使

$$\widehat{t_it_j}=\varphi_{ij}\ (\widehat{t_it_j}\text{表 } t_i \text{ 到 } t_j \text{ 的球面距离}) \tag{7}$$

对一切 $i\neq j$ 成立的充分必要条件是：

（ⅰ）$(n+1)$ 阶对称阵 $A=\begin{bmatrix}1 & & \cos\varphi_{ij}\\ & 1 & \\ & & \ddots \\ \cos\varphi_{ij} & & & 1\end{bmatrix}$ 是正半定的；

（ⅱ）$\det A=0$.

（证明见文献[3].）

§3 预给内角的单形之嵌入定理

下面我们着手证明本文之主要结果

定理 1 给了 $\binom{n+1}{2}$ 个实数 $0<\theta_{ij}<\pi(1\leqslant i<j\leqslant n+1)$，并令 $i\neq j$ 时 $\theta_{ji}=\theta_{ij}$，则当且仅当三个条件

（ⅰ）$(n+1)$ 阶对称阵 $A=\begin{bmatrix}1 & & -\cos\theta_{ij}\\ & 1 & \\ & & \ddots \\ -\cos\theta_{ij} & & & 1\end{bmatrix}$ 是正半定的；

（ⅱ）$\det A=0$；

（ⅲ）对一切 $i\neq j$，A 的代数余子式 $A_{ij}>0$. 同时满足时，在 $\boldsymbol{E}^n$ 中存在单形，它的任意两个侧面 f_i, f_j 所成之内角

$$\widehat{f_if_j}=\theta_{ij}. \tag{8}$$

证 先证条件之必要性，如果已有了 n 维单形 Ω 使 $\widehat{f_if_j}=\theta_{ij}$；取 Ω 的内切球，不失一般性，设它为单位球，诸切点 t_i 在该球上的球面距离应为

$$\varphi_{ij}=\pi-\theta_{ij}. \tag{9}$$

设以 $t_1, t_2, \cdots, t_{n+1}$ 为顶点之单形为 T，则 T 之外心——Ω 之内心必在 T 内部(引理 2). 从而球面上任两切点 t_i, t_j 都被其余各点所生成之径面分隔，由引理 3 知

$$A_{ij}>0$$

对一切 $i\neq j$ 成立，这里

$$A=\begin{bmatrix}1 & & \cos\varphi_{ij}\\ & 1 & \\ & & \ddots \\ \cos\varphi_{ij} & & & 1\end{bmatrix}=\begin{bmatrix}1 & & -\cos\theta_{ij}\\ & 1 & \\ & & \ddots \\ -\cos\theta_{ij} & & & 1\end{bmatrix}, \tag{10}$$

---

① $S_{m,r}$ 表曲率半径为 r 的 m 维球面型空间.

而 A_{ij} 为其代数余子式. 亦即(ⅲ)成立. 根据引理 4,(ⅰ)和(ⅱ)也成立,条件的必要性证毕.

下面往证条件之充分性:

既然(ⅰ)(ⅱ)成立,令 $\varphi_{ij}=\pi-\theta_{ij}$,由引理 4,可以将预给了球面距离 φ_{ij} $(1\leqslant i<j\leqslant n+1)$ 的诸点 $t_1,t_2,\cdots,t_{n+1}$ 嵌入于 $\boldsymbol{E}^n$ 中的某个单位球面 $S_{n-1,1}$ 上,我们首先证明,点集 $T=\{t_1,\cdots,t_{n+1}\}$ 生成 $\boldsymbol{E}^n$ 中的非退化单形,此单形亦记为 T.

设 t_i,t_j 的欧氏距离为 ρ_{ij},则

$$\rho_{ij}=2\sin\frac{\varphi_{ij}}{2}(i,j=1,2,\cdots,n+1).$$

下面的镶边行列式

$$D(T)=\begin{vmatrix} 0 & 1\cdots 1 \\ \begin{matrix}1\\ \vdots\\ 1\end{matrix} & \rho_{ij}^2 \end{vmatrix}$$

通常称为 T 的 Cayley-Menger 行列式,它与 T 的 n 维体积 $V(T)$ 有下列关系(见文献[3]或文献[8]):

$$(V(T))^2=\frac{(-1)^{n+1}}{2^n(n!)^2}D(T).$$

利用 $\rho_{ij}^2=2(1-\cos\varphi_{ij})$ 及行列式性质可算出

$$D(T)=(-2)^n\begin{vmatrix} 0 & 1\cdots 1 \\ \begin{matrix}1\\ \vdots\\ 1\end{matrix} & \cos\varphi_{ij} \end{vmatrix}=(-2)^n\begin{vmatrix} 0 & 1\cdots 1 \\ \begin{matrix}1\\ \vdots\\ 1\end{matrix} & A \end{vmatrix}=(-2)^n\left(-\sum_{i,j=1}^{n+1}A_{ij}\right).$$

由条件(ⅲ)知 $D(T)\neq 0$,于是 $V(T)\neq 0$,即 T 非退化. 以 T 为坐标单形建立重心坐标系,过点 t_i 作 $S_{n-1,1}$ 之切超平面 F_i,F_i 在此坐标系中之方程为

$$c_{i1}x_1+c_{i2}x_2+\cdots+c_{i,n+1}x_{n+1}=0(1\leqslant i\leqslant n+1), \tag{11}$$

此处系数

$$c_{ij}=\begin{cases}0 & (i=j),\\ \cos^2\dfrac{\theta_{ij}}{2} & (i\neq j).\end{cases} \tag{12}$$

事实上,只要应用直角坐标系与重心坐标系的变换公式,即可算出诸等式(12).

令

$$C=(c_{ij})_{i,j=1}^{n+1}=\begin{vmatrix} 0 & & \\ & & \cos^2\dfrac{\theta_{ij}}{2} \\ & 0 & \\ & & \ddots \\ \cos^2\dfrac{\theta_{ij}}{2} & & 0 \end{vmatrix} \tag{13}$$

并以 C_{ij} 表相应的代数余子式,考虑除 F_i 之外的其余几个切超平面的公共点 p_i. p_i 的重心坐标可以由方程组(11)(去掉第 i 个)解出:

$$(p_i)=(C_{i1},C_{i2},\cdots,C_{i,n+1}) \tag{14}$$

而(p_i)的各分量之和为

$$\sum_{i=1}^{n+1} C_{ij} = \begin{vmatrix} c_{11} & c_{12} & \cdots & c_{1,n+1} \\ c_{21} & c_{22} & \cdots & c_{2,n+1} \\ \cdots & \cdots & \cdots & \\ 1, & 1, & \cdots, 1 & \\ \cdots & \cdots & & \\ c_{n+1}, & 1, & \cdots, c_{n+1,n+1} & \end{vmatrix} (\text{第}\ i\ \text{行}). \tag{15}$$

再应用(12)及半角公式即得

$$\sum_{i=1}^{n+1} C_{ij} = \left(-\frac{1}{2}\right)^n \begin{vmatrix} a_{11} & a_{12} & \cdots & a_{1,n+1} \\ a_{21} & a_{22} & \cdots & a_{2,n+1} \\ \cdots & \cdots & \cdots & \\ 1 & 1 & \cdots & 1 \\ \cdots & \cdots & \cdots & \\ a_{n+1,1} & & \cdots & a_{n+1,n+1} \end{vmatrix} (\text{第}\ i\ \text{行}), \tag{16}$$

这里

$$a_{ij} = \begin{cases} 1 & (i=j), \\ -1\cos\theta_{ij} & (i\neq j). \end{cases}$$

参照(10)就有

$$\sum_{i=1}^{n+1} C_{ij} = \left(-\frac{1}{2}\right)^n \sum_{i=1}^{n+1} A_{ij}. \tag{17}$$

于是由条件(ⅲ)可知

$$\operatorname{sgn}\left(\sum_{j=1}^{n+1}\right) = (-1)^n, \tag{18}$$

亦即对 $i=1,2,\cdots,n+1$,(p_i)的分量之和具有相同的符号.

又由于 $c_{ij}\geqslant 0(i,j=1,2,\cdots,n+1)$,故由(令$|C|=\det C$)

$$\sum_{j=1}^{n+1} C_{ij}(p_j) = (0,0,\cdots,0,\underset{(\text{第}i\text{列})}{C},0,\cdots,0) \tag{19}$$

知顶点 t_i 的重心坐标可表为诸 p_j 的重心坐标的非负系数的线性组合,从而诸 t_i 在$\{p_1,p_2,\cdots,p_{n+1}\}$所生成的单形的内部或边界上,于是单形 T 的内点也都是单形

$$Q=p_1p_2\cdots p_{n+1}$$

的内点,由引理 3,T 外心 O 应当在 T 的内部,于是 O 也在 Ω 的内部,即 T 的外接球 $S_{n-1,1}$ 正是 Ω 的内切球. 此时 Ω 之诸内角

$$\widehat{F_iF_j}=\pi-\varphi_{ij}=\theta_{ij}(1\leqslant i<j\leqslant n+1).$$

故 Ω 正是我们所要的单形. 定理 1 证毕.

§4　一个应用

大家都知道,欧氏平面上任一三角形内角和为 π,所以一个三角形的三个角如果分别不

超过另一三角形的对应角的话，这两个三角形必然相似.

但是，在高维情形，类似的"内角和定理"不再成立了. 例如：我们很容易构造出两个四面体，它们的 6 个二面角之和是不相等的.

那么，我们能不能把一个四面体的各个二面角都稍稍减少一点，作出一个新的四面体呢？直观上像是不可能的. 下面，我们应用定理 1 给这个直观猜测的最普遍情形以严格证明.

定理 2　设 Ω 和 Ω^* 是 $\boldsymbol{E}^n$ 中的两个单形，以 θ_{ij}，θ_{ij}^* 分别记它们的诸内二面角. 如果对一切 $i\neq j(i,j=1,2,\cdots,n+1)$ 都有 $\theta_{ij}\leqslant\theta_{ij}^*$，则必有

$$\theta_{ij}=\theta_{ij}^*\ (i\neq j,i,j=1,2,\cdots,n+1).$$

即 Ω 与 Ω^* 相似.

证　考虑两个 $n+1$ 阶对称阵：

$$A=\begin{bmatrix}1 & & & \\ & & -\cos\theta_{ij} & \\ & 1 & & \\ & & \ddots & \\ -\cos\theta_{ij} & & & \\ & & & 1\end{bmatrix}=(a_{ij})_{i,j=1}^{n+1}$$

和

$$B=\begin{bmatrix}1 & & & \\ & & -\cos\theta_{ij}^* & \\ & 1 & & \\ & & \ddots & \\ -\cos\theta_{ij}^* & & & \\ & & & 1\end{bmatrix}=(b_{ij})_{i,j=1}^{n+1}.$$

再作一个含未知数 λ 的方程

$$\det(A+\lambda B)=0,\tag{20}$$

并令 A_{ij} 和 B_{ij} 表 A 和 B 的对应的代数余子式. 又令 $|A|=\det A$，$|B|=\det B$，将(20)展开得

$$|B|\lambda^{n+1}+\Big(\sum_{i,j=1}^{n+1}a_{ij}B_{ij}\Big)\lambda^n+\cdots+\Big(\sum_{i,j=1}^{n+1}b_{ij}A_{ij}\Big)\lambda+|A|=0.\tag{21}$$

由于 A,B 都是正半定的(定理 1 的必要条件(ⅰ))，方程(21)的根都是非正的，故(21)的任意两个系数不可能反号，特别是应当有

$$\Big(\sum_{i,j=1}^{n+1}a_{ij}B_{ij}\Big)\Big(\sum_{i,j=1}^{n+1}b_{ij}A_{ij}\Big)\geqslant 0.\tag{22}$$

但是由定理 1 必要条件(ⅱ)有

$$\begin{cases}\displaystyle\sum_{i,j=1}^{n+1}a_{ij}A_{ij}=(n+1)\det A=0,\\ \displaystyle\sum_{i,j=1}^{n+1}b_{ij}B_{ij}=(n+1)\det B=0.\end{cases}\tag{23}$$

另一方面从条件 $\theta_{ij}\leqslant\theta_{ij}^*$ 可以推出 $-\cos\theta_{ij}\leqslant-\cos\theta_{ij}^*$ (注意到 $\cos\theta$ 在 $[0,\pi]$ 上是单调递减的)，即

$$a_{ij}\leqslant b_{ij}.\tag{24}$$

再由定理 1 必要条件(ⅲ)$A_{ij}>0, B_{ij}>0$ 可知,除非对一切 i,j 有 $a_{ij}=b_{ij}$,由(24)及(23)将导出

$$\sum_{i,j=1}^{n+1} a_{ij}B_{ij} < \sum_{i,j=1}^{n+1} b_{ij}B_{ij} = 0$$

和

$$\sum_{i,j=1}^{n+1} b_{ij}A_{ij} > \sum_{i,j=1}^{n+1} a_{ij}A_{ij} = 0,$$

即 $\sum_{i,j=1}^{n+1} a_{ij}B_{ij}$ 与 $\sum_{i,j=1}^{n+1} b_{ij}A_{ij}$ 符号相反,此与(22)矛盾. 于是 $a_{ij}=b_{ij}$ 对一切 $i,j=1,\cdots,n+1$ 成立. 在这里我们又运用了文献[9, §§ 2～3]中的代数技巧. 定理 2 证毕.

参考文献

[1]Menger K. Untersuchungen über allgemeine Metrik. Math. Ann. ,1928(100):75 - 163.

[2]杨路,张景中. 单纯形构造定理的一个证明. 数学的实践与认识,1980(1):43 - 45.

[3]Blumenthal L M. Theory and Applications of Distance Geometry. 2nd ed. New York,1970.

[4]张景中,杨路. 有限点集在伪欧空间的等长嵌入. 数学学报,1980,23(4):481 - 487.

[5]Schoenberg I J. Linkages and Distance Geometry. Nederl,Akad. Wetensch Proc. Ser. A Math. ,1969(31):43 - 52.

[6]Morgan C L. Embedding Metric Spaces in Euclidean Space. J. Geometry,1974(5):101 - 107.

[7]Blumenthal L M,et al. Distribution of Points in n-Space. Amer. Math. Monthly. 1943(50):181 - 185.

[8]杨路,张景中. 关于有限点集的一类几何不等式. 数学学报,1980,23(5):740 - 749.

[9]杨路,张景中. Neuberg-Pedoe 不等式的高维推广及应用. 数学学报,1981,24(3):401 - 408.

Metric Spaces Which Cannot Be Isometrically Embedded in Hilbert Space

Yang Lu and Zhong Jing-zhong

(*University of Science and Technology of China*)

Let $A_1A_2A_3A_4$ be a planar convex quadrangle with diagonals A_1A_3 and A_2A_4. Is there a quadrangle $B_1B_2B_3B_4$ in Euclidean space such that $A_1A_3<B_1B_3$, $A_2A_4<B_2B_4$ but $A_iA_j>B_iB_j$ for other edges?

The answer is "no". It seems to be obvious but the proof is more difficult. In this paper we shall solve similar more complicated problems by using a higher dimensional geometric inequality which is a generalisation of the well-known Pedoe inequality (Proc. Cambridge Philos. Soc. 38(1942), 397—398) and an interesting result by L. M. Blumenthal and B. E. Gillam (Amer. Math. Monthly 50(1943), 181—185).

1. Definitions and main result

DEFINITION 1 Let $G=\{A_1, A_2, \cdots, A_{n+2}\}$ be an $(n+2)$-tuple in E^n. An edge A_iA_j of G is called "red" or "blue" if there exists uniquely a hyperplane $\pi_{ij}(G)$ containing $G\backslash\{A_i, A_j\}$ such that A_i and A_j lie to the opposite sides or the same side of $\pi_{ij}(G)$, respectively.

Some edges, of course, may be neither red nor blue.

DEFINITION 2 Let G be an $(n+2)$-tuple in E^n, (M, d) a semimetric space. A mapping $f: G\to(M, d)$, satisfying

(ⅰ) $|A_i-A_j|\leqslant d(f(A_i), f(A_j))$ if A_iA_j is a red edge of G,

(ⅱ) $|A_i-A_j|\geqslant d(f(A_i), f(A_j))$ if A_iA_j is a blue edge of G,

and the strict inequality holding at least for one edge red or blue, is called a "skew mapping" of G into (M, d). $f(G)$ is called a "skew image" of G, and G is called a "skew inverse image" of $f(G)$.

The following theorem gives a geometric condition under which a metric space (M, d) cannot be isometrically embedded in Hilbert space.

THEOREM 1. *If a metric space* (M, d) *contains a finite subset* R *which has a skew*

本文刊于《Bull. Austral. Math. Soc.》,第3卷,1984年.

inverse image in Euclidean space. then (M,d) cannot be isometrically embedded in Hilbert space l^2.

We shall prove this assertion in Section 3. Furthermore, its converse theorem is true for separable metric spaces. In fact, the authors have proved in [6] that a separable metric space which cannot be isometrically embedded in l^2 must contain a finite subset which has a skew inverse image in Euclidean space.

The proof [6] of the converse theorem, however, is very long and much more difficult than Theorem 1 itself so we need not repeat it here. The purpose of this note is only to prove Theorem 1 which is sufficient to answer the type of problems analogous to the one posed at the beginning of the present paper.

2. Notations and lemmas

Let $G=\{A_1,A_2,\cdots,A_{n+2}\}$ and $R=\{B_1,B_2,\cdots,B_{n+2}\}$ be two $(n+2)$-tuples in E^{n+1}, $a_{ij}=|A_i-A_j|$, $b_{ij}=|B_i-B_j|$, $(i,j=1,2,\cdots,n+2)$. By A, B denote the values of the determinants of the following two bordered matrices. respectively:

$$A=\begin{vmatrix} 0 & 1 & 1 & \cdots & 1 & 1 \\ 1 & & & & & \\ 1 & & & -\frac{1}{2}a_{ij}^2 & & \\ \vdots & & & & & \\ 1 & & & & & \end{vmatrix},\quad B=\begin{vmatrix} 0 & 1 & 1 & \cdots & 1 & 1 \\ 1 & & & & & \\ 1 & & & -\frac{1}{2}b_{ij}^2 & & \\ \vdots & & & & & \\ 1 & & & & & \end{vmatrix}. \tag{1}$$

By A_{ij} and B_{ij} denote the cofactors of $-\frac{1}{2}a_{ij}^2$ in A and $-\frac{1}{2}b_{ij}^2$ in B, $(i,j=1,2,\cdots,n+2)$, respectively.

LEMMA 1

$$\sum_{i=1}^{n+2}\sum_{j=1}^{n+2} a_{ij}^2 B_{ij} \geqslant 0,\ \sum_{i=1}^{n+2}\sum_{j=1}^{n+2} b_{ij}^2 A_{ij} \geqslant 0. \tag{2}$$

Proof. If G and R span two non-degenerate simplices in E^{n+1}, denoting by $V(G)$ and $V(R)$ the volumes of G and R, we have([4], P. 204, Theorem 1, or [5])

$$\sum_{i=1}^{n+2}\sum_{j=1}^{n+2} a_{ij}^2 B_{ij} \geqslant 2(n+1)[(n+1)!]^2 V(G)^{2/(n+1)} V(R)^{2-2/(n+1)}. \tag{3}$$

This is a generalisation of the Neuberg-Pedoe inequality which is the case $n=1$ in (3).

It is obvious by continuity that (3) holds still when G or R is degenerate; hence

$$\sum_{i=1}^{n+2}\sum_{j=1}^{n+2} a_{ij}^2 B_{ij} \geqslant 0,$$

analogously

$$\sum_{i=1}^{n+2}\sum_{j=1}^{n+2} b_{ij}^2 A_{ij} \geqslant 0.$$

LEMMA 2 *If $G=\{A_1, A_2, \cdots, A_{n+2}\}$ is an $(n+2)$-tuple in E^n and some cofactor A_{ij} in A is non-vanishing, then A_i and A_j lie to the opposite sides or the same side of the hyperpane $\pi_{ij}(G)$ when $A_{ij}<0$ or $A_{ij}>0$.*

This lemma is due to Blumenthal and Gillam([2], P. 183, Theorem 3. 1). There are merely a few differences of notation between the two statements.

LEMMA 3 *Let $G=\{A_1, A_2, \cdots, A_{n+2}\}$ be an $(n+2)$-tuple in E^n. If an edge A_iA_j is red or blue, then the corresponding cofactor A_{ij} is non-vanishing.*

Proof. We apply the following algebraic identity (4) which is very useful in distance geometry([1], § 41, P. 100). Let D be a symmetric determinant, D_{ii}, D_{jj} and D_{ij} be the corresponding cofactors in D, and D_{jj}^{ii} be the sub-determinant obtained by deleting the ith row, the ith column, the jth row and the jth column from D. Then, for $i \neq j$,

$$D_{ii}D_{jj}-D_{ij}^2=D \cdot D_{jj}^{ii}. \tag{4}$$

Now we apply this well-known identity to determinant A. It has been shown([4], P. 206, (1. 10)) that

$$A=-((n+1)!\ V(G))^2 \tag{5}$$

where $V(G)$ denotes the $(n+1)$-dimensional volume of the simplex spanned by G. Since G is an $(n+2)$-tuple in E^n, this simplex must be degenerate; hence $V(G)=0$ and so $A=0$. It follows that

$$A_{ii}A_{jj}-A_{ij}^2=0. \tag{6}$$

Suppose $A_{ij}=0$ for a certain i and a certain j, then either $A_{ii}=0$ or $A_{jj}=0$. Hence either A_j or A_i lies in the hyperplane $\pi_{ij}(G)$. (Since, by analogue with (5) we have $A_{ii}=-(n!\ V(G\backslash\{A_i\}))^2$, $A_{ii}=0$ implies that the simplex spanned by $G\backslash\{A_i\}$ is degenerate and the points of $G\backslash\{A_i\}$ including A_j lie in the same hyperplane which is just $\pi_{ij}(G)$.)

But, in this case, according to Definition 1, the edge A_iA_j is neither red nor blue, contradicting the hypothesis, and Lemma 3 has been proved.

3. Proof of Theorem 1

We use reduction to absurdity. Suppose a metric space (M, d) has been isometrically embedded in l^2 and there exists a finite subset R of M with a skew inverse image G in Euclidean space. From this we conclude that there exists $G=\{A_1, A_2, \cdots, A_{n+2}\}$ in E^n and $R=\{B_1, B_2, \cdots, B_{n+2}\}$ in l^2 such that

(ⅰ) $|A_i-A_j| \leqslant |B_i-B_j|$ if A_iA_j is red,

(ⅱ) $|A_i-A_j| \geqslant |B_i-B_j|$ if A_iA_j is blue,

and the strict inequality holds at least for one edge A_iA_j is red or blue.

Clearly, $G \subset E^n \subset E^{n+1}$ and $R \subset E^{n+1}$ because the widest position occupied by $n+2$ points of l^2 is only $(n+1)$-dimensional. We use the same notation as in Lemma 1: $a_{ij}=|A_i-A_j|$,

$b_{ij}=|B_i-B_j|$, and so on.

Since $G\subset E^n$ implies $A=0$ (by formula (5)), by simple calculation we have

$$\sum_{i=1}^{n+2}\sum_{j=1}^{n+2}a_{ij}^2A_{ij}=0, \tag{7}$$

and applying Lemma 1 we obtain

$$\sum_{i=1}^{n+2}\sum_{j=1}^{n+2}b_{ij}^2A_{ij}\geqslant 0=\sum_{i=1}^{n+2}\sum_{j=1}^{n+2}a_{ij}^2A_{ij}=0;$$

that is

$$\sum_{i=1}^{n+2}\sum_{j=1}^{n+2}(b_{ij}^2-a_{ij}^2)A_{ij}\geqslant 0, \tag{8}$$

First it is easy to verify that every term of the left side of (8) is non-positive:

when $A_{ij}=0$, $(b_{ij}^2-a_{ij}^2)A_{ij}=0$ and when $A_{ij}>0$, by Lemma 2 we know that A_iA_j is blue and by hypothesis $a_{ij}\geqslant b_{ij}$ so we have $(b_{ij}^2-a_{ij}^2)A_{ij}\leqslant 0$;

when $A_{ij}<0$, A_iA_j is red and by hypothesis $a_{ij}\leqslant b_{ij}$ and we have $(b_{ij}^2-a_{ij}^2)A_{ij}\leqslant 0$.

Then, according to the hypothesis of Theorem 1 and Definition 2, there exists at least one red or blue edge A_iA_j such that $a_{ij}\neq b_{ij}$. By Lemma 3 there exists at least one non-vanishing term of the left side of (8). We obtain

$$\sum_{i=1}^{n+2}\sum_{j=1}^{n+2}(b_{ij}^2-a_{ij}^2)A_{ij}<0, \tag{9}$$

which contradicts (8). This contradiction shows that (M, d) cannot be isometrically embedded in l^2 and the proof of Theorem 1 is complete.

4. A type of problem involving two metric point sets

Now let us answer the quadrangles problem which was posed at the beginning of the paper. Clearly, the mapping $A_1A_2A_3A_4\rightarrow B_1B_2B_3B_4$ is a skew mapping. According to Theorem 1, it is not possible to realize such a quadrangle in Euclidean space.

Of course, Theorem 1 may be applied to solve more complicated problems. For example: let Ω be a convex 6-faced polyhedron with vertices A_1, A_2, A_3, A_4, A_5 in E^3, such that Ω can be dissected into two tetrahedrons $A_1A_2A_3A_4$ and $A_1A_2A_3A_5$. Is there a 5-tuple $\Omega^*=\{B_1, B_2, B_3, B_4, B_5\}$ in E^4 such that $A_1A_2<B_1B_2$, $A_2A_3<B_2B_3$, $A_3A_1<B_3B_1$, $A_4A_5<B_4B_5$ but $A_iA_j>B_iB_j$ for other edges?

It can be seen easily that $A_1A_2, A_2A_3, A_3A_1, A_4A_5$ are red edges of Ω and other edges of Ω are blue. The mapping $A_1A_2A_3A_4A_5\rightarrow B_1B_2B_3B_4B_5$, therefore, is a skew mapping. By Theorem 1 we can assert that it is impossible to realize such a 5-tuple Ω^* in E^4.

There are a variety of conditions, each of which is necessary and sufficient to embed isometrically a metric space in Euclidean or Hilbert space; nevertheless, it is usually difficult to decide practically whether some given metric point set is embeddable or not.

Inequalities involving two metric point sets are often of great use for our work.

References

[1]L M Blumenthal. Theory and applications of distance geometry. Chelsea, New York, 1970.
[2]L M Blumenthal, B E Gillam. Distribution of points in n-space. Amer. Math. Monthly, 1943(50):181 - 185.
[3]D Pedoe. An inequality for two triangles. Proc. Cambridge Philos. Soc. ,1942(38):397 - 398.
[4]Yang Lu, Zhang Jingzhong. A generalisation to several dimensions of the Neuberg-Pedoe inequality, with applications. Bull. Austral. Math. Soc. ,1983(27):203 - 214.
[5]Yang Lu, Zhang Jingzhong. A high-dimensional extension of the Neuberg-Pedoe inequality its application(Chinese). Acta Math. Sinica. ,1981(24):401 - 408.
[6]Yang Lu, Zhang Jingzhong. A geometric criterion of metric embedding and the skew mapping, submitted.

关于空间曲线的 Johnson 猜想

杨　路　张景中
（中国科学技术大学）

一、引　　言

本文旨在对 Johnson 所提出的一个问题以及该问题的离散形式以肯定的回答.

数年前，Johnson[1]提出了涉及曲线大范围性质的一个猜想：设 $\boldsymbol{r}(s)$ 是一条闭曲线的自然参数表示，曲线长度为 L. 又设 p 是一个确定的正数，$0<p<L$. 考虑曲线上那些长度为 p 的弧所张的弦 $|\boldsymbol{r}(s+p)-\boldsymbol{r}(s)|$，令

$$d(p)=\min_{s}|\boldsymbol{r}(s+p)-\boldsymbol{r}(s)|. \tag{1.1}$$

则有

$$d(p)\leqslant\frac{L}{\pi}\sin\frac{\pi p}{L}, \tag{1.2}$$

等号成立仅当这曲线是一个圆.

Johnson 猜想是 Herda 问题的自然推广. 在(1.2)式中取 $p=\frac{L}{2}$ 可得不等式

$$d\left(\frac{L}{2}\right)<\frac{L}{\pi}. \tag{1.3}$$

这首先是被 Herda[2]所猜测而后为许多作者所证实的. Herda 将 $d\left(\frac{L}{2}\right)$ 叫做曲线的“伪直径”并将他的猜想叙述为：闭曲线的伪直径不超过 L/π，且仅当曲线是圆时其伪直径等于 L/π.

Herda 问题在各个不同的方向上被推广了[3～6]. 他本人已将这一结果推广到满足一定条件的闭曲面[7].

Alexander[8]提出了 Herda 问题和 Johnson 问题的离散形式. 他借助于度量平均的方法解决了离散的 Herda 问题并自信其方法对于 Johnson 猜想也能奏效，但最终未能给出相应的证明，无论是离散形式的或者连续形式的.

本文将对离散和连续两方面的问题分别给出解答，而不采取文献[8]中所建议的将离散结果连续化的步骤.

本文刊于《科学通报》，第 6 期，1984 年.

二、离散形式的 Johnson 猜想

定义 1　由欧氏空间中 n 个点构成的点列$\{\boldsymbol{r}_0,\boldsymbol{r}_1,\cdots,\boldsymbol{r}_{n-1}\}$叫做一个 n 链,如果$|\boldsymbol{r}_{i+1}-\boldsymbol{r}_i|=1$. 这里约定对一切整数 i 有 $\boldsymbol{r}_{i+n}=\boldsymbol{r}_i$. 一个 n 链叫做凸的,如果它构成平面凸 n 角形.

定义 2　对于欧氏空间中给定的 m 个项数相同的点列

$$\mathfrak{S}_k=\{\boldsymbol{r}_0^k,\boldsymbol{r}_1^k,\cdots,\boldsymbol{r}_{n-1}^k\},k=1,2,\cdots,m, \tag{2.1}$$

如果能找到一个点列 $\overline{\mathfrak{S}}=\{\overline{\boldsymbol{r}}_0,\overline{\boldsymbol{r}}_1,\cdots,\overline{\boldsymbol{r}}_{n-1}\}$,使

$$|\overline{\boldsymbol{r}}_i-\overline{\boldsymbol{r}}_j|^2=\frac{1}{m}\sum_{k=1}^{m}|\boldsymbol{r}_i^k-\boldsymbol{r}_j^k|^2, \tag{2.2}$$

对 $i,j=0,1,\cdots,n-1$ 都成立,我们就说 $\overline{\mathfrak{S}}$ 是这组点列$\{\mathfrak{S}_1,\mathfrak{S}_2,\cdots,\mathfrak{S}_m)$的"度量平均".

下面这个引理是距离几何中的已知结果:

引理 1[8~9]　对于欧氏空间中给定的 m 个项数相同的点列,它们的度量平均一定在欧氏空间中存在,但可能具有较高的维数.

离散形式的 Johnson 猜想,Alexander[8]是这样的提法:设$\mathfrak{S}=\{\boldsymbol{r}_0,\boldsymbol{r}_1,\cdots,\boldsymbol{r}_{n-1})$是一个 n 链;p 是一个确定的自然数,$1<p<n$. 令

$$d(p)=\min_i|\boldsymbol{r}_i-\boldsymbol{r}_{i+p}|. \tag{2.3}$$

则有

$$d(p)\leqslant\sin\frac{\pi p}{n}\Big/\sin\frac{\pi}{n}, \tag{2.4}$$

等号成立的充要条件是 $\boldsymbol{r}_0,\boldsymbol{r}_1,\cdots,\boldsymbol{r}_{n-1}$ 构成平面正多角形.

为证明此猜想,下述的已知结果是有用的:

引理 2(本文作者,见文献[10]中 Corollary 1.)　对于欧氏空间中给定的 m 个项数均为 n 的点列$\mathfrak{S}_1,\mathfrak{S}_2,\cdots,\mathfrak{S}_m$ 及其度量平均 $\overline{\mathfrak{S}}$,将 $\overline{\mathfrak{S}}$ 和每个$\mathfrak{S}_k$ 所张单形的$(n-1)$维体积分别记为 $\overline{V}$ 和 V_k(当然,其中某些单形可能是退化的,这时所对应的体积为零),则有

$$\overline{V}^{\frac{2}{n-2}}\geqslant\frac{1}{m}\sum_{k=1}^{m}V_k^{\frac{2}{n-1}}, \tag{2.5}$$

等号成立的充要条件是:这些点列中的每一个同其它任一个在欧几里得意义下是相似对应的.

我们还需要 Sallee 的一个近期结果:

引理 3[8,11]　对于一个非凸的 n 链 $\boldsymbol{r}_0,\boldsymbol{r}_1,\cdots,\boldsymbol{r}_{n-1}$,必存在一个凸的 n 链 $\boldsymbol{r}_0^*,\boldsymbol{r}_1^*,\cdots,\boldsymbol{r}_{n-1}^*$ 使得

$$|\boldsymbol{r}_i-\boldsymbol{r}_j|<|\boldsymbol{r}_i^*-\boldsymbol{r}_j^*| \tag{2.6}$$

对一切使$|i=j|\neq1(\bmod n)$的 i,j 成立.

上面这个引理叫做"弦的伸展定理".

现在我们来建立一个比(2.4)式更强的命题:

定理 1　设$\mathfrak{S}=\{\boldsymbol{r}_0,\boldsymbol{r}_1,\cdots,\boldsymbol{r}_{n-1}\}$是一个 n 链;p 是一个确定的自然数,$1<p<n$,令

$$\bar{d}(p,\mathfrak{S})=\left(\frac{1}{n}\sum_{i=0}^{n-1}|\boldsymbol{r}_i-\boldsymbol{r}_{i+p}|^2\right)^{\frac{1}{2}}. \tag{2.7}$$

则

$$\bar{d}(p,\mathfrak{S})\leqslant\sin\frac{\pi p}{n}\Big/\sin\frac{\pi}{n}, \tag{2.8}$$

等号成立的充要条件是$\mathfrak{S}$构成平面正多角形.

证 由紧致性可知在所有的 n 链中必存在一个 n 链$\mathfrak{S}_1$使得

$$\bar{d}(p,\mathfrak{S}_1)=\max_{\mathfrak{S}}\bar{d}(p,\mathfrak{S}). \tag{2.9}$$

设$\mathfrak{S}_1=\{\boldsymbol{r}_0,\boldsymbol{r}_1,\cdots,\boldsymbol{r}_{n-1}\}$. 易知$\mathfrak{S}_1$ 一定是凸链,否则由引理 3,它不能使 $\bar{d}(p)$取到最大值,与所设矛盾. 今另构造$(n-1)$个凸链如下:

$$\mathfrak{S}_2=\{\boldsymbol{r}_1,\boldsymbol{r}_2,\cdots,\boldsymbol{r}_{n-1},\boldsymbol{r}_0\},$$
$$\mathfrak{S}_3=\{\boldsymbol{r}_2,\boldsymbol{r}_3,\cdots,\boldsymbol{r}_0,\boldsymbol{r}_1\},$$
$$\cdots\cdots$$
$$\mathfrak{S}_n=\{\boldsymbol{r}_{n-1},\boldsymbol{r}_0,\cdots,\boldsymbol{r}_{n-3},\boldsymbol{r}_{n-2}\}.$$

对诸点列$\{\mathfrak{S}_1,\mathfrak{S}_2,\cdots,\mathfrak{S}_n\}$作度量平均,设为

$$\bar{\mathfrak{S}}=\{\bar{\boldsymbol{r}}_0,\bar{\boldsymbol{r}}_1,\cdots,\bar{\boldsymbol{r}}_{n-1}\}.$$

由定义有

$$|\bar{\boldsymbol{r}}_i-\bar{\boldsymbol{r}}_j|^2=\frac{1}{n}\sum_{k=0}^{n-1}|\boldsymbol{r}_{i+k}-\boldsymbol{r}_{j+k}|^2. \tag{2.10}$$

特别有

$$|\bar{\boldsymbol{r}}_i-\bar{\boldsymbol{r}}_{i+p}|^2=\frac{1}{n}\sum_{k=0}^{n-1}|\boldsymbol{r}_{i+k}-\boldsymbol{r}_{i+k+p}|^2=(\bar{d}(p,\mathfrak{S}))^2; \tag{2.11}$$

$$|\bar{\boldsymbol{r}}_i-\bar{\boldsymbol{r}}_{i+1}|^2=\frac{1}{n}\sum_{k=0}^{n-1}|\boldsymbol{r}_{i+k}-\boldsymbol{r}_{i+k+1}|^2=1. \tag{2.12}$$

可见 $\bar{\mathfrak{S}}$ 也是一个 n 链,并有

$$\bar{d}(p,\bar{\mathfrak{S}}))=|\bar{\boldsymbol{r}}_i-\bar{\boldsymbol{r}}_{i+p}|=\bar{d}(p,\mathfrak{S}_1). \tag{2.13}$$

前面已经讲过$\mathfrak{S}_1$ 构成平面凸多角形,从而$\mathfrak{S}_2,\mathfrak{S}_3,\cdots,\mathfrak{S}_n$ 也都是. 如果能证明这些多角形中的每一个同其它任何一个都是相似对应的,那么$\mathfrak{S}_1$ 的 n 个内角均相等,从而$\mathfrak{S}_1$ 是一个正则多角形,我们的目的就达到了.

用反证法. 如果上述的两两相似对应不全成立,由引理 2,这时有

$$\bar{V}^{\frac{2}{n-1}}>\frac{1}{n}\sum_{k=1}^{n}V_k^{\frac{2}{n-1}}, \tag{2.14}$$

即 $\bar{\mathfrak{S}}$ 具有$(n-1)$维正体积. 当 $n\geqslant 4$ 时 $\bar{\mathfrak{S}}$ 不在一个平面上,自然更不会是凸链. 于是由引理 3 必存在凸链$\mathfrak{S}^*=\{r_0^*,r_1^*,\cdots,r_{n-1}^*\}$使

$$|\bar{\boldsymbol{r}}_i-\bar{\boldsymbol{r}}_{i+p}|<|\boldsymbol{r}_i^*-\boldsymbol{r}_{i+p}^*| \tag{2.15}$$

对一切 i 成立. 从而有

$$\bar{d}(p,\mathfrak{S}_1)=\bar{d}(p,\bar{\mathfrak{S}})>\bar{d}(p,\mathfrak{S}^*). \tag{2.16}$$

这说明 $\overline{d}(p,\mathfrak{S}_1)<\max\limits_{\mathfrak{S}}\overline{d}(p,\mathfrak{S})$，与所设矛盾. 故 $\mathfrak{S}_1$ 必须是正则多角形. 而对正则多角形显然有 $\overline{d}(p,\mathfrak{S}_1)=\sin\frac{\pi p}{n}\Big/\sin\frac{\pi}{n}$. 定理 1 证毕.

注意到 $d(p)\leqslant\overline{d}(p)$，从定理 1 立即得到

系 1 离散形式的 Johnson 猜想(2.4)式是真的.

三、Fourier 级数的应用

为了证明引言中提出的 Johnson 猜想(1.2)式，可以采用与证明(2.4)式平行的推导过程. 连续曲线的度量平均将在 Lie 群上进行，这个度量平均将是 Hilbert 空间中的螺线. 然后运用 von Neumann[12] 和 Schoenberg 关于 Hilbert 空间的螺线和螺旋函数的解析理论来获得所要的结果. 但那样做必将涉及若干较为深远的概念. 我们这里给出的证明是比较直接的.

定理 2 设 C 是一条周长为 L 的空间闭曲线，其自然参数表示为 $\boldsymbol{r}=\boldsymbol{r}(s)$. 考虑 C 的那些长度为 p 的弧所张的弦 $|\boldsymbol{r}(s+p)-\boldsymbol{r}(s)|$，这里 p 是一个确定的正数并有 $0<p<L$. 将所有这些弦的长度 $|\boldsymbol{r}(s+p)-\boldsymbol{r}(s)|$ 对 s 作二阶平均

$$F(p)=\left(\frac{1}{L}\int_0^L|\boldsymbol{r}(s+p)-\boldsymbol{r}(s)|^2ds\right)^{\frac{1}{2}}. \tag{3.1}$$

则有

$$F(p)\leqslant\frac{L}{\pi}\sin\frac{\pi p}{L}, \tag{3.2}$$

而且等号仅当 C 是圆时成立.

证 不妨设 $L=2\pi$，这时我们只需证

$$F(p)\leqslant 2\sin\frac{p}{2}. \tag{3.3}$$

设 C 之重心为坐标原点，将 $\boldsymbol{r}(s)$ 展为富氏级数

$$\boldsymbol{r}(s)=\sum_{k=1}^{\infty}\boldsymbol{a}_k\cos ks+\boldsymbol{b}_k\sin ks, \tag{3.4}$$

这里缺零次项是因已设 C 的重心就是原点.

$$\begin{aligned}\boldsymbol{r}(s+p)-\boldsymbol{r}(s)&=\sum_{k=1}^{\infty}\boldsymbol{a}_k(\cos k(s+p)-\cos ks)+\boldsymbol{b}_k(\sin k(s+p)-\sin ks)\\&=\sum_{k=1}^{\infty}2\sin\frac{kp}{2}\left(\boldsymbol{b}_k\cos k\left(s+\frac{p}{2}\right)-\boldsymbol{a}_k\sin k\left(s+\frac{p}{2}\right)\right).\end{aligned} \tag{3.5}$$

由 Parseval 等式

$$\frac{1}{n}\int_0^{2\pi}|\boldsymbol{r}(s+p)-\boldsymbol{r}(s)|^2\mathrm{d}s=4\sum_{k=1}^{\infty}(|\boldsymbol{a}_k|^2+|\boldsymbol{b}_k|^2)\sin^2\frac{kp}{2}. \tag{3.6}$$

故

$$F^2(p)=2\sum_{k=1}^{\infty}(|\boldsymbol{a}_k|^2+|\boldsymbol{b}_k|^2)\sin^2\frac{kp}{2}. \tag{3.7}$$

$$\because \left|\sin\frac{kp}{2}\right| \leqslant k\sin\frac{p}{2}(0<p<2\pi;k=1,2,3,\cdots), \tag{3.8}$$

$$\therefore F^2(p)=2\sum_{k=1}^{\infty}k^2(|\boldsymbol{a}_k|^2+|\boldsymbol{b}_k|^2)\sin^2\left(\frac{p}{2}\right). \tag{3.9}$$

另一方面,对(3.4)式的导级数用 Parseval 等式有

$$\sum_{k=1}^{\infty}k^2(|\boldsymbol{a}_k|^2+|\boldsymbol{b}_k|^2)=\frac{1}{n}\int_0^{2\pi}|\boldsymbol{r}'(s)|^2ds=2. \tag{3.10}$$

将(3.10)式代入(3.9)式中得到

$$F(p)\leqslant 2\sin\frac{p}{2},$$

此即本定理所求证的不等式(3.3).

下面证明若上式等号成立,C 必是一个圆. 注意到(3.8)式中的等号只对 $k=1$ 成立,可知(3.9)式的等号成立当且仅当

$$|\boldsymbol{a}_2|=|\boldsymbol{b}_2|=|\boldsymbol{a}_3|=|\boldsymbol{b}_3|=\cdots=0. \tag{3.11}$$

这时有

$$\boldsymbol{r}(s)=\boldsymbol{a}_1\cos s+\boldsymbol{b}_1\sin s, \tag{3.12}$$

$$\boldsymbol{r}'(s)=\boldsymbol{b}_1\cos s-\boldsymbol{a}_1\sin s, \tag{3.13}$$

$$|\boldsymbol{r}'(s)|^2=|\boldsymbol{a}_1|^2\sin^2 s+|\boldsymbol{b}_1|^2\cos^2 s-\boldsymbol{a}_1\cdot\boldsymbol{b}_1\sin 2s. \tag{3.14}$$

微分之

$$(|\boldsymbol{a}_1|^2-|\boldsymbol{b}_1|^2)\sin 2s-2\boldsymbol{a}_1\cdot\boldsymbol{b}_1\cos 2s=0.$$

即

$$(|\boldsymbol{a}_1|^2-|\boldsymbol{b}_1|^2)\mathrm{tg}\,2s=2\boldsymbol{a}_1\cdot\boldsymbol{b}_1 \tag{3.15}$$

对一切 s 成立. 这导致

$$|\boldsymbol{a}_1|^2-|\boldsymbol{b}_1|^2=0,\boldsymbol{a}_1\cdot\boldsymbol{b}_1=0. \tag{3.16}$$

即$(\boldsymbol{a}_1,\boldsymbol{b}_1)$组成直角标架,于是方程(3.12)说明 $\boldsymbol{r}(s)$表示一个圆. 定理 2 证毕.

注意到(1.1)式定义的 $d(p)$不大于 $F(p)$,即有

系 2 Johnson 的闭曲线猜想(1.2)式是真的.

参考文献

[1]Herda. H. A. M. Monthly,1974(81):146 - 149.

[2]A. M. Monthly,1971(78):888 - 889.

[3]Witsenhausen H S. Proc. A. M. S. ,1972(35):240 - 241.

[4]Chakerian C D. A. M. Monthly,1974(81):153 - 155.

[5]Lutwak Erwin. Bull. Lond. M. S. ,1980(12):189 - 195.

[6]Falconer,K J. ,J. Lond. Math. Soc. ,1977,16(2):536 - 538.

[7]Herda,H,Collog. Math. ,1976,36(2):235 - 236.

[8]Alexander R. Lecture Notes in Math. ,1975(490):57 - 65.

[9]Pac. J. of Math. ,1979,85(1):1 - 9.

[10]Yang Lu,Zhang Jingzhong. Kexue Tongbao,1982,27(7):699－703.
[11]Sallee G T. Geomtriae Dedicata,1974(2):311－317.
[12]von Neumann J,Schoenberg I J. Trans. A. M. S. ,1941(50):226－251.

A Problem of Stolarsky's

Zhang Jing-zhong (张景中) and Yang Lu(杨　路)

(Institute of Mathematical Sciences, Chengdu Branch, Academia Sinica)

A few years ago, at the conference on metric geometry held in Michigen, K. B. Stolarsky posed a problem (cf. [1], problem 1) which challenges some classical methods of distance geometry.

Let T be a tetrahedron whose base is a triangle with edge-lengths a, b, c, and A, B, C be the lengths of edges opposite a, b, c, respectively. Stolarsky conjectured that the following inequality holds:

$$\frac{BC}{bc}+\frac{CA}{ca}+\frac{AB}{ab}\geqslant 1, \tag{1}$$

i. e.

$$aBC+bCA+cAB\geqslant abc. \tag{1*}$$

As he pointed out, without loss of generality, it is enough to verify the assertion for degenerate tetrahedrons (i. e. planar quadrangles).

That conjecture has been proved by M. S. Klamkin. Then, the new problem of Stolarsky's is whether one can write down a "one line proof" by rearranging the equation $D=0$, where D is the Cayley-Menger determinant for the vertices of planar quadrangle T. He showed[1] that the answer is "yes" if $a=b=c=1$, but it seems to be more difficult to find a solution in general case.

In this paper, an affirmative answer to Stolarsky's problem is given. We give a proof of (1) only by expanding and rearranging the Cayley-Menger determinant of T.

In accordance with the notations in [1], let D be the Cayley-Menger determinant (cf. [2], § 40, pp. 97—99) of the planar quadrangle T:

$$D=\begin{vmatrix} 0 & 1 & 1 & 1 & 1 \\ 1 & 0 & A^2 & B^2 & C^2 \\ 1 & A^2 & 0 & c^2 & b^2 \\ 1 & B^2 & c^2 & 0 & a^2 \\ 1 & C^2 & b^2 & a^2 & 0 \end{vmatrix}, \tag{2}$$

本文刊于《科学通报》,第 30 卷第 11 期,1985 年.

which is a homogeneous polynomial with six variables. Stolarsky wanted to deduce (1) or (1*) only from $D=0$; it is not allowed to use other geometric propositions except the triangle inequality as has already been used in [1] for the particular case of $a=b=c=1$.

Let us expand the determinant D and rearrange its terms as follows. Some of the steps written down by us need not require complicated calculations, but depend on some elementary algebraic techniques which are explained in the Annonote.

$$
\begin{aligned}
D &= -2a^2b^2c^2-2a^2B^2C^2-2b^2C^2A^2-2c^2A^2B^2+2(a^2+A^2)(b^2B^2+c^2C^2-a^2A^2) \\
&\quad +2(b^2+B^2)(c^2C^2+a^2A^2-b^2B^2)+2(c^2+C^2)(a^2A^2+b^2B^2-c^2C^2) \\
&= 4a^2b^2c^2-a^2\left\{(B^2+c^2-A^2)(b^2+C^2-A^2)+\begin{vmatrix} 0 & 1 & 1 & 1 \\ 1 & 0 & A^2 & B^2 \\ 1 & A^2 & 0 & c^2 \\ 1 & C^2 & b^2 & a^2 \end{vmatrix}\right\} \\
&\quad -b^2\left\{(C^2+a^2-B^2)(c^2+A^2-B^2)+\begin{vmatrix} 0 & 1 & 1 & 1 \\ 1 & 0 & A^2 & B^2 \\ 1 & B^2 & c^2 & 0 \\ 1 & C^2 & b^2 & a^2 \end{vmatrix}\right\} \\
&\quad -c^2\left\{(A^2+b^2-C^2)(a^2+B^2-C^2)+\begin{vmatrix} 0 & 1 & 1 & 1 \\ 1 & 0 & A^2 & C^2 \\ 1 & B^2 & c^2 & a^2 \\ 1 & C^2 & b^2 & 0 \end{vmatrix}\right\},
\end{aligned} \tag{3}
$$

where

$$
\begin{aligned}
\begin{vmatrix} 0 & 1 & 1 & 1 \\ 1 & 0 & A^2 & B^2 \\ 1 & A^2 & 0 & c^2 \\ 1 & C^2 & b^2 & a^2 \end{vmatrix} &= \pm\left[\begin{vmatrix} 0 & 1 & 1 & 1 \\ 1 & 0 & A^2 & B^2 \\ 1 & A^2 & 0 & c^2 \\ 1 & C^2 & b^2 & a^2 \end{vmatrix}^2\right]^{\frac{1}{2}} \\
&= \pm[(2A^2B^2+2c^2A^2+2c^2B^2-A^4-B^4-c^4) \\
&\quad \times(2C^2A^2+2b^2C^2+2b^2A^2-C^4-A^4-b^4)-2A^2D]^{\frac{1}{2}};
\end{aligned} \tag{4}
$$

$$
\begin{aligned}
\begin{vmatrix} 0 & 1 & 1 & 1 \\ 1 & 0 & A^2 & B^2 \\ 1 & B^2 & c^2 & 0 \\ 1 & C^2 & b^2 & a^2 \end{vmatrix} &= \pm\left[\begin{vmatrix} 0 & 1 & 1 & 1 \\ 1 & 0 & A^2 & B^2 \\ 1 & B^2 & c^2 & 0 \\ 1 & C^2 & b^2 & a^2 \end{vmatrix}^2\right]^{\frac{1}{2}} \\
&= \pm[(2B^2C^2+2a^2B^2+2a^2C^2-B^4-C^4-a^4) \\
&\quad \times(2A^2B^2+2c^2A^2+2c^2B^2-A^4-B^4-c^4)-2B^2D]^{\frac{1}{2}};
\end{aligned} \tag{5}
$$

$$
\begin{vmatrix} 0 & 1 & 1 & 1 \\ 1 & 0 & A^2 & C^2 \\ 1 & B^2 & c^2 & a^2 \\ 1 & C^2 & b^2 & 0 \end{vmatrix} = \pm\left[\begin{vmatrix} 0 & 1 & 1 & 1 \\ 1 & 0 & A^2 & C^2 \\ 1 & B^2 & c^2 & a^2 \\ 1 & C^2 & b^2 & 0 \end{vmatrix}^2\right]^{\frac{1}{2}}
$$

$$=\pm[(2C^2A^2+2b^2C^2+2b^2A^2-C^4-A^4-b^4)$$
$$\times(2B^2C^2+2a^2B^2+2a^2C^2-B^4-C^4-a^4)-2C^2D]^{\frac{1}{2}}; \quad (6)$$

Substituting (4—6) into (3) and putting $D=0$, we have

$$\begin{aligned}4a^2b^2c^2=&a^2[(B^2+c^2-A^2)(b^2+C^2-A^2)\pm(2A^2B^2+2c^2A^2+2c^2B^2\\&-A^4-B^4-c^4)^{\frac{1}{2}}\times(2C^2A^2+2b^2C^2+2b^2A^2-C^4-A^4-b^4)^{\frac{1}{2}}]\\&+b^2[(C^2+a^2-B^2)(c^2+A^2-B^2)\pm(2B^2C^2+2a^2B^2+2a^2C^2-B^4-C^4-a^4)^{\frac{1}{2}}\\&\times(2A^2B^2+2c^2A^2+2c^2B^2-A^4-B^4-c^4)^{\frac{1}{2}}]\\&+c^2[(A^2+b^2-C^2)(a^2+B^2-C^2)\pm(2C^2A^2+2b^2C^2+2b^2A^2-C^4-A^4-b^4)^{\frac{1}{2}}\\&\times(2B^2C^2+2a^2B^2+2a^2C^2-B^4-C^4-a^4)^{\frac{1}{2}}]\\\leqslant&a^2[(B^2+c^2-A^2)^2+(2A^2B^2+2c^2A^2+2c^2B^2-A^4-B^4-c^4)]^{\frac{1}{2}}\\&\times[(b^2+C^2-A^2)^2+(2C^2A^2+2b^2C^2+2b^2A^2-C^4-A^4-b^4)]^{\frac{1}{2}}\\&+b^2[(C^2+a^2-B^2)^2+(2B^2C^2+2a^2B^2+2a^2C^2-B^4-C^4-a^4)]^{\frac{1}{2}}\\&\times[(c^2+A^2-B^2)^2+(2A^2B^2+2c^2A^2+2c^2B^2-A^4-B^4-c^4)]^{\frac{1}{2}}\\&+c^2[(A^2+b^2-C^2)^2+(2C^2A^2+2b^2C^2+2b^2A^2-C^4-A^4-b^4)]^{\frac{1}{2}}\\&\times[(a^2+B^2-C^2)^2+(2B^2C^2+2a^2B^2+2a^2C^2-B^4-C^4-a^4)]^{\frac{1}{2}}\\=&4a^2BcbC+4b^2CacA+4c^2AbaB\\=&4abc(aBC+bCA+cAB).\end{aligned} \quad (7)$$

i. e.

$$abc\leqslant aBC+bCA+cAB.$$

And this is the proof.

Annonote. We may calculate by the following steps:

(i) $$D=\begin{vmatrix}0&1&1&1&1\\1&0&A^2&B^2&C^2\\1&A^2&0&c^2&0\\1&B^2&c^2&0&0\\1&C^2&b^2&a^2&0\end{vmatrix}+\begin{vmatrix}0&1&1&1&0\\1&0&A^2&B^2&0\\1&A^2&0&c^2&b^2\\1&B^2&c^2&0&a^2\\1&C^2&b^2&a^2&0\end{vmatrix}=D_1+D_2.$$

(ii) $$D_1=\begin{vmatrix}0&1&1&1&1\\1&0&A^2&B^2&C^2\\1&A^2&0&0&0\\1&B^2&c^2&0&0\\1&C^2&b^2&a^2&0\end{vmatrix}+\begin{vmatrix}0&1&1&0&1\\1&0&A^2&0&C^2\\1&A^2&0&c^2&0\\1&B^2&c^2&0&0\\1&C^2&b^2&0&0\end{vmatrix}=D_3+D_4.$$

(iii) $$D_2=-a^2\cdot\begin{vmatrix}0&1&1&1\\1&0&A^2&B^2\\1&A^2&0&c^2\\1&C^2&b^2&a^2\end{vmatrix}+b^2\begin{vmatrix}0&1&1&1\\1&0&A^2&B^2\\1&B^2&c^2&0\\1&C^2&b^2&a^2\end{vmatrix}.$$

(iv) $$D_4=\begin{vmatrix}0&1&1&0&1\\1&0&A^2&0&C^2\\1&A^2&0&c^2&0\\1&B^2&c^2&0&a^2\\1&C^2&b^2&0&0\end{vmatrix}-\begin{vmatrix}0&1&1&0&0\\1&0&A^2&0&0\\1&A^2&0&c^2&0\\1&B^2&c^2&0&a^2\\1&C^2&b^2&0&0\end{vmatrix}$$

$$=-c^2\begin{vmatrix}0&1&1&1\\1&0&A^2&C^2\\1&B^2&c^2&a^2\\1&C^2&b^2&0\end{vmatrix}-a^2c^2(A^2+C^2-b^2).$$

(v) Transform D_3 by subtracting the 5th column from the 2nd to 4th columns, the 5th row from the 2nd row and the 3rd row from the 4th, 5th rows.

$$D_3=\begin{vmatrix}0&0&0&0&1\\0&-2C^2&A^2-C^2-b^2&B^2-C^2-a^2&C^2\\1&A^2&0&0&0\\0&B^2-A^2&c^2&0&0\\0&C^2-A^2&b^2&a^2&0\end{vmatrix}$$

$$=\begin{vmatrix}2C^2&C^2-A^2+b^2&C^2-B^2+a^2\\B^2-A^2&c^2&0\\C^2-A^2&b^2&a^2\end{vmatrix}$$

$$=2a^2c^2C^2+b^2(C^2-B^2+a^2)(B^2-A^2)-c^2(C^2-B^2+a^2)(C^2-A^2)$$
$$-a^2(C^2-A^2+b^2)(B^2-A^2)$$
$$=2a^2c^2C^2+E_1+E_2+E_3.$$

(ⅳ) $E_1=-b^2(C^2-B^2+a^2)(A^2-B^2+c^2)+b^2c^2(C^2-B^2+a^2)$,

$E_2=-c^2(A^2-C^2+b^2)(B^2-C^2+a^2)+2a^2c^2(A^2-C^2)+b^2c^2(B^2-C^2+a^2)$,

$E_3=-a^2(C^2-A^2+b^2)(B^2-A^2+c^2)+a^2c^2(C^2-A^2+b^2)$,

$E_3+E_1+E_2=-a^2(B^2+c^2-A^2)(b^2+C^2-A^2)-b^2(C^2+a^2-B^2)(c^2+A^2-B^2)$
$-c^2(A^2+b^2-C^2)(a^2+B^2-C^2)+a^2c^2(A^2-C^2)+3a^2b^2c^2$.

(ⅶ) Substituting (vi) into (v), the result with (iv) into (ii), and the last result with (iii) into (i), we obtain (3).

(ⅷ) the proof of (4—6). D_{ij} denote the cofactor of the ith row and jth column of D, and $D_{ii,jj}$ the sub-determinant obtained by deleting the ith row, the ith column, the jth row and the jth column from D. Then for $i\neq j$, the following algebraic identity[2,3] holds

$$D_{ii}D_{jj}-D_{ij}D_{ji}=D_{ii,ij}D.$$

From this equality, putting $i=4$ and $j=5$ one can obtain (4); $i=3$ and $j=5$ obtain (5); $i=3$ and $j=4$ obtain (6).

(ix) By Cauchy inequality or in another way we have (7).

References

[1]Guy R K. Problems, The geometry of metric and linear spaces (Ed. L. M. Kelly). Lecture Notes in Math., Springer-Verlag, 1975(490): 233 - 244.

[2]Blumenthal L M. Theory and Applications of Distance Geometry. New York, 1970.

[3]杨路,张景中. 关于有限点集的一类几何不等式. 数学学报,1980,23(5):740 - 749.

Embedding of a Homeomorphism in a Flow and Asymptotic Embedding

Zhang Jing-zhong (张景中) and Yang Lu(杨　路)
(*University of Science and Technology of China*)

ABSTRACT

To embed a self-homeomorphism on a manifold in a flow, especially, to prove the uniqueness of the embedding, is rather difficult. In this paper, we introduce concepts of "asymptotic embedding" and "the flow asymptotic to a given flow"and find the connection between the two concepts. As a conclusion, we obtain a result about the uniqueness of embedding applicable to a class of fairly general cases.

INTRODUCTION

Let M be an n-dimensional C^r-manifold with or without a boundary. By a C^k-flow($0\leqslant k\leqslant r$) on M we mean a C^k-mapping $\varphi: R\times M\to M$ satisfying

1)$\varphi(0,x)=x, \forall x\in M$;

2)$\varphi(t_1+t_2,x)=\varphi(t_1,\varphi(t_2,x))$; $\forall\ t_1,t_2\in R, x\in M.$ (0. 1)

If φ is defined only on $R^+\times M$ ($R^+=[0,+\infty)$) such that 2) holds for $t_1,t_2\in R^+, x\in M$, then φ is called a semi-flow.

It is obvious that $\varphi(t,x)$ is a homeomorphism for each t when φ is a flow. This is not a necessary case when φ is a semi-flow. If φ is a semi-flow and the mapping $\varphi(t,x): M\to M$ is a homeomorphism for each $t\in R^+$, we call φ an invertible semi-flow.

Then let $f: M\to M$ be a mapping from M to itself. If there exists a real number $\lambda\geqslant 0$ such that

$$f(x)=\varphi(\lambda,x), \forall x\in M, \tag{0. 2}$$

we say that f can be embedded in flow (or semi-flow) φ and call λ the embedding index of f in φ.

本文刊于《中国科学(A辑)》,第28卷第7期,1985年.

What mapping can be embedded in a flow or a semi-flow? In what sense is the embedding unique? These problems have been considered by some researchers of functional equations for many years though they did not use the terminologies "flow" and "embedding". About the early work in this respect readers are referred to Abel[1], Schröder[2], Bödewadt[3], Koenigs[4], Kneser[5] and others.

In the investigations of dynamic systems theory which have been prospering in recent years, it is thought that the problem of embedding a homeomorphism or a mapping in a flow or a semi-flow is very interesting and significant. As Fort[6] has put it, however, embedding problem is rather difficult. In fact, most of the results were given for 1-dimensional case. For instance, Sternberg[7] and P. F. Lam[8-10] discussed the embedding of a self-homeomorphism on an interval in a flow. Z. S. Zhang[11] gave a necessary and sufficient condition for embedding a self-homeomorphism of a circle in a flow. In higher dimensions, Palis[12] and Sternberg[13] obtained some results under rather strong conditions. It is usually required that the discussed homeomorphism has and only has some hyperbolic fixed points.

If the mapping f may be embedded in a flow or a semi-flow, such an embedding usually has great arbitrariness. This can be clearly seen in [3] and [7]. For example, the self-homeomorphism $f:R\to R$, $f(x)=x+1$, of course, can be embedded in the flow $\varphi(t,x)=x+t$, but it can be embedded in the flow

$$\Phi(t,x)=G^{-1}(G(x)+T) \tag{0.3}$$

as well, where $G(x)=x+g(x)$ and $g(x)$ is a periodic continuous function with period 1 and the Lipschitz constant less than 1. It is easy to verify that $\Phi(t,x)$ is really a flow and

$$\Phi(1,x)=x+1.$$

It is a knotty problem to eliminate this arbitrariness. Early in 1944. it is proved for 1-dimensional cases that the self-homeomorphism of an interval can be embedded in a flow[3]. (In fact. this was predicted by G. H. Hardy in 1924)[14]. The discussion about the uniqueness, however, lasted up to recent years[8]. In order to eliminate this arbitrariness, usually, the mapping f and the considered flow must be subject to some stronger smoothness condition or other limit conditions, as presented in [8-10]. By a different method from them, the authors[15] obtained some results on the uniqueness of embedding a self-homeomorphism of an interval in a flow. In this paper, the results of [15] will be generalized to the cases of higher dimensions.

In higher-dimensional cases, the uniqueness of embedding cannot be assured, even under some very strong conditions. For example, the linear transformation on the complex plane, $f(z)=cz\,(0<c<1)$, can be embedded not only in flow

$$\varphi(t,z)=c^t z,$$

but also in the more general flow $\Phi_k(t,z)=e^{2k\pi i t}c^t z$, where k is any integer. Clearly, Φ_k is analytic.

Therefore, it is necessary to restrict the considered flow with some suitable conditions to avoid the indefiniteness. The concept of "asymptotic embedding" will be introduced for this purpose.

It should be pointed out that the flows and homeomorphisms discussed in this paper, which seem to be subject to more restrictions, are far more extensive than those in references concerning the uniqueness of embedding, such as [5,7—10,13].

Ⅰ. PRELIMINARIES

Let M be an n-dimensional C^r-manifold ($0\leqslant r\leqslant\infty$) and φ be a flow or an invertible semi-flow on M. In the latter case, we define $x=\varphi(-t,y)$ when $y=\varphi(t,x)$. In this way, 2) in (0.1) holds when both sides of it have definitions. The subset of M

$$\varphi_{x_0}=\{x\mid x=\varphi(t,x_0),t\in R,\varphi(t,x_0)\text{ have definition}\},\tag{1.1}$$

is called an orbit of φ passing through x_0.

For $x_0\in M$, if there exists a neighborhood B_{x_0} of x_0 and a number $t_0>0$ such that $B_{x_0}\cap\varphi(t,B_{x_0})=\phi$ when $|t|\geqslant t_0$, then we call x_0 a wandering point of φ. Here $\varphi(t,B_{x_0})$ denotes the image set of B_{x_0} under the mapping $\varphi(t,x)$.

Let Q denote the set of all wandering points of φ. Then the restriction of φ on $R\times Q$ is a flow or an invertible semi-flow which has no periodic orbit. For convenience we introduce

Definition 1.1. A subset $W(\subset M)$ which intersects exactly each orbit of φ at a unique point is called a cross-set of φ.

Definition 1.2. If W is a cross-set of φ, the mapping

$$J_\omega:M\to W\quad(J_\omega(x)=W\cap\varphi_x)\tag{1.2}$$

is called the orbit projection of φ with respect to W. If J_ω is continuous, we call W a section of φ.

Definition 1.3. If W is a cross-set of φ and Q is the wandering set of φ, then for every $x\in Q$, there exists a unique real number $t=t(x)$ such that

$$x=\varphi(t,J_\omega(x)).\tag{1.3}$$

We call the number t defined by (1.3) the time-index of φ with respect to W, and denote it by $t=T_\omega(x)$.

It is easily known that if $J_\omega(x)$ is continuous in Q, then $T_\omega(x)$ is also continuous.

A continuous flow may have neither section nor cross-set such that J_ω is continuous in Q. However, we can suitably narrow the definition range of φ, that is to say, We may consider the restriction φ_1 of φ on a submanifold $M_1\subset M$, such that φ_1 has a cross-set W and J_ω is continuous in Q_1, the wandering set of φ_1. This will facilitate our discussion.

Ⅱ. THE DEFINITION OF ASYMPTOTIC EMBEDDING AND THE STATEMENT OF MAIN THEOREM

Suppose that $A \subset Q \subset M$ and the restriction of the flow or invertible semi-flow φ on A is still a flow or an invertible semi-flow, and that the restriction of φ on A has a section W. J_ω and T_ω are defined as above. These notations are used throughout this paper.

If a homeomorphism, f on M can be embedded in φ, it is obvious that

$$\begin{aligned} &1.\ J_\omega(f^n(x)) = J_\omega(x), \qquad && x \in M, \\ &2.\ T_\omega(f^n(x)) = ns + T_\omega(x), \qquad && x \in Q, \end{aligned} \tag{2.1}$$

where n is any integer and s is the embedding index of f in φ. We may reduce (2.1) to the following limit equalities:

$$\begin{aligned} &\text{i)} \lim_{n\to\infty} J_\omega(f^n(x)) = J_\omega(x); \\ &\text{ii)} \text{For } s>0,\ \lim_{n\to+\infty} T_\omega(f(x)) = +\infty \text{ and } \lim_{T_\omega(x)\to+\infty} \{T_\omega(f(x)) - T_\omega(x)\} = s. \end{aligned} \tag{2.2}$$

This induces us to introduce the concept of "asymptotic embedding". As a preparation, we introduce the definition of "asymptotically co-orbital". Let B denote a submanifold of M throughout this paper.

Definition 2.1. Let f be a continuous self-mapping of $B \subset M$. If in each Compact subset B_1 of B, the formula $f^n(x) \in A$ holds uniformly for every $x \in B_1$ and sufficiently large n, the following

$$\lim_{n\to+\infty} J_\omega(f^n(x)) = H(x), x \in B_1 \tag{2.3}$$

is uniformly valid, and each equipotential of $H: B \to M$ (i. e. the set $\{x \mid H(x) = x_0\}$ for some $x_0 \in M$) is a 1-dimensional connected submanifold, then we say that f is "asymptotically co-orbital" with respect to φ. Mapping H is called the "orbit decomposition" of f with respect to φ.

Definition 2.2. Suppose that f on B is asymptotically co-orbital with respect to φ. If there exists a real number $s \geqslant 0$ such that, in each compact subset B_1 of B,

$$\lim_{n\to+\infty} (T_\omega(f^{n+1}(x)) - T_\omega(f^n(x))) = s \tag{2.4}$$

holds uniformly for every $x \in B_1$, then we say that f can be "asymptotically embedded" in φ, and call s the "asymptotic embedding index" of f in φ.

It can be easily seen that whether f can be asymptotically embedded in φ has nothing to do with the choice of the section W. Moreover, H in (2.3) and s in (2.4) are independent of the choice of the section W.

Definition 2.3. Let $f'(x) = \Phi(t, x)$ be a flow or a semi-flow on B. If for each positive number $\delta > 0$, f^δ can be asymptotically embedded in φ, each, f^δ has a nonvanishing asymptotic embedding index $s(\delta)$ and for each compact subset B_1 of B and each closed interval I on R^+, it holds uniformly with respect to (δ, x) on $I\times B_1$ that

$$\lim_{n\to+\infty}(T_\omega(f^{(n+1)\delta}(x))-T_\omega(f^{n\delta}(x)))=s(\delta). \tag{2.5}$$

then we say that Φ is a flow (or semi-flow) asymptotic to φ.

It is quite difficult, in general, to find a flow or a semi-flow in which the given mapping f can be embedded. But, it is easier to find a flow or semi-flow in which the given mapping can be asymptotically embedded. If we can find a close connection between embedding and asymptotic embedding, a great convenience for the research of embedding will be provided.

Clearly, the more extensive the mapping f is, the more empty the conclusion will be. Now let f be a self-homeomorphism on $B\subset M$ for which there exists a compact subset E of B such that

$$B=\bigcup_{k=-\infty}^{\infty} f^k(E)$$

and

$$\bigcup_{k=-n}^{n} f^k(E)\subset \operatorname{int}\bigcup_{k=-(n+1)}^{n+1} f^k(E);$$

in this case, we say that B can be compactly spanned by f and E is a compactly spanning set of B by f.

As to the relation between embedding and asymptotic embedding, the main result of this paper is the following.

Theorem 2.1. *Let φ be a continuous flow on M, Q the wandering set of φ and $A\subset Q\subset M$. Suppose that φ has a section W on A, f is a self-homeomorphism of $B\subset M$ and B is compactly spanned by f, and that f can be asymptotically embedded in φ (on B) with asymptotic embedding index $s>0$ and H is the orbit decomposition of f to φ. Then, f can be embedded in a flow asymptotic to φ if and only if in each compact subset of*

$$B^*=\{(x,y)\mid (x,y)\in B\times B, H(x)=H(y)\},$$

there holds

$$\lim_{n\to+\infty}(T_\omega(f^n(y))-T_\omega(f^n(x)))=T(x,y) \tag{2.6}$$

and $T(x,y)\neq 0$ when $x\neq y, (x,y)\in B^*$.

Ⅲ. SEVERAL LEMMAS

Lemma 3.1. *Function $T(x,y)$ determined by (2.6) has the following properties:*

i) $T(x,y_1)-T(x,y_2)=T(y_2,y_1)$,

ii) $T(x,y)=-T(y,x)$,

iii) $T(x,x)=0$,

iv) $T(x,f^n(x))=ns$, (*for any integer n.*) (3.1)

Proof. By (2.6), it holds true obviously for i), ii) and iii). Now we proceed to prove iv). Since f can be asymptotically embedded in φ, using (2.4) and noticing $H(x)=$

$H(f(x))$ we have

$$T(x,f^n(x)) = \lim_{m\to+\infty}(T_\omega(f^m(f^n(x))) - T_\omega(f^m(x)))$$
$$= \lim_{m\to+\infty}\sum_{k=1}^{n}(T_\omega(f^{m+k-1}(f(x))) - T_\omega(f^{m+k-1}(x)))$$
$$= n\lim_{m\to+\infty}(T_\omega(f^m(f(x))) - T_\omega(f^m(x))) = ns.$$

Lemma 3.2. *If $\Phi(t,x)$ is a continuous flow on B, and c, s are positive numbers such that*

$$f(x)=\Phi(s,x),\ g(x)=\Phi(c,x),$$

then

$$g^n(x)=f^{[nc/s]}\left(\Phi\left(\left\{\frac{nc}{s}\right\}s,x\right)\right), \tag{3.2}$$

where $[t]$ and $\{t\}$ denote the integral and decimal parts of t respectively, and when x varies within a certain compact subset of B it follows that

i) $\Phi\left(\left\{\frac{nc}{s}\right\}s,x\right)$ *varies within a certain compact subset of B.*

ii) $(y,z)=\left(\Phi\left(\left\{\frac{nc}{s}\right\}s,x\right),\Phi\left(\left\{\frac{nc}{s}\right\}s+c,x\right)\right)$ *varies within one compact subset of $B \times B$.*

iii) *If $y=\Phi(t,x)$ and t is bounded, then (x,y) varies within are compact subset of $B \times B$.*

Proof. By the definition of flow, 2) of (0.1), we have

$$g^n(x)=\Phi(nc,x)=\Phi\left(\left[\frac{nc}{s}\right]s+\left\{\frac{nc}{s}\right\}s,x\right)$$
$$=\Phi\left(\left[\frac{nc}{s}\right]s,\Phi\left(\left\{\frac{nc}{s}\right\}s,x\right)\right)=f^{[nc/s]}\left(\Phi\left(\left\{\frac{nc}{s}\right\}s,x\right)\right)$$

then (3.2) is true. By the continuity of Φ, the boundedness of $\left\{\frac{nc}{s}\right\}s$ and the compactness of the closed intervals, we know that i), ii), and iii) are true.

Lemma 3.3. *If $\Phi(t,x)$ is a flow asymptotic to φ on B, both Φ and φ are continuous,*

$$f(x)=\Phi(t_0,x),\ g(x)=\Phi(t_1,x),\ (t_0>0,t_1>0),$$

the orbit decompositions of f and g with respect to φ are H_f and H_g respectively, and the asymptotic embedding indexes of f and g with respect to φ are $s(t_0)$ and $s(t_1)$ respectively, then we have:

i) *each equipotential of H_f is an orbit of Φ;*

ii) $H_f=H_g=\lim_{t\to+\infty}J_\omega(\Phi(t,x))$;

iii) $s(t_0):s(t_1)=t_0:t_1$, *and hence Φ has no periodic orbit.*

Proof. Let $L(x_0)$ be an orbit of Φ passing through x_0. We turn to prove that H_f takes a constant value on $L(x_0)$. Consider the subset of $L(x_0)$

$$E=\left\{x \mid x=\Phi\left(\frac{k}{m}t_0, x_0,\right)\right\},$$

where m is a positive integer, and k an integer. Clearly, E is dense in $L(x_0)$, so what we need to prove is that H_f takes a constant value on E. In fact, setting $h(x)=\Phi\left(\frac{t_0}{m}, x\right)$ for each $\frac{k}{m}$ so that $h^m(x)=\Phi(t_0, x)=f(x)$. Since h can be asymptotically embedded in φ, we know that $\lim\limits_{n\to+\infty} J_\omega(h^n(x))$ exists. Hence

$$\begin{aligned} H_f\left(\Phi\left(\frac{k}{m}t_0, x_0\right)\right) &= \lim_{n\to+\infty} J_\omega\left(f^n\left(\Phi\left(\frac{k}{m}\right)t_0, x_0\right)\right)\right) \\ &= \lim_{n\to+\infty} J_\omega\left(h^{mn}\left(\Phi\left(\frac{k}{m}\right)t_0, x_0\right)\right)\right) \\ &= \lim_{n\to+\infty} J_\omega(h^{mn+k}(x_0)) \\ &= \lim_{n\to+\infty} J_\omega(h^n(x_0)) = \lim_{n\to+\infty} J_\omega(h^{mn}(x_0)) \\ &= \lim_{n\to+\infty} J_\omega(f^n(x_0)) = H_f(x_0). \end{aligned}$$

It can be seen that H_f takes a constant value on $L(x_0)$. If $\Gamma(x_0)$ denotes the equipotential of H_f passing through x_0, then $\Gamma(x_0)$ contains $L(x_0)$ entirely. Since $\Gamma(x_0)$ is a 1-dimensional connected submanifold, we shall have $L(x_0)=\Gamma(x_0)$ if $L(x_0)$ is homeomorphic to a circle. If $L(x_0)$ is an isolated point or homeomorphic to a line and $\Gamma(x_0)\neq L(x_0)$, we must have

$$\lim_{t\to+\infty} \Phi(t, x_0)=e\in\Gamma(x_0),$$

or

$$\lim_{t\to-\infty} \Phi(t, x_0)=e\in\Gamma(x_0),$$

so e belongs to B and is a fixed point of Φ. Therefore

$$\lim_{n\to+\infty} T_\omega(f^{n+1}(e))-T_\omega(f^n(e)))=\lim_{t\to+\infty}(T_\omega(e)-T_\omega(e))=0,$$

which contradicts the fact that the asymptotic embedding index of f is larger than zero. Thus ⅰ) is proved.

Then let us turn to prove ⅱ). We shall prove that for any $x_0\in B$,

$$H_f(x_0)=H_g(x_0)=\lim_{t\to+\infty} J_\omega(\Phi(t, x_0)).$$

Since H_f takes a constant value on $L(x_0)$, $\Phi(t, x_0)$ varies within a compact set when $0\leqslant t\leqslant t_0$ and f is asymptotically co-orbital to φ, it holds uniformly that

$$\lim_{\substack{n\to+\infty \\ 0\leqslant t\leqslant t_0}} J_\omega(f^n(\Phi(t, x_0)))=H_f(\Phi(t, x_0))=H_f(x_0).$$

On the other hand,

$$\begin{aligned} \lim_{t\to+\infty} J_\omega(\Phi(t, x_0)) &= \lim_{t\to+\infty} J_\omega\left(\Phi\left(\left[\frac{t}{t_0}\right]t_0+\left\{\frac{t}{t_0}\right\}t_0, x_0\right)\right) \\ &= \lim_{t\to+\infty} J_\omega\left(f^{[t/t_0]}\left(\Phi\left(\left\{\frac{t}{t_0}\right\}t_0, x_0\right)\right)\right) \end{aligned}$$

$$= \lim_{\substack{n\to+\infty \\ 0\leq t\leq t_0}} J_\omega(f^n(\Phi(t,x_0))).$$

so that

$$\lim_{t\to+\infty} J_\omega(f^n(\Phi(t,x_0)))=H_f(x_0).$$

Analogously, the limit equals $H_g(x_0)$ as well, and thus ii) is proved.

Now let us go to prove iii). By the uniform convergence of (2.5) of Definition 2.3, we know that $s(t)$ is continuous with respect to t. Hence we need only prove $s(t_0):s(t_1)=t_0:t_1$, for the rational numbers t_0 and t_1. Setting $t_0:t_1=m:k$ and

$$\Phi\left(\frac{t_0}{m},x\right)=\Phi\left(\frac{t_1}{k},x\right)=h(x),$$

from iv) of Lemma 3.1 we conclude that

$$s(t_0)=T(x,f(x))=T(x,h^m(x))=mT(x,h(x)),$$

$$s(t_1)=T(x,g(x))=T(x,h^k(x))=kT(x,h(x)).$$

Since $s(t_0)$ and $s(t_1)$ are positive, iii) is proved. This completes the proof of Lemma 3.3.

Ⅳ. PROOF OF THEOREM 2.1

We shall first prove the sufficiency of condition (2.6). For any $x_0\in B$, since f can be asymptotically embedded in φ, the points satisfying the equality $H(x)=H(x_0)$ form a 1-dimensional connected submanifold $\Gamma(x_0)$. We show that $T(x_0,y)$ is injective for $y\in\Gamma(x_0)$. In fact, let y_1,y_2 belong to $\Gamma(x_0)$ and $y_1\neq y_2$. By i) of Lemma 3.1. we have

$$T(x_0,y_1)-T(x_0,y_2)=T(y_2,y_1)\neq 0,$$

and by iv), $T(x,f^n(x))=ns$, it follows that

$$\lim_{n\to+\infty} T(x,f^n(x))=+\infty,$$

$$\lim_{n\to+\infty} T(x,f^{-n}(x))=-\infty,$$

and so $T(x_0,y)$ is injective with respect to y on $\Gamma(x_0)$, and continuously takes all real numbers. Here, the continuity is assured by the uniform convergence of (2.6) on a compact subset of B^*. Therefore, solving the system of equations

$$\begin{cases} T(x_0,y)=t, & (t\in R), \\ H(y)=H(x_0), & (x_0\in B), \end{cases} \tag{4.1}$$

we obtain the unique solution $\Phi: R\times B\to B$,

$$y=\Phi(t,x_0),\ x_0\in B,\ t\in R. \tag{4.2}$$

It is easy to show that $\Phi(t,x)$ is continuous with respect to t. Since B is compactly spanned by f, we can demonstrate that $\Phi(t,x)$ is continuous with respect to (t,x). In fact, let x_k be a sequence in B, $x_k\to x^*\in B$ and $\{t_k\}$ a real number sequence, $t_k\to t^*$, and set $y_k=\Phi(t_k, x_k)$, then $T(x_k,y_k)=t_k$. We need prove that $y_k\to y^*=\Phi(t^*,x^*)$. If $\{y_k\}$ has a limit point $\tilde{y}\in B$, by continuity of $T(x,y)$ and H, it is obvious that $H(\tilde{y})=H(x^*)$ and $T(\tilde{y},x^*)=$

t^*, and hence $\tilde{y}=\Phi(t^*, x^*)=y^*$ by the definition of Φ. It remains to prove that $\{y_k\}$ has no limit point except those in B and contains no divergent subsequence. Now let us show that $\{y_k\}$ is contained by a compact subset of B.

Since B is compactly spanned by f, there exist a compact subset E of B and a natural number m such that

$$x^* \in \operatorname{int} E_m, E_m=\bigcup_{n=-m}^{m} f^n(E)$$

and hence $x_k \in E_n$ for sufficiently large k. Also, because

$$B=\bigcup_{n=-\infty}^{\infty} f^n(E),$$

there exist $z_k \in E_m$ and a natural number m_k such that

$$y_k=f^{m}k(z_k)$$

for every $x_k \in E_m$, and hence

$$T(y_k, x_k)+T(x_k, z_k)=t_k+T(x_k, z_k)=T(y_k, z_k).$$

Number $T(x_k, z_k)$ is bounded due to the compactness of E_m and t_k is bounded due to the fact $t_k \to t^*$, so is $T(y_k, z_k)=m_k s$, or m_k. Thus, there exists N such that

$$\{y_k\} \subset \bigcup_{n=-N}^{N} f^n(E),$$

i. e., it is contained by some compact subset of B. The continuity of $\Phi(t,x)$ with respect to (t,x) is proved.

We then prove that $\Phi(t,x)$ is indeed a flow on B. Since $T(x,x)=0$, we have $\Phi(0,x)=x$. It is necessary to show

$$\Phi(t_1+t_2, x)=\Phi(t_1,(t_2, x)). \tag{4. 3}$$

In fact, we know by the definition of Φ, (4. 1), that (4. 3) is equivalent to

$$T(x, \Phi(t_1, \Phi(t_2, x)))=t_1+t_2. \tag{4. 4}$$

According to Lemma 3. 1. we have

$$\begin{aligned} T(x,\Phi(t_1,\Phi(t_2,x)))&=T(\Phi(t_2,x),\Phi(t_1,\Phi(t_2,x)))-T(\Phi(t_2,x),x)\\ &=T(z,\Phi(t_1,z))+T(x,\Phi(t_2,x))=t_1+t_2, \end{aligned}$$

and hence $\Phi(t,x)$ is a flow indeed. On the other hand, according to Lemma 3. 1, there exists $T(x,f(x))=s$, that is to say, $\Phi(s,x)=f(x)$ so $f(x)$ can be embedded in Φ. In this case, the embedding index of f in Φ is equal to that in φ.

Now we prove that Φ is asymptotic to φ. First, we shall verify that, for any $\delta>0$, the mapping $f^{\delta}(x)=\Phi(\delta,x)$ is asymptotically co-orbital to φ.

By Lemma 3. 2, if x varies within a compact subset of B, then so does $\Phi\left(\left\{\frac{n\delta}{s}\right\}s,x\right)$.

Since f is asymptotically co-orbital to φ, when x varies within one compact subset of B it holds uniformly that

$$\lim_{n\to+\infty} J_{\omega}(f^{n\delta}(x))=\lim_{n\to+\infty} J_{\omega}\left(f^{[n\delta/s]}\left(\Phi\left(\left\{\frac{n\delta}{s}\right\}s,x\right)\right)\right)$$

$$= \lim_{\substack{n\to+\infty \\ 0\leqslant t\leqslant s}} J_{\omega}(f^{[n\delta/s]}(\Phi(t,x)))=H(\Phi(t,x))=H(x).$$

This means that $f^{\delta}(x)$ is asymptotically co-orbital to φ. It should be noticed that the fact $H(x)=H(\Phi(t,x))$ used in the foregoing deducing is determined by the definition of Φ, (4.1).

In order to show that Φ is asymptotic to φ, we need also prove the uniform convergence of (2.5) with respect to (δ,x) on $I\times B_1$, where B_1 is any compact subset of B.

By Lemma 3.2 and Condition (2.6), we can easily see that it holds uniformly for $(\delta, x)\in I\times B_1$ that

$$\begin{aligned}
&\lim_{n\to+\infty}(T_{\omega}(f^{(n+1)\delta}(x))-T_{\omega}(f^{n\delta}(x))) \\
&\quad=\lim_{n\to+\infty}\left(T_{\omega}\left(f^{[n\delta/s]}\left(\Phi\left(\left\{\frac{n\delta}{s}\right\}s+\delta,x\right)\right)\right)-T_{\omega}\left(f^{[n\delta/s]}\left(\Phi\left(\left\{\frac{n\delta}{s}\right\}s,x\right)\right)\right)\right)-\delta+\delta \\
&\quad=\lim_{n\to+\infty}\left(T_{\omega}\left(f^{[n\delta/s]}\left(\Phi\left(\left\{\frac{n\delta}{s}\right\}s+\delta,x\right)\right)\right)\right)-T_{\omega}\left(f^{[n\delta/s]}\left(\Phi\left(\left\{\frac{n\delta}{s}\right\}s,x\right)\right)\right) \\
&\qquad-T\left(\Phi\left(\left\{\frac{n\delta}{s}\right\}s,x\right),\Phi\left(\left\{\frac{n\delta}{s}\right\}s+\delta,x\right)\right)\Big)+\delta \\
&\quad=\delta.
\end{aligned}$$

Here, we have used $T(\Phi(t,x),\Phi(t+\delta,x))=\delta$ and the conclusion that $(\Phi(t+\delta,x),\Phi(t,x))$ belongs to a compact subset of $B\times B$ when t and x belong to one compact subset of I and one of B respectively. Therefore, Φ is asymptotic to φ so that the sufficiency of Condition (2.6) in the theorem has been proved.

Now let us prove the necessity of this condition. That is to say, supposing that f has been embedded in a flow asymptotic to φ, $f=\Phi(t_0,x)$, we should prove that (2.6) holds uniformly in each compact subset of

$$B^{*}=\{(x,y)\mid(x,y)\in B\times B, H(x)=H(y)\}$$

and that $T(x,y)\neq 0$ when $x\neq y$.

From Lemma 3.3, each equipotential of $H(x)$ is just an orbit of Φ. Therefore when $(x,y)\in B^{*}$, both x and y are on the same orbit, so there exists a real number $\delta=\delta(x,y)$ such that

$$y=\Phi(\delta,x)=g^{\delta}(x),(x,y)\in B^{*}.$$

Now the question is whether $\delta(x,y)$ is bounded or not when (x,y) varies within one compact subset of B^{*}. We assert its boundedness. If not, then we can find a sequence $\{(x_n, y_n)\}$ in B^{*} such that $x_n\to x^{*}\in B$, $y_n\to y^{*}\in B$ and $\delta(x_n,y_n)\to+\infty$. Clearly $H(x^{*})=H(y^{*})$. Since Φ has no periodic orbit, there is a certain number t^{*} such that $y^{*}=\Phi(t^{*}, x^{*})$. By continuity, $\delta(x_n,y_n)\to t^{*}$, which contradicts the fact that $\delta(x_n,y_n)\to+\infty$. Hence $\delta(x,y)$ is bounded when (x,y) varies within one compact subset of B^{*}.

When (x,y) varies as above and $\delta(x,y)\neq 0$ (i. e. $x\neq y$), by Lemma 3.2 and Condition (2.5), it holds uniformly that

$$
\begin{aligned}
&\lim_{n\to+\infty}(T_{\omega}(f^{n}(y))-T_{\omega}(f^{n}(x)))\\
&=\lim_{n\to+\infty}(T_{\omega}(f^{n}(\Phi(\delta,x)))-T_{\omega}(f^{n}(x)))\\
&=\lim_{n\to+\infty}\left(T_{\omega}\left(g^{\left(\left[\frac{nt_0}{\delta}\right]+1\right)\delta}\left(\Phi\left(\left\{\frac{nt_0}{\delta}\right\}\delta,x\right)\right)\right)\right.\\
&\quad\left.-T_{w}\left(g^{\left[\frac{nt_0}{\delta}\right]\delta}\left(\Phi\left(\left\{\frac{nt_0}{\delta}\right\}\delta,x\right)\right)\right)\right)-s(\delta)+s(\delta)\\
&=s(\delta)
\end{aligned}
$$

where $s(\delta)$ is the asymptotic embedding index of g^{δ} in φ. From Lemma 3. 3, we have $s(\delta)\neq 0$ for $\delta\neq 0$. When $\delta=0$, it is obvious that $T_{\omega}(f^{n}(y))-T_{\omega}(f^{n}(x))=0$. This assures the uniform convergence of (2. 6) in each compact subset of B^{*}. The necessity of Condition (2. 6) has been proved. Thus the proof of Theorem 2. 1 is completed.

Ⅴ. UNIQUENESS OF EMBEDDING

If f can be embedded in both Φ and Φ^{*} which are flows asymptotic to φ, what relations are there between Φ and Φ^{*}? About this we have

Theorem 5. 1. *If f can be embedded in both Φ and Φ^{*} which are the flows asymptotic to φ on B, then there exists a constant α such that*

$$\Phi(t,x)=\Phi^{*}(\alpha t,x),x\in B. \tag{5. 1}$$

Proof. First. from Lemma 3. 3, the orbits of Φ are identical with those of Φ^{*}. That is to say, for each $x_0\in B$, the equipotential of H passing through x_0, $\Gamma(x_0)$, is the orbit of both Φ and Φ^{*}, where H is the orbit decomposition of f with respect to φ.

Suppose that the embedding indexes of f in Φ and Φ^{*} are δ and δ^{*} respectively, and that the asymptotic embedding index of f in φ is s. We are ready to prove

$$\Phi(t,x)=\Phi^{*}\left(\frac{\delta^{*}t}{\delta},x\right),x\in B. \tag{5. 2}$$

Clearly, if (5. 2) is true, the theorem is proved by setting $\alpha=\delta^{*}/\delta$.

Because of continuity, we need to derive (5. 2) only for a dense subset of R. We shall first show that (5. 2) is true when m is any positive integer and $t=\dfrac{\delta}{m}$. Setting

$$g_{*}(x)=\Phi^{*}\left(\frac{\delta^{*}}{m},x\right),$$

we have $g_{*}^{m}(x)=f(x)$. It follows from Theorem 2. 4 that

$$T(x,y)=\lim_{n\to+\infty}(T_{\omega}(f^{n}(y))-T_{\omega}(f^{n}(x))).$$

From Lemma 3. 1, we have

$$T(x,g_{*}(x))=\frac{1}{m}T(x,f(x))=\frac{s}{m}.$$

Analogously, setting $g(x)=\Phi\left(\dfrac{\delta}{m},x\right)$, we obtain

$$T(x,g(x))=\frac{1}{m}T(x,f(x))=\frac{s}{m}.$$

Therefore, both $g_*(x)$ and $g(x)$ satisfy

$$\begin{cases}T(x,y)=\dfrac{s}{m}, \\ H(y)=H(x),\end{cases} \quad (y=g_* \text{ and } y=g). \tag{5.3}$$

Since $H(y)=H(x)$ for the fixed x, y must vary on the orbit $\Gamma(x_0)$ and since $T(x,y)$ is invertible on $\Gamma(x_0)$, (5.3) has only one solution, hence $g=g_*$. Moreover,

$$g^k(x)=g_*^k(x), \text{ i. e. }, \Phi\left(\frac{k\delta}{m},x\right)=\Phi^*\left(\frac{k\delta^*}{m},x\right),$$

so (5.2) is true on a dense subset of R and thus Theorem 5.1 is proved.

On the other hand, we can find two flows Φ and Φ^* asymptotic to one flow. except the identity mapping. there does not exist a mapping which can be embedded in both Φ and Φ^* even though Φ and Φ^* are commutative, i. e.,

$$\Phi(t_1,\Phi^*(t_2,x))=\Phi^*(t_2,\Phi(t_1,x)). \tag{5.4}$$

Such an example will be given as follows.

We define a flow φ on $R\times R$:

$$\varphi(t,X)=(p(t,X),q(t,X)), \tag{5.5}$$

where $X=(x,y)$, $(x>0, y>0)$ and φ is determined by the system of equations

$$\begin{cases}\dfrac{1+\frac{1}{x}}{1+\frac{1}{y}}=\dfrac{1+\frac{1}{p}}{1+\frac{1}{q}}, \\ \log(x^2+y^2)=\log(p^2+q^2)-2t.\end{cases} \tag{5.6}$$

It is easy to verify that $\varphi=(p,q)$ is a flow on $R\times R$. Considering one section

$$W=\{(x,y)\mid x^2+y^2=1; x>0, y>0\}, \tag{5.7}$$

we know that $J_\omega(x,y)=(u,v)$ is determined by

$$\begin{cases}u^2+v^2=1, \\ \dfrac{1+\frac{1}{x}}{1+\frac{1}{y}}=\dfrac{1+\frac{1}{u}}{1+\frac{1}{v}},\end{cases} \tag{5.8}$$

and

$$T_\omega(x,y;u,v)=\frac{1}{2}\log\frac{u^2+v^2}{x^2+y^2}. \tag{5.9}$$

Then we define two flows Φ and Φ^*:

$$\Phi(t,X)=(r(t,X),s(t,X)), X\in \mathbf{R}\times\mathbf{R}, \tag{5.10}$$

where (r,s) is determined by

$$\begin{cases}r(t,X)=x, \\ s(t,X)=e^t y,\end{cases} \tag{5.11}$$

and $\Phi^*(t,X)=(r^*(t,X),s^*(t,X))$, where

$$\begin{cases} r^*(t,X)=e^t x, \\ s^*(t,X)=y. \end{cases} \tag{5.12}$$

It is easy to verify that Φ and Φ^* are commutative and can be asymptotically embedded in φ, but they are completely different from each other, that is, there is no mapping which can be embedded in both Φ and Φ^*, except the identity mapping.

Ⅵ. REMARKS

1. In this paper, we have only considered the embedding problem on the wandering set. The embedding problem about the mapping restricted within the set of periodic points is easier to be treated. As for other non-wandering points, in the systems not too complicated, they are often the limit points of the wandering set and periodic point set. Therefore, a general problem could be divided into several parts to be treated separately, and then the joining of these parts would be considered. Nevertheless, the problem of joining is rather difficult. The cases considered in this paper, however, have included all those cases considered in the references concerning the uniqueness of embedding.

2. The requirement of "asymptotic embedding" is not harsh. For example. when f is a mapping on an interval, it can be asymptotically embedded in flow $\varphi(t,x)=c^t x$, $(0<c<1)$, provided that $f(0)=0$ and $0<f'(0)<1$. We could show that for a class of higher-dimensional self-homeomorphisms, rather extensive and important to application, there are flows in which the given self-homeomorphisms can be asymptotically embedded.

3. Many problems remain to be considered, such as what relations are there between the topological structure of the orbits and that of the sections of a flow? and how can we say about the topological properties of the cross-sets of a flow without any section?

Furthermore, since the above-mentioned method is not restricted by "hyperbolic condition", it is applicable to more cases. In each case, how do we use Theorem 2.4 to obtain some more concrete results? There are a lot of work to do. At least, in the neighborhood of an isolated fixed point, in the neighborhood of the fixed points which form a lower-dimensional submanifold or in the neighborhood of a periodic orbit, we can use Theorem 2.4 to make a more meticulous analysis as the authors did in the rear part of [15].

References

[1]Abel N. Works. Posthumous Paper, 1881(2): 36 - 39.
[2]Schröder E. Math. Ann., 1871(3): 296 - 322.
[3]Bödewadt U T. Zeistc. Math., 1944, 49(9): 497 - 516.

[4]Koenigs G. Ann. de l'Ecole Norm Sup (3),1884:3-41.
[5]Kneser H. J. Reine Angew Math. ,1950(187):56-67.
[6]Fort M K Jr. ,Proc. Amer. Math. Soc. ,1955(6):960-967.
[7]Sternberg S. Duke Math J. ,1957(24):97-102.
[8]Lam P F. J. Diff. Equations,1978(30):31-40.
[9]Topological Dynamics:An International Symposium. New York,1968:319-333.
[10]Colloq. Math. ,1976(35):275-287.
[11]张筑生. 数学学报,1981,24(6):953-957.
[12]Palis J. Topology. 1969(8):385-405.
[13]Sternberg S. Amer. J. Math. ,1957(79):809-823,1958(80):623-632,1959(81):578-604.
[14]Hardy G H. Orders of Infinity. Cambridge,1924:31.
[15]张景中,杨路. 北京大学学报(自然科学版),1982(6):23-44.

An I. F. of Order 2^n to Compute Complex Roots by $n+1$ Informations

Yang Lu　Zhang Jing-zhong

(*Department of Mathematics*)

Abstract

In this paper the authors suggest a Several Points I. F. to compute the complex roots of polynomials and other analytic functions. It needs only $n+1$ informations, but the convergence rate is of order 2^n and the efficiency is higher than usual iterative programs.

One of the authors gave in [1] an I. F. in order to compute the real roots of transcendental equations. It only needs to use $n+1$ informations, but the convergence rate is of order 2^n and the efficiency is higher than usual iterative programs.

In September 1983 when the Conference DD4 was being held in Beijing, Professor M. Shub, on knowing the result of [1], asked authors whether this I. F. given in [1] can be used to compute the complex roots of polynomials. In fact, by the proof used in [1], one still finds it difficult to answer this question directly. In this paper, we shall show that the I. F. proposed in [1] is appliable not only to the computation of polynomials complex roots but also to the computation of those of analytic functions as well. Therefore this is an affirmative answer to the question of Professor Shub's.

1.

Let $f(z)$ be an analytic function defined on a region D in complex plane.

By referring to [2], if a complex sequence $\{z_m\}$ is defined by

$$Z_{m+1}=\varphi(Z_m,\omega_1(Z_m),\omega_2(Z_m),\cdots,\omega_k(Z_m)) \tag{1}$$

where $\omega_j\,(j=1,2,\cdots,k)$ are the functions dependent only on f and f', then φ is called a "Several Points I. F.". If there exists a root $\alpha\in D$ of f and α neighborhood U of α so that

本文刊于《中国科学技术大学学报》,第15卷第3期,1985年.

the sequence $\{Z_m\}$ defined by (1) converges to α and there exist a positive number $\tau>1$ and a positive number C dependent only on f and U so that

$$|\varphi(Z_0)-\alpha|\leqslant C|Z_0-\alpha|^{\tau},\text{for every } Z_0\in U, \tag{2}$$

then we say that the Several Points I. F. φ, with respect to root α of f, has a convergence rate of order τ. If φ with respect to each regular zero α(ie, $f(\alpha)=0$, $f'(\alpha)\neq 0$) of every analytic function has a convergence rate of order τ, we call φ an I. F. of order τ(for analytic functions).

To compute the ω_j's in(1), the values of f and its derivatives of some orders must be used. The number of these values is called"the number of informations"used to this I. F. For example, Newton method is an I. F. of order two and by two informations. It was given in [2]that I. F. of order $n+1$ by n informations.

An I. F. by $n+1$ informations was given in [1], which was proved being an I. F. of order 2^n for real functions which have continuous derivatives of order $n+1$, as follows.

Let $\omega_0=Z_0$, ω_j, be defined for $j=0,1,\cdots,m$ by using only the values of $f(\omega_0)$, $f(\omega_1)$, $\cdots$, $f(\omega_{j-1})$ and, $f'(\omega_0)$. To define ω_{m+1} we require another information $f(\omega_m)$.

Let $H_m(Z)$ be the interpolation polynomial of f by base points $\{\omega\}=\{Z_0, \omega_0, \omega_1, \cdots, \omega_m\}$. That is to say if ω_j appeared in $\{\omega\}$, s_j times, then $H_m(Z)$ would be a polynomial with a degree less than or equal to $m+1$ such as

$$H_m^{(s)}(\omega_f)=f^{(s)}(\omega_j),(s=0,1,\cdots,s_j-1). \tag{3}$$

Such an H_m, clearly, is defined by $m+2$ informations with respect to f. From $\omega_0=Z_0$ we know $H_m'(\omega_0)=f'(\omega_0)$. If ω_0, ω_1, $\cdots$, ω_m are distinct from one another, then the $m+1$ informations other than $f'(\omega_0)$ should be $f(\omega_j)(j=0,1,\cdots,m)$. Also, let

$$\begin{cases}\omega_{m+1}=\omega_m-\dfrac{f(\omega_m)}{H_m'(\omega_m)},(m=0,1,\cdots,n-1),\\ \varphi(Z_0)=\omega_m(Z_0).\end{cases} \tag{4}$$

Then, only $n+1$ informations are needed to obtain $Z_1=\varphi(Z^0)$. We should require in(4) $H_m'(\omega_m)\neq 0$ which will be proved for a neighborhood small enough of a simply root α of f. And it will be shown that when z_0 is approaching α sufficiently, the ω_j's defined by (3) and (4) are distinct from one another, hence the used $n+1$ informations include only one value of f' and others are the values of f.

2.

Let D be a bounded and simply connected region in complex plane and the boundary of D, ∂D, is piecewise smooth. Let $f(z)$ be all analytic function on $\overline{D}=D\cup\partial D$, $\alpha\in D$ such as $f(\alpha)=0$, $L>0$ be the length of ∂D and

$$\left.\begin{aligned} M&=\max_{z\in D}\{|f(Z)|\},\\ \Delta&=\min_{z\in D}\{|f'(Z)|\}>0,\end{aligned}\right\} \tag{5}$$

For $\delta>0$, we put

$$\left.\begin{aligned} B_\delta&=\left\{Z\,\middle|\,|Z-\alpha|<\delta\right\},\\ d_\delta&=\inf\left\{|Z_1-Z_2|\,\middle|\,Z_1\in\partial D, Z_2\in B_\delta\right\},\\ d&=\inf\left\{|Z-\alpha|\,\middle|\,Z\in\partial D\right\}=\delta+d_\delta.\end{aligned}\right\} \tag{6}$$

By(3),(4) and above notations we have

Theorem. For a given natural number $n\geqslant 1$, a given analytic function f on D and its a regular zero $\alpha\in D$, there exist numbers $\delta>0$ and $C>0$ such as, for any $\omega_0=z_0\in B_\delta, \omega_1, \omega_2, \cdots, \omega_n$ which may be defined by (3) and (4) satisfy

1° There is $0<q<1$ such as $|\omega_j-\alpha|\leqslant q|\omega_{j-1}-\alpha|$ $(j=1,2,\cdots,n)$, hence the sequence $\{z_{m+1}=\varphi(z_m)\}$ defined by (4) converges to α;

2° $|\varphi(z_0)-\alpha|\leqslant C|z_0-\alpha|^{2^n}$, hence φ is an I. F. of order 2^n.

Proof. We use the induction. When $n=1$, the I. F. defined by (3) and (4) is just Newton method, so the conclusion of the theorem is true to $n=1$, obviously. Let us suppose the conclusion is true to $n\leqslant l$, and go to prove this also to be true to $n=l+1$.

By inductive hypothesis, the $\omega_1, \omega_2, \cdots, \omega_l$ were defined by (3) and (4) for $\omega_0=z_0\in B_\delta$ and

$$|\omega_j-\alpha|\leqslant q|\omega_{j-1}-\alpha|, (1\leqslant j\leqslant l),$$

where $0<q<1$ and δ is a positive number related to l. Let δ be small enough so that $d_\delta>0$, hence $\omega_j\in B_\delta\subset D$.

First, let us prove that δ can become so small that $H_l'(\omega_l)\neq 0$, hence ω_{l+1} can be defined by (4). By Lagrange's or Hermite's interpolation formula we have

$$f(Z)=H_l(Z)+\frac{1}{2\pi i}\int_{\partial D}\frac{A_l(Z)}{A_l(\xi)}\cdot\frac{f(\xi)}{(\xi-Z)}d\xi, \tag{7}$$

where $z\in D, i=\sqrt{-1}$ and

$$A_l(Z)=(Z-Z_0)\prod_{j=0}^{l}(Z-\omega_j). \tag{8}$$

Deriving from (7) and letting $z=\omega_l$ we obtain

$$f'(\omega_l)=H'_l(\omega_l)+\frac{1}{2\pi i}\int_{\partial D}\frac{A_{l-1}(\omega_l)f(\xi)d\xi}{A_l(\xi)(\xi-\omega_l)}, \tag{9}$$

hence

$$\begin{aligned}\left|H_l'(\omega_l)\right|&\geqslant\left|f'(\omega_l)\right|-\frac{LM}{2\pi}\cdot\frac{2(\delta)^{l+1}}{d_\delta^{l+2}}\\ &\geqslant\Delta-\frac{LM}{2\pi d_\delta}\cdot\left(\frac{2\delta}{d_\delta}\right)^{l+1}.\end{aligned} \tag{10}$$

Since $d_\delta \to d > 0$ when $\delta \to 0$, it is shown that the right end of (10) is positive when $\delta > 0$ is so small that ω_{l+1} can be defined by (4).

To prove the conclusion 1° and 2°, we take $H_l(\omega_l) + H'_l(\omega_l)(z-\omega_l)$ as the interpolation polynomial of H_l by base points $\{\omega_l, \omega_l\}$. From (7) we have

$$H_l(z) = H_l(\omega_l) + H'_l(\omega_l)(z-\omega_l) + \frac{1}{2\pi i}\int_{\partial D} \frac{(z-\omega_l)^2 H_l(\xi)}{(\xi-\omega_l)^2(\xi-z)} d\xi, \tag{11}$$

Substituting (11) into (7) and applying (3) and (4) we obtain

$$f(z) = H'_l(\omega_l)(z-\omega_{l+1}) + \frac{1}{2\pi i}\int_{\partial D} \frac{(z-\omega_l)^2 H_l(\xi) d\xi}{(\xi-\omega_l)^2(\xi-z)} + \frac{1}{2\pi i}\int_{\partial D} \frac{A_l(z) f(\xi)}{A_l(\xi)(\xi-z)} d\xi, \tag{12}$$

Putting $z=\alpha$ in (12) we have

$$(\omega_{l+1}-\alpha) = \frac{\alpha-\omega_l}{2\pi i H'_l(\omega_l)}\left((\alpha-\omega_l)\int_{\partial D} \frac{H_l(\xi) d\xi}{(\xi-\omega_l)^2(\xi-\alpha)} + A_{l-1}(\alpha)\int_{\partial D} \frac{f(\xi) d\xi}{A_l(\xi)(\xi-\alpha)}\right). \tag{13}$$

To estimate the right end of (13) we need to estimate the integral in (7). It is obvious that

$$\left|\int_{\partial D} \frac{A_l(z) f(\xi)}{A_l(\xi)(\xi-z)} d\xi\right| \leqslant \frac{(|z-\alpha|+\delta)^{l+2} ML}{(d-\delta)^{l+2}(d-|z-\alpha|)}, (\delta < d, \ |z-\alpha| < d). \tag{14}$$

By (7) and (14), the estimation for H_l is obtained:

$$|H_l(z) - f(z)| \leqslant \frac{(|z-\alpha|+\delta)^{l+2} ML}{2\pi(d-\delta)^{l+2}(d-|z-\alpha|)}, (\delta < d, \ |z-\alpha| < d). \tag{15}$$

Let δ be small enough, denotting $|z-\alpha| = 2\delta$ by $\partial B_{2\delta}$, we have

$$\begin{aligned} \left|\int_{\partial D} \frac{(H_l(\xi) - f(\xi)) d\xi}{(\xi-\omega_l)^2(\xi-\alpha)}\right| &= \left|\int_{\partial D_{2\delta}} \frac{(H_l(\xi) - f(\xi)) d\xi}{(\xi-\omega_l)^2(\xi-\alpha)}\right| \\ &\leqslant 4\pi\delta \cdot \frac{(3\delta)^{l+2} ML}{2\pi(d-\delta)^{l+2}(d-2\delta)} \cdot \frac{1}{\delta^2 \cdot 2\delta} \\ &= \frac{9ML(3\delta)^l}{(d-\delta)^{l+2}(d-2\delta)}. \end{aligned} \tag{16}$$

Applying (16) and (10) to (13) we obtain

$$\begin{aligned} |\omega_{l+1}-\alpha| \leqslant & \frac{|\alpha-\omega_l|}{2\pi} \cdot \left(\Delta - \frac{LM}{2\pi d_\delta} \cdot \left(\frac{2\delta}{d_\delta}\right)^{l+1}\right)^{-1} \\ & \times \left\{\frac{3ML(3\delta)^{l+1}}{(d-\delta)^{l+2}(d-2\delta)} + \frac{ML\delta}{(d-\delta)^2 d} + \frac{ML\delta^{l+1}}{(d-\delta)^{l+2} d}\right\}. \end{aligned} \tag{17}$$

It can be seen that despite how small q is, we can find δ so small that

$$|\omega_{l+1}-\alpha| \leqslant q|\omega_l-\alpha|, \tag{18}$$

hence conclusion 1° is true.

Now we go to prove conclusion 2°. By induction hypothesis, it holds for $z_0 \in B_\delta$ that

$$|\omega_j-\alpha| \leqslant C_j |z_0-\alpha|^{2^j}, (1 \leqslant j \leqslant l) \tag{19}$$

where C s are some positive numbers dependent only on f, α, δ and j. Let the two ends of (13) be divided by $(z_0-\alpha)^{2^{l+1}}$; by using (10), (16) and (19) we have

$$\left|\frac{\omega_{l+1}-\alpha}{(z_0-\alpha)^{2^{l+1}}}\right|\leqslant\frac{1}{|2\pi H'_l(\omega_l)|}\left\{\left|\frac{\omega_l-\alpha}{(z_0-\alpha)^{2l}}\right|^2\cdot\left|\int_{\partial D}\frac{H_l(\xi)d\xi}{(\xi-\omega_l)^2(\xi-\alpha)}\right|\right.$$
$$\left.+\left|\prod_{j=0}^{l}\frac{\omega_j-\alpha}{(z_0-\alpha)^{2^j}}\right|\cdot\left|\int_{\partial D}\frac{f(\xi)d\xi}{A_l(\xi)(\xi-\alpha)}\right|\right\}$$
$$\leqslant\frac{(d-\delta)^{l+2}}{2\pi\Delta(d-\delta)^{l+2}-LM(2\delta)^{l+1}}$$
$$\times\left\{C_l^2\cdot\left(\frac{9ML(3\delta)^l}{(d-\delta)^{l+2}(d-2\delta)}+\frac{ML}{(d-\delta)^2d}\right)+C_1C_2\cdots C_l\cdot\frac{ML}{(d-\delta)^{l+2}d}\right\}$$
$$=\frac{ML}{2\pi\Delta(d-\delta)^{l+2}-LM(2\delta)^{l+1}}\cdot\left\{\frac{C_1C_2\cdots C_l}{d}+\right.$$
$$\left.C_l^2\left(\frac{(d-\delta)^l}{d}+\frac{9(3\delta)^l}{d-2\delta}\right)\right\}. \tag{20}$$

By the induction, therefore, conclusion 2° is true, so is the theorem.

3.

For the convenience of using, we shall give some estimations for δ and C, they are more concrete and simple.

Let $q<1$ be a prescribed positive number. To make 1° true, we ask how small δ must be. By (10) and (17), it is sufficient that

$$\left.\begin{array}{l}
1)\quad \Delta-\dfrac{LM}{2\pi\delta}\cdot\left(\dfrac{2\delta}{d_\delta}\right)^{l+1}>0,\ (l\geqslant 0),\\
2)\quad \dfrac{ML\delta}{(d-\delta)^2}\left(\dfrac{9(3\delta)^l}{(d-\delta)^l(d-2\delta)}+\dfrac{1}{d}+\dfrac{\delta^l}{(d-\delta)^l d}\right)\\
\qquad\leqslant 2\pi q\left(\Delta-\dfrac{LM(2\delta)^{l+1}}{2\pi(d-\delta)^{l+2}}\right),\ (l\geqslant 0),\\
3)\quad d-2\delta>0.
\end{array}\right\} \tag{21}$$

First, $\delta\leqslant\frac{d}{4}$ implies 3) and the following inequality is stronger than 1) and 2).

$$\frac{ML\delta}{(d-\delta)^2}\left(\frac{9}{d-2\delta}+\frac{1}{d}+\frac{1}{d}+2q\right)\leqslant 2\pi q\Delta. \tag{22}$$

By noticing $d-\delta\geqslant\frac{3}{4}d$, etc., (22) may be substituted by another which is stronger than (22).

$$\frac{16ML\delta}{9d^2}\left(\frac{10}{d}+q\right)\leqslant\pi q\Delta. \tag{23}$$

It follows

Remark 1. If δ satisfies

$$0<\delta\leqslant\min\left\{\frac{d}{4},\frac{9\pi q\Delta d^3}{16ML(10+qd)}\right\}, \tag{24}$$

then conclusion 1° of the theorem holds.

Noticing

$$\Delta \leqslant | f'(\alpha) | = \left| \frac{1}{2\pi} \int_{\partial D} \frac{f(\xi)}{(\xi-\alpha)^2} d\xi \right| \leqslant \frac{LM}{2\pi d^2}, \tag{25}$$

we have

$$\frac{9\pi q \Delta d^\delta}{16ML(10+qd)} \leqslant \frac{9qd}{320} < \frac{d}{4}, \tag{26}$$

hence (24) may be substituted by

$$0 < \delta \leqslant \frac{9\pi q \Delta d^3}{16ML(10+qd)}, \tag{27}$$

which is a simple condition, especially, this δ is independent of n.

(27) would be improved if we discuss it in detail.

Remark 2. If (2) is replaced by a limit form

$$\lim_{z_0 \to \alpha} \frac{\varphi(z_0 - \alpha)}{(z_0 - \alpha)^\tau} = C^*, \tag{2}^*$$

where τ is a positive integer, the conclusion 2° of the theorem may be correspondingly altered as follows:

2°*
$$\lim_{z_0 \to \alpha} \frac{\varphi(z_0) - \alpha}{(z_0 - \alpha)^{2^n}} = C_n^*, \tag{28}$$

where C_n^* is inductively given by

$$C_{l+1}^* = \frac{1}{f'(\alpha)} \left(\frac{f''(\alpha)}{2} \cdot (C_l^*)^2 + (-1)^l \cdot \frac{f^{(l+2)}(\alpha)}{(l+2)!} \cdot C_1^* C_2^* \cdots C_l^* \right), \tag{29}$$

which may be obtained by dividing (13) by $(z_0 - \alpha)^{2^{l+1}}$, letting z_0 tend to α and an induction with $C_0^* = 1$. (Compare [1, (15)].)

We wonder whether there is such an I. F. of order 2^n by $n+1$ informations, in which complex-computation may be avoided if $f(z)$ takes real values on real axis. By applying some modification of the method to [3], perhaps, it is possible to solve this problem.

References

[1]Zhang Jingzhong. 2^n-order I. F. Requiring $n+1$ Informations for Solving Transcendental Equations. Math. Numer. Sinica(Chinese, English Summary), 1980(4): 350 - 355.

[2]Ralston A, Wilf H S. Mathematical methods for Digital Computers, V. H. John Wiley & Sons Inc., 1968: 203.

[3] Zhang Jingzhong, Yang Lu. 2^n-order Cut-factor Method to Find the Roots of Polynomials. Math. Numer. Sinica(Chinese, English summary), 1982(4): 417 - 426.

伪对称集与有关的几何不等式

杨 路 张景中
(中国科学院成都分院)

§0 引 言

N个点在空间E^n中怎样分布算是"对称"的？例如在E^3中，当$N\neq 4,6,8,12,20$时，N个点不可能是某一正多面体的全部顶点，它们的分布能否具有某种对称性？这是在考虑某些几何不等式等号成立的条件时要产生的问题.有一大类涉及某个点集$\mathfrak{S}$的几何不等式[1]，其等号成立的条件都是:$\mathfrak{S}$关于其重心的惯量椭球是一个球.作者把具有这种性质的点集叫做"惯量等轴"，这一术语反映了点集在某种意义上的对称性.另外有许多几何不等式，它们的等号成立的条件是点集在不同意义或不同程度上的某种"对称性".

本文引进了E^n-伪对称性的概念，这是比"惯量等轴"更强的对称条件，并建立了有关的几何不等式.然后讨论了伪对称有限集的代数特征.文中将离散结果应用于紧致曲面，获得了超球面区别于其它曲面的一个新的特征.

设$\mathscr{F}$是E^n中的紧致曲面，F表$\mathscr{F}$的面积.令

$$M_r(\mathscr{F})=\frac{1}{F^2}\int_{\mathscr{F}}\int_{\mathscr{F}}|x-y|^r d\sigma(x)d\sigma(y).$$

表$\mathscr{F}$的"弦幂平均"，作者证明了不等式

$$M_r(\mathscr{F})\geqslant\frac{n+1}{n}M_2^2(\mathscr{F}),$$

等号成立当且仅当$\mathscr{F}$是一个超球面.

§1 伪对称集与离散的几何不等式

定义1 设$\mathfrak{S}$是n维欧氏空间中的一个点集.我们说$\mathfrak{S}$是E^n-伪对称的，如果$\mathfrak{S}$的凸包是n维的，而且

(ⅰ) $\mathfrak{S}$中所有的点都分布在E^n中的某一个球面S^{n-1}上；

(ⅱ) 球面S^{n-1}的中心O恰好是集$\mathfrak{S}$的重心；

(ⅲ) $\mathfrak{S}$关于O的惯量椭球是一个球.

据此定义，下列两个结论是显然的：

本文刊于《数学学报》，第28卷第6期，1986年.

引理 1 一个 n 维正多面体的全部顶点之集是 E^n -伪对称的. 一个 $n-1$ 维球面上所有点之集是 E^n -伪对称的.

引理 2 如果$\mathfrak{S}$和$\mathfrak{S}^*$都是 E^n -伪对称的,而且$\mathfrak{S}$和$\mathfrak{S}^*$分布在同一个球面上,则它们的并集$\mathfrak{S}\cup\mathfrak{S}^*$也是 E^n -伪对称的.

下面将给出伪对称集的一个几何特征.

设 $\mathfrak{S}=\{x_1,x_2,\cdots,x_N\}$是 E^n 的有限子集,令$|x_i-x_j|$表示两点 x_i,x_j 之间的欧氏距离,并令 M_r 表这些距离的 r 次幂的平均值

$$M_r(\mathfrak{S})=\frac{2}{N(N-1)}\sum_{1\leqslant i<j\leqslant N}|x_i-x_j|^r. \tag{1}$$

注意到$|x_i-x_i|=0$,也可将(1)式写成

$$M_r(\mathfrak{S})=\frac{1}{N(N-1)}\sum_{i=1}^{N}\sum_{j=1}^{N}|x_i-x_j|^r.$$

设 G 为点集$\mathfrak{S}$的重心,令

$$a_i=|x_i-G|\ (i=1,\cdots,N) \tag{2}$$

表示 x_i 到重心 G 的距离. 又令

$$L(\mathfrak{S})=\frac{2}{N(N-1)}\sum_{1\leqslant k<l\leqslant N}(a_k^2-a_l^2)^2 \tag{3}$$

表示这些$(a_k^2-a_l^2)^2$ 的平均值.

我们将建立下述的

定理 1 对于 E^n 中任何一个由 N 个点($N>n$)组成的有限点集$\mathfrak{S}$有不等式

$$M_4(\mathfrak{S})\geqslant\frac{N-1}{N}\ \frac{n+1}{n}M_2^2(\mathfrak{S})+L(\mathfrak{S}) \tag{4}$$

成立,这里等号成立的充分必要条件是:$\mathfrak{S}$关于其重心的惯量椭球是一个球.

这个定理的证明主要依赖于下面将要叙述的引理 3,它是一个更为普遍的已知结果的一个特款. 设以 N_1 表示$\mathfrak{S}$中各点间两两距离的平方和,令 N_2 表示以$\mathfrak{S}$中的点为顶点的所有三角形面积的平方和. 则有

引理 3(见文献[1]中的(1.5)式) 对于 E^n 中的任何一个 N 元素的有限点集$\mathfrak{S}$有不等式

$$N_1^2\geqslant\frac{8n}{n-1}NN_2 \tag{5}$$

成立,而等号成立的充分必要条件是:$\mathfrak{S}$关于其重心的惯量椭球是一个球.

定理 1 的证明 令 $a_{ij}=|x_i-x_j|$,并令 $I(x_i)$表示$\mathfrak{S}$关于点 x_i 的转动惯量,即

$$I(x_i)=\sum_{j=1}^{N}a_{ij}^2,i=1,\cdots,N. \tag{6}$$

又令 $I(G)$表$\mathfrak{S}$关于重心 G 的转动惯量. 根据熟知的事实有

$$I(x_i)=I(G)+N|x_i-G|^2. \tag{7}$$

于是

$$\sum_{1\leqslant k<l\leqslant N}(I(x_k)-I(x_j))^2=N^2,\quad \sum_{1\leqslant k<l\leqslant N}(a_k^2-a_l^2)^2=\frac{N^3(N-1)}{2}L(\mathfrak{S}). \tag{8}$$

又

$$
\begin{aligned}
\left(\sum_{1\leqslant i<j\leqslant N} a_{ij}^2\right)^2 &= \frac{1}{4}\left(\sum_{i=1}^{N}\left(\sum_{j=1}^{N} a_{lj}^2\right)\right)^2 \\
&= \frac{N}{4}\left(\sum_{i=1}^{N}\left(\sum_{j=1}^{N} a_{ij}^2\right)^2\right) - \frac{1}{4}\sum_{1\leqslant k<l\leqslant N}\left(\sum_{j=1}^{N} a_{kj}^2 - \sum_{j=1}^{N} a_{lj}^2\right)^2 \\
&= \frac{N}{4}\left(\sum_{i=1}^{N}\left(\sum_{j=1}^{N} a_{ij}^2\right)^2\right) - \frac{1}{4}\sum_{1\leqslant k<l\leqslant N}(I(x_k)-I(x_l))^2.
\end{aligned} \tag{9}
$$

现在令 Δ_{ijk} 表示以 x_i, x_j, x_k 为顶点的三角形的面积. 由熟知的面积公式我们有

$$
\begin{aligned}
16\sum_{1\leqslant i<j<k\leqslant N}\Delta_{ijk}^2 &= \sum_{1\leqslant i<j<k\leqslant N}(2a_{ij}^2a_{jk}^2 + 2a_{jk}^2a_{ki}^2 + 2a_{ki}^2a_{ij}^2 - a_{ij}^4 - a_{jk}^4 - a_{ki}^4) \\
&= 2\sum_{i=1}^{N}\left(\sum_{1\leqslant k<l\leqslant N} a_{ik}^2a_{il}^2\right) - (N-2)\sum_{1\leqslant i<j\leqslant N} a_{ij}^4.
\end{aligned}
$$

即

$$
2\sum_{i=1}^{N}\left(\sum_{1\leqslant k<l\leqslant N} - a_{ik}^2a_{il}^2\right) = (N-2)\sum_{1\leqslant i<j\leqslant N} a_{ij}^4 + 16N_2. \tag{10}
$$

将(8)和(10)代入(9),整理后得到

$$
N_1^2 = \frac{N^2}{4}\sum_{1\leqslant i\leqslant j\leqslant N} a_{ij}^4 + 4NN_2 - \frac{N^3(N-1)}{8}L(\mathfrak{S}). \tag{11}
$$

将(5)代入(11),整理后得到

$$
M_4(\mathfrak{S}) \geqslant \frac{N-1}{N}\,\frac{n+1}{n}M_2^2(\mathfrak{S}) + L(\mathfrak{S}).
$$

此即我们要证明的(4). 至于等号成立的条件,由于推导过程中不等号只出现于(5),故(5)等号成立的条件也就是(4)等号成立的条件. 这条件是:$\mathfrak{S}$关于其重心的惯量椭球是一个球.

定理 1a 对于 E^n 中任一个由 N 个点($N>n$)组成的有限点集$\mathfrak{S}$有不等式

$$
M_4(\mathfrak{S}) \geqslant \frac{N-1}{N}\,\frac{n+1}{n}M_2^2(\mathfrak{S}) \tag{12}
$$

成立,等号成立当且仅当$\mathfrak{S}$是 E^n -伪对称的.

证 (12)成立的充分必要条件是(4)的等号成立,而且 $L(\mathfrak{S})=0$. 显然 $L(\mathfrak{S})=0$ 等价于条件(ⅰ)和(ⅱ),而(4)的等号成立等价于条件(ⅲ).

这个定理建立了各点相互距离的四次幂平均同二次幂平均之间的一个不等式关系,而等号成立的条件恰好是前面所定义的伪对称性.

§2 应用于紧致曲面

设 $\mathscr{F}$ 是 E^n 中的紧致曲面,令 F 表 $\mathscr{F}$ 的面积. 取 $\mathscr{F}$ 的重心为坐标原点. 我们采用下面的量 $L(\mathscr{F})$ 来刻划 $\mathscr{F}$ 与超球面之间的偏离

$$
L(\mathscr{F}) = \frac{1}{F^2}\int_{\mathscr{F}}\int_{\mathscr{F}}(|x|^2 - |y|^2)^2 d\sigma(x)d\sigma(y). \tag{13}
$$

又令 $M_r(\mathscr{F})$ 表 $\mathscr{F}$ 的 r 阶弦幂平均如引言所述. 则由定理 1 取极限立即得到

定理 2 对 E^n 中任一个紧致曲面 $\mathscr{F}$ 有

$$M_4(\mathscr{F})\geqslant\frac{n+1}{n}M_2^2(\mathscr{F})+L(\mathscr{F}) \tag{14}$$

成立.

作为这定理的一个推论我们可以得到,使一个超球面区别于其它曲面的一个新的特征

定理 2a 对 E^n 中任意一个紧致曲面 $\mathscr{F}$ 有

$$M_4(\mathscr{F})\geqslant\frac{n+1}{n}M_2^2(\mathscr{F}), \tag{15}$$

等号当且仅当 $\mathscr{F}$ 是一个超球面时成立.

应该注意到,这里涉及的弦幂积分是二重曲面积分,它不同于经典积分几何中的弦幂积分 I_r 或 J_r. 在那里 J_r 是凸体上的二重积分,而 $I_r=\frac{1}{2}r(r-1)J_{r-n-1}$. 关于凸体的弦幂积分 I_r(或 J_r)之间的不等式一直是有兴趣的课题[2~4]. 另一方面,目前我们还不知道怎样把本文的不等式推广到一般的 M_r 上去.

§3　伪对称有限点集的代数特征

首先考虑满足定义 1 的部分条件——条件(ⅰ)和(ⅱ)的点集. 这类点集引起兴趣[5~6]是因为它能使某些泛函取到极值. $\mathfrak{S}$及有关记号如 §1 所述. 一般将 N 阶方阵(a_{ij}^2)叫做$\mathfrak{S}$的"平方距离阵",其重要程度可能不亚于 Cayley-Menger 阵[7].

定义 2 有限点集$\mathfrak{S}$的平方距离阵的特征多项式 $f(\lambda)=\det(a_{ij}^2-\lambda E)$ 叫做$\mathfrak{S}$的特征多项式. 又将镶边行列式

$$g(\lambda)=\begin{vmatrix} 0 & 1 & 1 & \cdots & 1 \\ 1 & -\lambda & a_{12}^2 & \cdots & a_{1N}^2 \\ 1 & a_{21}^2 & -\lambda & \cdots & \\ 1 & \vdots & & \ddots & \vdots \\ \vdots & \vdots & & & \vdots \\ 1 & a_{N1}^2 & & & -\lambda \end{vmatrix} \tag{16}$$

叫做$\mathfrak{S}$的次特征多项式.

引理 4 设$\mathfrak{S}$是满足条件(ⅰ)和(ⅱ)的 N 个点的集,则其特征多项式与次特征多项式有关系为

$$Nf(\lambda)=(\lambda-c)g(\lambda). \tag{17}$$

这里 $c=2NR^2$, R 是$\mathfrak{S}$所在球面的半径.

这引理的推导从略. 由之可直接推出

系 1 对于满足条件(ⅰ)和(ⅱ)的 N 个点的集$\mathfrak{S}$,它的次特征多项式 $g(\lambda)$ 的根都是特征多项式 $f(\lambda)$ 的根,此外 $f(\lambda)$ 还有一根为 $2NR^2$.

为弄清伪对称集的代数特征还需要下述的

引理 5 如果有限点集$\mathfrak{S}$的凸包是 n 维的,而且$\mathfrak{S}$关于其重心的惯量椭球是一个球,则$\mathfrak{S}$

的次特征多项式只有 n 个非零根，这 n 个非零根是相重的(n 重根).

这个结果是已知的，请参看文献[8]中的推论 4.1. 下面将建立本节的主要定理.

定理 3 设$\mathfrak{S}$是 N 个点的集($N>n$)，则$\mathfrak{S}$为 E^n -伪对称集的充分必要条件是：$\mathfrak{S}$的特征多项式与次特征多项式之间的关系为：

$$Nf(\lambda)=(\lambda-c)g(\lambda),$$

而且 $g(\lambda)$只有 n 个非零根，这 n 个非零根相重.

证 由引理 4 和引理 5 知，这条件的必要性是显然的. 下面证条件的充分性.

设 $g(\lambda)$的 n 个非零根为 n 重根 λ_0. 按 $g(\lambda)$的定义将多项式展开. 注意到 $g(\lambda)$的其它根都是零，由根与系数的关系容易算出

$$n\lambda_0=-\frac{2}{N}\sum_{1\leqslant i<j\leqslant N}a_{ij}^2. \tag{18}$$

其次考虑 $f(\lambda)$的根，由假设 $f(\lambda)$除拥有 $g(\lambda)$的全部根外还另有一根 c. 注意到平方距离阵的迹必然是零，可知 $f(\lambda)$所有根之和为零. 于是有

$$c=-n\lambda_0. \tag{19}$$

将多项式 $f(\lambda)$按定义展开，注意到它的 n 个根是 λ_0，一个根是 $-n\lambda_0$，而其余的根是零. 由根与系数的关系容易算出

$$\frac{n(n+1)}{2}\lambda_0^2=\sum_{1\leqslant i<j\leqslant N}a_{ij}^4. \tag{20}$$

将(18)与(20)比较就得

$$M_4(\mathfrak{S})=\frac{N-1}{N}\frac{n+1}{n}\cdot M_2^2(\mathfrak{S}).$$

由定理 1 知，$\mathfrak{S}$是 E^n -伪对称的. 定理 3 证毕.

参考文献

[1]杨路，张景中. 关于有限点集的一类几何不等式. 数学学报，1980，23(5)：740－749.

[2]Santaló L A. Integral Geometry and Geometric Probability. Addison-Wesley Pub.，1976.

[3]吴大任. On the relations between the integrals for the power of chords of a convex set. 第一次全国微分几何学术讨论会，1983.

[4]任德麟. The unified inequalities governing the integrals for the power of chords. 第二次国际双微讨论会(DD2)，1981.

[5]Alexander R，Stolarsky K B. Extremal problems of distance geometry related to energy integrals. Trans. Amer. Math. Soc.，1974(193)：1－31.

[6]Björk G. Distributions of positive mass which maximize a certain generalized energy integral，Ark. Mat.，1956(3)：255－269.

[7]Blumenthal L M. Theory and Applications of Distance Geometry. New York，1970.

[8]张景中，杨路. 有限点集在伪欧空间的等长嵌入. 数学学报，1981，24(4)：481－487.

线段上连续自映射嵌入半流的充分必要条件

杨　路　张景中

（中国科学院成都分院）

关于离散动力系统的研究，近期已不限于考察自同胚的迭代，而且也注意于连续自映射迭代生成的半动力系统. 例如，P. Collet 等的书[1]，Block、熊金城、周作领等作者的近期工作[2~5]，以及张筑生的博士论文[6]. 对应于自同胚嵌入流的问题[7~9]，提出连续自映射嵌入半流的问题，是十分自然的事.

设 M 是拓扑空间，$\mathbf{R}^+=[0,+\infty)$，如果连续映射 $\varphi:\mathbf{R}^+\times M\to M$ 满足

1° $\varphi(0,x)=x(\forall x\in M)$；

2° $\varphi(s+t,x)=\varphi(s,\varphi(t,x))(\forall s,t\in\mathbf{R}^+,x\in M)$.

则称 φ 是 M 上的半流（C^0 半流）.

设 f 是 M 上的自映射，φ 是 M 上的半流. 如果有 $s\geqslant 0$，使 $\varphi(s,x)=f(x)(\forall x\in M)$，则称 f 可以嵌入半流 φ. 对于 M 是线段的情形，同胚嵌入流的问题早已有丰富的成果[8~12]. 但连续自映射嵌入半流的条件，尚属未知.

本文找到了线段上连续自映射嵌入 C^0 半流的充分必要条件；这条件说明，线段上的半流一般虽不可逆，其构造却出人意料地特殊.

定理 1　线段 I 上的连续自映射 f 可嵌入 C^0 半流的充要条件是

1）　f 在 I 上单调不减；并且

2）　如果 $x_1\neq x_2$ 而 $f(x_1)=f(x_2)=x^*$，则 x^* 是 f 的不动点.

换言之，f 除了在不动点的某邻域可能取常值之外，在其它点都是严格递增的.

为了证明定理 1，我们需要证明几个引理，引理中的 I 均指开的、闭的或半开半闭区间.

引理 1　若连续映射 $f:I\to I$ 可嵌入 I 上的 C^0 半流，则 f 的一切周期点都是不动点.

证　记 $f^t(x)=\varphi(t,x)$ 为 f 所嵌入的半流，并不妨设 $f(x)=\varphi(1,x)=f^1(x)$. 用反证法：设 x_0 是 f 的 k 阶周期点，$k>1$. 考虑点集

$$J=\{y\,|\,y=f^t(x_0),t\geqslant 0\}.$$

显然 $J\subset I$，且对一切 $t\geqslant 0$，有 $f^t(J)\subset J$.

对任一 $y\in J$，总有

$$f^k(y)=f^{k+t}(x_0)=f^t(f^k(x_0))=f^t(x_0)=y.$$

因而对 $y_1,y_2\in J$，只要 $f(y_1)=f(y_2)$，总有

$$y_1=f^{k-1}(f(y_1))=f^{k-1}(f(y_2))=y_2,$$

本文刊于《数学学报》，第 29 卷第 2 期，1986 年.

从而知 f 在 J 上是严格单调的. 于是 $f^{\frac{1}{2}}$ 也在 J 上严格单调. 由 $f^{\frac{1}{2}}(J)\subset J$,故 $f^{\frac{1}{2}}\circ f^{\frac{1}{2}}=f$ 在 J 上严格递增,从而 f 在 J 上只能有周期为 1 的周期点,此与 $x_0\in J$ 矛盾. 引理 1 证毕.

引理 2 若连续映射 $f:I\to I$ 可嵌入 I 上的半流 $\varphi(t,x)=f^t(x)$,则对 $\forall x_0\in I,t>0$,由 $f^t(x_0)=x_0$ 知 $f(x_0)=x_0$,反之亦然.

证 由引理 1 可知,若 $f^t(x_0)=x_0$,则必有 $f^{\frac{t}{2^n}}(x_0)=x_0(n=1,2,3,\cdots)$. 不妨设 $f^1(x)=f(x)$,又令 $\dfrac{1}{t}$ 之二进表示为

$$\frac{1}{t}\sum_{n=0}^{\infty}\frac{a_n}{2^n}\quad(a_0\text{ 为非负整数,其它 }a_n=1\text{ 或 }0),$$

于是对任意正整数 m,有

$$\varphi\Big(\sum_{n=0}^{m}\frac{a_nt}{2^n},x_0\Big)=x_0,$$

由连续性,令 $m\to+\infty$ 即得 $f(x_0)=\varphi(1,x_0)=x$. 交换 f 与 f^t,可知由 $f(x_0)=x_0$ 能推知 $f^t(x_0)=x_0$. 引理 2 证毕.

引理 3 若连续映射 $f:I\to I$ 可嵌入 I 上的 C^0 半流,则对任意 $x_0\in I$,当 $f(x_0)>x_0(<x_0)$时,$f^t(x_0)$关于 t 单调非减(非增);如果令 e 表示大于(或小于)x_0 的 f 的最小(大)不动点[①],则当 $f^t(x_0)<e$(或$>e$)时,$f^t(x_0)$关于 t 严格递增(递减),而且

$$\lim_{t\to+\infty}f^t(x_0)=e.$$

证 由 $f^1(x_0)>x_0(<x_0)$,若有 $t_1>0$ 使 $f^{t_1}(x_0)<x_0(>x_0)$,则有 $t_2>0$ 使 $f^{t_2}(x_0)=x_0$,由引理 2 知 $f(x_0)=x_0$,此与所设 $f(x_0)>x_0(<x_0)$矛盾. 故对一切 $t>0$,有 $f^t(x_0)>x_0$ $(<x_0)$.

下面指出,若 t 在$[0,t^*)$上变化时恒有 $x_0\leqslant f^t(x_0)<e$ $(x_0\geqslant f^t(x_0)>e)$,则 $f^t(x_0)$关于 $t\in[0,t^*]$严格递增(递减). 事实上,若不然,则有 $t_2>t_1$,使 $f^{t_2}(x_0)=f^{t_1}(x_0)\in[x_0,e)$(或$(e,x_0]$),令 $y_0=f^{t_1}(x_0)$,$s=t_2-t_1>0$,则有 $f^s(y_0)=y_0$,从而由引理 2 知 $f(y_0)=y_0$,此与$[x_0,e)$($(e,x_0]$)内无不动点之假设矛盾.

如果对某个 t^* 有 $f^{t^*}(x_0)=e$,可断言对一切 $t\geqslant t^*$ 有 $f^t(x_0)=e$. 事实上,由 $f(e)=e$,故对一切 $t\geqslant0$ 有 $f^{t^*+t}(x_0)=f^t(e)=e$. 于是当 $t\in[t^*,+\infty)$时 $f^t(x_0)$取常值 e. 故 $f^t(x_0)$的单调性证毕. 且有 $\lim\limits_{t\to+\infty}f^t(x_0)=z_0\leqslant e$(或$\geqslant e$). 显然,$f(z_0)=\lim\limits_{t\to+\infty}f^{t+1}(x_0)=z_0$,但$[x_0,e)$上($(e,x_0]$上)无不动点,故 $z_0=e$. 引理 3 证毕.

下面往证定理 1 所述条件之必要性:

设 f 的全体不动点之集为 E,则 $I\backslash E$ 由一些互不相交的区间 Δ 构成. $f(x)-x$ 在每个 Δ 上不变号,且当 $x\in\Delta$ 时 $f(x)\in\bar{\Delta}$,这是由引理 3 结论显然可以推出的. 现在考虑 Δ 内任两点 $x_2>x_1$,则当 $f(x)-x$ 在 Δ 上为正(负)时,有 $t^*>0$,使 $f^{t^*}(x_1)=x_2(f^{t^*}(x_2)=x_1)$,(因 $\lim\limits_{t\to+\infty}f^t(x)$是 Δ 的端点)从而必有

$$f(x_2)=f[f^{t^*}(x_1)]\geqslant f(x_1),$$

---

① 对某些 x_0,这样的不动点可能不存在. 对于这种 x_0,令 e 表 I 的右(左)端点.

$$(f(x_1)=f[f^{t*}(x_2)]\leqslant f(x_2)),$$

这里等式仅当 $f(x_1)$(或 $f(x_2)$)是 f 的不动点时成立. 条件的必要性证毕.

下面往证条件的充分性. 为此我们还要一个引理:

引理 4 设 $f:[a,b]\to[a,b]$连续,$f(a)=a$,$f(b)=b$,当 $x\in(a,b)$时 $f(x)>x$. 有 $c\in(a,b]$使 f 在$[a,c]$上严格递增且$[c,b]$上恒为常值. 则 f 可嵌入 C^0 半流.

证 当 $c=b$ 时,f 是$[a,b]$的自同胚,问题早已解决,见文献[10]或文献[8]. 这里只就 $c<b$ 的情形来证明. 考虑连续映射 $g:[0,1]\to[0,1]$,

$$g(x)=\begin{cases}2x & \left(x\in\left[0,\dfrac{1}{2}\right]\right),\\ 1 & \left(x\in\left[\dfrac{1}{2},1\right]\right),\end{cases}$$

易验证,$g(x)$可嵌入半流 $\varphi:[0,+\infty)\times[0,1]\to[0,1]$,

$$\varphi(t,x)=\begin{cases}2^t x & \left(x\in\left[0,\dfrac{1}{2^t}\right]\right),\\ 1 & \left(x\in\left[\dfrac{1}{2^t},1\right]\right),\end{cases}$$

于是,我们只要证明:存在同胚 $h:[a,b]\to[0,1]$,使 $f=h^{-1}\circ g\circ h$,即下列交换图成立:

$$\begin{array}{ccc}[a,b] & \xrightarrow{f} & [a,b]\\ {\scriptstyle h}\downarrow & & \downarrow{\scriptstyle h}\\ [0,1] & \xrightarrow{g} & [0,1].\end{array}$$

下面着手构造这样的 h:

1° 对 $x\in[c,b]$,取 $h(x)$为任一个满足 $h(b)=1$,$h(c)=\frac{1}{2}$的严格递增连续函数,例如

$$h(x)=\frac{1}{2}\left[\frac{x-c}{b-c}+1\right]\quad(c\leqslant x\leqslant b),$$

2° 记 $x_0=c$,$x_{n+1}=f^{-1}(x_n)$①,显然有 $x_n\to a(n\to+\infty)$,且 $x_{n+1}<x_n$. 在$[x_1,x_0]$上取 $h(x)$为任一个满足 $h(x_0)=\frac{1}{2}$,$h(x_1)=\frac{1}{4}$的严格递增连续函数.

3° 若已确定了 $h(x)$在$[x_n,x_{n-1}]$($n\geqslant1$)上的定义,且

$$h(x_n)=\frac{1}{2^{n+1}},h(x_{n-1})=\frac{1}{2^n},$$

$h(x)$在$[x_n,x_{n-1}]$上严格递增,则我们用等式

$$h(x)=g^{-1}\circ h\circ f(x)\quad(x\in[x_{n+1},x_n])$$

来确定 $h(x)$在$[x_{n+1},x_n]$上的值. 易验证 h 在诸 x_k 处取值是确定的.

4° 最后,令 $h(a)=0$.

这样定义的 h,确是$[a,b]$到$[0,1]$的同胚,而且显然满足 $h^{-1}\circ g\circ h=f$,从而 f 可在$[a,b]$

---

① 在$[a,c]$上 f 的反函数存在并用 f^{-1}表示.

上嵌入 C^0 半流.

下面我们来完成定理 1 的条件的充分性证明. 不妨假定 I 的两端都是 f 的不动点. 否则,我们可以先对 f 的定义域开拓,使两端为不动点,嵌入某个流之后再取其限制,如作者在[5]中定理 7 的证明中所做的那样. 在此假定之下,$I\backslash E$ 的每个构成区间的两端点都是 f 的不动点. 显然,f 只能够在一个端点附近取常值. 如 $f(x)-x$ 在构成区间 Δ 内为正,由引理 4,它可以在 Δ 上嵌入 C^0 半流. 若 $f(x)-x$ 在 Δ 内为负,而 Δ 两端点为 $v>u$,作变换

$$\tilde{f}(x)=u+v-f(u+v-x),(x\in\Delta),$$

则 $\tilde{f}$ 满足引理 4 之条件,从而可在 Δ 上嵌入 C^0 半流,f 也就可以嵌入了.

剩下的事,不过是把 f 在诸 Δ 上嵌入的 C^0 半流凑成 I 上的 C^0 半流. 这当然是显然可行的. 定理 1 证毕.

定理 1 多少有点出乎意料:可嵌入 C^0 半流的 C^0 映射居然如此特殊! 这是因为,半流定义是很强的要求. 我们猜想,如把半流定义中的要求 1°去掉——姑且把去掉了 1°之后的 φ 叫做拟半流吧——,可嵌入的函数,将要包括一些非单调函数. 此外,作为定理 1 应用的例,显然有

推论 线段上的解析自映射 f 如果不能嵌入 C^0 流而能嵌入 C^0 半流,则 f 恒等于常数.

参考文献

[1] Collet P, et al. Iterated maps on the interval as dynamical systems. Birkhäuser Boston, 1980.

[2] Block L. Continuous maps of the interval with finite nonwandering set. Trans, Amer. Math. Soc., 1978(240): 221 - 230.

[3] 熊金城. 线段连续自映射的迭代极限点. 中国科学技术大学学报, 1981, 11(3): 113.

[4] 周作领, 刘旺金. 线段自映射拓扑熵为零的一个充分条件. 数学进展, 1982, 11(3): 216 - 219.

[5] 张景中, 杨路. 论逐段单调连续函数的迭代根. 数学学报, 1983, 26(4): 399 - 412.

[6] 张筑生. 微分半动力系统的不变集. 中国科学, (待发表).

[7] 张筑生. 圆周上自同胚的嵌入流与变换群的作用. 数学学报, 1981, 24(6): 954 - 957.

[8] Fort M K. The embedding of homeomorphisms in flows. Proceedings of Amer. Math. Soc., 1955(6): 960 - 967.

[9] 张景中, 杨路. 单变元实迭代半群的存在唯一准则. 北京大学学报(自然科学版), 1982(6): 23 - 45.

[10] Boedewadt U T. Zur Iteration reeller Funktionen. Zeitsc. Math., 1944, 49(3): 497 - 516.

[11] Sternberg S. Local C^n transformations of the real line. Duke Math. J., 1957(24): 97 - 102.

[12] Ping-Fun Lam. Embedding a differentiable homeomophism in a flow. J. Diff. Equations, 1978(30): 31 - 40

The Second Type of Feigenbaum's Functional Equations

Yang Lu(杨　路) and Zhang Jing-zhong(张景中)

(Institute of Mathematical Sciences, Academia Sinica, Chengdu)

ABSTRACT

Recently the topics concerning the Feigenbaum phenomenon have become important for research of dynamical systems. Among these topics, a key problem is to observe the existence and behavior of solutions to the Feigenbaum functional equation. This paper raises another type of Feigenbaum's functional equations which would play the same role as the original equation posed by Feigenbaum, and, it is more convenient for research because this type has better geometrical intuitiveness. The concrete connection between the two types of equations is shown. A feasible method is given to construct the continuous valley-unimodal solutions of the second type of Feigenbaum functional equations. The existence of its C^r solutions is also shown. Finally, some further problems are posed.

Ⅰ. INTRODUCTION

The remarkable Feigenbaum phenomenon in the research of dynamical systems originated from the population equation

$$x_{n+1}=x_n(a-bx_n) \tag{1.1}$$

or its equivalent form

$$x_{n+1}=1-\mu x_n^2 (0<\mu<2, x\in[-1,1]). \tag{1.2}$$

This can be also due to the iterations of the function

$$f_\mu(x)=1-\mu x^2 (0<\mu<2, x\in[-1,1]). \tag{1.3}$$

As usual, if x_0 satisfies

$$f^n(x_0)=x_0 (\text{and } f^k(x_0)\neq x_0 \text{ for } k=1,2,\cdots,n-1), \tag{1.4}$$

then we call x_0 an n-periodic point of f. Furthermore, if

本文刊于《中国科学》(A辑)》,第29卷第12期,1986年.

$$(f^n(x))'_{x=x_0}=0, \tag{1.5}$$

we call x_0 a superstable periodic point. Mathematicians and other scientists have paid great attention to the problem of whether a superstable periodic point exists for a certain mapping or not.

A few years ago, Feigenbaum[1] observed the iterations of the mapping family (1.3). He discovered that the dynamical properties of f_μ have some catastrophes at a series of critical μ values. When μ increased to μ_n ($\mu_n>\mu_{n-1}, n=1,2,\cdots$), there appeared the superstable 2^n-periodic point of f_{μ_n}. It is more interesting that $\mu_n\to\mu_\infty$ and

$$(\mu_\infty-\mu_n)^{\frac{1}{n}}\to\delta^{-1}, \tag{1.6}$$

with

$$\delta=4.669201609\cdots \tag{1.7}$$

which is a universal constant available not only for (1.3) but also for many physical problems[2] concerning the bifurcation of periodic orbit.

For explaining this phenomenon, Feigenbaum posed some hypotheses proved in some particular cases[3,4]. Among them an important key hypothesis is the existence of solutions to the functional equation

$$\begin{cases} g(x)=-\dfrac{1}{\lambda}g(g(-\lambda x)), \\ g(0)=1, -1\leqslant g(x)\leqslant 1, \end{cases} \tag{1.8}$$

($\lambda\in(0,1)$ to be determined, $x\in[-1,1]$),

where $g(x)$ satisfies certain conditions. For example, he supposed

$$g(x)=1-a|x|^{1+\varepsilon}+o(|x|^{1+\varepsilon}), (a>0), \tag{1.9}$$

and $g(x)=g(-x)$, $g'(x)$ is continuous, $g(x)$ is unimodal, and so on.

Eqs. (1.8) have become a remarkable subject to research. It was proved[4] for ε small enough that (1.8) has a solution $g(x)$ satisfying (1.9). Furthermore, it was shown[5] that (1.8) has a concave C^2 solution for some number λ_0, where $\lambda_0^2\in[0.152, 0.165)$. The proofs given in [4] and [5] are non-constructive. A numerical solution was given in [6] by computer. As to whether (1.8) has any other solutions, how the solutions are constructed and what properties they have, these problems remain to be investigated.

In this paper the authors raise another type of Feigenbaum's functional equations, give out all continuous valley-unimodal solutions and find a non-harsh condition sufficient for a continuous solution to become a C^1 solution. It was also pointed out that the problem to find the even peak-unimodal solutions can be due to the valley-unimodal solutions to the new type of Feigenbaum's equations. This will give a feasible method to construct the even peak-unimodal solutions for all kinds of (1.8).

We suggest the second type of Feigenbaum's equations,

$$\begin{cases} \varphi(x)=\dfrac{1}{\lambda}\varphi(\varphi(\lambda x)), \\ \varphi(0)=1, 0\leqslant\varphi(x)\leqslant 1, \end{cases} \tag{1.10}$$

$(\lambda \in (0,1)$ to be determined, $x \in [0,1])$,

because (1.10) is simpler than (1.8), has better geometrical intuitiveness and would play the same role as (1.8) in the research of the Feigenbaum phenomenon. The following theorem shows the close connection between (1.10) and (1.8).

Theorem 1.1. *Between the solutions of* (1.8) *and* (1.10), *there are the following connections*:

(i) *if an even function* $g(x)$ *is a continuous peak-unimodal solution of* (1.8), *then*

$$\varphi(x) = |g(x)|, (x \in [0,1]) \tag{1.11}$$

is a continuous valley-unimodal solution of (1.10);

(ii) *if* $\varphi(x)$ *is a continuous valley-unimodal solution of* (1.10), *then* $\varphi(x)$ *takes its minimum* $\varphi(\alpha)=0$ *at some* $\alpha \in (0,1)$ *and the even function*

$$g(x) = (\mathrm{sgn}(\alpha - |x|))(\varphi(|x|) \tag{1.12}$$

is a continuous peak-unimodal solution of (1.8).

Proof. (i) It is easy to check directly.

(ii) Let φ take its minimum at α. It follows from (1.10)

$$\varphi(\alpha) = \frac{1}{\lambda}\varphi(\varphi(\lambda\alpha)) \geqslant \frac{1}{\lambda}\varphi(\alpha), \tag{1.13}$$

that is, $\varphi(\alpha)=0$. On the other hand, we can assert

$$\lambda < \alpha, \tag{1.14}$$

or otherwise put $x_0 = \lambda^{-1}\alpha$,

then

$$\varphi(x_0) = \frac{1}{\lambda}\varphi(\varphi(\alpha)) = \frac{1}{\lambda}\varphi(0) = \frac{1}{\lambda} > 1,$$

which contradicts (1.10).

By (1.14), We can derive in a counter direction to the reasoning of (i) to complete the proof of (ii), and so it is omitted here.

Remark 1.1. If $g(x)$ satisfies (1.8)(or(1.10)), and $h(x)$ defined on $\mathbf{R}$(or $\mathbf{R}^+$) is a strictly monotonic continuous function such that

$$h(-\lambda x) = -\lambda h(x) \ (\text{or } h(\lambda x) = \lambda h(x)),$$

let

$$g^*(x) = h^{-1} \circ g \circ h(x).$$

Then $g^*(x)$ satisfies (1.8) (or(1.10)) as well.

Remark 1.2. If $g(x)$ satisfies (1.8)(or(1.10)), and let $\tau > 0$ and

$$h(x) = (\mathrm{sgn}\, x)|x|^{\frac{1}{\tau}},$$

$$g^*(x) = h^{-1} \circ g \circ h(x),$$

then $g^*(x)$ satisfies (1.8)(or(1.10)) by exchanging λ for λ^τ.

By using the transformation given in Remark 1.2, the $\varphi(x)$ given in (1.11) can be rubbed smooth at $x=\alpha$, and so the C^k solutions of (1.10) are obtained from those of (1.8).

Conversely, the $g(x)$ given in (1.12) can be rubbed smooth at $x=0$, and so C^k solutions of (1.8) can be obtained from those of (1.10). The details are omitted here.

Perhaps, the Feigenbaum phenomenon is based on a more extensive mathematical foundation since the solutions of (1.8) are more than those given in [4−6]. It requires to investigate further whether this opinion is correct.

Ⅱ. THE METHOD OF CONSTRUCTING VALLEY-UNIMODAL CONTINUOUS SOLUTIONS

Let $\varphi(x)$ be a valley-unimodal continuous solution of (1.10). Theorem 1.1 shows that there is $\alpha\in(0,1)$ such that φ takes its minimum $\varphi(\alpha)=0$ at $x=\alpha$. It follows clearly from (1.14)

$$\varphi(1)=\lambda<\alpha=\varphi(\lambda\alpha). \tag{2.1}$$

In this case, φ has a unique fixed point $\beta=\varphi(\beta)$ in $(0,\alpha)$. As a preparation, we shall prove a result stronger than (2.1).

Lemma 2.1. *Let $\varphi(x)$ be a valley-unimodal continuous solution of (1.10), which takes $\varphi(\alpha)=0$ as its minimum at $x=\alpha$. Then*

(ⅰ) *$\varphi(x)$ has a unique fixed point $\beta=\varphi(\beta)$ in $[0,1]$, and*

$$\lambda<\beta<\alpha, \tag{2.2}$$

(ⅱ) *$\psi(x)=\varphi(\lambda x)$ has a unique periodic point*

$$\alpha=\varphi(\lambda\alpha). \tag{2.3}$$

Proof. (ⅰ) By (2.1), we know that the maximum of φ at $[\alpha,1]$ is $\lambda<\alpha$, so φ has no fixed point on $[\alpha,1]$, that is, φ has a unique fixed point $\beta\in(0,\alpha)$. Let us go to prove $\lambda<\beta$ by reduction to absurdity. If $\lambda\geq\beta$, then $\frac{\beta}{\lambda}\in[0,1]$, hence (by (1.10))

$$\varphi\left(\frac{\beta}{\lambda}\right)=\frac{1}{\lambda}\varphi\left(\varphi\left(\lambda\cdot\frac{\beta}{\lambda}\right)\right)=\frac{\beta}{\lambda}, \tag{2.4}$$

that is, $x_0=\frac{\beta}{\lambda}$ is a fixed point of φ. The uniqueness of the fixed point of φ shows $\frac{\beta}{\lambda}=\beta$, so $\lambda=1$, which contradicts $\lambda\in(0,1)$.

(ⅱ) The $\psi(x)$ has a unique fixed point $\alpha=\psi(\alpha)$ since it is strictly decreasing on $[0,1]$, and its other periodic points must be 2-periodic points. If $x_0=\psi^2(x_0)$, it follows from (1.10)

$$\lambda\varphi(x_0)=\varphi(\varphi(\lambda x_0))=\varphi(\psi(x_0)), \tag{2.5}$$

$$\lambda^2\varphi(x_0)=\lambda\varphi(\psi(x_0))=\varphi(\psi^2(x_0))=\varphi(x_0).$$

Since $\lambda^2<1$, then $\varphi(x_0)=0$ and $x_0=\alpha$. Lemma 2.1 is proved.

Lemma 2.2. *Let φ_1,φ_2 be two valley-unimodal continuous solutions of (1.10). If*

$$\left.\begin{array}{l}1^\circ\ \varphi_1(\lambda)=\varphi_2(\lambda)=a,\\ 2^\circ\ \varphi_1(x)=\varphi_2(x)\ \textit{for}\ x\in[0,a]\end{array}\right\}, \tag{2.6}$$

then $\varphi_1(x)=\varphi_2(x)$ *on the whole* $[0,1]$.

Proof. Because of (2.2), it follows $\lambda<\beta_i$, and because of decreasing φ_i, we have

$$a=\varphi_i(\lambda)>\varphi_i(\beta_i)=\beta_i>\lambda, \tag{2.7}$$

where β_i is a fixed point of φ_i (from $a>\beta_i$ and condition 2 we know $\beta_1=\beta_2$), and so φ_1 equals φ_2 on $[0,\lambda]$. If we let

$$\psi(x)=\varphi_1(\lambda x)=\varphi_2(\lambda x),(x\in[0,1]), \tag{2.8}$$

then $\psi(x)$ has a unique fixed point α on $[0,1]$. Clearly,

$$\varphi_1(\alpha)=\varphi_2(\alpha)=0. \tag{2.9}$$

Observe the two series $\{\psi^{2k}(0)\}$ and $\{\psi^{2k+1}(0)\}$, $(k=0,1,2,\cdots)$. The series $\psi^{2k}(0)$ is increasing and $\psi^{2k+1}(0)$ decreasing because $\psi^2(x)$ is an increasing function, $\psi^2(0)>0$ and $\psi^2(1)<1$. By (ii) of Lemma 2.1, the two series have the common limit α so that

$$[0,1]=\{\alpha\}\cup\left(\bigcup_{k=0}^{\infty}\Delta k\right), \tag{2.10}$$

where $\Delta k=[\psi^k(0);\psi^{k+2}(0)]$, and $[X;Y]$ denotes the interval with the endpoints X and Y.

It suffices to prove $\varphi_1(x)=\varphi_2(x)$ on every Δk. We make use of the induction on k. For $k=0$, $\Delta_0=[0,\alpha]$, from 2 in (2.6), there holds $\varphi_1(x)=\varphi_2(x)$ on Δ_0. If

$$\varphi_1(x)=\varphi_2(x),(x\in\Delta k), \tag{2.11}$$

mapping Δ_{k+1} onto Δ_k by ψ^{-1} and using (1.10), we obtain

$$\varphi_1(x)=\lambda\varphi_1(\psi^{-1}(x))=\lambda\varphi_2(\psi^{-1}(x))=\varphi_2(x). \tag{2.12}$$

By the induction and (2.10), we have proved $\varphi_1(x)$ equals $\varphi_2(x)$ on $[0,\alpha)\cup(\alpha,1]$. Since $\varphi_1(\alpha)=\varphi_2(\alpha)$ clearly, Lemma 2.2 is proved.

Theorem 2.1. *Let* λ *and* a *be real numbers satisfying* $0<\lambda<a<1$ *and* $\varphi_0(x)$ *be a continuous and strictly decreasing function on* $[0,1]$. *If*

(i) $\varphi_0(0)=1,\varphi_0(\lambda)=a,\varphi_0(a)=\lambda^2$, (2.13)

(ii) $\psi(x)=\varphi_0(\lambda x)$ *has a unique periodic point* α *on* $[0,1]$, *then there exists a unique continuous function* $\varphi(x)$ *satisfying*

$$\begin{cases}\lambda\varphi(x)=\varphi(\varphi(\lambda x)), & x\in[0,1],\\ \varphi(x)=\varphi_0(x), & x\in[0,a],\end{cases} \tag{2.14}$$

and $\varphi(x)$ *is a valley-unimodal function with minimum* $\varphi(\alpha)=0$.

Conversely, if φ_0 *is the restriction on* $[0,\varphi(\lambda)]$ *of a valley-unimodal solution to* (1.10), *then* (i) *and* (ii) *must hold.*

Proof. The $\psi^2(x)$ is strictly increasing on $[0,1]$ since $\psi(x)$ is strictly decreasing. Also,

$$\psi^2(0)=\varphi_0(\lambda)=a>0,\psi^2(1)=\varphi_0(\lambda a)<1,$$

then the series $\psi^{2k}(0)$ is strictly increasing and $\psi^{2k+1}(0)$ is strictly decreasing $(k=0,1,2,\cdots)$. By the condition (ii), the two series have the common limit $\alpha=\psi(a)$. Let

$$\Delta_k=[\psi^k(0);\psi^{k+2}(0)],(k=0,1,\cdots). \tag{2.15}$$

The φ is defined on Δ_k by induction as follows.

The function φ_0 has been defined on Δ_0. If for $k=0,1,\cdots,m$, the continuous and strictly monotonic functions φ_k have been defined such that

$$\lambda\varphi_{k-1}(\psi^{-1}(x))=\varphi_k(x) \text{ for } x\in\Delta k, \tag{2.16}$$

by setting

$$\lambda\varphi_m(\psi^{-1}(x))=\varphi_{m+1}(x) \text{ for } x\in\Delta_{m+1}, \tag{2.17}$$

we can define the continuous and monotonic functions φ_k for every Δ_k, satisfying

$$\lambda\varphi_k(x)=\varphi_{k+1}(\psi(x)),(k=0,1,2,\cdots). \tag{2.18}$$

Putting $x=\psi^k(0)$, we obtain

$$\lambda\varphi_k(\psi^k(0))=\varphi_{k+1}(\psi^{k+1}(0)). \tag{2.19}$$

Hence

$$\varphi_k(\psi^k(0))=\lambda^k\varphi(0)=\lambda^k. \tag{2.20}$$

Putting $x=\psi^{k+2}(0)$, we obtain

$$\varphi_k(\psi^{k+2}(0))=\lambda^k\varphi(\psi^2(0))=\lambda^k\varphi(a)=\lambda^{k+2}. \tag{2.21}$$

By combining (2.20) and (2.21), we have

$$\varphi_k(\psi^{k+2}(0))=\varphi_{k+2}(\psi^{k+2}(0))=\lambda^{k+2}. \tag{2.22}$$

In other words, φ_k and φ_{k+2} have the same value at the common endpoint of the intervals Δ_k and Δ_{k+2}. We can set, therefore,

$$\varphi(x)=\begin{cases}\varphi_k(x) & \text{for} \quad x\in\Delta_k,\\ 0 & \text{for} \quad x=\alpha.\end{cases} \tag{2.23}$$

The φ is a continuous unimodal function defined on $[0,1]$ because of $0<\lambda<1$, the strict monotonicity of φ_k on Δ_k, (2.21) and (2.22). Also, the $\varphi(x)$ satisfies (1.10) because of (2.18).

As to the uniqueness of $\varphi(x)$, it follows from Lemma 2.2. The conditions (i) and (ii) must be satisfied by the continuous valley-unimodal solutions to (1.10), which is a trivial fact. Theorem 2.1 is proved.

The following lemma provides another clue for constructing the continuous valley-unimodal solutions to (1.10).

Lemma 2.3. *Let φ_1,φ_2 be two continuous valley-unimodal solutions to* (1.10). *If*

$$\varphi_1(x)=\varphi_2(x)$$

for $x\in[\lambda,1]$, then the equality holds on the whole $[0,1]$.

Proof. By (2.1), there is a number $\alpha\in(\lambda,1)$ such that $\varphi_i(\alpha)=0,(i=1,2)$. Let

$$\begin{aligned}\varphi_+(x)=\varphi_1(x)=\varphi_2(x),(x\in[\alpha,1]),\\ \varphi_-(x)=\varphi_1(x)=\varphi_2(x),(x\in[\lambda,\alpha]).\end{aligned} \tag{2.24}$$

Then the inverse functions of φ_+ and φ_-, φ_+^{-1} and φ_-^{-1}, both are uniquely determined.

Denote Δ_k by $[\lambda^{k+1},\lambda^k]$. It is sufficient to prove $\varphi_1=\varphi_2$ on every $\Delta_k(k=0,1,2,\cdots)$.

This holds for $k=0$ clearly. If it holds on Δ_k for $0\leqslant k\leqslant m$ that

$$\varphi_1(x)=\varphi_2(x)=\varphi(x),$$

where $\varphi(x)$ denote the common values of φ_1 and φ_2 on $[\lambda^{m+1},1]$, then we know from (1.10) that

$$\lambda\varphi\left(\frac{x}{\lambda}\right)=\lambda\varphi_i\left(\frac{x}{\lambda}\right)=\varphi_i(\varphi_i(x)),(i=1,2,x\in\Delta_{m+1}),\tag{2.25}$$

since $\frac{x}{\lambda}\in[\lambda^{m+1},1]$. It holds $\varphi_i(x)\geqslant\varphi_i(\lambda)\geqslant\varphi_-(\lambda)>\lambda$ because $x\leqslant\lambda$. By (2.24), if $\varphi_i(x)\in[\alpha,1]$, then (2.25) can be written as

$$\lambda\varphi\left(\frac{x}{\lambda}\right)=\varphi_+(\varphi_i(x)),\tag{2.26}$$

hence

$$\varphi_i(x)=\varphi_+^{-1}\left(\lambda\varphi\left(\frac{x}{\lambda}\right)\right),(x\in\Delta_{m+1}).\tag{2.27}$$

if $\varphi_i(x)\in[\lambda,\alpha]$, then (2.25) can be written as

$$\lambda\varphi\left(\frac{x}{\lambda}\right)=\varphi_-(\varphi_i(x))\tag{2.28}$$

hence

$$\varphi_i(x)=\varphi_-^{-1}\left(\lambda\varphi\left(\frac{x}{\lambda}\right)\right),(x\in\Delta_{m+1}).\tag{2.99}$$

The equality $\varphi_1(x)=\varphi_2(x)$ holds, therefore, in both cases for $x\in\Delta_{m+1}$. By induction, $\varphi_1(x)=\varphi_2(x)$ on every Δ_k. Lemma 2.3 is proved. Thus, we have obtained another method to construct the continuous valley-unimodal solutions to (1.10):

Theorem 2.2. *Let λ and α be real numbers satisfying $0<\lambda<\alpha<1$ and $\varphi_0(x)$ be a continuous function on $[\lambda,1]$. If*

i) $\varphi_0(1)=\lambda,\varphi_0(\alpha)=0,\varphi_0(\varphi_0(\lambda))=\lambda^2,\varphi_0(\lambda)>\lambda,$ (2.30)

ii) *φ_0 is strictly increasing on $[\lambda,\alpha]$ and strictly decreasing on $[\alpha,1]$, then there exists a unique valley-unimodal function $\varphi(x)$ satisfying*

$$\begin{cases}\lambda\varphi(x)=\varphi(\varphi(\lambda x)),x\in[0,1],\\ \varphi(x)=\varphi_0(x), \qquad x\in[\lambda,1].\end{cases}\tag{2.31}$$

Conversely, if φ_0 is the restriction on $[\lambda,1]$ of a continuous valley unimodal solution to (1.10), then i) *and* ii) *must hold.*

Proof. Here we give only the procedure for constructing $\varphi(x)$. As to the details for proving, readers may refer to the method of the proof to Theorem 2.1 and use Lemma 2.3. We construct $\varphi(x)$ by the following method. Set

$$\left.\begin{aligned}\varphi_+(x)=\varphi_0(x),(x\in[\alpha,1]),\\ \varphi_-(x)=\varphi_0(x),(x\in[\lambda,\alpha]),\end{aligned}\right\}\tag{2.32}$$

$$\Delta_k=[\lambda^{k+1},\lambda^k],(k=0,1,2,\cdots).\tag{2.33}$$

Then, φ_0 is defined on Δ_0. For $x\in\Delta_1=[\lambda^2,\lambda]$, we set

$$\varphi_1(x)=\begin{cases}\varphi_+^{-1}\left(\lambda\varphi_0\left(\frac{x}{\lambda}\right)\right),(x\in[\lambda^2,\lambda\alpha]),\\ \varphi_-^{-1}\left(\lambda\varphi_0\left(\frac{x}{\lambda}\right)\right),(x\in[\lambda\alpha,\lambda]),\end{cases} \tag{2.34}$$

and then give the inductive definition

$$\varphi_{k+1}(x)=\varphi_+^{-1}\left(\lambda\varphi_k\left(\frac{x}{\lambda}\right)\right),(x\in\Delta_{k+1}). \tag{2.35}$$

Finally, we define

$$\varphi(x)=\begin{cases}1, & (x=0),\\ \varphi_k(x), & (x\in\Delta_k).\end{cases} \tag{2.36}$$

The definition is acceptable and $\varphi(x)$ is a continuous valley-unimodal function because of ⅰ) and ⅱ). The other details are omitted.

So far we can construct all the continuous valley-unimodal solutions to (1.10) feasibly. The next section will answer whether the unimodal C^1 solutions can be induced along this clue.

Ⅲ. The Method of Constructing Continuously Differentiable Solutions

Using the constructing method given in Theorem 2.1 and restricting the initial function $\varphi_0(x)$ suitably, We can obtain the C^1 solutions to (1.10).

Theorem 3.1. *If $\varphi_0(x)$ is defined on $[0,a]$, $0<\lambda<a<1$, such that*

1° $$\varphi_0(0)=1,\varphi_0(\lambda)=a,\varphi_0(a)=\lambda^2, \tag{3.1}$$

2° *$\psi(x)=\varphi_0(\lambda x)$ has a unique periodic point $\alpha=\psi(\alpha)$,*

3° *$\varphi_0(x)=g(x^\tau)$, where $\tau>0$, $g(x)$ is continuously differentiable on $[0,a]$ and $g'(x)<0$,*

4° $$\varphi_0'(\lambda)\cdot\varphi_0'(a)=\lambda^{1-\tau}, \tag{3.2}$$

5° $$|\varphi_0'(\lambda\alpha)|>1, \tag{3.3}$$

then the solution $\varphi(x)$ of Eq. (1.10), which is constructed from φ_0 by the method described in the proof of Theorem 2.1, is continuously differentiable on $(0,1]$, $\varphi'(x)<0$ for $x\in(0,\alpha)$ and $\varphi'(x)>0$ for $x\in(\alpha,1]$.

Proof. Referring to (2.15)—(2.18), we proceed to prove that φ_k and φ_{k+2} have the same derivative at $\psi^{k+2}(0)$, the common endpoint of Δ_k and Δ_{k+2}. This can be verified directly for $k=0$ as follows: For the equations

$$\left.\begin{aligned}\varphi_1(\psi(x))=\lambda\varphi_0(x),\\ \varphi_2(\psi^2(x))=\lambda^2\varphi_0(x),\end{aligned}\right\}x\in[0,a], \tag{3.4}$$

deriving from x and by $\psi(x)=\varphi_0(\lambda x)$ and $\varphi_0(x)=g(x^\tau)$, we have

$$\left.\begin{aligned}&\varphi_1'(\psi(x))\cdot(\lambda x)^{\tau-1}g'((\lambda x)^{\tau})=x^{\tau-1}g'(x^{\tau}),\\&\varphi_2'(\psi^2(x))\cdot\varphi_0'(\lambda\psi(x))\cdot(\lambda x)^{\tau-1}g'((\lambda x)^{\tau})=x^{\tau-1}g'(x^{\tau}).\end{aligned}\right\}\tag{3.5}$$

Let $x\to 0$, we obtain

$$\left.\begin{aligned}&\varphi_1'(1)=\lambda^{1-\tau},\\&\varphi_2'(\psi^2(0))\cdot\varphi_0'(\lambda)\cdot\lambda^{\tau-1}.\end{aligned}\right\}\tag{3.6}$$

It follows from (3. 2) (note $\psi^2(0)=a$)

$$\varphi_2'(\psi^2(0))=\varphi_0'(\psi^2(0)).\tag{3.7}$$

Now, suppose

$$\varphi_k'(\psi^{k+2}(0))=\varphi_{k+2}'(\psi^{k+2}(0)).\tag{3.8}$$

Let us go to prove that

$$\varphi_{k+1}'(\psi^{k+3}(0))=\varphi_{k+3}'(\psi^{k+3}(0)).\tag{3.9}$$

In fact, it holds from (2. 18) that

$$\varphi_{k+1}(\psi(x))=\lambda\varphi_k(x),(\text{for } x\in[\psi^k(0);\psi^{k+2}(0)]).\tag{3.10}$$

From x, we derive

$$\varphi_{k+1}'(\psi(x))\cdot\psi'(x)=\lambda\varphi'(x).\tag{3.11}$$

Setting $x=\psi^{k+2}(0)$, we have

$$\varphi_{k+1}'(\psi^{k+3}(0))\cdot\psi'(\psi^{k+2}(0))=\lambda\varphi_k'(\psi^{k+2}(0)).\tag{3.12}$$

Substituting $k+2$ for k in (3. 10), deriving from x and setting $x=\psi^{k+2}(0)$, we obtain

$$\varphi_{k+3}'(\psi^{k+3}(0))\cdot\psi'(\psi^{k+2}(0))=\lambda\varphi_{k+2}'(\psi^{k+2}(0)).\tag{3.13}$$

The $\psi'(\psi^{k+2}(0))$ is non-vanishing because $g(x)<0$. Then, it follows from the inductive assumption that

$$\varphi_{k+1}'(\psi^{k+3}(0))=\varphi_{k+3}'(\psi^{k+3}(0)).\tag{3.14}$$

We have proved, therefore, the continuity of $\varphi(x)$ on $(0,\alpha)\cup(\alpha,1]$. By (3. 6), (3. 11) and $g(x)<0$, the $\varphi'(x)$ is non-vanishing except for $x=0$ and $x=\alpha$, that is, $\varphi'(x)<0$ for $x\in(0,\alpha)$ and $\varphi'(x)>0$ for $(\alpha,1]$.

Finally, from

$$\varphi(\psi^k(x))=\lambda^k\varphi_0(x),\tag{3.15}$$

we derive

$$\varphi'(\psi^k(x))\cdot\psi'(x)\cdot\psi'(\psi(x))\cdots\psi'(\psi^{k-1}(x))=\lambda^k\varphi_0'(x).\tag{3.16}$$

The $\psi'(x)$ (that is, $\lambda\varphi_0'(\lambda x)$) and $\varphi_0'(x)$ would tend to the same non-vanishing constant, or have the same order of infinitesimals as $x\to 0$. Also, the $\psi'(\psi^j(x))$ has a non-vanishing infimum for $j\geqslant 1$ and tends to the limit value, $\psi'(\alpha)=\lambda\varphi_0'(\lambda\alpha)$, as $j\to+\infty$. By combining (3. 3), it follows uniformly that

$$\lim_{k\to+\infty}\varphi'(\psi^k(x))=0,(x\in[0,1]),\tag{3.17}$$

which implies that $\varphi'(x)$ is continuous at $x=\alpha$ and $\varphi'(\alpha)=0$. Theorem 3. 1 is proved.

In this theorem, we only show the continuity of $\varphi'(x)$ on $(0,1]$ because we did not assume the existence of $\varphi_0'(0)$. We can assure the continuity of $\varphi'(x)$ at $x=0$ if $\tau\geqslant 1$;

furthermore, $\varphi'(0)\neq 0$ if $\tau=1$.

It is rather difficult to construct the continuously differentiable solutions satisfying $\varphi'(0)\neq 0$ by using the method used in Theorem 2. 2. If we permit $\varphi(0)=0$ or $-\infty$, the method analogous to Theorem 2. 2 is available. The details are omitted here.

We can use the method analogous to Theorem 3. 1 to obtain the C^k solutions to (1. 10) as well. It is difficult, however, to construct a C^2 solution $\varphi(x)$ with $\varphi''(\alpha)\neq 0$. We can see for this, by means of the following theorem.

Theorem 3. 2. *If $\varphi(x)$ is a continuous valley-unimodal solution such that*

ⅰ) $\varphi(\alpha)=0, 0<\alpha<1$,

ⅱ) *there exists a positive number σ satisfying*

$$\lim_{x\to\alpha}\frac{\varphi(x)}{|x-\alpha|^{\sigma}}=L>0, \tag{3.18}$$

then, by $\varphi_0(x)$ denoting the restriction on $[0,\varphi(\lambda)]$ of $\varphi(x)$ and setting $\psi(x)=\varphi_0(\lambda x)$ for $x\in[0,1]$, we have

$$\varphi(x)=\lim_{n\to\infty}\frac{L|\psi^n(x)-\alpha|^{\sigma}}{\lambda^n}. \tag{3.19}$$

Proof. By means of a corollary from (1. 10),

$$\lambda^n\varphi(x)=\varphi(\psi^n(x)), \tag{3.20}$$

and combining (3. 18) with (3. 20), we obtain

$$\begin{aligned}\varphi(x)&=\frac{\varphi(\psi^n(x))}{\lambda^n}=\lim_{n\to+\infty}\frac{\varphi(\psi^n(x))}{|\psi^n(x)-\alpha|^{\sigma}}\cdot\frac{|\psi^n(x)-\alpha|^{\sigma}}{\lambda^n}\\&=\lim_{n\to+\infty}\frac{L|\psi^n(x)-\alpha|^{\sigma}}{\lambda^n}.\end{aligned} \tag{3.21}$$

Theorem 3. 2 is proved.

This tells us that if $\varphi(x)$ is a C^2 function satisfying $\varphi''(\alpha)\neq 0$, then (3. 18) holds for $\sigma=2$, so the limit function $\varphi(x)$ determined by (3. 19) must coincide with the function constructed in Theorem 2. 1. It does not hold in general. This problem is rather complicated, which we shall discuss in another paper.

There are many problems about Eq. (1. 10), such as, whether there exists any convex C^2 solution or analytic solution, whether they exist uniquely, how we can classify them and what solutions have close connection with the Feigenbaum phenomenon.

Reference

[1]Feigenbaum M J. J. Stat. Phys. ,1978(19):25 - 52.

[2]郝柏林. 物理学进展,1983(3):329 - 416.

[3]Feigenbaum M J. J. Stat. Phys. ,1979(21):669 - 706.
[4]Collet P,Eckmann J-P,Lanford O E. Ⅲ,Commun. Math. Phys. ,1980(76):211 -254.
[5]Campanino M Epstein H. ibid. 1981(79):261 - 302.
[6]Lanford O E. Ⅲ,Mathematical problems in Theoretical physics,Proceedings. Lausanne, 1979,Springer-Verlag. 1980.

度量嵌入的几何判准与歪曲映像

张景中　杨　路

（中国科学院成都分院）

Banach 和 Mazur[1] 早就指出：可分度量空间总可以保距地嵌入区间上连续函数空间 C. 又由 Urysohn 的熟知定理，可分度量空间可拓扑嵌入 Hilbert 空间. 随后许多学者对什么样的度量空间能保距地嵌入 Hilbert 空间感兴趣.

在[2]中指出，可分度量空间能否保距嵌入 l^2，决定于其有限子集是否都能保距嵌入于欧氏空间. 至于度量嵌入（即保距嵌入）欧氏和 Hilbert 空间的条件，则先后在[2～4]中以不同形式被得到[5,6]. 这些条件（包括作者在[20]中给出的一个充要条件）实质上都是代数的，用起来有时不大方便. 甚至对直观性很强的几何问题，也难于直接判断.（实例见最后一节）

为处理这类问题，本文给出可分度量空间保距嵌入于 l^2 的几何判准法. 我们先给出欧氏空间一类有限子集到某度量空间的“歪曲映象”（Skew mapping）的概念. 然后证明：可分度量空间可保距嵌入 l^2 的充要条件是，它不含有欧氏空间有限子集的歪曲象.

最后给出这一几何判准的应用实例. 度量嵌入的其它应用，见[7～9]及[21].

§1. 定义及主要结果

欧氏空间 E^n 的有限点集 $\mathfrak{S}=\{A_1,A_2,\cdots,A_m\}$ 叫做一个“m 点形”. 每个 A_i 叫做 $\mathfrak{S}$ 的顶点，线段 A_iA_j 叫做 $\mathfrak{S}$ 的边 $(i,j=1,2,\cdots,m)$.

定义 1　设 $(n+2)$ 点形 $\mathfrak{S}\subset E^n$. $\mathfrak{S}$ 的一条边 A_iA_j 叫做是“红”的，如果有唯一的超平面 $\pi_{ij}(\mathfrak{S})\supset\mathfrak{S}\backslash\{A_i,A_j\}$，使得 A_i 和 A_j 位于 $\pi_{ij}(\mathfrak{S})$ 的异侧；一条边 A_iA_j 叫做是“兰”的，如果有唯一的超平面 $\pi_{ij}(\mathfrak{S})\supset\mathfrak{S}\backslash\{A_i,A_j\}$，使得 A_i 和 A_j 位于 $\pi_{ij}(\mathfrak{S})$ 的同侧. $\mathfrak{S}$ 的某些边可以是无色的：非红亦非兰. 例如，A_i 或 A_j 在 $\pi_{ij}(\mathfrak{S})$ 上，或 $\mathfrak{S}\backslash\{A_i,A_j\}$ 之维数低于 $n-1$ 因而包含它的超平面不唯一的时候.

例如，平面凸四角形具有兰边与红对角线.

定义 2　设 $(n+2)$ 点形 $\mathfrak{S}\subset E^n(M,d)$ 是一个半度量空间. 映射 $f:\mathfrak{S}\to(M,d)$ 叫做一个“歪曲映象”，如果它满足下述条件：

$|A_i-A_j|\leqslant d(f(A_i),f(A_j))$，当 A_iA_j 是 $\mathfrak{S}$ 的红边；

$|A_i-A_j|\geqslant d(f(A_i),f(A_j))$，当 A_iA_j 是 $\mathfrak{S}$ 的兰边；

并且至少有 $\mathfrak{S}$ 的某一条红边或兰边使得 $|A_i-A_j|\neq d(f(A_i),f(A_j))$. 这时 $f(\mathfrak{S})$ 叫做 $\mathfrak{S}$ 的一

本文刊于《数学学报》，第 29 卷第 5 期，1986 年.

个"歪曲象",$\mathfrak{S}$叫做 $f(\mathfrak{S})$的"歪曲原象".

定理 1 若度量空间(M,d)含有一个有穷子集 $\mathcal{R}$,$\mathcal{R}$ 在欧氏空间中有一个歪曲原象,则(M,d)不能保距地嵌入 Hilbert 空间.

定理 2 如果可分度量空间(M,d)不能保距地嵌入于 l^2,则(M,d)必含有有限子集 $\mathcal{R}$,$\mathcal{R}$ 在欧氏空间有一个歪曲原象.

换言之,可分度量空间可以保距嵌入 l^2 的充要条件是:它不含有欧氏空间有限子集的歪曲象.

§2. 不可嵌入条件的充分性

设$\mathfrak{S}=\{A_1,A_2,\cdots,A_{n+2}\}$和$\mathcal{R}=\{B_1,B_2,\cdots,B_{n+2}\}$是 $n+1$ 维欧氏空间中的两个子集. 令 $a_{ij}=|A_i-A_j|$,$b_{ij}=|B_i-B_j|$$(i,j=1,2,\cdots,n+2)$,又令

$$A=\begin{vmatrix} 0 & 1\cdots 1 \\ 1 & \\ \vdots & -\frac{1}{2}a_{ij}^2 \\ 1 & \end{vmatrix},B=\begin{vmatrix} 0 & 1\cdots 1 \\ 1 & \\ \vdots & -\frac{1}{2}b_{ij}^2 \\ 1 & \end{vmatrix},$$

并以 A_{ij},B_{ij} 分别记$-\frac{1}{2}a_{ij}^2$在A 中和$-\frac{1}{2}b_{ij}^2$在B 中的代数余子式,于是有

引理 1
$$\sum_{i=1}^{n+2}\sum_{j=1}^{n+2}a_{ij}^2B_{ij}\geqslant 0,\sum_{i=1}^{n+2}\sum_{j=1}^{n+2}b_{ij}^2A_{ij}\geqslant 0. \tag{1.1}$$
(这是作者[16]或[17]中结果的明显推论).

引理 2[10] 若$\mathfrak{S}\subset E^n$ 而且A 的一个代数余子式$A_{ij}\neq 0$,则 $A_{ij}>0$ 时点 A_i,A_j 位于超平面 $\pi_{ij}(\mathfrak{S})$之同侧;$A_{ij}<0$ 时 A_i,A_j 位于 $\pi_{ij}(\mathfrak{S})$的异侧.

引理 3 设$\mathfrak{S}=\{A_1,A_2,\cdots,A_{n+2}\}\subset E^n$ 的某一条边 A_iA_j 是红的或兰的,则对应的$A_{ij}\neq 0$.

证 以 A_{jj}^{ii} 表示从A 中去掉$-\frac{1}{2}a_{ij}^2$和$-\frac{1}{2}a_{ji}^2$所在的两行两列后剩下的子行列式,则有([18,19])

$$A_{ii}A_{jj}-A_{ij}^2=A\cdot A_{jj}^{ii}.$$

但对 E^n 中的$n+2$ 点形总有 $A=0$(见[18,19]),故得 $A_{ii}A_{jj}-A_{ij}^2=0$. 若 $A_{ij}=0$,则 A_{ii}和A_{jj}中至少有一个是 0,即 A_i 与A_j 二者至少有一个点落在超平面 $\pi_{ij}(\mathfrak{S})$上. 按定义,这时 A_iA_j 是无色的,与所设矛盾. 证毕.

定理 1 的证明 用反证法. 假定(M,d)已保距嵌入 l^2 而且有子集$\mathcal{R}\subset M$,$\mathcal{R}$在 E^n 中有歪曲原象$\mathfrak{S}=\{A_1,A_2,\cdots,A_{n+2}\}$,$\mathcal{R}$ 在 l^2 中之嵌入象设为

$$\mathcal{R}^*=\{B_1,B_2,\cdots,B_{n+2}\}.$$

由歪曲映象定义不妨认为

$$|A_i-A_j|\leqslant|B_i-B_j|,\text{当 } A_iA_j \text{ 红},$$

$$|A_i - A_j| \geqslant |B_i - B_j|, \text{当} A_iA_j \text{兰},$$

而且至少有$\mathfrak{S}$的一条红边或兰边 A_iA_j 使 $|A_i - A_j| \neq |B_i - B_j|$. 沿用引理 1 之记号，注意到 $\mathfrak{S} \subset E^n \subset E^{n+1}$，而 E^n 中 $n+2$ 个点的 Cayle-Menger 行列式为 0[5]，变换之得

$$\sum_{i=1}^{n+2} \sum_{j=1}^{n+2} a_{ij}^2 A_{ij} = 0. \tag{2.1}$$

结合引理 1(注意 $\mathscr{R}^* \subset E^{n+1}$) 得

$$\sum_{i=1}^{n+2} \sum_{j=1}^{n+2} b_{ij}^2 A_{ij} \geqslant 0 = \sum_{i=1}^{n+2} \sum_{j=1}^{n+2} a_{ij}^2 A_{ij},$$

即

$$\sum_{i=1}^{n+2} \sum_{j=1}^{n+2} (b_{ij}^2 - a_{ij}^2) A_{ij} \geqslant 0. \tag{2.2}$$

在(2.2) 左端的每一项中，当 $A_{ij} > 0$ 时由引理 2，A_iA_j 是兰边，按定理所设 $a_{ij} \geqslant b_{ij}$，故 $(b_{ij}^2 - a_{ij}^2) A_{ij} \leqslant 0$. 当 $A_{ij} < 0$ 时 A_iA_j 是红边，故 $a_{ij} \leqslant b_{ij}$，亦有$(b_{ij}^2 - a_{ij}^2) A_{ij} \leqslant 0$. 但至少有一条红边或兰边使 $a_{ij} \neq b_{ij}$，由引理 3，可知(2.2) 左端至少有一项为负，于是

$$\sum_{i=1}^{n+2} \sum_{j=1}^{n+2} (b_{ij}^2 - a_{ij}^2) A_{ij} < 0. \tag{2.3}$$

与(2.2)矛盾. 证毕.

§ 3. 不可嵌入条件的必要性

引进下述几个引理：

引理 4 设 $\mathscr{R} = \{B_1, B_2, \cdots, B_{n+2}\} \subset (M, d)$. 若 $\mathscr{R}$ 任一个$(n+1)$点子集均可保距嵌入 E^n，任一 n 点子集均可保距嵌入 E^{n-2}，则 $\mathscr{R}$ 可保距嵌入 E^n.

证 根据距离几何中关于欧氏空间合同阶的定理([5], § 38, p. 95)：半度量空间可保距嵌入 E^m 的充要条件是，它的任何一个$(m+3)$点子集都可保距嵌入 E^m. 因此，只要证明 $\mathscr{R}$ 的任一个$(n+1)$点子集都可保距嵌入于 E^{n-2} 即可. 由于 E^n 是 n 维线性空间，这是显然的. 证毕.

在以下的引理 5 到引理 9 中，使用如下记号：

ⅰ) $\pi = E^{n-1}$，为 E^n 中定向超平面，单位法向量为 $\bar{e}$，

ⅱ) $\mathscr{D} = \{A_1, A_2, \cdots, A_n\} \subset E^{n-1}$，$\mathscr{D}$ 为非退化单形，

ⅲ) $\mathscr{D}$ 中过 A_k 的高线之垂足为 H_k，

ⅳ) $h_k \bar{e}_k = A_k - H_k$，$h_k > 0$，$|\bar{e}_k| = 1$，

ⅴ) $\mathscr{D}_k = \mathscr{D} \backslash A_k$，$\pi_k$ 为 $\mathscr{D}_k$ 在 E^{n-1}中生成的 $n-2$ 维超平面，其法向量为 $\bar{e}_k$， (3.1)

ⅵ) A, A_{n+1}, A_{n+2}都是 E^n 中的点，

ⅶ) $\mathfrak{S} = \{A_1, A_2, \cdots, A_{n+2}\}$.

必要时再引入其它记号.

引理 5 若 $A_{n+1} \notin \pi$，且有实数组$\{x_1, \cdots, x_{n+1}\}$使

$$\begin{cases}1^\circ & x_1+x_2+\cdots+x_{n+1}=1,\\ 2^\circ & x_1A_1+x_2A_2+\cdots+x_{n+1}A_{n+1}=A_{n+2},\end{cases}\tag{3.2}$$

则当 $x_k>0$ 时，A_kA_{n+2}是$\mathfrak{S}$的兰边；$x_k<0$ 时，A_kA_{n+2}是$\mathfrak{S}$的红边；$x_k=0$ 时，A_kA_{n+2}无色.

证 作单形$\mathfrak{S}_{n+2}=\mathfrak{S}\backslash A_{n+2}=\mathscr{D}\cup\{A_{n+1}\}$. 设$\mathfrak{S}_{n+2}$的过 A_k 的高线为 A_kH，令 $\bar{h}=A_k-H$，由(3.2)得

$$A_{n+2}=x_k\bar{h}+x_kH+\sum_{\substack{j=1\\ j\neq k}}^{n+1}x_jA_j.\tag{3.3}$$

取 $\bar{h}$ 为$\mathfrak{S}\backslash\{A_k,A_{n+2}\}$所决定的$(n-1)$维超平面 $\pi_{k,n+2}$之法向量，则有

$$A_{n+2}-x_k\bar{h}=A^*\in\mathfrak{S}_{k,n+2}.$$

由

$$\bar{h}\cdot(A_{n+2}-A^*)=x_k|\bar{h}|^2$$

即得所要之结论. 证毕.

定义 3 若 $A\in\pi$，$H\in\pi_k$，使

$$A=H+t\bar{e}_k,\tag{3.4}$$

这里点 H 和实数 t 显然由 A 唯一确定. 对给定的 θ，$0<|\theta|<\frac{\pi}{2}$，令

$$A^*=A^*(\theta,k)=H+t\bar{e}_k\cos\theta+t\bar{e}\sin\theta,\tag{3.5}$$

称 A^* 为 A 的 $\theta-k$ 扰动.

这里 $\pi,\pi_k,\bar{e}_k,\bar{e}$ 如(3.1)所述. 以下引理 6 至引理 9 都是显然的，证明略去.

引理 6 若 $A\in\pi$，A^* 是 A 的 $\theta-k$ 扰动，则

1° $|A^*-A_j|=|A-A_j|$，$1\leqslant j\leqslant n$，$j\neq k$；

2° 若 A，A_k 在 π_k 的同侧，则$|A^*-A_k|>|A-A_k|$；　　(3.6)

3° 若 A，A_k 在 π_k 的异侧，则$|A^*-A_k|<|A-A_k|$.

引理 7 若 $A_{n+1}\in\pi$，$A_{n+2}\in\pi$，A_{n+1}^*和 A_{n+2}^*分别为 A_{n+1}，A_{n+2}的 $\theta-k$ 扰动，则 A_{n+1}^*，A_{n+2}^*和 $\mathscr{D}_k$ 中$(n-1)$个点同在 E^n 的某个 $n-1$ 维超平面上. 从而在 E^n 的 $n+2$ 点形$\{A_1,A_2,\cdots,A_n,A_{n+1}^*,A_{n+2}^*\}$中，$A_kA_{n+1}^*$，$A_kA_{n+2}^*$无色.

以下为方便，记 $\mathfrak{S}^*=\{A_1,\cdots,A_{n+1},A_{n+2}^*\}$，$\mathfrak{S}^{**}=\{A_1,\cdots,A_n,A_{n+1}^*,A_{n+2}^*\}$. 而 A_{n+1}^*，A_{n+2}^*由上下文给出.

引理 8 设 $A_{n+2}\in\pi$，$A_{n+1}\in\pi$，A_{n+2}^*为 A_{n+2}的 $\theta-k$ 扰动，则有 $\varepsilon>0$，使当$|\theta|<\varepsilon$ 时：若 A_{n+2}和 A_k 在 π_k 之同侧，则 $A_kA_{n+2}^*$是$\mathfrak{S}^*$的兰边；若 A_{n+2}和 A_k 在 π_k 之异侧，则 $A_kA_{n+2}^*$是$\mathfrak{S}^*$的红边.

引理 9 设 $A_{n+1}\in\pi$，$A_{n+2}\in\pi$，$A_{n+2}\notin\pi_k$ 而 $A_{n+1}\notin\pi_j$，A_{n+2}^*为 A_{n+2}之 $\theta-k$ 扰动，A_{n+1}^*为 A_{n+1}之 $\varphi-j$ 扰动，则有 $\varepsilon>0$，使当$|\theta|<\varepsilon$ 且$|\varphi|<|\varepsilon\theta|$时有：若 A_{n+2}^*，A_{n+1}^*在 π 之同侧而 A_{n+1}，A_{n+2}不在 π 之同侧或 A_{n+2}^*，A_{n+1}^*在 π 之异侧而 A_{n+1}，A_{n+2}不在 π 之异侧，则当 A_{n+2}与 A_k 在 π_k 之同侧(异侧)时，$A_kA_{n+2}^*$为$\mathfrak{S}^{**}$之兰边(红边)；当 A_{n+1}与 A_j 在 π_j 之同侧(异侧)时，$A_jA_{n+1}^*$为$\mathfrak{S}^{**}$之兰边(红边).

定理 2 之证明 以下保距嵌入简称嵌入. 由(M,d)不能嵌入 l^2，则有$(n+2)$点子集 $\mathscr{R}=$

$\{B_1,\cdots,B_{n+2}\}$不能嵌入E^{n+1}. 不妨设$n\geqslant 2$是这种整数中最小者，于是$\mathscr{R}$的$(n+1)$点子集均可嵌入E^n，由引理4，$\mathscr{R}$中至少有一个n点子集不能嵌入E^{n-2}. 设此n点子集为$\{B_1,\cdots,B_n\}$，并设φ_i是$\mathscr{R}\backslash B_{n+i}$到$E^{n+1}$的一个嵌入$(i=1,2)$. 记$\varphi_1(B_j)=A_j\,(j\neq n+1)$，$\varphi_2(B_j)=\widetilde{A}_j\,(j\neq n+2)$，则有$E^{n+1}$到自身的运动$T$，使

$$T(\widetilde{A}_i)=A_i,\ i=1,2,\cdots,n+1.$$

从而$T\circ\varphi_2$也是$\mathscr{R}\backslash B_{n+2}$到$E^{n+1}$的嵌入. 现在得到了$E^{n+1}$中的点集$\mathfrak{S}=\{A_1,A_2,\cdots,A_{n+2}\}$，满足

$$|A_i-A_j|=d(B_i,B_j),\ 2\leqslant i+j\leqslant 2n+2. \tag{3.7}$$

因$\{B_1,\cdots,B_n\}=\mathscr{D}$不能嵌入$E^{n-2}$，故$\mathscr{D}$生成$(n-1)$维超平面$\pi=E^{n-1}$. 设$\pi$在$E^{n+1}$中法空间的一组单位正交基为$\bar{e}_1,\bar{e}_2$，则有唯一的$\{x_i\}$，$\{y_i\}$，使

$$\begin{cases}A_{n+1}=x_1A_1+x_2A_2+\cdots+x_nA_n+x_{n+1}\bar{e}_1+x_{n+2}\bar{e}_2,\\ A_{n+2}=y_1A_1+y_2A_2+\cdots+y_nA_n+y_{n+1}\bar{e}_1+y_{n+2}\bar{e}_2,\\ x_1+\cdots+x_n=y_1+\cdots+y_n=1.\end{cases} \tag{3.8}$$

再记

$$s=\sqrt{x_{n+1}^2+x_{n+2}^2},\ t=\sqrt{y_{n+1}^2+y_{n+2}^2},\ l=\left|\sum_{i=1}^{n}(x_i-y_i)A_i\right|. \tag{3.9}$$

我们断言，如果

$$l^2+(s-t)^2\leqslant d^2(B_{n+1},B_{n+2})\leqslant l^2+(s+t)^2, \tag{3.10}$$

则$\mathscr{R}$必可嵌入E^{n+1}. 事实上，令

$$\begin{aligned}A_{n+1}^{\theta}&=\sum_{i=1}^{n}x_iA_i+\bar{e}_1 s\cos\theta+\bar{e}_2 s\sin\theta,\\ A_{n+2}^{*}&=\sum_{i=1}^{n}y_iA_i+\bar{e}_1 t,\end{aligned} \tag{3.11}$$

则θ由0变到π时，$|A_{n+1}^{\theta}-A_{n+2}^{*}|^2$由$l^2+(s-t)^2$变到$l^2+(s+t)^2$，故有$\theta=\alpha\in[0,\pi]$使

$$|A_{n+1}^{\alpha}-A_{n+2}^{*}|^2=d^2(B_{n+1},B_{n+2}). \tag{3.12}$$

注意到(3.7)，并计算$|A_i-A_{n+1}^{\alpha}|$和$|A_i-A_{n+2}^{*}|$，可知$\{A_1,\cdots,A_n,A_{n+1}^{\alpha},A_{n+2}^{*}\}$恰为$\mathscr{R}$在$E^{n+1}$中的嵌入象. 此与假设矛盾，故以下恒设(3.10)不成立，即

$$d^2(B_{n+1},B_{n+2})>l^2+(s+t)^2\ (\text{或}\ d^2(B_{n+1},B_{n+2})<l^2+(s-t)^2). \tag{3.13}$$

以下分三种情形讨论：

Ⅰ　$st>0$时，令

$$\begin{aligned}&A_{n+1}^{*}=\sum_{i=1}^{n}x_iA_i+s\bar{e}_1,\ A_{n+2}^{*}=\sum_{i=1}^{n}y_iA_i+\delta t\bar{e}_1,\\ &\delta=\begin{cases}-1, \text{当}\ d^2(B_{n+1},B_{n+2})>l^2+(s+t)^2,\\ 1, \text{当}\ d^2(B_{n+1},B_{n+2})<l^2+(s-t)^2.\end{cases}\end{aligned} \tag{3.14}$$

易验证$\mathfrak{S}^{**}$是$\mathscr{R}$的歪曲原象.

Ⅱ　$st=0$但$s+t>0$时，不妨设$s>0$，$t=0$，即$A_{n+1}\notin\pi$而$A_{n+2}\in\pi$. 在(3.8)中令$s\bar{e}=x_{n+1}\bar{e}_1+x_{n+2}\bar{e}_2$，则

$$A_{n+1}=\sum_{i=1}^{n}x_iA_i+s\bar{e},\sum_{i=1}^{n}x_i=1,$$
$$A_{n+2}=\sum_{i=1}^{n}y_iA_i,\qquad\sum_{i=1}^{n}y_i=1.\tag{3.15}$$

从而$\mathfrak{S}$在由 π 和 $\bar{e}$ 生成的空间 E^n 中.设对某个 k 有 $y_k>0$,作 A_{n+2} 的 $\theta-k$ 扰动 A_{n+2}^*:

$$A_{n+2}^*=H_{n+2}+\beta\bar{e}_k\cos\theta+\beta\bar{e}\sin\theta,\tag{3.16}$$

这里 $A_{n+2}=H_{n+2}+\beta\bar{e}_k$.由 $y_k>0$ 知,A_{n+2} 和 A_k 在 π_k 之同侧.由引理 8,$|\theta|$ 足够小时,$A_kA_{n+2}^*$为$\mathfrak{S}^*$之兰边.又由引理 6

$$|A_{n+2}^*-A_i|=|A_{n+2}-A_i|,i\neq k,1\leqslant i\leqslant n,$$
$$|A_{n+2}^*-A_k|>|A_{n+2}-A_k|.\tag{3.17}$$

当 $d^2(B_{n+1},B_{n+2})>l^2+(s+t)^2$ 时,取 θ 使$\beta\sin\theta<0$,即使 $A_{n+1}A_{n+2}^*$是$\mathfrak{S}^*$之红边.$|\theta|$足够小时$|A_{n+1}-A_{n+2}^*|^2$ 与 $l^2+s^2=|A_{n+1}-A_{n+2}|^2$ 充分接近,以致

$$d(B_{n+1},B_{n+2})>|A_{n+1}-A_{n+2}^*|.\tag{3.18}$$

由(3.7),(3.17),(3.18),知$\mathfrak{S}^*$是$\mathscr{R}$的歪曲原象.

当 $d^2(B_{n+1},B_{n+2})<l^2+(s-t)^2$ 时,取 θ 使$\beta\sin\theta>0$,使得 $A_{n+1}A_{n+2}^*$是$\mathfrak{S}^*$的兰边.$|\theta|$足够小时有

$$d(B_{n+1},B_{n+2})<|A_{n+1}-A_{n+2}^*|,\tag{3.19}$$

则易证$\mathfrak{S}^*$也是$\mathscr{R}$的歪曲原象.

Ⅲ　$s=t=0$ 时,由(3.8),

$$A_{n+1}=\sum_{i=1}^{n}x_iA_i,A_{n+2}=\sum_{i=1}^{n}y_iA_i,\ |A_{n+1}-A_{n+2}|^2=l^2,$$
$$\sum_{i=1}^{n}x_i=1,\sum_{i=1}^{n}y_i=1.\tag{3.20}$$

可以分四种特款来讨论:

1°　$d^2(B_{n+1},B_{n+2})>l^2$,且有 $1\leqslant k\leqslant n$ 使 $x_ky_k<0$,此时 A_{n+1},A_{n+2}在 π_k 之异侧,A_{n+1},A_{n+2}的 $\theta-k$ 扰动 A_{n+1}^*,A_{n+2}^*在 π 的异侧,即 $A_{n+1}^*A_{n+2}^*$是$\mathfrak{S}^{**}$之红边.$|\theta|$足够小时有

$$d(B_{n+1},B_{n+2})>|A_{n+1}^*-A_{n+2}^*|.\tag{3.21}$$

又由引理 6 知

$$|A_{n+j}^*-A_i|=|A_{n+j}-A_i|,1\leqslant i\leqslant n,i\neq k,j=1,2.\tag{3.22}$$

由引理 7,$A_kA_{n+1}^*$和 $A_kA_{n+2}^*$无色.综合(3.21),(3.22),(3.7),即知$\mathfrak{S}^{**}$是$\mathscr{R}$的歪曲原象.

2°　$d^2(B_{n+1},B_{n+2})>l^2$,但对一切 $1\leqslant i\leqslant n$ 均有 $x_iy_i\geqslant0$.设 $y_k>0,x_k\geqslant0$,又设 $x_j>0$,则 $A_{n+2}\notin\pi_k,A_{n+1}\notin\pi_j$.作 A_{n+2}的 $\theta-k$ 扰动和 A_{n+1}的 $\varphi-j$ 扰动 A_{n+1}^*和 A_{n+2}^*,并使 A_{n+1}^*,A_{n+2}^*在 π 的异侧.又由 $x_ky_k\geqslant0$ 知 A_{n+1},A_{n+2}不在 π_k 之异侧.由 $y_k>0$,A_{n+2}与 A_k 在 π_k 之同侧.由 $x_j>0$,A_{n+1}与 A_j 在 π_j 同侧.由引理 9,当$|\theta|$和$\left|\dfrac{\varphi}{\theta}\right|$足够小时,$A_jA_{n+1}^*,A_kA_{n+2}^*$均为$\mathfrak{S}^{**}$之兰边.再由引理 6,有

$$
\begin{aligned}
&|A_{n+1}^{*}-A_i|=|A_{n+1}-A_i|,1\leqslant i\leqslant n,i\neq j,\\
&|A_{n+2}^{*}-A_i|=|A_{n+2}-A_i|,1\leqslant i\leqslant n,i\neq k,\\
&|A_{n+2}^{*}-A_k|>|A_{n+2}-A_k|,|A_{n+1}^{*}-A_j|>|A_{n+1}-A_j|.
\end{aligned}
\tag{3.23}
$$

注意到 $A_{n+1}^{*}A_{n+2}^{*}$ 为 $\mathfrak{S}^{**}$ 之红边且当 $|\theta|,|\varphi|$ 足够小时

$$
d(B_{n+1},B_{n+2})>|A_{n+1}^{*}-A_{n+2}^{*}|,
\tag{3.24}
$$

综合(3.24),(3.23),(3.7),知 $\mathfrak{S}^{**}$ 是 $\mathscr{R}$ 的歪曲原象.

3° $d^2(B_{n+1},B_{n+2})<l^2$,且有 $1\leqslant k\leqslant n$ 使 $x_ky_k>0$. 证法类似于 1°,从略.

4° $d^2(B_{n+1},B_{n+2})<l^2$,且对一切 $1\leqslant i\leqslant n$ 有 $x_iy_i\leqslant 0$. 证法类似于 2°,从略,

综合以上各款,定理 2 证毕.

§4. 例

本文结果特别适用于离散型度量空间.关于连续型度量空间,近来研究工作较多.例如,完备的、凸的并且外凸的度量空间保距嵌入 l^2 的问题,最终归结为对"Euclid 四点性质"的研究[11~14].

对于更一般的度量空间的嵌入问题,只考虑四点性质是不够的,而我们的判准法,适用于包括四点形在内的多点形的考虑.以下仅举四例:

1. 欧氏空间中能否有这样两个四角形:$A_1A_2A_3A_4$ 和 $B_1B_2B_3B_4$,前者为平面凸四角形,它的两条对角线 A_1A_3,A_2A_4 分别比 B_1B_3,B_2B_4 较短,而其余四边 A_1A_2,A_2A_3,A_3A_4,A_4A_1 分别比 B_1B_2,B_2B_3,B_3B_4,B_4B_1 较长?按定义,A_1A_3 和 A_2A_4 是红的而其余四边是兰的.又按定义,$A_i\to B_i(i=1,2,3,4)$ 是歪曲映象.由定理 1,四点形 $B_1B_2B_3B_4$ 不可能在 l^2 中实现.

2. 设 A_1,A_2,A_3,A_4,A_5 是 E^3 中凸六面体 Ω 的顶点,且 Ω 可剖分为两个四面体 $A_1A_2A_3A_4$,$A_1A_2A_3A_5$.问:在欧氏空间中能否找到五点形 $B_1B_2B_3B_4B_5$,使得 $A_1A_2<B_1B_2$,$A_2A_3<B_2B_3$,$A_3A_1<B_3B_1$,$A_4A_5<B_4B_5$,而 Ω 的其余各棱 A_iA_j 均大于对应的 B_iB_j?

按定义,A_1A_2,A_2A_3,A_3A_1,A_4A_5 是红的,其余的 A_iA_j 是兰的.而 $A_i\to B_i$ 是歪曲映象,故五点形 $B_1B_2B_3B_4B_5$ 不可能在 l^2 中实现.

3. 在仿射平面上引入范数 $\|(x,y)\|_p=(x^p+y^p)^{1/p}$ 而得到空间 $l^p(p\geqslant 1)$,于是诱导出距离 $\|z_1-z_2\|_p$.熟知的一个事实是:l_2^p 中四点形不一定都能保距地嵌入 l^2.我们的结果有助于举出具体例子.当 $1\leqslant p<2$ 时考虑四点 $(0,0)$,$(1,0)$,$(1,1)$,$(0,1)$,将它们的欧氏距离与 l_2^p 距离比较,易知这是歪曲映象,故它们不能嵌入 l^2.当 $p>2$ 时,考虑 $(1,0)$,$(0,1)$,$(-1,0)$,$(0,-1)$,得到类似的结论.

4. 设 Ω 为处处具有负 Gauss 曲率的完备曲面上的单连通区域.$B_1B_2B_3B_4$ 是 Ω 中测地四角形,B_4 在测地三角形 $B_1B_2B_3$ 内部,从而

$$
\angle B_1B_4B_2+\angle B_2B_4B_3+\angle B_3B_4B_1=2\pi.
\tag{4.1}
$$

如果这个测地四角形充分小,已经证明它不能保距嵌入 l^2([15],Ch. 11),大范围情形又如何呢?

在欧氏平面 E^2 中作四点形 $A_1A_2A_3A_4$,使有 $A_iA_4=B_iB_4(i=1,2,3)$,且

$$\angle A_iA_4A_j=\angle B_iB_4B_j(i,j=1,2,3).$$

由(4.1),A_4 在$\triangle A_1A_2A_3$ 内.按定义 A_iA_4 为兰边而 A_iA_j 为红边.

在曲面 Ω 上和平面 E^2 上分别计算各自的测地线 B_iB_j 和 $A_iA_j(i,j=1,2,3)$的长度,由负曲率曲面的性质易知

$$A_1A_2<B_1B_2,A_2A_3<B_2B_3,A_3A_1<B_3B_1.$$

按定义,$A_i\to B_i(i=1,2,3,4)$是歪曲映象.由定理1,测地四角形 $B_1B_2B_3B_4$ 不能嵌入于 l^2.

对高维负 Gauss 曲率的超曲面,不难建立类似的结果.事实上,本节所有例子均可推广到较高的维数.

参考文献

[1]Banach S. Théorie des Opérations Linéaires. Warsaw,1932.

[2]Schoenberg,I J. Remarks to M. Fréchet's article,Sur la définition axiomatique d'une classe d'espaces vectoriels distancies applicables vectoriellement sur l'espace de Hilbert. Ann. of Math. ,1935(36):724-732.

[3]Menger K. Untersuchungen über allgemeine Metrik. Math. Ann. ,1928(100):75-163.

[4]Wilson W A. On the imbedding of metric sets in Euclidean Space. Amer. J. Math. ,1935(57):322-326.

[5] Blumenthal L M. Theory and Applications of Distance Geometry. Chelsea, New York,1970.

[6]Morgan C L. Embedding metric spaces in Euclidean space. J. Geometry,1974(5):101-107.

[7]Alexander R. Metric embedding techniques applied to geometric inequalities,in Lecture Notes in Math. Springer-Verlag,1975(490):57-65.

[8]Generalized sums of distance,Metric Averaging in Euclidean and Hilbert spaces. Pac. J. Math. ,1979(85):1-9.

[9]Alexander R,Stolarsky K B. Extremal problems of distance geometry related to energy interals. Trans. Amer. Math. Soc. ,1974(193):1-31.

[10]Blumental L M,Gillam B E. Distribution of points in n-space. Amer. Math. Monthly,1943(50):181-185.

[11]Valentine J E. On criteria of Blumenthal for inner-product spaces. Fund. Math. ,1971(72):265-269.

[12]Andalafte E Z,Freese R W. An equilateral four-point property which characterizes generalized Euclidean spaces. J. Geometry,1983(20):151-154.

[13]Freese R W,Andalafte E Z. A new class of four-point properties which characterizes Euclidean spaces. J. Geometry,1982(18):43-53.

[14]Kelly L M. On the equilateral feeble four-point properties. in Lecture Notes in Math. , Springer-Verlag,1975(490):13-16.

[15]Blumenthal L M, Menger K. Studies in Geometry. W. H. Freeman and Company, 1970.
[16]Yang Lu, Zhang Jingzhong. A generalisation to several dimensions of the Neuberg-Pedoe inequality with applications. Bull. Austral. Math. Soc. , 1983(27): 203 - 214.
[17]杨路，张景中. Neuberg-Pedoe 不等式的高维推广及其应用. 数学学报，1981(24): 401 -408.
[18]杨路，张景中. 抽象距离空间的秩的概念. 中国科技大学学报，1980，10(4): 52 - 65.
[19]杨路，张景中. 关于有限点集的一类几何不等式. 数学学报，1980，23(5): 740 - 749.
[20]张景中，杨路. 有限点集在伪欧空间的等长嵌入. 数学学报，1981，24(4): 481 - 487.
[21]杨路，张景中. 关于空间曲线的 Johnson 猜想. 科学通报，1984，29(6): 329 - 333.

Average Distance Constants of Some Compact Convex Spaces

Yang Lu　Zhang Jing-zhong

(Chengdu Branch, Academia Sinica)

Abstract

This paper is inspired by the research of the so-called "average distance property" which interests many authors nowadays. Let (X, d) be a compact connected metric space. Then there is a uniquely determined constant $a(X, d)$ with the following property: For each positive integer n, and for all $x_1, \cdots, x_n \in X$, there exists $y \in X$ for which

$$\frac{1}{n}\sum_{j=1}^{n} d(x_j, y) = a(X, d).$$

In this article, an explicit expression of $a(X, d)$ is discussed for the compact convex subsets of a class of symmetric spaces including Banach spaces and Lobachevsky spaces. It is given for such a compact convex subset X which is considered as a subspace (X, d) that

$$a(X, d) = \min_{x \in X} \max_{y \in X} d(x, y).$$

This is useful to evaluate the values of $a(X, d)$ for some concrete examples.

Key words: average distance constant, straight symmetric space, normal segment

1 Introduction

Recently a result concerning the compact connected metric spaces interests many authors,[5,6,7,9,2], that is

Theorem A Let (X, d) be a compact connected metric space. Then there is a uniquely determined constant $a(X, d)$ with the following property: For each positive integer n, and for all $x_1, \cdots, x_n \in x$, there exists $y \in X$ for which

本文刊于《中国科学技术大学学报》,第 17 卷第 1 期,1987 年.

$$\frac{1}{n}\sum_{j=1}^{n}d(x_j,y)=a(X,d). \tag{1}$$

This is the so-called "average distance property". The corresponding constant is known as "average distance constant"(a. d. c.) of the space. It is a surprising fact that though the class of compact connected metric spaces is wide, the "average distance property" owned by the class is quite strong. Stadje[6] remarks that "this property of compact connected metric spaces is nontrivial even in the simplest examples".

It is rather difficult to find an explicit expression for a. d. c. for a given space. As far as we know, there are only a small number of results for a few spaces such as the spheres in E^n. No other explicit expression for a. d. c. are known even for the planar polygons.

In this paper the authors show that

$$a(X,d)=\min_{x\in X}\max_{y\in X}d(x,y) \tag{2}$$

for some compact convex spaces including the compact convex subsets of Banach spaces and Lobachevsky spaces. In this way, for example, the a. d. c. of a planar convex N-polygon can be evaluated in a time of $O(N)$.

2 The Statement of Results

Definition 1 A metric space (M,d) is called a straight symmetric space, provided

i) An isometry $f_p:M\to M$ is defined for each $p\in M$, with $f_p(x)=x$ if and only if $x=p$;

ii) A subset $L(x,y)$ called "a straight line" is defined for $x\in M$, $y\in M$ and $x\neq y$, such that $x,y\in L(x,y)\subset M$ and $L(x,y)$ is the image of the whole real axis under all isometry;

iii) Every straight line L to which the point p belongs is an invariant set with respect to f_p.

Obviously, Banach spaces and Lobachevsky spaces are straight symmetric, for which f_p is a symmetry with respect to p.

Definition 2 A compact connected subset of a straight line in a straight symmetric space is called "a normal segment" and denoted by $[x,y]$ where x and y are its endpoints.

Definition 3 A subset X of a straight symmetric space is "convex" if every normal segment $[x,y]$ with endpoints $x\in X$ and $y\in X$ is contained by X.

Busemann pointed out that the class of straight spaces contains more members than Banach spaces and hyperbolic spaces. Other interesting examples of such spaces can be found in [1]. It is, therefore, certainly worthwhile to observe them in general.

A main result of this paper is following.

Theorem 1 Let X be a compact convex subset of a straight symmetric space (M,d). Then, the a. d. c. of the subspace (X,d) is given by

$$a(X,d)=\min_{x\in X}\max_{y\in X}d(x,y).$$

Since Banach spaces are straight symmetric, it follows that

Corollary 1 Let X be a compact convex subset of a Banach space and d be the derived metric by the norm. Then the a. d. c. of the subspace (X, d) is given by

$$a(X,d)=\min_{x\in X}\max_{y\in X} d(x,y).$$

In terms of geometry, that is

Corollary 2 Let X be a compact convex subset of a Banach space and d be the derived metric by the norm. Then $a(X,d)$ equals the radius of the smallest Banach sphere enclosing X and having the centre contained by X.

It is easy to show for a compact convex subset X of Euclidean space that the centre of the smallest sphere enclosing X (namely, the spanning sphere of X) must belong to X, so we have

Corollary 3 The a. d. c. of a compact convex subset of Euclidean space is equal to the radius of the smallest sphere enclosing the subset itself.

Recently, an algorithm was given in [4], spending a time of $O(N)$ to find the spanning circle of a planar N-polygon, so we have

Corollary 4 The a. d. c. of a planar convex N-polygon can be evaluated in a time of $O(N)$.

On the other hand, by the well-known Jung's theorem:[3] "A subset of E^n with diameter D must be enclosed by a sphere with radius

$$\sqrt{\frac{n}{2(n+1)}}D,"$$

and by Corollary 3, we obtain

Corollary 5 Let X be a compact convex subset of E^n. Then

$$a(X,d)\leqslant\sqrt{\frac{n}{2(n+1)}}D(x,d), \tag{3}$$

where $D(X,d)$ is the diameter of X. This result can be found in [7].

Since the n-dimensional hyperbolic space H^n is straight symmetric, we have

Corollary 6 Let X be a compact convex subset of H^n and d be the metric in H^n. Then,

$$a(X,d)=\min_{x\in X}\max_{y\in X} d(x,y),$$

Corollary 7 The a. d. c. of a compact convex subset of a hyperbolic space equals the radius of the smallest sphere enclosing the subset itself.

On the other hand, since the authors have proved in [8] that "A subset of H^n with diameter D must be enclosed by a sphere with diameter

$$\frac{2}{\sqrt{-K}}\mathrm{sh}^{-1}\left(\sqrt{\frac{2n}{n+1}}\mathrm{sh}\,\sqrt{-K}\frac{D}{2}\right), \tag{4}$$

where $K<0$ is the curvature of the space", it follows that

Corollary 8 Let X be a compact convex subset of H^n. Then

$$a(X,d)\leqslant\frac{1}{\sqrt{-K}}\mathrm{sh}^{-1}\left(\sqrt{\frac{2n}{n+1}}\mathrm{sh}\left(\frac{1}{2}\sqrt{-K}D(x,d)\right)\right), \tag{5}$$

where $K<0$ is the curvature of the space and $D(X,d)$ is the diameter of X.

The proof of Theorem 1 is postponed till Section Ⅳ.

3 Lemmas

In order to prove Theorem 1, we need the following Lemmas.

Lemma 1 Let $[x_1,x_2]$ be a normal segment in a straight symmetric space (M,d) and x be the mid-point of $[x_1,x_2]$. Then it holds for all $y\in M$ that

$$d(x_1,y)+d(x_2,y)\geqslant 2d(x,y). \tag{6}$$

Proof Since (M,d) is straight symmetric, there is an isometry described in Definition 1, $f_x:M\to M$, (namely, the symmetry with respect to x) such that

$$x_2=f_x(x_1), \tag{7}$$

put

$$y'=f_x(y), \tag{8}$$

then

$$d(x,y)=d(x,y')=\frac{1}{2}d(y,y'), \tag{9}$$

and

$$d(x_2,y)=d(x_1,y'). \tag{10}$$

Thus

$$\begin{aligned} d(x_1,y)+d(x_2,y)&=d(x_1,y)+d(x_1,y') \\ &\geqslant d(y,y') \\ &=2d(x,y). \end{aligned}$$

Lemma 2 Let $[x_1,x_2]$ be a normal segment in a straight symmetric space and $x\in[x_1,x_2]$ such that

$$d(x_1,x):d(x_2,x)=m, \tag{11}$$

where m is a positive integer. Then it holds for all $y\in M$ that

$$d(x_1,y)+md(x_2,y)\geqslant(m+1)d(x,y). \tag{12}$$

Proof Make use of induction on m. The proposition holds for $m=1$ because of Lemma 1.

We take a point $x'\in[x_1,x]\subset[x_1,x_2]$ such that

$$d(x_1,x'):d(x',x):d(x,x_2)=(m-1):1:1. \tag{13}$$

If it holds for any $y\in X$ that

$$d(x_1,y)+(m-1)d(x,y)\geqslant md(x',y), \tag{14}$$

by combining it with the following inequality from Lemma 1,

$$md(x',y)+md(x_2,y)\geqslant 2md(x,y), \tag{15}$$

we have

$$d(x_1,y)+md(x_2,y)\geqslant(m+1)d(x,y).$$

The induction is completed.

Lemma 3 Let X be a compact convex subset of a straight symmetric space (M,d). Given n points $x_1,\cdots,x_n\in X$. Then there exists a point $x^n\in X$ such that

$$\sum_{j=1}^{n}d(x_j,y)\geqslant nd(x^n,y) \tag{16}$$

holds for all $y \in X$.

Proof Let x^2 be the mid-point of $[x_1, x_2]$. Take $x^j \in [x^{j-1}, x_j] \subset X$ for $j=3,4,\cdots,n$, successively, such that

$$d(x_j, x^j) : d(x^{j-1}, x^j) = j-1, \tag{17}$$

the convexity of X implies the existence of these normal segments and all x^j.

Make use of induction on n. The proposition holds for $n=2$ clearly.

Assume

$$\sum_{j=1}^{n-1} d(x_j, y) \geqslant (n-1) d(x^{n-1}, y), \tag{18}$$

then by combining it with the following inequality from Lemma 2,

$$d(x_n, y) + (n-1) d(x^{n-1}, y) \geqslant n d(x^n, y), \tag{19}$$

we have

$$\sum_{j=1}^{n} d(x_j, y) \geqslant nd(\lambda^n, y).$$

The induction is completed.

4 Proof of Theorem 1

Proof of Theorem 1 Let X be a compact convex subset of a straight symmetric space (M, d). Given n points $x_1, \cdots, x_n \in X$. Take a point $x^* \in X$ satisfying

$$\max_{y \in X} d(x^*, y) = \min_{x \in X} \max_{y \in X} d(x, y), \tag{20}$$

Then it holds for $j=1,\cdots,n$ that

$$d(x^*, x_j) \leqslant \max_{y \in X}(x^*, y) = \min_{x \in X} \max_{y \in X} d(x, y), \tag{21}$$

hence

$$\sum_{j=1}^{n} d(x^*, x_j) \leqslant n(\min_{x \in X} \max_{y \in X} d(x, y)). \tag{22}$$

On the other hand, by Lemma 3, there exists a point $X^n \in X$ for which

$$\sum_{j=1}^{n} d(x_j, y) \geqslant nd(x^n, y)$$

holds for all $y \in X$. Since

$$\max_{y \in X} d(x^n, y) > \min_{x \in X} \max_{y \in X} d(x, y), \tag{23}$$

there exists a point $y^* \in X$ satisfying

$$d(x^n, y^*) \geqslant \min_{x \in X} \max_{y \in X} d(x, y), \tag{24}$$

hence

$$\sum_{j=1}^{n} d(x_j, y^*) \geqslant n(\min_{x \in X} \max_{y \in X} d(x, y)). \tag{25}$$

Because (22), (25) and the connectivity of X, there exists a point $\bar{y} \in X$ such that

$$\sum_{j=1}^{n} d(x_j, \bar{y}) = n(\min_{x \in X} \max_{y \in X} d(x, y)), \tag{26}$$

By the uniqueness of a. d. c. , the proof of Theorem 1 is completed.

5 Remark

The expression (2) would hold for the compact convex subsets of more metric spaces which need not be straight if we restrict these subsets to suitable size. For example, we have a similar result on spherical spaces as follows.

Theorem 2 Let X be a compact convex subset of spherical space S^n with curvature r^{-2} and d be the metric of S^n. Then the a. d. c. of the subspace (X, d) is given by

$$a(X, d) = \min_{x \in X} \max_{y \in X} d(x, y),$$

if

$$\max_{x, y \in X} d(x, y) \leqslant \frac{1}{2}\pi r, \tag{27}$$

The proof of Theorem 2 is analogous to that of Theorem 1 since the reasoning as Lemmas 1 - 3 is available.

References

[1]Busemann H. The geometry of geodesics, New York: Academic Press, 1955: 345 - 347.

[2]Gross O. The rendezvous value of a metric space. Ann. of Math. Studies, 1964(52): 49 -53.

[3]Hadwiger H, et al. Combinatorial geometry in the plane, Holt, Rinehartand Winston, 1964: 46.

[4]Megiddo N. Linear-time algorithms for linear programming in R^3 and related problems. SIAM J. Comput. 1983(12): 759 - 776.

[5]Morris s A, Nickolas P. On the average distance property of compact connected metric spaces. Arch. Math. , 1983(40): 456 - 463.

[6]Stadje W. A property of compact connected spaces. Arch. Math. , 1981(36): 275 - 280.

[7]Strantzen J. An average distance result in Euclidean n-space. Bull. Austral. Math. Soec. , 1982(26): 321 - 330.

[8] Yang Lu, Zhang Jingzhong. Spanning-radius of a compact set of hyperbolic space. Scientia Sinica(A), Vol. XXVI(1983).

[9]Yost D. Average distances in compact connected spaces. Bull. Austral. Math. Soc. , 1982 (26): 331 - 342.

逐段单调连续函数嵌入拟半流问题

张景中　杨　路

（中国科学院成都分院）

设 M 是度量空间，$R^+=[0,+\infty)$. 连续映射

$$\varphi: M\times R^+\to M \tag{1}$$

$(y=\varphi(x,t);y,x\in M,t\in R^+)$若满足

$$\begin{cases}1^\circ & \varphi(x,0)=x,\ \forall x\in M,\\ 2^\circ & \varphi(\varphi(x,t_1),t_2)=\varphi(x,t_1+t_2),\ \forall x\in M,\ \forall t_1,t_2\in R^+,\end{cases} \tag{2}$$

则称其为 M 上的连续半流.

设 F 是 M 上的连续自映射. 如果有 M 上的连续半流 φ，使

$$F(x)=\varphi(x,1),\ \forall x\in M, \tag{3}$$

则说 F 可以嵌入连续半流 φ.

在定义(1)，(2)中，把 R^+ 换成 $R=(-\infty,+\infty)$，则 φ 叫做 M 上的连续流. 显然，若 φ 为流，φ 限制于 $M\times R^+$ 上为半流，这时称(3)中的 F 可嵌入流. 易知可嵌入连续流的映射必为同胚. 关于线段上或圆周上的同胚可嵌入连续流的条件的研究，已有相当完善的结果，例如[1—3]，但映射嵌入半流的条件，目前所知甚少.

若 F 可嵌入半流 φ，$F(x)=\varphi(x,1)$，我们令

$$f(x)=\varphi\left(x,\frac{1}{n}\right),$$

则由(2)得

$$f^n(x)=F(x), \tag{4}$$

这里 $f^n(x)=f\circ f^{n-1}(x)$，$f^0(x)=x$. 通常把满足(4)的 f 叫做 F 的 n 次迭代根. 关于映射的迭代根的存在与计算的问题，很早就引起人们的兴趣[4],[5]，至今仍不断有新的工作发表[6],[7]. 多数作者的工作，限于 F 为线段上严格递增连续自映射这一最简单情形.

以下设 F 为$[a,b]$上的连续自映射. 如果有 $a=c_0<c_1<\cdots<c_N<c_{N+1}=b$，使 F 在$[c_i, c_{i+1}]$上严格单调$(i=0,1,\cdots,N)$，且 F 在 $c_1,c_2,\cdots,c_N$ 处取极大或极小，则称 F 为$[a,b]$上的逐段严格单调连续自映射. 所有这样的 F 之集记作 $S_{[a,b]}$.

对于 $F\in S_{[a,b]}$，作者[8]给出了 F 对所有正整数 n 均有 n 次连续迭代根的充要条件. 这自然引出下列问题：对 $F\in S_{[a,b]}$，若 F 对所有正整数 n 有 n 次连续迭代根，是否它就能嵌入$[a,b]$上的连续半流？

本文刊于《数学学报》，第 30 卷第 1 期，1987 年.

下面将应用[8]的一些结论,通过简单的讨论指出:对 $F\in S_{[a,b]}$,当且仅当 F 在$[a,b]$上严格递增时,即 $N=0$ 时,它才能嵌入连续半流. 这样看来,$S_{[a,b]}$ 中映射嵌入连续半流的问题,其内容是平凡的.

注意到(4)的成立仅依赖于(2)中的 2°,这启发我们减弱对半流 φ 的要求而引出"拟半流"的概念:如果(1)中的连续映射 φ 满足(2)中的条件 2°,我们称 φ 为 M 上的连续拟半流. 显然,只要 F 能嵌入连续拟半流,它就一定有任意 n 次的连续迭代根.

自然会问:对 $F\in S_{[a,b]}$,F 何时可以嵌入连续的拟半流? 对此,本文将给出一个完全的回答:

定理 2 设 $F\in S_{[a,b]}$,以 A,B 分别记 F 在$[a,b]$上的下确界和上确界,则 F 可以嵌入$[a,b]$上的连续拟半流的充要条件是:存在$[a,b]$的子区间$[\bar{a},\bar{b}]$,使 F 在$[\bar{a},\bar{b}]$上严格递增,且 $\bar{a}\leqslant F(\bar{a})=A<B=F(\bar{b})\leqslant\bar{b}$.

为证明此定理,需要引用[8]中一些概念和结论,并建立几个简单的引理.

终本文以 $\varphi(x,t)$ 记$[a,b]$上的某个连续拟半流. 有时为了强调 t 固定而 x 为主变元,也写为 $\varphi_t(x)=\varphi(x,t)$.

按[8],我们称 $x_0\in(a,b)$ 为 F 的一个"单调点",如果有 $\varepsilon>0$ 使 F 在$[x_0-\varepsilon,x_0+\varepsilon]$上严格单调的话. 显然,$[a,b]$上的连续自映射 F 属于集 $S_{[a,b]}$ 的充要条件是 F 的非单调点为有限个. 这时,以 $N(F)$ 记 F 在(a,b)内的非单调点的个数,U_t 记 $\varphi_t(x)$ 的全体单调点之集,并以 A_t,B_t 分别记 $\varphi_t(x)$ 在$[a,b]$上的下确界和上确界.

根据拟半流之定义,并注意到 $f\circ g$ 的单调点之集是 g 的单调点之集的子集,便有下述引理.

引理 1 若 $0\leqslant t_1<t_2$,则 $U_{t_1}\supset U_{t_2}$,且 $A_{t_1}\leqslant A_{t_2}\leqslant B_{t_2}\leqslant B_{t_1}$.

引理 1 证明略去.

引理 2 若 $F(x)=\varphi_1(x)\in S_{[a,b]}$,则对一切 $t>0$,$U_t=U_1$,$\varphi_t\in S_{[a,b]}$.

证 由于此时(4)对一切正整数 n 有连续解 f,由[8]之定理 1(或 §8 之 2°),可知必有 $N(F)=N(F^2)=\cdots=N(F^{2^n})=N(\varphi,_n)$. 同理,由 $\varphi_{\frac{1}{2}}\circ\varphi_{\frac{1}{2}}=F$ 知 $\varphi_{\frac{1}{2}}\in S_{[a,b]}$,由方程

$$f^n(x)=\varphi_{\frac{1}{2}}(x)$$

对一切正整数 n 有连续解知 $N(\varphi_{\frac{1}{2}})=N(F)$. 于是推出

$$N(\varphi,_{-n})=N(F)=N(\varphi,_n),n=0,1,2,\cdots.$$

再用引理 1,即得引理 2 之结论.

引理 3 若 $F(x)=\varphi_1(x)\in S_{[a,b]}$,则有$[a,b]$的子区间$[\bar{a},\bar{b}]$,使对一切 $t>0$,$\varphi_t(x)$在$[\bar{a},\bar{b}]$严格递增,且 $\bar{a}\leqslant A_t<B_t\leqslant\bar{b}$.

证 由引理 2 可知,对任一 $t>0$,$\varphi_t\in S_{[a,b]}$. 由于 φ_t 对一切正整数 n 有 n 次连续迭代根,故根据[8]中 §8 之 2°,3°,4°,5°可知,有$[a,b]$的子区间$[\bar{a}_t,\bar{b}_t]$,使 φ_t 在$[\bar{a}_t,\bar{b}_t]$上严格递增且在 $\bar{a}_t,\bar{b}_t$ 处取到极值,且 $\bar{a}_t\leqslant A_t<B_t\leqslant\bar{b}_t$. 由引理 2,对一切 $t>0$,φ_t 在$[a,b]$上的极值点集都相同,故对 $t_1\neq t_2$,$(\bar{a}_{t_1},\bar{b}_{t_1})$和$(\bar{a}_{t_2},\bar{b}_{t_2})$或者重合,或者不相交. 由引理 1,$(A_{t_1},B_{t_1})$与$(A_{t_2},B_{t_2})$不可能不相交,故$(\bar{a}_{t_1},\bar{b}_{t_1})$与$(\bar{a}_{t_2},\bar{b}_{t_2})$必重合. 引理 3 证毕.

由此立刻得到 $S_{[a,b]}$ 中的映射嵌入连续半流的充要条件:

定理 1 若 $F\in S_{[a,b]}$，则 F 可以嵌入连续半流的充要条件是 F 在 $[a,b]$ 上严格递增.

证 条件的充分性早已是周知的，例如见[1]. 下面证明必要性. 若 F 可嵌入连续半流 $\varphi(x,t)$，则 $\varphi(x,0)=x$，从而 $A_0=a,B_0=b$. 因连续半流也是连续拟半流，由引理 3，有 $[a,b]$ 的子区间 $[\bar{a},\bar{b}]$ 使 F 在 $[\bar{a},\bar{b}]$ 上严格增且 $\bar{a}\leqslant A_t<B_t\leqslant\bar{b}$. 令 $t\to 0$ 得 $\bar{a}\leqslant A_0<B_0\leqslant\bar{b}$，亦即 $a=\bar{a},b=\bar{b}$，即 F 在 $[a,b]$ 上严格递增. 证毕.

下面我们退而考虑 $S_{[a,b]}$ 中映射嵌入连续拟半流的问题，先证明定理 2 中条件的必要性.

引理 4 若 $F\in S_{[a,b]}$，F 可嵌入拟半流中 $\varphi(x,t)$，则有 $[a,b]$ 的子区间 $[\bar{a},\bar{b}]$，使 F 在 $[\bar{a},\bar{b}]$ 上严格递增，且对一切 $x\in[a,b]$，有 $\bar{a}\leqslant F(\bar{a})\leqslant F(x)\leqslant F(\bar{b})\leqslant\bar{b}$.

证 由拟半流之定义，对 $0\leqslant t\leqslant 1$ 有

$$F(x)=\varphi(x,1)=\varphi_{1-t}\circ\varphi_t(x).$$

由引理 3，当 $x\in[a,b]$ 时，有 $\varphi_t(x)\in[\bar{a},\bar{b}]$（这里 $[\bar{a},\bar{b}]$ 如引理 3 所述，亦即本引理中将证明的断言中所指者），$\varphi_{1-t}(x)$ 在 $[\bar{a},\bar{b}]$ 上严格递增，因而

$$\varphi_{1-t}(\bar{a})\leqslant F(x)\leqslant\varphi_{1-t}(\bar{b}),x\in[a,b].$$

令 $t\to 0$ 得 $F(\bar{a})\leqslant F(x)\leqslant F(\bar{b})$. 证毕.

由引理 4，定理 2 之必要性得证，下面的引理给出定理 2 中充分性之证明.

引理 5 若 F 为 $[a,b]$ 上的连续自映射，F 在 $[a,b]$ 的某个子区间 $[\bar{a},\bar{b}]$ 上严格递增，且有 $\bar{a}\leqslant F(\bar{a})\leqslant F(x)\leqslant F(\bar{b})\leqslant\bar{b}$ 对一切 $x\in[a,b]$ 成立，则 F 可嵌入 $[a,b]$ 上的某连续拟半流.

证 记 F 在 $[\bar{a},\bar{b}]$ 上的限制为 G. 由熟知的结果，知 G 可嵌入 $[\bar{a},\bar{b}]$ 上的连续半流 $\psi(x,t)=\psi_t(x)$. 显然 $\psi(\bar{a},t)$ 当 t 增加时不减，而 $\psi(\bar{b},t)$ 当 t 增加时不增. $\psi_t(x)$ 有定义于 $I_t=[\psi_t(\bar{a}),\psi_t(\bar{b})]$ 的反函数 $\psi_{-t}(x)$. 当 $0\leqslant t\leqslant 1$ 时，$\psi_{-t}(x)$ 的定义域包含了 $[F(\bar{a}),F(\bar{b})]=[\psi_1(\bar{a}),\psi_1(\bar{b})]\subset[\bar{a},\bar{b}]$，因而我们可用下列定义把 ψ 开拓成为定义于 $[a,b]\times[0,+\infty)$ 上的 φ：

$$\varphi(x,t)=\psi_{t-n}(F^n(x)),n=1,2,3,\cdots,t\in[n-1,n)x\in[a,b],\tag{5}$$

这里 $(t-n)\in(0,1]$ 而 $F^n(x)\in[F(\bar{a}),F(\bar{b})]$，故定义是合理的. 当 $(t-n)\to 0^-$ 时有 $\varphi(x,t)\to F^n(x)=\varphi(x,n)$，故 $\varphi(x,t)$ 在 $[a,b]\times[0,+\infty)$ 上连续. 剩下的是验证等式

$$\varphi(x,t_1+t_2)=\varphi(\varphi(x,t_1),t_2).$$

我们分两种情形来讨论.

ⅰ) $n-1\leqslant t_1<n,m-1\leqslant t_2<m$，且 $m+n-1\leqslant t_1+t_2<m+n$.

ⅱ) $n-1\leqslant t_1<n,m-1\leqslant t_2<m$，且 $m+n-2\leqslant t_1+t_2<m+n-1$.

在情形ⅰ)，有

$$\begin{aligned}\varphi(x,t_1+t_2)&=\psi_{t_1+t_2-(m+n)}\circ F^{m+n}(x)\\&=\psi_{t_1+t_2-(m+n)}\circ F^m\circ F^n(x)\\&=\psi_{t_1+t_2-(m+n)}\circ\psi_m\circ F^n(x)\\&=\psi_{t_1+t_2-(m+n)}\circ\psi_{m+n-(t_1+t_2)}\circ\psi_{t_1+t_2-n}\circ F^n(x)\\&=\psi_{t_1+t_2-n}\circ F^n(x)\\&=\psi_{t_2-m}\circ\psi_{m-t_2}\circ\psi_{t_1+t_2-n}\circ F^n(x)\end{aligned}$$

$$
\begin{aligned}
&=\psi_{t_2-m}\circ\psi_{m-t_2}\circ\psi_{t_1+t_2-n}\circ\psi_{n-t_1}\circ\psi_{t_1-n}\circ F^n(x)\\
&=\psi_{t_2-m}\circ\psi_m\circ\psi_{t_1-m}\circ F^n(x)\\
&=\psi_{t_2-m}\circ F^m\circ\psi_{t_1-n}\circ F^n(x)\\
&=\varphi_{t_2}\circ\varphi_{t_1}(x).
\end{aligned}
$$

在情形ⅱ),当 $m>1$ 时,有

$$
\begin{aligned}
\varphi(x,t_1+t_2)&=\psi_{t_1+t_2-(m+n-1)}\circ F^{m+n-1}(x)\\
&=\psi_{t_1+t_2-(m+n-1)}\circ F^{m-1}\circ F^n(x)\\
&=\psi_{t_1+t_2-(m+n-1)}\circ\psi_{m-1}\circ F^n(x)\\
&=\psi_{t_1+t_2-(m+n-1)}\circ\psi_{(m+n-1)-(t_1+t_2)}\circ\psi_{t_1+t_2-n}\circ F^n(x)\\
&=\psi_{t_1+t_2-n}\circ F^n(x)\\
&=\psi_{t_2-m}\circ\psi_{m-t_2}\circ\psi_{t_1+t_2-n}\circ F^n(x)\\
&=\psi_{t_2-m}\circ\psi_{m+t_1-n}\circ F^n(x)\\
&=\psi_{t_2-m}\circ\psi_m\circ\psi_{t_1-n}\circ F^n(x)\\
&=\psi_{t_2-m}\circ F^m\circ\psi_{t_{1-n}}\circ F^n(x)\\
&=\varphi_{t_2}\circ\varphi_{t_1}(x).
\end{aligned}
$$

当 $m=1$ 时,有

$$
\begin{aligned}
\varphi(x,t_1+t_2)&=\psi_{t_1+t_2-n}\circ F^n(x)\\
&=\psi_{t_2-m}\circ\psi_{m-t_2}\circ\psi_{t_2}\circ\psi_{t_{1-n}}\circ F^n(x)\\
&=\psi_{t_2-m}\circ\psi_m\circ\psi_{t_{1-n}}\circ F^n(x)\\
&=\psi_{t_2-m}\circ F^m\circ\psi_{t_{1-n}}\circ F^n(x)\\
&=\varphi_{t_2}\circ\varphi_{t_1}(x).
\end{aligned}
$$

引理 5 证毕.

由引理 5 即得定理 2 中条件之充分性证明.

本文的讨论并不排斥 a,b 为 ∞ 的情形.例如,把定理 2 用于 $(-\infty,+\infty)$ 上,F 为多项式的情形,即可断言:若 F 为奇次多项式,当且仅当 $F'\circ F\geqslant 0$ 时 F 可嵌入连续拟半流,这时,F 其实也可以嵌入连续半流,因它在 $[-\infty,+\infty]$ 上严格增,若 F 的次数是正的偶数,则 $F'\circ F\geqslant 0$ 仅仅是 F 可嵌入连续拟半流的必要条件.如果要 F 可嵌入连续拟半流,还应当要求:当 F 的最高次项系数为正时,F 的最右极小值点应当是它的最小值点,当 F 的最高次项系数为负时,F 的最左极大值点应当是它的最大值点.证明从略.

附带指出,[8]中结尾时,关于"偶次多项式 F 在 $(-\infty,+\infty)$ 上有所有 n 次连续迭代根的充要条件为 $F'\circ F\geqslant 0$"这一断言是不确切的.事实上,$F'\circ F\geqslant 0$ 是必要条件而非充分条件.有趣的是,$F'\circ F>0$ 是充分条件而非必要条件.不过,很容易应用[8]中§8 的 1°~5°彻底弄清楚这一问题.兹将结论叙述于下:

若 F 为非常值的偶次多项式,则 F 在 $(-\infty,+\infty)$ 上对所有正整数 n 有 n 次连续迭代根的充分必要条件是:$F'\circ F\geqslant 0$,并且

1° 当 F 的最高次项系数为正时,若 x_0 是 F 的最右极小值点而 x_1 是 F 的最小值点,则当 $F(x_1)=x_0$ 时必有 $F(x_0)=x_0$.

2° 当 F 的最高次项系数为负时，若 x_0 是 F 的最左极大值点而 x_1 是 F 的最大值点，则当 $F(x_1)=x_0$ 时必有 $F(x_0)=x_0$.

对作者在[8]中所举的这一例题上的疏忽，广西大学麦结华同志曾赐函指出，我们在此谨致谢忱，并向编者及读者表示歉意.

参考文献

[1]Fort M K. The embedding of homeomorphisms in flows. Proc. of Amer. Math. Soc. , 1955(6):960 - 967.

[2]Lam P F. Embedding homeomorphisms in differential flows. Colloq. Math. ,1976(35): 275 - 287.

[3]张筑生. 圆周上自同胚的嵌入流与变换群的作用. 数学学报，1981(24):953 - 957.

[4]Schröder E. Ueber Iterate Funktionen. Math. Ann. ,1871(3):296 - 322.

[5]Boedewadt U T. Zur Iteration realler Funktionen. Zeitsc. Math. ,1944,49(3):497 - 516.

[6]Rice R E. Iterative square roots of Čebyšev Polynomials. Stochastica,1979,3(2):1 - 14.

[7]C C Cowen. Iteration and the Solution of functional equations for functions analytic in the unit disk. Trans. of Amer. Math. Soc. ,1981,265(1):69 - 95.

[8]张景中，杨路. 论逐段单调连续函数的迭代根. 数学学报，1983,26(4):398 - 412.

[9]杨路，张景中. 线段上连续自映射嵌入半流的充分必要条件. 数学学报，1986,29(2): 180 -183.

度量和与 Alexander 对称化

杨　路　张景中

（中国科学院成都分院）

提　要

近期 R. Alexander 在度量几何研究中提出了度量和与对称化的基本概念. 本文作了进一步工作，建立了度量和与对称化不增高空间维数的一个充要条件；并举例说明其应用.

§1. 引　言

Alexander[1,2,3]建议在度量几何中应用度量加或度量平均的方法. 他特别指出，这一方法与 Sallee[4]的"弦的伸展定理"相结合，用以解决某些几何极值问题是卓有成效的.

设欧氏空间中有 m 个点列，每个点列的项数均为 N：

$$\mathfrak{S}=\{r_1^k, r_2^k, \cdots, r_N^k\}, k=1,2,\cdots,m, \tag{1.1}$$

如果能找到一个点列

$$\mathfrak{S}=\{r_1, r_2, \cdots, r_N\}$$

使

$$| r_i - r_j |^2 = \sum_{k=1}^{m} | r^k - r_j^k |^2 \tag{1.2}$$

对 $i,j=1,2,\cdots,N$ 都成立，我们就说点列 $\mathfrak{S}$ 是这组点列 $\{\mathfrak{S}_1, \mathfrak{S}_2, \cdots, \mathfrak{S}_m\}$ 的"度量和"，记为

$$\mathfrak{S}=\mathfrak{S}_1+\mathfrak{S}_2+\cdots+\mathfrak{S}_m. \tag{1.3}$$

显然，满足条件(1.2)的 $\mathfrak{S}$ 不是唯一的，而是组成一个合同类. 类中的每一个点列都是 $\{\mathfrak{S}_1, \mathfrak{S}_2, \cdots, \mathfrak{S}_m\}$ 的度量和.

并且，我们又把满足条件

$$| \bar{r}_i - \bar{r}_j |^2 = \frac{1}{m} \sum_{k=1}^{m} | r_i^k - r_j^k |^2 \tag{1.4}$$

的点列 $\overline{\mathfrak{S}}=\{\bar{r}_1, \bar{r}_2, \cdots, \bar{r}_N\}$ 叫做这组点列 $\{\mathfrak{S}_1, \mathfrak{S}_2, \cdots, \mathfrak{S}_m\}$ 的"度量平均". 记为

$$\overline{\mathfrak{S}}=\frac{1}{m}(\mathfrak{S}_1+\mathfrak{S}_2+\cdots+\mathfrak{S}_m). \tag{1.5}$$

进一步考虑连续型度量和或度量平均. 设

本文刊于《数学年刊(A 辑)》，第 8 卷第 2 期，1987 年.

$$\mathfrak{G}(t)=\{r_1(t),r_2(t),\cdots,r_N(t)\},\alpha<t<\beta \tag{1.6}$$

是由点列构成的单参数族. 如果在 Hilbert 空间 l^2 中能找到一个点列 $\mathfrak{G}=\{r_1,r_2,\cdots,r_N\}$ 使得

$$|r_i-r_j|^2=\int_\alpha^\beta |r_i(t)-r_j(t)|^2dt \tag{1.7}$$

对 $i,j=1,2,\cdots,N$ 都成立,我们就说点列 $\mathfrak{G}$ 是点列族$\{\mathfrak{G}(t)|\alpha<t<\beta\}$的度量和,记为

$$\mathfrak{G}=\int_\alpha^\beta \mathfrak{G}(t)dt. \tag{1.8}$$

类似地,把满足条件

$$|\bar{r}_i-\bar{r}_j|^2=\frac{1}{\beta-\alpha}\int_\alpha^\beta |r_i(t)-r_j(t)|^2dt \tag{1.9}$$

的点列 $\overline{\mathfrak{G}}=\{\bar{r}_1,\bar{r}_2,\cdots,\bar{r}_N\}$叫做$\{\mathfrak{G}(t)|\alpha<t<\beta\}$的度量平均,记为

$$\overline{\mathfrak{G}}=\frac{1}{\beta-\alpha}\int_\alpha^\beta \mathfrak{G}(t)dt. \tag{1.10}$$

下列事实是已知的[1,3]:对于欧氏空间中任给的一族项数相同的点列,它们的度量和或度量平均一定在有限维或无穷维的欧氏空间(后者指 Hilbert 空间)中存在.

一个相当重要然而尚未解决的问题是:如果所给的一族点列每一个都在 E^n 内,什么条件能够保证它们的度量和也在 E^n 中? 粗略言之,度量和不增高空间维数的条件是什么?

Alexander 的文章提醒我们去注意这一问题. 事实上,如果度量和不增高空间维数,则往往难于直接应用 Sallee 的弦的伸展定理来达到所要的目的. 在[3]中正好遇上了这样的困难,以致妨碍了这方法的进一步运用.

本文将针对这一问题,给出度量和不增高空间维数的一个充要条件,即文中定理 1(离散型)和定理 4(连续型),并据此建立了所谓"对称化"(定义见后文)维数不增高的条件,应用此条件对 Johnson 的一个猜想给出了证明.

§2. 作为预备工具的离散模型

记号 conv $\mathfrak{G}$ 表示点列 $\mathfrak{G}$ 的凸包,这凸包的维数 dim(conv $\mathfrak{G}$)当然也就是 $\mathfrak{G}$ 所占有的欧氏空间的维数.

定理 1 设 $\mathfrak{G}=\mathfrak{G}_1+\mathfrak{G}_2+\cdots+\mathfrak{G}_m$,令

$$n=\max_k\{\dim(\operatorname{conv}\mathfrak{G}_k)\}, \tag{2.1}$$

并设每个点列 $\mathfrak{G}_k\subset E^n(k=1,2,\cdots,m)$的第一点都落在原点,则 $\dim(\operatorname{conv}\mathfrak{G})=n$ 的充要条件是:存在一个 $\mathfrak{G}_{k_*}\in\{\mathfrak{G}_1,\mathfrak{G}_2,\cdots,\mathfrak{G}_m\}$和一组线性映射 $L_k:E^n\to E^n$ 使得

$$L_k(\mathfrak{G}_{k_0})=\mathfrak{G}_k,k=1,2,\cdots,m. \tag{2.2}$$

将关系(2.2)表述得详细些就是:如果设 $\mathfrak{G}_k=\{r_1^k,r_2^k,\cdots,r_N^k\}$,则有

$$L_k(r_i^{k_0})=r_i^k,i=1,2,\cdots,N,k=1,2,\cdots,m. \tag{2.3}$$

证　充分性　可设

$$r_i^k=(x_{i1}^k x_{i2}^k\cdots r_{in}^k)^\tau(\tau\text{ 表转置}), \tag{2.4}$$

并令

$$Q_k=(r_1^k r_2^k \cdots r_N^k),k=1,2,\cdots,m, \tag{2.5}$$

均表示相应的 $n\times N$ 矩阵. 又 Q_k 的行向量可记为

$$V_l^k=(x_{1l}^k x_{2l}^k \cdots x_{nl}^k),l=1,2,\cdots,n. \tag{2.6}$$

条件(2.2)或(2.3)相当于,对每个 $k=1\ 2,\cdots,m$ 存在着 $n\times n$ 实矩阵 M_k 使得

$$M_k Q_{k_0}=Q_k. \tag{2.7}$$

由(2.1)知道①

$$\max_k\{\text{rank}(Q_k)\}=n. \tag{2.8}$$

由(2.7)和(2.8)必须有

$$\text{rank}(Q_{k_0})=n. \tag{2.9}$$

现在令

$$r_i^*=(x_{i1}^1\cdots x_{in}^1 x_{i1}^2\cdots x_{in}^2\cdots x_{i1}^m\cdots x_{in}^m)^\tau, \tag{2.10}$$

每个 r_i^* 可视为空间 E^{mn} 中的一个点,这样构成的点列

$$\mathfrak{G}^*=\{r_1^*,r_2^*,\cdots,r_N^*\} \tag{2.11}$$

当然含有原点($r_1^*=◎$),并易验证满足条件

$$|r_i^*-r_j^*|^2=\sum_{k=1}^m |r_i^k-r_j^k|^2,$$

即

$$\mathfrak{G}^*=\mathfrak{G}_1+\mathfrak{G}_2+\cdots+\mathfrak{G}_m. \tag{2.12}$$

又令

$$Q=(r_1^* r_2^* \cdots r_N^*) \tag{2.13}$$

表示相应的 $N\times mn$ 矩阵. 显然,Q 的全部行向量就是前面(2.6)所定义的 V_l^k.

根据(2.7),Q_k 的每一行向量都可以表为 Q_{k_0} 的行向量的线性组合,即每一个 V_l^k 都可以表示为 $V_1^{k_0},V_2^{k_0},\cdots,V_n^{k_0}$ 的线性组合. 于是

$$\text{rank}(Q)=\text{rank}(Q_{k_0})=n,$$

而②

$$\dim(\text{conv}\ \mathfrak{G}^*)=\text{rank}(Q)=n.$$

必要性　由(2.1)可知在诸 $\mathfrak{G}_k$ 中至少有一个占据空间的维数是 n,不妨设其为 $\mathfrak{G}_{k_*}$,即

$$\dim(\text{conv}\ \mathfrak{G}_{k_*})=n. \tag{2.14}$$

然后仍按(2.4),(2.10),(2.13)作出矩阵 Q. 由假设 $\mathfrak{G}^*$ 占据空间维数不超过 n,故有

$$\text{rank}(Q)=n. \tag{2.15}$$

这样 Q 的全部行向量 $V_l^k(k=1,2,\cdots,m;l=1,2,\cdots,n)$ 的最大无关组只包括 n 个向量;根据(2.14),我们可以取 $V_l^{k_0}(l=1,2,\cdots,n)$ 作为这样一个极大线性无关组,其余的 V_l^k 都可以通过它们线性表出;这等价于存在线性映射 $L_k:E^n\to E^n$ 使得 $L_k(\mathfrak{G}_{k_0})=\mathfrak{G}_k$. 证毕.

系 1　将定理 1 叙述中之度量和代之以度量平均,全部结论仍然成立.

---

① 因 $\mathfrak{G}_k$ 含有原点,它占有空间的维数应等于它的秩.

② 因 $\mathfrak{G}^*$ 含有原点,它占有空间的维数应等于它的秩.

系 2 设 $\dim(\text{conv}\,\mathfrak{G}_k)=n(k=1,2,\cdots,m)$，即每个 $\mathfrak{G}_k$ 都占有 n 维空间，则其度量和(或度量平均)也占有 n 维空间的充要条件是：所有 $\mathfrak{G}_k$ 属于同一个仿射类. 详细地说：存在一系列仿射变换 $A_{hk}:E^n\to E^n$ 使得

$$A_{hk}(r_i^h)=r_i^k;i=1,2,\cdots,N;h,k=1,2,\cdots,m. \tag{2.16}$$

系 3 如果所有 $\mathfrak{G}_k(k=1,2,\cdots,m)$ 属于同一个仿射类，则其度量和 $\mathfrak{G}$ 及度量平均也都属于这个仿射类.

证 设 $\dim(\text{conv}\,\mathfrak{G}_k)=n(k=1,2,\cdots,m)$. 由诸 $\mathfrak{G}_k$ 之仿射等价性按系 2 有

$$\dim(\text{conv}\,\mathfrak{G})=n,$$

于是

$$\dim(\text{conv}(\mathfrak{G}+\mathfrak{G}))=n. \tag{2.17}$$

(因为 $\mathfrak{G}+\mathfrak{G}$ 显然相似对应于 $\mathfrak{G}$，故占有相同维数空间). 而(2.17)可写为

$$\dim(\text{conv}(\mathfrak{G}_1+\mathfrak{G}_2+\cdots+\mathfrak{G}_m+\mathfrak{G}))=n. \tag{2.18}$$

这说明度量和 $\mathfrak{G}_1+\mathfrak{G}_2+\cdots+\mathfrak{G}_m+\mathfrak{G}$ 不增高空间维数. 由系 2 知 $\mathfrak{G}$ 应与所有 $\mathfrak{G}_k$ 属于同一仿射类. 至于 $\overline{\mathfrak{G}}$，它显然是同 $\mathfrak{G}$ 相似对应的. 证毕.

定义 将一个多角形用它的顶点按邻接次序构成的点列 $\mathfrak{G}_1=\{r_1,r_2,\cdots,r_N\}$ 来表示. 然后另外构造 $(N-1)$ 个点列如下：

$$\left.\begin{aligned}&\mathfrak{G}_2=\{r_2,r_3,\cdots,r_N,r_1\},\\&\mathfrak{G}_3=\{r_3,r_4,\cdots,r_1,r_2\},\\&\cdots\cdots\\&\mathfrak{G}_N=\{r_N,r_1,\cdots,r_{N-2},r_{N-1}\}.\end{aligned}\right\} \tag{2.19}$$

我们将这些 $\mathfrak{G}_k(k=1,2,\cdots,N)$ 的度量平均

$$\overline{\mathfrak{G}}=\frac{1}{N}(\mathfrak{G}_1+\mathfrak{G}_2+\cdots+\mathfrak{G}_N)$$

所表示的多角形叫做是原多角形的“对称化”(为了与 Steiner 对称化相区别，可以将此处的对称化称为 Alexander 对称化).

在不致引起混淆时，可以对点列及其代表的多角形用相同记号表示，就说多角形 $\overline{\mathfrak{G}}$ 是多角形 $\mathfrak{G}_1$ 的对称化.

定理 2 设 $\mathfrak{G}_1$ 是一个多角形[①]，$\overline{\mathfrak{G}}$ 是它的对称化. $\overline{\mathfrak{G}}$ 为平面多角形的充要条件是：$\mathfrak{G}_1$ 是一个仿射正多角形.

证 充分性：$\mathfrak{G}_1$ 是一个仿射正多角形，显而易见 $\mathfrak{G}_1$ 与(2.19)中的任何一个 $\mathfrak{G}_k$ 都是仿射等价的. 由系 1(或系 2)知 $\overline{\mathfrak{G}}$ 与 $\mathfrak{G}_1$ 应占有同维数空间，即 $\overline{\mathfrak{G}}$ 是平面多角形.

必要性 如果 $\overline{\mathfrak{G}}$ 是平面多角形，因为度量平均不能降低空间维数，$\dim(\text{conv}\,\mathfrak{G}_1)\leqslant\dim(\text{conv}\,\overline{\mathfrak{G}})$；又据约定 $\mathfrak{G}_1$ 不是那种全部顶点在一直线上的多角形，故有 $\dim(\text{conv}\,\mathfrak{G}_1)\geqslant 2$. 于是由系 2 和系 3，$\overline{\mathfrak{G}}$ 和所有 $\mathfrak{G}_k$ 均属于同一个仿射类.

令 $\overline{\mathfrak{G}}=(\bar{r}_1,\bar{r}_2,\cdots,\bar{r}_N)$，由于 $\overline{\mathfrak{G}}$ 是 $\mathfrak{G}_1$ 的对称化，由(2.19)不难验证

① 这里，不考虑那种全部顶点在一直线上的退化多角形.

$$|\bar{r}_i-\bar{r}_j|=|\bar{r}_{i+p}-\bar{r}_{j+p}|, \tag{2.20}$$

如果令 $\bar{r}_{i+N}=\bar{r}_i$，这等式对一切自然数 p 成立. 这说明 $\overline{\mathfrak{G}}$ 的对称群包含了一个 N 阶循环群作为子群，对于一个平面 N 角形来讲，这恰好是它成为正多角形的条件.

既然 $\overline{\mathfrak{G}}$ 是正多角形，与之仿射对应的 $\overline{\mathfrak{G}}_1$ 当然就是仿射正多角形了. 证毕.

将此结果应用于等边多角形可以得到：

定理 3　设 $\mathfrak{G}_1$ 是一个等边多角形(边数大于 4)，$\overline{\mathfrak{G}}$ 是它的对称化. $\overline{\mathfrak{G}}$ 为一平面多角形的充要条件是：$\mathfrak{G}_1$ 是一个平面正多角形.

证　不妨设 $\mathfrak{G}_1$ 的各边长皆为 1，即

$$|r_i-r_{i+1}|=1, i=1,2,\cdots,N. \tag{2.21}$$

由定理 2 已经知道，$\overline{\mathfrak{G}}$ 为平面多角形的充要条件是 $\mathfrak{G}_1$ 为仿射正多角形；为证明本定理，只需证明边数大于 4 的等边仿射正多角形一定是平面正多角形就够了.

现在设 $\mathfrak{G}_1$ 是一个仿射正多角形(自然是平面凸多角形). 令 θ_i 表示顶点在 r_i 的 $\mathfrak{G}_1$ 的内角；又令 Δ_i 表示三角形 $r_{i-1}r_ir_{i+1}$ 的面积，则因 $\mathfrak{G}_1$ 是仿射正多角形而有

$$\Delta_1=\Delta_2=\cdots=\Delta_N. \tag{2.22}$$

再结合(2.21)就得到

$$\sin\theta_1=\sin\theta_2=\cdots=\sin\theta_N. \tag{2.23}$$

这说明 $\mathfrak{G}_1$ 的所有内角至多能取某两个值，不妨令其为 θ 和 $\pi-\theta$.

如果 $\mathfrak{G}_1$ 的所有内角相等，据题设它又是等边的，当然就是正多角形，定理已得证.

如果 $\mathfrak{G}_1$ 的内角确实取得了两个不同的值 θ 和 $\pi-\theta\left(设\ \theta>\frac{\pi}{2}\right)$，不妨设 $\theta_2=\theta,\theta_3=\pi-\theta$，这时 $r_1r_2r_3r_4$ 组成一个菱形. 至于 θ_4，有两种可能取值：或是 $\theta_4=\theta$，这时 $r_2r_3r_4r_5$ 也组成菱形，故 r_5 重合于 r_1，即 $\mathfrak{G}_1$ 的边数为 4，与题设矛盾；或是 $\theta_4=\pi-\theta<\theta$，这时 $\mathfrak{G}_1$ 的一边 r_4r_5 进入菱形 $r_1r_2r_3r_4$ 内部，与 $\mathfrak{G}_1$ 的凸性矛盾. 证毕.

下面将举例说明，如何应用本节所述的几个离散型的定理来处理某些离散型的几何极值问题. 为此还将引用 Sallee 的一个近期结果：

定理 S(G. T. Sallee[4])　设 $\mathfrak{G}=\{r_1,r_2,\cdots,r_N\}$ 是一个多角形. 如果 $\mathfrak{G}$ 不是一个平面多角形，则必存在一个平面凸多角形 $\mathfrak{G}^*=\{r_1^*,r_2^*,\cdots,r_N^*\}$，使得

$$|r_i-r_{i+1}|=|r_i^*-r_{i+1}^*|,$$

并且

$$|r_i-r_j|<|r_i^*-r_j^*| \tag{2.24}$$

对一切 $|i-j|\neq 0,1 \pmod N$ 的 i,j 成立.

例 1　将多角形 $\mathfrak{G}$ 的顶点 r_i 和 r_{i+p} 决定的对角线叫做“跨度为 p 之对角线”. 将 $\mathfrak{G}$ 的所有跨度为 p 的对角线的平方和记为 $\sum_p(\mathfrak{G})$，即

$$\left.\begin{aligned}\sum\nolimits_p(\mathfrak{G})&=\sum_{i=1}^{N}|r_i-r_{i+p}|^2, 1\leqslant p<\frac{N}{2};\\ \sum\nolimits_p(\mathfrak{G})&=\frac{1}{2}\sum_{i=1}^{N}|r_i-r_{i+p}|^2, p=\frac{N}{2}.\end{aligned}\right\} \tag{2.25}$$

求证比值$\sum_p(\mathfrak{G})/\sum_1(\mathfrak{G})$当且仅当$\mathfrak{G}$为仿射正多角形时取到最大值.

证 首先,对一个固定的p,由紧致性可知必存在一个N角形$\mathfrak{G}_1$使

$$\sum_p(\mathfrak{G}_1)/\sum_1(\mathfrak{G}_1)=\max_{\mathfrak{G}}\left\{\sum_p(\mathfrak{G})/\sum_1(\mathfrak{G})\right\}. \tag{2.26}$$

如果$\mathfrak{G}_1$不是仿射正多角形,作它的对称化$\overline{\mathfrak{G}}$,由定理2,$\overline{\mathfrak{G}}$不是平面正多角形.对$\overline{\mathfrak{G}}$运用定理S(弦的伸展定理)可得平面凸多角形$\mathfrak{G}^*$使

$$\sum_1(\mathfrak{G}^*)=\sum_1(\overline{\mathfrak{G}}),$$

$$\sum_p(\mathfrak{G}^*)>\sum_p(\overline{\mathfrak{G}}),1<p\leqslant\frac{N}{2}. \tag{2.27}$$

另一方面由对称化的定义可得

$$\sum_1(\overline{\mathfrak{G}})=\sum_1(\mathfrak{G}_1),\sum_p(\overline{\mathfrak{G}})=\sum_p(\mathfrak{G}_1). \tag{2.28}$$

结合(2.27)就有

$$\sum_p(\mathfrak{G}^*)/\sum_1(\mathfrak{G}^*)>\sum_p(\mathfrak{G}_1)/\sum_1(\mathfrak{G}_1). \tag{2.29}$$

此与(2.26)矛盾.故$\mathfrak{G}_1$必须是仿射正多角形.

而且,这时$\mathfrak{G}_1$的对称化$\overline{\mathfrak{G}}$是正多角形,对正多角形而言比值$\sum_p\Big/\sum_1$是很容易计算的,经简单计算我们有

$$\max_{\mathfrak{G}}\left\{\sum_p(\mathfrak{G})/\sum_1(\mathfrak{G})\right\}=\begin{cases}\left(\sin\dfrac{p\pi}{N}\Big/\sin\dfrac{\pi}{N}\right)^2, & \text{当 } 1<p<\dfrac{N}{2}\text{ 时},\\[2ex] \dfrac{1}{2}\sin^2\dfrac{\pi}{N}, & \text{当 } p=\dfrac{N}{2}\text{ 时},\end{cases} \tag{2.30}$$

例 2 设多角形$\mathfrak{G}$的顶点的重心为r_0,诸顶点到r_0的距离平方和

$$I(\mathfrak{G})=\sum_{k=1}^{N}|r_k-r_0|^2, \tag{2.31}$$

叫做$\mathfrak{G}$的中心矩.$\mathfrak{G}$各边平方和仍然如例1那样用$\sum_1(\mathfrak{G})$表示.B. H. Neumann[5]曾经证明

$$\sum_1\geqslant2\left(1-\cos\frac{2\pi}{N}\right)I \tag{2.32}$$

这个结果很容易从(2.30)推出.因为熟知有

$$I=\frac{1}{N}\sum_{1\leqslant i<j\leqslant N}|r_i-r_j|^2 \tag{2.33}$$

从而

$$\frac{I}{\sum_1}=\frac{1}{N}\sum_{p=1}^{\left[\frac{N}{2}\right]}\left(\frac{\sum_p}{\sum_1}\right)\leqslant\frac{1}{2N}\sum_{p=1}^{N}\left(\frac{\sin\frac{p\pi}{N}}{\sin\frac{\pi}{N}}\right)^2=\frac{1}{4\sin^2\frac{\pi}{N}}. \tag{2.34}$$

即$\sum_1\geqslant2\left(1-\cos\dfrac{2\pi}{N}\right)I$,等号成立的充要条件是:$\mathfrak{G}$是一个仿射正多角形.Neumann也给出了一个等号成立的充要条件,二者的等价性似乎并非一目了然的.

例 3 每边长为1的等边N角形,它的中心矩不超过$N/4\sin^4\dfrac{\pi}{N}$.这可以从(2.32)直接

推出. 广而言之,总周长为 L 的等边 N 角形,其中心矩不超过 $\dfrac{L^2}{4N\sin^2\frac{\pi}{N}}$. 即

$$I\leqslant\frac{L^2}{4N\sin^2\frac{\pi}{N}},\tag{2.35}$$

等号成立的充要条件是,所讨论的多角形是平面正 N 角形.

例 4 设 $\mathfrak{G}$ 是每边长为 $\frac{L}{N}$ 的等边 N 角形,令

$$d(p,\mathfrak{G})=\min_i|r_i-r_{i+p}|\tag{2.36}$$

表示 $\mathfrak{G}$ 的跨度为 p 的弦的最小值. [3]中猜想有

$$d(p,\mathfrak{G})\leqslant\frac{L}{N}\sin\frac{\pi p}{N}/\sin\frac{\pi}{N},\tag{2.37}$$

等号当且仅当 $\mathfrak{G}$ 为平面正 N 角形$\left(\text{每边长为}\frac{L}{N}\right)$时成立. 这是 Johnson 的一个猜想[6]的离散形式. [3]中建议用度量平均的方法来处理,但未能完成其证明. 现证明如下:

将 $\mathfrak{G}$ 的所有跨度为 p 的弦的二阶平均记为

$$\overline{d}(p,\mathfrak{G})=\left(\frac{1}{N}\sum_{i=1}^{N}|r_i-r_{i+p}|^2\right)^{\frac{1}{2}}.\tag{2.38}$$

在所有边长为 $\frac{L}{N}$ 的等边 N 角形中,由紧致性,必存在一个 $\mathfrak{G}_1$ 使得

$$\overline{d}(p,\mathfrak{G}_1)=\max_{\mathfrak{G}}\{\overline{d}(p,\mathfrak{G})\}.\tag{2.39}$$

然后作 $\mathfrak{G}_1$ 的对称化 $\overline{\mathfrak{G}}$. 如果 $\mathfrak{G}_1$ 不是平面正 N 角形(又非菱形),则由定理 3,$\overline{\mathfrak{G}}$ 必定不是平面多角形. 于是由定理 S(弦的伸展定理)一定存在一个平面凸 N 角形 $\mathfrak{G}^*$,它的每边长为 $\frac{L}{N}$ 而其余各对角线都严格大于 $\overline{\mathfrak{G}}$ 中对应的对角线;从而有

$$\overline{d}(p,\mathfrak{G}_1)=\overline{d}(p,\overline{\mathfrak{G}})<\overline{d}(p,\mathfrak{G}^*).\tag{2.40}$$

这说明 $\overline{d}(p,\mathfrak{G}_1)<\max\limits_{\mathfrak{G}}\{\overline{d}(p,\mathfrak{G})\}$,与(2.39)矛盾. 故 $\mathfrak{G}_1$ 必须是平面正 N 角形或菱形,对这两种情况显然都有(经直接计算)

$$\overline{d}(p,\mathfrak{G}_1)=\frac{L}{N}\sin\frac{\pi p}{N}\Big/\sin\frac{\pi}{N}.\tag{2.41}$$

又因显然有 $d(p,\mathfrak{G})\leqslant\overline{d}(p,\mathfrak{G})$,故所要之结果得证. 事实上可以证明更强的结论

$$\overline{d}(p,\mathfrak{G})\leqslant\frac{L}{N}\sin\frac{\pi p}{N}\Big/\sin\frac{\pi}{N},$$

当且仅当 $\mathfrak{G}$ 是一个平面正 N 角形或者菱形时等号成立.

§3. 连续型度量和与对称化

为了处理连续型度量和(其定义见(1.7))与度量平均(其定义见(1.9)),需要事先作一些准备. 一个项数为 d 的点列

$$\mathfrak{G}=\{r_1,r_2,\cdots,r_d\},$$

通常叫做是“占有最广位置”的，如果其凸包 conv $\mathfrak{G}$ 是一个$(d-1)$维的非退化单形.

定义 一个实函数 $V(\mathfrak{G})=V(r_1,r_2,\cdots,r_d)$如下：当 $\mathfrak{G}$ 占有最广位置时，令

$$V(\mathfrak{G})=\text{conv }\mathfrak{G}\text{ 的}(d-1)\text{维体积};$$

否则令 $V(\mathfrak{G})=0$. 这样定义的 V 显然是 d 个变元$(r_1,r_2,\cdots,r_d)$的连续函数.

引理 1 对于一族依赖于连续单参数的点列 $\mathfrak{G}(t)=\{r_1(t),r_2(t),\cdots,r_d(t)\}$，$\alpha<t<\beta$，我们有

$$V^{\frac{2}{d-1}}\left(\int_\alpha^\beta \mathfrak{G}(t)dt\right)\geqslant\int_\alpha^\beta V^{\frac{2}{d-1}}(\mathfrak{G}(t))dt. \tag{3.1}$$

这里的记号 $\int_\alpha^\beta \mathfrak{G}(t)dt$ 表族$\{\mathfrak{G}(t)\}$的度量和.

证 首先，对于两个点列的度量和 $\mathfrak{G}_1+\mathfrak{G}_2$，如果 $\mathfrak{G}_1$，$\mathfrak{G}_2$ 占有最广位置的话，文[10]的 Th. 1* 证明了

$$V^{\frac{2}{d-1}}(\mathfrak{G}_1+\mathfrak{G}_2)\geqslant V^{\frac{2}{d-1}}(\mathfrak{G}_1)+V^{\frac{2}{d-1}}(\mathfrak{G}_2); \tag{3.2}$$

由连续性，当 $\mathfrak{G}_1$ 或 $\mathfrak{G}_2$ 退化(不占有最广位置)时(3.2)式仍然成立. 进而可以推出

$$V^{\frac{2}{d-1}}\left(\sum_{k=1}^m \mathfrak{G}_k\right)\geqslant\sum_{k=1}^m V^{\frac{2}{d-1}}(\mathfrak{G}_k). \tag{3.3}$$

然后通过一般的极限过程就可以得到

$$V^{\frac{2}{d-1}}\left(\int_\alpha^\beta \mathfrak{G}(t)dt\right)\geqslant\int_\alpha^\beta V^{\frac{2}{d-1}}(\mathfrak{G}(t))dt,$$

证毕.

定理 4 设 $\mathfrak{G}=\int_\alpha^\beta \mathfrak{G}(t)dt$，令

$$n=\sup_{\alpha<t<\beta}\{\dim(\text{conv }\mathfrak{G}(t))\}, \tag{3.4}$$

并设 $\forall t\in(\alpha,\beta)$，点列 $\mathfrak{G}(t)$的第一点都落在原点，则 $\dim(\text{conv }\mathfrak{G})=n$ 的充要条件是：存在着一个点列 $\mathfrak{G}(t_1)\in\{\mathfrak{G}(t)\mid\alpha<t<\beta\}$和一族线性映射 $L_t:E^n\to E^n$ 使得

$$L_t(\mathfrak{G}(t_1))=\mathfrak{G}(t),\alpha<t<\beta. \tag{3.5}$$

将条件(3.5)表述得详细些就是：若令

$$\mathfrak{G}(t)=\{r_1(t),r_2(t),\cdots,r_N(t)\},$$

则有

$$L_t(r_i(t_1))=r_i(t),i=1,2,\cdots,N,\alpha<t<\beta. \tag{3.6}$$

证 充分性的证明可以由定理 1 取极限而得到，事属显然. 下面是必要性的证明.

由于维数 dim 只取离散值，故(3.4)中的上确界必能取到，即存在 $t_1\in(\alpha,\beta)$使

$$\dim(\text{conv }\mathfrak{G}(t_1))=n. \tag{3.7}$$

假如有一个数 $t_2\in(\alpha,\beta)$使得把 $\mathfrak{G}(t_1)$映为 $\mathfrak{G}(t_2)$的线性映射不存在，我们将证明这种情况下有 $\dim(\text{conv }\mathfrak{G})>n$.

由于 $\dim(\text{conv}(\mathfrak{G}(t_2)))\leqslant\dim(\text{conv}(\mathfrak{G}(t_1)))=n$，我们可以找到 $\mathfrak{G}(t_1)$一个含$(n+2)$项的子列

$$\mathfrak{G}'(t_1)=\{r_{i_1}(t_1),r_{i_2}(t_1),\cdots,r_{i_{n+1}}(t_1)\}, \tag{3.8}$$

它与$\mathfrak{S}(t_1)$占有相同维数空间($\dim(\mathrm{conv}\ \mathfrak{S}'(t_1))=n$),并且使得,把$\mathfrak{S}'(t_1)$映为$\mathfrak{S}(t_2)$中相应子列

$$\mathfrak{S}'(t_2)=\{r_{i_1}(t_2),r_{i_2}(t_2),\cdots,r_{i_{n+1}}(t_2)\} \tag{3.9}$$

的线性映射$L:E^n\rightarrow E^n$不存在.

对$\mathfrak{S}'(t_1)$和$\mathfrak{S}'(t_2)$应用定理1,得到

$$\dim(\mathrm{conv}(\mathfrak{S}'(t_1)+\mathfrak{S}'(t_2)))>n. \tag{3.10}$$

但是,$(n+2)$个点占有的最广位置不过是$(n+1)$维空间,故$\mathfrak{S}'(t_1)+\mathfrak{S}'(t_2)$占有最广位置,即其凸包$\mathrm{conv}(\mathfrak{S}'(t_1)+\mathfrak{S}'(t_2))$是$E^{n+1}$中非退化单形而有

$$V(\mathfrak{S}'(t_1)+\mathfrak{S}'(t_2))>0, \tag{3.11}$$

这里V是前面定义过的$(n+2)$元实函数.

现在取一个足够小的正数$\delta>0$,使得区间$\Delta_1=[t_1-\delta,t_1+\delta]$和$\Delta_2=[t_2-\delta,t_2+\delta]$都含在$(\alpha,\beta)$中而且$\Delta_1\cap\Delta_2=\varnothing$. 并令$\Delta_3=(\alpha,\beta)\backslash\{\Delta_1\cup\Delta_2\}$.

又置

$$\mathfrak{S}'(t)=\{r_{i_1}(t),r_{i_2}(t),\cdots,r_{i_{n+1}}\} \tag{3.12}$$

作其度量和$\mathfrak{S}'=\int_\alpha^\beta\mathfrak{S}'(t)dt$,将$\mathfrak{S}'$一分为三:

$$\mathfrak{S}'=\mathfrak{S}_1+\mathfrak{S}_2+\mathfrak{S}_3, \tag{3.13}$$

其中

$$\mathfrak{S}_k=\int_{\Delta_k}\mathfrak{S}'(t)dt,k=1,2,3.$$

经过换元

$$\mathfrak{S}_1+\mathfrak{S}_2=\int_{t_1-\delta}^{t_1+\delta}(\mathfrak{S}'(t)+\mathfrak{S}'(t+t_2-t_1))dt. \tag{3.14}$$

对(3.14)应用引理1(在(3.1)中置$d=n+2$)有

$$V^{\frac{2}{n+1}}(\mathfrak{S}_1+\mathfrak{S}_2)\geqslant\int_{t_1-\delta}^{t_1+\delta}V^{\frac{2}{n+1}}(\mathfrak{S}'(t)+\mathfrak{S}'(t+t_2-t_1))dt. \tag{3.15}$$

右端积分号下的被积函数

$$\varphi(t)=V^{\frac{2}{n+1}}(\mathfrak{S}'(t)+\mathfrak{S}'(t+t_2-t_1))$$

是t的非负连续函数,而且由(3.11)有

$$\varphi(t_1)=V^{\frac{2}{n+1}}(\mathfrak{S}'(t_1)+\mathfrak{S}'(t_2))>0, \tag{3.16}$$

故其积分必取正值. 于是由(3.15)得到

$$V(\mathfrak{S}_1+\mathfrak{S}_2)>0, \tag{3.17}$$

即$\mathrm{conv}(\mathfrak{S}_1+\mathfrak{S}_2)$是$E^{n+1}$中非退化单形,也即

$$\dim(\mathrm{conv}(\mathfrak{S}_1+\mathfrak{S}_2))=n+1.$$

从而$\dim(\mathrm{conv}\ \mathfrak{S})\geqslant\dim(\mathrm{conv}\ \mathfrak{S}')\geqslant\dim(\mathrm{conv}(\mathfrak{S}_1+\mathfrak{S}_2))>n$. 证毕.

作为定理4的推论,前节中的系1,2,3都可以平行推广到连续型,就不必一一证明了. 今后在应用中,可对连续型度量和或度量平均直接引用系1,2,3.

定义 设C是一条可求长的闭曲线,其参数表示为$r=r(t)(0\leqslant t\leqslant T)$. 我们把Hilbert空间中一条闭曲线$\overline{C}:\overline{r}=\overline{r}(t)(0\leqslant t<T)$叫做曲线$C$按参数$t$的"对称化",如果有关系

$$|\bar{r}(t_1)-\bar{r}(t_2)|^2=\frac{1}{T}\int_0^T|r(t_1+t)-r(t_2+t)|^2dt \tag{3.18}$$

对区间$[0,T]$中的一切t_1,t_2都成立(注意:将$r(t)$视为以T为周期的函数,于是它对一切$t\in(-\infty,+\infty)$都有意义).

这样的对称化的存在性是早已解决了的问题[1,3].此外从(3.18)可以看到,弦长$|\bar{r}(t_1)-\bar{r}(t_2)|$只依赖于$t_1-t_2$.这样的曲线$\bar{C}$正是 von Neumann[7]和 Sehoenberg 用分析工具研究过的 Hilbert 空间中的闭螺旋线,其对称群是一个单参数连续群.

定理 5 设C是一条可求长的平面闭曲线,其参数表示为$r=r(t)(0\leqslant t\leqslant T)$.又令$\bar{C}$:$\bar{r}=\bar{r}(t)$是$C$按参数$t$的对称化,则$\bar{C}$为平面曲线的充要条件是:

(ⅰ)C是一个椭圆;

(ⅱ)参数t是C的“仿射弧长”的线性函数.

证 必要性 设$\bar{C}$是平面曲线,我们来证明,对每一个固定的u值,$r(t)\mapsto r(t+u)$都是曲线C到自身的一个仿射对应.如果不是这样,则能找到C上的4个不共线的点$r(t_1)$,$r(t_2)$,$r(t_3)$,$r(t_4)$和某个数$u_0\in[0,T]$,使得$r(t_k)\mapsto r(t_k+u_0)$(其中$k=1,2,3,4$)不是线性对应.那么由定理4,点列族

$$\mathfrak{G}(u)=\{r(t_1+u),r(t_2+u),r(t_3+u),r(t_4+u)\}\quad(0\leqslant u\leqslant T) \tag{3.19}$$

的度量平均

$$\bar{\mathfrak{G}}=\frac{1}{T}\int_0^T\mathfrak{G}(u)du$$

肯定不是平面四角形.从另一方面看,曲线$\bar{C}$上的点列$\{\bar{r}(t_1),\bar{r}(t_2),\bar{r}(t_3),\bar{r}(t_4)\}$合同于$\bar{\mathfrak{G}}$,(这是根据(3.18)),故$\bar{C}$不可能是平面曲线.这与所设矛盾.由此可见,对每一个固定的u值,$r(t)\mapsto r(t+u)$都是曲线C到自身的一个仿射对应.再由系3通过极限过程容易证明,映射$r(t)\mapsto\bar{r}(t)$也是一个仿射对应,这对应将曲线C仿射地映到其对称化曲线$\bar{C}$上.

前面已经讲过,闭曲线的对称化一定是 von Neumann[7]意义下的闭螺旋线.而平面上的闭螺旋线只有圆,因而$\bar{C}$必定是一个圆.既然已经证明C是$\bar{C}$的仿射象,故C是椭圆,(ⅰ)成立.

如果一条可求长曲线$\bar{r}=\bar{r}(t)$的弦长$|\bar{r}(t_1)-\bar{r}(t_2)|$只依赖于参数值的差$t_1-t_2$,这曲线(已知是一螺旋线)的弧长必然是参数$t$的线性函数,这是[7]中 Theorem 4 结论的一部分.

这样,我们这个定理中的对称化曲线$\bar{C}$:$\bar{r}=\bar{r}(t)$的参数t应当是$\bar{C}$的弧长s的线性函数.已知$\bar{C}$是一个圆,一个圆的孤长s为其仿射弧长λ的线性函数(按仿射弧长的定义,见[8,9].对于足够光滑的曲线,$\lambda=\int_0^s k^{\frac{1}{3}}(s)ds$,其中$k(s)$是曲线的曲率函数).于是,$\bar{C}$的仿射弧长$\lambda$是参数$t$的线性函数.

又在仿射对应下,曲线的仿射弧长只改变一个常数倍.既然C是$\bar{C}$的仿射象,$r(t)\mapsto\bar{r}(t)$是仿射对应,故C的仿射弧长也应为t的线性函数.(ⅱ)成立.

充分性比较显然,故略去.证毕.

定理 6 设C是一条可求长的平面闭曲线,其弧长参数表示为$r=r(s)(0\leqslant s\leqslant L)$.又令$\bar{C}$:$\bar{r}=\bar{r}(s)$是$C$按参数$s$的对称化,则$\bar{C}$为平面曲线的充要条件是:$C$是一个圆.

证 充分性十分显然,一个圆的按弧长参数的对称化必合同于自身.

必要性 如果 $\overline{C}$ 是平面曲线,由定理 4(只用到必要性部分),C 必是一个椭圆而且参数 s 是仿射弧长 λ 的线性函数. 注意到 s 是弧长参数. 我们有(令 $k(s)$ 表 C 的曲率函数)

$$\int_0^s k^{\frac{1}{3}}(s)ds = \lambda = as + b. \tag{3.20}$$

微分之得到 $k(s)=a^3$,即 C 是一个圆. 证毕.

这个定理对于应用而言是比较方便的. 以 Herda 在[6]中提及(后来 Alexander 又在[3]中提到)的 Johnson 的一个猜想为例,我们将用对称化的方法借助定理 6 进行处理.

例 5(Johnson 猜想) 设 $r(s)$是闭曲线 C 的自然参数表示,曲线长度为 L、又设 σ 是一个确定的正数,$0<\sigma<L$. 考虑曲线上那些长度为 p 的弧所张的弦的最小值. 令

$$f(\sigma,C)=\min_s|r(s+\sigma)-r(s)|, \tag{3.21}$$

则有

$$f(\sigma,C)\leqslant\frac{L}{\pi}\sin\frac{\pi\sigma}{L}, \tag{3.22}$$

等号成立当且仅当 C 是一个圆.

为证明此命题,我们可以引用下面的

定理 $\mathbf{S}^*$ (G. T. Sallee[4]) 设 $C:r=r(s)$是一条可求长闭曲线,$0\leqslant s\leqslant L$. 如果 C 不是平面凸曲线,则必存在一条平面凸闭曲线 $C^*:r^*=r^*(s)$,$0\leqslant s\leqslant L$,使得 C^* 与 C 有相等的长度 L,而且

$$|r(s_1)-r(s_2)|<|r^*(s_1)-r^*(s_2)| \tag{3.23}$$

对一切 $s_1\neq s_2$,$0\leqslant s_1<L$,$0\leqslant s_2<L$ 都成立.

应该指出,Sallee 只对(3.23)证明了不等号$\leqslant$成立,但不难证明,代之以严格的不等号$<$时,定理也是真的.

例 5 的证明不等式(3.22)可以由这问题的离散模型,例 4 的(2.37)通过极限过程得到. 剩下的问题是求证等号成立的条件.

圆能够使(3.22)取得等号,这无论是由直接计算或通过极限过程都是显而易见的. 下面我们假定 C 不是圆,往证 $f(\sigma,C)$不能取到最大值.

按 C 的弧长参数 s 作对称化 $\overline{C}:\overline{r}=\overline{r}(s)$,由定理 6,$\overline{C}$ 不是平面曲线. 再由定理 S^*,存在着一条与 $\overline{C}$ 等长的曲线 $C^*:r^*=r^*(s)$,使得

$$|\overline{r}(s+\sigma)-\overline{r}(s)|<|r^*(s+\sigma)-r^*(s)| \tag{3.24}$$

对一切 $\sigma\in(0,L)$都成立. 注意到$|\overline{r}(s+\sigma)-\overline{r}(s)|$仅与 σ 有关而与 s 无关,故有

$$|\overline{r}(s+\sigma)-\overline{r}(s)|<\min_s|r^*(s+\sigma)-r^*(s)|=f(\sigma,C^*). \tag{3.25}$$

仍参照对称化的定义(3.18),我们有

$$\begin{aligned} f(\sigma,C)&=\min_s|r(s+\sigma)-r(s)| \\ &\leqslant\left(\frac{1}{L}\int_0^L|r(s+\sigma)-r(s)|^2ds\right)^{\frac{1}{2}} \\ &=|\overline{r}(s+\sigma)-\overline{r}(s)|<f(\sigma,C^*). \end{aligned} \tag{3.26}$$

这说明 $f(\sigma,C)<f(\sigma,C^*)\leqslant\frac{L}{\pi}\sin\frac{\pi\sigma}{L}$,即 C 不能使(3.22)取等号. 例 5 证毕.

作者曾在[11]中用 Fourier 展开的方法给出一个别证. 但此处的证法阐明对称化概念的效用.

参考文献

[1]Alexander R. Pac. J. of Math. ,1979,85(1):1 - 9.
[2]Alexander R. Geometriae Dedicada,1977(6):87 - 94.
[3] Alexander R. Metric embedding techniques applied to geometric inequalities, in the geometry of metric and linear spaces,Ed. L. M. Kelly,Springer-Verlag,1975:57 - 65.
[4]Sallee G T. Geometriae Dedicada,1974(2):311 - 317.
[5]Neumann B H. J. London Math. Soc. ,1941(16):230 - 245.
[6]Herda H. Amer. Math. Monthly,1974(81):146 - 149.
[7]von Neumann J,Schoenberg I J. Trans. Amer. Math. Soc. ,1941(50):226 - 251.
[8]苏步青. 仿射微分几何,北京:科学出版社,1982.
[9]Blashke W. Vorlesungen uber Differentialgeometrie Ⅱ,Chelsea,New York,1967.
[10]Yang Lu,Zhang Jingzhong. Kexue Tongbao,1982,27(2):699 - 703.
[11]杨路,张景中. 关于空间曲线 Johnson 的猜想. 科学通报,1984(6):329 - 333.

高维映射嵌入多参流与渐近嵌入多参流

武　河　　　　　　张景中　杨　路
（中国科学技术大学数学系，合肥）（中国科学院成都分院数理研究室）

摘　要

本文引入了多参流的概念，讨论了多参流与我们通常研究的单参流的一些关系，并利用正规流与渐近嵌入的思想给出了一类高维收缩映射嵌入多参流的充要条件.

设 M 为 n 维 $C^r(0\leqslant r<+\infty)$ 流形，所谓 M 上的 C^k-m 参流 $(0\leqslant k\leqslant r, 1\leqslant m\leqslant n)\ \varphi$ 是指这样一个 C^k 映射：

$$\varphi = M\times\mathbf{R}^m\to M$$

满足

(1) $\varphi(x,0)=x\quad(\forall x\in M)$,

(2) $\varphi(\varphi(x,t_1),t_2)=\varphi(x,t_1+t_2)\quad(\forall x\in M,\forall t_1,t_2\in\mathbf{R}^m)$,

如果 φ 仅在 $M\times\mathbf{R}^{+m}(\mathbf{R}^{+m}=[0,+\infty)^m)$ 上有定义且使(2)对 $t_1,t_2\in\mathbf{R}^{+m}$，$x\in M$ 时成立，我们称 φ 为 M 上的 C^k-m 参半流.

若 φ 为流，可以证明对任一固定的 $t\in\mathbf{R}^m$，$\varphi(x,t)$ 为同胚. φ 为半流时则不一定如此.

设 f: $M\to M$ 为 M 上的映射，若有 $\lambda\in\mathbf{R}^{+m}$，使

$$f(x)=\varphi(x,\lambda)\ (\forall x\in M),$$

则说 f 可以嵌入流(或半流) φ，称 λ 为 f 关于 φ 的嵌入指数.

什么样的映射可以嵌入流或半流呢？这种嵌入在什么意义上是唯一的？这些问题很早就为一些函数方程的研究者所关心，这方面最早的工作可以追溯到 Abel[1]. 近年来在动力系统理论的研究中，人们对同胚嵌入流或连续映射嵌入半流的问题仍感兴趣，但是正如 Fort 在文献[2]中所指出的那样，嵌入问题是一个相当困难的问题，因而关于这个问题的多数结果都是在一维情形下得到的，如 Fort[2]，Sternberg[3]，Lam[4~6] 都讨论了线段上自同胚嵌入流的问题. 张筑生[7]给出了圆周上自同胚嵌入流的充要条件. 关于高维情形，麦结华[8]给出了带及圆柱面的自同胚与平移的拓扑共轭(从而可以嵌入连续流)的充要条件. Palis[9]，Sternberg[10]在相当强的条件下得到了一些高维的结果，他们通常要求所讨论的同胚有且仅有双曲不动点. 张景中、杨路[11,12]不要求所讨论的同胚一定具有双曲不动点，引入了标准流与渐近嵌入的思想，取得了一些相当普遍的结果. 以上所讨论的流均为单参流. 我们知道，单参流一般可以看作一阶常微分方程(组)的解曲线，流的参数在这里即为时间参数. 但高阶常

本文刊于《中国科学(A辑)》，第1期，1988年.

微分方程(组)的解可以是一个多参曲线族,有时会是多参流[13]. 因而多参流的嵌入问题是一个有意义的问题,可惜一直未见有人考虑. 我们在这篇文章里讨论了多参流与单参流的一些关系,并且利用文献[11,12]的思想,通过引入标准流得出了一类在不动点收缩的同胚可以嵌入多参流的充要条件,并且证明了在一定的限制下所嵌入的流是唯一的. 在一个方面拓广和发展了这个古老、困难而又有意义的领域.

一、多参流与单参流的一些关系

在引言中我们给出了多参流的定义,一个最自然的问题就是:多参流与单参流的关系是什么? 具体地说,一个多参流中的参数间满足怎样的关系才能使之成为单参流呢? 为了行文方便,现给出两个定义:

定义 1.1 设 φ 为 n 维 C^r 流形 M 上的 C^k-m 流($0\leqslant k\leqslant r, 1\leqslant m\leqslant n$),如果存在 m 个 C^k 映射:

$$f_j:\mathbf{R}\to\mathbf{R}(j=1,2,\cdots,m),$$

使得 $\varphi_1(x,s)=\varphi\left(x,\begin{pmatrix}f_1(s)\\ \vdots\\ f_m(s)\end{pmatrix}\right)$ $(\forall x\in M, \forall s\in\mathbf{R})$ 为一单参流,则称 φ 可以限制成单参流.

定义 1.2 设 $\varphi_1,\cdots,\varphi_n$ 为 n 维 C^r 流形 M 上的 C^k 单参流,如果存在 M 上的 C^k-m 参流 φ,使得 φ 可以分别限制成 $\varphi_1,\cdots,\varphi_n$,则称 φ 为 $\varphi_1,\cdots,\varphi_n$ 生成的 n 参流.

有了这两个定义,就可以问一个多参流最多能限制成多少个本质不同的单参流呢? 反之,给定 n 个不同的单参流能生成 n 参流吗?

首先我们有

定理 1.1 设 φ 为 n 维 C^r 流形 M 上的 C^k-m 流($0\leqslant k\leqslant r, 1\leqslant m\leqslant n$),且 $\varphi(x,T)=x\Leftrightarrow T=0$,则 φ 可以限制成单参流当且仅当 φ 的参数 $T=\begin{pmatrix}t_1\\ \vdots\\ t_m\end{pmatrix}$ 可以表示为 $T=\begin{pmatrix}a_1 & & 0\\ & \ddots & \\ 0 & & a_m\end{pmatrix}\begin{pmatrix}s\\ \vdots\\ s\end{pmatrix}$,其中 $\begin{pmatrix}a_1 & & 0\\ & \ddots & \\ 0 & & a_m\end{pmatrix}$ 为常矩形,$s\in\mathbf{R}$.

证 ($\Rightarrow$)设 φ 可以限制成单参流,即存在 m 个 C^k 映射:$f_j:\mathbf{R}\to\mathbf{R}(j=1,2,\cdots,m)$使得

$$\varphi_1(x,s)=\varphi\left(x,\begin{pmatrix}f_1(s)\\ \vdots\\ f_m(s)\end{pmatrix}\right)\quad(\forall x\in M, \forall s\in\mathbf{R})$$

为一单参流.

由流的定义我们可以推出

$$\varphi\left(x,\begin{pmatrix}f_1(s+t)\\ \vdots\\ f_m(s+t)\end{pmatrix}\right)=\varphi\left(x,\begin{pmatrix}f_1(s)+f_1(t)\\ \vdots\\ f_m(s)+f_m(t)\end{pmatrix}\right)\quad(\forall x\in M, \forall s,t\in\mathbf{R}),$$

又因为 $\varphi(x,T)=x \Leftrightarrow T=0$.

所以 $f_j(s+t)=f_j(s)+f_j(t)(j=1,2,\cdots,m)$，由 f_j 的连续性知 $f_j(s)=a_j s$，a_j 为常数. 必要性证毕.

(⇐)显然.

此外，如果我们给定了 n 维流形 M 上的两个单参流 $\varphi_1(x,s)$ 和 $\varphi_2(x,t)$，它们能生成 M 上的双参流吗？首先当 φ_1,φ_2 可交换时，我们说 φ_1 和 φ_2 可以生成双参流. 事实上我们只要令

$$\varphi: M\times\mathbf{R}^2\to M$$

为 $\varphi\left(x,\begin{pmatrix}s\\t\end{pmatrix}\right)=\varphi_2(\varphi_1(x,s),t)=\varphi_1(\varphi_2(x,t),s)$，容易验证 φ 为 φ_1,φ_2 生成的双参流. 当然还可以有其它的定义方式，比如：

$$\bar{\varphi}\left(x,\begin{pmatrix}s\\t\end{pmatrix}\right)=\varphi_2(\varphi_1(x,\alpha s+\beta t),as+bt)$$

$$\left(\forall x\in M,\forall s,t\in\mathbf{R},\alpha,\beta,a,b\in\mathbf{R}\text{ 为常量且}\begin{vmatrix}\alpha & a\\ \beta & b\end{vmatrix}\neq 0\right)$$

也为 φ_1,φ_2 生成的双参流. 容易看出 φ 与 $\bar{\varphi}$ 只差参数间的一个可逆线性变换. 我们称两个只差参数间的一个可逆线性变换的流为等价的. 显然上述 φ 与 $\bar{\varphi}$ 是等价的. 因而自然想到，φ_1，φ_2 生成的双参流在这种等价意义下唯一吗？

定理 1.2　如果 φ_1,φ_2 为 n 维流形 M 上的 C^k 单参流，且 φ_1,φ_2 可交换，则 φ_1,φ_2 可以生成 M 上的 C^k 双参流. 进一步，如果 φ_1 与 φ_2 不等价，则它们生成的双参流在等价意义下是唯一的.

证　φ_1,φ_2 可以生成双参流已明. 下证在等价意义下所生成的双参流的唯一性.

设 φ_1,φ_2 生成的双参流为 $\varphi\left(x,\begin{pmatrix}s\\t\end{pmatrix}\right): M\times\mathbf{R}^2\to M$，由生成的定义及定理 1.1 知，必存在常量 $a,b,\alpha,\beta\in\mathbf{R}$，使得：

$$\varphi\left(x,\begin{pmatrix}a & s\\ b & s\end{pmatrix}\right)=\varphi_1(x,s),$$

$$\varphi\left(x,\begin{pmatrix}\alpha & t\\ \beta & t\end{pmatrix}\right)=\varphi_2(x,t)(\forall x\in M,\forall s,t\in\mathbf{R})$$

成立.

因为 φ_1 与 φ_2 不等价，所以 $\begin{vmatrix}a & \alpha\\ b & \beta\end{vmatrix}\neq 0$，不妨记 $\begin{pmatrix}a & \alpha\\ b & \beta\end{pmatrix}^{-1}=\begin{pmatrix}\lambda & q\\ \mu & r\end{pmatrix}$，则

$$\begin{aligned}\varphi\left(x,\begin{pmatrix}s\\t\end{pmatrix}\right)&=\varphi\left(x,\begin{pmatrix}a & \alpha\\ b & \beta\end{pmatrix}\begin{pmatrix}\lambda & q\\ \mu & r\end{pmatrix}\begin{pmatrix}s\\t\end{pmatrix}\right)\\&=\varphi\left(x,\begin{pmatrix}a & \alpha\\ b & \beta\end{pmatrix}\begin{pmatrix}\lambda s+qt\\ \mu s+rt\end{pmatrix}\right)\\&=\varphi_1(\varphi_2(x,\mu s+rt),\lambda s+qt).\end{aligned}$$

又由于$\begin{pmatrix}\lambda & q\\ \mu & r\end{pmatrix}\neq 0$,故$\varphi\left(x,\begin{pmatrix}s\\t\end{pmatrix}\right)$与$\varphi_1(\varphi_2(x,t),s)$是等价的. 所以$\varphi\left(x,\begin{pmatrix}s\\t\end{pmatrix}\right)$在等价意义下是唯一的. 证毕.

对于k个单参流,我们也有同样的结论:

定理 1.3 如果$\varphi_1,\varphi_2,\cdots,\varphi_k$为$n$维流形$M$上的$k$个$C^r$单参流,且它们两两可换,则它们可以生成$M$上的$C^r_{-k}$参流. 进一步,如果$\{\varphi_j:j=1,\cdots,k\}$两两不等价,则它们生成的$k$参流在等价意义下是唯一的.

证 类似于定理(1.4).

有了这个定理,我们还可以问一个有趣的问题,即给定一个单参流,问与它可交换并且不等价的单参流有多少,进而就可以研究单参流向多参流的拓广问题. 这里我们给出一个简单的例子.

例:$\varphi_1(x,s)=\begin{pmatrix}\lambda^s & s\lambda^{s-1}\\ 0 & \lambda^s\end{pmatrix}x\quad(\lambda\in\mathbf{R}^+,s\in\mathbf{R},x\in\mathbf{R}^2)$,通过计算可找到一类与$\varphi_1$可交换的$\mathbf{R}^2$上的单参流具有形式

$$\varphi_2(x,t)=\begin{pmatrix}\mu^s & at\mu^{t-1}\\ & \mu^t\end{pmatrix}x\quad(\lambda\in\mathbf{R}^+,t\in\mathbf{R},x\in\mathbf{R}^2).$$

这样我们就知道了φ_1可以拓广成多参流,并且可以定出它的结构.

二、n 参正规流及渐近嵌入的定义

定义 2.1 设M为n维流形,$\mathbf{R}^n$为n维实空间,定义连续映射

$$G:M\times\mathbf{R}^n\to M$$

满足 (1)$G(x,0)=x\quad(\forall x\in M)$,

(2)$G(G(x,T_1),T_2)=G(x,T_1+T_2)\quad(\forall x\in M,\forall T_1,T_2\in\mathbf{R}^n)$,

(3)$G(x,\mathbf{R}^n)=M\quad(\forall x\in M)$,

则称G为M上的n参流或M上的实变换群. 如果G还满足

(4)$G(x,T)=x\Leftrightarrow T=0$,

则称G为M上的n参正规流.

例:(1)$G_0(x,T)=x+T(x\in\mathbf{R}^2,T\in\mathbf{R}^2)$,

(2)$G_1(x,T)=G_1\left[\begin{pmatrix}x_1\\y_1\end{pmatrix},\begin{pmatrix}s\\t\end{pmatrix}\right]=\begin{pmatrix}a^s & \\ & b^t\end{pmatrix}\begin{pmatrix}x_1\\y_1\end{pmatrix}$

$$\left(0<a,b<1;T=\begin{pmatrix}s\\t\end{pmatrix}\in\mathbf{R}^2;x=\begin{pmatrix}x_1\\y_1\end{pmatrix}\in\mathbf{R}^{+2}\right),$$

(3)$G_2(x,T)=\begin{pmatrix}a^s & sa^{s-1}\\ & a^s\end{pmatrix}\begin{pmatrix}b^t & \\ & b^t\end{pmatrix}\begin{pmatrix}x_1\\y_1\end{pmatrix}\left(0<a,b<1;\begin{pmatrix}x_1\\x_1\end{pmatrix}\in(0,+\infty)^2\right)$,

(4)$G_3(z,T)=a^se^{it}z\left(T=\begin{pmatrix}s\\t\end{pmatrix}\in\mathbf{R}^2;z\in\mathbf{C}-复数域\right)$.

容易验证 G_0, G_1, G_2 均为双参正规流，而 G_3 仅为双参流.

引理 2.1　若 $G(x,T)$ 为 M 上的 n 参正规流，则对任给的 $x,y\in M$，有唯一的 $T=T(x,y)$ 满足

$$G(x,T)=y$$

且 $T=T(x;y)$ 连续，关于 x,y 均为同胚映射.

证　任给 $x,y\in M$，因为 $G(x,\mathbf{R}^n)=M$，故有 $T\in\mathbf{R}^n$，使得

$$G(x,T)=y.$$

我们说这个 T 是唯一的，如若不然，假设还存在 $S\in\mathbf{R}^n$，使得

$$G(x,S)=y,$$

则有

$$G(x,T)=y=G(x,S),$$

故得

$$G(x,T-s)=x.$$

由定义 2.1 的(4)知 $T-S=0$，即 $T=S$. 所以存在唯一的 $T=T(x,y)$，使 $G(x,T)=y$，即

$$G(x,T(x,y))=y,$$

$T(x,y)$ 的连续性显然. 下证 $T(x,y)$ 关于 x,y 的同胚性. $\forall x_0\in M$，我们说 $T(x_0,\cdot):M\to\mathbf{R}^n$ 为同胚映射.

事实上，任给 $y_1,y_2\in M$，$y_1\neq y_2$，假设

$$T(x_0,y_1)=T(x_0,y_2),$$

则 $y_1=G(x_0,T(x_0,y_1))=G(x_0,T(x_0,y_2))=y_2$，矛盾. 又因为 $G(x_0,\mathbf{R}^n)=M$，故 $T(x_0,M)=\mathbf{R}^n$. 至此知 $T(x_0,\cdot)$ 既为单射又为满射，加之连续性我们知道 $T(x_0,\cdot):M\to\mathbf{R}^n$ 为同胚映射. 同理 $T(\cdot,y_0):M\to\mathbf{R}^n$ 亦为同胚映射. 证毕.

推论 2.1　对于引理 2.1 中的 $T(x,y)$，有下面等式成立：

$$T(x,y)=T(x_0,y)-T(x_0,x)\quad(\forall x_0\in M).$$

证　令 $T_x=T(x_0,x)$，$T_y=(x_0,y)$，则

$$\begin{aligned}G(x,T_y-T_x)&=G(G(x_0,T(x_0,x)),T_y-T_x)\\&=G(G(x_0,T_x),T_y-T_x)\\&=G(x_0,T_y)=y.\end{aligned}$$

所以

$$T(x,y)=T_y-T_x,$$

即

$$T(x,y)=T(x_0,y)-T(x_0,x).$$ 证毕.

定义 2.2　若 $F\in C^0(M,M)$，使得

$$\lim_{m\to+\infty}T(F^n(x),F^{n+1}(x))=\alpha_F=\begin{pmatrix}\alpha_1\\\vdots\\\alpha_n\end{pmatrix}$$

关于 x 在 M 上局部一致地成立，则称 F 可以渐近嵌入于 G. α_F 称为 F 关于 G 的渐近嵌入指数.

定义 2.3　若 M 上的 n 参半流 $F(x,T)$ 中的每个 $F(x,T_0)$(其中 T_0 的每个分量 $\geqslant 0$，这时记为 $T_0\geqslant 0$)都可以渐近嵌入于 G，并且 $T_0\geqslant 0$. 当且仅当 $F(x,T_0)$ 关于 G 的渐近嵌入指数 $\alpha(T_0)\geqslant 0$，并满足 T_0 与 $\alpha(T_0)$ 的分量中的 0 的个数相同，则 F 叫做渐近嵌入于 G 的 n 参半流.

定义 2.4 设 M 为 n 维流形，$p\overline{\in}M$，$p\in\partial M$（∂M 记为 M 的边界），并且 $M\cup\{p\}$ 为闭的，F 为 M 上的连续映射，且有

$$\lim_{x\to p}F(x)=p,\quad \lim_{n\to+\infty}F^n(x)=p\quad(\forall x\in M),$$

则称 F 在 p 点为收缩的.

定义 2.5. 设 $G(x,T)$ 为 M 上的 n 参正规流，任给 $x\in M$，任给 $\mathbf{R}^{+n}$ 中的向量列 $T_m=\begin{bmatrix}t_m^1\\ \vdots\\ t_m^n\end{bmatrix}$（当 $m\to+\infty$ 时，$t_m^i\to+\infty$，$i=1,2,\cdots,n$），局部一致地有

$$\lim_{m\to+\infty}G(x,T_m)=q.$$

当存在 $j_0\in\{1,2,\cdots,n\}$ 使 $t_m^{j_0}=0(m=1,2,\cdots)$ 且 $t_m^i\to+\infty(m\to+\infty,j\neq j_0)$ 时，$G(x,T_m)$ 局部一致地收敛到 M 中一点，则称 $G(x,T)$ 为在 q 点正向收缩的 n 参正规流.

推论 2.2 设 $G(x,T)$ 为 M 上的在 q 点正向收缩的 n 参正规流，则任给 $T_0\in\mathbf{R}^{+n}$ 且 $T_0>0$，$G(x,T_0)$ 为在 q 点收缩的映射.

证 由定义 2.4 及 2.5 显然.

推论 2.3 设 $G(x,T)$ 为 M 上的在 q 点正向收缩的 n 参正规流，则

$$G(x,\mathbf{R}^{+n})=M_x\subset M\cup\{q\}\text{为一有界闭域},$$

$$G(x,\mathbf{R}^{+k}\times\{0\}\times\mathbf{R}^{+(n-k-1)})=M_{x,k}\subset M\text{ 为一有界闭域}.$$

证 由定义 2.5 显然.

三、主要定理及其证明

为了行文方便，我们先对二维情形给出我们的定理.

定理 3.1 设 M 为 $\mathbf{R}^2$ 中一单连通域，$F\in C^\circ(M,M)$ 且在 0 点收缩，G 为 M 上的在 0 点正向收缩的 2 参正规流，F 可以渐近嵌入于 G，渐近嵌入指数 $\alpha_F>0$（指每一分量 >0）且 $F(x)\in G(x,\mathbf{R}^{+2})$，则 F 可以嵌入一个渐近于 G 的 2 参半流 $F(x,T)$ 且 $F(x,\mathbf{R}^{+2})=G(x,\mathbf{R}^{+2})$ 的充要条件为：

$$\Phi_n(x,y)=T(F^n(x),F^n(y)).$$

在 $y\in G(x,\mathbf{R}^{+2})$ 上局部一致收敛于关于 x,y 均为同胚的映射 $\Phi(x,y)$，且 $\Phi(x,G(x,\mathbf{R}^{+2}))=\mathbf{R}^{+2}$.

证 先证充分性.

因为 $\Phi_n(x,y)$ 在 $y\in G(x,\mathbf{R}^{+2})$ 上局部一致收敛到关于 x,y 均为同胚的映射 $\Phi(x,y)$.

由推论 2.1 知

$$T(x,y)=T(x_0,y)-T(x_0,x),$$

取定 x_0，即有

$$\Phi(x,y)=\Phi(x_0,y)-\Phi(x_0,x).$$

记 $\varphi(x)=\Phi(x_0,x)$，由条件知 $\varphi(x)$ 为同胚映射，即 φ^{-1} 存在且连续，又因为 $\Phi(x,G(x,\mathbf{R}^{+2}))=\mathbf{R}^{+2}$，故唯一地有

$$y=\varphi^{-1}(\varphi(x)+T)=F(x,T)\quad(T\in\mathbf{R}^{+2}),$$

可以验证 $F(x,T)$ 为半流，这里略.

由 $F(x)\in G(x,\mathbf{R}^{+2})$，知 $\Phi_n(x,F(x))$ 收敛，对照渐近嵌入指数的定义知 $\Phi(x,F(x))=\alpha_F$，所以 $F(x)=F(x,\alpha_F)$，即 F 可以嵌入半流 $F(x,T)$.

因为任给 $x\in M$，任给 $s\in\mathbf{R}^{+2}$，有

$$y=G(x,s)=\varphi^{-1}(\varphi(x)+\Phi(x,y))=F(x,\Phi(x,y)),$$

而 $\Phi(x,y)\in\mathbf{R}^{+2}$，则 $G(x,\mathbf{R}^{+2})\subset F(x,\mathbf{R}^{+2})$.

又因为 $\Phi(x,G(x,\mathbf{R}^{+2}))=\mathbf{R}^{+2}$，所以有

$$F(x,\mathbf{R}^{+2})\subset G(x,\mathbf{R}^{+2}).$$

故 $F(x,\mathbf{R}^{+2})=G(x,\mathbf{R}^{+2})$.

因为 $G(x,s)$ 在 0 点正向收缩，故由推论 2.3 知 $G(x,\mathbf{R}^{+2})=M_x\subset M\cup\{0\}$ 为一有界闭区域. 所以 $F(x,\mathbf{R}^{+2})=M_x$ 为一有界闭区域.

由于 $\forall s\in\mathbf{R}^{2+}$，$\forall x\in M$，有

$$y=G(x,s)=F(x,\Phi(x,y)),$$

不妨记 $\Phi(x,y)=H(x,s)$，显然连续，故有 $H(x,\mathbf{R}^{+2})=\mathbf{R}^{+2}$. 又 $H(x,0)=\Phi(x,G(x,0))=\Phi(x,x)=0$，故有

$$H(x,\mathbf{R}^{+}\times\{0\})=\mathbf{R}^{+}\times\{0\}\quad(\text{或}\{0\}\times\mathbf{R}^{+}),$$
$$H(x,\{0\}\times\mathbf{R}^{+})=\{0\}\times\mathbf{R}^{+}\quad(\text{或 }\mathbf{R}^{+}\times\{0\}).$$

不妨设　$H(x,\mathbf{R}^{+}\times\{0\})=\mathbf{R}^{+}\times\{0\}$,

　　　　$H(x,\{0\}\times\mathbf{R}^{+})=\{0\}\times\mathbf{R}^{+}$.

由此可得

$$G(x,\mathbf{R}^{+}\times\{0\})=F(x,H(x,\mathbf{R}^{+}\times\{0\}))=F(x,\mathbf{R}^{+}\times\{0\}),$$

同理 $G(x,\{0\}\times\mathbf{R}^{+})=F(x,\{0\}\times\mathbf{R}^{+})$.

由推论 2.3 知 $G\left(x,\begin{pmatrix}\mathbf{R}^{+}\\0\end{pmatrix}\right)$ 为 M 中有界闭域，则 $F\left(x,\begin{pmatrix}\mathbf{R}^{+}\\0\end{pmatrix}\right)=G\left(x,\begin{pmatrix}\mathbf{R}^{+}\\0\end{pmatrix}\right)$ 亦为 M 中一有界闭域. 进一步 $F\left(x,\begin{pmatrix}\mathbf{R}^{+}\\t\end{pmatrix}\right)(\forall t\in\mathbf{R}^{+})$ 亦为 M 中一有界闭域. 更进一步 $F\left(x,\begin{pmatrix}\mathbf{R}^{+}\\I\end{pmatrix}\right)$ ($I\subset\mathbf{R}^{+}$ 为一有界闭域) 亦为 M 中一有界闭域.

同理，$F\left(x,\begin{pmatrix}I\\\mathbf{R}^{+}\end{pmatrix}\right)$ ($I\subset\mathbf{R}^{+}$ 为一有界闭域) 为 M 中一有界闭域.

下面证明：$\forall T>0$，$F(x,T)$ 可以渐近嵌入于 G，即证明 $T[F(x,nT),F(x,nT+T)]$ 局部一致收敛. 事实上：

$$\begin{aligned}&T[F(x,nT),F(x,nT+T)]\\&=T\left[F\left(x,n\begin{pmatrix}s\\t\end{pmatrix}\right),F\left(x,n\begin{pmatrix}s\\t\end{pmatrix}+T\right)\right]\quad\left(T=\begin{pmatrix}s\\t\end{pmatrix}>0\right)\\&=T\left[F\left(x,k_n\alpha_F+n\begin{pmatrix}s\\t\end{pmatrix}-k_n\alpha_F\right),F\left(x,k_n\alpha_F+n\begin{pmatrix}s\\t\end{pmatrix}+T-k_n\alpha_F\right)\right],\end{aligned}$$

记 $\alpha_F=\begin{pmatrix}a\\b\end{pmatrix}$ 且设若 $\frac{s}{a}\geqslant\frac{t}{b}$ 取 $k_n=\left[\frac{nt}{b}\right]$，否则 $\frac{s}{a}<\frac{t}{b}$ 时取 $k_n=\left[\frac{ns}{a}\right]$. 这里不妨设 $\frac{s}{a}\geqslant\frac{t}{b}$，即取

$k_n=\left[\frac{nt}{b}\right]$，记 $\beta_n=\begin{pmatrix}s_n\\t_n\end{pmatrix}=n\begin{pmatrix}s\\t\end{pmatrix}-k_n\alpha_F$，则 $0\leqslant t_n\leqslant b$，则

$$\lim_{n\to+\infty}T[F(x,nT),F(x,nT+T)]$$
$$=\lim_{n\to+\infty}T[F(x,k_n\alpha_F+\beta_n),F(x,k_n\alpha_F+\beta_n+T)]$$
$$=\lim_{n\to+\infty}T[F^{k_n}(F(x,\beta_n)),F^{k_n}(F(x,\beta_n+T))]$$
$$=\lim_{n\to+\infty}\Phi_{k_n}[F(x,\beta_n),F(x,\beta_n+T)]$$
$$=\lim_{\substack{n\to+\infty\\ y\in M_x}}\Phi_{k_n}[y,F(y,T)]$$，其中 M_x 为 M 中某有界闭域（由上面的讨论及 β_n 的构成知）
$$=\Phi[y,F(y,T)]\quad(y\in M_x)$$
$$=T.$$

若 $T\in\mathbf{R}^{+2}$，且仅有一分量为 0，则由推论 2.3 直接有

$$\lim_{n\to+\infty}T[F(x,nT),F(x,nT+T)]$$
$$=T(y,F(y,T))\quad y\in M$$
$$=T(y,G(y,\widetilde{T}))\quad \widetilde{T}\text{ 仅有一分量为 }0$$
$$=\widetilde{T},$$

至此已证明 $F(x,T)$ 可以正向渐近嵌入于 G.

唯一性：若 F 可以同时嵌入正向渐近嵌入于 G 的 2 参半流 $F(x,T)$ 和 $\widetilde{F}(x,T)$，且 $F(x,\alpha)=\widetilde{F}(x,\alpha)=F(x)$，则 $\forall\, T>0$，

$$\lim_{n\to+\infty}T[F^n(x),F^n(F(x,T))]=\Phi(x,F(x,T))=T.$$

同理 $\Phi(x,\widetilde{F}(x,T))=T$.

由 Φ 的同胚性知
$$F(x,T)=\widetilde{F}(x,T),$$
故
$$F(x,T)=\widetilde{F}(x,T)\quad(x\in M,T\in\mathbf{R}^{+2}).$$

下证必要性：

设 F 已嵌入渐近于 G 的半流 $F(x,T)$，要证的是 $\Phi_n(x,y)=T(F^n(x),F^n(y))$ 在 $y\in G(x,\mathbf{R}^{+2})$ 上局部一致收敛于关于 x,y 均为同胚的映射，且 $\Phi(x,G(x,\mathbf{R}^{+2}))=\mathbf{R}^{+2}$.

因为
$$\lim_{n\to+\infty}F^n(x)=0\quad(\forall\, x\in M),$$
$$F(x,\mathbf{R}^{+2})=G(x,\mathbf{R}^{+2}),$$

故 $\forall\, T\in\mathbf{R}^{+2}$，由定义 $\forall\, y=F(x,T)\in G(x,\mathbf{R}^{+2})$，有

$$\lim_{n\to+\infty}T_G[F^n(x),F^n(F(x,T))]$$
$$=\lim_{n\to+\infty}T_G[F^n(x),F(F^n(x),T)]$$
$$=\alpha(T)\in\mathbf{R}^{+2}.\tag{Δ}$$

（因为 $F(x,T)$ 可以渐近嵌入于 G，知（Δ）式局部一致成立，其中 $\alpha(T)$ 为 $F(x,T)$ 关于 G 的渐近嵌入指数）.

我们可以给出 $\alpha(T)$ 的结构表达.

（Ⅰ）因为 $F(x,\mathbf{R}^{+2})=G(x,\mathbf{R}^{+2})$，

所以
$$\lim_{x\to0}F(x,s)=\lim_{x\to0}G(x,T)=0\in\mathbf{R}^2$$
（由 G 在 0 点的收缩性）.

由(Δ)式及流的定义知：

$$\alpha(T_1+T_2)=\alpha(T_1)+\alpha(T_2)\quad(T_1,T_2\in\mathbf{R}^{+2}),$$

故有

$$\alpha(T)=AT\quad(A\text{ 为 2 阶常方阵}).$$

(Ⅱ)由于$\alpha(T)\in\mathbf{R}^{+2}$且保持$T$中分量的0的个数，故$A$可逆且$A$中元素均为非负元素，则知$\alpha(T)$关于$T$为同胚且$\alpha(\mathbf{R}^{+2})=\mathbf{R}^{+2}$.

同样，由$\alpha(T)$保持T中分量的0的个数这个性质，我们可以推出：任给$s,t\in\mathbf{R}^{+}$，s,t不全为0，有$F\left(x,\begin{pmatrix}s\\0\end{pmatrix}\right)\neq F\left(x,\begin{pmatrix}0\\t\end{pmatrix}\right)$，故有$F(x,S)\neq F(x,T)\quad(\forall S,T\in\mathbf{R}^{+2},S\neq T)$. 因而引理2.1对半流$F(x,T)$也成立，即$\forall y\in G(x,\mathbf{R}^{+2})$，关于半流的$T_F(x,y)$存在，且关于$x,y$均为同胚映射. 即有

$$\alpha(T)=AT=AT_F(x,F(x,T))=AT_F(x,y).\qquad(*)$$

故由(Δ)式和(*)式知$\Phi_n(x,y)$在$y\in G(x,\mathbf{R}^{+2})$上局部一致收敛到关于$x,y$均为同胚的映射.

从(Ⅱ)知$\alpha(\mathbf{R}^{+2})=\mathbf{R}^{+2}$，即$\Phi(x,G(x,\mathbf{R}^{+2}))=\mathbf{R}^{+2}$，必要性证毕.

对于n维情形有着完全相同的结果，下面只给出定理的叙述，不加以证明.

定理 3.2 设M为$\mathbf{R}^n$中一连通域，$F\in C^{\circ}(M,M)$且在0点收缩，G为M上的在0点正向收缩的n参正规流，F可以渐近嵌入于G，渐近嵌入指数$\alpha_F>0$(指每一分量>0)且$F(x)\subset G(x,\mathbf{R}^{+2})$，则$F$可以嵌入一个渐近于$G$的$n$参半流$F(x,T)$且$F(x,\mathbf{R}^{+n})=G(x,\mathbf{R}^{+n})$的充要条件为：

$$\Phi_n(x,y)=T(F^n(x),F^n(y)).$$

在$y\in G(x,\mathbf{R}^{+n})$上局部一致收敛于关于$x,y$均为同胚的映射$\Phi(x,y)$且$\Phi(x,G(x,\mathbf{R}^{+n}))=\mathbf{R}^{+n}$，以及$\Phi(x,G(x,\cdot))$把$\mathbf{R}^{+n}$中的$n$条棱一一对应地变为$\mathbf{R}^{+n}$中的$n$条棱(这里的棱指$\overbrace{\{0\}\times\cdots\times\{0\}}^{i-1}\times\mathbf{R}^{+}\times\overbrace{\{0\}\times\cdots\times\{0\}}^{n-i}$).

推论 3.1[11] 设G为区间(a,b)上的单参正规流，$F(x)$在(a,b)上连续、递增，没有不动点，$F(b)=b$，F把(a,b)变为(a,b)且F可以渐近嵌入于G，渐近嵌入指数为正，则F可以嵌入唯一的渐近于G的单参半流的充要条件是：二元函数列

$$\Phi_n(x,y)=T_G[F^n(x),F^n(y)](n=1,2,\cdots)$$

在区域$D:\begin{cases}a<x<b\\x\leqslant y<b\end{cases}$上收敛到连续、严格单调的函数$\Phi(x,y)$.

四、定理 3.1 的应用

Sternberg[10]讨论了高维同胚与对角型线性映射的共轭问题，事实上这可以视作高维单参流的嵌入问题，也可以看成高维多参流的嵌入问题，其中最典型的一种情形如下：

定理[10] 4.1 设$\widetilde{M}$为n维欧氏空间中原点的某个邻域且不含原点，F是$\widetilde{M}\cup\{0\}$上的C^k同胚且保持原点不动，并且具有非0的Jocobian，如果有：

(1)$J(F)$的特征值$s_1,s_2,\cdots,s_n$满足

$$|s_i|<1(i=1,2,\cdots,n),$$

(2)F 的线性项没有多重基本因子(multiple elementary divisors),并且

$$s_i \neq s_1^{i_1} s_2^{i_2} \cdots s_n^{i_n} \quad \left(\forall \text{ 非负整数 } i_j \text{ 使} \sum_{j=1}^{n} i_j > 1 \right),$$

则当 k 充分大时,存在一个 C^k 同胚 R 使得 RFR^{-1} 为线性的,即

$$RFR^{-1}(x) = \begin{pmatrix} s_1 & & & \\ & s_2 & & \\ & & \ddots & \\ & & & s_n \end{pmatrix} x \quad (x \in \widetilde{M}).$$

在我们的定理 3.1 中只要取 $M=\widetilde{M}, G(x,T)=R^{-1}\begin{pmatrix} s_1^{t_1} & & \\ & \ddots & \\ & & s_n^{t_n} \end{pmatrix} R(x)$ (这里不妨设 $s_j > 0, x \in M$),显然 F 可以嵌入于 G,当然渐近嵌入于 G. 很容易看出 F 满足定理 3.1 中的充分条件,即 F 可以嵌入一个 n 参流,由唯一性知这个 n 参流就是 G. 可见满足定理 4.1 条件的必然也满足定理 3.1 的条件,并且在定理 3.1 中指出了可嵌入流的唯一性.

此外,我们并不像 Sternberg[10] 所做的那样,要求映射不动点为双曲的,事实上有很多在不动点非双曲的映射可以嵌入单参或多参流. 有兴趣的读者可以在一维情形找出具体例子,由于篇幅问题,这里不再赘述.

参考文献

[1]Abel N. Works,V. 2. Posthumous paper,1881:36 - 39.

[2]Fort M K Jr. Proc. Amer. Math. Soc. ,1955(6):960 - 967.

[3]Sternberg S. Duke Math. J. ,1957(24):97 - 102

[4]Lam P F. J. Diff. Equations,1978(30):31 - 40

[5]Topological Dynamics:An International Symposium,New York,1968:319 - 333.

[6]Colloq. Math. ,1976(35):275 - 287.

[7]张筑生. 数学学报,1981(24):953 - 957.

[8]麦结华. 中国科学 A 辑,1985(1):17 - 23.

[9]Palis J. Topology,1969(8):385 - 405.

[10]Sternberg S. Amer. J. Math. ,1957(79):809 - 823;1958(80):623 - 632;1959(81):578 - 604.

[11]张景中,杨路. 北京大学学报,1982(6):23 - 44.

[12]中国科学 A 辑,1985(1):32 - 43.

[13]Gottschalk W H, Hedlund G A. Topological Dynamies, Amer. Math. Soc. Colloq. Publications,Vel. XXXY1. 11.

The Criterion Algorithm of Relation of Implication between Periodic Orbits(I)

Zhang Jing-zhong Yang Lu Zhang Lei

(*Chengdu Branch, Academia Sinica, Chengdu*)

Abstract

In recent years, there is a wide interest in Sarkovskii's theorem and the related study. According to Sarkovskii's theorem, if the continuous self-mapf of the closed interval has a 3-periodic orbit, then f must has an n-periodic orbit for any positive integer n. But f can not has all n-periodic orbits for some n.

For example, let

$$f(x)=\begin{cases} x+\frac{1}{2} & \left(x\in\left[0,\frac{1}{2}\right]\right) \\ 2(1-x) & \left(x\in\left(\frac{1}{2},1\right]\right) \end{cases}$$

Evidently f has only one kind of 3-periodic orbit in the two kinds of 3-periodic orbits. This explains that it isn't far enough to uncover the relation between periodic orbits by information which Sarkovskii's theorem has offered. In this paper, we raise the concept of type of periodic orbits, and give a feasible algorithm which decides the relation of implication between two periodic orbits.

Ⅰ. Introduction

In recent years, there is extensive interest in Sarkovskii's theorem and the related topics. A. N. Sarkovskii's[1] suggested arranging all natural numbers in the following order:

$\lhd : 3\lhd 5\lhd 7\lhd 9\lhd \cdots\lhd 2n-1\lhd 2n+1\lhd \cdots;$

$\lhd 6\lhd 10\lhd 14\lhd \cdots\lhd 2\cdot(2n-1)\lhd 2\cdot(2n+1)\lhd \cdots;$

$\lhd 12\lhd 20\lhd 28\lhd \cdots\lhd 4\cdot(2n-1)\lhd 4\cdot(2n+1)\lhd \cdots;$

本文刊于《Applied Mathematics and Mechanics》,第11卷第11期,1989年.

$\cdots;\lhd 2^k\cdot 3\lhd 2^k\cdot 5\lhd 2^k\cdot 7\lhd\cdots\lhd 2^k\cdot(2n-1)\lhd 2^k\cdot(2n+1)\lhd$
$\cdots;\cdots 2^m\lhd 2^{m-1}\lhd\cdots\lhd 16\lhd 8\lhd 4\lhd 2\lhd 1$

and proved the following:

Sarkovskii's theorem Let f be a continuous self-mapping of the closed interval. If f has an n-periodic point, then f must have an m-periodic point for any m satisfying $n\lhd m$.

Let f be a continuous mapping. For any positive integer n, we define f^n, inductively by $f^0=Id$(identity mapping)and $f^n=f\circ f^{n-1}$. If there exists a point x_0 satisfying

$$\begin{cases} f^n(x_0)=x_0 \\ f^k(x_0)\neq x_0 \quad (k=1,2,\cdots,n-1) \end{cases} \tag{1.1}$$

then x_0 is called an n-periodic point of f and $\{x_0, x_1=f(x_0),\cdots,x_{n-1}=f(x_{n-2})\}$ or $\{f^k(x_0);k=0,1,\cdots,n-1\}$ is called an n-periodic orbit of f.

For any fixed positive integer n, there may be different kinds of n-periodic orbits. This can be explained by introducing the concept of types of periodic orbits.

Arrange n elements of an n-periodic orbit $\{f^k(x_0);k=0,1,\cdots,n-1\}$ into the following increasing sequence

$$z_1<z_2<\cdots<z_n. \tag{1.2}$$

If $f(z_i)=z_j$, then let $a_i=j$ for $i=1,2,\cdots,n$. The ordered set $(a_1,\cdots,a_n)$of positive integers is called the type of the n-periodic orbit $\{f^k(x_0);k=0,1,\cdots,n-1\}$.

For example, there are two 4-periodic orbits:

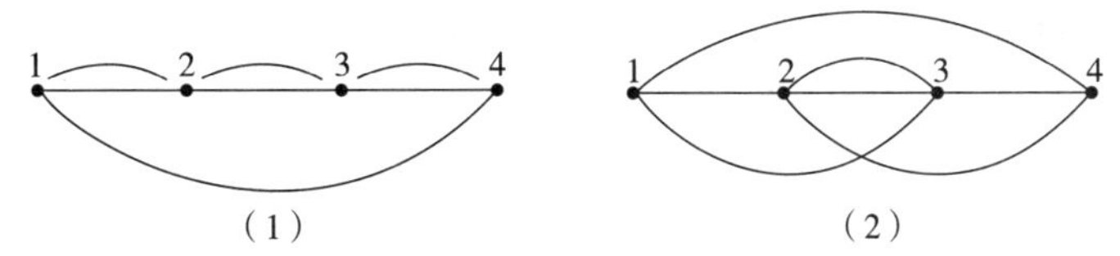

(1) (2)

where the type of (1) is (4,1,2,3), and the type of (2) is (4,3,1,2).

Let n be the period of some periodic orbits. It can be seen easily that there is only one type (1) or (2,1) for $n=1$ or 2, respectively. When $n=3$, there are two types (2,3,1) and (3,1,2). When $n=4$, there are six types, besides the two types mentioned above, there remain (3,1,4,2),(3,4,2,1),(2,3,4,1) and (2,4,1,3).

Generally speaking, there are $(n-1)!$ types for all the n-periodic orbits.

By Sarkovskii's theorem, if the continuous self-mapping f of a closed interval has a 3-periodic orbit, then f must have an n-periodic orbit for any positive integer n. But f need't have all types of n-periodic orbits.

Example Let

$$f(x)=\begin{cases} x+\frac{1}{2} & \left(x\in\left[0,\frac{1}{2}\right]\right) \\ 2(1-x) & \left(x\in\left(\frac{1}{2},1\right]\right) \end{cases} \tag{1.3}$$

Evidently, f has a 3-periodic orbit $\{0,1/2,1\}$ (with type(2,3,1)). But f has only one type (3,4,2,1) among the six types of 4-periodic orbits. Likewise, f has also only one type (3,5,4,2,1) or (4,6,5,3,2,1) among the twenty-four types of 5-periodic orbits or one hundred and twenty types of 6-periodic orbits, respectively.

Does f have only one type for any n-periodic orbit? No, in fact f has two types of 7-periodic orbits, three types of 8-periodic orbits and thirteen types of 12-periodic orbits. All these results were obtained by computer. It is very complicated to obtain the above results in the usual ways.

A natural problem is that if f has a periodic orbit with some type, what type of periodic orbits should f have?

If the existence of any periodic orbit with type A implies the existence of a periodic orbit with type B for every continuous self-mapping of a closed interval, we say "type A implies type B" and denote this by $A \lhd B$.

For instance, in the previous example we have $(2,3,1) \lhd (3,4,2,1)$ and so on.

Then, given both different types A and B. Type A implies type B or type B implies type A or they imply each other or not. which is right at all? For a certain type, say, A, how do we determine the types implied by type A?

It isn't enough to answer the above questions only by the information which Sarkovskii's theorem has offered. This needs more deep and careful study.

The implication between types may be discussed in a more general view. The concepts of s-minimum periodic orbit and simple periodic orbit were introduced in [5] and [6], and the following result was given in [6]:

If a continuous self-mapping of a closed interval has n-periodic orbit, then it must have a simple n-periodic orbit.

These results in [5] and [6] may be regarded as the beginning for exploring the implication between types of periodic orbits.

It seems quite difficult to find out, for the problem of implication between types, a perfect answer clear at one glance as Sarkovskii's theorem. In this paper, however, a feasible algorithm is given to determine whether a type implies another type or not. The results offered by computer are helpful to us in further research on periodic orbits.

The following paragraphs introduce the principle of algorithm, the design of program, the computed results and the related conjectures.

Ⅱ. The Principle of Algorithm

We consider the general cases.

Let $S_n = \{x_1 < x_2 < \cdots < x_n\} \subset R^1$ and $C(S_n)$ denote the space of mappings from S_n into itself. For $f \in C(S_n)$, we define the piecewise linear extension $\bar{f}$ of f as follows

$$\begin{cases} \bar{f}(x_i)=f(x_i) & (i=1,2,\cdots,n) \\ \bar{f} \text{ is linear on } [x_i,x_{i+1}] & (i=1,2,\cdots,n-1) \end{cases}$$

and let $L(S_n)$ denote the set consisting of all such $\bar{f}$.

For some $f\in C(S_n)$, S_n may be an n-periodic orbit of f. We define

$C_p(S_n)=\{(f\in C(S_n)S_n$ is a periodic orbit of $f\}$

$L_p(S_n)=\{\bar{f}\in L(S_n)\mid f\in C_p(S_n)\}$.

In the following, we shall explain that all types of m-periodic orbits of $\bar{f}$ can be determined by finite steps for any $\bar{f}\in L(S_n)$ and show that the $\bar{f}$ has a periodic orbit with type B if $A\lhd B$ for any type B, where $f\in C_p(S_n)$ and A is the type of f.

For any $f\in C(S_n)$, let $a_i=j$ if $f(x_i)=x_j$. The n-tuple$(a_1,a_2,\cdots,a_n)$ of positive integers is called the type of f on S_n, denoted by

$$A_f=(a_1,a_2,\cdots,a_n) \tag{2.1}$$

When $f\in C_p(S_n)$, A_f is also the type of n-periodic orbit S_n of f.

Obviously, if $A_f=A_g$ for $f,g\in C(S_n)$, then $\bar{f}$ and $\bar{g}$ are orientation-preserving topologically conjugate, and they have the periodic orbits with the same type. Thus, we may discuss for any n-tuple in place of S_n. Without loss of generality, let $S_n=\{1,2,\cdots,n\}$. Evidently,

$$A_f=(f(1),f(2),\cdots,f(n)) \tag{2.2}$$

We use the usual method of symbolic dynamical systems. Let interval $[1,n]$ be divided into the $n-1$ parts:

$$\Delta_i=(i,i+1) \quad (i=1,2.\cdots,n-1)$$

and let

$$\begin{cases} u_i=\min\{f(i),f(i+1)\} \\ v_i=\max\{f(i),f(i+1)\} & (i=1,2,\cdots,n-1) \\ D_i=\mathrm{sgn}(f(i+1)-f(i)) \end{cases}$$

When $v_i>u_i$, $\bar{f}$ maps Δ_i into (u_i,v_i). When $u_i=v_i$, $\bar{f}$ maps Δ_i into the point $f(i)$. Besides, when $D_i=1$, $\bar{f}$ is strictly increasing on Δ_i, when $D_i=-1$, $\bar{f}$ is strictly decreasing on Δ_i; when $D_i=0$, $\bar{f}$ is a constant on Δ_i.

Consider infinite sequences formed by $1,2,\cdots$and $n-1$: $I=i_0i_1i_2\cdots i_k\cdots(1\leqslant i_k\leqslant n-1)$. Let M be the set of all I.

Definition 1 Let $I=i_0i_1i_2\cdots i_k\cdots\in M$, I is called f-admissible, if

$$u_{i_k}\leqslant i_{k+1}<v_{i_k} \quad (k=0,1,2,\cdots) \tag{2.3}$$

where u_{i_k} and v_{i_k} were defined as above. By M_f we denote the set which consists of all f-admissible sequences in M.

For $I=\{i_k\}\in M_f$, by definition it always holds

$$D_{i_k}\neq 0 \tag{2.4}$$

So, we may introduce order relation on M_f.

Definition 2 Let $I, J\in M_f, I\neq J$ and

$$I=i_0 i_1 i_2\cdots i_k\cdots$$

$$J=j_0 j_1 j_2\cdots j_k\cdots$$

We say $I<J$ if one of the following two conditions holds:

(ⅰ) $i_0<j_0$

(ⅱ) there exists k such that $i_l=j_l$ for $l=0,1,\cdots,k-1$, and

$$D_{i_0}D_{i_1}\cdots D_{i_{k-1}}\cdot i_k<D_{i_0}\cdot D_{i_1}\cdots D_{i_{k-1}}\cdot j_k \tag{2.5}$$

According to the above definition, obviously M_f is a total ordering set

For any $x\in[1,n]$, if there exists a positive integer k such that $\overline{f}^k(x)$ is an integer, then x is called a trivial point of f. Evidently we need only consider those periodic orbits consisting of nontrivial points and integers respectively for clarifying the types of periodic orbits of f.

Let us give a corresponding relation between elements of M_f and the set of non-trivial points of $[1,n]$.

Definition 3 Let $x\in[1,n]$ be a non-trivial point and

$$i_k=[\overline{f}^k(x)] \quad (k=0,1,2,\cdots) \tag{2.6}$$

where $[\cdot]$ denotes the maximum integer not more than x (that is to say, $\overline{f}^k(x)\in\varDelta_{i_k}$), and $I(x)=i_0\cdots i_k\cdots$. The sequence $I(x)$ is called the itinerary of x under $\overline{f}$.

The following lemmas can be easily proven.

Lemma 1 For any non-trivial point x, there is always $I(x)\in M_f$.

Lemma 2 Let x, y be two non-trivial points, then $I(x)\leqslant I(y)$ if $x<y$.

Lemma 3 If non-integer x is an m-periodic point of $\overline{f}$, then $I(x)$ must be a repetitive sequence and the length of its minimum repetitive path must be a divisor of m.

Definition 4 Let $x\in[1,n]$. If there exists $J\in M_f, J=j_0 j_1 j_2\cdots j_k\cdots$ such that

$$\overline{f}^k(x)\in\overline{\varDelta}_{j_k} \tag{2.7}$$

($\overline{\varDelta}$ denotes the closure of $\varDelta$), then we say that x mates with J, denoted by $x\wedge J$.

Evidently, x may mate with different J's, and reversely, J may mate with different x's as well.

But we still have

Lemma 4 Let $x_1\wedge J_1, x_2\wedge J_2$. If $J_1<J_2$ and $x_1\neq x_2$, then $x_1<x_2$.

Lemma 5 For any repetitive sequence $J\in M_f$, there exists a periodic point x of $\overline{f}$ satisfying:

(ⅰ) If x is trivial, then $x\wedge J$. If x is non-trivial, then $I(x)=J$.

(ⅱ) The period of x is just the length of the minimum repetitive path of J.

(ⅲ) The type of periodic orbit which is generated by x can be determined uniquely by J in finite steps.

Above several lemmas can be proven by known methods. Here, only lemma 5(ⅲ) need to be explained as follows. Let J_k denote the sequence obtained by omitting the first k elements of J for $k=0,1,2,\cdots$. Obviously, $\bar{f}^k(x) \wedge J_k$. Because the period of x is exactly equal to the length of the minimum repetitive path of J, all J_k's are pairwise different for $k=0,1,2,\cdots,m-1$. By definition 2 we can define the order between J_k's. Since we may arrange the order of $\{\bar{f}^k(x)\}_{k=0}^{m-1}$ from the order of J_k's according to lemma 4, the type of periodic orbit generated by x is determined uniquely.

Q. E. D.

In order to find out the types of all periodic orbits of $\bar{f}$, we need consider all sequences of the minimum repetitive paths with length m in M_f. These sequences determine the types of some m-periodic orbits of f. Is this enough to solve our problem? In other words, need we consider the sequences of the minimum repetitive paths with length d for every divisor d of m so as not to omit some types of the m-periodic orbits of f? In fact, this is not necessary.

Lemma 6 Let $x\in[1,n]$ be a non-trivial m-periodic point of f. d is the length of the minimum repetitive path of $I(x)$. If $d<m$ then there exists an integer $x^*\in[1,n]$ such that x^* is also an m-periodic point of f, and the type of periodic orbit generated by x^* is the same as the type of periodic orbit generated by x.

Proof Let $y=\bar{f}^d(x)$, then $I(x)=I(y)$. This shows that $\bar{f}^k(x)$ and $\bar{f}^k(y)$ are in the same interval Δ_i for any nonnegative integer k. So, $\bar{f}$ is linear on $[\bar{f}^k(x);\bar{f}^k(y)]$, where Δ_i is defined as previous, and $[\alpha,\beta]$ or $[\beta,\alpha]$ is denoted by $[\alpha;\beta]$. Suppose that $F=\bar{f}^m$, $g=\bar{f}^d$ and $m=d\cdot l$, then $F=g^l$. Let $I(x)=i_0 i_1\cdots i_k\cdots$. Because x and y are the fixed points of f, F is identity mapping on $[x,y]$. By $I(x)\in M_f$ and the definition of $\bar{f}$, we have $|\bar{f}'|\geqslant 1$ on $\Delta_{i_0}, \Delta_{i_1}, \cdots, \Delta_{i_{d-1}}$. Since $F'=(\bar{f}^m)'=1$, there must be $|\bar{f}'|=1$ on each Δ_{i_k}. Thus, $\bar{f}$ maps exactly Δ_{i_0} into itself. Because of $g(x)=y$, x and y belong to the same l-periodic orbit of g. Obviously, there must be $l=2$. So, $g(y)=x$.

For convenience sake, suppose that $x<y$. We shall point out that if x^* is the left endpoint of Δ_{i_0}, then x^* is also an m-periodic point of f and the type of periodic orbit generated by x^* is the same as the type of periodic orbit generated by x.

Since g is linear on Δ_{i_0}, $g(x)=y$, $g(y)=x$ and g maps the integer x^* into an integer, we obtain $g(i_0)=i_0+1$ and $g(i_0+1)=i_0$. Thus i_0 and i_0+1 belong to the same periodic orbit on $\bar{f}$ and their period is a divisor of m. It can be seen that $\Delta_{i_0}, \Delta_{i_1}, \Delta_{i_2}, \cdots, \Delta_{i_{d-1}}$ are pairwise different since f maps exactly Δ_{i_k} into $\Delta_{i_{k+1}}$; otherwise, the length of the minimum repetitive path must be smaller than d. Since i_0 and i_0+1 can be mapped into each endpoint of all Δ_{i_k} under the iteration of f, the number of these endpoints is $d+1$ at least. Thus $t\geqslant d+1>m/2$. This explains that $t=n$ and further, $\overline{\Delta_{i_0}}, \overline{\Delta_{i_1}}, \cdots, \overline{\Delta_{i_{d+1}}}$ are pairwise disjoint.

Now, we have obtained two m-periodic orbits:

$$\{x, \bar{f}(x), \cdots\cdots, \bar{f}^{m-1}(x)\}$$
$$\{x^*, \bar{f}(x^*), \cdots\cdots, \bar{f}^{m-1}(x^*)\},$$

We shall show that the two periodic orbits have the same type. Obviously, one needs only to show the following fact.

For any positive integer $0\leqslant k<l\leqslant m-1$, if $\bar{f}^k(x)<\bar{f}^l(x)$, then $\bar{f}^k(x^*)<\bar{f}^l(x^*)$; if $\bar{f}^l(x)<\bar{f}^k(x)$, then $\bar{f}^l(x^*)<\bar{f}^k(x^*)$.

Since x and x^* belong to the same interval $\overline{\Delta}_{i_0}$, $\bar{f}^k(x^*)$, $\bar{f}^k(x)$ and $\bar{f}^l(x^*)\bar{f}^l(x)$ belong to the same interval $\overline{\Delta}_{i_k}$ and $\overline{\Delta}_{i_l}$ respectively. When $\overline{\Delta}_{i_k}$ is different from $\overline{\Delta}_{i_l}$, the desired fact follows. When $\overline{\Delta}_{i_k}=\overline{\Delta}_{i_l}$ (i. e. $l=k+d$), there must be $\bar{f}^l(x)=\bar{f}^k(y)$. Let $y^*=i_0+1=\bar{f}^d(x^*)$. Then $\bar{f}^l(x^*)=\bar{f}^k(y^*)$. By $x^*<x<y<y^*$ and the monotonicity of $\bar{f}$ on $\overline{\Delta}_{i_s}$, we may assert that $\bar{f}^k(x)<\bar{f}^k(y)$ implies $\bar{f}^k(x^*)<\bar{f}^k(y^*)$, and $\bar{f}^k(y)<\bar{f}^k(x)$ implies $\bar{f}^k(y^*)<\bar{f}^k(x^*)$.

This shows that the previous two periodic orbits have the same type. Therefore, we may obtain the following conclusion.

Theorem 1 For the given $f\in C(S_n)$ and a positive integer m, there exists a feasible algorithm which may determine the types of all m-periodic orbits of $\bar{f}$.

In fact, by lemma 5 and lemma 6 we need only list the sequences of all the minimum repetitive paths with length m in M_f and determine the types of the m-periodic orbits which these sequences correspond to, adding the types of the m-periodic orbits of f.

In order to solve completely our problem to find out the types of periodic orbits implied in a given type of periodic orbits, we still need a lemma.

Lemma 7 Let $f\in C(S_n)$, $S_n\subset[a,b]$ and $\varphi(x)$ be a continuous extension of f on $[a,b]$. Then, for any m-periodic orbit of $\tilde{f}\in L(S_n)$, there exists an m-periodic orbit of φ such that the two periodic orbits have the same type.

This lemma seems obvious, but its proof needs a lot of explanations. So it is put in the appendix.

Now we may obtain the following theorem from lemma 7.

Theorem 2 Let $f\in C(S_n)$, with type A, and m be a positive integer. There exists a feasible algorithm, for any type B with $A\lhd B$, which can determine type B of m-periodic orbits.

In particular, for the given type A and type B of the two periodic orbits, we may determine whether $A\lhd B$ or $B\lhd A$ or both hold or not.

Appendix Proof of Lemma 7

This section will give a proof of lemma 7 which is raised in section Ⅱ.

We first do some preliminary work.

Definition 0. 1 Let $f:\mathbf{R}\to\mathbf{R}$ be a continuous mapping from real line to itself, $I=[a,b]\subset\mathbf{R}$. f is called the pseudo-increasing(pseudo-decreasing), if, f satisfies $f(a)<f(x)<f(b)$ $(f(a)>f(x)>f(b))$ for any $x\in$ int I. I is called a pseudo-increasing interval (pseudo-decreasing interval) of f.

Definition 0. 2 Let I and J be two non-trivial closed subintervals of $\mathbf{R}$. If max $I\leqslant$ min

J, we say that I is smaller than J, denoted as $I<J$.

Lemma 0.1 Let $f: I \to \mathbf{R}$ be a continuous mapping from closed interval to real line. If there exist closed subintervals $I_1, I_2, \cdots, I_{n-1}$, of I satisfying $I_0 \xrightarrow{f} I_1 \to \cdots \to I_{n-1} \to I_0$ ($I \xrightarrow{f} J$ denotes $f(I) \supset J$), then there exists $x_0 \in I_0$ such that $f^n(x_0)=x_0$, $f^i(x_0) \in I_i$ for $i=0,1,\cdots,n-1$.

The proof of the above lemma is trivial.

Lemma 0.2 Let $f: I \to \mathbf{R}$ be a continuous mapping from closed interval to real line. $I_1<I_2<\cdots<I_n$ are the closed subintervals of I, $J=[a,b] \subset I$ and $f(J) \supset \bigcup_{i=1}^{n} I_i$. Then, when $f(a)<f(b)$ $(f(a)>f(b))$, there exist $n+1$ pseudo-increasing (pseudo-decreasing) intervals $J', J_1<J_2<\cdots<J_n$. $(J_1>J_2>\cdots>J_n)$ such that $\bigcup_{i=1}^{n} J_i \subset J' \subset J$, $f(J')=[f(a); f(b)]$ and $f(J_i)=J_i$ for $i=1,2,\cdots,n$.

Proof Let $J'=[a',b']$ and $I_i=[a_i,b_i]$, $i=1,2,\cdots,n-1$. We discuss the following two cases.

(1) $f(a)<f(b)$

Let $a'=\max\{x \in J \mid f(x)=f(a)\}$ and $b'=\min\{a<x\leqslant b \mid f(x)=f(b)\}$.

Obviously, $J'=[a',b'] \subset J$ is a pseudo-increasing interval of f, and $f(J')=[f(a),f(b)]$.

Let

$$c_1=\min\{x \in J' \mid f(x)=a_1\}$$
$$d_1=\min\{c_1<x\leqslant b' \mid f(x)=b_1\}$$
$$c_i=\min\{d_{i-1}\leqslant x\leqslant b' \mid f(x)=a_i\}$$
$$d_i=\min\{c_i<x\leqslant b' \mid f(x)=b_i\}$$
$$(i=2,3,\cdots,n)$$

Since $a_i<b_i$ for $1\leqslant i\leqslant n$, by the definition of J', there must exist a pseudo-increasing interval $J_i \subset [c_i,d_i] \subset J'$ such that $f(J_i)=I_i$ for $i=1,2,\cdots,n$. Obviously $J_i<J_{i+1}$ by $[c_i, d_i]<[c_{i+1},d_{i+1}]$ for $i=1,2,\cdots,n-1$.

(2) $f(a)>f(b)$

The definition of a' and b' is the same as in (1). Then J' is a pseudo-decreasing interval of f and $f(J')=[f(b),f(a)]$.

Let

$$d_1=\max\{x \in J' \mid f(x)=a_1\}$$
$$c_1=\max\{a'\leqslant x<d_1 \mid f(x)=b_1\}$$
$$d_i=\max\{a'\leqslant x\leqslant c_{i-1} \mid f(x)=a_i\}$$
$$c_i=\max\{a'\leqslant x<d_i \mid f(x)=b_i\}$$
$$(i=2,3,\cdots,n)$$

Since $a_i<b_i$, by the selection of J', there must exist pseudo-decreasing interval $J_i \subset$

$[c_i, d_i] \subset J'$ such that $f(J_i) = I_i$ for $i = 1, 2, \cdots, n$. Obviously $J_{i+1} < J_i$ by $[c_{i+1}, d_{i+1}] < [c_i, d_i]$ for $i = 1, 2, \cdots, n-1$. Q. E. D

Theorem (Lemma 7 in Ⅱ) Let $f \in C(S_n)$, $S_n \subset [a, b]$ and continuous function $\varphi(x)$ be an extension of f on $[a, b]$. Then, for any m-periodic orbit $\{x_0, x_1, \cdots, x_{m-1}\}$ of $\bar{f} \in L(S_n)$, there exists an m-periodic orbit $\{z_0, z_1, \cdots, z_{m-1}\}$ of φ such that the two periodic orbits have the same type.

Proof If S_n contains a periodic orbit whose type is the same as that of $\{x_0, x_1, \cdots, x_{m-1}\}$, then the conclusion follows obviously. In the following, we don't consider this case. Thus, we may assume that x_0 is a non-trivial point of f.

By lemma 6, m is the length of the minimum repetitive path of $I(x_0)$.

Let

$$I(x_0) = I_0 I_1 \cdots I_{m-1} I_0 I_1 \cdots I_{m-1} \cdots\cdots$$

Then, the first m elements of $I(x_0)$ satisfy the relation as follows:

$$I_0 \xrightarrow{\bar{f}} I_1 \to \cdots \to I_{m-1} \to I_0 \tag{A. 1}$$

and $\bar{f}^i(x_0) \in \operatorname{int} l_i$, for $i = 0, 1, \cdots, m-1$.

Since φ is a continuous extension of f. by lemma 0. 2 there exists the closed interval $J_i (\subset J_i)$ satisfying the following condition:

If $\bar{f}$ increases (decreases) on I_i, then φ is pseudo-increasing (pseudo-decreasing) on J_i and $\varphi(J_i) = \bar{f}(J_i)$.

Obviously,

$$J_0 \xrightarrow{\varphi} J_1 \to \cdots \to J_{m-1} \to J_0 \tag{A. 2}$$

and {A. 2} is the prime repetitive loop with length m.

If $I_0, I_1, \cdots, I_{m-1}$ are pairwise different, then, by the selection of J_i and lemma 0. 1, the m-periodic orbit which is obtained by (A. 2) has the same type as that of periodic orbit $\{x_0, x_1, \cdots, x_{m-1}\}$. The desired conclusion is obtained.

If $\{I_i\}_{i=0}^{m-1}$ have the two same elements at least, without loss of generality, we may assume that I_0 arises two times at least in (A. 1) (not containing the least I_0 in (A. 1). Let $I_{s1}, \cdots, I_{st}$ be all different elements arranged after I_0 in (A. 1), and $I_{S1} < I_{S2} < \cdots < I_{St}$ $(t \geq 1)$, where $I_0 \to I_{st}$, is a subpath in (A. 1) for $i = 1, 2, \cdots, t$.

If $\bar{f}$ increases (decreases) on I_0, then φ is pseudo-increasing (pseudo-decreasing). By lemma 0. 2, there must exist $(I_{Si}^{(0)})_{i=1}^{t}$ and $(J_{Si}^{(0)})_{i=1}^{t}$ such that

$$I_{S_1}^{(0)} < I_{S_2}^{(0)} < \cdots < I_{S_t}^{(0)}$$
$$(>)(>)(>)$$
$$J_{S_1}^{(0)} < J_{S_2}^{(0)} < \cdots < J_{S_t}^{(0)}$$
$$(>)(>)(>)$$

$I_{S_i}^{(0)} \subset I_0$, $J_{S_i}^{(0)} \subset J_0$, $\bar{f}(I_{S_i}^{(0)}) = I_{S_i}$ and $\varphi(J_{S_i}^{(0)}) = J_{S_i}$. And $\bar{f}$ increases (decreases) on $I_{S_i}^{(0)}$; φ is pseudo-increasing(pseudo-decreasing)on $J_{S_i}^{(0)}$ for $i=1,2,\cdots,t$.

Now, we use $I_{S_i}^{(0)}\to I_{S_i}$ and $J_{S_i}^{(0)}\to J_{S_i}$ in place of $I_0\to I_{S_i}$ and $J_0\to J_{S_i}$ in(A. 1)and(A. 2) respectively, $i=1,2,\cdots,t$. We obtain to new prime repetitive loops (A. 3) and (A. 4) with length m-respectively from (A. 1) and (A. 2):

$$I'_0\xrightarrow{\bar{f}}I'_1\to\cdots\to I'_{m-1}\to I'_0 \tag{A. 3}$$

satisfying $\bar{f}^i(x_0)\in \text{int } I'_i, \bar{f}^{k_i}(\partial I'_i)\subset S_n$ for some positive integer $k_i(i=0,1,\cdots,m-1)$, where $\partial I'_i$ denotes the endpoints of interval I'_i.

$$J'_0\xrightarrow{\varphi}J'_1\to\cdots\to J'_{m-1}\to J'_0 \tag{A. 4}$$

satisfying $\varphi^{l_i}(\partial J'_i)\subset S_n$ for some positive integer $l_i(i=0,1,\cdots,m-1)$.

Obviously, (A. 3) and (A. 4) satisfy the relation as follows:

If $I'_i<I'_j$ then $J'_j<J'_j$ $(0\leqslant i,j\leqslant m-1)$.

According to the previous discussion, we know that (A. 3) and (A. 4) are compared with (A. 1) and (A. 2) respectively, satisfying (*) the number of different elements in the latter is smaller than that in the former.

In the following, we continue to handle the repetitive elements of (A. 3) and (A. 4) by the same method. After handling for finite times, we may obtain two prime repetitive loops with length m as follows by(*):

$$\begin{aligned} &I_0^*\xrightarrow{\bar{f}}I_1^*\to\cdots\to I_{m-1}^*\to I_0^* \\ &J_0^*\xrightarrow{\varphi}J_1^*\to\cdots\to J_{m-1}^*\to J_0^* \end{aligned} \tag{A. 5}$$

satisfying

ⅰ) $\text{int } I_i^*\cap\text{int } I_j^*=\varnothing$, $\text{int } J_i^*\cap\text{int } J_j^*=\varnothing$. for $0\leqslant i,j\leqslant m-1, i\neq j$.

ⅱ) If $I_i^*<I_j^*$, then $J_i^*<J_j^*$, $0\leqslant i,j\leqslant m-1$.

ⅲ) For $0\leqslant i\leqslant m-1$, there exist k_i^* and l_i^* such that $J_i^{k^*}(\partial I_i^*)\subset S_n$ and $\varphi^{l_i^*}(\partial J_i^*)\subset S_n$.

ⅳ) $\bar{f}^i(x_0)\in\text{int } I_i^*$, for $0\leqslant i\leqslant m-1$.

By (A. 5) and lemma 0. 1, there exists an m-periodic point z_0 of φ such that $z_i=\varphi_i(z_0)\in\text{int } J_i^*$ for $i=0,1,\cdots,m-1$. By ⅰ)—ⅳ)above, we may infer that the periodic orbit $\{z_0, z_1,\cdots,z_{m-1}\}$ of φ and the periodic orbit $\{x_0,x_1,\cdots,x_{m-1}\}$ of f have the same type.

Q. E. D.

References

[1]Sarkovskii A N. Ukr. Mat. Z. ,1964(16):6 - 71.
[2]Stefan P. Comm. Math. Phys. ,1977(54):237 - 248.
[3]Zhang Jingzhong, Yang Lu. Adv. in Math. ,1987,16(1):33 - 48. (in Chinese)
[4]Coppel W A. Math. Proc. Camb. Phil. Soc. ,1983(93):397 - 408.
[5]Block L,D Hart. Ergod. Th. and Dynam. Sys,1983(3):533 - 539.

The Criterion Algorithm of Relation of Implication between Periodic Orbits(II)

Zhang Jing-zhong Yang Lu Zhang Lei
(Chengdu Branch, Academia Sinica, Chengdu)

Abstract

In recent years there is a wide interest in Sarkovskii's theorem and related study. According to Sarkovskii's theorem, if the continuous self-map f of the closed interval has a 3-periodic orbit then f must has an n-periodic orbit for any positive integer n. But f can not have all n-periodic orbits for some n.

Example . Let

$$f(x)=\begin{cases} x+1/2 & x\in[0,1/2] \\ 2(1-x) & x\in[1/2,1] \end{cases}$$

Evidently, f has only one kind of 3-periodic orbit in the two kinds of 3-periodic orbits which explains that it isn't far enough to uncover relation between periodic orbits by the information which Sarkovskii's theorem has offered. In this paper, we raise the concept of type of periodic orbits and give a feasible algorithm which decides the relation of implication between the two kinds of periodic orbits.

Let f be a continuous self-mapping of the closed interval I. If there exist the pairwise different points $x_0, x_1, \cdots, x_{n-1}$, on I such that $f(x_{n-1})=x_0$, $f(x_k)=x_{k+1}$, for $k=0,1,2,\cdots,n-2$, then the set $\{x_0, x_1, \cdots, x_{n-1}\}$ is called an n-periodic orbit of f. Let $x_0, x_1, \cdots, x_{n-1}$ be arranged into $z_1<z_2<\cdots<z_n$ in increasing order. If $f(z_i)=z_j$, then let $a_i=j$ for $i=1,2,\cdots,n$. The n-tuple$(a_1, a_2, \cdots, a_n)$of the positive integers is called the type of the n-periodic orbit. Obviously there exist $(n-1)!$ types for all n-periodic orbits.

If a continuous function with a periodic orbit with type A must have a periodic orbit with type B, then we say type A implies type B, denoted by $A \lhd B$. Given concretely two types A and B, type A implies type B, or type B implies type A, or they imply each other or not. Which is the right at all? This is a quite interesting question in one-dimensional

本文刊于《Applied Mathematics and Mechanices》,第 11 卷第 2 期,1990 年.

dynamic system.

The first part of this paper has given a feasible criterion algorithm which may determine the implication between any two types by computer and the algorithmic principle which may determine if one type implies another type. In the second part, we shall give a program according to the principle, and list some concrete computed results.

Ⅲ. Explanation of the Program

We have written an algorithmic program in BASIC language, which may be executed by CASIO-PB 700 mini-microcomputer. Of course, the program may be also executed by any microcomputer(only revising some symbols when necessary). The program is listed in the following. We have recorded the tape so that it can be used directly in the same type of microcomputer by the profession who are interested in this problem.

The beginning of the program shows a menu. Please input choicely one of 0, 1 and 2 (6th line). Inputting 0, i. e. choosing LIST, the machine will output one repetitive path of each sequence consisting of the minimum repetitive path with length k in M_f, the arranging order of the elements in the periodic orbit determined corresponding to each sequence and the type corresponding to each periodic orbit(where the meaning of K will be explained below). Inputting 1, i. e. choosing NO SAME, the repetitive types obtained from the above will be omitted. Inputting 2, i. e. choosing CHECK, the machine will check if f implies some periodic orbit of the given type.

After choosing the menu, the screen will show input of the number N of elements in the domain of f then the K, the period of all possible periodic orbits which you want to list or the period of the given periodic orbit which you want to check if f implies.

Choosing $P=2$, the screen will show input of{C!(J); $J=1,2,\cdots,K$}, the type of some periodic orbit, so as to check if f implies it. Then, the type of f, {A!(J,0); $J=1,2,\cdots,N$) is asked to be input. By this time, the input of all information is finished.

The machine will print immediately N. K and{A! (J); $J=1,2,\cdots,N$) ending with" * OMP"(When choosing $P=2$, if f implies the checked type{C!(J); $J=1,2,\cdots,K$}, it will still print the{C! (J); $J=1,2,\cdots,K$}, ending with" * * * CK"). Then, it will print the sup and inf of the interval which $\overline{f}$ maps Δ_i into, for $i=1,2,\cdots,N-1$. For instance, 2—(2, 4) implies that $\overline{f}(\Delta_2)$ covers three intervals of Δ_2, Δ_3 and Δ_4

At last, the computed results will be printed. When choosing $P=0$ (LIST), every possible repetitive path with length K will be printed, ending with" * ". Then, the ordinal symbols of the periodic points corresponding to each repetitive path with length K will be printed, ending with"---ODER". Then, the type of periodic orbit corresponding to each repetitive path with length K will be printed, ending with" * * * * MP". When choosing $P=1$, the repetitive types will be omitted and won't be printed(For this, in the 13rd line of the program, we introduce D!(K, 100) which deposits the found types, where D!(K, 1 00) deposits one hundred types at most by the restriction of depositing capacity of mini-microcomputer). When choosing $P=2$, only the minimum repetitive path corresponding to the type of a given periodic orbit, the ordinal symbols of periodic points in the periodic orbit corresponding to the minimum repetitive path and the type of the periodic orbit corresponding to the minimum repetitive path will be printed. If f does imply the checked type, YES will be printed at last, and NO otherwise.

In our program, the types of trivial periodic orbits are not listed specially. It is not necessary of course to check the implication between periodic orbits.

In the following, we shall explain the meaning of some variants and lines in our program.

The variant symbol "!" denotes the semi-precision in order to take up smaller interval storage.

From the 25th line to the 28th line, the divisors B!(E, 3) of K are determined so as to omit the repetitive paths with length K, which consist of the repetitive paths with shorter length.

From the 75th Line to the 625th line, the repetitive paths with length K are found, i. e. all the K-repetitive sequences are found in M from 700th line to the 740th line, the sequences of the minimum repetitive paths with length smaller than K are omitted in the above K-repetitive sequences. From the 745th line to the 775th line, the repetitive K-repetitive sequences are omitted. From the 860th line to the 935th line, the elements in the K-periodic orbit corresponding to each K-repetitive path are given the symbols {B!(L, 2); $L=1, 2, \cdots, K$} in increasing order. From the 965th line to the 975th line, the type of the K-periodic orbit corresponding to each K-repetitive path is printed.

From the 1000th line to the 1050th line, choosing $P=1$, the repetitive types are omitted. Choosing $P=2$, each found type will be compared with the given type C!(J). If both are not the same, one will continue to find the type. If both are the same, one will print YES and end.

This program should have a great room for improvement. In particular, when choosing $P=2$ (CHECK), the more repetitive paths which need not be checked may be omitted beforehand by the features of the type of given periodic orbit. Thus, the computed quantity may be reduced.

The following is the outline of flowchart:

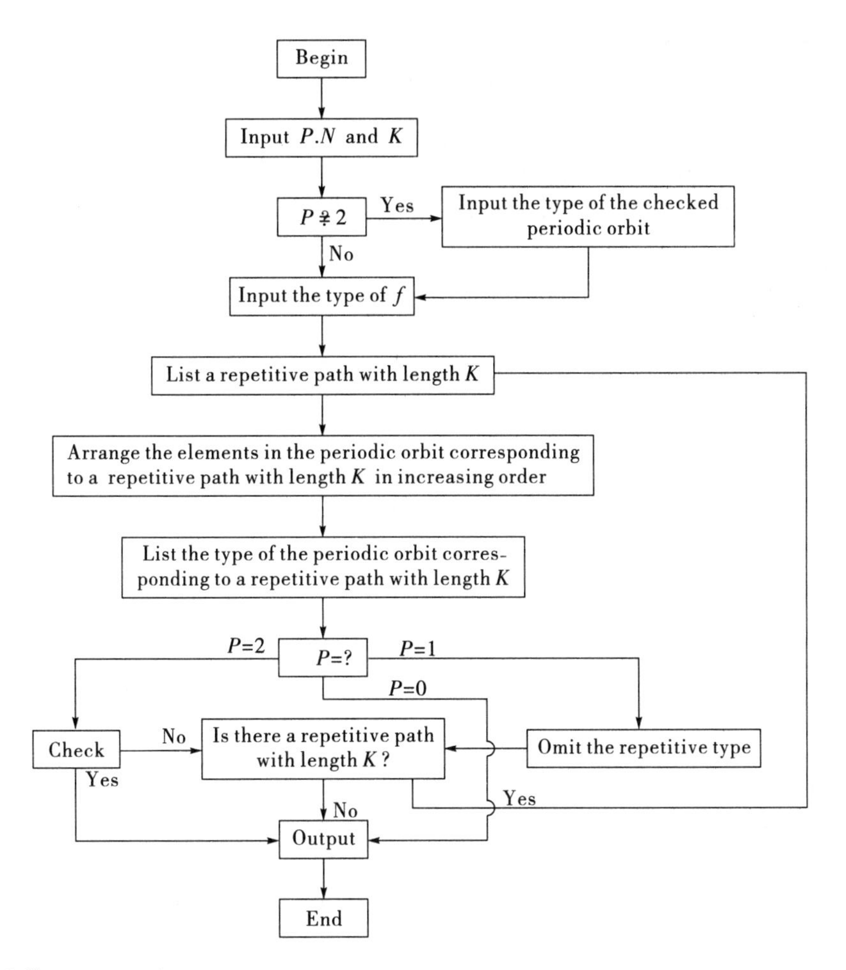

The following is the program:

```
5   ERASE A!. B!. C!. D!
6   INPUT"0:LIST 1:NO SAME 2:CHECK P=" :P
10  INPUT"N=";N. "K=";K
12  LPRINT"N=";N. "K=";K.
13  IF P=1 THEN DIM D! (K,100)
20  DIM A! (N,3). B! (K,6). C! (K)
21  IF P=2 THEN FOR J=1 TO K:INPUT"C!(J)=";C!(J):NEXT J:GOTO 995
25  E=0:I=0
26  FOR J=1 TO INT(K/2)
27  IF K MOD J=0 THEN E=E+1:B!(E,3) =J
28  NEXT
29  FOR J=1 TO N
30  INPUT"A !(J,0)";A!(J,0)
35  IF J=1 THEN GOTO 60
```

```
40  L=J-1:U=A!(L,0):V=A!(J,0):A!(L,2) =SGN(V-U)
43  IF U>V THEN W=U:U=V:V=W
45 A!(L,3)=V-1:A!(L,1)=U
47  IF U=V THEN A!(L,3)=0:A!(L,1)=0
48  IF J=N THEN A!(N,3)=A!(N-1,3):A! (N,1)=A!(N-1,1)
50  NEXT J
51  FOR J=1 TO N
52  LPRINT A!(J,0):". ";
53  IF J=N THEN LPRINT" * OMP"
54  NEXT J
55  FOR J=1 TO K
60  B!(J,0)=0
70  NEXT J
71  FOR J=1 TO N-1
T3  LPRINT J;"-(";A!(J,1):". ";A!(J,3):")"
74  NEXT J
T5  X=1:R=0
80  B!(1,0)=1:Z=1
85  IF  B!(1,0)=N-1 THEN END
90  IF  B!(1,0)>A!(Z,3)THEN Q=0:GOTO 120
95  IF  A!(Z,1)<=B!(1,0)THEN R=B!(1,0) :Q=1:GOTO 120
100  IF A!(Z,1)>B!(1,0)THEN R=A!(Z,1): Q=1:GOTO 120
105  IF Z>A!(B!(X-1,0),3)THEN B!(X,0)= 0:Q=-1:GOTO 120
120  X=X+Q:IF X=1 THEN Z=B!(1,0)+1: B!(1,0)=Z:GOTO 85
125  IF Q<1 THEN B!(X,0)=B!(X,0)+1
130  IF B!(X,0)=0 THEN B!(X,0)=R
135  Z=B!(X,0)
136  IF Z>A!(B!(X-1,0),3)THEN Q=-1:B! (X,0)=0:COTO 120
140  IF X=K THEN GOTO 600
150  GOTO 85
600  IF Z>A!(B!(K-1,0),3)THEN Q=-1: B!(X,0)=0:GOTO 120
610  IF B!(1,0)<A!(Z,1)THEN Z=Z+1: GOTO 600
620  IF B!(1,0)>A!(Z,3)THEN Z=Z+1: GOTO 600
625  B!(K,0)=Z
626  GOTO 700
630  FOR J=1 TO K:IF P=1 THEN GOTO 660
640  LPRINT B!(1,0);". ";
```

```
650  IF J=K THEN LPRINT" * "
660  NEXT J
661  IF P=3 THEN GOTO 950
665  GOTO 860
670  Z=Z+1:GOTO 600
680  PRINT E.
700  IF E=0 THEN GOTO 745
705  FOR J=1 TO E
710  FOR F=1+B!(J,3)TO K
720  H=(F MOD B!(J,3)):IF H=0 THEN H=B!(J,3)
730  IF(B!(H,0)-B!(F,0)) ^ 2>0 THEN GOTO  740
735  IF F=K THEN GOTO 670
736  NEXT F
740  NEXT J
745  G=0
750  FOR J=2 TO K
760  IF B!(J,0)=B!(1,0)THEN G=G+1:B! (G,4)=J
770  NEXT J
775  IF G=0 THEN GOTO 630
780  FOR J=1 TO G
790  FOR J=1 TO K-2
800  W=(B!(J,4)+F) MOD K:IF W=0 THEN W=K
810  IF B!(1+F,0)<B!(W,0)THEN GOTO  840
820  IF B!(1+F,0)>B!(W,0)THEN GOTO  670
830  NEXT F
840  NEXT J
850  GOTO 630
860  FOR J=0 TO K
865  B!(J,2)=J
870  NEXT J
875  FOR J=1 TO K
880  H=2 ^ J:IF H>=2 * K THEN GOTO 950
885  FOR G=0 TO INT(K/H)+1
890  U=G * H+1:V=U+H/2
895  IF V>K THEN GOTO 940
896  IF U>=G * H+H THEN GOTO 935
897  IF V>G * H+H THEN GOTO 935
```

```
898  IF V>K THEN GOTO 935
900  X0=B!(U,2):Y0=B!(V,2)
910  FOR W=0 TO K-1
911  B!(0,6)=1
912  B!(0,0)=B!(K,0)
915  X1=X0+W:IF X1>K THEN X1=X1-K
917  Y1=Y0+W:IF Y1>K THEN Y1=Y-K
918  IF W>0 THEN B!(W,6)=B!(W-1,6)*A! (B!(X1-1,0),2)
921  U1=B!(X1,0)*B!(W,6):V1=B!(Y1,0)*B! (W,6)
922  IF U1=V1 THEN GOTO 930
923  IF U1<V1 THEN U=U+1:GOTO 896
924  FOR L=V TO U+1 STEP-1
925  B!(L,2)=B!(L-1,2)
926  NEXT L
927  B!(U,2)=Y0
928  U=U+1:V=V+1
929  GOTO 896
930  NEXT W
935  NEXT G
940  NEXT J
950  IF P=2 THEN GOTO 1300
951  FOR J=1 TO K
955  B!(B!(J,2),5)=J:IF P=1 THEN GOTO 960
956  LPRINT B!(J,2); ".";
957  IF J=K THEN LPRUNT "…ODER"
960  NEXT J
963  IF P=1 THEN GOTO 1000
965  FOR J=1 TO K
966  L=B!(J,2)+1:IF L>K THEN L=L-K
967  IF P=2 THEN GOTO 1200
970  LPRINT B!(L,5);",";
972  IF J=K THEN LPRINT"****MP"
975  NEXT J
977  IF P=3 THEN LPRINT"YES".
978  IF P=3 THEN END
980  IF P=2 THEN P=3:GOTO 950
981  IF P=2 THEN END
```

```
990   GOTO 670
995   FOR J=1 TO K
996   LPRINT C!(J);". ";
997   IF J=K THEN LPRINT" * * * CK".
998   NEXT J
999   GOTO 25
1000   FOR J=1 TO K
1005   L=B!(J,2)+1:IF L>K THEN L=L-K
1010   D!(J,I)=B!(L,5)
1015   NEXT J
1020   IF I=0 THEN I=I+1:GOTO 965
1025   FOR Q=0 TO I-1
1030   FOR J=1 TO K
1035   IF(D!(J,Q)-D!(J,I)) ^2>0 THEN GOTO 1045
1040   IF J=K THEN GOTO 670
1044   NEXT J
1045   NEXT Q
1050   I=I+1:GOTO 965
1200   IF B!(L,5)=C!(J)THEN GOTO 975
1210   GOTO 670
1300   FOR J=1 TO K
1810   B!(B!(J,2),5)=J
1320   NEXT J
1830   GOTO 965
```

Ⅳ. Some Computed Results

As is well-known, a 3-periodic orbit must imply an n-periodic orbit for any positive integer n. But what types of periodic orbits does a 3-periodic orbit imply after all? The following lists the types of the K-periodic orbits' which are implied by the type (2,3,1) for $K=4,5,\cdots,12$.

$K=4$(one repetitive path, one type)

(3,4,2,1)

$K=5$(two repetitive paths, one type)

(3,5,4,2,1)

$K=6$(two repetitive paths, one type)

(4,6,5,3,2,1)

$K=7$(four repetitive paths, two types)

(4,7,6,5,3,2,1)

(4,5,7,6,3,2,1)

$K=8$(five repetitive paths, three types)

(5,8,7,6,4,3,2,1)

(5,6,8,7,4,3,2,1)

(4,5,8,7,6,3,2,1)

$K=9$(eight repetitive paths, four types)

(5,9,8,7,6,4,3,2,1)

(5,6.8,9,7,4,3,2,1)

(5,7,9,8,6,4,3,2,1)

(4,6,9,8,7,5,3,2,1)

$K=10$(eleven repetitive paths, six types)

(6,10,9,8,7,5,4,3,2,1)

(6,8,10,9,7,5,4,3,2,1)

(4,6,10,9,8,7,5,3,2,1)

(5,6,7,10,9,8,4,3,2,1)

(5,6,9,10,8,7,4,3,2,1)

(6,7,9,10,8,5,4,3,2,1)

$K=11$(eighteen repetitive paths, nine types)

(6,11,10,9,8,7,5,4,3,2,1)

(6,7,11,10,9,8,5,4,3,2,1)

(6,9,11,10,8,7,5,4,3,2,1)

(4,7,11,10,9,8.6,5,3,2,1)

(5,6,7,11,10,9,8,4,3,2,1)

(5,7,10,11,9,8,6,4,3,2,1)

(5,7,8,11,10,9,6,4,3,2.1)

(6,8,10,11,9,7,5,4,3,2,1)

(6,7,8,10,11,9,5,4,3,2,1)

$K=12$(twenty-five repetitive paths, thirteen types)

(7,12,11,10,9,8,6,5,4,3,2,1)

(7,8,12,11,10,9,6,5,4,3,2,1)

(6,7,12,11,10,9,8,5,4,3,2,1)

(4,7,12,11,10,9,8,6,5,3,2,1)

(7,10,12,11,9,8,6,5,4,3,2.1)

(6,7,8,12,11,10,9,5,4,3,2,1)

(5,6,8,12,11,10,9,7,4,3,2,1)

(5,7,11,1 2,10,9,8,6,4,3,2,1)

(5,7,9,12,11,10,8,6,4,3,2,1)

(7,9,11,12,10,8,6,5,4,3,2,1)

(6,7,9,11,12,10,8,5,4,3,2. 1)

(6,7,8,10,12,1 1,9,5,4,3,2,1)

(7,8,9,11,12,10,6,5,4,3,2,1)

Here, two points should be pointed out. First the orientation reversing topological conjugacy of the type (2,3,1) is the type (3,1,2),which is obtained by 4 minus 2,3 and 1 in order and reversing the order of the computed results. If we construct the orientation reversing topological conjugacy of each K-periodic type of orbit which the type(2,3,1) implies,i. e. by $K+1$ minus the components of the K-periodic type of orbit in order,and then reversing their order we can obtain each K-periodic type of orbit which the type(3,1,2)implies. For instance,we know immediately(3,1,2)$\lhd$(4. 3,1,2) from(2,3,1)$\lhd$(3,4,2,1)and(3,1,2,)$\lhd$(5,4,2,1,3)from(2,3,1)$\lhd$(3,5,4,2,1) etc.

Next,let $f\in C_p(S_n)$. The type of f is called the unimodal if $\overline{f}(x)$ is a unimodal function. By the unimodality of the type (2,3,1),the types which the type (2,3,1) implies must be unimodal as well. Thus, it's enough to write only the left elements of the unimodal type. For instance,(3,5,4,2,1) may be denoted simply by(3,…),(4,5,7,6,3,2,1,) by (4,5,…) and (5,6,7,10,9,8,4. 3,2,1,) by (5,6,7,…). So long as we know the number of elements in a unimodal type, we can write other components of the type according to the above simple mark. The above two explanations can bring us the convenience for the discussion of type.

In addition,we still compute the implication between the six kinds of 4-periodic orbits and the twenty-four kinds of 5-periodic orbits respectively.

The following is the implication between the six kinds of 4-periodic orbits.

$$(3,1,4,2)\lhd\begin{cases}(4,1,2,3)\lhd(4,3,1,2)\\(2,3,4,1)\lhd(3,4,2,1)\end{cases}$$

$$(2,4,1,3)\lhd\begin{cases}(4,3,1,2)\\(3,4,2,1)\end{cases}$$

As to the implication between twenty-four kinds of 5-periodic orbits, we have computed twelve. The information of the other twelve may be obtained by the orientation reversing topological conjugacy(In the types of 4-periodic orbits,the types(3,1,4,2)and(2,4,1,3)are the orientation reversing topologically self-conjugate. Obviously,there does not exist the orientation reversing topologically self-conjugate for any periodic orbit with odd period). We list them concretely as follows(simply,(2,4,1,5,3)denoted by(24153)):

$$
(2\ 4\ 1\ 5\ 3)\lhd\begin{cases}\boxed{\begin{Bmatrix}(3\ 1\ 4\ 5\ 2)\\(4\ 1\ 2\ 5\ 3)\end{Bmatrix}\lhd\begin{Bmatrix}(2\ 3\ 4\ 5\ 1)\lhd(2\ 4\ 5\ 3\ 1)\lhd(3\ 5\ 4\ 2\ 1)\\(5\ 1\ 2\ 3\ 4)\lhd(5\ 3\ 1\ 2\ 4)\lhd(5\ 4\ 2\ 1\ 3)\end{Bmatrix}}^{*}\\[2ex](2\ 5\ 1\ 3\ 4)\lhd\boxed{\begin{Bmatrix}(2\ 5\ 4\ 1\ 3)\lhd(3\ 4\ 5\ 1\ 2)\lhd(3\ 5\ 4\ 2\ 1)\\(3\ 5\ 2\ 1\ 4)\lhd(4\ 5\ 1\ 2\ 3)\lhd(5\ 4\ 2\ 1\ 3)\end{Bmatrix}}^{**}\end{cases}
$$

$$
(3\ 4\ 2\ 5\ 1)\lhd(4\ 3\ 1\ 5\ 2)\lhd\boxed{\begin{Bmatrix}(3\ 1\ 4\ 5\ 2)\\(4\ 1\ 2\ 5\ 3)\end{Bmatrix}\lhd\begin{Bmatrix}(2\ 3\ 4\ 5\ 1)\\(5\ 1\ 2\ 3\ 4)\end{Bmatrix}\lhd\begin{Bmatrix}(2\ 4\ 5\ 3\ 1)\\(5\ 3\ 1\ 2\ 4)\end{Bmatrix}\lhd\begin{Bmatrix}(3\ 5\ 4\ 2\ 1)\\(5\ 4\ 2\ 1\ 3)\end{Bmatrix}}^{*}
$$

$$
(4\ 5\ 2\ 3\ 1)\lhd(5\ 4\ 1\ 3\ 2)\lhd(5\ 4\ 2\ 1\ 3)
$$

The groups of the orientation reversing topological conjugacy about the above three groups are listed as follows respectively:

$$
(3\ 1\ 5\ 2\ 4)\lhd\begin{cases}\boxed{}^{*}\\(2\ 3\ 5\ 1\ 4)\lhd\boxed{}^{**}\end{cases}
$$

$$
(5\ 1\ 4\ 2\ 3)\lhd(4\ 1\ 5\ 3\ 2)\lhd\boxed{}^{*}
$$

$$
(5\ 3\ 4\ 1\ 2)\lhd(4\ 3\ 5\ 2\ 1)\lhd(3\ 5\ 4\ 2\ 1)
$$

where $\boxed{}^{*}$ and $\boxed{}^{**}$ denote respectively two constant groups under the orientation reversing topological conjugacy, and their concrete contents are given when $\boxed{}$ appears first.

From the above-mentioned, it may be seen that there exists the relation not to be implied mutually between two different types, but there does exist not the mutual implication. In fact, we have the following conclusion.

Proposition Let $f\in C_p(S_n^{(0)})$, $g\in C_p(S_m^{(0)})$. If $A_f\lhd A_g$, and $A_g\lhd A_f$, then $A_f=A_g$.

Proof Let $S_n^{(i)}=\{x_1^{(i)}<x_2^{(i)}<\cdots<x_n^{(i)}\}$ and $S_m^{(i)}=\{y_1^{(i)}<y_2^{(i)}<\cdots<y_m^{(i)}\}$. Assume that $A_f\neq A_g$. Without loss of generality, let $\bar{f}$ be increase on $[x_1^{(0)},x_2^{(0)}]$, $n\leqslant m$. Let $[x_1^{(0)},c]$ be the maximum interval on which f is increasing (obviously, $c\in S_n^{(0)}$).

By $A_f\lhd A_g$, there must exist $g_1\in C_p(S_m^{(1)})$ such that $A_{g_1}=A$, $g_1(y_i^{(1)})=\bar{f}(y_i^{(1)})$, $1\leqslant i\leqslant m$, and $x_1^{(0)}<y_1^{(0)}$.

(1) Since $A_f\neq A_g$, $A_{g_1}=A_g\lhd A_f$, and $\bar{f}$ is a continuous extension of g_1, there must exist the $f_1\in C_p(S_n^{(1)})$ by lemma 6 and lemma 7 such that $A_{f_1}=A_f$, $f_1(x_i^{(1)})=\bar{f}(x_i^{(1)})$, $1\leqslant i\leqslant n$, and $y_1^{(1)}<x_1^{(1)}<c$.

(2) Since $A_{f_1}=A_f\lhd A_g$, for the same reason, there must exist the $g_2\in C_p(S_m^{(2)})$ such that $A_{g_2}=A_g$, $g_2(y_i^{(2)})=\bar{f}(y_i^{(2)})$, $1\leqslant i\leqslant m$ and $x_1^{(1)}<y_1^{(2)}<c$.

In the following, we apply repeatedly the methods of (1) and (2) so that we obtain two sequences of $\{x_1^{(i)}\}$ and $\{y_1^{(i)}\}$ satisfying the condition that

$$
\left\{\mathrm{ord}_{\bar{f}}(x_1^{(i)})\right\}_{i=1}^{\infty}\ \text{are the set of } n\text{-periodic orbits of the same type,}\qquad (*_1)
$$

$\left\{\text{ord}_{\bar{f}}(y_1^{(i)})\right\}_{i=1}^{\infty}$ are the set of m-periodic orbits of the same type. and

$$x_1^{(i-1)}<y_1^{(i)}<x_1^{(i)}<c, \quad i=1,2,\cdots$$

Thus $\lim\limits_{i\to\infty}x_1^{(i)}=x_0;\lim\limits_{i\to\infty}y_1^{(i)}=x_0$ $(*_2)$

and $\bar{f}^n(x_0)=x_0\leqslant c$

According to the choice of the $\{x_1^{(i)}\}$ and $x_1^{(1)}>x_1^{(0)}$ we may infer that $x_0\notin S_n^{(0)}$. By the continuity of $\bar{f}$, there must exist a closed interval $I\subset(x_1^{(0)},c)/S_n^{(0)}$ such that $\bar{f}^i$ is linear and strictly monotone on I for $i=1,2,\cdots,n$. By $(*1)$ and $(*2)$, there exists an i such that $J=[x_1^{(i)},x_1^{(i+1)}]\subset I$ satisfying $\bar{f}^n(J)=J$ and $\bar{f}^n(z)=z$ for any $z\in J$. But $y_1^{(i+1)}\in J$ implies that $\bar{f}^n(y_1^{(i+1)})=y_1^{(i+1)}$ and m is a divisor of n. Thus $m=n$ by our assumption $n\leqslant m$. Obviously, $\bar{f}^{j-k}$ is linear and strictly monotone on $[\bar{f}^k(x_1^{(i)}),\bar{f}^k(x_1^{(i+1)})]$for any k and j satisfying $k<j$, $0\leqslant k,j\leqslant n$. Thus, when $\bar{f}^j(x_1^{(i)})<\bar{f}^k(x_1^{(i)})$ $\bar{f}^j(x_1^{(i)})>\bar{f}^k(x_1^{(i)})$, we infer that $\bar{f}^j(y_1^{(i+1)})<\bar{f}^k(y_1^{(i+1)})$ $\bar{f}^j(y_1^{(i+1)})>f^k(y_1^{(i+1)})$ from $\bar{f}^j(x_1^{(i+1)})<\bar{f}^k(x_1^{(i+1)})$ $(\bar{f}^j(x_1^{(i+1)})>\bar{f}^k(x_1^{(i+1)}))$ by $(*1)$. This explains that $\text{orb}_{\bar{f}}(x_1^{(i)})$ and $\text{orb}_{\bar{f}}(y_1^{(i+1)})$ are two periodic orbits of the same type-contradicting $A_f\neq A_g$.

The initial computed results of this paragraph lead us to raise some questions:

(1)What types of periodic orbits do the type of a 3-periodic orbit imply? or otherwise, what types of periodic orbits imply the types of 3-periodic orbits?

(2)Let the number of the types of k-periodic orbits which a 3-periodic orbit implies be T_k. How do we compute T_k or estimate the order of T_k with respect to k?

(3)For the given k, there always exist some types of k-periodic orbits which are not implied by other types of k-periodic orbits(when $k=4$, there are two such types, (3,1,4,2) and(2,4,1,3); when $k=5$, there are six such types which may be seen from the previous computed results). These types of periodic orbits are called no harm"types of primitive periodic orbits". What character do they have? How many types of the primitive periodic orbits with period k are there?

(4)For the given k, the type of periodic orbit with period k, which implies most of the types of periodic orbits with period k may be called no harm"type of complex periodic orbit". Then what character does the type of complex periodic orbit have?

(5)Does there exist the polynomial algorithm for determining the implication between two types?

We think that it is significant but quite thorny to clarify thoroughly the mutual relation between the types. We hope that the exploration of this aspect can arouse the interest of the profession.

References

[1]Sarkovskii A N. Coexistence of cycles of a continuous map of the line into itself. Ukr. Math. Zh. ,1964(16):61 - 71. (in Russian)

[2]Stefan P. A theorem of Sarkovskii on the coexistence of periodic orbits of continuous endomorphisms of the real line. Comm. Math. Phys. ,1977(54):237 - 248.

[3]Zhang Jingzhong, Yang Lu. Some theorems on the Sarkovskii order. Adv. in Math. , 1987,16(1):33 - 48. (in Chinese)

[4]Coppel W A. Sarkovskii-minimal orbits. Math. Proc. Camb. phil. Soc. , 1983(93): 397 - 408.

[5]Block L, D Hart. Stratification of the space of unimodal interval maps. Ergod, Th. and Dynam. Sys. ,1983(3):533 - 539.

定理机械化证明的数值并行法及单点例证法原理概述

张景中　杨　路

（中科院成都分院数理科学研究室）

摘要　本文浅近地介绍以检验数值实例为基本手段的两种方法——洪加威提出单点例证法和张景中，杨路提出的数值并行法以及这两种方法与吴文俊数学机械化理论的关系.

用机械的方法证明数学定理，曾是数百年前一些卓有远见的数学家的美妙幻想.由于吴文俊教授近十年来的杰出工作[1-7]，这一美妙幻想已在电子计算机帮助下成功地变为现实.

在吴文俊数学机械化理论与方法的影响之下，这一研究领域在我国日趋活跃，近年来提出的通过数值实例的检验来证明几何定理的思想与方法，尤其引起广泛的兴趣甚至惊讶！

这里浅近地介绍一下以检验数值实例为基本手段的两种方法——洪加威提出单点例证法[8]和张景中、杨路提出的数值并行法[11-12]，以及这两种方法与吴文俊数学机械化理论的关系.

一、最初的起点

一个极为简单的事实，等式

$$(x+1)(x-1)=x^2-1 \tag{1}$$

是一个恒等式.把左端展开，移项，合并，便可证明.

但也可以用数值实验的方法证明.取 $x=0$，两端都是 -1；取 $x=1$，两端都是 0；取 $x=2$，两端都是 3.这就证明了(1)是恒等式.

道理很简单，如果它不是恒等式，就是一个不高于二次的一元代数方程.这种方程至多有两个根.现在已有 $x=0,1,2$ 三个根了，那表明它不是方程而是恒等式.

推广到高次，便是

命题 A.　设 $f(x)$ 与 $g(x)$ 都是不超过 n 次的多项式.如果有 $n+1$ 个不同的数 $a_0, a_1, \cdots, a_n$ 使 $f(a_k)=g(a_k)(k=0,1,2,\cdots,n)$，则等式

$$f(x)=g(x) \tag{2}$$

是恒等式.

通俗地说：要问一个给定的不高于 n 次的一元代数等式是不是恒等式，只要分别用 $n+1$ 个数值代替变元检验，即可得出结论.

本文刊于《数学的实践与认识》，第 1 期，1989 年.

换句话,举足够多的数值例子可以证明一元代数恒等式,其基本依据是:n 次代数方程至多有 n 个根.

现在换一个角度来看等式(1),如果有人说,只要取一个较大的 x 代入,比如 $x=6$,就足以证明它是恒等式!这能行吗?

初看似乎令人吃惊,细想也就不奇怪了.在(1)式中,左端展开后最多 4 项,每项系数的绝对值至多为 1.整理、合并之后,系数的绝对值是不大于 5 的正整数或 0.如果(1)不是恒等式,把它整理之后,应当是一个方程,即

$$ax^2+bx+c=0. \tag{3}$$

这里,a,b,c 不全为 0,都是绝对值不大于 5 的整数.如果取 $x=6$ 时,有

$$a\times 6^2+b\times 6+c=0, \tag{4}$$

可知 $a=b=c=0$.

若 $a=0$,由 $6b+c=0$ 得 $6|b|=|c|$.

当 $b=0$ 时得 $c=0$.当 $b\neq 0$ 时得 $|c|=6|b|\geqslant 6$,这与 $|c|\leqslant 5$ 矛盾.

若 $a\neq 0$,由 $36a+6b+c=0$ 得 $36|a|\leqslant 6|b|+|c|$,即

$$3b\leqslant|36a|\leqslant|6b|+|c|\leqslant 35. \tag{5}$$

仍然矛盾.

这证明了(1)是恒等式.

这种办法也可以推广到高次.更具体地有:

命题 B　设 $f(x)$ 与 $g(x)$ 都是不超过 n 次的多项式.如果知道 $f(x)-g(x)$ 的标准展开式中系数绝对值最大者不大于 L,非 0 系数绝对值最小者不小于 $S>0$.设

$$|\hat{x}|=p\geqslant\frac{L}{S}+2,$$

则

$$S\leqslant|f(\hat{x})-g(\hat{x})|\leqslant Sp^{n+1}. \tag{6}$$

证　设

$$f(x)-g(x)=c_0x^k++c_1x^{k-1}+\cdots+c_k, \tag{7}$$

这里 $0\leqslant k\leqslant n$,而 $c_0\neq 0$.若 $k=0$,显然有(6)成立.

以下设 $1\leqslant k\leqslant n$.这时

$$\begin{aligned}|f(\hat{x})-g(\hat{x})|&\geqslant|c_0\hat{x}^k|-|c_1x^{k-1}+c_2x^{k-2}+\cdots+c_k|\\&\geqslant Sp^k-L(p^{k-1}+p^{k-2}+\cdots+p+1)\\&\geqslant\frac{Sp^k}{p-1}\left(p-1-\frac{L}{S}\left(1-\frac{1}{p^k}\right)\right)\geqslant S\left(p-1-\frac{L}{S}\right)\\&\geqslant S.\end{aligned} \tag{8}$$

另一方面,显然有 $|f(\hat{x})-g(\hat{x})|\leqslant L\dfrac{p^{n+1}-1}{p-1}\leqslant Sp^{n+1}$.

不等式(6)表明,对 $|\hat{x}|\geqslant\dfrac{L}{S}+2$,$\hat{x}$ 不可能是方程 $f(x)-g(x)=0$ 的根.如果计算表明居然有 $f(\hat{x})=g(\hat{x})$,即可断言 $f(x)=g(x)$ 是恒等式(至于不等式(6)的右端,后面将用到).

通俗地说：举一个足够大的数值例子，即可以证明一条一元代数恒等式. 其基本依据是：代数方程的根的绝对值，不超过其绝对值最大的系数与最高次项系数之比的绝对值加1（在命题B中，采用了略强的条件，是为了得到(6)中确定的正的下界，这在后面将用到）.

命题A与命题B都不是什么新鲜事. 但是，数学里的许多有用的方法，往往发源于极其平凡的朴素思想.

着眼于多项式根的界限，从命题B起步，可以引出洪加威的单点例证法.

着眼于多项式根的个数，从命题A起步，张景中、杨路提出了数值并行法.

两种方法的思想，大体上都是在1984年产生的. 这时，吴文俊的工作已在世界上有了广泛的影响. 吴先生的理论与方法揭示了这一领域的光明前景，吸引了他们着手这方面的研究. 在洪加威的著名论文[8]发表前约一年，他与本文作者曾就此不止一次交换过各自的想法. 但都没有改变自己的着眼之点.

二、自然的推广

一粒小小的种子，能变成枝繁叶茂的树木. 因为种子里蕴含了树木的信息，但种子还不是树. 要成为树，需要成长的时间与其它条件.

从前面所说的命题A与命题B起步，发展到定理机械化证明的例证法与并行法，需要进一步的工作.

首先要做的，是从一元推广到多元.

命题A的推广是：

定理A 设 $f(x_1,x_2,\cdots,x_m)$ 是 $x_1,x_2,\cdots,x_m$ 的多项式，它关于 x_k 的次数不大于 n_k、对应于 $k=1,2,\cdots,m$，取数组 $a_{k,l}(l=0,1,2,\cdots,n_k)$ 使得 $l_1\neq l_2$ 时，有 $a_{k,l_1}\neq a_{k,l_2}$. 如果对任一组 $\{l_1,l_2,\cdots,l_m,0\leqslant l_k\leqslant n_k\}$ 有

$$f(a_{1,l_1},a_{2,l_2}\cdots,a_{m,l_m})=0, \tag{9}$$

则 $f(x_1,x_2,\cdots,x_m)$ 是恒为0的多项式.

证 对 m 作数学归纳. $m=1$ 时即命题A. 设要证的命题对 $m=j$ 已真，往证它对 $m=j+1$ 也真. 这时把 $f(x_1,x_2,\cdots,x_{j+1})$ 写成

$$c_0x^n+c_1x^{n-1}+\cdots+c_n. \tag{10}$$

这里 $x=x_{j+1}$，$n=n_{j+1}$，而 $c_k=c_k(x_1,x_2,\cdots,x_j)$ 是关于 x_k 次数不大于 n_k 的多项式，$k=1,2,\cdots,j$.

取定 $\hat{x}_k=a_{k,l_k}(k=1,2,\cdots,j,0\leqslant l_k\leqslant n_k)$，则 $f(\hat{x}_1,\hat{x}_l,\cdots,\hat{x}_j,x)$ 是 x 的不超过 n 次的多项式. 对于 x 的 $n+1$ 个不同的值 $a_{j+1,0},a_{j+1,1},\cdots a_{j+1,n}$ 总有 $f(\hat{x}_1,\hat{x}_2,\cdots,\hat{x}_j,a_{j+1,t})=0$，$t=0,1,\cdots,n$. 由命题A可知 $f(\hat{x}_1,\hat{x}_2,\cdots,\hat{x}_l,x)$ 是恒0多项式. 这表明 $c_k(\hat{x}_1,\hat{x}_2,\cdots,\hat{x}_j)=0$. 又因 $\hat{x}_k$ 可以是 $a_{k,0},a_{k,1},\cdots,a_{k,nk}$ 中任一个，由归纳前提可知 $c_k(x_1,\cdots,x_j)$ 是恒0多项式. 由数学归纳法，命题得证.

为便于理解，不妨看一个 $m=2$ 的特例. 要证明等式

$$(x+y)(x-y)=x^2-y^2 \tag{11}$$

是恒等式,注意到它关于 x,y 的次数都不大于 2,故可取 $x=0,1,2$ 和 $y=0,1,2$,分别组成变元(x,y)的 9 组值$((0,0),(0,1),(0,2);(1,0),(1,1),(1,2);(2,0),(2,1)(2,2))$代入验算即可.

一般说来,在定理 A 的条件下,需要验算的变元数值共有$(n_1+1)(n_2+1)\cdots(n_m+1)$组.

作者曾把定理 A 在 $m=2$ 时的特款选作 1985 年度数学专业研究生入学考试的高等代数试题,竟无人做出.可见,这个重要的事实在大学代数课程中似乎没有提及.

类似地,命题 B 也可以推广到多元.

定理 B 设 $f(x_1,x_2,\cdots,x_m)$是 $x_1,x_2,\cdots,x_m$ 的多项式,它关于 x_k 的次数不大于 n_k,$1\leqslant k\leqslant m$. 又设它的标准展式中系数的绝对值最大者不大于 L,非 0 系数绝对值最小者不小于 $S>0$. 如果变元的一组值 $\hat{x}_1,\hat{x}_2,\cdots,\hat{x}_m$ 满足

$$\begin{cases}|\hat{x}_1|=p_1\geqslant\dfrac{L}{S}+2,\\|\hat{x}_2|=p_2\geqslant p_1^{n_1+1}+2,\\|\hat{x}_3|=p_3\geqslant p_2^{n_2+1}+2,\\\cdots\cdots\\|\hat{x}_m|=p_m\geqslant p_m^{n_{m-1}+1}+2,\end{cases}\tag{12}$$

则有

$$|f(\hat{x}_1,\hat{x}_2,\cdots,\hat{x}_m)|\geqslant S>0.\tag{13}$$

证 对 m 作数学归纳. 当 $m=1$ 时,命题 B 成立. 现在设命题对 $m-1$ 真,往证它对 m 亦真. 记 $g(x_1,x_2,\cdots,x_m)=f(\hat{x}_1,\hat{x}_2,\cdots,\hat{x}_m)$,则 g 是 $x_2,x_3,\cdots,x_m$ 这$(m-1)$个变元的多项式,它的系数具有形式为

$$c(\hat{x}_1)=c_0\hat{x}_1^n+c_1\hat{x}_1^{n-1}+\cdots+c_n(0\leqslant n\leqslant n_1),\tag{14}$$

这些多项式 $c(x)$是不全为 0 多项式,系数 c_i 的绝对值最大不超过 L,非 0 者绝对值最小不小于 S,由命题 B 可知,当 $c(x)$非恒 0 时

$$S\leqslant c(\hat{x}_1)\leqslant Sp_1^{n_1+1}=L_1.\tag{15}$$

而由(12)知$|\hat{x}_2|=p_2\geqslant p_1^{n_1+1}+2=\dfrac{L_1}{S}+2$. 由归纳前提可知$|g(\hat{x}_2,\hat{x}_3,\cdots,\hat{x}_m)|=|f(\hat{x}_1,\hat{x}_2,\cdots,\hat{x}_m)|\geqslant S$. 证毕.

定理 B 告诉我们一桩似乎出人意料的事实:即使要验证一个多元代数恒等定,也可以只用一个数值例子. 不过这个例子要涉及很大的变元数值. 大到什么程度呢? 从(12)易于看出. 大致上有

$$|\hat{x}_m|>\left(\frac{L}{S}\right)^{(n_1+1)(n_2+1)\cdots(n_{m-1}+1)}.\tag{16}$$

这表明,运算涉及的有效数字之长正比于

$$\ln|\hat{x}_m|>(n_1+1)(n_2+1)\cdots(n_m+1)\ln\frac{L}{S}.\tag{17}$$

作一次乘法,运算工作量正比于有效数字长度的平方,即$(\ln|\hat{x}_m|)^2$. 如果采用精度倍增的模方法,可把工作量降到正比于 $\ln|\hat{x}_m|$,但要附加一些转换工作量.

如果用定理 A 的结果,验算时涉及变元的有效数字之长仅为 $\ln\max\{n_j+1\}=Q$. 但类似的验算工作要进行 $(n_1+1)(n_2+1)\cdots(n_m+1)$ 次,作这么多次乘法的工作量正比于

$$(n_1+1)(n_2+1)\cdots(n_m+1)=Q. \tag{18}$$

一般地说来,这当然要比 $(\ln|\hat{x}_m|)^2$ 小得多. 即使应用定理 B 时采取了精度倍增技术,一般说来仍不如用定理 A 合算. 这是因为

(1)精度倍增技术要附加转换工作量和必要的空间;

(2)从证明可看出,定理 A 适用于一般域上的多项式,定理 B 仅能用于赋范域;

(3)定理 B 要求检验多个例子,每个例子都很容易算,这本质上适于高度并行化;

(4)在恒等式不成立的情形,检验多个小例子尤其合算,因一旦有一例失败,即可断言恒等式不成立而停机,而按定理 A 检验一个大例子,必须进行到底才见分晓.

但是,用一个特例就可检验一条普遍的命题,这种观念的确是美妙而吸引人的. 基于定理 B 的"单点例证法"发表后,在有关研究领域的国内外同行中引起了很大的兴趣. 而另一方面,基于定理 A 的"数值并行法"(也曾被叫做多点例证法)却在计算机上取得了更为成功的实现.

我们把基于定理 A 的检验恒等式方法叫"数值并行法",一方面是为了强调它的适于高度并行计算的特点;另一方面,这一名称比"多点例证法"有更多的内涵,它可以包括另一些类似的但并非多点例证的方法.(第四节将提及). 此外,叫"多点例证法"易产生误解,以为它是"单点例证法"的重复操作或推广. 事实上,两者之间并无依赖关系. 它们一开始所依赖的原理就不相同.

三、可靠的基础——吴 Ritt 除法

到现在为止,我们只说明了验算一个"大"例子或多个"小"例子可以检验一个代数等式是不是恒等式. 这和大家感兴趣的几何定理的机器证明仍有相当大的距离!

通常初等几何中的定理,如果在前提和结论中不涉及不等式,总可以用坐标法化成这样的问题:假设已知有一些代数等式,求证某一个代数等式成立. 以熟知的西姆松(Simson)定理为例:

西姆松定理 在 $\triangle ABC$ 的外接圆上任取一点 P,自 P 向 BC,CA,AB 引垂线,垂足顺次为 R、S、T,则 R、S、T 三点在一直线上.

这个定理涉及 7 个点:A,B,C,P,R,S,T,设它们的笛卡儿坐标为:$A(x_1,y_1)$,$B(x_2,y_2)$,$C(x_3,y_3)$,$P(x_4,y_4)$,$R(x_5,y_5)$,$S(x_6,y_6)$,$T(x_7,y_7)$为了减少变量,不妨设 $\triangle ABC$ 外接圆心为原点,设圆半径为 1,则可取 $P=(1,0)$再由 R 在 BC 上等,得

$$\begin{cases}x_5=\lambda x_2+(1-\lambda)x_3,y_5=\lambda y_2+(1-\lambda)y_3,\\x_6=\mu x_3+(1-\mu)x_1,y_6=\mu y_3+(1-\mu)y_1,\\x_7=\rho x_1+(1-\rho)x_2,y_7=\rho y_1+(1-\rho)y_2.\end{cases} \tag{19}$$

这样可以用 λ,μ,ρ 三个参数代替 x_5,x_6,x_7 和 y_5,y_6,y_7 这六个变量. 这时,9 个变量 $x_1,y_1,x_2,y_2,x_3,x_3,\lambda,\mu,\rho$ 应当满足 6 个等式:

$$\begin{cases}A_1\equiv x_1^2+y_1^2=1,\\ A_2\equiv x_2^2+y_2^2=1,\\ A_3\equiv x_3^2+y_3^2=1,\\ A_4^*\equiv(1-\lambda x_2-(1-\lambda)x_3)(x_2-x_3)-(\lambda y_2+(1-\lambda)y_3)(y_2-y_3)=0,\\ A_5^*\equiv(1-\mu x_3-(1-\mu)x_1)(x_3-x_1)-(\mu y_3+(1-\mu)y_1)(y_3-y_1)=0,\\ A_6^*\equiv(1-\rho x_1-(1-\rho)x_2)(x_1-x_2)-(\rho y_1+(1-\rho)y_2)(y_1-y_2)=0,\end{cases}\tag{20}$$

这里,A_1 表示 A 在单位圆上,A_4^* 表示 $PR\perp BC$,等等.要证的结论是 R,S,T 共直线,即

$$G\equiv(x_5-x_6)(y_5-y_7)-(x_5-x_7)(y_5-y_6)=0.$$

对(20)进行约化,利用 A_1,A_2,A_3 可把 A_4^*,A_5^*,A_6^* 变为:

$$\begin{cases}A_4\equiv 2(x_2x_5+y_2y_3-1)\lambda-(x_2x_3+y_2y_3-1)+(x_2-x_3)=0,\\ A_5\equiv 2(x_3x_1+y_3y_1-1)\mu-(x_3x_1+y_3y_1-1)+(x_3-x_1)=0,\\ A_6\equiv 2(x_1x_2+y_1y_2-1)\rho-(x_1x_2+y_1y_2-1)+(x_1-x_2)=0.\end{cases}\tag{21}$$

而结论 G 中则可消去 $x_5,y_5\cdots$ 而换成(19)中诸右端,整理之,得:

$$G\equiv(\lambda\rho+\lambda\mu+\mu\rho-\lambda-\mu-\rho+1)[x_1(y_2-y_3)+x_2(y_3-y_1)+x_3(y_1-y_2)]=0.\tag{22}$$

这里,G 是一个含有 9 个变元的多项式.但它的 9 个变元之间有 $A_1\sim A_6$ 这 6 个方程联系着,所以不能用上一节所说的验算一个或若干个例子的办法判定它是不是恒为 0.一般说来,利用这 6 个方程可以消去 6 个变元,剩下 3 个自由变元.从几何上看,不妨取 x_1,x_2,x_3 为自由变元.如果能够消去另外 6 个非自由变元而得到一个仅含 x_1,x_2,x_3 的多项式 $\Phi(x_1,x_2,x_3)$ 就可从 $\Phi(x_1,x_2,x_3)$ 出发用 x_1,x_2,x_3 的数值来检验了.吴-Ritt 方程,正好提供了消去非自由变元的可能性.

定理 W(吴文俊-Ritt) 设 f 和 g 都是关于 $x_1,x_2,\cdots,x_n,y$ 的多项式

$$\begin{cases}f=c_0y^m+c_1y^{m-1}+\cdots+c_m,\\ g=b_0y^l+b_1y^{l-1}+\cdots+b_l.\end{cases}\tag{23}$$

这里,$m\geqslant1,l\geqslant1$,诸 c_k,b_j 是 $x_1,x_2,\cdots,x_n$ 的多项式,且 c_0,b_0 不恒为 0,则可以用机械的方法确定出多项式 $Q(x_1,\cdots,x_n,y)$ 与 $f_1(x_1,\cdots,x_n,y)$,使得:

1.如果一组 $\hat{x}_1,\hat{x}_2,\cdots,\hat{x}_n,\hat{y}$ 满足

$$g(\hat{x}_1,\hat{x}_2,\cdots,\hat{x}_n,\hat{y})=0,$$

则有

$$Q_1(\hat{x}_1,\hat{x}_2,\cdots,\hat{x}_m,\hat{y})f(\hat{x}_1,\hat{x}_2,\cdots,\hat{x}_m,\hat{y})=f_1(\hat{x}_1,\cdots,\hat{x}_m,\hat{y}),\tag{24}$$

2.f_1 关于 y 的次数小于 m.

这里不再详述这个定理的证法,只指出:利用综合除法可得多项式 P,Q,f 使

$$P\cdot g+Q_1\cdot f=f_1\tag{25}$$

就足够了.

继续用定理 W,再一次降低 f_1 中 y 的次数,最后可以把 y 从 f 中消去,即找到一个多项式 $Q=Q_1$、$Q_2\cdots Q_n$ 和 $x_1,x_2,\cdots,x_n$ 的多项式 F,使对任一组满足 $g(\hat{x}_1,\cdots,\hat{x}_n,\hat{y})=0$ 的 $\hat{x}_1$,$\hat{x}_2,\cdots\hat{x}_n,\hat{y}$,有

$$Q(\hat{x}_1,\hat{x}_2,\cdots,\hat{x}_n,\hat{y})f(\hat{x}_1,\hat{x}_2,\cdots,\hat{x}_n,\hat{y})=F(\hat{x}_1,\cdots,\hat{x}_n).\tag{26}$$

更一般地，在代数化了的几何命题中，如果假设条件为（设已按吴-Ritt 方程整理为不可约升列）

$$g_j(u_1,u_2,\cdots,u_s;x_1,x_2,\cdots,x_j)=0(j=1,2,\cdots,t), \tag{27}$$

而结论为

$$f(u_1,u_2,\cdots,u_s;x_1,x_2,\cdots,x_t)=0. \tag{28}$$

我们就可以利用(27)顺次从 f 中消去变元 $x_t,x_{t-1},\cdots,x_1$. 在消去过程中得到多项式 $Q(u_1,\cdots,u_s;x_1,\cdots,x_t)$ 和 $\Phi(u_1,u_2,\cdots,u_s)$，使对任一组满足(27)的变元 $\hat{u}_1,\hat{u}_2,\cdots,\hat{u}_s,\hat{x}_1,\hat{x}_2,\cdots,\hat{x}_t$，有（参看[1]中 132 页引理 4).

$$Q(\hat{u}_1,\cdots,\hat{u}_s,\hat{x}_1,\cdots,\hat{x}_t)f(\hat{u}_1,\cdots,\hat{u}_s,\hat{x}_1,\cdots,\hat{x}_t)=\Phi(\hat{u}_1,\cdots,\hat{u}_s). \tag{29}$$

但是我们实际上可以来个"引而不发"——应用 Q 与 Φ 可以机械地得到这一结果，而并不真地去做逐步消去诸变元 $x_1,\cdots,x_t$ 的工作. 办法是用数值检验法直接判定 $\Phi(u_1,u_2,\cdots,u_t)$ 是不是恒为 0 的多项式. 若 $\Phi\equiv 0$，我们就可以断言：对于满足假设条件(27)的 $\hat{u}_1,\cdots,\hat{u}_s,\hat{x}_1,\cdots,\hat{x}_t$，只要 $Q(\hat{u}_1,\cdots,\hat{u}_s,\hat{x}_1,\cdots,\hat{x}_t)\neq 0$（这就是著名的吴氏非退化条件），则 $f(\hat{u}_1,\cdots,\hat{u}_s,\hat{x}_1,\cdots,\hat{x}_t)=0$. 简言之：要证的命题一般成立.

数值检验的办法，原则上不难说清楚：取一组 $\hat{u}_1,\cdots,\hat{u}_s$ 代入(27)，解出对应的 $\hat{x}_1,\cdots,\hat{x}_t$ 代入 f. 若 $f=0$，则 $\Phi=0$. 如果求解过程中途不能进行，则表明非退条件 $Q\neq 0$ 不满足，从(29)可知仍有 $\Phi(\hat{u}_1,\cdots,\hat{u}_s)=0$.

如果出现某一组 $\hat{u}_1,\cdots,\hat{u}_s$ 及对应的 $\hat{x}_1,\cdots,\hat{x}_t$，使得 $f\neq 0$，则要检验的命题不真. 这时即可停机，而不必继续检验了.

当然，也可以不用求解的办法，而用综合除法消去部分或全体的 $x_1,\cdots,x_t$，得到 $\Phi(\hat{u}_1,\cdots,\hat{u}_s)$.

这里有两个技术问题，但并不难处理：

(1)如果用单点例证法，应当知道 Φ 关于变元 $u_1,\cdots,u_s$ 的次数和系数绝对值的界限才知如何取诸 $\hat{u}_j$. 如果用数值并行法，也要知道 Φ 关于各变元次数的界限才知道算多少组. 追踪 Φ 的构造过程，这些问题都不难解决. 当然，次数的界限估计要比系数界限估计容易很多.

(2)在计算机上作数值计算，会有误差. 如何知道最终算出的 $f(\hat{u}_1,\cdots,\hat{u}_s,\hat{x}_1,\cdots,\hat{x}_t)$ 是真正的 0，还是很小的数呢？洪加威为此写了文[9]. 但用了吴氏理论的结果(29)，这个问题十分容易解决：如果假设条件(27)中都是整系数多项式（大量几何命题都是如此！），不难发现(29)中的 Q,f,Φ 也都是整系数多项式. 只要 $\hat{u}_1,\hat{u}_2\cdots,\hat{u}_s$ 取整数值，假定 $\Phi(\hat{u}_1,\cdots,\hat{u}_s)\neq 0$ 时必有

$$|\Phi(\hat{u}_1,\cdots,\hat{u}_s)|\geqslant 1, \tag{30}$$

也就是

$$|f(\hat{u}_1,\cdots,\hat{u}_s,\hat{x}_1,\cdots,\hat{x}_t)|\geqslant\frac{1}{|Q(\hat{u}_1,\cdots,\hat{u}_s,\hat{x}_1,\cdots,\hat{x}_t)|}>0, \tag{31}$$

而 $|Q(\hat{u}_1,\cdots,\hat{u}_s,\hat{x}_1,\cdots,\hat{x}_t)|$ 是可以估出的！这样，只要数值计算表明 $|f(\hat{u}_1,\cdots,\hat{u}_s,\hat{x}_1,\cdots,\hat{x}_t)|<|Q(\hat{u}_1,\cdots,\hat{u}_s,\hat{x}_1,\cdots,\hat{x}_t)|^{-1}$，即可断言 $\Phi(\hat{u}_1,\cdots,\hat{u}_s)=0$. 非退化条件 $Q\neq 0$ 满足时即知 $f=0$.

很清楚,有了吴-Ritt除法,数值检算的来龙去脉相当简单.顺便说一句,这里实际上已用更便捷的方法获得了比[8,9]文中更为一般的结果.

现在,回过头来看看本节一开始写出的西姆松定理,要用数值并行法证明它,必须估计消去了λ,μ,ρ和y_1,y_2,y_3之后得到的$\Phi(x_1,x_2,x_3)$关于自由变元x_1,x_2,x_3的次数.老老实实按吴-Ritt法做,估出的次数使得我们不得不做数万次的数值检验.下面我们使用一种变换技巧使数值检验的必需次数大大降低.

作变换:

$$\begin{cases}u_k=x_k+iy_k,\\ v_k=x_k-iy_k,\end{cases}\quad \begin{cases}x_k=\dfrac{1}{2}(u_k+v_k),\\ y_k=\dfrac{1}{2i}(u_k-v_k),\end{cases}\quad k=1,2,3;i=\sqrt{-1}. \tag{32}$$

则诸A_j及G成为u_k,v_k,λ,μ,ρ的多项式.把u_1,u_2,u_3看成自由变元,我们来看消去另外6个变元后得到的$\varphi(u_1,u_2,u_3)$中u_1,u_2,u_3的次数是多少.

设f是$u_1,u_2,u_3,v_1,v_2,v_3,\lambda,\mu,\rho$的多项式.如果$f$关于这些变元的次数顺次不超过$k_1,k_2,k_3\cdots,k_9$,则记作

$$N(f)\leqslant(k_1,k_2,k_3,k_4,k_5,k_6,k_7,k_8,k_9).$$

由(22)及(32)可以看出:

$$N(G)\leqslant(1,1,1,1,1,1,1,1,1), \tag{33}$$

这是因为,形如x_ky_k的项,才含有u_k,v_k的二次项.

利用条件A_6从G中消去ρ得g_1.在消去过程中要用A_6中ρ的系数多项式乘G,因而要把A_6中诸变量次数与G中对应次数相加.因

$$N(A_6)\leqslant(1,1,0,1,1,0,0,0,1), \tag{34}$$

故得

$$N(g_1)\leqslant(2,2,1,2,2,1,1,1,0). \tag{35}$$

类似地,用条件A_5消去μ得g_2,用条件A_4消去λ得g_3.由于

$$\begin{cases}N(A_5)\leqslant(1,0,1,1,0,1,0,1,0),\\ N(A_4)\leqslant(0,1,1,0,1,1,1,0,0),\end{cases} \tag{36}$$

故得

$$N(g_2)\leqslant(3,2,2,3,2,2,1,0,0), \tag{37}$$

$$N(g_3)\leqslant(3,3,3,3,3,3,0,0,0), \tag{38}$$

注意到,经过代换之后,A_1,A_2,A_3成为

$$\begin{cases}A_1\equiv u_1v_1=1,\\ A_2\equiv u_2v_2=1,\\ A_3\equiv u_3v_3=1,\end{cases} \tag{39}$$

把u_3^6乘g_3,即可消去v_3,得g_4,故

$$N(g_4)\leqslant(3,3,6,3,3,0,0,0,0). \tag{40}$$

依次再消去v_2,v_1之后,得到一个仅含u_1,u_2,u_3的多项式$\varphi(u_1,u_2,u_3)$.而φ关于每个变元的次数不超过6.一般而言,检验φ是否恒0多项式时,要用$7\times7\times7=343$组变元值代入.这

相当于在单位圆上取 7 个不同的点，再从其中任取三点（可重复）作为 A,B,C 来检验此命题，不过这 7 个点中可以有复坐标点. 如果 7 个点中有一个是 $P=(1,0)$，则含有 P 点的 $\{A,B,C\}$ 组显然使命题之结论成立，故实际上只需考虑由其余 6 点中任取 3 点的组合，而且 3 点中有两点相同时也不必检验. 3 个点之间的顺序也与命题结论是否成立无关. 故要检验的组数实际上只有 $\frac{1}{6}(6\times5\times4)=20$(组). 显然，这在微机上是轻而易举的. 在 PB－700 袖珍机运行仅 40 秒.

应当指出，变换(32)的引入不过是虚拟的讨论，目的是估计试验的组数. 估计完成之后，具体试验仍可以设定诸 x_k 值代入而不必真的作变换. 从而可以不涉及复运算.

由于几何定理中圆方程是常见的，这种降次变换技巧对我们的算法十分重要. 对它详细的讨论与有关证明，请参看[12].

给定 x_1,x_2,x_3 求解 y_1,y_2,y_3 时，解并不唯一. 例如，$x_1=\frac{1}{2}$ 时将有 $y_1=\pm\frac{\sqrt{3}}{2}$，如何选取符号呢？其实，符号可以任取，因为 $\Phi(x_1,x_2,x_3)$ 中 y_1 并不出现，因而把 $\frac{\sqrt{3}}{2}$ 和 $-\frac{\sqrt{3}}{2}$ 代入时，只要有一个结果使 $\Phi(x_1,x_2,x_3)$ 为 0 就足够了.

如果假设条件中只涉及四次以下的方程，则数值计算可以带分母及根式进行，这样，得到的结果是准确的. 另外，复数变元的出现也并不影响我们的结论，因为定理 A、定理 B、定理 W 都不排斥复数运算.

四、可喜的前景

数值并行方法用于定理的机器证明，还有许多变化与发展：

1. 单点例证法与数值并行法可以结合使用. 对某些变元用单点例证法，另一些变元用数值并行法.

2. 数值并行法与吴－Ritt 除法可以结合使用——先消去一些变元，再对剩下的约束变元用数值并行法，这样可对不同的问题找寻组合优化的证法.

3. 既然吴－Ritt 除法可以把“条件等式”问题化为“自由变元恒等式”问题，那么，任一种检验代数恒等式的数值方法都可以和吴－Ritt 除法结合起来形成几何定理机械化证明的数值算法. 这已超出了例证法的范围，但仍可以用数值并行法. 例如，检验一个代数恒等式，可以通过计算一阶偏导数而实现，可以通过计算某一点的台劳系数而实现，可以通过计算对某一组点的差分而实现. 这样的每个想法都可引出有前景的研究.

4. 数值并行法还可以用来证明几何不等式[10]. 基本想法是：若 $f(x)$ 是某紧致流形上的 C^∞ 函数. 如果对某点 x_0 有 $f(x_0)>0$，则在 x_0 邻域有 $f(x)>0$，如果 $f(x_0)=0$，则可用展开成台劳级数的方法检验 x_0 邻域是否有 $f(x)\geqslant0$. 这样一来，在每个点 x_0 邻域，不等式 $f(x)\geqslant0$ 是否成立都是可以机械判定的. 用一下有限元覆盖定理即可得出断言：可以经过有限步骤检验来判定紧流形上关于 C^∞ 函数的不等式. 这种检验，本质上是并行的.

最后，让我们看看数值并行法在哲学上意味着什么.

数值并行法把一个一般性的断言化为众多的实例来加以检验. 这些实例检验工作,可以由不同的人,在不同的机器上(当然也可以用纸笔),在不同的年代进行. 他们的大量经验结合起来,便可以肯定一条定理成立,这不正是人类使用了数千年甚至更久的经验归纳法吗?除了数学之外,这种经验归纳法已被各种学科承认. 定理机器证明的数值并行法,用演绎的逻辑推出:在一定条件与范围之内,经验归纳法对数学研究也是行之有效的!

定理机器证明的数值并行法,也从一个角度回答了"为什么经验归纳法在各学科的研究中获得了如此之大的成功?"这一哲学问题.

当所研究的规律的数学形式是可用数值并行法处理的命题时,经验归纳法将是可靠的.

当然,我们不能设想数值并行法适用于一切问题,例如哥德巴赫问题. 但它的范围,也不限于欧氏几何. 我们已用它验证了一些非欧几何定理,其中包括新发现的定理.

单点例证法显示了从特殊事例到一般命题的联系,数值并行法则揭示了大量经验中蕴涵着客观规律的奥秘. 两者辉映成趣. 但它们又同时在吴氏机械化数学理论中找到了根据,这又从一个方面展示了这个新开垦的领域的丰富内涵与美好前景!

参考文献

[1]吴文俊. 几何定理机器证明的基本原理(初等几何部分). 北京:科学出版社,1984.
[2]吴文俊. 初等几何判定问题与机械化证明. 中国科学,1977(6):507-516.
[3]吴文俊. 走向几何的机械化——评 Hilbert 的名著《几何基础》. 数学物理学报,1982(2):125-138.
[4]Wu Wen-tsun. Some Recent Advances in Mechanical Theorem Proving of Geometries. Automated. Theorem proving:After 25 Years,AMS,1983:235-242.
[5]Basic Principles of Mechanical Theorem Proving in Elementary Geometries. 系统科学与数学,1984(4):207-235.
[6]A Mechanization Method of Geometry I. 数学季刊,1986(1):1-14.
[7]吴文俊. 几何学的机械化. 数学学报,1986(3):204-216.
[8]洪加威. 能用例证法来证明几何定理吗? 中国科学,1986(3):234-242.
[9]洪加威. 近似计算有效数字的增长不超过几何级数. 中国科学,1986(3):225-233.
[10]陶懋颀、张景中、杨路. 用多点例证法证明一个几何不等式,(待发表).
[11]邓米克. 证明构造性几何定理的数值并行法,(待发表).
[12]Zhang Jingzhong,Yang Lu,Deng Mike. The Parallel Numerical Method of Mechanical theorem Proving(to appear).

动力系统中的分形集

杨　路　张景中　曾振柄
（中国科学院成都分院数理科学研究室）

一、引言 fractal 和自相似

函数迭代的研究，可以追溯到一百多年以前，其中对后来影响较深者，有 E. Schröder[2]，N. H. Abel[3] 和 Charles Babbage[4]，文献[1]的附录中介绍了他们的工作. 但是由于迭代运算与代数运算的迥然不同，研究工作艰难曲折[5~10]. 直到 1974 年 Li and Yorke 发表了 *Period three implies chaos* 和 Sharkovskiĭ 工作的重新发现，才带来迭代论研究的一系列突破性进展，迭代论中最基本的问题诸如迭代根和嵌入流、周期轨、动力系统的稳定性、分歧现象和浑沌的研究，相继获得重要的发现. 这些深刻的结果令人振奋，激起数学家们更大的兴趣. 同时也为科学界其他分支所注视.

可以说，迭代论即动力系统研究的每一重大进展都归功于计算机所提供的便利. 由于把一函数 $f(x)$ 的 n 次迭代 $f^n(x)$ 表示成 x, n 的分析式 $F(n, x)$ 一般是不可能的，这样就很难由 $f(x)$ 的性质推断 $f^n(x)$ 的演变. 就是说给定变换方式和初始状态，虽然其未来的每一时刻都是确定的，但我们却不能跨越时间而预知它. 计算机高速度数值运算的能力，使数学家们看到了原来不可知的未来世界的奇妙景象. 而动力系统的 fractal 几何就是其中之一[11~13].

传统几何例如欧氏几何. 微分几何或微分拓扑的研究对象都是非常规则的图形，如圆、直线、可微曲面、流形等等. 而对于另外的不规则图形，则没有进行深入的研究. 例如，可填满正方形的 Peano 曲线，处处连续但处处不可微的 Weierstrass 函数的图象，它们被发现后，至今已超过一百年了，但是直到 B. B. Mandelbrot 提出 fractal 几何之前，却一直扮演着“反例”的角色，人们总是小心翼翼地回避它们，没有人去征服这些陷阱. 实际上，就是在普通的空间诸如欧氏空间或者现实自然界中，极不规则图形是普遍存在的，而规则图形都是理想化的结果. 这也是 fractal 几何得以迅速应用的原因之一. 在已经发表的学术论文中，我们可以知道 fractal 集的模型已用于各种自然现象和科学实验现象的描述，如粒子的 Brown 运动、流体的湍流和渗透扩散、植物生长、海岸线和地表面的形状、天空中飘忽变幻的浮云、降雨区的边界、宇宙中星系的分布等，甚至于股票市场价格的波动. 在数学纯理论的许多领域，例如数论和非线性微分方程中，也出现 fractal 集，奇异吸引子就是典型的 fractal 集.

英文 fractal 一词是 1975 年 Mandelbrot 由拉丁文 fractus 引入的，原义是“破碎、断片”.

本文刊于《数学进展》，第 19 卷第 2 期，1990 年.

Mandelbrot 在 1977 年对 fractal 集给出过一个“tentative definition”：A fractal is a set whose Hausdorff-Besicovitch dimension strictly exceeds its topological dimension. 这个定义不能表达 fractal 集的根本特征，正如 Mandelbrot 所说：But I like this definition less and less, and take it less and less seriously. 因为按照这一定义，一些极不规则的图形，其几何性质完全与分形集的相同，却因其 Hausdorff 维数是整数而被排除在 fractal 的门外. Mandelbrot 认为，fractal 概念的本质在于其几何特征，而维数则是外在性质(见[14]). 根据 J. Falconer 的著作[15]可知 fractal 集的几何特征主要在两个方面，其一是这种集合在其几乎每一点的每一个邻域里，点的分布是零落散乱、疏稠无规的；其二是这种集合在其几乎每一点是没有切线的. 它们的确切含义我们将在本节予以扼要地说明. 据此，我们认为用“分形”一词翻译英文的 fractal，或径采用 fractal，而不使用“分数(维)”，以相应于 fractal 和 fractional 之原义的区别. 而在本文中，我们总使用 fractal 以求明了.

本节内容包括 Hausdorff 维数，稠密度和可切性，fractal s-集，自相似(self-similar) fractals 及其 Hausdorff 维数的计算，Weierstrass 函数形成的 fractal 集. 以及随机 fractals，这里主要介绍 Brown 运动.

Hausdorff 维数可以对于一般的距离空间的任意子集来定义，此处考虑 n 维欧氏空间 R^n，其一子集 U 的直径 $|U|$ 如通常所指

$$|U|=\sup\{|x-y|: x,y\in U\}.$$

一子集族 $\{U_i\}$ 称为另一集合 E 的一个 δ-覆盖，如果

$$E\subset\bigcup_i U_i \quad 而且\ 0<|U_i|\leqslant\delta, 对每一\ i.$$

设 $E\subset R^n$ 是任一子集，$s\geqslant 0$ 是一非负实数，则对于 $\forall\delta>0$，定义

$$H_\delta^s(E)=\inf\left\{\sum_{i=1}^{\infty}|U_i|^s, \{U_i\}_{i=1}^{\infty}\ 为\ E\ 的\ \delta\text{-覆盖}\right\}.$$

易见对固定的 s，H_δ^s 随着 δ 减小而增大，所以下述极限存在或为 ∞：

$$H^s(E)=\lim_{\delta\to 0}H_\delta^s(E),$$

当 s 是某一整数 n 时，$H^\delta(E)$ 相当于 E 的 n 维体积(Lebesgue 测度) $L^n(E)$，相差一个只与 n 有关的常数因子

$$c_n=\pi^{\frac{1}{2}n}/2^n\left(\frac{1}{2}n\right)!$$

例如，若 E 是一有界的光滑曲面 F，就有

$$H^s(F)=\infty, 0\leqslant s<2,$$

$$H^2(F)=\frac{\pi}{4}\text{Area of } F,$$

$$H^s(F)=0, 2<s<\infty.$$

对于 R^n 中的任意集合 E，照样存在一个唯一的临界值 $s_0=\dim E$，满足

$$H^s(F)=\infty, 若\ 0\leqslant s<\dim E,$$

$$0\leqslant H^s(F)<\infty, 若\ s=\dim E,$$

$$H^s(F)=0, 若\ \dim E<s<\infty.$$

这是因为若 $s<t$，对 $\forall\delta$ 都有，

$$H_\delta^s(E)\geqslant\delta^{s-t}H_\delta^t(E).$$

它说明不仅 $H^s(E)$ 随 s 的增加而不增，而若 H^t(且)是正数，则 $H^s(E)=\infty$.

处于分界处的数 $s=\dim E$ 称为集合 E 的 Hausdorff 维数，或者称为 Hausdorff-Besicovitch 维数，后者在 Hausdorff 测度及其几何的研究领域中作出了主要贡献. 集合 E 的 Hausdorff s-测度就是极限 $H^s(E)$.

应该指出，把 $H_\delta^s(E)$ 定义中的 δ-可数覆盖 $\{U_i\}$ 改为 V_i 均是凸的或小球或方体，或均是开的或闭的，所得 $\dim E$ 和 $H^{\dim E}(E)$ 均不改变.

R^s 的一子集 E 称为是整数维或者非整数维根据 $\dim E$ 是否整数而定. 非整数维(non-integer dimension)的使用是为防止 fractional dimension 和 fractal dimension 的混淆. 非整数维集的经典例子 Cantor 集，易计算标准 Cantor 集的 Hausdorff 维数是 $\frac{\log 2}{\log 3}$，两个 Cantor 集在平面上的乘积 $\dim=\frac{\log 4}{\log 3}$.

Hausdorff 维数是描述一集合之规模的整体性尺度，其值也基本上反映了集合内部点的分布的规则或不规则的程度. 更加精细的两个尺度是集合在每一点的稠密性和可切性，它们都是局部的量. 设 $E\subset R^s$ 是一 s-子集，即 $\dim E=s$，$\forall x\in R^s$，E 在 x 点的稠度和疏度(上下稠密度，upper and lower density)定义为

$$\overline{D}^s(E,x)=\overline{\lim_{r\to 0}}\frac{H^s(E\cap B_r(x))}{(2r)^s},$$

$$\underline{D}^s(E,x)=\underline{\lim_{r\to 0}}\frac{H^s(E\cap B_r(x))}{(2r)^s}.$$

$x\in E$ 称为是规则点，如果 $\overline{D}^s(E,x)=D^s(E,x)$，不然就叫极不规则点(regular and irregular). E 本身叫规则的，如果它的 H^δ-几乎所有点是规则的，叫做不规则的，如果它的 H^s-几乎所有点是不规则的. H^s-几乎所有点是指除开一个 $H^s(E')=0$ 的集合 E'，R^s 中任一集合可分成两个集的并

$$E=E_r\cup E_{ir},$$

其中 E_r 和 E_{ir} 分别由 E 的规则点和不规则点作成. 可证明它们分别是规则集和不规则集. 集合 E 在一点 x 沿一方向有切线或没有，其严格意义也是由 Hausdorff 测度来定义的. 这里不再叙述详细过程. 下述结果均可从[15]中找到它们的证明.

设 $E\subset R^s$ 是一 s-集，则其 Hausdorff 维数与规则性和可切性的关系为

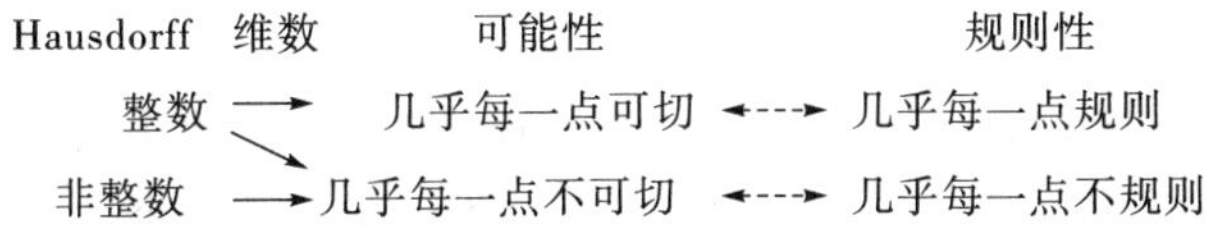

Hausdorff 维数	可能性		规则性
整数 → (↘)	几乎每一点可切	←---→	几乎每一点规则
非整数 →	几乎每一点不可切	←---→	几乎每一点不规则

Hausdorff 维数为整数的不规则集的存在性可用下例说明(见[15]).

例　R^2 上的不规则 1-集.

设 f 是定义于$[0,1)$上的实函数，$x\in[0,1)$的像 $f(x)$是

$$f(.x_1x_2\cdots)=.y_1y_2\cdots,$$

其中$.x_1x_2\cdots$和$.y_1y_2\cdots$是4进展开(设每一x_i后不全是3),且$y_i=5-x_i(\mathrm{mod}\ 4)$.

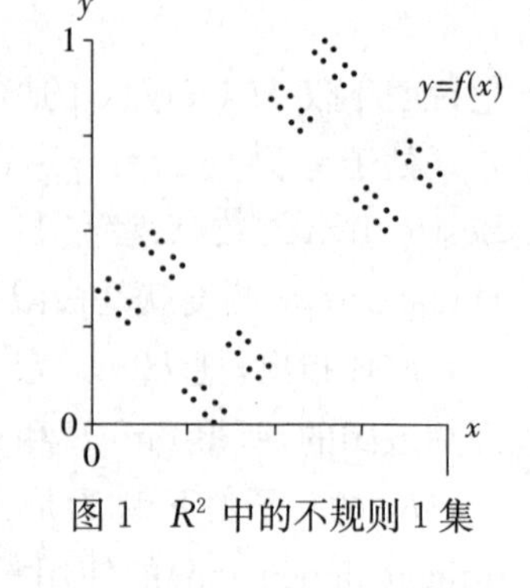

图1 R^2 中的不规则1集

集合E定义为f的图象,如图1,

$$E=\{(x,f(x)):0\leqslant x<1\}.$$

因为E在z轴上的投影为$[0,1)$,所以

$$\dim E\geqslant 1.$$

又因为$\forall k$,可用4^k个边长为4^{-k}的正方形覆盖E,所以

$$H_{s}^{1-k}(E)\leqslant 4^k\cdot(\sqrt{2}\cdot 4^{-4})^1=\sqrt{2},$$

$$H^1(E)\leqslant\sqrt{2},$$

这说明

$$\dim E\leqslant 1.$$

以上说明E的Hausdorff维数为整数1.

为证明E的不规则性,将E投影到与x轴夹角为$\pm 45°$的两直线上,易见其Lebesgue测度均为0.关于投影与规则性的关系,有定理如下:

定理 R^2上一个1-集E不规则当且仅当它在两个或以上的不同方向上的投影的Lebesgue测度为0(参见[15]).

R^n中一个整数维集是否规则也有类似结果.

定理 设E是R^s中一个k-集,k是整数.那么E为规则的导致$L^k(\mathrm{proj}_nE)>0$对几乎所有的$\Pi\in G_{n,k}$;E为不规则的导致$L^k(\mathrm{proj}_nE)=0$,对几乎一切的$\Pi\in G_{n,k}$. $G_{n,k}$表示R^n的所有k维子空间构成的Grassmann流形(见[16],[15]).

另外,在维数不超过1的时候,规则性与完全不连通性(totally disconnection)紧密相关:

定理 设$E\subset R^n$,$s=\dim E\leqslant 1$,则E不规则当且仅当E是完全不连通的(见[15],[17]).

许多经典的fractal集在几何结构上都具有自相似的特征.一个图形称为是自相似(self-similar)的,如果它的每一局部都可以放大为整个图形. Cantor集是自相似的fractal集,但Koch雪花曲线则是一自相似fractal曲线,它的构造为:从一个正三角形开始,将每一直线段三等分,将中间一份换成与它能形成指向外侧的小正三角形的折线段;然后再将新图形的每一直线段三等分,而将中间一份换成折线段;…如此继续,得到的极限曲线是连续的但每一点都不可切的,如图2,它是德国数学家H. von Koch于1904年提出的.可以计算其Hausdorff维数是$\dfrac{\log 4}{\log 3}$.

尽管Koch雪花曲线在本世纪初就为人们所知,但却一直被作为"反例"看待. 1982年,Mandelhrot把这种用递归步骤产生自相似分形集的方法一般化(见[12]).一个源图(generator)是由若干线段组成的,其中有两个点为特定点;每一次迭代把上一次迭代图的每一直线段改换为源图的相似像,且源图中两特定点的像就是这一直线段的两个端点.这实际上是几何变换(相似)的迭代,极限图形就是这一几何变换的不变集.这种分形集的Hausdorff

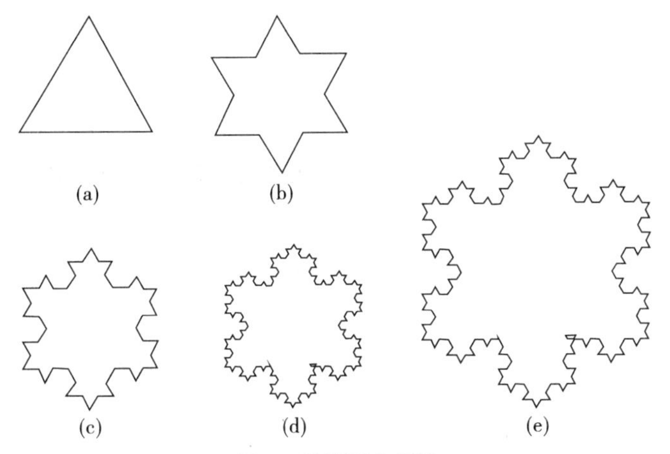

图 2 科赫雪花曲线

维数可以从几何变换的比例因子 $r_1, r_2, \cdots, r_m$ 求得. 如果每一直线段的变换像与另外直线段的变换像不相交,则下述方程的解 s 就是迭代的极限图形的 Hausdorff 维数,如果有相交则它是一个上界

$$\sum_{i=1}^{\infty} r_i^s = 1.$$

Hutchinson[18] 研究了更为广泛的问题. 设 $\{\psi_1, \psi_2, \cdots, \psi_m\}$ 是 $R^n \to R^n$ 的收缩映射族,即

$$|\psi_i(x) - \psi_i(y)| \leqslant c_i |x-y|, \forall x, y \in R,$$
$$c_i < 1, i = 1, 2, \cdots, m.$$

则有

定理 存在唯一的非空紧集 E 满足

$$E = \bigcup_{i=1}^{m} \psi_i(E).$$

并且, $\forall$ 紧集 F,在 Hausdorff 距离的拓扑意义下

$$\psi^K(K) \to E.$$

这里映射 ψ 是 R^n 的子集之间的集值映射, $\forall F \subset R^n$,

$$\psi(F) = \bigcup_{i=1}^{m} \psi_i(F).$$

S. Hayashi[19] 中将自相似 fractal 集作为 Taski 映射的不动点,也有类似定理. 下面的定理用以估计前一不变集的 Hausdorff 维数(见[20],[15]).

定理 若 $\{\psi_i\}_{i=1}^m$ 是 $R^n \to R^n$ 的收缩,满足

$q_i |x-y| \leqslant |\psi_i(x) - \psi_i(y)| \leqslant r_i |x-y|, i = 1, 2, \cdots, m. x, y \in \bar{V}$, V 是 E 的某一开集,满足当 $i \neq j$ 时

$$\psi_i(V) \cap \psi_j(V) = \varnothing, i, j = 1, 2, \cdots, m.$$

则 $\dim E \in [s, t]$,其中 s, t 满足

$$\sum_{i=1}^{m} q_i^s = 1 = \sum_{i=1}^{m} r_i^t.$$

若对某些 $i \neq j$, $\psi_i(V) \cap \psi_j(V) \neq \varnothing$,则 $\dim E \leqslant t$ 成立.

上一定理的一个特殊情况是$\{\psi_i\}_{i=1}^m$为一列相变换，这时$0<q_i=r_i<1$. 自相似集的严格定义就是E对某一列收缩相似$\{\psi_i\}_{i=1}^m$满足

$$E=\bigcup_{i=1}^m \psi_i(E),\psi_i(E)\cap\psi_j(E)=\varnothing,1\leqslant i<j\leqslant m.$$

一个特殊的情况是Mandelbrot递归方法生成的自相似分形集，当源图(generator)的每个直线段等长时，则相似比$r_1=r_2=\cdots=r_m=r$，这时分形集的Hausdorff维数s为

$$s=\frac{\log m}{\log\dfrac{1}{r}}.$$

当源图形是非退化的折线段时，这个值严格大于1. 这时设折线段为$p_0p_1\cdots p_m$，则相似比

$$r=\frac{|\overrightarrow{p_0p_1}|}{|\overrightarrow{p_0p_m}|}\geqslant\frac{|\overrightarrow{p_0p_1}|}{\sum\limits_{i=0}^{m-1}|\overrightarrow{p_ip_{i+1}}|}=\frac{1}{m}.$$

除非$p_0p_1\cdots p_m$退化为直线段，严格不等号成立.

动力系统中许多极限集合都是具有类似于自相似几何特性的fractal集. 例如Hénon吸引子，Julia集，它们都不是严格意义下的自相似集，但具有自相似几何特性的精髓. 这就是迭式构造，或者叫全息结构，即每一片断都含有整体的全部信息. 自然数的šharkovskiĭ序就是这样一种迭式构造：$3\triangleleft 5\triangleleft\cdots\triangleleft 2\times3\triangleleft 2\times5\triangleleft\cdots\triangleleft\cdots\triangleleft 2^n\triangleleft\cdots\triangleleft 2\triangleleft 1$.

如果将片断理解为每一点的后继，则每一片断在一"相似"变换(除以因子2^k)之后又得到完整的šharkovskiĭ序，除非这片断是有限的. 激光全息照片是一种自相似图形，植物的生长也表现出某种自相似性(生物全息律，见[21]). 球面上复解析动力系统中的Julia集许多情况下是拟自相似的，其严格意义将在后面说明. 自仿射分形集(self-affine fractal set)也是一种迭式构造，类似于自相似集，E称为自仿射的，如果有一组仿射变换$\varnothing_1,\varnothing_2,\cdots,\varnothing_m$使$E=\bigcup\limits_{i=1}^m\varnothing_i(E)$而且$\varnothing_i\{(E)\cap\varnothing_j(E)=\varnothing,1\leqslant i<j\leqslant m$. 见Mandelbrot[22-24].

如果在用递归方法构造自相似集或者一般的迭式分形集的过程中引入随机变量，使迭代中ψ_i的出现按一定的概率分布，则得到的将是统计自相似集(statistically self-similar set)或统计迭式集. 例如图3所示的图形，设迭代时添进反射作为随机变量. 其Hausdorff维数的计算方法一样. 这里是$\left(\frac{1}{2}\right)^s+\left(\frac{1}{2}\right)^s+\left(\frac{1}{2}a\right)^s=1$的解$s$.

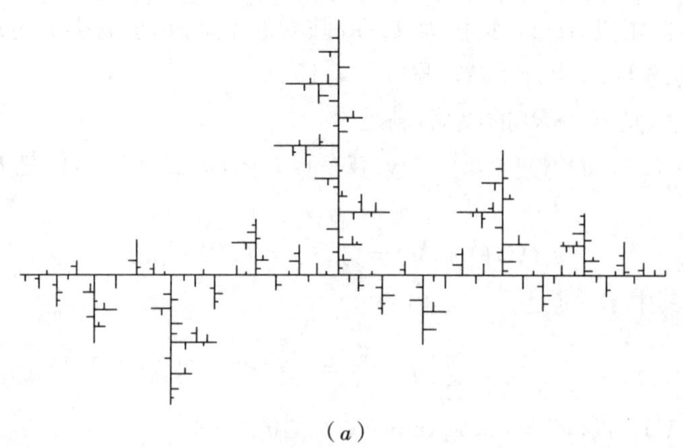

(*a*)

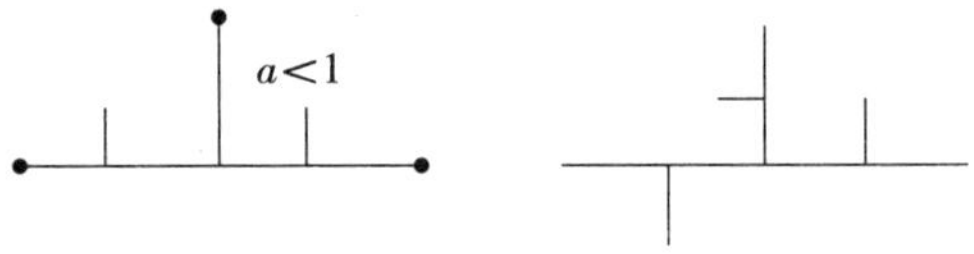

a<1

(b)

图3 维数是 s,其中 $2\left(\frac{1}{2}\right)^s+\left(\frac{1}{2}a\right)^s=1$.

R. Brown 在 1827 年观察到悬浮在液体中的粒子处于不停地运动中,所沿路径极不规则. Wiener 过程是 Brown 运动的一个严格的统计学模型,它实际上是统计自仿射分形的一个恰当例子. 设 $n\geqslant 1$,Ω 表示所有连续路径 $\omega:[0,\infty)\to R^n$,将 $\omega(0)$ 置于原点,而 $\omega(t)$ 表示沿 ω 路径运动的粒子在时刻 t 的位置. 则可证明存在定义于 Ω 的一大类子集上的概率测度 $p(p(\Omega)=1)$ 满足

(a)$\omega(t_2)-\omega(t_1)$ 与 $\omega(t_4)-\omega(t_3)$ 无关,若 $t_1<t_2<t_3<t_4$;

(b) $p\{\omega:|\omega(t+h)-\omega(t)|\leqslant\rho\}$

$=\mathrm{c}h^{-\frac{n}{2}}\int_0^{\rho}r^{n-1}e^{-\frac{r}{2}\frac{2}{h}}dr$,c 是使 $p(\Omega)=1$ 的规范化常数.

由(b)可知,路径 $\omega(t)$ 和 $\frac{1}{\sqrt{b}}\omega(bt)$ 在分布上是相同的,对 $\forall b>0$. 而 $\omega(t)$ 到 $\frac{1}{\sqrt{b}}\omega(bt)$ 的变换是仿射变换([15],[22]).

分形集可以分为两类:确定性的和随机性的(deterministic fractals and statistical(or random)fractals). 前者是按一定基本规则精确构成的,自相似集属于这一类,Cantor 集,Koch 曲线以外还有 sierpenski gaskets,Havlin carpets 等,Pythagoras[107] 树花等,但也有一些不属于自相似或迭式的,例如 Weierstrass 函数的图象. 随机 fractal 集的构造中引入了随机因素,这使得它们更加与物理现象相接近,因而得到许多学者的研究,这些模型,有如 self-avoid-ing random walk,lattice animals,random percolation,Ising model,irreversible growth model. 仅不可逆生长模型,就有多种,如 Eden(伊甸园)model,DLA 过程(diffusion limited particle aggregation,diffusion limited cluster aggregation),ballistic model,化学模型. M. Kolb 还制作了一个大约 25 分钟长的影片介绍这些理论模型[26]. 这类研究论文能从有关刊物上广泛见到、并有若干论文辑([27],[28]),但理论性的研究有待深入.

这一节结束之前,介绍 Weierstrass 型函数图象的 Hausdorff 维数的计算. 最著名的例子是

$$f(x)=\sum_{n=1}^{\infty}\lambda^{(s-2)n}\sin(\lambda^n x),1<s<2,\lambda>1.$$

它是 Weierstrass 提出的连续但处处不可微的例子. 易证明曲线 $\Gamma=\{(x,f(x)),0\leqslant x\leqslant 1\}$ 的 Hausdorff 维数 $\leqslant s$,等号的严格证明尚未见到. 另一个例子是 Mandelbrot[11] 提出的,它是 Weierstrass 函数的变形

$$f(x)=\sum_{n=-\infty}^{+\infty}\lambda^{(s-2)n}(1-\cos\lambda^n x),1<s<2,\lambda>1,$$

这一函数的 Hausdorff 维数 $=s$,它满足

$$f(\lambda x)=\lambda^{2-s}f(x).$$

如果把 Weierstrass 函数中的 $\sin x$ 换成如下定义的分段函数 $g(x)$

$$g(4k+x)=\begin{cases}x, & 0\leqslant x<1,\\ 2-x, & 1\leqslant x<3,\\ x-4, & 3\leqslant x<4,\end{cases}$$

则所得函数仍是连续但处处不可微的，其图象具有 Hausdorff 维数 $\dim=s$.

下面定理在确定 Lipschitz 函数的 Hausdorff 维数的上界时是很方便的([15]).

定理　设 $f:[0,1]\to\mathbf{R}$ 满足

$$|f(x+h)-f(x)|\leqslant ch^{2-s},$$

对一切 x 和 $0<h\leqslant h_0$，其中 c,h_0 是正常数，则 $\dim\Gamma\leqslant s,\Gamma=\{(x,f(x)),0\leqslant x\leqslant 1\}$.

关于 Weierstrass 函数的分形性，详见 Berry&Lewis[25]. 对多变元的情况，Ausloos & Berman 近来的工作[117]可以参考.

二、奇异吸引子的几何

计算机方法在近十多年来动力系统的研究中起到了极大的推动作用. 许多重要现象的发现，特别是对于简单系统的复杂行为的观察，无不首先起因于计算机上的工作. 最早而且最有名的例子无疑是 Lorenz 吸引子([29],1963).

考虑位于两个平行的光滑平面间的具有一致深度的液体层，设上下两表面的温度差是常值. 这一系统的偏微分方程在 1962 年已由 Saltzmann[30] 所给出，Lorenz 通过 Galerkin 方法将它变为下面的三阶常微分方程系统：

$$\dot{x}=\sigma(-x+y),$$
$$\dot{y}=\rho x-y-xz,$$
$$\dot{z}=-\beta z+xy.$$

式中 β 是与上下平面间无量纲距离有关的常数，σ 是 prandtl 数，ρ 是系统某一临界值的 Rayleigh 商. 现在设 β,σ 为固定值而让 ρ 变化. Lorenz 证明存在闭的简单连通区域 $D\subset R^3$，含有原点，而 D 边界上的向量均指向内侧. 所以存在含于 D 的吸引集合 A

$$A=\bigcap_{t\geqslant 0}\varnothing_t(D).$$

当 $\rho<1$ 时原点是全局吸引的，所有轨道都趋于它. 当 $\rho>1$ 时，系统有两个非平凡的不动点 p_+,p_-：

$$p_{\pm}=(\pm\sqrt{\beta(\rho-1)},\rho-1),$$

如果 $\sigma<\beta+1$，则这两个不动点总是稳定的. 如果 $\sigma>\beta+1$，则当 $1<p<\rho_h$ 时两不动点稳定，而当 $p>\rho_h$ 时是不稳定的，式中

$$\rho_h=\sigma(\sigma+\beta+3)/(\sigma-\beta-1).$$

此即系统当 $p=\rho_h$ 时在 $p_{\pm}$ 处产生 Hopf 分歧. 现在考虑 p_+ 和 p_- 的不稳定流形，由于系统线性化的特征值是

$$\lambda=-(\sigma+\beta+1),\lambda=\pm i\sqrt{2\sigma(\sigma+1)(\sigma-\beta-1)},$$

可知 $W^n(p_+),W^n(p_-)$ 是二维流形，而原点 p 的不稳定流形 $W^n(p)$ 是一维的. 显然它们都包

含在 A 中. 所以这时 A 的结构很复杂.

又因为系统的散度为 $-\sigma-\beta-1<0$，说明吸引集合 A 具有零体积. Lotenz 研究了当 $\sigma=10,\beta=\frac{8}{3},\rho=28$（这时 $\rho_h\approx24.74$）时系统的行为，通过计算机观察到吸引集 A 的许多“奇异”性质.

（1）A 中任何轨道都是不稳定的.

（2）轨道对初值有敏感依赖性，即任意两轨道不论其初始点如何靠近都将随 t 的增加而截然不同.

（3）A 含有不可数无穷多个在 A 中稠密的轨道.

（4）A 的周期轨是稠密的.

下面的图 4 是由计算机绘制的 Lorenz 方程的一个非周期轨，它基本上刻画了 Lorenz 吸引子的形状，Hüseyin Koeak[31] 系统地介绍了使用计算机研究动力系统的方法. 由于 Lorenz 方程是一个经典的例子，许多地方已有专门介绍，可以参看文献[32,33,34].

另一个有影响的 Hénon 吸引子，它是 1976 年法国天文学家 M. Hénon[35] 发现的，考虑二次映射的标准形式：

$$x_{n+1}=y_n,$$
$$y_{n+1}=bx_n+ay_n-y_n^2.$$

对于 $a=3.1678,b=0.3$，Hénon 的数值研究发现了系统的一个吸引集，它可以看成原点的不稳定流形的极限集. 如图 5，Hénon 吸引子清楚地显示出某种自相似性和 Cantor 形截面. 这里将不再深入介绍这些熟知的结果.

“奇异吸引子”一直没有得到严格的定义，即使是“吸引子”，它的涵义在各种文献中也不一致[36~42]. 下面是 David Ruelle[38] 的关于吸引集. 吸引子的定义：

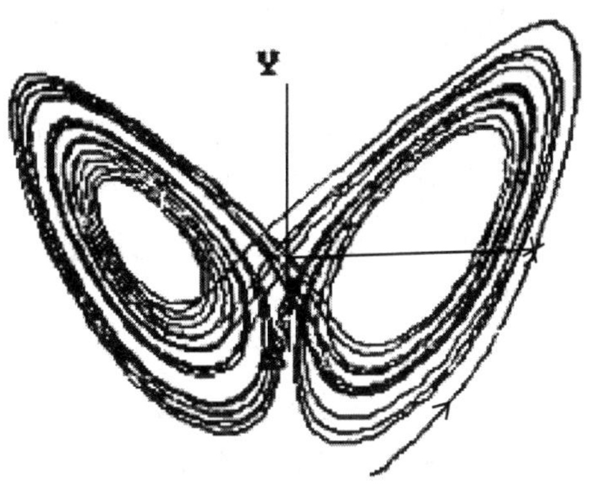

图 4　Lorenz 方程的一个非周期轨

设 M 是有限维流形，$f:M\to M$ 是连续映射. M 的一非空子集 Λ 称为是 f 的吸引集（attracting set），如果 Λ 有紧致邻域 U 满足 $\Lambda=\bigcap_{k\geqslant0}f^kU$，而且对于充分大的 n 均有 $f^nU\subset U$. 这个邻域 U 称为吸引集 Λ 的一个基本邻域. 开集 $W=\bigcup_{n\geqslant0}f^{-n}U$ 称为 Λ 的吸引域或者吸附域（basin of attraction），它与 U 的选择没有关系，$W=\{x\in M,f^nx\to\Lambda,n\to\infty\}$.

吸引子（attractor）是链回归意义下吸引集的不可约分解. 任给 $a,b\in M,a\succ b$ 的意义是 $\forall\varepsilon>0$，有一个从 a 到 b 的 ε-伪轨，$\succ$ 称为链回归走向，或者 a 走向 b. 链回归等价 $a\sim b$ 指 $a\succ b$ 而且 $b\succ a$. 这样每一点 $a\in M$ 都按链回归等价意义对应于 M 的一个子集$[a]$，称为等价类，吸引集的链回归等价类的划分获得吸引子. 显然，吸引集中至少含有一个吸引子.

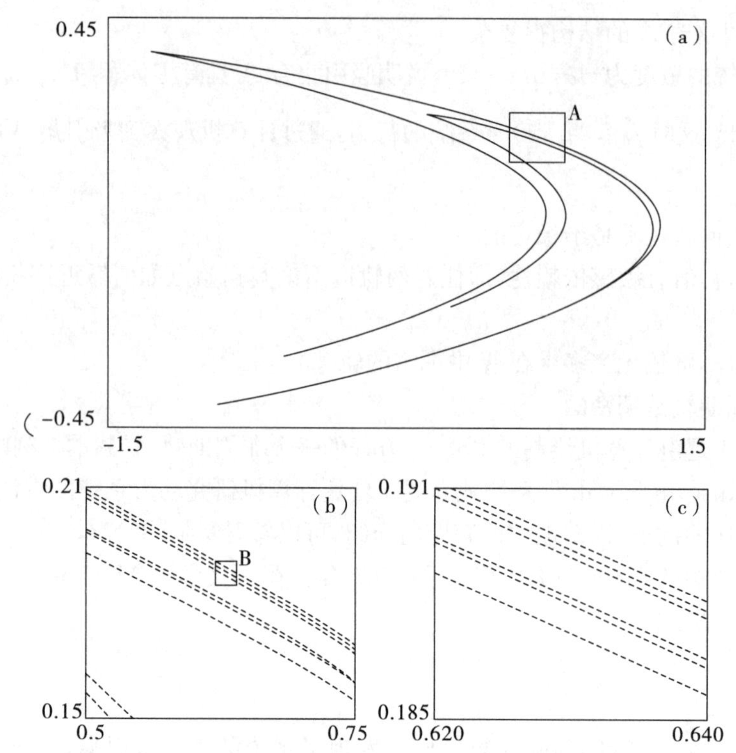

图 5 Hénon 映射$(x,y)\to(1+y-ax^2,bx)$，$a=1.4$，$b=0.3$，(a)，(0.631，0.189)的次 100 000 迭代，(b)，(a)中区域 A 的放大，100000 次迭代，(c)，(b)中区域 B 的放大，1000000 次迭代.

定理 一吸引集 Λ 是一个吸引子当且仅当它是链回归的.

David Ruelle 同时考虑了按此方式定义的吸引子在 f 的小扰动上具有某种稳定性(详见[38]，[42]). 从上述意义下考虑，由计算机产生的吸引子确实也是"理论上的"吸引子.

关于吸引子的"奇异"性，按 J. Guck-enheimer and P. Holmes[32] 的定义，是根据其中是否含有横截同宿轨而言的. 熟知的结果是横截同宿轨导致某种类型的转移或马蹄结构，也就产生 Li-Yorke 意义下的浑沌不变集，参见[43]～[46]. 这是就吸引子上的动力系统而考虑的，本节的主要内容则是对于奇异吸引子的几何结构加以探讨.

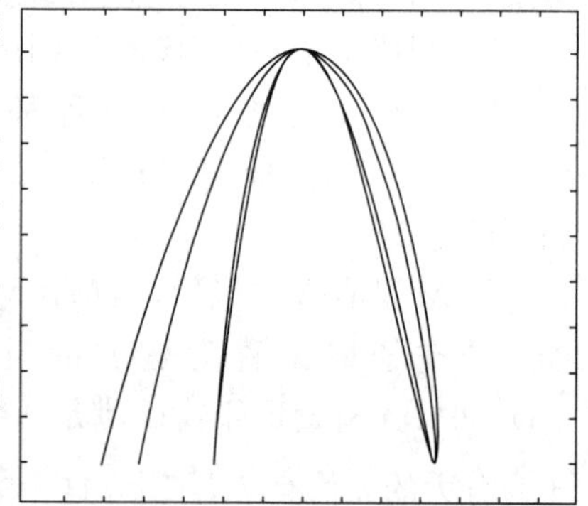

图 6 原点的不稳定流形的前几圈，横轴从-0.1到 1.1，纵轴从-0.1到 1.1.

1. Lauwerier 映射[47～49,51]

这一映射是马蹄映射的变型，定义于单位正方形$[0,1]\times[0,1]\to[0,1]\times[0,1]$：

$$\begin{cases} x\to bx(1-2y)+y, \\ y\to 4y(1-y) \qquad (0<b\leqslant 1), \end{cases}$$

它的两个不动点(0,0),$\left(\frac{3}{4+2b},\frac{3}{4}\right)$都是双曲的,它们的稳定流形分别是线段 $y=0$ 和 $y=\frac{3}{4}$,不稳定流形也可以用分析表达式写出.

当 $0<b<1$ 时,作变换

$$x=\frac{1}{2}-\frac{1-b}{2b}u,$$
$$y=\frac{1}{2}-\frac{1}{2}v,$$

则 Lauwerier 映射写为

$$\begin{cases}u_{n+1}=b(1+u_n)v_n\\ v_{n+1}=2v_n^2-1,\end{cases}$$

则

$$\begin{cases}u_n=bv_n+b^2v_{n-1}v_{n-2}+b^3v_{n-1}v_{n-2}v_{n-3}+\cdots+b^nv_{n-1}v_{n-2}\cdots v_0+b^nv_{n-1}v_{n-2}\cdots v_0\cdot u_0,\\ v_n=\cos(2^n\cos^{-1}v_0).\end{cases}$$

注意到有

$$v_0v_1v_2\cdots v_{m-1}=\frac{2^{-m}\sin(2^mz)}{\sin z},$$

式中 $z=\cos^{-1}v_0$.

所以 u_n,v_n 可以写成下面的级数

$$\begin{cases}u_n=\sum_{k=1}^{n}\left(\frac{b}{2}\right)^k\frac{\sin(2^kz)}{\sin(2^{n-k}z)}+\left(\frac{b}{2}\right)^n\frac{\sin(2^nz)}{\sin z}\cdot u_0,\\ v_n=\cos(2^nz),\end{cases}$$

对于一初始点(u_0,v_0),从上式可以看出 $n\to\infty$时,因$\left(\frac{b}{2}\right)^n\to 0$,$(u_n,v_n)$实际上只与 v_0 有关.

若引入参数 t 使 $z=2^{-n}t$,则得到 Lauwerier 映射的不变曲线

$$\begin{cases}u(t)=\sum_{k=1}^{\infty}\left(\frac{b}{2}\right)^k\frac{\sin t}{\sin(2^{-k}t)},\\ v(t)=\cos t.\end{cases}$$

用原来的变量 x,y 写成

$$\begin{cases}x=\frac{1}{2}-\frac{1}{2}\left(\frac{1}{b}-1\right)\sum_{k=1}^{\infty}b^k\phi_k(t),\\ y=\sin^2\left(\frac{t}{2}\right),t\geqslant 0.\end{cases}$$

式中函数

$$\phi_k(t)=\frac{\sin t}{2^k\sin(2^{-k}t)},k=1,2,\cdots.$$

可以证明这一曲线就是(0,0)的不稳定流形,记为 J_0,因为

(1)Lauwerier 逆映射下$(x(t),y(t))\in J_0\to\left(x\left(\frac{t}{2}\right),y\left(\frac{t}{2}\right)\right)$,

(2)相应于参数值 $t_n=2^n\pi, n=1,2,3,\cdots$,

$x_n=\left(1-\frac{b}{2}\right)b^{n-1}, y_n=0$,而$(x_n, y_n)\to(0,0)$.

另一不动点$\left(\frac{3}{4+2b}, \frac{3}{4}\right)$的不稳定流形 J_1 可用相同的方法得到,

$$\begin{cases} x=\frac{1}{2}-\frac{1}{2}\left(\frac{1}{b}-1\right)\sum_{k=1}^{\infty}\psi_k(t)\left(\frac{b}{2}\right)^k, \\ y=\sin^2\left(\frac{\pi}{3}+\frac{t}{2}\right), -\infty<t<\infty, \end{cases}$$

式中

$$\psi_k(t)=\frac{\sin\left(\frac{2}{3}\pi+t\right)}{(-2)^k\sin\left(\frac{2}{3}\pi+\frac{t}{(-2)^k}\right)}.$$

Lauwerier 映射的不稳定流形的极限集(闭包)是一个典型的奇异吸引子. 现在考虑它与水平线的截面,将看到它是一个 Cantor 型集. 取固定值 $\theta, 0<\theta<1$,设

$$t_n=(n+\theta)\cdot 2\pi, n=0,1,2,\cdots,$$

则 J_0 与线段 $y=\sin^2\pi\theta$ 的截面是$(x(t_n), y(t_n))$,即

$$x_n=\frac{1}{2}-\frac{1}{2}\left(\frac{1}{b}-1\right)\Sigma\left(\frac{b}{2}\right)^k=\frac{\sin 2\pi\theta}{\sin(2^{-k}(n+\theta)\cdot 2\pi)}, y_n=\sin^2\pi\theta,$$

$$n=0,1,2,\cdots.$$

设 θ 的二进展开为

$$\theta=.b_{-1}b_{-2}b_{-3}\cdots$$

对每一实数 $\beta, 0<\beta<1$,引进一个整数序列 $N(\beta, n)$ 与之对应,其方法是,若 β 的二进展开为

$$\beta=.b_0b_1b_2\cdots$$

则整数列 $N(\beta, n)$ 由下面的二进式表达

$$b_0, b_1b_0, b_2b_1b_0, \cdots, b_nb_{n-1}\cdots b_1b_0, \cdots.$$

于是 β 指定了 J_0 与 $y=\sin^2\pi\theta$ 的截面的一个子集.

$$x_\beta(n)=x((N(\beta, n)+\theta)2\pi), y_\beta(n)=\sin^2\pi\theta, n=0,1,2,\cdots$$

用 $b_i(-\infty<i<\infty)$ 可将 $x_\beta(n)$ 表达为

$$\begin{aligned} x_\beta(n) = {} & \frac{1}{2}-\frac{1}{2}\left(\frac{1}{b}-1\right)\sum_{k=1}^{n+1}\left(\frac{b}{2}\right)^k\frac{\sin 2\pi\theta}{\sin(.b_{k-1}b_{k-2}\cdots b_0b_{-1}b_{-2}\cdots)2\pi} \\ & -\frac{1}{2}\left(\frac{1}{b}-1\right)\sum_{k=n+2}^{\infty}\left(\frac{b}{2}\right)^k\frac{\sin 2\pi\theta}{\sin 2^{-k+n+1}(.b_nb_{n-1}\cdots b_0b_{-1}b_{-2}\cdots)2\pi} \end{aligned}$$

$$n=0,1,2,\cdots$$

可以看出此序列当 $n\to\infty$ 时收敛于点 $x(\beta)$.

$$x(\beta)=\frac{1}{2}-\frac{1}{2}\left(\frac{1}{b}-1\right)\sum_{k=1}^{\infty}\left(\frac{b}{2}\right)^k\frac{\sin 2\pi\theta}{\sin(.b_{k-1}b_{k-2}\cdots b_1b_0b_{-1}b_{-2}\cdots)2\pi}$$

这个点位于水平线 $y=\sin^2\pi\theta$. 对于无理数 β,因为序列 $N(\beta, n)$ 是无限的,所得 $x(\beta)$ 是 J_0 与

水平线 $y=\sin^2\pi\theta$ 截面的极限点；对于有理数 β，点 $x(\beta)$ 是 $x(\beta_n)$ 的极限点，其中 β_n 为趋近于 β 的无理数列，所以 $\forall\beta$，$0<\beta<1$，$x(\beta)$ 总是 $\bar{J}_0$ 与 $y=\sin^2\pi\theta$ 截面的极限点. 现在证明 J_0 与 $y=\sin^2\pi\theta$ 截面上的每一点也是 $\bar{J}_0$ 与 $y=\sin^2\pi\theta$ 截面的极限点. 任一整数 $n=0,1,2,\cdots$ 可写为二进表达式，设

$$\begin{aligned}n+\theta&=b_m'\cdots b_2'b_1'b_0'.\,b_{-1}b_{-2}b_{-3}\cdots\\&=\cdots0\cdots0b_m'\cdots b_2'b_1'b_0'.\,b_{-1}b_{-2}b_{-3}\cdots,\end{aligned}$$

那么对应的点是

$$\begin{aligned}x(n)&=\frac{1}{2}-\frac{1}{2}\left(\frac{1}{b}-1\right)\sum_{k=1}^{\infty}\left(\frac{b}{2}\right)^k\frac{\sin2\pi\theta}{\sin(.\,b_{k-1}'b_{k-2}'\cdots b_1'b_0'.\,b_{-1}b_{-2}\cdots)2\pi}\\&=x(\beta),\end{aligned}$$

其中 $\beta=\beta(n)$ 由下面二进式定出

$$\beta=.\,b_0'b_1'b_2'\cdots b_m'0\cdots$$

而 $x(\beta)$ 是 $\bar{J}_0\cap\{(x,y):0\leqslant x\leqslant1,y=\sin^2\pi\theta\}$ 的极限点，所以 (x_n,y_n) 也是，以上说明 $\bar{J}_0\cap\{(x,y):0\leqslant x\leqslant1,y=\sin^2\pi\theta\}$ 的每一点都是其自身的极限点，即不稳定流形的闭包的每一截面是一不可数完全集(perfect set). 当 $0<b<\frac{1}{2}$ 时不难证明这个截面的测度是 0，所以它是 Cantor 型的. 注意 $\bar{J}_0=\bar{J}_1$. 图 6 是由计算机绘出的奇异吸引子，$b=\frac{1}{3}$.

Lauwerier 映射吸引子的维数可以由 Lyapunov 数确定. Lyapunov 数分别是 $\lambda_1=2$，$\lambda_2=\frac{b}{2}$，对应的 Lyapunov 指数

$$\sigma_1=\log2,\sigma_2=\log\left(\frac{b}{2}\right).$$

因而吸引子的 Lyapunov 维数[50]

$$d_L=1+\frac{\log2}{\log\frac{2}{b}}.$$

根据 Kaplan-Yorke[52-52′]，有下面的谨慎猜测：

$$d_H=d_L.$$

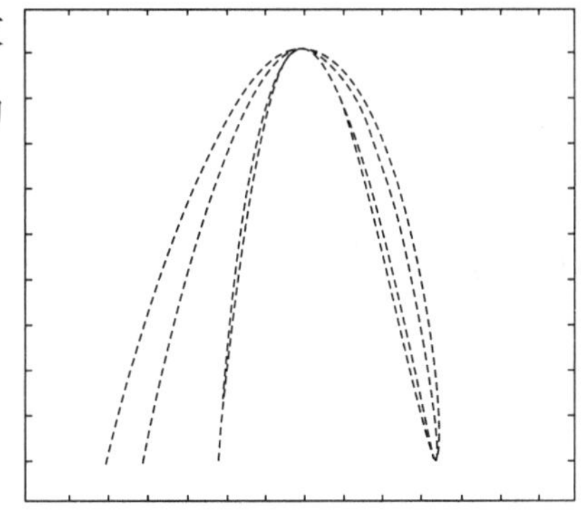

图 7　原点的不稳定流形上的一个轨道，坐标尺度如图 6

式中 d_H 为吸引子的 Hausdorff 维数. 从 Hausdorff 维数和集合的不规则性的关系知道 Lauwerier 映射的奇异吸引集是一个 fractal 集.

值得指出，前面的研究方法适用于广泛的一类映射，例如

$$\begin{cases}x_{n+1}=bx_n+(1-b)y_n,\\y_{n+1}=f(y_n).\end{cases}$$

其中 $f(0)=0$，$0<b<1$. 鞍点 $(0,0)$ 的不稳定流形是

$$\begin{cases}x=\left(\frac{1}{b}-1\right)\sum\limits_{k=1}^{\infty}b^k\phi\left(\frac{t}{a^k}\right),\\y=\phi(t).\end{cases}$$

$\phi(t)$是满足一定条件的周期函数.

对许多映射的奇异吸引子的研究表明奇异吸引子一般是具有分数 Hausdorff 维的 fractal 集. 吸引域的边界也显示出 fractal 结构,此处介绍另一个二维映射.

2. Grebogi-Ott-Yorke 映射[53,54]

$$\begin{cases}\theta_{n+1}=2\theta_n,\text{mod }2\pi,\\ z_{n+1}=az_n+z_n^2+\beta\cos\theta_\pi.\end{cases}$$

这一映射有两个不动点$(0,z_b)$,$(0,z_c)$,它们分别在吸引域边界和吸引子上,其中 $z_b=\frac{1}{2}(1-a)-\frac{1}{2}\sqrt{(1-a)^2-4\beta}$, $z_c=\frac{1}{2}(1-a)+\frac{1}{2}\sqrt{(1-a)^2-4\beta}$. 图 7 是 $\alpha=0.5$, $\beta=0.04$ 时的吸引子和吸附界,即吸引域的边界. 因为对充分大的 $|z|$, $z_{n+1}>z_n^2$,所以 z 的轨道趋近于 $z=+\infty$,说明 $z=+\infty$是 Grebogi-Ott-Yorke 映射的一个吸引集. 又因带形区域$|z|\leqslant z_c=0.1$上此映射映到自身,而 θ 相当于符号动力系统的转移,它是 Li-Yorke 意义下浑沌的[45,46],所以这一区域中的吸引集是浑沌的. 对 z_n 加入估计可知吸附界的上半部分位于带形$\frac{1}{2}\leqslant z\leqslant z_c$,映射限制于此带形区域时在 z 方向是扩张的,即$\frac{\partial}{\partial z_n}(\alpha z_n+z_n^2+\beta\cos\theta_n)>1$,所以每一竖线段映为另一更长的线段且穿过带形区域,因而对每一 θ,有唯一的值 $z(\theta)$使得点$(\theta,z(\theta))$在映射下恒留于带形区域,所得曲线$\{(\theta,z(\theta)),\theta=\theta \text{ mod } 2\pi\}$是 Grebogi-Ott-Yorke 映射的不变集,它是 $z=\infty$吸引点和有界吸引集的吸附域(basins of attractors)的分水岭,即吸附界. 下半平面的吸附界是前一吸附界曲线的逆象之一,另一逆象是其自身. 前面的讨论说明每一吸附界同胚于圆周. 以下将证明它们是 Weierstrass 型连续但处处不可微函数的图象,因而是 fractal 集.

由 $z(\theta)$的定义知道 $z(\theta)$是周期 2π 单值函数,

$$z(2\theta)=\alpha z(\theta)+z(\theta)^2+\beta\cos\theta,$$

代入 $\theta=0$,得

$$z_\pm(0)=\frac{1}{2}(1-\alpha)\pm\frac{1}{2}\sqrt{(1-\alpha)^2-4\beta}.$$

$(0,z_+(0))$位于上半平面的吸附界,$(0,z_-(0))$是上半平面的另一不动点. 现在求$(0,z_+(0))$的两个逆象$(\pi,z_+(\pi))$,$(\pi,z_-(\pi))$,由

$$z_+(0)=\alpha z_\pm(\pi)+z_\pm(\pi)^2-\beta,$$

得 $z_+(\pi)$,$z_-(\pi)$的值,

$$z_\pm(\pi)=\frac{1}{2}(-\alpha\pm\sqrt{\alpha^2+4[\beta+z_+(0)]}.$$

显然 $z_+(\pi)>0$, $z_-(\pi)<0$,即$(\pi,z_+(\pi))$和$(\pi,z_-(\pi))$分别位于上半平面和下半平面,它们都在各半平面的吸附界上.

下一步通过$(\pi,z_+(\pi))$的两个逆象求得吸附界的另外四点$\left(\frac{\pi}{2},z_+\left(\frac{\pi}{2}\right)\right)$,

$\left(\frac{3\pi}{2},z_+\left(\frac{3\pi}{2}\right)\right),\left(\frac{\pi}{2},z_-\left(\frac{\pi}{2}\right)\right),\left(\frac{3\pi}{2},z_-\left(\frac{3\pi}{2}\right)\right)$. 此过程可以写为：

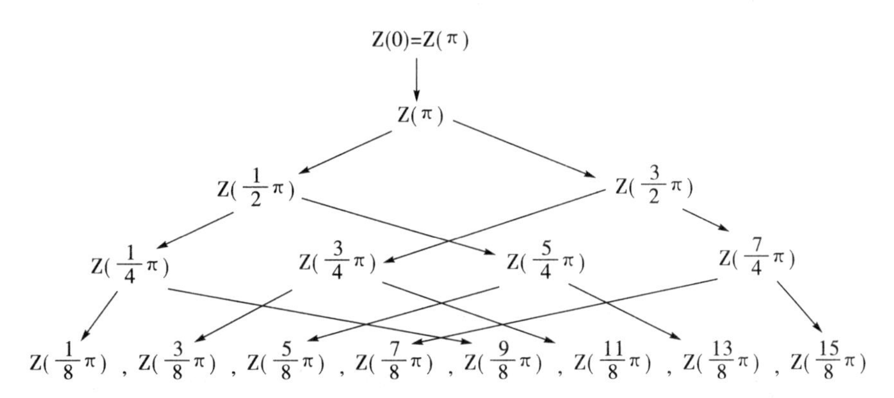

这样对一切有理数 r，$0\leqslant r\leqslant 1$，可以求出 $z_\pm(r\cdot 2\pi)$ 的值，将其连续化到区间 $[0,2\pi]$，即得到两个吸附界曲线 $z_+(\theta)$，$z_-(\theta)$. 计算 12 步后得到的吸附界曲线如图 8.

现在利用扰动方法讨论 $|\beta|\ll 1$ 时吸附界曲线的微分性质. 设 $z(\theta)$ 写为 β 的级数

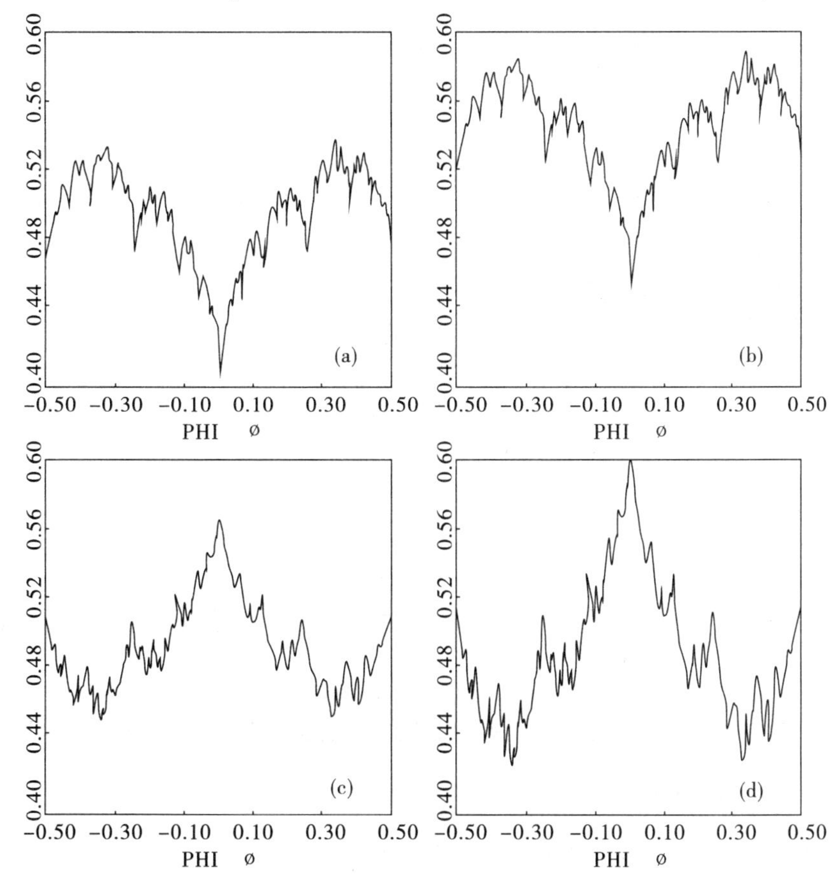

图 8　分形边缘 $\Lambda(Z>0)_t$ 参数为：$\alpha=0.5$，(a)$\beta=0.04$，(b)0.06，(c)-0.04，(d)，-0.06，纵轴为 $\phi=\theta/2\pi$

$$z(\theta)=z_0(\theta)+\beta z_1(\theta)+\beta^2 z_2(\theta)+\cdots\cdots,$$

则将 $z(2\theta)$ 和 $z(\theta)$ 代入关系

$$z(2\theta)=\alpha z(\theta)+z(\theta)^2+\beta\cos\theta,$$

比较两边 $1,\beta,\beta^2,\cdots,\beta^n,\cdots$ 的系数,有关系式

$$z_0(\theta)=1-\alpha,$$
$$z_1(2\theta)=\lambda z_1(\theta)+\cos\theta,$$
$$z_n(2\theta)=\lambda z_n(\theta)+\sum_{\substack{i+j=n\\ i,j\geqslant 1}} z_i(\theta)z_j(\theta),n=2,3,\cdots,$$

式中 $\lambda=2-\alpha$.

由 z_1 的函数方程求得

$$z_1(\theta)=-\sum\lambda^{-(k+1)}\cos 2^k\theta.$$

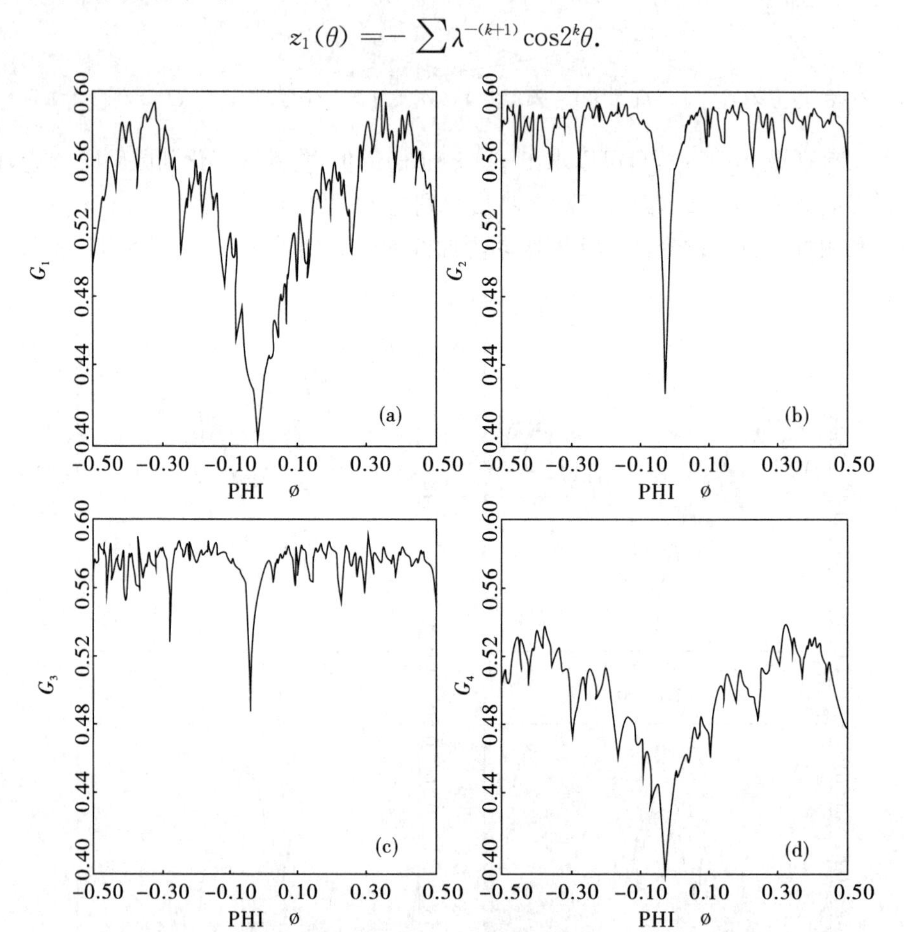

图 9 Weierstrass 函数

当 $1<\lambda<2$ 时这一函数是 Weierstrass 函数,连续但无处可微,其图象的 Hausdorff 维数是 $d=2-\frac{\ln\lambda}{\ln 2}$. 当 $|\beta|$ 相当小时,$z(\theta)$ 的性质取决于 $z_1(\theta)$.

设 $z_1(\theta),\cdots,z_{n-1}(\theta)$ 均已表示为 Fourier 级数,则

$$\sum_{\substack{i+j=n\\ i,j\geqslant 1}} z_i(\theta)z_j(\theta)=\sum_{k=0}^{\infty}\alpha_k^{(n)}\cos k\theta$$

常数 $a_k^{(n)}$ 与 $a_k^{(n-1)},a_k^{(n-2)},\cdots,k=0,1,\cdots,\infty$ 有关. 可设对于固定的 $n,a_k^{(n)}\to 0$ 当 $k\to\infty$. 于是

可以从函数方程中求得如下的 $z_n(\theta)$ 的 Fourier 级数

$$z_n(\theta)=\frac{a_0^{(n)}}{1-\lambda}-\frac{1}{\lambda}\sum_{p=1}^{\infty}a_{2p-1}^{(n)}\cos(2p-1)\theta$$

$$-\frac{1}{\lambda}\sum_{q=1}^{\infty}\sum_{p=1}^{\infty}\frac{1}{\lambda^p}\Big(\sum_{m=0}^{p}\lambda^m a_{(2q-1)2m}^{(n)}\Big)\cos(2q-1)2^p\theta,$$

它也是连续但无处可微的函数. 图 9 是按扰动方法得到的函数 $z_1(\theta)$, $z_2(\theta)$, $z_3(\theta)$ 以及它们的合成 $z(\theta)\approx z_0(\theta)+\beta z_1(\theta)+\beta^2 z_2(\theta)+\beta^3 z_3(\theta)$ 的图象.

现在计算 Grebogi-Ott-Yorke 映射的浑沌吸引集的吸附界曲线的 Lyapunov 指数和 fractal 维数. 对于初始值 (z_0,θ_0), 有

$$z_{n+1}=az_n+z_n^2+\beta\cos 2^n\theta,$$

于是沿 z 轴的 Lyapunov 指数,

$$\sigma_z=\lim_{N\to\infty}\frac{1}{N}\sum_{n=1}^{N}\ln|\alpha+2z_n|$$

可以写为积分形式, 利用 $z(\theta)$ 的解 $z(\theta)=\sum_{n=1}^{\infty}\beta^n z_n(\theta)$,

$$\sigma_z=\langle\ln(\alpha+2z)\rangle$$

$$=\ln\lambda+\Big\langle\ln\Big[1+\frac{2}{\lambda}(\beta z_1+\beta^2 z_2+\beta^3 z_3+\cdots)\Big]\Big\rangle,$$

式中括号

$$\langle\cdot\rangle=\frac{1}{2\pi}\int_0^{2\pi}(\cdot)d\theta,$$

将上述积分展开到 β^3 阶有

$$\sigma_z\approx\ln\lambda+\frac{2\beta}{\lambda}\langle z_1\rangle+\beta^2\Big(\frac{2}{\lambda}\langle z_2\rangle-\frac{2}{\lambda^2}\langle z_1^2\rangle\Big)+\beta^3\Big(\frac{2}{\lambda}\langle z_3\rangle-\Big(\frac{2}{\lambda}\Big)^2\langle zz_2\rangle+\frac{8}{3\lambda^3}\langle z_1^3\rangle\Big).$$

计算 $\langle\cdot\rangle$ 项,

$$\langle z_1\rangle=0,$$

$$\langle z_1^2\rangle=\frac{1}{2(\lambda^2-1)},\ \langle z_1^3\rangle=-\frac{\lambda^3+2}{4\lambda(\lambda^3-1)},$$

$$\langle z_2\rangle=\frac{1}{1-\lambda}\langle z_1^2\rangle=\frac{-1}{2(\lambda-1)(\lambda^2-1)},$$

$$\langle zz_2\rangle=\frac{\lambda^3+2}{4(\lambda^2-1)(\lambda^3-1)},$$

$$\langle z_3\rangle=\frac{2}{1-\lambda}\langle zz_2\rangle=\frac{-(\lambda^3+2)}{2(\lambda-1)(\lambda^2-1)(\lambda^3-1)}.$$

由以上结果

$$\sigma_z\approx\ln\lambda-L\beta^2-M\beta^3,$$

其中

$$L=\frac{2\lambda-1}{\lambda^2(\lambda-1)(\lambda^2-1)},\ M=\frac{(\lambda^3+2)(8\lambda^3-5\lambda^2-2\lambda+2)}{3\lambda^4(\lambda-1)(\lambda^2-1)(\lambda^3-1)}.$$

为求吸附界的 fractal 维数 d, 先计算它沿 z 方向的维数. 在 $\theta=\theta_0$ 处沿 z-方向的长为 L

的线段在逆映射之下缩短为$\frac{L}{\delta_0}$，其中$\delta_0=a+2z(\theta_0)$，作用n次后长度为$\frac{L}{\Delta(\theta_0)}$，$\Delta(\theta_0)=\delta_0\delta_1\cdots\delta_{n-1}$. 而$\theta$方向上长度缩减为原来的$\frac{1}{2^n}$. 若以直径为$\varepsilon_n=\frac{1}{2^n}$的方形覆盖吸附界，所需的数目最小约为

$$N(\varepsilon_n)=\frac{L}{\Delta(\theta_0)\varepsilon_n}.$$

所以 fractal 维数

$$d_z=\lim_{\varepsilon\to 0}\frac{\ln N(\varepsilon)}{\ln\frac{1}{\varepsilon}}=\lim_{n\to\infty}\frac{\ln\left(2^n\cdot\frac{L}{\Delta(\theta_0)}\right)}{\ln 2^n}=1-\lim_{n\to\infty}\frac{1}{n\ln 2}\sum_{i=0}^{n-1}\ln\delta_i$$

$$=1-\lim_{n\to\infty}\frac{1}{n\ln 2}\sum_{i=0}^{n-1}\ln|\alpha+2z_i|,$$

比较 Lyapunov 指数σ_z的定义，得

$$d_z=1-\frac{\sigma_z}{\ln 2}$$

因为曲线$z(\theta)$在任一点的坡度为90°，即其线性化向量平行于z轴，维数

$$d=1+d_z=2-\frac{\sigma_z}{\ln 2}.$$

可计算每一曲线$z_1(\theta)$，$z_2(\theta)$，…的 fractal 维数，它们都是

$$d=2-\frac{\ln\lambda}{\ln 2},$$

注意到$z(\theta)=\sum_{n=0}^{\infty}\beta^n z_n(\theta)$的维数却与此不同.

很多时候吸附界(basin boundary)是光滑曲线，但更多的情况下它是 fractal 集，例如含参数动力系统$f:C\to C, z\mapsto z^2+p$，唯在$p=0$时它的吸附界是单位圆周，p为非 0 的小数时吸附界是一 fractal 拟圆周(quasicircle，指起伏波动不大的闭曲线)，而p较大时则为完全不连通集，见[55]，[56]. 这些集合就是 Julia 集，我们将在后文深入讨论.

按 Mac Donald etc[57]的分类，fractal 吸附界“曲线”有如下三种：

(ⅰ)局部不连通(locally disconnected). Van der Por 系统在一定的参数范围内具有一个 Cantor 集为其吸附界(Levi[58]，Cartwright and Little wood[59])，Lorenz[60]系统在一定的参数之下也具有这种吸附界(Kaplan and Yorke[61]). 另外的例子，如下面写出的环域映射和一维 logistic 映射

$$\begin{cases}\theta_{n+1}=\theta_n+a\sin 2\theta_n-b\sin 4\theta_n-x_n\sin\theta_n,\\ x_{n+1}=-J_0\cos\theta_n,\end{cases}$$

$$x_{n+1}=\lambda-x_n^2,\quad -2\leqslant x\leqslant 2.$$

图 10 是前一映射的吸引域图，吸引集分别是不动点$(0,-J_0)$，(π,J_0).

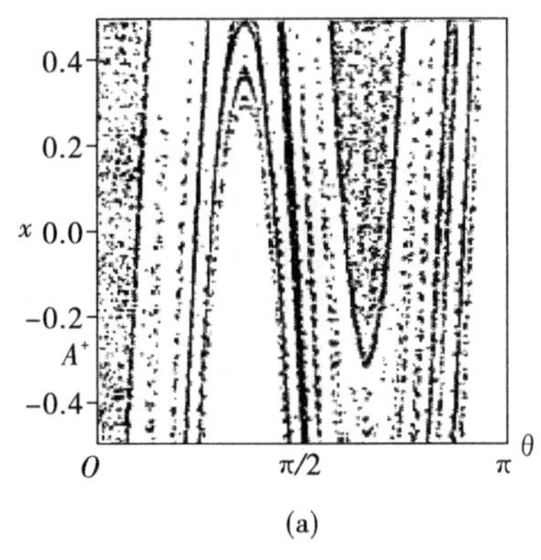

(a)

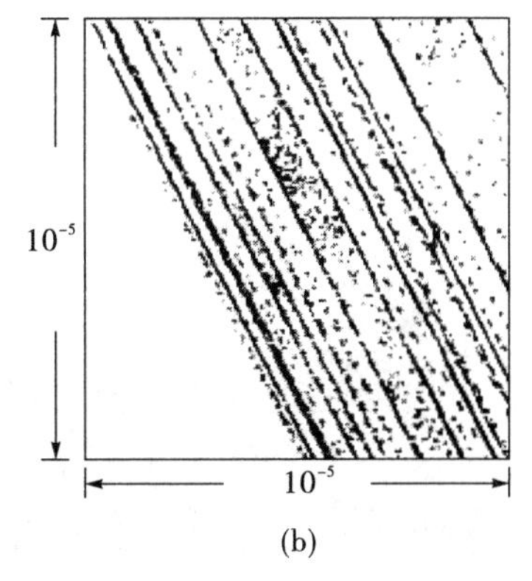

(b)

图 10 (a)两个吸引子 A^-(黑区)和 A^+(白区)的吸引区域；

(b) 图(a)的局部放大，范围：$1.92200\leqslant\theta\leqslant1.92201$，$-0.50000\leqslant x\leqslant-0.49999$，扩大 100000 倍.

(ⅱ)局部连通但不是拟圆周(quasicircle). 一局部连通曲线称为是拟圆周，如果存在常数$k>0$，满足对任意 $\varepsilon>0$，任意距离$<\varepsilon$ 的两点可用直径$<k\varepsilon$ 的连通段相连. 其直观意义是曲线的坡度为有限的，坡度是前一定义中的 k 的下界，显然它与坐标系无关. Grebogi-Ott-Yorke 映射的吸附界就是一个坡度$=+\infty$的曲线，尽管它同胚于圆周. 类似于此但较简单的映射[61,62]：

$$\begin{cases}\theta_{n+1}=\lambda\,\theta_n \bmod 2\pi,\\ x_{n+1}=\mu x_n+\cos\theta_n,\end{cases}$$

其中 λ 为整数. 当 $\lambda>\mu>1$ 时，映射的两个吸引集分别是 $x=-\infty$和 $x=+\infty$，吸引域的界线可以表示为

$$x=-\sum_{n=1}^{\infty}\mu^{-n}\cos(\lambda^{-n}\theta),$$

这是一个 Weierstrass 型曲线. 图 11 是这一曲线和它分成的吸附域.

(ⅲ)局部连通而且为拟圆周，具有这种吸附界的映射如

$$f:C\to C,z\mapsto z^2+p,$$

$$\begin{cases}x_{n+1}=x_n^2-y_n^2+\dfrac{1}{2}x_n-\dfrac{1}{2}y_n,\\ y_{n+1}=2x_ny_n+\dfrac{1}{2}x_n+\dfrac{1}{2}y_n,\end{cases}$$

它们的吸引集都是原点和无穷远处，其吸附界曲线如图 12.

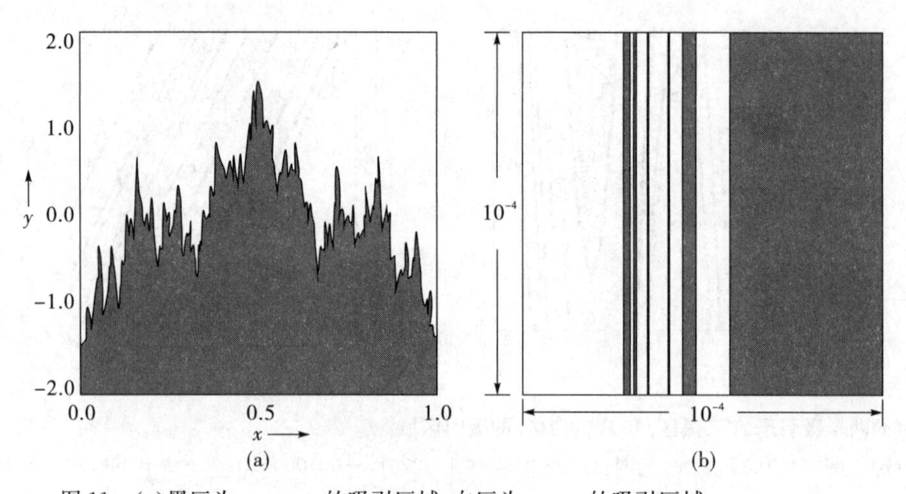

图 11 (a)黑区为 $y=-\infty$ 的吸引区域,白区为 $x=\infty$ 的吸引区域;

(b)图(a)的局部放大,范围 $0.38317\leqslant x\leqslant 0.38327$, $0.4450\leqslant y\leqslant 0.4451$.

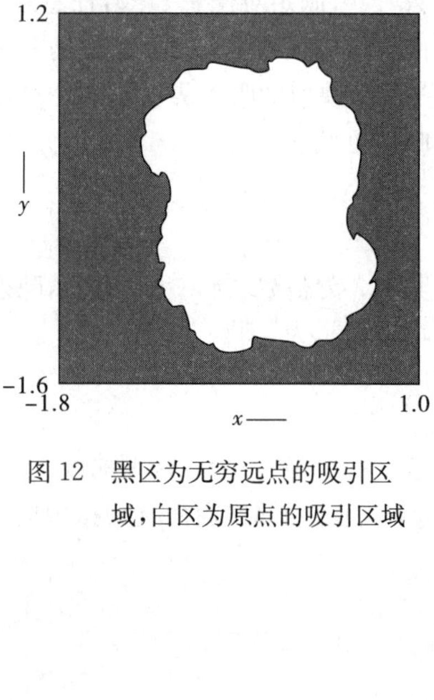

吸附界作为不同吸引子吸引区域的分水岭,它的分形性表明了动力系统长期行为对于初值的敏感依赖性,即终极状态关于一些初始点的不确定性. 点 x 称为 ε -初值不确定的,假如小球 $B(x,\varepsilon)$ 在迭代下不能保持处于某同一个吸引子的吸引域. 设很多点均匀地散布于相空间,则 $\forall\,\varepsilon>0$,可以统计这些点中属于 ε -不确定的部分,设它占总数的 $f(\varepsilon)$, $f(\varepsilon)$ 与 ε 有某种"比例"关系. 很多情况下有

$$f(\varepsilon)\sim\varepsilon^{a},a\text{ 为常数}.$$

这时此常数 a 称为系统的不确定度,它与吸附界的维数 d 有如下关系:

$$a=D-d,$$

此处 D 是相空间的维数,而 d 是容度维数,即

$$d=\lim_{\delta\to 0}\frac{\ln N(\delta)}{\ln\frac{1}{\delta}},$$

图 12 黑区为无穷远点的吸引区域,白区为原点的吸引区域

$N(\delta)$ 是用直径为 δ 的 D 维方体覆盖指定点集所需要的最少数目. a,d 的关系式可略加如下说明:对设定的初值误差 $\varepsilon>0$,可用 $N(\delta)\sim\varepsilon^{-d}$ 个直径为 ε 的 D 方体完全覆盖吸附界,这个覆盖的总体积约为 $\left(\frac{1}{\sqrt{D}}\varepsilon\right)^{D}\cdot N(\varepsilon)\sim\varepsilon^{D-d}$,注意到这一覆盖也包含了所有 $\frac{1}{2}\varepsilon$ 一初始值不确定的点,所以

$$f\left(\frac{1}{2}\varepsilon\right)\sim\varepsilon^{D-d}.$$

同样,Lyapunov 指数以及由它得到的 Lyapunov 维数也是刻划吸附界的 fractal 性的量度. 这些量将在后面专门介绍.

三、解析动力系统中的 fractals

动力系统中的 fractal 集,有一大类来源于复平面上解析映射的迭代.

$$T: z \mapsto f(z),$$

它所显示出的性质惊人地复杂而且漂亮. P. Fatou 和 Gaston Julia 在第一次世界大战期间开创了这一方向的研究,针对于 $F(z)=p(z)/q(z)$ 即有理映射的情形. 他们的工作[63~67]成为此领域的经典内容. 但后来将近半个世纪这方面的研究似乎间断了. 由于个人计算机的使用重新唤起对它的兴趣,现在对于有理函数的迭代的研究十分活跃,收获也极为丰富. 如 Sullivan[68,69],Douady,Hubbard[70],Mandelbrot[12,71],Herman[72] 等等. P. Blanchard[73] 系统地总结了直到近几年来这一问题的进展. 我们将在这一节的前一部分介绍由复平面上解析映射迭代所产生的 fractal 集,包括指数解析映射 $z \mapsto e^z$ 的动力系统. 后一部分的内容是与 fractal 集有关的函数系迭代方法.

1. Julia 集 as a fractal

设 $F: C \to C$ 是解析函数,则根据 F 的迭代可把复平面 C 分成两部分:Fatou 集和 Julia 集. 其定义是

$$\mathscr{F}(F)=\{z, z\in C, \text{满足对某一开邻域} U, z\in U, \text{使} F^n|_U (n=1,2,\cdots)$$
$$\text{是正规族或等价地说} F^n|_{\overline{U}} (n=1,2,\cdots) \text{是同等连续的}\}$$
$$\mathscr{J}(F)=C \backslash \mathscr{F}(F)$$

按照定义,Fatou 集 $\mathscr{F}(F)$ 是开集而 Julia 集 $\mathscr{J}(F)$ 是闭集. 这里将无穷远点∞也包含在 C 中,所以迭代实际上是在 Riemann 球面上进行的. 为了书写便利,我们还将用 $J(F)$ 表示映射的 Julia 集. 下面叙述的 Julia 集的一个等价说法对于确定给定映射 F 的 Julia 集,特别是用计算机方法时,是非常有用的.

定理 $J(F)$ 等于所有排斥性周期点之集合的闭包.

F 的一个 n 周期点 z_0 按照 $(F^n)'(z_0)$ 的绝对值大小分为吸性、中性、斥性的. 具体地,设 $\lambda=(F^n)'(z_0)$.

(ⅰ)若 $0\leqslant|\lambda|<1$,则 z_0 称为 F 的吸性周期点或稳定周期点(attracting,or stable),特别当 $\lambda=0$ 时称为超吸性或超稳定的(super—).

(ⅱ)若 $|\lambda|=1$,则 z_0 称为 F 的中性周期点(neutral).

(ⅲ)若 $|\lambda|>1$,则 z_0 称为 F 的斥性或不稳定周期点(repelling,or unstable).

有理函数 R 有如下的性质:

(1) $J(R)$ 是 R 的非空的完全不变集,即 $R^{-1}(J)=J=R(J)$.

(2) $\mathrm{Int}\, J(R)=\varnothing$,除非 $J(R)=C$.

(3) $J(R)$ 是完全集(perfect).

(4) $\forall z \in J(R), \overline{\bigcup_{n\geqslant 0} R^{-n}(z)} = J(R)$.

这里我们将展开对 Julia 集几何结构的讨论. 有理映射 R 的 Julia 集除有可能 $J(R)=C$

以外，可以分成三类，它们是光滑的或 fractal Jordan 曲线，或完全不连通集. 在最后一种情况下，结合前面列举的性质(3)，知 $J(R)$是一个 Cantor 集，这时动力系统 $R|_J$ 相当于符号动力系统(d 符号序列的单边转移). 起决定作用的是临界点的轨道和无穷远点的稳定集. 在引出有关定义之前我们先列出几个著名的映射以及其 Julia 集.

例 1　(Lattès'example)映射 R：

$$R(z)=\frac{(z^2+1)^2}{4z(z^2-1)}$$

的 Julia 集是整个复平面[74]. 证明亦可见[73]. 这一映射有一个在 C 中稠密的轨道. Fatou[64~66]得到下面的一个等价条件.

定理　$J(F)=C$ 当且仅当有理映射 F 有一个在 C 中稠密的轨道.

Herman[72]利用 Sullivan 的非游荡域定理[68]给出了构造具有稠密轨道之映射的一般方法.

例 2　(Douady's Rabbit)映射 R_p：

$$R_p: z \mapsto z^2+p, p\in C.$$

随着 p 的变化，这一映射的 Julia 集 $J(R_p)$变幻万端，呈现出美妙的图象. 当 $p=0$ 时 $J(R_p)$就是单位圆周，而 p 在 $p=0$ 附近时，$J(R_p)$是 Hausdorff 维数大于 1 的 fractal 拟圆周，在 p 充分大时 $J(R_p)$是 Cantor 型 fractal 集，见[55,56]. Douady 兔子是 p 满足 $p^3+2p^2+p+1=0$，$\mathrm{Im}(p)>0$ 时的 Julia 集 $J(R_p)$，如图 13，这一 Julia 集是连通的，而 Fatou 集则由无限多个简单连通区域组成. p 的渐近值是 $p\approx-0.12256117+0.74486177i$，图 14 是取 $p=-0.1226+0.7449\,i$ 时由计算机绘出的，参见[75,70,47]. 另一个著名的情形是圣马克吸引子[76,71]，在 R_p 中取 $p=-\frac{3}{4}$所得. 图 15 也是由计算机绘出的，$J(R_p)$是 fractal Jordan 曲线，其 Hausdorff 维数大于 1. 图 16 是上述映射在 $p=-0.3125$ 时 Julia 集 $J(R_p)$的图形，这也是一个 fractal 拟圆周，见[77]，根据 Ahlfors[78]的定义，这里拟圆周指单位圆周的拟共形同胚象.

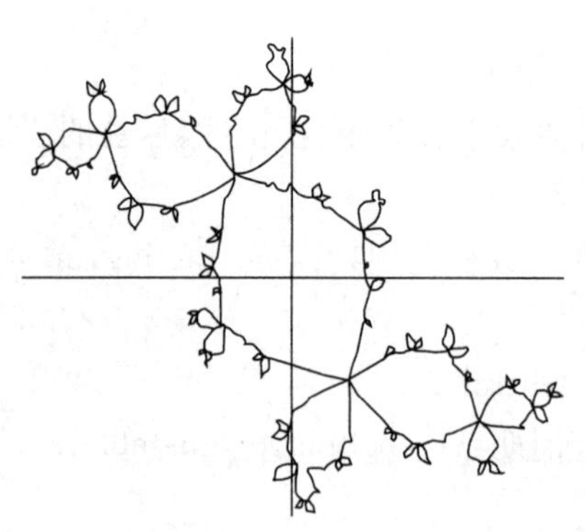

图 13　Douady 兔子，$z\mapsto z^2+c$ 的 Julia 集，这里 $c^3+2c^2+c+1=0$，$\mathrm{Im}(c)>0$.

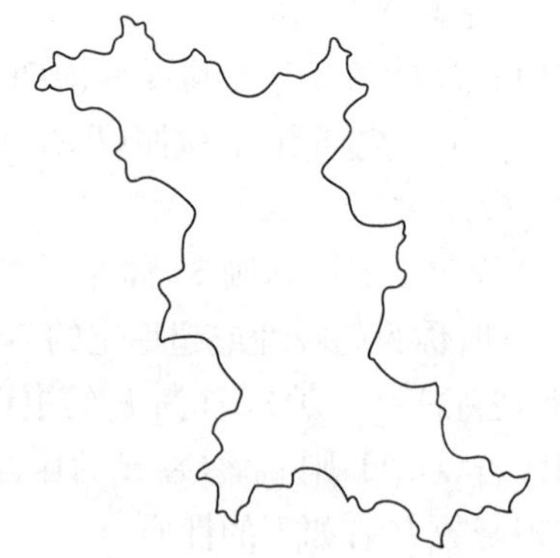

图 14　圆周在共形映射多次迭代所得的 Douady 兔子的近似

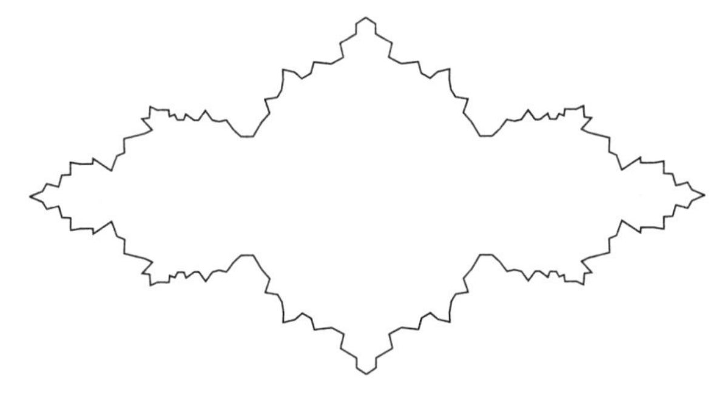

图 15　San Marco 吸引子

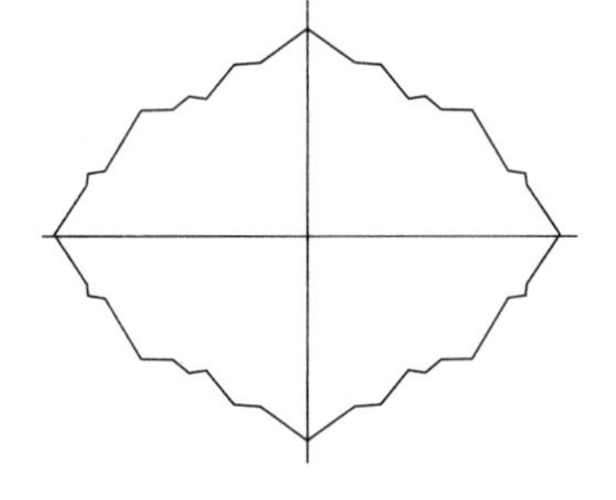

图 16　多项式的 $P(z)=z^2-0.3125$ 的 Julia 集

另外的一些情形是 $J(R_p)$ 为完全不连通因而是 Cantor 型 fractal 集. 图 17(a)是 $R:z\mapsto z^2+0.3$ 时的图形,图 17(b)(c)是前一图形的局部的放大. 这预示着 Julia 集可能是类似于自相似集的 fractal.

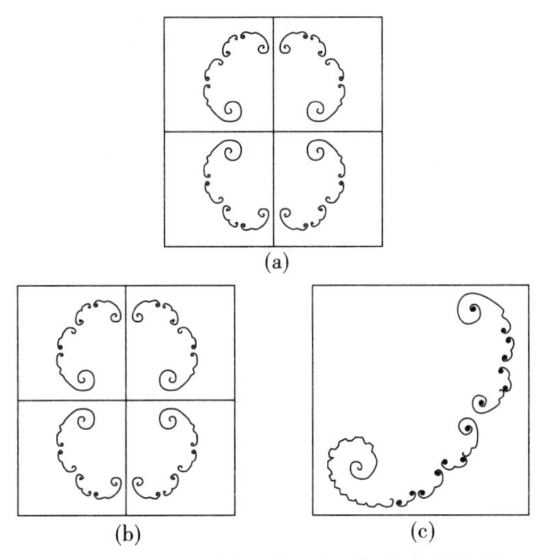

图 17　$p(z)=z^2+3$ 的 Julia 集的分形特征

例 3　Julia 集的另外一些例子. Julia 集与多项式零点的迭代算法有密切关系. 一个纯迭代算法指的是下面的一个有理映射

$$T:P\times C\rightarrow C,$$

其中 P 是 C 上的所有多项式的集合. 例如 Newton 算法

$$T_f(z)=z-\frac{f(z)}{f'(z)}.$$

一个值得研究的问题是对于某个给定的 f,C 上的哪些点 z 在 T_f 的迭代下收敛到 f 的一个零点 z_0 即

$$z_k=T_f^k(z)\rightarrow z_0\quad 当\ k\rightarrow\infty,$$

这些点实际上构成有理函数 $T_f:C\rightarrow C$ 的 Fatou 集,而 Julia 集就是在 T_f 迭代下不收敛到 f 任一零点的点的集合. 显然,我们希望发现这样的一个算法 T,它使任意的 $f\in P$,或者在弱一点的范围 $f\in P_d$(所有次数$\leqslant d$ 的多项式的集合),Julia 集 $J(T_f)$是零测度的. 曾猜想[79],不存在在整个 C 上一般收敛的纯迭代算法. Curt McMullen 最近证明了这一猜测的正确性

(见[80,81]):

定理　设$d>3$,$T:P_d\times C\to C$是关于f的系数和z的有理运算定义的算法,那么不存在全测度的开集$U\subset P_d\times C$使得$\forall(f,z)\in U$,当$k\to\infty$时$T_f^k(z)=z_k$收敛到f的一个根.

在$d=2$时Newton方法是一般收敛的,而$d=3$时McMullen找到了一般收敛的纯迭代算法.图18是Newton算法用于三次多项式$z^3-1=0$时产生的有理函数$T(z)=\frac{2z^3+1}{3z^2}$的Julia集,这一集合具有二维零测度.即几乎任何初始值的轨道收敛于方程的根.有关Newton迭代法的动力系,可参见Cosnard and Masse[101],Hurley and others[102~104],Saari and Urenko[105].

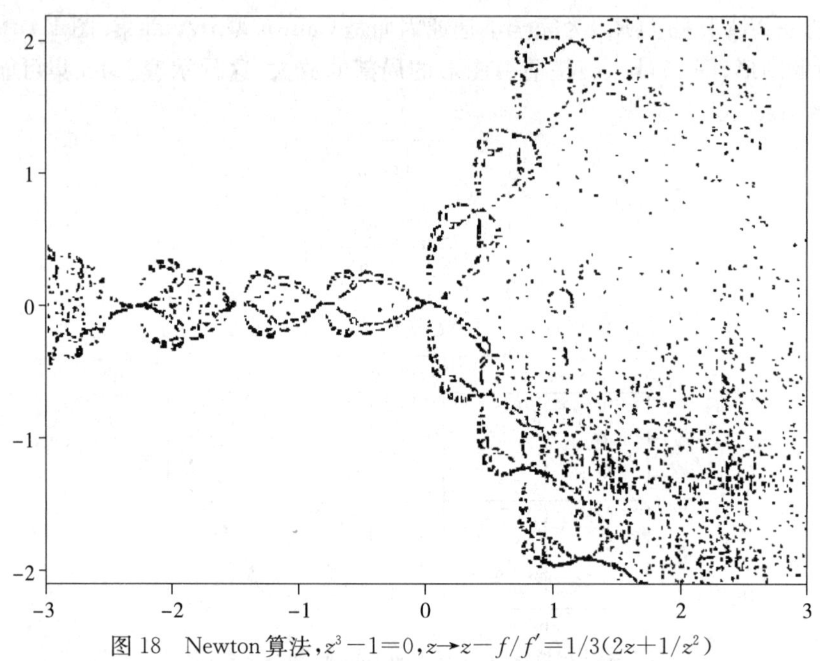

图18　Newton算法,$z^3-1=0$,$z\to z-f/f'=1/3(2z+1/z^2)$

定义　点$c\in C$称为是有理映射$R:C\to C$的临界点,如果$R'(c)=0$.R的所有临界点的集合记为$K(R)$,或简记为K.K'表示所有有限临界点的集合.

记号$A(z)$表示含有z的使R^n,$n=1,2,\cdots$限制于其上为正规族的最大域,称为z的直接稳定集.$O^+(z)$的正向轨道$\{R^z(z),n=1,2,\cdots\}$.$\forall z\in C$,$z\notin A(\infty)$相当于$O^+(z)$包含于一个有界集中.

定理

1.$K'(p)\cap A(\infty)=\varnothing$当且仅当Julia集$J(p)$是连通集,对于复平面上的多项式$p$.

2.$K'(p)\subset A(\infty)$,则Julia集$J(p)$是完全不连通的,而且$p|J$同构于d符号单边转移,d是多项式p的度数(degree).

由上一定理易知任一二次多项式的Julia集或者是连通集或者是完全不连通集.利用拓扑共轭可将二次映射变为标准形式$p(z)=\lambda z+z^2$或者$p(z)=z^2-\mu$.下一定理说明Julia集是一fractal曲线,因为任何规则曲线或可求长曲线的Hausdorff维数$=1$,如[15].

定理　$1<\dim J(p_\lambda)<2$,对于$0<|\lambda|<1$,$p_\lambda=\lambda z+z^2$;进一步地有:$J(p_\lambda)$是一个拟

圆周.

有关证明可见[82,83].

无论是完全不连通的 Julia 集(这一情形下因而就是 Cantor 型 fractal 集,有时称为"Cantor 尘")还是拟圆周形状的连通的 Julia 集,都是某种意义上的自相似 fractal 集. 拟自相似概念是自相似的推广,它包括了自仿射. 其严格定义如下:

定义 距离空间(X,d)到自身的函数 f 称为是拟等距(quasi-isometry),假如存在正常数 k_0,使得

$$\frac{1}{k_0}d(x,y)<d(f(x),f(y))<k_0 d(x,y)$$

对一切 $x,y\in X$ 成立. 集合 $S\subset C$ 称为是拟自相似的(quasi-selfsimilar),如果对任意的 $x\in S$ 和足够小的实数 r,集合$\varnothing_r(S\cap B(x,r))$可由一拟等距映为 S,此处$\varnothing_r$ 表示放大$\frac{1}{r}$倍. 因此复平面或者 Euclid 空间的一子集是拟自相似集的含义就是这一集合上的任何一小片在放大为一定尺寸后就与这集合的整体是拟等距的. Sullivan[77,84]的下述定理指出 Julia 集在一定条件下的自相似特征.

定理 如果 $J\cap\overline{O^+(K)}=\varnothing$,则 J 是拟自相似的. 这个结论对于一般的有理映射成立.

关于 Julia 集的 Hausdorff 维数,有 Ruelle[85]的结果.

定理 在任一只含有限制于其 Julia 集上为扩张的有理映射的邻域中,Julia 集的 Hausdorff 维数是映射诸系数的实解析函数.

R 在其 Julia 集 $J(R)$是扩张指$\forall z\in J(R)$,有 n 使$|(R^n)'(z)|>1$. 可以证明这时 n 可取为一个与 z 无关的常数.

作为这一段的结束,我们讨论下面映射的 Julia 集的 Hausdorff 维数的求法

$$f(z)=z^Q+p,Q\text{ 为整数},p\in C.$$

Lucy Garnett[86]用数值方法研究了 $Q=2,p\in R$ 且 $p\approx 0$ 的情形,他得到

$$\dim J(f_p)=1+a_2p^2+\cdots$$

Widom,Bensimon,Kadanoff and Shenker[55]用扰动方法研究了 $z\mapsto z^Q+p$ 当 p 较小和较大时的 Hausdorff 维数,并提出一种用于估计的数值方法.

$p=0$ 时,映射 $z\mapsto z^Q$ 的 Julia 集是单位圆周,设 p 充分小使 $z\mapsto z^Q+p$ 的 Julia 集与圆周同胚,则可以将 Julia 集参数化为

$$z(t)=e^{2\pi it}(1+pU_1(t)+p^2U_2(t)+\cdots),U_m(t)=U_m(t+1),$$

因为 Julia 集在 f 的不变性,$z(t)$应满足

$$f(z(t))=z(Qt).$$

(参见[87]). 把 $z(t)$代入上一方程并比较展开后 p 的各阶系数,得下列函数方程

$$U_1(Qt)-QU_1(t)=e^{-2\pi iQt},$$

$$U_2(Qt)-QU_2(t)=\frac{1}{2}Q(Q-1)\cdot U_1^2(t),$$

$$\cdots$$

$$U_m(Qt)-QU_m(t)=\sum_{\substack{\sum_{j=1}^{Q} i_j=m\\ 0<i_j<m}} U_{i_1}(t)U_{i_2}(t)\cdots U_{i_a}(t),$$

$$\cdots$$

$$U_0(t)=1.$$

由以上方程可以依次求出函数 $U_1(t),U_2(t),\cdots,U_m(t),\cdots$. 如令 $\varphi(t)=\frac{-1}{Q}\sum_{l=0}^{\infty}Q^{-l}\cdot e^{-2\pi iQ^l t}$,则

$$U_1(t)=\varnothing(Qt),$$

$$U_2(t)=\frac{1}{2}(Q(Q-1)\sum_{l_1,l_2=1}^{\infty}Q^{-(l_1+l_2)}\varnothing((Q^{l_1}+Q^{l_2})t),$$

$$\cdots$$

利用 $z(t)$的表达式可计算 Julia 集的 Hausdorff 维数. 根据 Ruelle[88] 的一个定理;

定理 若 D_0 是 f 的 Julia 集的 Hausdorff 维数,则级数

$$\zeta(u)=\exp\sum_{n=1}^{\infty}A_n(D_0)\frac{u^n}{n}$$

收敛于一个亚纯函数,$u=1$ 是其简单极点,而且单位圆盘内 $\zeta(u)$没有其它极点和零点. 式中 $A_n(D)$的定义是

$$A_n(D)=\sum_{z\in Fixf^n}\left|\frac{df^n}{dz}\right|^{-D},$$

由 Ruelle 的定理立即得到 $D_0=\dim J(f)$的必要条件是

$$\lim A_n(D_0)=1,$$

这一公式用于计算 $\dim J(f)$是非常方便的,如果我们能够求出 $A_n(D)$的表达式. 这里,$\mathrm{Fix}f^n$ 可以根据关系 $f(z(t))=z(Qt)$写为

$$\mathrm{Fix}f^n=\left\{z(t_j):t_j=\frac{j}{Q^n-1},j=0,1,\cdots,Q^n-2\right\},$$

于是

$$\left.\frac{df^n}{dz}\right|_{z=z(tj)}=\prod_{m=0}^{n-1}f'(z(Q^m t_j))=Q^n\Big[\prod_{m=0}^{n-1}z(Q^m t_j)\Big]^{Q-1},$$

$$A_n(D)=Q^{-nD}(Q^n-1)\left\langle\prod_{m=0}^{n-1}|z(Q^m t_j)|^{-(Q-1)D}\right\rangle_n,$$

这里记号$\langle\cdot\rangle_n$ 表示

$$\langle a(t)\rangle_n=\frac{1}{Q^n-1}\sum_{j=0}^{Q^n-2}a(t_j).$$

把 $z(t)$ 的表达式代入前面的 $A_n(D)$ 可以得到 p 的幂级数：

$$A_n(D)=Q^{-nD}(Q^n-1)\left\{1+p\,\overline{p}\frac{D^2n}{4}+\delta_{Q,2}(p^2\,\overline{p}+\overline{p}^2p)\right.$$

$$\left.\times\left[\left(\frac{D^2}{4}\frac{n}{2^n-1}\right)+\left(\frac{D^3}{16}+\frac{D^2}{8}\right)n\right]+O(p^4n^2)\right\}$$

将此式代入 $\lim\limits_{z\to\infty}A_n(D_0)=1$ 得到

$$D_0=1+\frac{p\,\overline{p}}{4\ln Q}+\delta_{Q,2}\frac{3(p^2\,\overline{p}+\overline{p}^2p)}{16\ln Q}.$$

对于较大的 p，$z\mapsto z^Q+p$ 的 Julia 集 $J(f_Q)$ 集结于 f 的 Q 个不稳定不动点 $z\approx(-p)^{\frac{1}{Q}}$ 附近，可以得到估计

$$A_n(D)\sim Q^{n(1-D)}|p|^{-\frac{nD}{Q}},$$

所以有

$$D_0=\frac{Q\ln Q}{\ln|p|}.$$

2. 整函数的 Julia 集

复平面上整数迭代的研究是 Fatou[67] 1926 年开始的. Misiurewicz[89] 于 1980 年解决了 Fatou 在六十年前提出的一个问题，他证明指数映射 $z\mapsto e^z$ 的 Julia 集 $J(\exp)=C$，从而激起了研究者们在这方面的兴趣. 现在已经有很多重要的发现，特别是超越整函数的 Julia 集与多项式或有理映射的 Julia 集的不相同性质. 下面的定理(Devaney and Tangerman[90])是一个显著的代表.

定理 设 E 是临界有限的整超越函数，并且具有一个有一渐近方向和双曲性的指数片 D，则 Julia 集

$$J(E)=\text{closure}\{z,E^n(z)\to\infty\}.$$

在解释有关定义之前，我们先指出：指数映射 e^z，三角函数 $\sin z$，$\cos z$ 都属于上一定理所要求的范围；对于多项式 $p(z)$，迭代下趋于∞处的点一定不在 $J(p)$ 中.

一整超越函数 E 称为临界有限的(critically finite)，如果它只有有限多个临界值和渐近值. 设这些临界值和渐近值(有限的)为开圆盘 D 所包含，$\Gamma=C\setminus D$，[90]中证明：$E^{-1}(\Gamma)$ 的每一个分支 T 是闭包含有∞的圆盘，而且 $E:T\to\Gamma$ 是一个通用覆盖. 每个分支都被称为是指数片(exponential tract). 关于指数片的渐近方向和双曲性的完全定义，可见[90].

Devaney[91]研究了指数映射 $z\overset{E_\lambda}{\mapsto}\lambda e^z$ 的 Julia 集的几何结构，有

定理 1. 若$\lambda>\dfrac{1}{e}$，则 $J(E_\lambda)=C$；

2. 若 $0<\lambda<\dfrac{1}{e}$，则 $J(E_\lambda)$ 是曲线的 Cantor 集，位于竖直线 $x=p$ 的右半平面，p 为 E_λ 的较大不动点.

曲线的 Cantor 集形状如图 19. 它可以看成一 Cantor 集和闭半线$[0,\infty)$的积，称为 Cantor 束. 这是因为其中所有曲线沿着同一方向趋于∞. 下一定理说明整超越函数 Julia 集的 fractal 性质. 设 T 是 $E^{-1}(\Gamma)$ 的任一分支，即指数片，则 $J_T(E)=\{z\in J(E)\,|\,E^n(z)\in T$ for *all n*}是 $J(E)$的闭不变子集.

定理[90] 设 E 是临界有限的整超越函数,具有指数片 T 和 $E|_T$ 的渐近方向 θ^*,则 $J(E)$的子集 $J_T(E)$包含有一个 Cantor 束(bouquet).

关于 Julia 集的 Hausdorff 维数,可参看 McMullen 最近的工作[92];Hausdorff 测度,可见 Eremenko and Ljubic[93]. Julia 集上的动力行为和整函数迭代动力系统的稳定性,参见[94～97]. 最近的工作还可参看[98～100].

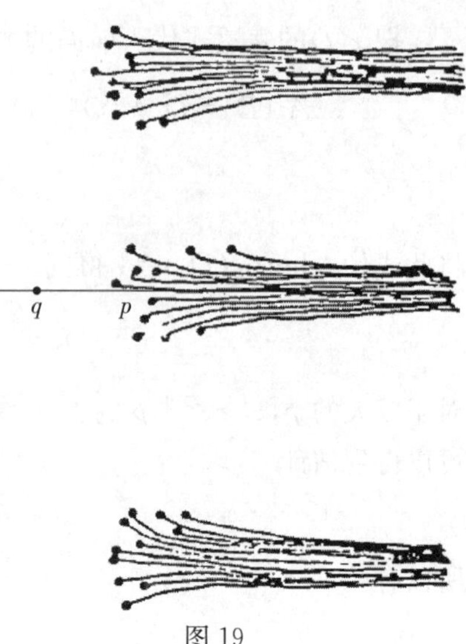

图 19

3. Mandebrot 集

二次多项式是最简单的有理映射,但它的迭代却已表现出极为丰富的性质. 我们已经知道,它的 Julia 集是一个 Cantor 尘(dust)或者连通集,例如 fractal 拟圆周,这取决于二次多项式的系数. 由于每个二次映射解析共轭于

$$f_c(z)=z^2+c,c\in C.$$

所以,计算临界 $z=0$ 的轨道 $O^+(0)=\{c,c^2+c,(c^2+c)^2+c,\cdots\}$是否收敛于$\infty$即可断言 $J(f_c)$是不连通还是连通. Mandelbrot 集定义为由复平面的使 $J(f_c)$为连通集的点 c 的集合. Mandelbrot 首先在计算机上研究了这一集合,它所显示的结构之复杂远非人之所料. 这是一个边界为 fractal 的自相似集. 图 20 是计算机绘出的 Mandelbrot 集.

Mandelbrot[14]使用人们较为熟悉的语言对 Mandelbrot 集作了形象的描述. 一个原子是复平面上的一个最大区域(开集),其中任何两个 c 值的 $J(f_c)$是同胚的,原子中有一些是心形而另一些是近于圆盘形的. Mandelbrot 集的原子有可数无限多个. 核是原子中的超稳定的 c 值. 一分子由可数无限多个原子结合而成,其中仅有一个是心形而其余为近于圆盘形,分子中的每两个原子都由一个只与有限多个其它原子相交的连通线相连. Mandelbrot 集中分子有无限多个,其中之一是大陆而其余皆为小岛. 任何两个分子可用 Mandelbrot 集中的一连通线路联结,这线路相交于无限多个其它的分子,沿着这线路不属于分子的内容的点形成一个 Cantor 集. 而任何两个分子之间都有一个光滑同胚.

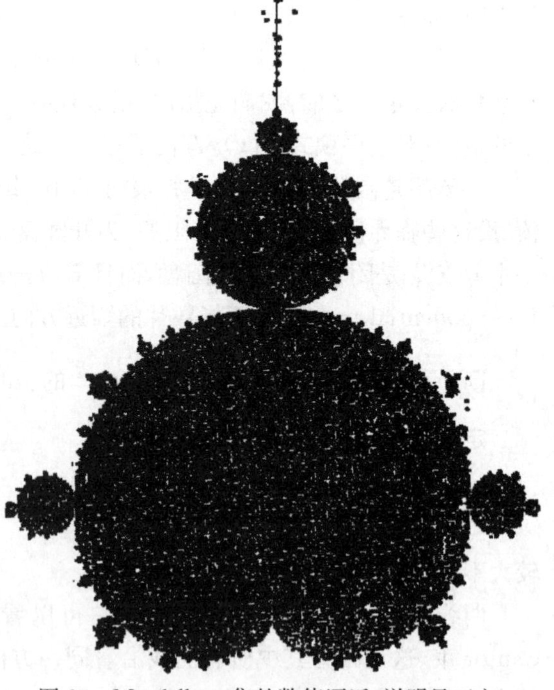
图 20 Mandelbrot 集的数值逼近,说明见正文

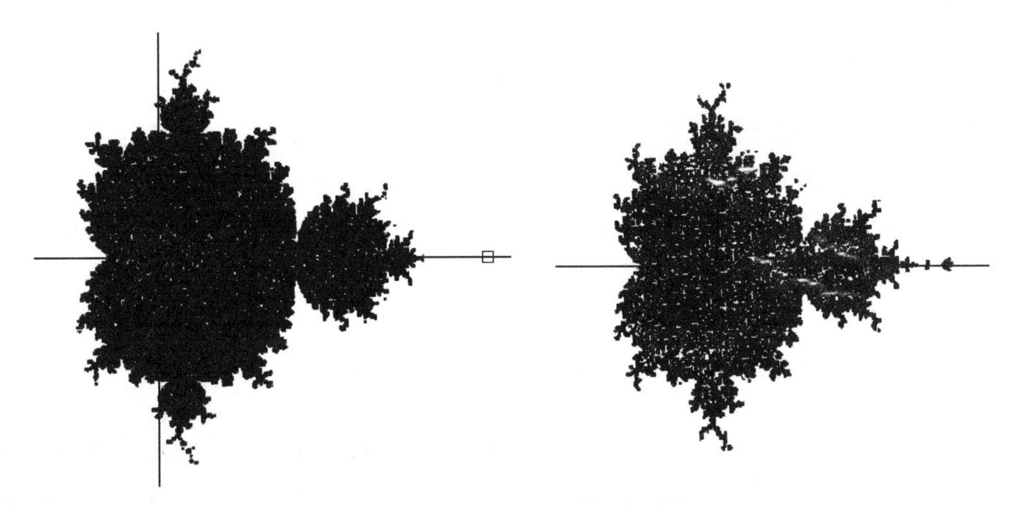

图 21　Mandelbrot 集及其局部放大

Mandelbrot 集的连通性与它的计算机图形的直观显示是相反的，这一性质是 Douady and Hubbard[70] 证明的.

定理　Mandelbrot 集是连通的.

现在尚不清楚 Mandelbrot 集的局部连通性.

Mandelbrot 集边界的维数，也是值得研究的问题之一.

Mandelbrot 集的概念还可推广到其它的含参函数族[132]或带有参数的函数系[108]的迭代上. 我们将在下一段介绍这样的一个例子. 图 21 是 $f_c(z)=z^2+c$ 的 Mandelbrot 集的"自相似性"(类似于 Šharkovskiĭ 序自相似性)的直观说明. 将前一图的小方形放大得到后一图形，这一过程可以无限制地重复下去.

4. 函数系的迭代

迭代函数系(iterated function systems)是 Barnsley and Demko 在[20]中作为产生和分类 fractal 集的统一方式而引入的. Julia 集和其它很多一些 fractal 集可以看成某些迭代函数系的吸引集.

设 $w_i:K\to K$ 是紧度量空间到自身的连续函数，$i=1,2,\cdots,d$，则可定义集值函数 w: $2^K\to 2^K$

$$w:2^K\to 2^K, w(X)=\bigcup_{i=1}^{d} w_i(X)$$

对 $\forall x\in K$，$w(x)$ 表示 $w\{x\}$. 称 $\{K,w_1,w_2,\cdots,w_d\}$ 为一个迭代函数系. x 在 $\{K,w_i\}$ 之下的吸引集定义为

$$A(x)=\lim_{n\to\infty} w^n(x).$$

其严格意义就是 $a\in A(x)$ 当且仅当对 a 的任意邻域 N，有无限多个 n 使 $N\cap w^n(x)\neq\varnothing$.

如果 $A(x)$ 与 x 的选择无关，就把 $A=A(x)$ 称为迭代函数系 $\{K,w_i\}$ 的吸引集，它是 K 的紧子集，且在 w 下不变，即 $w(A)=A$.

迭代函数系 $\{K,w\}$ $(w=\{w_1,w_2,\cdots,w_d\})$ 的逆定义为 K 上的集值映射

$$w^{-1}(X)=\{x\in K, w(x)\in X\}.$$

$\{K,w\}$的排斥集(repeller)指$\{K,w^{-1}\}$的吸引集(attractor).

若 $f:C\to C$ 是有理映射,可将它考虑为 Riemann 球面 $\overline{C}=C\cup\{\infty\}$上的函数,后一空间是紧的,则$\{\overline{C},f\}$的逆是迭代函数系$\{\overline{C};f_1^{-1},f_2^{-1},\cdots,f_d^{-1}\}$,其中 $f_i^{-1}(i=1,2,\cdots d, d=\text{degree}(f))$是 f 的逆的 d 个分支.因为 f 的吸性不动点至多有(2^d-2)个,所以,几乎对所有的 $z\in\overline{C}$, $A_{f^{-1}}(z)=A_{f^{-1}}$与 z 的选择无关,可证明

$$A_{f^{-1}}=J(f).$$

这说明映射 f 的 Julia 集就是迭代函数系$\{\overline{C},f\}$的排斥集,或者说$\{\overline{C};f_1^{-1},f_2^{-1},\cdots,f_d^{-1}\}$的吸引集.

迭代 Riemann 面(iterated Riemann Surface)是另一种重要的迭代函数系.设$R(w,z)$是关于复变元 $w,z\in C$ 的复系数多项式,w,z 的度数 degree $w=d$,degree $z=e\geqslant 1$,则由 $w_1(z)$,$i=1,2,\cdots,d$ 定义的迭代函数系$\{\overline{C},w_i\}$称为 $R(w,z)$生成的 iterated Riemann Surface.其中 $w_i(z)(i=1,2,\cdots,d)$是

$$R(w,z)=0, z\in\overline{C}$$

的 d 个解.考虑下面的两个例子:

例 1　设 $a,b,c\in C$ 是不共线三点,$R(w,z)$如下:

$$R(w,z)=(2w-z-a)(2w-z-b)(2w-z-c),$$

则$\forall x\in C, A(x)=A$ 是以 a,b,c 为顶点的 Sierpinski 三角形,如图 22.

例 2　设 $a,b,c\in C$ 是不共线三点,由下述 $R(w,z)$

$$R(w,z)=(w-3z+2a)(w-3z+2b)(2w-3z+c)$$

生成的迭代函数系的吸引集是$\{\infty\}$,而排斥集为一如图 23 的 Cantor 树(tree).

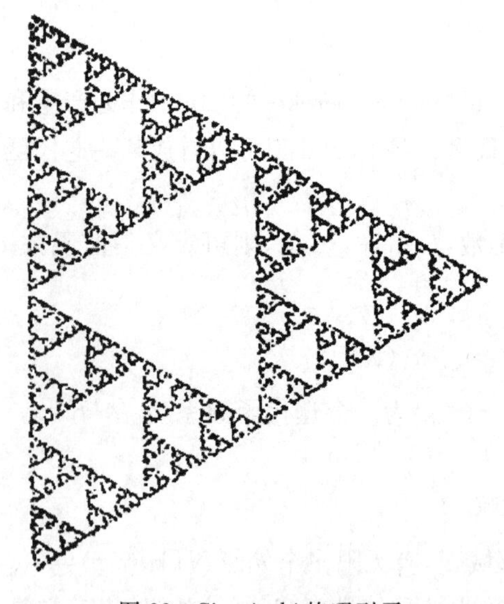

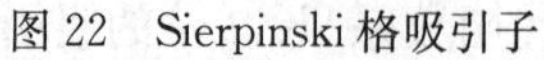

图 22　Sierpinski 格吸引子

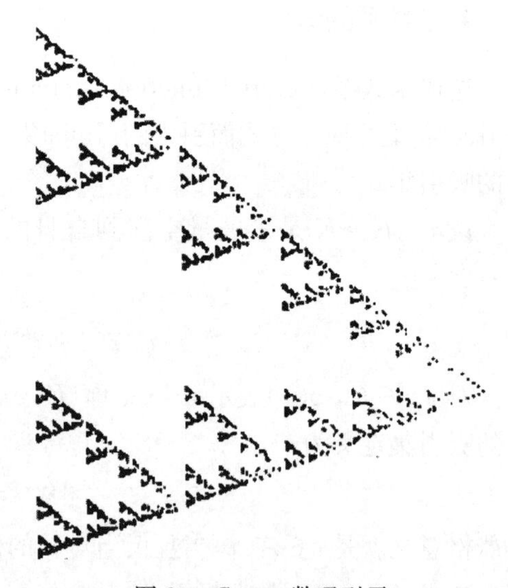

图 23　Cantor 数吸引子

对具有双曲性的迭代函数系,其吸收集的结构和 fractal 维数我们了解得较为深入.$\{K, w_i:i=1,2,3,\cdots,d\}$是双曲的是指存在常数 $0\leqslant s<1$,满足

$$|w_i(x)-w_i(y)|<s|x-y| \text{ for all } x,y\in K, i=1,2,,\cdots,d,$$

定理 设$\{K;w_1,w_2,\cdots,w_d\}$是双曲迭代函数系，则它有唯一的吸引集A，而且存在Cantor集到A上的一个连续映射.

定理 设$\{K;w_1,w_2,\cdots,w_d\}$是双曲迭代函数系，且$w_i(A)\cap w_j(A)=\varnothing, 1\leqslant i\neq j\leqslant d$，$A$是上一定理得到的吸引集，则$w^{-1}|A\approx\sigma|_{\sum d}$.

此处$\sum_d$是d个符号的单边无穷序列集，σ为转移映射，下面关于吸引集的Hausdorff维数的结论，在第一节中已介绍过类似公式.

定理 在上一定理的条件下，若K是R^n的子集，且常数$0<s_i\leqslant\bar{s}_i<I$使

$$s_i|x-y|\leqslant|w_i(x)-w_i(y)|\leqslant\bar{s}_i|x-y|$$
$$\text{for all } x,y\in A, i=1,2,\cdots,d.$$

则A的Hausdorff维数界于

$$\min\{n,l\}\leqslant\dim A\leqslant u,$$

其中l,u是下述方程的正数解

$$\sum_{i=1}^{d}s_i^l=1=\sum_{i=1}^{d}\bar{s}_i^n.$$

若对某些$i\neq j, w_i(A)\cap w_j(A)\neq\varnothing$，则上界的估计依然成立.

以上只是简略地介绍了迭代函数的理论，它的另一个应用是重建fractal，即用迭代函数系渐近由其它方法产生的fractal集，详见文献[20]，[106].

下面考虑几个具体的迭代函数系，它们的吸引集合相当于Julia集的推广.

Pythagoras树花（blossom of the Pythagoras tree）这个问题源于荷兰工程师A. Bosman，他在二次世界大战期间研究了下面的过程. 从以原点为中心边长为2的正方形的顶边出发，作一等腰直角三角形，然后以此三角形的直角边为底分别向外作小正方形；从目前的所有最小三角形的顶边开始重复上述步骤，这一程序的极限图形叫做Pythagoras树，如图24.

Pythagoras树的构造过程可以看成是复平面上的含有两个映射的迭代函数系$\{c;A,B\}$，其中

$$A: z\mapsto 1+2i-\frac{1+i}{2}z,$$

$$B: z\mapsto -1+2i-\frac{1-i}{2}z.$$

于是Pythagoras树$=\lim\limits_{n\to\infty}\bigcup\limits_{k=1}^{n}w^k(Q)$，这里$w=\{A,B\}$，而$Q$表示前面的正方形. 变换$A$相当于以$1+i$为中心(不动点)旋转$-\frac{3}{4}\pi$后再缩小成$\frac{1}{\sqrt{2}}$倍，$B$相当于以$-1+i$为中心旋转$+\frac{3}{4}\pi$后再缩小成$\frac{1}{\sqrt{2}}$倍.

H. A. Lauwerier[107]研究了这一迭代函数系的不变曲线(单位区间的某个连续映射像)，它位于Pythagoras树中. 被称为Pythagoras树花. 推广到一般的相似变换，通过适当的坐标变换，则A,B可写为标准形式

$$A: z \mapsto 1+az,$$
$$B: z \mapsto 1+bz,$$

其中 $a,b\in C$,不变曲线经由下面方式得到:

$\forall r\in[0,1]$,设 r 的二进展开式为 $0,r_1r_2\cdots r_n\cdots$,它所对应的点可由 $\cdots w_n\cdots w_2w_1(0)$ 确定,其中 w_i 取映射 A 或 B 视 $r_i=0$ 或 1 而定. 写得较为形式化就是定义

$$s_k=(1-r_k)a+r_kb, k=0,$$
$$z(r)=c_0+c_1+c_2+\cdots+c_k+\cdots,$$

而 $c_0=1$,且归纳地定义

$$c_k=s_kc_{k-1}, k=1,2,\cdots$$

易证当 $|a|<1$, $|b|<1$ 时 $z(r)$ 是收敛的. 实际上这时 $\{C;A,B\}$ 是一个双曲迭代系,从而由前面已经叙述过的定理,$\{z(r),r\in[0,1]\}$ 就是迭代系的吸引集 Lauwerier 将 Pythagoras 树花与 Julia 集进行了比较,分析了它们的相同之处,例如一定参数范围内极限集合的完全不连通性等. 图 25 是由计算机绘出的 Pythagoras 树花.

下面的例子着重考虑函数系迭代动力系统与参数的关系问题,即 Mandelbrot 集的推广. 复平面上两个伸缩比例相同的相似变换

$$T_+z=sz+1,$$
$$T_-z=sz-1,$$

其中 $s\in C$. 它们构成最简单的迭代函数系,其吸引集合可以写成

$$A(s)=\{\pm1\pm s\pm s^2\pm s^3\pm\cdots,\text{对一切由}+,-\text{构成的序列}\}.$$

依次可以很方便地用小计算机打印出 $A(s)$ 的图像*. 对于 s 不同的值,$A(s)$ 可能是连通的或不连通的. 例如当 $s\in\mathbf{R}$ 且 $0<s<\frac{1}{2}$,由迭代函数系的一般结论,$A(s)$ 是 Cantor 集. Barnsley and Harrington[108] 考虑参数 s 的集合

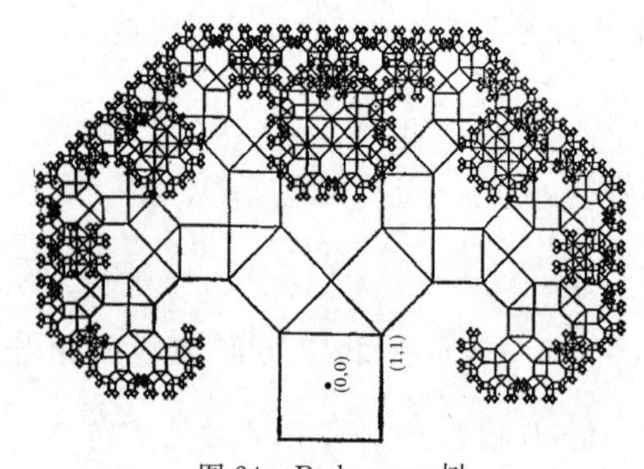

图 24 Pythagoras 树

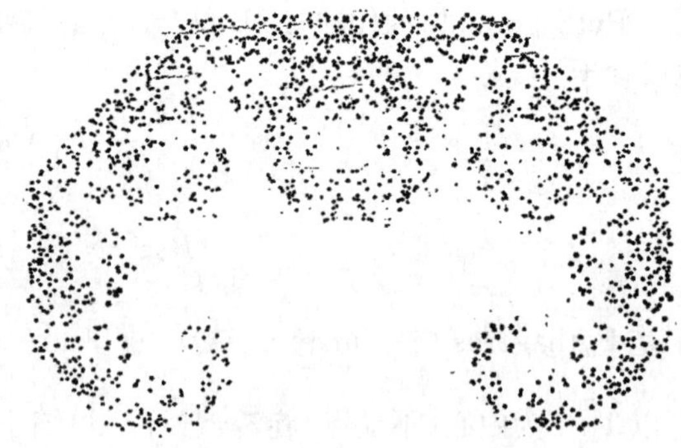
图 25 Pythagoras 数的极限

图 26 是这一点集的图象,图 27 是其一局部的放大,它显示出 ∂D 即边界是较为复杂的.

* 这种用+−号序列对应极限集的方法可应用于研究 $z\mapsto z^2-\mu$ 的 Julia 集 as 吸引集 of $\{\overline{C},\sqrt{2+\mu},-\sqrt{2+\mu}\}$,见[118].

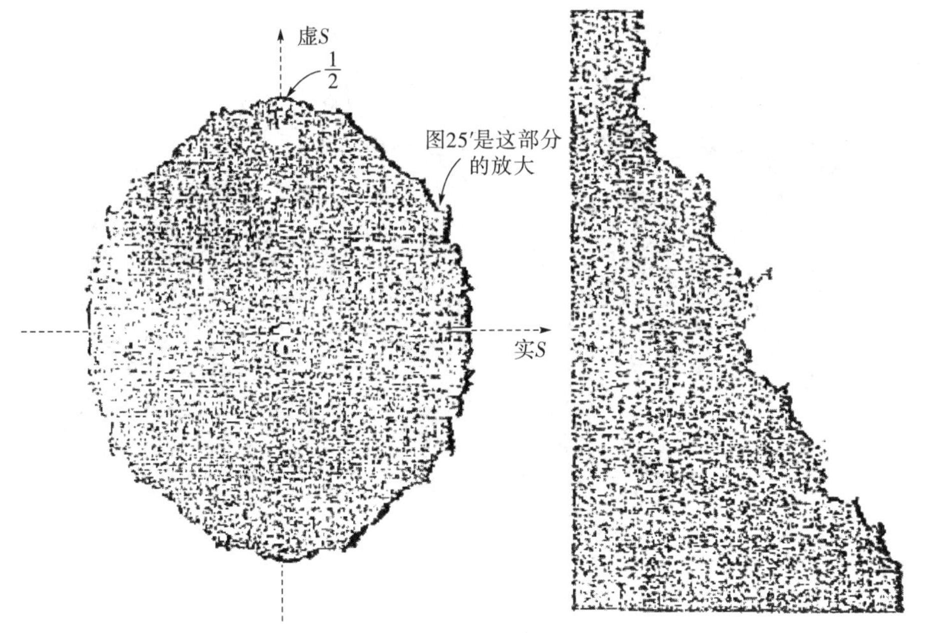

图 26　Mandelbrot 集 D

图 27　集 D 的部分放大，其中 0.49≤Res≤0.55，0.35≤Im s≤0.45.

关于∂D 的位置，有

$$(1)\partial D\subset\left\{s, s\in C, \frac{1}{2}\leqslant|s|\leqslant\frac{1}{\sqrt{2}}\right\};$$

$$(2)I=\{0.5\leqslant s\leqslant 0.53\}\supset\partial D.$$

前一个结论，因 $\dim A(s)\leqslant\dfrac{\log\frac{1}{2}}{\mathrm{los}|s|}$，所以 $|s|<\dfrac{1}{2}$ 时

$$\dim A(s)<1.$$

$A(s)$是完全不连通的. $\left\{s\in C, |s|<\dfrac{1}{2}\right\}\subset D$；而另一半即当$\dfrac{1}{\sqrt{2}}<|s|<1$ 时 $A(s)$连通可应用下一定理：

(3)$A(s)$为完全不连通或连通，取决于 $T_+(A(s))$与 $T_-(A(s))$相交为空集或不空.

D 是 Mandelbrot 集的类推，但有些性质似乎两者相差甚远. 如连通性，Barnsley and Harrington 根据计算机结果猜测∂D 是不连通的. 但 Mandelbrot 集的边界∂M 的连通性已由 Douady and Habbard[109] 所证明(亦见[110～111]).

最后特别指出，在线性映射迭代系$\{C; sz+1, sz-1\}$中取 $s=\dfrac{1}{2}(1+i)$则得到一个龙曲线(dragon curve)，这是一个著名的 fractal 集，如见[13，112，113]，它上面的动力系统的有关研究，近期有[114～116]的工作，那里的龙曲线或龙区域(domain)都是以递归的方法建立的. 图 28 中 $A(s)$称为双生龙曲线.

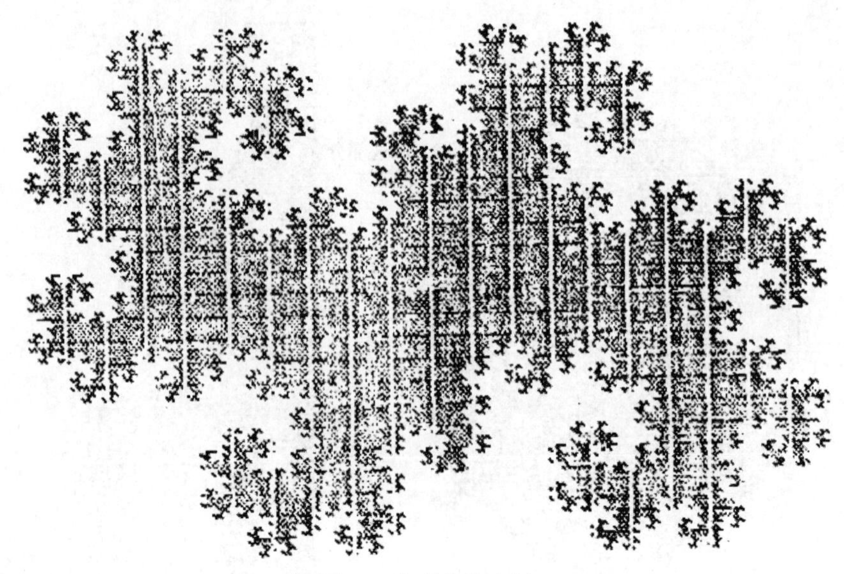

图 28　龙形曲线之例

四、Fractals 的量度

Hausdorff 维数是定量地描述一个点集规则或不规则的几何尺度,同时其整数部分反映出图形的空间规模. 但也不是一贯如此的,例如存在整数 Hausdorff 维数的不规则集. 对于由动力系统(迭代)产生的 fractal 集,我们常可以从不同角度作数量刻划. 这些量可以分为两大类:一种是静态(static)性的,例如 Hausdorff 维数,它基本上只研究 fractal 集图形本身的几何性质,其结果大体上可说明动力系在其上的独立参与的变量的数目,另一种是动态(dynamic)性的,它们主要根据 fractal 集上面动力系的性质而得到,其结果说明了初值条件在单位时间内的信息平均损失量,诸如距离熵(metric entropy),或者 Lyapunov 指数等. 这一节将讨论几种主要的量度方法,它们之间的关系并指出一些问题.

Hausdorff 维数无疑是最重要的量度之一,它的定义在第一节就已经叙述过. 设任一集合 $V\subset R^n$,则

$$
\begin{aligned}
d_{\mathrm{II}}(V) &= \inf\left\{a,\lim_{a\to 0}\inf\sum_j(\operatorname{diam} c_j)^a=0\right\}\\
&= \sup\left\{a,\lim_{a\to 0}\inf\sum_j(\operatorname{diam} c_j)^a=\infty\right\}
\end{aligned}
$$

式中 $\inf\sum\limits_j(\operatorname{diam} C_j)^a=\inf\left\{\sum\limits_j(\operatorname{diam} C_j)^a,V\subset\bigcup\limits_{j=1}^{\infty}C_j\right.$,and $\left.0<\operatorname{diam} C_j<\varepsilon,j=1,2,\cdots\right\}$ 是与 V,ε 有关的量.

给定一函数 $f:M\to M$,它的迭代动力系产生的 fractal 集,例如奇异吸引集,其 Hausdorff 维数一般很难通过理论方法或计算机方法求得. 如果在 $d_H(V)$ 的定义中限制 $\dim C_j=\varepsilon$,即

$$
\inf\sum_j(\operatorname{diam} C_j)^a=\inf\left\{\sum_j(\operatorname{diam} C_j)^a,V\subset\bigcup_{j=1}^{\infty}C_j,\ \text{and}\ \operatorname{diam} C_j=\varepsilon,j=1,2,\cdots,N_\varepsilon\right\}
$$

则所求出的值记为 $d_0(V)$

$$d_0(V) = \inf\left\{a, \lim_{a\to 0}\inf\sum_j (\mathrm{diam}\ C_j)^a = 0\right\}$$

称为 V 的容度(capacity),若以 $N(\varepsilon)$表示以直径为 ε 的小球覆盖 V 时所需的最少数目,则

$$\inf\sum_j (\mathrm{diam}\ C_j)^a = N(\varepsilon)\cdot\varepsilon^a,$$

从而

$$d_0(V) = \lim_{\varepsilon\to 0}\frac{\log N(\varepsilon)}{\log\frac{1}{\varepsilon}}.$$

但是一般情况下一个集合的容度和 Hausdorff 维数的值是不相同的,而且由于容度在变量代换之下不能保持为常数,所以用它来代替 Hausdorff 维数显然是不合适的.我们希望"维数"是某些变换之下的不变量,例如同胚或者共轭. Ott, Withers and Yorke[119] 讨论了 Hausdorff 维数和信息维数在一定条件变换下的不变性.

定理 设 $v\subset R^n$, $F: R^n\to R^n$ 是可射映射,则

$$d_H F(v) = d_H V.$$

例 1 考虑[0,1]中有理数的集合,由容度的定义易知 $d_0(Q)=1$,但已知任何可数集的 Hausdorff 维数等于 0,见[15,120].

2. Cantor 集 C_β 由每一次保留一区间两端的$\frac{1}{2^\beta}$而得到,则易得 C_β 的 Hausdorff 维数 $d_H(C_\beta)$=容度 $d_0(C_\beta)=\frac{1}{\beta}$. 若作变量代换 $F: x\mapsto(-\log x)^{-a}$,则可分析 $F(C_\beta)$的容度 $d_0(F(C_\beta))$,有

$$d_0 F(C_\beta) \geqslant \frac{1}{1+\alpha}.$$

所以,只要选 α,β 使 $\beta>1+\alpha>1$ 就有

$$d_0 F(C_\beta) > d_0 C_\beta.$$

而 Hausdorff 按前面的定理仍有

$$d_H F(C_\beta) = \frac{1}{\beta}.$$

信息维数的定义源于 Balatoni and Renyi[121],

$$d_I(\mu) = \lim_{\varepsilon\to 0}\frac{\sum_{i=1}^{N(\varepsilon)} P_i \ln P_i}{\ln\varepsilon}.$$

其中测度 μ 的支承集由 $N(\varepsilon)$个边长为 ε 的小方体覆盖,p_i 是第 i 个小方体的测度(μ)是动力系的概率不变测度).

Farmer, Ott and Yorke[50], Grassberger, Procaccia, Hentschel[122~124] 定义的 Renyi 维数,是信息维数和容度的推广.设 μ 是不变概率测度,则

$$d_q(\mu) = \frac{1}{q-1}\lim_{\varepsilon\to 0}\frac{\ln\sum_{i=1}^{N(\varepsilon)} p_i^q}{\ln\varepsilon},$$

对 $q\neq 1$ 均有定义,注意到上式中若 $q=0$,

$$d_0(\mu)=d_0(V)=\lim_{\varepsilon\to 0}\frac{\ln N(\varepsilon)}{\ln\frac{1}{\varepsilon}}.$$

Hentschel and Procaccia[124]证明信息维数可由 Renyi 维数在 $q\to 1$ 时取得

$$d_I(\mu)=\lim_{q\to 1}d_q(\mu).$$

同容度类似，当 $q\neq 1$ 时，维数 $d_q(\mu)$ 在变量代换下不能保持，见[119].

定理 设 μ 是某动力系统在不变集 V 上的不变概率测度，$F:V\subset R^n\to R^n$ 可微而且 $|F'|,\frac{1}{|F'|}$ 均有界，则

$$d_I(\mu)=d_I(v),$$

其中 $v(w)=\mu(F^{-1}(w))$.

另外还有一些试图统一维数的工作，例如 R. Badii and A. Politi[125]. 设 $V\subset R^n$ 是有界集合，考虑一个参考点 x 和其它 $(n-1)$ 个从 V 中随机取出的点，以 $\delta(n)$ 表示点 x 到其它 $n-1$ 个点的最近距离，$p(\delta,n)$ 表示 n 个点之间 nn 个距离的概率分布，则定义

$$M_r(n)=\int_0^\infty\delta^\gamma p(\delta,n)d\delta.$$

由 Badii and Politi[126,127]，当 n 充分大时

$$M_r(n)\approx n^{-\frac{\gamma}{D(\gamma)}},\text{for some }D(\gamma).$$

于是定义维数函数

$$D(\gamma)=-\lim_{a\to\infty}\frac{\gamma\ln n}{\ln M_\gamma(n)}.$$

它可以看成 $(-\infty,+\infty)$ 到自身的映射. Badii and Politi 证明 $D(\gamma)$ 对 γ 是单调不减的.

定理 $\inf\{\gamma,D(\gamma)=\gamma\}=d_0$,

$D[(1-q)d_q]=d_q$,

$D(0)=d_I$.

Badii-Politi 维数函数的最直接的用途是为利用计算机求奇异吸引子的容度和信息维数提供一种方法. Badii and Politi 计算了 Hénon 映射的吸引子，Sinai[126]映射，Zaslavskij[129,125]映射，Lorenz 吸引子的维数函数对一些不同的 γ 的取值，他们还计算了流(微分方程)，例如 Roessler 双曲浑沌系统[130]，Navier-Stokes 方程的吸引子的 $D(\gamma)$ 函数.

对一指定动力系统的奇异吸引子或者它的其它 fractal 集，其维数函数的性质与下面的量关系极大

$$\lambda=D'(d_0).$$

一个重要的结论是

定理 总有 $0\leqslant\lambda\leqslant 1$.

$\lambda=D'(d_0)$ 表现了维数函数 $D(\gamma)$ 在不动点附近的变化快慢程度，称为 fractal 集的均匀度 (uniformity factor)(见 Badii and Politi[125,126,131]).

Renyi 维数和 Badii-Politi 维数，包括容度和信息维数，都主要适用于通过计算机进行的

数值研究.

Hausdorff 维数的计算方法,对于具有迭式结构(例如自相似或者回归 fractal)的集合,已有卓有成效的工作,有如 Bedfold[133],Hotchinson[18] 等. 但是对于由一般非线性映射迭代动力系统产生的 fractal 集,这些结果都难以应用,所以关于动力系 fractal 集的 Hausdorff 维数的结论和计算方法实际上几乎没有.

一个在理论上意义重大而便于计算的量是 Lyapunov 指数和 Lyapunov 维数. Lyapunov 指数反映轨道的扩张和收缩率,考虑由可微映射 $f:U\to E$ 生成的离散系统,开集 $U\subset E$,E 为欧氏空间或 Banach 空间. $\forall X\in U$,当 $\varepsilon>0$ 很小时,$B(x,\varepsilon)$在 f 的作用下可以得到一系列的椭球

$$B(x,\varepsilon),f(B(x,\varepsilon)),f^2(B(x,\varepsilon)),\cdots,f^n(B(x,\varepsilon)).$$

只要 ε 充分小总可以使这个序列达到任意指定的长度,设半轴的长度是按下面的方式变化的:

$$\varepsilon_i^{(n)}=\varepsilon\cdot e^{A_i(x,n)},n\to\infty.$$

则平均值

$$\lambda_i=\lim_{a\to\infty}\frac{1}{n}\int_U d\mu(x).A_i(x,n)\quad(\mu(U)=1),$$

就是 $f:U\to E$ 的 Lyapunov 指数. 这是一个直观但不很严格的说法. 较为正式的理论可见[134~139].

定义 设 $f:U\to E$ 是 C^1 的且切映射是紧的,$K\subset E$ 满足 $f(K)\subset K$,$x\in K$. 在 x 点 f 的 Lyapunov 指数是

$$\limsup_{a\to 0}\frac{1}{n}\log\|(D_X f^n)v\|,0\neq V\in E$$

这一定义的建立需要一系列有关存在性的证明. 例如存在 K 的全概率子集 K_0 使 $\forall x\in K_0$,极限

$$\lim_{a\to+\infty}\frac{1}{n}\log\|(D_x f^n)v\|$$

存在(有限或$-\infty$)对一切 $v\neq 0$,而且所有的极限构成一序列

$$\lambda_1(x)>\lambda_2(x)>\cdots$$

以及进一步地 E 关于 $\lambda_1(x),\lambda_2(x),\cdots$ 的分解,即存在于空间 $E_1(x),E_2(x),\cdots,F_1(x),F_2(x),\cdots$,使

$$E=E_1(x)\oplus E_2(x)\oplus\cdots\oplus E_j(x)\oplus F_j(x)\text{for all }j,$$

$$\lim\frac{1}{n}\log\|(D_x f^n)v\|=\lambda_j(x),\text{for all }v\in E_j(x)\text{and all }j.$$

当 x 是不动点的时候,$\lambda_1(x),\lambda_2(x),\cdots$是线性映射 $D_x f$ 的谱分解中诸数的绝对值之对数、对有限维空间 Lyapunov 指数的数目等于空间维数.

根据 Lyapunov 谱型可以对系统的动力行为获得大致的认识. 例如三维空间里,吸引集按其 Lyapunov 指数的符号分为:(+,0,-),奇异吸引子;(0,0,-),环面;(0,-,-),极限环;(-,-,-),不动点等. Lorenz 吸引子属于(+,0,-)一类,参见 A. Wolf[140,141].

由小椭球 $f^n B(x,\varepsilon)$ 半轴长变化，不难利用计算机求流和离散动力系统的 Lyapunov 指数. 细节性的材料，包括算法的 FORTRAN 程序，Wolf, Swift, Swinney and Vastano 的[142]可作参考. 这里指出 Gram-Schmit 规划化的必要性，它是为避免计算机将很小的量处理为 0(沿负 Lyapunov 指数方向的半轴长)以及避免小椭球在按指数速度扭曲时其它方向的半轴被沿最大 Lyapunov 指数方向的半轴所掩没或吞并.

因为 Lyapunov 指数的理论已经颇为成熟，而且这个方向上已有不少专门著作和经典工作，所以我们将不再深入介绍. 作为这一节的结束，以下简略地考虑 Lyapunov 指数和 Hausdorff 维数以及其它量度(例如拓扑熵)的联系，以及另一个关于 fractal 维数和浑沌发生机制的问题.

动力系统的奇异吸引子的 Lyapunov 指数虽然便于研究，但从几何角度来看，我们仍希望对于它的 fractal 维数特别是 Hausdorff 维数有所了解. 自然首先很关心 Hausdorff 维数与 Lyapunov 指数有什么关系. 这方面有著名的 Kaplan and Yorke conjecture. 设 $\lambda_1 \geqslant \lambda_2 \geqslant \cdots$ 是 Lyapunov 指数的由大到小的排列，若 $\lambda_1 \geqslant \lambda_2 \geqslant \cdots \geqslant \lambda_j \geqslant 0 > \lambda_{j+1} > \cdots$，则相应的 Lyapunov 维数定义为

$$d_L = j + \frac{\sum_{i=1}^{j} \lambda_i}{|\lambda_{j+1}|}.$$

F. Ledrappier[143]证明

$$d_H \leqslant d_L.$$

如同 J. L. Kaplan and J. A. Yorke[60,61]，自然有猜想

$$d_H \overset{?}{=} d_L.$$

在二维情况下，Young[114]实际上肯定地证明了它. 这个结果叙述为严格的定理，可以写为

定理 设 $f: M \to M$ 是 C^2 微分同胚，而 M 是一紧致曲面，μ 是其上的遍历的 Borel 概率测度. 设 Lyapunov 指数 $\lambda_1 \geqslant \lambda_2$，则

$$d_h = h_\mu(f) \cdot \left(\frac{1}{\lambda_1} - \frac{1}{\lambda_2}\right).$$

这一结论中的 $h_\mu(f)$ 称为 f 的 μ -熵. 有关熵的理论参考 P. Walters[145]，Ya. G. Sinai[146]，F. Tekens[147]和 F. Ledrappier and L. —S. Young[148]. Young 的上一定理在用于 Kaplan-Yorke 猜想时，有 $\lambda_1 > 0 > \lambda_2$，$h_\mu(f) = \lambda_1$.

对于高维的情形，已经有许多例子说明 Kaplan-Yorke 猜想不成立(见[149～152]). 同时也有一些工作，如 Mori[153,154]，试图修改 Lyapunov 维数的定义以保持类似于 Kaplan-Yorke 的关系成立.

接下来考虑另外一个问题. Lyapunov 指数是描述奇异吸引集上动力行为的数量指标，而 fractal 维数则是刻画奇异吸引集几何结构的尺度. 因为吸引集的几何结构从根本上来说取决于动力系统的方程(流或映射)，那么，反过来，我们是否能够根据吸引集的几何结构对它上面的动力行为作某些断言呢？即由 dimension of attractor 来推断 chaotic dynamics. 从某种意义上说，这是 Kaplan-Yorke 猜想的反问题，那里主要的目的是由 dynamic behavior 推断 geometric structure.

目前这个问题的研究甚少，且主要限于计算机研究. A. K. Agarwal，K. Banerjee and J. K. Bhattacharjee[152]研究了下列几种含参系统在由倍周期分歧达到浑沌状态之临界参数处系统的奇异吸引子的 fractal 维数，这些系统包括 Lorenz 系统，描述 thermohaliae 系统中对流和浑沌的 double-diffusive 方程([155,156])：

$$\begin{aligned}&\dot{x}=\sigma(-x+y+u),\\&\dot{y}=-xz+r_1(1+\varepsilon\cos\omega t)x-y,\\&\dot{z}=xy-bz,\\&\dot{u}=xv-sr_2x-us,\\&\dot{v}=-ux-bsv.\end{aligned}$$

Rössler 系统([130,157])：

$$\begin{aligned}&\dot{x}=-y-z,\\&\dot{y}=ex+y,\\&\dot{z}=f+xz-\mu z.\end{aligned}$$

结果发现，在倍周期分歧进入浑沌状态的临界参数(记为 c)，所对应的系统 R_c 的 Poincaré 映射吸引子的 fractal 维数，是一个与具体系统无关的常数

$$D(R_c)=0.54,$$

而且在临界参数附近的系统 R_r

$$D(R_r)-D(R_c)\propto|r-c|.$$

这几种系统都是二次的流. 考虑 Hénon 映射产生的离散动力系统

$$\begin{aligned}&x_{n+1}=1-cx_n^2+y_n,\\&y_{n+1}=\beta x_n.\end{aligned}$$

当参数 $\beta<1$ 时，上述系统可产生倍周期浑沌，如 $\beta=0.3$ 时 $c_\infty(\beta)=1.059$，这时 $D(R_c)=0.548$；如果 β 取别的值. 则 $c_\infty(\beta)$将随这变化，但 D 值即 fractal 维数保持为常值. 这说明 $D\approx0.54$ 很有可能是一个普适值. 这一点的确认还需要纯理论的证明.

另外，含参系统在沿其它方式进入浑沌的临界态或者别的突变处的 fractal 集的维数均有待进一步地研究.

五、一些问题

本文最后一部分的目的是提供一些与动力系统有关的 fractal 几何问题，并介绍它们的背景.

1. fractals 重构

我们已经知道 fractal 集可以由多种方式形成，它们基本上与迭代和递归过程有联系，而所形成的 fractals 也大多表现出某种自相似性或拟自相似性. 但是不论使用何种方法，所得到的 fractals 并不能概括所有的 fractal 集合，而是总有其特征. 现在经常用于生成 fractal 集的方法包括映射迭代，图形迭代，细胞自动机(cellular automata，见[164～165])，重整化群

(renormalization,见[166～169]).

fractals 重构问题广义而言是任给一个几何上认为是 fractal 的图形,能否以某个指定的方式生成它,狭义则是指能否通过映射迭代来实现这一 fractals 图形. 这里之所以特别提出映射迭代,是因为该过程可以在计算机上十分方便地完成. 这样,fractals 的重构问题可以理解为动力系统研究的某种逆问题.

问题 给定一 fractal 集,找出一动力系统其吸引集是此给定集合;或者更广泛地说,给出一测度,找出一动力系统其不变测度等于或接近于给定者.

对于自相似的 fractal,此问题已有

定理 设 A 是紧致自相似集,$A\subset R^n$,则存在线性映射 $L_1,L_2,\cdots,L_p:R^n\to R^n$,使迭代函数系$\{L_1,L_2,\cdots,L_p;K\}$的吸引集就是集合 A;这里 K 是包含 A 的某个球.

对于一般的 fractal,总可以用一系列迭代函数系吸引集来逼近. Barnsley, Ervin, Hardin and Lancaster 的下述拼贴定理是非常有用的. 见[170],[171],[158].

College Theorem 设 $A\subset R^n$ 是紧致集合. 若有压缩映射 $w_1,w_2,\cdots,w_p:K\to K$ 满足

$$h(A,\bigcup_{i=1}^{p} w_i(A))<\varepsilon,$$

则迭代函数系$\{w_1,w_2,\cdots,w_p;K\}$的吸引集 A^* 满足

$$h(A^*,A)<\frac{1}{1-s}\cdot\varepsilon.$$

此处 $h(\cdot,\cdot)$是 Hausdorff 距离,而 $0\leqslant s<1$ 是 $w_1,w_2,\cdots,w_p$ 的一致压缩常数,即

$$d(w_i(x),w_i(y))\leqslant s\cdot d(x,y),\text{for all } x,y\in K \text{ and } i=1,2,\cdots,p.$$

由上述定理可知,只要把集合 A 划分成充分小的片,就可以用线性映射构成的迭代函数系的吸引集来逼近 A. 也就是说,任意的 fractal 集总可以用一系列的自相似 fractals 来逼近.

fractals 重构问题有许多实际背景,特别是生长过程的模拟. 也许我们能够通过研究一植物种子的形态而用一动力系统实现此植物整个生长过程的模拟.(参阅[171],[158]).

2. 三次多项式的 Julia 集和 Mandelbrot 集

有理映射的迭代动力系统主要决定于其临界点的迭代. d 次复多项式的迭代,其分类随着 d 的增加而复杂化. 对于 $d=2$,即 $f_\lambda(z)=z^2-\lambda$,目前已有许多研究,而对于 $d\geqslant3$ 则还有很多的问题有待研究. 任一 d 次复多项式可以写成

$$p(z)=z^d+a_{d-2}a^{d-2}+\cdots+a_0,$$

其参数空间是$(a_{d-2},a_{d-3},\cdots,a_0)\in C^{d-1}$,

Mandelbrot 集可以推广为下述的 Connectedness-Locus:

$$L_d=\{\lambda\in C^{d-1}\mid K_{p\lambda}\text{是连通集}\}.$$

其中 K_p 定义为 $K_p=\{z\mid p^n(z)\not\to\infty\}$称为多项式的 filled-in Julia 集.

A. Douady and John H. Hubbard 证明

定理 Mandelbrot 集 L_2 是连通的.

最近,Bodil Branner[158] 证明

定理 Mandelbrot 集 L_3 是连通的.

于是自然有下述猜测

猜测 Mandelbrot 集 L_d 也是连通的.

顺便指出,Peitgen and Richter 在 1986 年出版了一本名为 *The Beauty of Fractals* 的书[159],含有大量精美的黑白及彩色插图,其中四维数集 L_3 显示出非常奇特的几何景观.

我们已经知道,二次多项式 $p_\lambda(z)=z^2+\lambda$ 的迭代结果依赖于其唯一的临界点 $z=0$, $p_\lambda(0)=\lambda, p_\lambda^2(0)=\lambda^2+\lambda,\cdots$,这一序列是否有界. 如果 0 吸引到$\infty$,则 filled-in Julia 集 K_λ 是 Cantor 集,如果 0 不吸引到∞,则 K_λ 是连通的. 但是对于三次多项式 $p_{a,b}(z)=z^3-3a^2z+b$,其临界点是 $a,-a$. 如果 $a,-a$ 均吸引到∞,则 $K_{a,b}$是 Cantor 集;如果 $a,-a$ 均不被吸引到∞,则 $K_{a,b}$是连通的. 而若 $a,-a$ 之中只有一点吸引到∞,则 $K_{a,b}$ 的变化变得复杂了. 这时 $K_{a,b}$是不连通的,但不一定是完全不连通的.

Paul Blanchard,Branner,B and Hubbard,J. 等人专门研究了三次多项式的迭代,他们使用的 Kneading sequence 有非常重要的作用.

设 p 是一多项式,定义逃逸函数 $h:C\to R^+U\{0\}$:

$$h(z)\lim_{K\to\infty}\frac{1}{d^K}\log_+|p^K(z)|,$$

此处是 d 是 p 的次数,而 $\log_+$ 是

$$\log_+(x)=\begin{cases}\log x, x\geqslant 1.\\ 0, \quad \text{otherwise}.\end{cases}$$

于是 h 是 C 上的连续函数,filled-in Julia 集 $K=h^{-1}(0)$.

现在继续讨论三次多项式,这里设其临界点为 c_1,c_2,而 $c_1\in w^s(\infty), c_2\in K$. 曲线 $L=h^{-1}(h(c_1))$(称为 level curve)是两个有界连通盘(disks)A,B 的边界,$A\cap B=\{c_1\}\subset L$. $A\cup B\subset U'=h^{-1}\{[0,h(p(c_1))]\}$,后者同胚于圆盘,设 $c_2\in B$. 则 Kneading sequence 是如下定义的:

$$K_i=\begin{cases}A \text{ if } p^i(c_2)\in A\\ B \text{ if } p^i(c_2)\in B\end{cases}, i=0,1,2,\cdots\cdots$$

[158]中 P. Blanchard 研究了 Kneading-sequence 是 $BBB\cdots, BABABA\cdots, BAAA\cdots$的三次多项式迭代. 下面叙述其结果.

定理 设 p 是三次多项式,其 Julia 集是不连通的,但不是完全不连通的,如果其 Kneading 序列是 $BBB\cdots$,则有同胚$\varnothing:\Sigma'\to\{K_p$ 的连通分支$\}$使$\varnothing\cdot\sigma=\bar{p}\cdot\varnothing$,又若 $s\in\Sigma'$ 是Σ_3 的元素,则$\varnothing(s)$是一点. 其中Σ'是由四个符号$\{1,2,3,B\}$构成的Σ_4 上转移的 σ 的不变子集,Σ_3 是由$\{1,2,3\}$三个符号构成的序列的集合. 而 $\bar{p}$ 是 p 在$\{K_p$ 的连通分支$\}$上的诱导.

定理 设 $A-B$ Kneading sequence 是 $BABABA\cdots$,记Σ'是$\Sigma_s=\Pi\{1,2,3,A,B\}$的$\sigma$-不变子空间,则有共轭$\varnothing:\Sigma'\to\{K_p$ 的连通分支$\}$. 若 $s\in\Sigma_5\cap\Pi\{1,2,3\}$,则$\varnothing(s)$是一点.

定理 设 $A-B$ Kneading sequence 是 $BAAA\cdots$,则映射 $p|_z$ 拓扑共轭于$\sigma^*|_{\Sigma_3/\sim}$,其中$\sim$是由 $2111\cdots\sim3111\cdots$生成的最小等价关系. $\sigma^*:\Sigma_3/\sim\to\Sigma_3/\sim$由$\sim$诱导.

下面三个问题对于仔细研究三次多项式的迭代和 Mandelbrot 集 L_3 是非常必要的.

问题(1)　$A-B$ Kneading sequence 可以不是周期的吗？如果可以，相应的动力系统如何？

(2)If the Kneading sequence is preperiodic, is the Julia set a Cantor set?

(3)For a given Kneading sequence, calculate the number of its Mandelbrot in each leaf of the trefoil clover.

第三个问题的背景参见[158]. 关于三次多项式的迭代，可参见[172～175].

3. 整超越函数的 Julia 集爆炸

考虑临界有限的整超越函数族的 Julia 集的变化. 当参数变化时，其 Julia 集可能发生爆炸，即由一个无处稠密的集合突然变化为整个复平面. Robert L. Devancy 等人在复指数函数族和复正弦函数族中发现这种现象. 这一点与多项式映射族有很大不同[参阅 176～178]. 关于正弦函数族 $s_\lambda(z)=\lambda\sin z$ 的结论如下：

定理　设 $s_\lambda(z)=\lambda\sin z$. 则在 1 的任意邻域 $U\subset C$，存在 λ，使 $J(s_{\lambda_0})=C$[参见 158].

问题　设 E_λ 是一整超越函数族，每一 E_λ 是临界有限的. 参数平面上什么样的分歧点可能产生 Julia 集爆炸？

已经知道 $s_\lambda(z)$ 中在 $\lambda=1$ 附近使 $J(s_\lambda)=C$ 的 λ 构成一个曲线，即 hair.

问题　对一般的整超越函数族，其情形如何？

4. Julia 集的准周期点的代数特征

Pierre Moussa 等人研究了整系数多项式和代数整系数多项式的 filled-in Julia 集内准周期点的代数特征. 结果说明准周期点与代数整数之间有紧密的联系.

定理　设 p 是一具有有理整系数的多项式. 若代数数 z 及其所共轭代数数均含于集合 K_p，则 z 是 p 的准周期点；反之，又是若 p 是单变元的，则 p 的任一准周期点(which must be contained in K_p)必是代数数[179].

这一定理的证明要引用 Kronecker 和 Fekete[180] 的一些经典结论.

定理　If the domain D has a transfinite diameter less than one, there is only a finite number of algebraic integers sitting in D together with their conjugates.

特别，对于 D 是单位闭圆盘，设代数整数 z 及其所有共轭均含于 D，则 $z=0$ 或 z 是单位根.

对于 $D=[-2,+2]$，同一条件之下，结论是 $z=2\cos\pi q$，此处 q 是有理数.

类似于 p 是有理整数为系数的多项式，前一定理的结论可以用适当的形式推广到代数整数为系数的多项式上.

Douady 和 Moussa 讨论了二次多项式族 $f_\lambda(z)=z^2+\lambda$ 的 Mandelbrot 集的类似问题. 有下述结果.

定理　若 $\lambda\in C$ 是一代数整数，λ 及其所有共轭均含于集合 M，则 $z=0$ 是 $f_\lambda(z)$ 的准周期点；反之亦然[158,181].

进一步的研究问题包括：

问题 如何将前面的结论推广到有理分式的迭代?

问题 对于 d 次多项式族的 Mandelbrot 集,其中所含有的代数数有何特点?

有关这类问题的研究,可参考文献[182～185].

参考文献

[1]张景中,熊金城.迭代论,上海科技出版社,将出版.

[2]Schroder E. Über Iterate Funktionen,Math. Ann. ,1871(5):295 - 322.

[3]Abel N H. Works V. 2,Posthumous paper,1881:36 - 39.

[4]Pubbey J M. The Math. Works of Charles Babbage,Cambridge Univ. Press,1978.

[5]张景中,杨路.北京大学学报(自然),1982(6):23 - 45.

[6]张景中,杨路.数学学报,1983,26(4):398 - 412.

[7]张景中,杨路.中国科学(A),1985(1):32 - 43.

[8]Lam P F. Colloq. Math. ,1976(35):275 - 287.

[9]Cowen C C. Tran. A. M. S. ,1981,265(1).

[10]杨路,张景中.数学学报,1986,29(2):180 - 183.

[11]Mandelbrot B B. Les objects fractals;forme,hasard et dimension,Paris: Flammarion,1975.

[12] Mandelbrot B B. Fractals, Form, Chance, and Dimension, San Francisco: W. H. Freeman & Co. 1977.

[13]Mandelbrot B B. The Fractal Geometry of Nature,San Francisco: W. H. Freeman & Co. 1982.

[14]Mandelbrot B B. On Fractal Geometry,and a Few of the Mathematical Questions It Has Raised. Proceeding of the International Congress of Mathematicians,1982.

[15] Falconer J K. The Geometry of Fractal Sets, Cambridge Univ. Press, New York (1985).

[16]Federer H. Tran. A. M. S. ,62,1947(62):114 - 192.

[17]Besicovitch A S. Math. Ann. ,1928(998):442 - 464.

[18]Hutchinson J E. Indiana Univ. Math. Journal,1981(30):713 - 747.

[19]Hayashi S. Publ. RIMS,Kyoto Univ. ,1985(21):1059 - 1066.

[20]Barnsley B M,Demko S. Proc Royal Soc. London,A399,1985:243 - 275.

[21]张颖清.自然杂志,4(1981),243 - 248;生物全息学研究,山东大学出版社(1985);又见自然杂志,第 10 卷第 9 期(1987).

[22] Mandelbrot B B. Self-affine Fractal Sets, Ⅰ, Ⅱ, Ⅲ, in Fractals in Physics, ed. by Pietronero,L. and Tosatti,E. ,Northholand,1986.

[23] Qiu, W, Chou W, Yin Y. Some Problems on Complex Analytic Dynamics, Fudan University,1988(全国第二届动力系统及其应用学术讨论会(杭州)资料).

[24]MacMullen C. Nogaya Math. J. ,1984(96):1 - 11.

[25]Berry M V,Lewis Z V. Proc. Royal Math. London,A370,1980:459 - 483.

[26]Kolb M. Film on aggregation Processes, in Fractals in Physics, ed. by Pictronero, L. and Tossatti, E. , Northholand, 1986.

[27]Pietronero L, Tossatti E. (ed), Fractals in Physics, North-Holland, 1986.

[28]Shlesinger. (ed.), Proceedings of a Symposium on Fractals in the Physical Sciences. JourRal of Statistical Physics, V. 36, nos. 5/6, 1984: 519－919.

[29]Lorenz E N. J. Atmos. Sci. , 1963(20): 130－141

[30]Salzman B. J. Atmos. Sci. , 1962(19): 239－341

[31] Kocak H. Diffrential and Difference Equations through Computer Experiments. Springer-Verlag, New York, 1996.

[32] Guckenheimer J, Holmes, P. Nonlinear Oscillations, Dynamical Systems, and Bifurcations of Vector Fields, Springer-Verlag, New York, 1983.

[33]Sparrows C. The Lorenz Equations, In Chaos, ed. by Holden. A. V. , Princeton Univ. , Press, Princeton, New Jersey, 1986.

[34]Bunimovlch L A. Statistical Properties of Lorenz Attractors, In Nolinear Dynamical and Turbulence, ed. by Barenblatt G I, looss G, Joseph D D. Pitman Advanced Publishing Program. Boston, London, Melbourne, 1983.

[35]Henon M. Comm. Math. Phys. , 1976(50): 69－77.

[36]Ruelle D, Takens F. Comm. Math. Phys. , 1971(20); 1971(23); 1981(82).

[37]Sinai J G, Vul, E, D. Physica, 2D, 1981.

[38]Ruelle D. Small Random Pertubations and the Definition of Atlractors. In Geometric Dynamics, ed. by Palis Jr J. Lecture Notes in Math. , 1007, Springer-Verlag, Berlin, Heidelberg, 1983.

[39]Thom R. Stabilite structurelle et morphogenese, W. A. Benjamin, Reading, Mass. , 1972.

[40]Smale S. Bull. A. M. S. , 1967(73): 748－817.

[41]Williams R F. Publ. Math. I. H. E. S. , 1974(43): 169－203.

[42] Ruelle D. Small random Pertubation of Dynamical Systems and the Definition of Attrattor, Comm. Math. Phys. , to appear.

[43]Smale S. Diffeomorphisms with Many Periodic Points. In Diffrential and Combinatorial Topology, Cairns, S. S. ed. , Princeton Univ. Press, Princeton, New Jersey.

[44]钱敏，严寅. 科学通报，1985，30(13)：961－965.

[45]周作领. 科学通报，1987，32(4)；数学学报，1987，30(2)：284－288

[46]曾振柄. 马蹄结构蕴涵 Li-Yorke 强浑沌，(待发表).

[47]Lauwerier H A. Two Dimensional Iterative Maps, In Chaos, ed. by Holden, A. V. , Princeton Univ. Press, Princeton, New Jersey, 1966.

[48]Lauwerier H A. Physica, 7D, 1986: 146－154.

[49]Lauwerier H A. J. of Appl. Math. , 1986.

[50]Farmer J D Ott E and Yorke J A. Physica, 7D, 1983: 153－180.

[51]Poincare H. J. de Math. Ser,1890,4(6):313－365.

[52]Fredericksön P,Kaplan J L,Yorke J A. J. Diff. Eq. ,1985(49):185.

[52′]Kaplan J L,Yorke J A. Chaotic Behavior of Multidimensional Difference Equations. In Lecture Notes of Math. ,730,Springer-Verlag,Berlin,1978.

[53]Grebogi C,Ott E,Yorke J A. ,Ergod. Th. & Oyanam. 8ys. ,1985(5):341－372.

[54]Yamaguchi Y. Physics Letiers,A,V. 117,n. g(Sept. 1986),450－458.

[55]Widom M,Bensimon D,Kadanoff L P,Shenker S J. J. of statistical Physics,1983,32(3).

[56]Collet P. J. Math. Pures et Appl. ,1984(63):391－406.

[57]McDonald S M,Grebogi C,Ott E,Yorke J A. Physica,17D,1985:125－153.

[58]Levi M. Qualitative Analysis of the Periodically Forced Relaxation oscillations. Mem. Amer. Math. Soc. 1981(32):244.

[59]Cartwright M L,Littlewood J E. Ann. Math. ,1951(54):1,also:J. London Math. Soc. ,1945(20):180.

[60]Kaplan J L,Yorke J A. Comm. Math. Phys. ,1979(67):93.

[61]Kaplan J L,Yorke J A. In Functional Differential Equations and Approximation of Fixed Points. Lecture Nores of Math. ,730,Springer-Verlag,New York,1979.

[62]Grebogi C. Ott E,Yorke J A. Phys. Rev. Lett. ,1983(50):935.

[63]Julia G. J. Math. ,1918(8):47－245.

[64]Fatou P. Bull. Soc. Math. France,1919(47):161－271.

[65]Fatou P. Bull. Soc. Math. France,1920(48):33－94.

[66]Faton J. Bull. Soc. Math. France,1920(48):208－314.

[67]Fatou J. Aeta Math. ,1926(47):337－370.

[68]Sullivan D. Quasiconformal Homeomorphisms and Dynamics,Ⅰ,preprint.

[69]吕以攀. 复解析动力系统理论基础. 全国复解析动力系统讲习班资料,山东省荷泽师专数学系印,1987 年 4 月.

[70] Sullivan D. Quasiconformal Homeomorphisms and Dynamics,Ⅲ: Topological Conjugacy Classes of Analytic Endomorphisms,preprint.

[70]Douady A,Hubbard J. C. R. Acad. Sci. Paris,1982(294):123－126.

[71]Mandelbrot B B. Ann. New York Aead. Sei. ,1980(357):249－259.

[72]Herman P R. Bull. Soc. Math. France,tome112(1984),fas. 1,93－142.

[73]Blanchard P. Ball. A. M. S. ,1984,11(1):85－142.

[74]Lattes S. Note aux. C. R. Acad. Sci. Paris,1918(166):26－28.

[75]Douady A. Systems dynamiques holomorphes,Seminaire Bourbaki,35° annee,1982/83(599):1982.

[76]Mandelbrot B B. Physica,D7,1983:224－239.

[77]Sullivan D. Conformal Dynamical Systems,preprint.

[78]Ahlfors L. Complex Analysis,McGraw-Hill,1979.
[79]Smale S. Bull. A. M. S. ,new series,1985(15):87 - 121.
[80]McMullen C. Ann. Math. 2nd series,V. 125,1987(3):467 - 494.
[81]Smale S. Algorithms for Solving Equations,Proceedings of the International Congress of Mathematicians,1986;又见:数学译林,1987,6(3):191 - 212.
[82]Bowen R. Publ. I. H. E. S. ,1980(50).
[83]Mane R,Sad P,Sullivan D. On the Dynamics of Rational Maps:submitted to J. of the Ecole Normal Superieur.
[84] Sullivan. D. Seminar on Conformal and Hyperbolic Geometry. I. H. E. S. seminar notes,march 1982.
[85]Ruelle D. Analytic Repellers. J. Ergod. Th. & Dynam. Sys. ,1982.
[86]Garnett L. Calculation of Hausdorff Dimension,in preparation.
[87]Brolin H. Ark. Mat. ,1965(6):103.
[88]Ruelle D. Submitted to J. Ergod. Th. & Dynam. Sys. ,1982(2).
[89]Misiurewicz M. Ergod. Th. & Dynam. Sys. ,1981(1):103 - 106.
[90]Devaney R L,Tangerman ,F. Ergod. Th. & Dynam. Sys. ,1986(6):489 - 503.
[91]Devaney R L. Bull. A. M. S. ,1984,11(1):167 - 171.
[92] McMullen C. Area and Hausdorff Dimension of Julia Sets of Entire Functions, preprint.
[93]Eremenko A,Ljubic M. UkrSSR. Acad. Sei. Kharkov,1984(6).
[94]Devaney R L,Krych M. Ergod,Th. & Dynam. Sys. ,1984(4):35 - 52.
[95]Ghys E. Goldberg L,Sullivan D. Ergot. Th. & Dynam. Sys. ,1985(5):330 - 335.
[96]Devaney R L. Proc. A. M. S. ,1985(94):545 - 548.
[97]Eremenko A,Ljubic M. UkrUSSR. Acad. Sci. Kharkov,1984(29).
[98]Baker I N. Proc. London Math. Soc. ,1984(49):563 - 576.
[99]Baker I N,Rippon P J. Ann. Acad. Sci. Feas. Ser. IA Math. ,1984(9):85 - 141.
[100]Goldberg L,Keen L. A Finiteness Theorem for a Dynamical Class of Entire Functions,to appear.
[101]Cosnard M,Masse C. C. R. Acad. Sci. Paris,1983(297):549 - 552.
[102]Hurley M. Attracting Orbits in Newton's Method. *Trans. A. M. S.* , to appear.
[103]Hurley M. Ergod. Th. & Dynam. Sys. ,1986(6):561 - 569.
[104]Hurley M,Martin C. SIAM. J. Math. Anal. ,1984(15):238 - 252.
[105]Saari D,Urenko,J. Amer. Math. Month. ,1985(91):3 - 17.
[106]Barnsley M F,Demko S G. Rational Approximation of Fractals,In Lecture Notes of Math. ,1105,73 - 88,Springer-Verlag.
[107]Lauwerier H A. The Pythagrass tree as a Julia set,preprint.
[108]Barnsley M F,Harrington A N. Physiea,15D,1985:421 - 432.

[109]Douady A,Hubbard J. Compets Rendus(Paris),1982(294):123-126.
[110]Douady A,Hubbard J. On the Dynamics of Polynomial like mapping,preprint(1984).
[111]Thurston. On the Dynamics of Iterated Maps,preprint.
[112]Knuth D. The Arts of Computing Programming,Ⅱ,section 4. 1,Addison Wesley,1969.
[113]Davis C,Knuth D. J. of Recreational Math. ,1970(3):66-81.
[114]Mitzutani M,Ito S. Dynamical Systems on Dragon Domains. In Dynamical Systems and Nonlinear Oscillations,ed. by Giko Ikegami,World Scientific 1986,Singapore.
[115]Mitzutani M,Ito S. Dynamical Systems on Dragon Domains,to appear.
[116]Mitzutani M,Ito S. A New Characterization of Dragon and Dynamical System,to appear.
[117]Ausloos M,Berman D H. Proc. R. Soc. London,A400,1985:331-350.
[118]Barnsley M F,Geronimo J S,Harrington A N. J. of Stat. Phys. ,1984(37):51-92.
[119]Ott E,Withers W D,Yorke J A. J. of Star. Phys. ,1984(36):687-697.
[120]Billingsley Ergodic Theory and Information,Wiley,New York,1965.
[121]Balatoni J,Renyi A. Publ. Math. Inst. Hung. Acad. Sai. ,1965(1):9. (English translation in The Selected Papers of A. Renyi,v. I,Akademia,Budapest,1976).
[122]Grassberger P. Phys. Lett. ,97A,1983:227,also Generilized Dimensions of Strange Attractors,Wuppertal,preprinting.
[123]Grassberger P,Procaccia I. Phys. Rev. Lett. ,1983(50):346.
[124]Hentschel H G E,Procaccia I. Physica,8D,1983:435.
[125]Badii R,Politi A. J. of Stal. Phys. ,1985(40):725-750.
[126]Badii R,Politi A. Phys. Rev. Lett. ,1984(52):1661.
[127]Badii R,Politi A. Phys. Leff. ,104A,1984:303.
[128]Sinai Y G. Russ. Math. ,Surveys,1972(4):21.
[129]Zaslavskij G M. Phys. Lett. ,69A,1978:145.
[130]Rossler O E. Zeitschriften Naturforschung,38A,1983:788.
[131]Badii R,Politi A. Numerical Investigation of nonuniform Fractals,In Fractals in Physics,ed. by Pietronero,L. and Tossatti,Elsevier Sci. Publ. ,1986:453-456.
[132]Cvitanovic P. Jenscn M H,Kadanoff L P,Procaccia I. Circle Maps in the Complex Plane,preprinting.
[133]Bedford T. J. London Math. Soc. ,1986,33(2):89-100.
[134]Bennetic G,Galgani L,Streicyn J-M. Lyapunov Characteristic exponents for smooth dynamical systems and for Hamiltonian systems;a method for computing all of them. Meccanica 15,1989:9-20.
[135]Lyapunov A M. Ann. Math. Study,1947(17).
[136]Oseledec V I. Trass. Moscow Math. Soc. ,1988(19):197.
[137]Mane R. Lyapunov Exponents and Stable Manifold for Compact Transformations,In

Lecture Notes in Math. , 1007: Geometric Dynamics, ed. by J. Paris Jr. , 522 - 577, Springer-Verlag, Berlin, Heidelberg, 1983.
[138]Ruelle D. Publ. , I. H. E. S. , 1979(50).
[139]Ruelle D. Ann. Math. 1982(115).
[140]Wolf A. Quantifying Chaos with Lyapunov Exponents, In Chaos, ed. by Holden, A. V. , 273 - 290, Princeton Univ. Press, Princeton, New Jersey, 1986.
[141]郝柏林. 物理学进展, 1983, 3(3): 329 - 416.
[142]Wolf A, Swift J B, Swinney H L, Vastano J A, Physica, 16D, 1985: 285 - 317.
[143]Ledrappier F, Comm. Math. Phys. , 1981(81): 229.
[144]Young L S. Ergod. Th. & Dynam. Sys. , 1982(2): 109.
[145]Watters D. An Introduction to Ergodic Theory, Springer-Verlag, New York. 1982.
[146]Sinai Y G. Dokl. Akad. Nauk SSSR, 1959(124): 768.
[147]Takens F. Invariants Related to Dimension and Entropy, In Atas. do 13 Coloquio Brasileiro de Mathematica, 1984.
[148]Ledrappier F, Young L S. The Metric Entropy of Diffeomorphisms, I and Ⅱ, Berkeley preprints, 1984.
[149]Alexander J C, Yorke J A. Fat Baker'S Transformation, Univ. Mariland prepriut, 1982.
[150]Grassberger P, Procaccia I. Physica, 9D, 1983: 189.
[151]Shtern V N. Phys. Lett. , 99A, 1983: 268.
[152]Agarwal A K, Banerjee K, Bhattacharjee J K. Phys. Left. , 119A, 1986(6): 280 - 282.
[153]Mori H. Progress in Theotic Phys. , 1980(65): 1044.
[154]Mori H, Fujisaka H. Progress in Theotic Phys. , 1980(63): 1931.
[155]Dacosta L N, Knobloch, E, Weiss N. J. Fluid Mechanics, 1982(109): 25.
[156]Agarwal A, Batarcharjee J K, Banerjee K. Phys. Lett. , A111, 1985: 329.
[157]Rossler O E. Phys. Lett. , A57, 1976: 196.
[158]Barnsley M F, Demko S G. Chaotic Dynamics and Fractals, Academic Press, 1986.
[159]Peitgen H O, Richter P H. The Beauty of Fractals, Springer-Verlag, 1986.
[160]李忠. 拟共形映射及其在黎曼曲面中的应用. 科学出版社, 1988.
[161]Codd E F. Cellular Automata, Academic Press, New York, 1968.
[162]Wilson S J. Discrete Appl. Math. , 1984(8): 91 - 99.
[163]Wilson S J. Physica, 10D, 1984: 69 - 74.
[164]Wolfram S. Rev. of Modern Phys. , 1983(65): 601 - 644.
[165]Wolfram S. Physica, 10D, 1984: 1 - 35.
[166]Burkhardt T W, van Leeuwan J M J. Real Space Renormaiization, v. 30, Springer-Verlag, Berlin, Heidelberg, New York, 1982.
[167]Derrida B, Itzykson C, Luck J M. Comm. Math. Phys. , 1984(94): 115 - 167.
[168]Derrida B, De Seze L, Itzykson C. J, Siat. Phys. , 1985(55): 559.

[169]Derrida B,Eckmann J P,Erzan A. J. Phys. ,A16,1983:893.
[170]Bafnsley M F,Ervin V,Hardin,D,Lancaster J. Solution of an Inverse Problem for Fractals and Other Sets,Geogia Teeh. preprint,1984.
[171]Demko S,Hedges L,Naylor B. Construction of fractal Objects with Iterated Function Systems,Proceedings of SIGRAPH 1985.
[172]Blanchard P. Symbols for Cubics and other Polynomials,preprint.
[173]Branner B,Hubbard J. Iteration of Cubic Polynomials, Ⅰ:The Global Structure of Parameter Space,personal communication to Paul Blanchard.
[174]Branner B,Hubbard J. Iteration of Cubic Polynomials, Ⅱ:Patterns and Parapatterns, personal communication to Paul Blanchard.
[175]Branner B. The Parameter Space for Complex Cubic Polynomials. Proceedings of the Conference on Chaotic Dynamics,Georgia Tech,1985.
[176]Devaney R. ,Structural Instability of exp(z),to appear in *Proc. A. M. S.*
[177]Devaney R. Bursts into Chaos. Phys. Lett. ,1984(104):385 - 387.
[178]Douady A, Hubbard J. Etude Dynamique des Polynomes Complex. Publ. Math. D'Orsay,preprint.
[179]Moussa P,Geronimo J S,Bessis D. Ensemblea de Julia et proprietes de localisation des famillcs iterees d'entiers algebraiques,C. R. Acad. Sci. Paris,299,ser. I(1 984),281 - 284.
[180]Fekete M. Math. Z. ,1923(17):228 - 249.
[181]Douady A. Private communicatiou to Pierre Moussa.
[182]Barnsley M F,Bessis. D,Moussa P. J. Math. Phys. ,1979(20):535 - 546.
[183]Moussa P,Anr. Inst. Hsnri Poincare,1983(38):309 - 347.
[184]Douady A,Hubbard J H. C. R. Acad. Sci,Paris,294,ser. I,1982:123 - 216.
[185]Douady A. Seminaire Bourbaki,n. 599,Asterisque,105 - 106(1983),39 - 63.

The Parallel Numerical Method of Mechanical Theorem Proving

Zhang Jing-zhong and Yang Lu
Institute of Mathematical Sciences, Chengdu Branch of Academia Sinica, China

Deng Mike
Department of Mathematics, University of Science and Technology of China, China

Abstract. In this paper, we present results of the work which allow us to prove geometry theorems by the *parallel numerical method* based on the *multi-instance numerical verification* of algebraic identity. The algebraic principle of the parallel numerical method is discussed and illustrated intuitively. The advantages of our method are given. It is acceptable on the *complexity of both memory and time*. It can be used to prove non-trivial geometric theorems by microcomputer, even by hand. We give some examples proved by parallel numerical method, including certain new unexpected results.

1. Introduction

In the recent decade, owing to Wu Wentsun's outstanding work, a great breakthrough was made in the field of mechanical theorem proving which has attracted increasingly world-wide attention. Deciding mechanically whether a proposition is true or false is now possible in some fields, particularly elementary geometry, a goal of many mathematicians since Descartes's age.

The realization of mechanical theorem proving depends on the high-speed calculation of computers. To raise the computer's calculating capacity, one important method is to develop the parallel machine and the parallel algorithm. Therefore, the study of the parallel algorithm of mechanical theorem proving has profound significance.

In this paper, we discuss the parallel numerical method of mechanical theorem proving based on the multi-instance numerical verification of algebraic identity. It is well known that there are many methods to decide whether an algebraic equality of univariable is identical. For instance, given a_i, b_k, to see if the equality

本文刊于《Theoretical Computer Science》, 1990 年.

$$(x+a_1)(x+a_2)(x+a_3)(x+a_4)=b_0x^4+b_1x^3+b_2x^2+b_3x+b_4 \quad (1)$$

is identical, we can use any one of the following methods:

(1) Formal calculation: expand the left side of the equality, merge similar terms, then compare the coefficients.

(2) Multi-instance numerical verification: take five distinct values $\hat{x}_1, \hat{x}_2, \hat{x}_3, \hat{x}_4, \hat{x}_5$ of x, substitute each of them for x in (1), then calculate (1), respectively. If for each $\hat{x}_j$ $(j=1,2,3,4,5)$, (1) holds, then (1) is identical, otherwise (1) is an algebraic equation, with degree not more than 4, it is impossible that (1) has five roots.

(3) Single-instance numerical verification: take a large enough positive number p and $\hat{x}$ such that $|\hat{x}|>p$. If for $x=\hat{x}$, (1) holds, then (1) is identical. Or else, by transporting and packing terms we infer the equation

$$c_0x^4+c_1x^3+c_2x^2+c_3x+c_4=0 \quad (2)$$

from (1). Assume $c_0\neq 0$. If $\hat{x}$ were the root of (2), when $|\hat{x}|\geqslant 1$, we should have

$$\begin{aligned}|\hat{x}| &= |(c_1x^3+c_2x^2+c_3x+c_4)/c_0x^3| \\ &\leqslant (|c_1|+|c_2|+|c_3|+|c_4|)/|c_0|. \quad (3)\end{aligned}$$

Now take $p=(|c_1|+|c_2|+|c_3|+|c_4|)/|c_0|$, $|\hat{x}|>p$ implies that $\hat{x}$ is not a root of (2).

The verification of an algebraic identity is intrinsically interrelated with the proof of equality theorem in elementary geometry, which is due to Wu Wentsun's work[5,6,11]. As a result, each of the three above verifying methods can be regarded as the basis of relevant ideas in mechanical theorem proving. Wu's method uses the method of formal calculation.

Along the lines of single-instance numerical verification (3), Hong Jiawei suggested the method of geometry theorem proving [2, 3]. His work aroused the interest of many colleagues. Further discussion of the computational complexity and the practical feasibility of his method is pending.

From the numerical verification (2), Zhang Jingzhong and Yang Lu proposed the parallel idea and the parallel numerical method of geometry theorem proving. In recent years, practical research on the feasibility of this method has been undertaken[1,4,12]. As the result indicates, it is acceptable on the complexity of both memory and time. The method can be used to prove non-trivial geometric theorems by microcomputer, even by hand. Moreover, this method can also be used to prove some quite difficult geometric inequality successfully[4].

With the aid of this method, the authors discovered certain kinds of new non-trivial theorems, including a number of somewhat unexpected results, e.g. Example 4.3 and 4.4 in the paper.

2. Elementary lemmas

To develop the multi-point numerical verification in Section 1 to the mechanical method for geometry theorem proving, we have to overcome two obstacles: one is to extend the case of univariable to that of multi-variable, the other is to turn the verification of geometric proposition into that of algebraic identity. The two following lemmas are introduced for these purposes.

At first, we give the following.

Definition 2.1. Suppose $S_1, S_2, \cdots, S_m$ are m subsets of the set K, S_j has t_j elements, where t_j is a positive integer, for $j=1,2,\cdots,m$. We call the Cartesian product of the above m subsets

$$S=S_1\times S_2\times\cdots\times S_m \tag{4}$$

the m-dimensional lattice array on K, the array $(t_1, t_2\cdots, t_m)$ the size of the lattice array. Clearly, S has $t_1t_2\cdots t_m$ elements.

Lemma 2.2. *Let $f(x_1, x_2, \cdots, x_m)$ be a polynomial in variables $x_1, x_2, \cdots, x_m$, with coefficients in the field K, the degree of f in x_k be not more than n_k, where n_k is a non-negative integer, for $k=1,2,\cdots,m$. If there exists one m-dimensional lattice array S on K with size $(n_1+1, n_2+1, \cdots, n_m+1)$, such that f vanishes over S, i. e. for any $(\hat{x}_1, \hat{x}_2, \cdots, \hat{x}_m)\in S$,*

$$f(\hat{x}_1,\hat{x}_2,\cdots,\hat{x}_m)=0, \tag{5}$$

then, f is identical to zero.

Proof. Induction on m. $m=1$, f is a polynomial in x_1, with degree not more than n_1. Correspondingly, one-dimensional lattice array is a subset of K with size (n_1+1), i. e. there are n_1+1 elements in S. If f is not identical to zero but it vanishes on S, then f has n_1+1 roots at least, which contradicts the fact that f has only n_1 roots.

Suppose that the theorem is true for $m=j$. Let us prove it is also true for $m=j+1$. Write $f(x_1,x_2,\cdots,x_j,x_{j+1})$ in the following form:

$$c_0x^n+c_1x^{n-1}+\cdots+c_n. \tag{6}$$

For convenience, let $x=x_{j+1}$, $n=n_{j+1}$, where $c_k=c_k(x_1,\cdots,x_j)$ is the polynomial in x_1, $x_2,\cdots,x_j$ with degree in x_k not more than n_k.

Suppose there is a $(j+1)$-dimensional lattice array $S=S_1\times S_2\times\cdots\times S_{j+1}$ on K with size $(n_1+1, n_2+1, \cdots, n_{j+1}+1)$, such that f vanishes over S. Take an element $(\hat{x}_1,\hat{x}_2,\cdots,\hat{x}_j)$ in $S^*=S_1\times S_2\times\cdots\times S_j$, $f(\hat{x}_1,\hat{x}_2,\cdots,\hat{x}_1,x)$ is a polynomial in x with degree not more than n. By induction hypothesis,

$$f(\hat{x}_1,\hat{x}_2,\cdots,\hat{x}_j,\hat{x})=0 \tag{7}$$

for any $\hat{x}\in S_{j+1}$. Since S_{j+1} has $n+1$ elements, $f(\hat{x}_1,\hat{x}_2,\cdots,\hat{x}_j,x)$ is identical to zero. Hence,

$$c_k(\hat{x}_1,\hat{x}_2,\cdots,\hat{x}_j)=0(k=0,1,2,\cdots,n) \tag{8}$$

for any element $(\hat{x}_1,\hat{x}_2,\cdots,\hat{x}_j)$ in S^*, where S^* is a j-dimensional lattice array with size $(n_1+1, n_2+1, \cdots, n_j+1)$. By induction hypothesis, we have that c_k is identical to zero, so is $f(x_1,x_2,\cdots,x_j,x_{j+1})$.

Lemma 2. 2 completes the extension of verifying algebraic equality of univariable to that of multi-variable. Now we set about the other task: to turn a geometrie proposition into an algebraic equality.

For any polynomial φ, we denote the degree of φ in y by $\deg(\varphi, y)$. We have the following.

Lemma 2. 3. *Let f and g be the polynomials in variables $u_1, u_2, \cdots, u_n, x$ over the field K:*

$$f = a_0 x^m + a_1 x^{m-1} + \cdots + a_m,$$

$$g = b_0 x^l + b_1 x^{l-1} + \cdots + b_l, \tag{9}$$

where $m \geqslant 1, l \geqslant 1, a_k, b_k$ are the polynomials in $u_1, u_2, \cdots, u_n$ over K, a_0, b_0 are not zero polynomial, there is a mechanical method to determine the polynomials P, Q and R in $u_1, u_2, \cdots, u_n, x$ over K. such that

$$Pf + Qg = R^l \tag{10}$$

and

$$\deg(P, x) \leqslant l-1, \quad \deg(Q, x) \leqslant m-1, \quad \deg(R, x) = 0. \tag{11}$$

Suppose $l \geqslant m$, and

$$A_0 = \deg(f, u), \quad A_1 = \deg(g, u_1) + (l-m+1)A_0, \tag{12}$$

$$A_{k+1} = 2A_k + A_{k-1},$$

then $\deg(R, u_1) \leqslant A_m$.

Proof. In fact, the process of constructing R is that of applying the division algorithm for f and g, which can be decomposed into a series of steps to eliminate the highest power in x. Let

$$g_1 = b_0 x^{l-m} f - a_0 g. \tag{13}$$

From(9), we infer $\deg(g, x) = l_1 < l$, and

$$\deg(g, u_l) \leqslant \max\{\deg(b_0 f, u_l), \deg(a_0 g, u_l)\}$$

$$\leqslant \deg(fg, u_l) \leqslant \deg(f, u_l) + \deg(g, u_l). \tag{14}$$

If $l_1 \geqslant m$, we replace g with g_1 in(9), repeat step(13), then we have g_2. Generally, if having

$$g_k = b_{0,k} x^{l_k} + b_{1,k} x^{l_k - 1} + \cdots + b_{l_k, k}, \tag{15}$$

and $l_k \geqslant m$, where $b_{0,k}$ is a polynomial in $u_1, u_1, u_2, \cdots, u_n$ which is not identical to zero, we take

$$g_{k+1} = b_{0,k} x^{l_k - m} f - a_0 g_k, \tag{16}$$

to obtain a series of $g_1, g_2, \cdots, g_k, \cdots$. By(9), (15), (16), it is easy to know that

$$\deg(g, x) > \deg(g_1, x) > \cdots > \deg(g_k, x) > \deg(g_{k+1}, x) > \cdots, \tag{17}$$

$$\deg(g_{k+1}, u_l) \leqslant \deg(g_k, u_l) + \deg(f, u_l). \tag{18}$$

If $\deg(g_{p_l - 1}, x) \geqslant m > \deg(g_{p_1}, x)$ holds for some $k = p_1$, stop this process. Using (18)

Remark: this result is well known, here we emphasize the estimation for $\deg(R, U_1)$.

and (14) repeatedly, we obtain

$$\deg(g_{p_1}, u_l) \leqslant \deg(g, u_l) + p_1 \deg(f, u_l). \tag{19}$$

Let $f_{-1}=g, f_0=f, f_1=g_{p_1}$, using (16) and (13) repeatedly, we obtain

$$f_1 = P_1 f_0 + Q_1 f_{-1}, \tag{20}$$

where

$$\deg(P_1, x) = l-m, \quad \deg(Q_1, x) = 0,$$
$$\deg(f_1, x) < \deg(f_0, x),$$
$$\deg(f_1, u_1) = \deg(g_{p_1}, u_1) \leqslant \deg(f_1, u_1) + p_1 \deg(f_0, u_1).$$
$$p_1 \leqslant l-m+1.$$

Suppose we have a set of f_j,

$$f_j = a_{0,j} x^{m_j} + a_{1,j} x^{m_j - 1} + \cdots + a_{m,j} \tag{22}$$

for $j=1,2,\cdots,k$, and polynomials P_j, Q_j in $k[u_1, u_2, \cdots, u_n, x]$ such that

$$f_j = P_j f_{j-1} + Q_j f_{j-2}, \tag{23}$$

where

$$\deg(P_i, x) = m_{i-2} - m_{i-1}, \quad \deg(Q_1, x) = 0 \quad (m_0 = m, m_1 = 1),$$
$$\deg(f_i, x) < \deg(f_{i-1}, x),$$
$$\deg(f_j, u_i) \leqslant \deg(f_{j-2}, u_i) + p_j \deg(f_{j-1}, u_i), \tag{24}$$
$$p_j \leqslant m_{j-2} - m_{j-1} + 1 (j=1,2,\cdots,k; i=1,2,\cdots,n).$$

If $\deg(k_k, x) > 0$, let $f=f_k, g=f_k$, in(9), repeating the process(13)-(16), we obtain f_{k+1}, satisfying(22), (23), (24), for $j=1,2,\cdots,k+1$, i. e. dividing f_{k-1} by f_k to obtain f_{k+1}, which is a polynomial in $u_1, u_2, \cdots, u_n, x$. In the above process, it is quite evident that (22), (23), (24) hold. By (20) and $f=f_0, g=f_1$, we have

$$f_1 = P_1 f + Q_1 g, \tag{25}$$

combining with

$$f_2 = P_2 f_1 + Q_2 f_0, \tag{26}$$

we obtain

$$f_2 = (P_2 P_1 + Q_2) f + P_2 Q_1 g. \tag{26}$$

Rewrite (26′) as

$$f_2 = \hat{P}_2 f + \hat{Q}_2 g, \tag{27}$$

where $\hat{P}_2 = P_2 P_1 + Q_2, \hat{Q}_2 = P_2 Q_1$, hence(note $\deg(Q_j, x)=0$)

$$\deg(\hat{P}_2, x) \leqslant \deg(P_1, x) + \deg(P_2, x) = l-m+m-m_1 = l-m_1,$$
$$\deg(\hat{Q}_2, x) \leqslant \deg(P_2, x) = m_0 - m_1.$$

In general, if we have

$$f_k = \hat{P}_k f + \hat{Q}_k g, \quad f_{k-1} = \hat{P}_{k-1} f + \hat{Q}_{k-1} g, \tag{28}$$

combining with

$$f_{k+1} = P_{k+1} f_k + Q_{k+1} f_{k-1} \tag{29}$$

we obtain

$$f_{k+1} = (P_{k+1} \hat{P}_k + Q_{k+1} \hat{P}_{k-1}) f + (P_{k+1} \hat{Q}_k + Q_{k+1} \hat{Q}_{k-1}) g. \tag{30}$$

Suppose inductively

$$\deg(\hat{P}_k, x) \leqslant l - m_{k-1}, \quad \deg(\hat{Q}_k, x) \leqslant m - m_{k-1}, \tag{31}$$

let

$$\hat{P}_{k+1} = P_{k+1}\hat{P}_k + Q_{k+1}\hat{P}_{k-1}, \quad \hat{Q}_{k+1} = P_{k+1}\hat{Q}_k + Q_{k+1}\hat{Q}_{k-1},$$

we have

$$f_{k+1} = \hat{P}_{k+1} f + \hat{Q}_{k+1} g, \tag{32}$$

and

$$\begin{aligned} \deg(\hat{P}_{k+1}, x) &\leqslant \deg(P_{k+1}, x) + \deg(\hat{P}_k, x) \leqslant l - m_k, \\ \deg(\hat{Q}_{k+1}, x) &\leqslant \deg(P_{k+1}, x) + \deg(\hat{Q}_k, x) \leqslant m - m_k. \end{aligned} \tag{33}$$

Since $\deg(f_{k+1}, x) = m_{k+1} < \deg(f_k, x) = m_k$, there is a positive integer s, such that $\deg(f_s, x) = 0$, where

$$f_s = \hat{P}_s f + \hat{Q}_s g, \tag{34}$$

and $\deg(\hat{P}_s, x) \leqslant l - m_{s-1} < l$, $\deg(\hat{Q}_s, x) \leqslant m - m_{s-1} < m$. Let $R = f_s$, $P = \hat{P}_s$, $Q = \hat{Q}_s$, we have(10),(11), which are just the first row conclusions of Lemma 2. 3.

Now let us prove $\deg(R, u_t) \leqslant A_m$, where A_m is given by (12). We will conclude that

$$\deg(f_k, u_t) \leqslant A_{m-m_k} \tag{35}$$

for $k = 0, 1, 2, \cdots, s$.

Obviously,(35) holds for $k = 1, 2, \cdots, t < s$. By the last formula in (24), we obtain

$$\begin{aligned} \deg(f_{t+1}, u_t) &\leqslant \deg(f_{t-1}, u_t) + (m_t - 1 - m_1 + 1)\deg(f_t, u_1) \\ &\leqslant A_{m-m_t} + (m_{t-1} - m_t + 1)A_{m-m_t+1}. \end{aligned} \tag{36}$$

Therefore, in order to prove (35), we need only prove

$$A_{m-m_{t-1}} + (m_{t-1} - m_t + 1)A_{m-m_{t+1}+1} \leqslant A_{m-m_t+1}. \tag{37}$$

Let $\alpha = m - m_{t-1}$, $\beta = m_{t-1} - m_t$, then α, β are positive integers, (37) can be replaced by

$$A_{\alpha+\beta+1} \geqslant (\beta+1)A_{\alpha+1} + A_\alpha. \tag{38}$$

We use induction on β, $\beta = 1$, (38) becomes

$$A_{\alpha+2} \geqslant 2A_{\alpha+1} + A_\alpha. \tag{39}$$

It holds because of (12). Suppose that (38) is true for $\beta = \omega$. For $\beta = \omega + 1$, we have

$$\begin{aligned} A_{\alpha+\omega+2} &= 2A_{\alpha+\omega+1} + A_{\alpha+\omega} \\ &\geqslant 2((\omega+1)A_{\alpha+1} + A_\alpha) + A_{\alpha+1} \\ &\geqslant (\omega+2)A_{\alpha+1} + A_\alpha. \end{aligned} \tag{40}$$

So by induction,(35) is true. Let $k = s$ in (35), since $m = \deg(f_s, x) = 0$, we obtain $\deg(R, u_1) \leqslant A_m$, and conclude Lemma 2. 3.

We will cite the following immediate corollaries without proofs. Both corollaries are the principal bases of the subsequent sections.

Corollary 2. 4. *Suppose f and g are as in Lemma 2. 3, there are Q in $K[u_1, u_2, \cdots, u_n, x]$, R in $K[u_1, u_2, \cdots, u_n]$ such that*

(1) $\deg(Q, x) \leqslant \deg(f, x) - 1$,

(2) $\deg(R, u_j) \leqslant A_m$, *where A_m is given by*(12),

(3) *If there is a set of* $\hat{u}_1,\cdots,\hat{u}_n,\hat{x}$ *such that* $f(\hat{u}_1,\cdots,\hat{u}_n,\hat{x})=0$, *we have*

$$Q(\hat{u}_1,\cdots,\hat{u}_n,\hat{x})g(\hat{u}_1,\cdots,\hat{u}_n,\hat{x})=R(\hat{u}_1,\cdots,\hat{u}_n). \tag{41}$$

Corollary 2.5. *In Lemma 2.3, if* $1\leqslant m\leqslant 2$, *the inequality* $\deg(R,u_j)\leqslant A_m$ *can be expressed simply*

$$\deg(R,u_j)\leqslant m\deg(g,u_j)+(ml-m+1)\deg(f,u_j). \tag{42}$$

3. The illustration of the method

In Euclidean geometry, Lobachevsky geometry and spherical geometry, the hypotheses of many theorems can be expressed by a set of algebraic equations as follows:

$$H_k^*: f_k(u_1,u_2,\cdots,u_n;x_1,x_2,\cdots,x_n)=0.\ (k=1,2,\cdots,s) \tag{43}$$

and the conclusion is an algebraic equation

$$C: G(u_1,u_2,\cdots,u_n,x_1,x_2,\cdots,x_n)=0. \tag{44}$$

By Wu-Ritt's algorithm, we can turn (43) into the equivalent "ascending set" [1]

$$H_k: F_k(u_1,u_2,\cdots,u_n,x_1,\cdots,x_k)=0. \tag{45}$$

Many theorems in elementary geometry with hypotheses can be changed easily into the ascending set. In the above equations, $u_1,u_2,\cdots,u_n$ are independent variables. each x_k is the dependent variable determined by the condition (43) or (45).

Applying repeatedly Corollary 2.5 of Lemma 2.3. in substance, applying Wu-Ritt's Well-ordering Theorem, we can mechanically obtain Q_k, which is a polynomial in $u_1,u_2,\cdots,u_n,x_1,\cdots,x_k (k=1,2,\cdots,s)$ and a polynomial R in $u_1,u_2,\cdots,u_n$, such that

(1) the degree of Q_k in x_k is less than that of F_k.

(2) if $\hat{u}_1,\cdots,\hat{u}_n,\hat{x}_1,\cdots,\hat{x}_k$, satisfy (43) (or (45)), they also satisfy

$$Q_1Q_2\cdots Q_nG=R. \tag{46}$$

First regarding F_s and G as f and g_s in x, as x in Lemma 2.3. we obtain Q_s and R_s which does not contain x_s. Next, regarding F_{s-1} and R_s as f and g, we obtain Q_{s-1} and R_{s-1} which does not contain x_{s-1} and x_s. So we obtain successively $Q_s,Q_{s-1},\cdots,Q_1$, and R which does not contain $x_s,x_{s-1},\cdots,x_1$.

If the polynomial R is identical to zero. (46) shows that if hypotheses (43) (or (45)) are satisfied and the non-degenerate condition

$$Q_1Q_2\cdots Q_s\neq 0 \tag{47}$$

holds, it must hold that $G=0$, i. e. the conclusion holds; i. e. the proposition holds generically. Wu's method is to decide whether a proposition true or false by calculating R.

However, according to Corollary 2.5 of Lemma 2.3, we can estimate the degree of R in each u_t without calculating R. We put

$$\deg(R)=(\deg(R,u_1),\deg(R,u_2),\cdots,\deg(R,u_n)). \tag{48}$$

By Lemma 2.2, to assert that R is identical to zero, we need only verify that R vanishes over a lattice array with size

$$(\deg(R,u_1)+1,\deg(R,u_2)+1,\cdots,\deg(R,u_n)+1). \tag{49}$$

We have more than one method using the left side of (46) to decide whether R is equal to zero without calculating R. We may take a set of the independent variables $\hat{u}_1, \hat{u}_2, \cdots, \hat{u}_n$, calculate the dependent variables $\hat{x}_1, \hat{x}_2, \cdots, \hat{x}_n$, from (45), then decide whether (44) holds after substituting them in (44). If $G=0$, we have $R=0$ (if $\hat{u}_1, \hat{u}_2, \cdots, \hat{u}_n$ cannot be determined, which indicates (47) is not satisfied, it still holds that $R=0$).

Another method is to take a set of $\hat{u}_1, \hat{u}_2, \cdots, \hat{u}_n$, substitute them in (44), (45), calculate R by division algorithm. then decide whether R is equal to zero. This process is similar to Wu's division method, but simpler than the latter since the value of each u was given numerically.

Summarily, our mechanical method for proving a theorem can be divided into three steps.

Step 1: Choose the appropriate coordinate system, independent and dependent variables, write down the set of algebraic equations which express the hypotheses and conclusion of the proposition, turn them into the corresponding ascending set.

Step 2: Estimate the upper bound for the degree of $R(u_1, u_2, \cdots, u_n)$ in each u_i obtained by eliminating the dependent variables to determine the size of the lattice array to be verified.

Step 3: Take the particular elements of the lattice array, calculate them one by one. If it fails for an instance, the proposition is false. Otherwise, the proposition is true generically.

Step 1: is similar to Wu's method. Step 2 can be done by hand, it is easy to give an algorithm for estimating upper bound. Usually, Step 3 is done by computer. Some of the following examples show that sometimes computing by hand or analyzing intuitively would be sufficient to verify certain of the non-trivial theorems.

We call the algorithm based on the above principle the parallel numerical method of mechanical theorem proving, because in Step 3, the verification for different elements of the lattice array can proceed in a parallel manner. The remarkable advantages of our method would be

(1) The design of program is simple so that it is easy to popularize.

(2) Different individuals, with different machines, in different times, can cooperate in verifying the same theorem. Of course, one can verify a theorem quickly on parallel computer, as the main work, verifying instances, can be done in highly parallel form.

(3) It is quite free to choose the elements of the lattice array. We can take the instances that are easy to verify.

(4) The numerical calculation needs less memory, so the verification can be done even on a pocket computer.

(5) If the proposition is false, the algorithm can stop when one instance is false. It is favourable particularly to disprove a conjecture.

(6) It is applicable to prove geometric theorems on a general field.

Remark 3.1 In general, many of the hypotheses of geometric propositions are linear or quadratic equations. It is uncomplicated to calculate the exact values of dependent variables after taking $\hat{u}_1, \hat{u}_2, \cdots, \hat{u}_n$. If there are equations with higher degrees in the hypotheses, we can only obtain the approximate values of $\hat{x}_j$, and thus the approximate value of $G(\hat{u}_1, \cdots, \hat{u}_n, \hat{x}_1, \cdots, \hat{x}_s)$. Is it exactly equal to zero?

Suppose F_1 and G are polynomials with integral coefficients, then R is also a polynomial with integral coefficients. If the values of the independent variables are integers, the value of R is also an integer. When $R(\hat{u}_1, \hat{u}_2, \cdots, \hat{u}_n) \neq 0$, we have

$$|R(\hat{u}_1, \hat{u}_2, \cdots, \hat{u}_n)| \geqslant 1. \tag{50}$$

By (46), we conclude

$$|G| = \frac{|R|}{|Q_1 Q_2 \cdots Q_s|} \geqslant \frac{1}{|Q_1 Q_2 \cdots Q_s|} > 0. \tag{51}$$

It is not hard to estimate the upper bound of the coefficients and degrees of Q_j from F_j and G. So, we can estimate the upper bound M of $|Q_1 Q_2 \cdots Q_s|$ for a particular set of $\hat{u}_1, \hat{u}_2, \cdots, \hat{u}_n, \hat{x}_1, \hat{x}_2, \cdots, \hat{x}_s$. If $|G| \neq 0$, then

$$|G| \geqslant \frac{1}{M}. \tag{52}$$

We can estimate G accurately so that the error is less than $1/(2M)$, the exact value of G equals zero only if the approximate value of $|G|$ is less than $1/(2M)$.

The pre-estimation for the coefficients and degrees of Q_j, and how to assure the precision of estimation for G, forms another subject for study.

Remark 3. 2. After taking the values of $\hat{u}_1, \cdots, \hat{u}_n$, the value of x_j may not be unique. At this moment, we can choose any value of x_j, since in the right side of (46), R contains only $u_1, u_2, \cdots, u_n$, so that for the same $\hat{u}_1, \hat{u}_2, \cdots, \hat{u}_n$, satisfying (45), the choice of values for $\hat{x}_1$ such that the value of R is fixed if (45) holds, does not matter.

4. Examples

The following examples show that we can prove non-trivial geometric theorems, including new theorems by the parallel numerical method.

Example 4. 1 (*Simson's Theorem*). Let P be a point on the circumscribed circle of triangle ABC. From P, perpendiculars are drawn to BA, CA, AB, the feet of the perpendiculars are R, S, T, respectively. Prove that R, S, T are collinear.

Suppose the Descartes coordinates of points A, B, C, P, R, S, T, are (x_i, y_i) $(i=1, \cdots, 7)$, respectively. Let the circumcenter of triangle ABC be the origin, the circumradius be 1, the coordinates of $P(x_4, y_4) = (1, 0)$. As R is in BC, S in CA, T in AB, we have

$(x_5, y_5) = \lambda(x_2, y_2) + (1-\lambda)(x_3, y_3)$,

$(x_6, y_6) = \mu(x_3, y_3) + (1-\mu)(x_1, y_1)$,

$(x_7, y_7) = \rho(x_1, y_1) + (1-\rho)(x_2, y_2)$.

Replacing the variables $x_5, x_6, x_7, y_5, y_6, y_7$ with the variables λ, μ, ρ, the hypotheses of the proposition are expressed as follows:

$H_1: x_1^2+y_1^2=1,$

$H_2: x_2^2+y_2^2=1,$

$H_3: x_3^2+y_3^2=1,$

$H_4: (1-\lambda x_2-(1-\lambda)x_3)(x_2-x_3)-(\lambda y_2+(1-\lambda)y_3)(y_2-y_3)=0,$

$H_5: (1-\mu x_3-(1-\mu)x_1)(x_3-x_1)-(\mu y_3+(1-\mu)y_1)(y_3-y_1)=0,$

$H_6: (1-\rho x_1-(1-\rho)x_2)(x_1-x_2)-(\rho y_1+(1-\rho)y_2)(y_1-y_2)=0.$

where H_1, H_2, H_3 indicate that A, B, C are on the unit circle, H_4, H_5, H_6 indicate that $PR\perp BC$, $PS\perp CA$, $PT\perp AB$. The conclusion to be proved is given by

$$(x_5-x_6)(y_5-y_7)=(x_5-x_7)(y_5-y_6),$$

i. e.

$$x_5y_6-x_6y_5+x_6y_7-x_7y_6+x_7y_5-x_5y_7=0.$$

Substituting λ, μ, ρ for $x_5, x_6, x_7, y_5, y_6, y_7$ in the last expression, we obtain

$$\begin{aligned} C: &(\lambda x_2+(1-\lambda)x_3)(\mu y_3+(1-\mu)y_1-\rho y_1-(1-\rho)y_2) \\ &+(\mu x_3+(1-\mu)x_1)(\rho y_1+(1-\rho)y_2-\lambda y_2-(1-\lambda)y_3) \\ &+(\rho x_1+(1-\rho)x_2)(\lambda y_2+(1-\lambda)y_3-\mu y_3-(1-\mu)y_1)=0. \end{aligned}$$

To reduce H_1, H_2, H_3, take an usual transformation

$$\begin{cases} u_k=x_k+iy_k, \\ v_k=x_k-iy_k, \end{cases} \quad \begin{cases} x_h=\dfrac{1}{2}(u_k+v_k), \\ y_k=\dfrac{1}{2i}(u_k-v_k), \end{cases} \quad i=\sqrt{-1}. \ (k=1,2,3).$$

Then, the hypothesis equations can be written as

$H_1: F_1(u_1, v_1)=u_1v_1-1=0,$

$H_2: F_2(u_2, v_2)=u_2v_2-1=0,$

$H_3: F_3(u_3, v_3)=u_3v_3-1=0,$

$H_4: F_4(u_2, u_3, v_2, v_3, \lambda)=0,$

$H_5: F_5(u_1, u_3, v_1, v_3, \mu)=0,$

$H_6: F_6(u_1, u_2, v_1, v_2, \rho)=0,$

the conclusion equation as

$$C: \quad G(u_1, u_2, u_3, v_1, v_2, v_3, \lambda, \mu, \rho)=0.$$

We need only estimate the degrees in each variable without writing the accurate expressions.

By a 9-dimensional vector we denote the degrees of a polynomial in $u_1, u_2, u_3, v_1, v_2, v_3, \lambda, \mu, \rho$. In practice,

$$\deg(G)\leqslant(1,1,1,1,1,1,1,1,1)$$

expresses that the degree of G in the nine variables is not more than 2, and

$$\deg(F_6)\leqslant(1,1,0,1,1,0,0,0,1).$$

By the Corollary 2. 5 of Lemma 2. 3, dividing G by F_6 (in ρ), we obtain R_6, such that

$$\det(R_6)\leqslant(2,2,1,2,2,1,1,1,0).$$

Since

$$\deg(F_5)\leqslant(1,0,1,1,0,1,0,1,0),$$

dividing R_6 by F_5 (in μ), we obtain R_5, such that

$$\deg(R_5)\leqslant(3,2,2,3,2,2,1,0,0),$$

dividing R_5 by F_4 (in λ), we obtain R_4, such that

$$\deg(R_4)\leqslant(3,3,3,3,3,3,0,0,0).$$

Regarding u_1, u_2, u_3 as independent variables, dividing R_4 by F_3, F_2, F_1 successively, we obtain R only in u_1, u_2, u_3:

$$\deg(R)\leqslant(6,6,6,0,0,0,0,0,0).$$

That means we need only verify the theorem on a lattice array with size (7, 7, 7). The geometric significance is: choose arbitrarily 7 points on the unit circle (may be imaginary point, e. g. $(\sqrt{3}, \sqrt{-2})$), take any 3 points as A, B, C, then verify $G=0$ is true or false. Suppose P is one of the 7 points. Obviously, if P coincides with one of A, B, C, then $G=0$. If two of A, B, C are coincident, also $G=0$. Deleting these trivial instances, the number of instances to be verified is

$$\frac{1}{6}(6\times5\times4)=20,$$

which is easy to do on a microcomputer.

In practice, we first choose u_1, u_2, u_3, then turn them into x_i, y_i for calculation, so that we can use the original equations without doing the transformation. Since it only involves rational operations, we can avoid the errors of values in the computer.

Example 4. 2 (*Ptolemy's Theorem*). A, B, C and D are four points on a circle. Show

$$AB\cdot CD\pm AD\cdot BC\pm AC\cdot BD=0.$$

Suppose the hypotheses are

$$H_i:\quad x_i^2+y_i^2=1\quad(i=1,2,3,4),$$

the conclusion is

$$\sqrt{[(x_1-x_2)^2+(y_1-y_2)^2][(x_3-y_4)^2+(x_4-y_4)^2]}$$
$$\pm\sqrt{[(x_1-x_4)^2+(y_1-y_4)^2][(x_2-x_3)^2+(y_2-y_3)^2]}$$
$$\pm\sqrt{[(x_1-x_3)^2+(y_1-y_3)^2][(x_2-x_4)^2+(y_2-y_4)^2]}=0.$$

By H_1-H_4, the conclusion becomes

$$\sqrt{(1-x_1x_2-y_1y_2)(1-x_3x_4-y_3y_4)}\pm\sqrt{(1-x_1x_4-y_1y_4)(1-x_2x_3-y_2y_3)}$$
$$\pm\sqrt{(1-x_1x_3-y_1y_3)(1-x_2x_4-y_2y_4)}=0.$$

After removing the signs of the square roots twice, we have

$$C:\quad G(x_1,x_2,x_3,x_4,y_1,y_2,y_3,y_4)=0,$$

the sum of degrees of in x_t, and y_t is not more than 2 for each fixed subscript i. Take the transformation

$$x_i=\frac{1}{2}(u_i+v_i),\quad y_i=\frac{1}{2i}(u_i-v_i),(i=1,2,3,4).$$

The hypotheses become

$$\mathrm{H_i}: \quad u_i v_i = 1, (i=1,2,3,4),$$

and the conclusion becomes

$$\mathrm{C}: \quad G^*(u_1, u_2, u_3, u_4, v_1, v_2, v_3, v_4) = 0,$$

where the degree in each variable is not more than 2. Substituting $t_i = u_i^{-1}$ in G^*, and eliminating denominators, we obtain

$$\Phi(u_1, u_2, u_3, u_4) = 0,$$

where $\deg(\Phi, u_i) \leqslant 4 (i=1,2,3,4)$. According to Lemma 2.2, all we have to do is verify the conclusion on a lattice array with size (5,5,5,5), e.g. take

$$u_i \in \{1, e^{(2\pi/5)i}, e^{(4\pi/5)i}, e^{-(2\pi/5)i}, e^{-(4\pi/5)^i}\},$$

where $i=1,2,3,4$, i.e. all u_i's are the vertices of a regular pentagon. We notice that $u_i \neq u_j$ for $i \neq j$. If two vertices of a quadrangle are coincident, the theorem holds obviously. Now, we need only discuss the cases in which u_1, u_2, u_3, u_4 are mutually distinct. In these cases, the validity of Ptolemy's Theorem turns into the validity of equality $a^2 = a+1$, where a is the ratio of the diagonal length to the side length of a regular pentagon.

Using one trivial instance we have proved the Ptolemy's Theorem! Of course, such examples are very rare. Usually, we need to calculate several numerical instances, as in Example 4.1.

Now, let us introduce two new theorems of spherical geometry which we found by means of the parallel numerical method.

Example 4.3. Suppose the area of a spherical triangle on a unit sphere is π. Show the spherical distance of any two midpoints of sides is $\frac{1}{2}\pi$, i.e. the midpoints of three sides form an equilateral right triangle.

Let A, B, C denote the three angles of the triangle and a, b, c denote the corresponding sides, respectively. Suppose the spherical distance of the two midpoints of a, b is m. By the Cosine Law, we have

$$\cos m = \cos C \sin \frac{1}{2}a \sin \frac{1}{2}b + \cos \frac{1}{2}a \cos \frac{1}{2}b,$$

the conclusion of the theorem, $m = \frac{1}{2}\pi$, is equivalent to

$$\cos C \sin \frac{1}{2}a \sin \frac{1}{2}b + \cos \frac{1}{2}a \cos \frac{1}{2}b = 0.$$

In order to eliminate the half angles, transpose the terms, then calculate the square. We obtain

$$\cos^2 C(1-\cos a)(1-\cos b) = (1+\cos a)(1+\cos b),$$

where $\cos a$ and $\cos b$ can be defined as follows

$$\cos A = \cos a \sin B \sin C - \cos B \cos C,$$
$$\cos B = \cos b \sin A \sin C - \cos A \cos C.$$

The hypothesis of the theorem, the area is π, is equivalent to $A+B+C=2\pi$. So, we can replace $\cos C$ and $\sin C$ with $\cos(A+B)$ and $-\sin(A+B)$ to eliminate angle C in the above equalities. Let

$$s_1=\cos a, s_2=\cos b,$$

the conclusion can be written as the polynomial equality

$$\mathrm{C}: G(\cos A, \cos B, \sin A, \sin B, s_1, s_2)=0.$$

It is easy to see, in each term of G, the sum of degrees in $\cos A$ and $\sin A$ and the sum of degrees in $\cos B$ and $\sin B$ are not more than 2.

The hypotheses (i. e. the two above cosine formulas concerning $\cos a$ and $\cos b$) can be written as follows

$$\mathrm{H}_1: \quad f_1(\cos A, \cos B, \sin A, \sin B, s_1, s_2)=0,$$
$$\mathrm{H}_2: \quad f_2(\cos A, \cos B, \sin A, \sin B, s_1, s_2)=0.$$

In each term of f_1, the sum of degrees in $\cos A$ and $\sin A$ is not more than 1, the sum of degrees in $\cos B$ and $\sin B$ is not more than 2, the degrees in s_1 and s_2 are not more than 1 and 0, respectively. And in each term of f_2, the sum of degrees in $\cos A$ and $\sin A$ is not more than 2, the sum of degrees in $\cos B$ and $\sin B$ is not more than 1, the degrees in s_1 and s_2 are not more than 0 and 1, respectively.

Leading into new variables

$$u_1=\mathrm{e}^{iA}, \quad u_2=\mathrm{e}^{iB}.$$

Substituting them in f_1, f_2, G and reducing the denominators, we obtain

$$\mathrm{H}_1: \quad f_1^*(u_1, u_2, s_1, s_2)=0,$$
$$\mathrm{H}_2: \quad f_2^*(u_1, u_2, s_1, s_2)=0,$$
$$\mathrm{C}: \quad G^*(u_1, u_2, s_1, s_2)=0,$$

where the degrees of f_1^*, f_2^* and G^* in variables are not more than (2, 4, 1, 0), (4, 2, 0, 1) and (4, 4, 1, 1), respectively. By Corollary 2. 5 of Lemma 2. 3, we know that if we divide G^* by f_1^*, f_2^* (in s_1, s_2), we obtain the polynomial equality

$$\varnothing(u_1, u_2)=0,$$

where the degrees of $\varnothing$ in u_1, u_2 are not more than 10. By Lemma 2. 2, regarding the symmetry, we need only verify 66 pairs of variables (u_1, u_2).

Example 4. 4. Given a spherical triangle whose perimeter is half of the length of the great circle. Prove the sum of the cosines of the three internal angles equals 1.

Let the lengths of sides be a, b, c, the corresponding internal angles be A, B, C, respectively. By the Cosine Law, we need only prove

$$\frac{\cos a-\cos b \cdot \cos c}{\sin b \cdot \sin c}+\frac{\cos b-\cos c \cdot \cos a}{\sin c \cdot \sin c}+\frac{\cos c-\cos a \cdot \cos b}{\sin a \cdot \sin b}=1.$$

Eliminating the denominators, we have a triangular polynomial equality

$$\sin a \cdot \sin b \cdot \sin c+\sin a \cdot \cos b \cdot \cos c+\cos a \cdot \sin b \cdot \cos c$$
$$+\cos a \cdot \cos b \cdot \sin c-\sin a \cdot \cos a-\sin b \cdot \cos b$$
$$-\sin c \cdot \cos c=0.$$

By the hypothesis, $a+b+c=\pi$, we can replace sin c and cos c with $\sin(a+b)$ and $-\cos(a+b)$. Eliminating c from the last expression, we obtain

$$G(\cos a, \sin a, \cos b, \sin b)=0,$$

where, in each term, the sum of degrees in cos a and sin a is not more than 2, similarly with cos b and sin b. Now lead into new variables

$$u_1=e^{ia}, u_2=e^{ib}.$$

After substituting and reducing the demoninators, we obtain the equality to be proved

$$\Phi(u_1, u_2)=0,$$

where the degrees of Φ in u_1, u_2, are not more than 4. So we need only verify the theorem on a lattice array with size(5,5), in addition to symmetry, we can conclude our theorem by calculating 15 instances.

This theorem is quite odd; it is hard to imagine that for some triangle with a certain perimeter, its sum of cosines of internal angles is invariant. There is no similar result in Euclidean or Lobachevsky geometry.

Example 4. 5(*Pappus' Theorem*). Given two distinct lines l, m. Let A_1, A_2, A_3 be on l, and B_1, B_2, B_3 be on m. By C_3, C_2, C_1 denote the intersection points of A_1B_2 and A_2B_1, A_1B_3 and A_3B_1, A_2B_3, and A_3B_2, respectively. Show C_1, C_2, and C_3 are collinear.

We only discuss the case in which l and m are not parallel. We can choose I as the X-axis, m as the Y-axis. Let the coordinates of A_1, A_2, A_3 be $(u_1,0), (u_2,0), (u_3,0)$, B_1, B_2, B_3 be $(0,v_1), (0,v_2), (0,v_3)$, and C_1, C_2, C_3 be $(x_1,y_1), (x_2,y_2), (x_3,y_3)$.

The hypotheses of the theorem are

$$\begin{aligned}
&H_1: v_2x_3+u_1y_3-u_1v_2=0,\\
&H_2: v_1x_3+u_2y_3-u_2v_1=0,\\
&H_3: v_3x_2+u_1y_2-u_1v_3=0,\\
&H_4: v_1x_2+u_3y_2-u_3v_1=0,\\
&H_5: v_3x_1+u_2y_1-u_2v_3=0,\\
&H_6: v_2x_1+u_3y_1-u_3v_2=0,
\end{aligned}$$

and the conclusion is

$$C: \begin{vmatrix} x_1 & y_1 & 1 \\ x_2 & y_2 & 1 \\ x_3 & y_3 & 1 \end{vmatrix}=0.$$

By H_1-H_6, the dependent variables can be expressed as follows

$$x_1=\frac{L_1}{L}, y_1=\frac{L_2}{L},$$

$$x_2=\frac{M_1}{M}, y_2=\frac{M_2}{M},$$

$$x_3=\frac{N_1}{N}, y_3=\frac{N_2}{N},$$

where L, L_1, L_2 are the polynomials in u_2, u_3, v_2, v_3, M, M_1, M_2 are the polynomials in u_1,

u_3, v_1, v_3, and N, N_1, N_2 are the polynomials in u_1, u_2, v_1, v_2; the degrees in each variable are not more than 1. Substituting into the conclusion and eliminating the denominators, we have

$$\Phi=\begin{vmatrix} L_1 & L_2 & L \\ M_1 & M_2 & M \\ N_1 & N_2 & N \end{vmatrix}=0,$$

where the degree of $\Phi=\Phi(u_1, u_2, u_3, v_1, v_2, v_3)$ in each variable is obviously not more than 2. So we can choose $\{0,1,2\}$ as the range of each variable. It is clear that $u_i=0$ implies that A_i is the origin, at this moment, there are two of C_1, C_2, C_3 coincident; obviously the theorem is true. Thus we can choose u_1, u_2, u_3 in $\{1,2\}$; this means that there are two of A_1, A_2, A_3 coincident, e. g. A_1 and A_2 are coincident, so all C_1, C_2, C_3 are on A_2B_3; the theorem is true again. We have turned the theorem into trivial cases.

Example 4. 6(*Morley's theorem*). Given a triangle, a trisectrix of each angle intersects the adjacent trisectrix of another angle. Such three intersection points form an equilateral triangle.

For convenience, we suppose the three angles are $3A$, $3B$, $3C$. By the Sine Law and Cosine Law, we write the conclusion of the theorem as follows

$$\begin{aligned}
&[\sin C\cdot\sin 3B\cdot\sin(A+B)\cdot\sin(B+C)]^2\\
&+[\sin B\cdot\sin 3C\cdot\sin(C+A)\cdot\sin(C+B)]^2\\
&-[\sin C\cdot\sin 3A\cdot\sin(C+A)\cdot\sin(A+B)]^2\\
&-[\sin A\cdot\sin 3C\cdot\sin(C+A)\cdot\sin(B+C)]^2\\
&-2\cos A\cdot\sin B\cdot\sin C\cdot\sin 3B\cdot\sin 3C\cdot\sin(C+A)\cdot\sin(A+B)\\
&\cdot\sin^2(B+C)+2\cos B\cdot\sin C\cdot\sin A\cdot\sin 3C\cdot\sin 3A\\
&\cdot\sin(B+C)\cdot\sin(A+B)\cdot\sin^2(C+A)=0.
\end{aligned}$$

As usual, we turn the trigonometric function into the exponential function. Let

$$u_1=e^{iA}, u_2=e^{iB}, y=e^{iC}, x=\sqrt{-3}.$$

The hypothesis, $A+B+C=\frac{1}{3}\pi$, can be expressed as follows:

$$H_1: x^2+3=0,$$

$$H_2: 1+x-2u_1u_2y=0,$$

and the conclusion as

$$C: G(u_1, u_2, y)=0.$$

It is easy to see all the degrees of G in three variables are not more than 10. According to Corollary 2. 5 of Lemma 2. 3, we can estimate immediately the equality

$$G_1(u_1, u_2, x)=0,$$

which is concluded by eliminating the dependent variable y (dividing G by H_2), where the degrees of G_1 in u_1, u_2 are not more than 20. If we divide G_1 by H_1 to eliminate x, the degrees in variables would increase to 40. But here, we take another way.

Let

$$G_1(u_1,u_2,x)=\Phi_1(u_1,u_2)+\Phi_2(u_1,u_2)x.$$

By H_1, we only need to decide whether $\Phi_1(u_1,u_2)$ and $\Phi_2(u_1,u_2)$ are identical to 0. By Lemma 2. 2, each of them should be verified on a lattice array of size(21,21). Considering the symmetry, we need only verify 210 pairs of variables (u_1,u_2).

Example 4. 7(*Butterfly Theorem*). From M, the midpoint of a chord PQ of a circle, draw other chords AB and CD, the chords AD and BC intersect PQ in X and Y, respectively, Show M is also the midpoint of XY.

We can estimate the degree of R, a polynomial only in independent variables u_1, u_2, u_3, in each u_1 not more than 8. Since A, C, Q are not coincident with $P(1,0)$, and distinct mutually. We need only verify $8\times7\times6=336$ instances.

Example 4. 8(*Feuerbach's Theorem*). The nine-point circle of a triangle is tangent with the circles which contact with all sides of the triangle.

We can estimate the degree of R, the last polynomial in two independent variables u_1 and u_2, in each u_i not more than 16, so we need verify $17\times17=289$ instances.

Appendix

The running time of the above examples on the PB－700 pocket computer(storage 4K (can be expanded to 16K), significant digit length 12, language BASIC)

Example 4. 1. 40 s, each instance 2s.

Example 4. 2. Need not run.

Example 4. 3. 150 s, each instance 3 s.

Example 4. 4. 11 s, each instance 1 s.

Example 4. 5. Need not run.

Example 4. 6. 23 min; for this example, we designed a program of doubling precision for the operation using a significant digit length of 17, so the calculation slows down.

Example 4. 7. 17 min.

Example 4. 8. 15 min; a program of doubling precision was used.

Acknowledgement

In this study we are inspired by Professor Wu Wentsun's related works. One of the authors has an exchange of views on the possibility of parallel numerical method with Professors Hong Jiawei and Hou Zhenting. We would like to express our thanks to them.

References

[1]Deng Mike. the parallel numerical method of constructive geometric theorem proving.

Sri, Bull China, to appear.

[2]Hong Jiawei. Can the single-instance method prove a geometric theorem? Sci. Sinica., 1986(29):234-242, (in Chinese).

[3]Hong Jiawei. The growth of significant digit length in approximate calculation is not faster than that of geometrical series. Sci. Sinica., 1986(29):225-233, (in Chinese).

[4]Tao Maoqi, Zhang Jingzhong, Yang Lu. Proving a geometric inequality by multi-instance method. Sci. Sinica. China, to appear.

[5]Wu Wen-tsun. Basic Principles of Mechanical Theorem-proving of Geometries(Part on Elementary Geometries. Science Press. China, 1984.

[6]Wu Wen-tsun. On the decision problem and the mechanization of theorem-proving in elementary geometry. Sci. Sinica, 1978(21): 159-172(re-published in Automated Theorem Proving: After 25 Years(Amer. Mathematical Soc. Providence, RI, 1983)213-234).

[7]Wu Wen-tsun. Toward mechanization of geometry—Some comments on Hilbert's "Grundlagen der Geometrie". Acta Math. Scientia, China, 1982(2):125-138.

[8]Wu Wen-tsun. Basic principles of nechanical theorem proving in elementary geometries. J. System sct Math. Sci., China, 1984(4):207-235.

[9]Wu Wen-tsun. A mechanization method of geometry 1 Elementary geometry. Chinese Quart. J. Math 1, 1986(1):1-14.

[10]Wu Wen-tsun. Mechanization of geometry. J, System Sci. Math Sci., 1986(6): 204-216.

[11]Wu Wen-tsun. Some recent advances in mechanical theorem proving of geometries, in Automated Theorem Proving After 25 Years(Amer. Mathematical Soc. Providence, RI. 1983:235-242,

[12]Zhang Jingzhong, Yang Lu. Principle of parallel numerical method and the single-instance method of mechanical theorem proving. Math. Practice Theory, China, 1989(1):34-43.

A Method to Overcome the Reducibility Difficulty in Mechanical Theorem Proving

Zhang Jing-zhong
International Centre for Theoretical Physics, Trieste, Italy

and Yang Lu
Department of Mathematics, La Trobe University,
Bundoora, Victoria 3083, Australia

ABSTRACT

In this paper, an algorithm is introduced to overcome the difficulty caused by the reducibility of some algebraic varieties which express the hypotheses of theorems to be proved. The problem arose as one used algebraic methods for automated theorem proving. Here we show that the difficulty could be overcome if one uses the mutual pseudo division to replace or complete the pseudo division in Wu's method which has had great success in this field.

1. INTRODUCTION

In 1977, Wu[1]suggested an algebraic method for automated theorem proving, which is greatly successful in some fields including elementary geometry. S. C. Chou[2]published in 1988 a monograph in which 512 non-trivial geometry theorems were proved mechanically by using Wu's method.

But, just as what has been pointed out in [1] and [2], the Wu method and other algebraic methods may be invalid for checking a proposition which involves a so-called reducible variety. This difficulty was much discussed and vigorously pursued([3]-[6]). Both Chou[2]and Wu [3]have proposed some ways to avoid or lighten the difficulty and have achieved some progress. But in general, it is still a big troublesome problem which

本文刊于《INTERNATIONAL CENTRE FOR THEORETICAL PHYSICS》, 1990 年.

restricts the existing algebraic methods to be used smoothly and efficiently for our purpose.

In this paper, an improved method will be proposed. It is based on the Wu method. The key of our new method is to replace or complete the pseudo division to reduce the independent unknowns in the Wu method by the mutual pseudo division. With the improved method, one can determine whether a proposition is generically true or false, and can further determine the sub-varieties on which the proposition is generically true or false. And then, of course, it is not necessary to do some difficult work of the Ritt decomposition algorithm which is the first step of the so-called "complete method of Wu-general cases" suggested in[2].

One new method has been proposed by Zhang Jingzhong, Yang Lu and Deng Mike in [7], where it was only the starting point of the parallel numerical method and not to be connected with the reducibility difficulty.

This work is organized as follows: some notations are introduced in Sec. 2, the new method is explained in Sec. 3 and last, some examples are discussed in Sec. 4.

2. NOTATIONS, SOME PRELIMINARY KNOWLEDGE

Let K be a field with characteristic zero, and $A=K[x_1,\cdots,x_m]$ be a polynomial ring with unknowns $x_1,\cdots,x_m$. The field of rational functions over K with unknowns $x_1,\cdots,x_m$ be denoted by $K(x_1,\cdots,x_m)$. We often use their abbreviation such as $x=x_1,\cdots x_m$, and $K(x)=K(x_1,\cdots,x_m)$, etc.

Suppose $f_1,\cdots,f_n$ be a set of polynomials in $K[x]$, then $K[x]/(f_1,\cdots,f_n)$ denote the quotient ring of $K[x]$ modulus $f_1,\cdots,f_n$.

Given an extension field of K, named E, and let E^m be the m-fold Cartesian product of E. A subset V of E^m is called an algebraic set or a variety, if V is the set of common zeros of all elements of a nonempty polynomial set S, i. e.,

$$V=\{(a_1,\cdots,a_m)\in E^m \mid f(a_1,\cdots,a_m)=0 \quad \text{for all} \quad f\in S\}.$$

We denote V by $V(S)$, or E-$V(S)$.

A nonempty variety is called reducible variety if it can be expressed as the union of two proper subsets, where each of them is a variety. Otherwise, it is called an irreducible one.

Let f be a polynomial. Denote the degree of f in the variable x_i by $deg(f,x_i)$. The class of f is the smallest integer c such that f is in $K[x_1,\cdots,x_c]$. We denote it by $class(f)$. If f is in K we define $class(f)=0$. Let $c=class(f)>0$ and $lv(f)$ denote the leading variable x_c of f. Considering f as a polynomial in x_c, let $lc(f)$ denote the leading coefficient of f which is also called the initial of f. Let $ld(f)$ denote the leading degree of f, i. e, $deg(f, x_c)$. A polynomial g is reduced with respect to f if $deg(g,x_c)<deg(f,x_c)$, where $c=class(f)>0$.

Let $C=f_1,f_2,\cdots,f_n$ be a sequence of polynomials in $K[x]$. We call it an ascending chain if either $s=1$ and $f_1 \neq 0$ or $s>1$ and satisfy that

(ⅰ) $0<class(f_i)<class(f_j)$ for $i<j$.

(ⅱ) The initials of the f_j are reduced with respect to f_i for $i<j$.

Suppose $C=f_1, \cdots, f_n$ is an ascending chain, not consisting of a constant. After a suitable renaming of the x_i, we may assume that $class(f_i)=i$ for $i=1,\cdots,n$, and distinguish the x_{a+j} by calling them u_j for $j=1,\cdots,d$ where $d \geqslant 0$ and $d+s=m$. We call d the dimension of variety $V(C)$, denote it by $d=dim\ V(C)$. If C_1 is also an ascending chain and $V(C_1) \subset V(C)$, we say $V(C_1)$ is a degenerate subvariety of $V(C)$ when $dim\ V(C_1)<dim\ V(C)$.

An ascending chain $f_1, \cdots, f_n$ is called irreducible if each f_i is irreducible on the quotient ring $K(u)[x_1, \cdots, x_{i-1}]/(f_1, \cdots, f_{i-1})$. Suppose that $f_1, \cdots, f_{k-1}$ $(0<k \leqslant 0)$ is irreducible, but $f_1, \cdots, f_k$ is reducible, then it is easy to show that there are polynomials g and h in $K[u,x]$ reduced with respect to $f_1, \cdots, f_k$ such that $class(g)=class(h)=class(f_k)$ and $gh \in$ the ideal $(f_1, \cdots, f_k)$ of A, i. e.,

$$gh=Q_1 f_1+\cdots+Q_k f_k \quad (Q_j \in A=K[u,x]).$$

Let

$$C_1=f_1, \cdots, f_{k-1}, g, f_{k+1}, \cdots, f_s$$
$$C_2=f_1, \cdots, f_{k-1}, g, f_{k+1}, \cdots, f_s.$$

Then we have two ascending chains C_1 and C_2, and both $V(C_1)$ and $V(C_2)$ are the non-degenerate subvariety of $V(C)$.

3. NEW METHOD AND ITS PRINCIPLE

In our method, the basic algebraic tool is still pseudo division. Now let us describe this operation.

Let A be a commutative ring, and

$$f=a_n v^n+\cdots+a_0$$
$$h=b_k v^k+\cdots+b_0 \quad (k>0, b_k \neq 0)$$

be two polynomials in $A[v]$, where v is a new indeterminate. Then the pseudo division proceeds as follows:

Pseudo Division First let $r=f$. Then repeat the following process until $m=deg(r, v)<k$: $r:=b_k{}^r-c_m v^{m-k}h$, where c_m is the leading coefficient of r. It is easy to see that m strictly decreases after each iteration. At the end, we have the pseudo remainder $prem(f,h,v)=r=r_0$ and the following formula,

$$\begin{cases} b_k^{\lambda} f=qh+r_0 \ (\lambda \leqslant n-k+1, \deg(r_0, v)<\deg(h,v)) \\ r_0=\text{prem}(f,h,v). \end{cases} \tag{3.1}$$

If there is an ascending chain

$$C: f_i(u_1, x_1, \cdots, x_i) \ (i=1,2,\cdots,s)$$

and a polynomial $g=g(u_1, x_1, \cdots, x_s)$, then we can do successive pseudo divisions:

$$\begin{cases} R_{s-1}=prem(g,f_s,x_s) \\ R_{s-2}=prem(R_{s-1},f_{s-1},x_{s-1}) \\ \cdots \\ R_0=prem(R_1,f_1,x_1). \end{cases} \tag{3.2}$$

Here R_0 is called the final remainder and is denoted by

$$R_0=prem(g,f_1,\cdots,f_s) \tag{3.3}$$

The Wu method is based on the The Remainder Formula of SPD (successive pseudo divisions): Let $f_1,\cdots,f_k$ and R_0 be the same as above. There are some non-negative integers $\lambda_1,\cdots,\lambda_n$ and polynomials $Q_1,\cdots,Q_s$ such that

(1) $I_1^{\lambda_1}\cdots I_s^{\lambda_s}g=Q_1f_1+\cdots+Q_sf_s+R_0$, where the I_j are the leading coefficients of f_j.

(2) $\deg(R_0,x_i)<\deg(f_i,x_i)$, for $i=1,\cdots,r$.

The Wu method told us: If $f_1=\cdots=f_k=0$ and $I_j\neq 0$ for $j=1,\cdots,s$, then $R_0\equiv 0$ implies $g=0$. When the ascending chain $f_1,\cdots,f_k$ are irreducible, $g=0$ also implies $R_0\equiv 0$. Here $R_0\equiv 0$ means that R_0 is identity with zero. But if $R_0\not\equiv 0$ and the ascending chain $f_1,\cdots,f_k$ are reducible, we could not know whether $g=0$ or not. This is the problem we want to solve.

Now let us introduce the mutual pseudo division or Euclidean pseudo division which is based on the pseudo division. Let f and h be the same as above, then the mutual pseudo division proceeds as follows:

Mutual Pseudo Division First let $r_{-1}=f$ and $r_0=h$, then repeat the following process until $deg(r,v)=0$ or $r=0$ (i.e., $r\in A$):

$$\begin{cases} r:=prem(r_{-1},r_0,v) \\ r_{-1}:=r_0 \\ r_0:=r. \end{cases} \tag{3.4}$$

Suppose $r_1,r_2,\cdots,r_{l+1}$ are the remainders which were given in the process (3.4). By (3.1). we have

$$J_lr_{l-1}=q_lr_l+r_{l+1}\,(l=0,\cdots\cdots,t) \tag{3.5}$$

where $J_l\in A$ and $J_l\neq 0$, and $q_l\in A[v]$. It is easy to know that $t<k$ and $deg(r_l,v)<deg(r_{l-1},v)$. By (3.4), the final remainder r_{l+1} is in A. We denote it by

$$r_{t+1}=Eprem(f,h,v) \tag{3.6}$$

We introduce polynomials P_l and Q_l as follows:

$$\begin{cases} P_t=J_{t_1}Q_t=-q_t \\ P_{l-1}=J_{l-1}Q_l \\ Q_{l-1}=P_l-q_{l-1}Q_l,\ l=t,t-1,\cdots,1. \end{cases}$$

Then we have

Lemma 1

$$P_lr_{l-1}+Q_lr_l=r_{l+1}\,(l=0,1,\cdots,t). \tag{3.8}$$

Proof When $l=t$, we know that (3.8) is true by setting $l=t$ in (3.5). For $l=t-1,\cdots,$

0, we use the backward mathematical induction.

Suppose (3.4) is true for $l=k+1$ and for $l=k$. We are going to prove that it is still true for $l=k-1$. In fact, taking $l=k-1$ in (3.5) and $l=k$ in (3.8), we get

$$J_{k-1}r_{k-2}=q_{k-1}r_{k-1}+r_k \tag{3.9}$$

$$P_k r_{k-1}+Q_k r_k=r_{t+1} \tag{3.10}$$

To reduce r_k from (3.9) and (3.10), we have

$$J_{k-1}Q_k r_{k-2}+(P_k-q_{k-1}Q_k)r_{k-1}=r_{t+1} \tag{3.11}$$

Noticed that $J_{k-1}Q_k=P_{k-1}$ and $P_k-q_{k-1}Q_k=Q_{k-1}$ by (3.7), we have proved Lemma 1.

Lemma 2 If $r_{t+1}=0$, then

$$Q_l r_t=(-1)^{t-l+1}J_l J_{l+1}\cdots J_t\cdot r_{l-1}\quad (l=0,1,\cdots,t) \tag{3.12}$$

Proof When $l=t$, we know that (3.12) is true by setting $l=t$ in (3.8) since $r_{t+1}=0$. For $l=t$ to 0 use again the backward mathematical induction, Suppose (3.12) is true for $l\geqslant k$. We are going to prove that it is still true for $l=k-1$. In fact, using (3.7) and the inductive assumption we obtain

$$\begin{aligned} Q_{k-1}r_t &=(P_k-q_{k-1}Q_k)r_t \\ &=(J_kQ_{k+1}-q_{k-1}Q_k)r_t \\ &=(-1)^{t-k+2}J_kJ_{k+1}\cdots J_t(r_k+q_{k-1}r_{k-1}) \\ &=(-1)^{t-k}J_{k-1}J_k\cdots J_t\cdot r_{k-2} \end{aligned} \tag{3.13}$$

and the lemma is proved.

For convenience, we introduce some notations to represent P_0, Q_0, r_t and $(-1)^tJ_0J_1\cdots J_t$ which has already been used:

$$\begin{cases} E_{pp}(f,h,v)=P_0 \\ E_{Pr}(f,h,v)=Q_0 \\ E_{pr}(f,h,v)=r_t \\ E_{pJ}(f,h,v)=(-1)^tJ_0J_1\cdots J_t. \end{cases}$$

Taking $l=0$ and $l=1$, in (3.7), we have

Corollary 1 If $Eprem(f,h,v)=0$, then

$$\begin{cases} E_{pp}(f,g,v)\cdot E_{pr}(f,h,v)=E_{pJ}(f,h,v)h \\ E_{pq}(f,h,v)\cdot E_{pr}(f,h,v)=E_{pJ}(f,h,v)(-f). \end{cases} \tag{3.15}$$

Let us now describe the successive mutual pseudo division (SMPD). Suppose we have an ascending chain

$$C: f_1(u,x_1),\cdots,f_k(u,x_1,\cdots,x_s)$$

and a polynomial $g(u,x_1,\cdots,x_s)$, Let:

$$\begin{cases} R_{s-1}=Eprem(g,f_s,x_s) \\ R_{s-2}=Eprem(R_{s-1},f_{s-1},x_{s-1}) \\ \cdots \\ R_0=Eprem(R_1,f_1,x_1). \end{cases} \tag{3.16}$$

The polynomial R_0 is called the final remainder of SMPD, and it is denoted by

$$R_0=Eprem(g,f_1,f_2,\cdots,f_s). \tag{3.17}$$

For simplicity, we also set

$$\begin{cases}U_i=E_{pp}(R_i,f_i,x_i)\\V_i=E_{pq}(R_i,f_i,x_i)(i=1,\cdots,s,R_s=g)\\W_i=E_{pr}(R_i,f_i,x_i).\end{cases} \tag{3.18}$$

Then we have

Corollary 2 (The Remainder Formula of SMPD) Let f_i,U_i,V_i and $R_0=Eprem(g,f_1,\cdots,f_s)$ be as above. Then

$$R_0=U_1\cdots U_sg+V_1f_1+U_1V_2f_2+U_1U_2V_3f_3+\cdots+U_1\cdots U_{s-1}V_sf_s. \tag{3.19}$$

Proof Taking the case $l=0$, in(3.8)of Lemma 1, and using(3.14), (3.18), we obtain

$$U_iR_i+V_if_i=R_{i-1}(i=1,\cdots,s:R=g) \tag{3.20}$$

when $i=1$, (3.20) becomes

$$R_0=U_1R_1+V_1f_1. \tag{3.21}$$

Replacing R_i *by* $U_{i+1}R_{i+1}+V_{i+1}f_{i+1}$ in(3.21)for $i=1,2,\cdots,s-1$, we get(3.19)in the final step.

We are now in a position to give our main results for solving the problem proposed in Section 1. These results are stated in the following two theorems.

Theorem 1 Let $C:f_1(u_g,x_1),\cdots,f_k(u_gx_1,\cdots,x_s)$ be an ascending chain, $g(u,x_1,\cdots,x_s)$ be a polynomial and $R_0=Eprem(g,f_1,\cdots,f_s)$. Then

(ⅰ) $g\neq0$ generically when $f_1=\cdots=f_s=0$ if R_0 is not the zero element of $A[u]$. In fact $g(u,x)=0$ can hold at most on the surface $\Gamma:R_0(u_1,\cdots,u_d)=0$

(ⅱ) $g=0$ when $f_1=\cdots=f_k=0$ if R_0 is the zero element of $A[u]$, under the further, assumption

$$U_1U_2\cdots U_s\neq0(U_i=E_{pp}(R_i,f_i,x_i)). \tag{3.22}$$

Theorem 1 is obvious by the Remainder Formula of SMPD(3.19). But we have to investigate further for the subsidiary condition(3.22). If $class(U_i)<i$ for $i=1,2,\cdots,s$, it is shown that the SMPD which has been done for $g,f_1,\cdots,f_s$ is just the same as SPD, and (3.22)must be the same as the Wu non-degenerate condition. For the general case, we have

Theorem 2 Suppose it is given an ascending chain $C=\{f_1(u,x_1),\cdots,f_k(u,x_1,\cdots,x_s)\}$and a polynomial g$(u,x_1,\cdots,x_s)$ Then there is a mechanical method to decompose $V(C)$ into some sub - variety

$$V(C)=V(C_1)\cup V(C_2)\cup\cdots\cup V(C_a)\cup V^* \tag{3.23}$$

such that

(ⅰ) V^* is a degenerate variety.

(ⅱ) The equality $g=0$ holds generically or fails generically on each $V(C_i)$, and we can determine mechanically which case appear for every $V(C_i)$ respectively.

(iii) For the ascending chain $C_i=\{f_1^{(i)},\cdots,f_s^{(i)}\}$, we have $class(f_j^{(i)})=j$ and $ld(f_j^{(i)})\leqslant ld(f_j)$. $(i=1,2,\cdots,\alpha,f=1,2,\cdots,s)$.

We need the simple

Lemma 3 Suppose it is given an ascending chain $C=\{f_1(u,x_1),\cdots,f_s(u,x_1,\cdots,x_s)\}$ and a polynomial $g(u,x_1,\cdots,x_s)$. If $class(g)=\alpha,ld(g)<ld(f_\alpha),g$ is reduced with respect to $f_1,\cdots,f_{\alpha-1}$ and $V(f_1,\cdots,f_s,g)$ is non-degenerate, then we can determine an ascending chain $f_1^*,\cdots,f_s^*$ such that

(i) $V(f_1^*,\cdots,f_s^*)=V(f_1,\cdots,f_s,g)$

(ii) $class(f_i^*)=class(f_i)$ and $ld(f_i^*)\leqslant ld(f_i)(i=1,2,\cdots,s)$

(iii) $ld(f_1^*)+\cdots+ed(f_s^*)<ld(f_1)+\cdots+ld(f_s)$.

Proof Put $N=ld(f_1)+\cdots+ld(f_s)$ and do induction for $N=1,2,\cdots$. When $N=1$, the proposition is true since $V(f_1,\cdots,f_s,g)$ must be degenerate.

Suppose the proposition is true for $N=1,\cdots,k-1$. Let us prove that it is still true for $N=k$. In fact, let $f_\alpha^*=g$ and $g^*=prem(f_\alpha,f_1,\cdots,f_{\alpha-1},g)$, then we have

$$V(f_1,\cdots,f_{\alpha-1},f_{\alpha+1}^*,\cdots,f_s,g^*)=V(f_1,\cdots,f_s,g) \tag{3.24}$$

because $ld(F_\alpha^*)<ld(f_\alpha)$, so

$$ld(f_1)+\cdots+ld(f_{\alpha-1})+ld(f_\alpha)+ld(f_{\alpha+1})+\cdots+ld(f_s))>k_i$$

by the inductive hypothesis we finish the proof.

The Proof of Theorem 2 Put $N=ld(f_1)+\cdots+ld(f_s)$ and do induction for $N=1,2,\cdots$ When $N=1$, it is obvious. Suppose the conclusion is true for $N=1,2,\cdots,k-1$, we are going to prove that it is still true for $N=k$. Let $g^*=Eprem(g,f_s,x_s)$ and $C^*=\{f_1,\cdots,f_{s-1}\}$. Since $ld(f_1)+\cdots+ld(f_{s-1})<k$, by the inductive hypothesis, we can decompose mechanically $V(C^*)$ into

$$V(C^*)=V(C_1^*)\cup V(C_2^*)\cup\cdots\cup V(C_\alpha^*)\cup V^* \tag{3.25}$$

such that (i), (ii) and (iii) are satisfied.

Let $C_i^*=\{f_{i,1}^*,f_{i,2}^*\cdots,f_{i,s-1}^*\}$. If $g^*\neq 0$ generically for some $V(C_i^*)$, then obviously $g\neq 0$ generically on $V(f_{i,1}^*,\cdots,f_{i,s-1}^*)$. So we need consider only $V(C_j^*)$ on which the equality $g^*=0$ holds generically. For simplicity, by our inductive assumption, it is enough to assume that $V(C_j^*)$ is just $V(f_1,\cdots,f_{s-1})$.

Notice that $g^*=0$ generically on $V(f_1,\cdots,f_{s-1})$ means that $R_{s-1}=Eprem(g,f_s,x_s)$ is the zero in the ring $K[u,x_1,\cdots,x_{s-1}]/(f_1\cdots,f_{s-1})$. Use Corollary 1 (and by (3.18)), we have

$$\begin{cases}U_s\cdot W_s=E_{pJ}(g,f_s,x_s)f_s\\ V_s\cdot W_s=-E_{pJ}(g,f_s,x_s)g.\end{cases} \tag{3.26}$$

Let $g^{**}=E_{pJ}(g,f_s,x_s)$, we can decompose $V(f_1,\cdots,f_{s-1})$ as (3.25), such that $g^{**}=0$ holds generically or fails generically on each $V(C_i^*)$. *If* $g^{**}\neq 0$ on some $V(C_i^*)$, we can decompose $V(f_{i,1}^*,\cdots,f_{i,s-1}^*,f_s)$ into $V(B_1)$ and $V(B_2)$, where

$$B_1: f_{i,1}^*, \cdots, f_{i,s-1}^*, W_s$$
$$B_2: f_{i,1}^*, \cdots, f_{i,s-1}^*, U_s.$$

For B_1, we have $g=0$ generically on $V(B_1)$ by(3.26). For B_2, since $class(w_s)=s$ we have

$$ld(f_{i,1}^*)+\cdots+ld(f_{i,s-1}^*)+ld(U_s)<k,$$

we are done by our inductive assumption.

Then we discuss the $V(C_j^*)$ on which $g^{**}=0$ holds generically. We still suppose $V(C_j^*)$ is just the $V(f_1, f_2, \cdots, f_{s-1})$.

By(3.14), $g^{**}=E_{pJ}(g, f_s, x_s)=(-1)^t J_0 J_1 \cdots J_t$, where $J_i \in K[u, x_1, \cdots, x_{s-1}]$. Of course we may suppose each J_i is reduced with respect to $f_j(f=1,2,\cdots,3-1)$ and $J_i \neq 0$. Since $g^{**}=0$ generically on $V(f_1, \cdots, f_{s-1})$ has been supposed, without considering degenerate sub-variaties, we can decompose $V(f_1, \cdots, f_{s-1})$ into $V(f_1, \cdots, f_{s-1}, J_j)$. Using Lemma 3, we can determine an ascending $f_1^*, \cdots, f_{s-1}^*$ such that

$$V(f_1^*, \cdots, f_{s-1}^*)=V(f_1, \cdots, f_{s-1}, J_i)$$

when $V(f_1, \cdots, f_{s-1}, J_i)$ is non-degenerate for each i respectively, and

$$ld(f_1^k)+\cdots+ed(f_{s-1}^*)<ld(f_1)+\cdots+ld(f_{s-1}).$$

This means that

$$ld(f_1^*)+\cdots+ld(f_{s-1}^*)+ld(f_s)<k$$

and we do not need to discuss further the sub-variety $V(f_1^*, \cdots, f_{s-1}^*, f_s)$, i. e., variety $V(f_1, \cdots f_s)$, by our inductive assumption.

The proof of Theorem 2 has given an algorithm inductively. But in practice, it may be better to take the following steps:

Suppose α is the least number among the i's such that $class(U_i)=i$. By the remainder formula of SMPD(3.19), we have

$$\begin{aligned} R_0 &= E_{\text{prem}}(R_{\alpha-1}, \cdots, f_{\alpha-1}) \\ &= U_1 U_2 \cdots U_{\alpha-1} R_{\alpha-1} + V_1 f_1 + \cdots + U_1 \cdots U_{\alpha-2} V_{\alpha-1} f_{\alpha-1}. \end{aligned} \tag{3.27}$$

So the assumption $f_1=\cdots=f_{\alpha-1}=0$ implies $R_{\alpha-1}=0$ under the non-degenerate condition $U_1 \cdots U_{\alpha-1} \neq 0$. It is shown that $R_{\alpha-1}$ is the zero element in the ring $K(u)[x_1, \cdots, x_{\alpha-1}]/(f_1, \cdots, f_{\alpha-1})$, and we have

$$\begin{cases} U_\alpha \cdot W_\alpha = E_{pJ}(R_\alpha, f_\alpha, x_\alpha) f_\alpha \\ V_\alpha \cdot W_\alpha = -E_{pJ}(R_\alpha, f_\alpha, x_\alpha) R_\alpha \end{cases} \tag{3.28}$$

in the ring $A^*[x_\alpha]$ by(3.12), where A^* is the ring $K(u)[x_1, \cdots, x_{\alpha-1}]/(f_1, \cdots, f_{\alpha-1})$.

Then the variety $V(C)=V(f_1, \cdots, f_s)$ can be decomposed into three: $V(C_1)$, $V(C_2)$ and $V(C_3)$ where

$$\begin{cases} C_1=\{f_1, \cdots, f_{\alpha-1}, W_\alpha, f_{\alpha+1}, \cdots, f_\alpha, E_{pJ}(R_\alpha, f_\alpha, x_\alpha) \neq 0\} \\ C_2=\{f_1, \cdots, f_{\alpha-1}, U_\alpha, f_{\alpha+1}, \cdots, f_\alpha, E_{pJ}(R_\alpha, f_\alpha, x_\alpha) \neq 0\} \\ C_3=\{f_1, \cdots, f_s, E_{pJ}(R_\alpha, f_\alpha, x_\alpha)=0\}. \end{cases} \tag{3.29}$$

The subsidiary condition $E_{pJ} \neq 0$ or $E_{pJ}=0$ may be the non-degenerate or degenerate condition respectively, merely. But it is possible that these subsidiary conditions could cause the new decomposition of some one among $f_1, \cdots, f_{\alpha-1}$ and so $V(C_3)$ is non-degenerate, as example 3 in the next section.

For $V(C_1)$, we need not get $Eprem(R_\alpha, W_\alpha, x_\alpha)$ again. If $class(U_\alpha) < i$ for $i = \alpha+1, \cdots s$. it is shown $g=0$ generically on $V(C_1)$. Otherwise, we use Theorem 2 and repeat the discussion(3. 27)-(3. 29) again.

Suppose the condition $E_{pJ}(R_\alpha, f_\alpha, x_\alpha) \neq 0$ has been reduced and $V(C_2)$ could be represented by $V(f_1^*, f_2^*, \cdots, f_\alpha^*, f_{\alpha+1}, \cdots, f_s)$. We have to get $Eprem(R_\alpha, f_\alpha^*, x_\alpha)$ and $Eprem(R_\alpha, f_1^*, \cdots, f_\alpha^*)$ again. Of course, repeat the above discussion.

If $V(C_3)$ is non-degenerate, we have to do all the things for it as for $V(C_2)$.

Notice that it may be much better to join the new method with Wu's, as follows: We may get $prem(g, f_1, \cdots, f_s)$ by the Wu method firstly. If $prem(g, f_1, \cdots, f_s)=0$, then it is finished. Otherwise, let

$$g := prem(g, f_1, \cdots, f_s)$$

and use a new method.

4. EXAMPLES

Example 1 Suppose

$$H: \begin{cases} f_1 = x_1 - u = 0 \\ f_2 = x_2^2 - 2x_1x_2 + u^2 = 0 \end{cases}$$

To check equality $g = x_2 - x_1 = 0$.

Solution Since $prem(g, f_1, f_2) = x_2 - u \neq 0$ we should use the new method to continue the work.

$$R_1 = Eprem(x_2 - u, f_2, x_2) = 2ux_1 - u^2$$

$$U_2 = x_2 - 2(x_1 - u), W_2 = x_2 - u$$

$$R_0 = Eprem(R_1, f_1, x_1) = 0$$

It is easy to check that $f_2 = U_2 W_2$ under the condition $f_1 = 0$, (where $E_{pJ} = 1$, so it is needed not to consider C_3), so that $\{f_1, f_2\}$ could be decomposed into

$$C_1 = \{f_1, W_2\}$$

$$C_2 = \{f_1, U_2\}$$

By our above discussion, we know that $g=0$ generically on $V(C_1)$. And $g=0$ generically on $V(C_2)$ too, by the Wu method, since

$$prem\{x_2 - u, f_1, U\} = 0.$$

Therefore, H implies $g=0$.

Example 2 Suppose

$$H: \begin{cases} f_1 = x_1^2 - ux_1 = 0 \\ f_2 = x_2^2 + x_1x_2 - u^2 = 0 \end{cases}$$

To check equality $g=x_1+x_2+1=0$.

Solution　Since $prem(g,f_1,f_2)=x_1+x_2+1\neq 0$, we could not make sure by SPD what the conclusion is. But we have

$$R_1=Eprem(g,f_2,x_2)=x_1+1-u^2$$

$$R_0=Eprem(R_1,f_1,x_1)=(u^2-1)(u^2-u-1)\neq 0$$

So the equality $g=0$ is false generically.

Example 3　Suppose

$$H:\begin{cases} f_1=x_1^2+ux_1=0 \\ f_2=x_2^2+(x_1^2-u)x_2+u^2x_1=0 \end{cases}$$

To check equality $g=x_2^2+x_1(1-u)x_2-ux_1^2=0$.

Solution　We have $prem(g,f_1,f_2)=(x_1+u)x_2\neq 0$. So more work should be done. Let

$$g^*=(x_1+u)x_2$$

We have

$$R_1=Eprem(g^*,f_2,x_2)=-u(x_1+u)x_1^2$$

$$U_2=x_2+(x_1^2-u),W_2=(x_1+u)x_2$$

$$R_0=Eprem(R_1,f_1,x_1)=0$$

By Theorem 2, U_2W_2 must equal to $E_{pJ}(g^*,f_2,x_2)\cdot f_2$ under the assumption $f_1=0$. It is easy to check that $U_2W_2=(x_1+u)f_2$. Then $V(f_1,f_2)$ can be decomposed into three parts $V(C_1)$, $V(C_2)$ and $V(C_3)$, where

$$C_1:\{f_1,W_2,(x_1+u\neq 0)\}\Rightarrow\{x_1,x_2\},$$

$$C_2:\{f_1,U_2,(x_1+u\neq 0)\}\Rightarrow\{x_1,x_2+(x_1^2-u)\}$$

$$C_3:\{f_1,f_2(x_1+u=0)\}\Rightarrow\{x_1+u,f_2\}$$

We have known that $g=0$ holds generically on $V(C_1)$. Since $Eprem(g^*,x_1,x_2+(x_1^2-u))\neq 0$, $g=0$ is false generically on $V(C_2)$. Finally, $g=0$ is true generically on $V(C_3)$ because $prem(g^*,x_1+u,f_2)=0$.

The following examples 4 and 5 are taken from [2]. Of course, we use the new method which is different from [2].

Example 4　On the two sides AC and BC of triangle ABC, two squares $ACDE$ and $BCFG$ are drawn. M is the midpoint of AB. Show $DF=2CM$.

Solution　Let $A=(u_1,0)$, $B=(u_2,u_3)$, $C=(0,0)$, $D=(0,u_1)$, $F=(x_1,x_2)$ and $M=(x_3,x_4)$, then the hypothesis can be represented by

$$H:\begin{cases} h_1=x_2^2+x_1^2-u_2^2-u_3^2=0(CF=BC) \\ h_2=u_3x_2+u_2x_1=0(CF\perp BC) \\ h_3=2x_3-u_2-u_1=0 \\ h_4=2x_4-u_3=0\ (M\text{ is the midpoint of }AB) \end{cases}$$

The conclusion we want is

$$g=4x_4^2+4x_3^2-x_2^2+2u_1x_2-x_1^2-u_1^2=0\ (DF=2CM).$$

The ascending chain obtained from H is

$$C:\begin{cases} f_1=(u_3^2+u_2^2)x_1^2-u_3^4-u_2^2u_3^2=0 \\ f_2=u_3x_2+u_2x_1=0 \\ f_3=2x_3-u_2-u_1=0 \\ f_4=2x_4-u_3=0. \end{cases}$$

Firstly, we get that $prem(g,f_1,\cdots,f_4)=2u_1u_2u_3(u_3-x_1)\neq0$. However we could not say that the proposition is false in general because the Wu method may be invalid when C is reducible.

Then let $g^*=x_1-u_3$ and we have

$$Eprem(g^*,f_1,\cdots,f_4)=0$$
$$U_1=(u_3^2+u_2^2)(x_1+u_3),W_1=(x_1-u_3).$$

It is easy to check that $U_1W_1=f_1$ and C can be decomposed into

$$C_1:(W_1,f_2,f_3,f_4)$$
$$C_2:(U_1,f_2,f_3,f_4).$$

We have known $g^*=0$ to be generically true on $V(C_1)$. But it is false generically on $V(V_2)$ since $Eprem(g^*,U_1,f_2,f_3,f_4)=Eprem(g^*,U_1)\neq0$. It is shown that as has been pointed out in [2], the conclusion is true for one subcase of the configuration (both the two squares are outward to $\triangle ABC$ or both inward to $\triangle ABC$), but not for the other. (One square is outward but the other is inward.)

Example 5 (The Butterfly Theorem) A,B,C and D are four points on the circle(O) and E is the intersection of AC and BD. Through E draw a line perpendicular to OE, meeting AD at F and BC at G. Show that $FE=GE$.

Solution Let $E=(0,0)$, $O=(u_1,0)$, $A=(u_2,u_3)$, $B=(x_1,u_4)$, $C=(x_3,x_2)$, $D=(x_3,x_4)$, $F=(0,x_6)$ and $G=(0,x_7)$. Then the hypothesis equations are:

$$H:\begin{cases} h_1=-x_1^2+2u_1x_1-u_4^2+u_3^2+u_2^2-2u_1u_2=0\ (OA=OB) \\ h_2=-x_3^2+2u_1x_3-x_2^2+u_3^2+u_2^2-2u_1u_2=0\ (OA=OC) \\ h_3=-u_3x_3+u_2x_2=0\ (C,A \text{ and } E \text{ are collinear}) \\ h_4=-x_5^2+2u_1x_5-x_4^2+u_3^2+u_2^2-2u_1u_2=0\ (OA=OD) \\ h_5=-u_4x_5+x_1x_4=0\ (D,B \text{ and E are collinear}) \\ h_6=(-x_5+u_2)x_6+u_3x_5-u_2x_4=0\ (F,A \text{ and } D \text{ are collinear}) \\ h_7=(-x_3+x_1)x_7+u_4x_3-x_1x_2=0\ (G,B \text{ and } C \text{ are collinear}) \end{cases}$$

The conclusion is $g=x_7+x_6=0$.

We obtain an ascending chain from H:

$$C:\begin{cases} f_1=x_1^2+2u_1x_1-u_4^2+u_3^2+u_2^2-2u_1u_2=0 \\ f_2=-(u_3^2+u_2^2)x_2^2+2u_1u_2u_3u_2+u_3^4+(u_2^2-2u_1u_2)u_3^2=0 \\ f_3=-u_3x_3+u_2x_2=0 \\ f_4=-(2u_1x_1+u_3^2+u_2^2-2u_1u_2)x_4^2+2u_1u_4x_1x_4+(u_3^2+u_2^2-2u_1u_2)u_4^2=0 \\ f_5=-u_4x_5+x_1x_4=0 \\ f_6=(-x_1x_4+u_2u_4)x_6+u_4u_3x_5-u_2u_4x_4=0 \\ f_7=(-u_2x_2+u_3x_1)x_7+u_4u_3u_3-u_3x_1x_2=0. \end{cases}$$

We have

$$prem(g,f_1,\cdots,f_7)=(u_3x_1-u_2x_4)[((x_1-u_2)x_2+u_3x_1)x_4-u_2u_4x_2]\neq 0,$$

then we could not make sure of the conclusion. Let $g^*=prem(g,f_1,\cdots,f_7)$, and then we get that:

$$R_4=Eprem(g^*,f_5,f_6,f_7)=g^*,$$

$$\begin{aligned} R_3 &=Eprem(R_4,f_4,x_4) \\ &=2u_3u_4^2(x_1-u_2)(u_3x_1-u_2u_4)x_1[(u_2^2+u_3^2)x_2+u_3(u_2^2+u_3^2+2u_1u_2)] \end{aligned}$$

$$\begin{aligned} W_4 &=E_{pr}(R_4,f_4,x_4) \\ &=(u_3x_1-u_2u_4)[((x_1-u_2)x_2+u_3x_1)x_4-u_2u_4x_2] \end{aligned}$$

$$\begin{aligned} U_4 &=E_{pp}(R_4,f_4,x_4) \\ &=(-(x_1^2+u_4^2)[(x_1-u_2)x_2+u_3x_1]x_4 \\ &\quad +[(2u_1u_4-u_2u_4)x_1^2-2u_1u_2u_4x_1-u_2u_4^3)x_2+2u_1u_3u_4x_1^2] \end{aligned}$$

$$R_2=Eprem(R_3,f_3,x_3)R_3$$

$$R_1=Eprem(R_2,f_2,x_2)=0$$

$$W_2=R_3, U_2=-x_2+u_3$$

$$R_0=Eprem(R_1,f_1,x_1)=0.$$

Using Theorem 2, we have

$$W_2U_2=2u_3u_4^2(x_1-u_2)(u_3x_1-u_2u_4)f_2$$

under condition H, and so f_2 can be decomposed into

$$\begin{aligned} f_2 &=(-x_2+u_3)((u_2^2+u_3^2)x_2+u_3(u_2^2+u_3^2-2u_1u_2)) \\ &=(-x_2+u_3)f_2^* \end{aligned}$$

under the non-degenerate condition $2u_3u_4^2(x_1-u_2)(u_3x_1-u_2u_4)\neq 0$. Then we decompose $V(C)$ into $V(C_1)$ and $V(C_2)$, where

$$C_1:\{f_1,f_2,f_3,f_4,f_5,f_6,f_7\}.$$

$$C_2:\{f_1,-x_2+u_3,f_2,f_3,f_4,f_5,f_6,f_7\}.$$

Since $Eprem(R_2,-x_2+u_3,f_1)\neq 0$, the conclusion $g=0$ is false generically on $V(C_2)$.

For $V(C_1)$, since $R_3=E_{pJ}(R_2,f_2,x_2)f_2^*=0$, we know that

$$W_4U_4=E_{pJ}(R_4,f_4,x_4)f_4$$

under the condition $f_1=f_2^*=f_3=0$. Then we can decompose f_4 into

$$f_4=f_4^*\cdot f_4^{**}$$

under non-degenerate condition

$$E_pJ(R_4,f_4,x_4)=u_2u_3(u_3x_1-u_2u_4)(x_1^2+u_4^2)\neq 0,$$

where

$$f_4^* = (2u_1x_1 + u_2^2 + u_3^2 - 2u_1u_2)x_4 + u_4(u_2^2 + u_3^2 - 2u_1u_2)$$

$$f_4^{**} = -x_4 + u_4.$$

Then we can decompose $V(C_1)$ into $V(C_1^*)$ and $V(C_2^*)$, where

$$C_1^*: (f_1, f_2^*, f_3, f_4^*, f_5, f_6, f_7)$$

$$C_2^*: (f_1, f_2^*, f_3, f^{**}, f_5, f_6, f_7).$$

Since $Eprem(R_4, f_2^*, f_3, f_4^{**}) \neq 0$ and $prem(g^*, f_1, f_3, f_4^*, f_5, f_6, f_7) = prem(R_4, f_1, f_2^*, f_3, f_4^*) = 0$, the conclusion is false generically on $V(C_2^*)$ but true on $V(C_1^*)$. It is shown that the conclusion $g=0$ is true generically if we suppose $-x_2+u_2 \neq 0$ and $-x_4+u_4 \neq 0$ to exclude the varieties $V(C_2)$ and $V(C_2^*)$.

As what has been pointed in [2] and [3], the situation like Example 4 and Example 5, i. e. the conclusion may be false in some cases, is caused by some ambiguity when using algebraic equations to encode certain geometric conditions. It is shown that our method could remove the ambiguity mechanically.

Acknowledgements

One of the authors (J. Z.) would like to thank professor Abdus Salam, the International Atomic Energy Agency and UNESCO for hospitality at the International Centre for Theoretical Physics, Trieste. The authors would like to extend their thanks to Professor A. Verjovsky for reading the paper carefully and for valuable suggestions.

References

[1] Wu Wen-tsun. On the decision problem and mechanization of theorem proving in elementary geometry. Scientia Sinica, 1978(21): 157 - 179.

[2] Shang-Ching Chou. Mechanical Geometry Theorem Proving. D. Reidel Publishing Company. 1988.

[3] Wu Wen-tsun. On reducibility problem in mechanical theorem proving of elementary geometries. Chinese Quarterly J. Math., 1987, 2(2): 1 - 19.

[4] B Kutzler, S Stifter. Automated geometry theorem proving using Buchberger's algorithm. preprint for SYMSAC, 1986.

[5] D Kapur. Ceometry theorem proving using Hilbert's Nullstellensatz. preprint for SYMSAC, 1986.

[6] Hu Sen, Wang Dongming. Fast factorization of polynomials over rational number field of its extension fields. Kexue Tongbao, 1986(31): 150 - 156.

[7] Zhang Jingzhong, Yang Lu and Deng Mike. The parallel numerical method of mechanical theorem proving. Research Report IMS - 30, Inst. Math. Sci. Chengdu, P. R. China, 1988.

What Can We Do with Only a Pair of Rusty Compasses?

Zhang Jing-zhong, Yang Lu and Hou Xiao-rong

ABSTRACT. Compasses are called rusty, if one can draw only the unit circle with them. We prove that from two points A and B, with only rusty compasses, one can draw the points of k-section of AB, and all the vertices of a regular n-gon which has a side AB, where k is any integer greater than 1, and $n=3,4,5,6,8,12,17,257\cdots$, etc. Generally, let A be $(0,0)$ and let B be $(\lambda,0)$, then one can draw all the points (λ_x,λ_y) where x and y are any elements in some regular 2^m-extension of the rational field, for $m=1,2,3,\cdots$

1. Introduction

A pair of compasses is called rusty if we can draw only the unit circle with it. What can we do with only a rusty compass? This is an old problem[1]. As we know, no one had given any solution exactly. Maybe many people think that it is hard to do some things with only a rusty compass.

Occasionally, a surprising discovery opens a new page on the problem. Several years ago, D. Pedoe[2] discovered an interesting fact: given two points A, B but without the segment AB, if $AB<2$, it is easy to find a point C such that $AC=BC=AB$, with only a rusty compass.

Figure 1 shows five unit circles: $\odot A$, $\odot B$, $\odot F$, $\odot D$ and $\odot E$. And the process of construction is that $F\in\odot A\cap\odot B$, $D\in\odot F\cap\odot A$, $E\in\odot F\cap\odot B$ and $C=(\odot D\cap\odot E)\backslash F$. Here we choose D and E such that both $A-F-D$ and $F-B-E$ are in counterclockwise order. So we have $AC=BC=AB$.

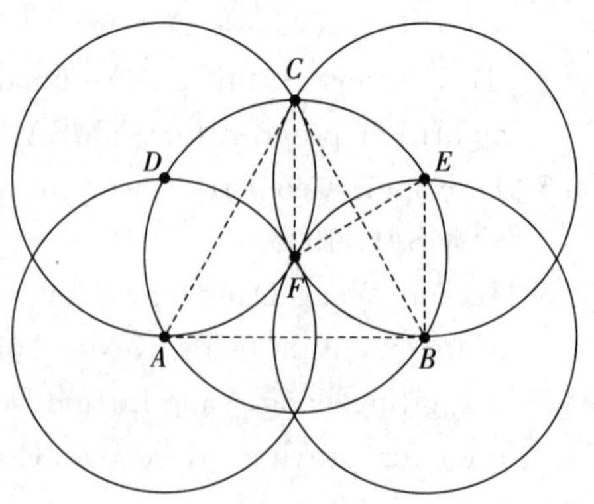

Fig. 1

The proof is simple. It is enough to point out that

本文刊于《Geometriae Dedicata》,1992 年.

$$\frac{1}{2}\angle ACB=\angle FCB=\frac{1}{2}\angle FEB=30°.$$

So $\angle A=\angle B=\angle C=60°$, That is all.

This discovery encouraged Pedoe to ask other things. He proposed two further problems more difficult and significant as follows:

PROBLEM 1. Given two points A and B but without segment AB, prove or disprove that the third vertex C of the regular $\triangle ABC$ can be found when $AB\geqslant 2$, using only a pair of rusty compasses[3].

PROBLEM 2. Given two points A and B but without segment AB, prove or disprove that the midpoint M of the segment AB can be found, using only a pair of rusty compasses[4].

J. Zhang, L. Yang and X. Hou have given not only positive solutions to Pedoe's two problems but also much further results[5]-[9]. An unexpected conclusion is that the effect of a rusty compass is equivalent to a ruler and a fine compass when the construction was started from only two points. But most of these works have not been published in English. Here we shall sum up all these results more clearly by a simpler method.

By the way, some geometers remarked that it is impossible to divide segments with rusty compasses[10].

For convenience, some notions will be used throughout the paper:

(1) UC(X) - the unit circle with centre X.

(2) RC - construction with only a rusty compass.

(3) $Z=X\nabla Y$ means that $\triangle XYZ$ is a regular triangle with vertices labelled $X-Y-Z$ in counterclockwise order.

(4) $Z=\square(WXY)$ means that WXYZ is a parallelogram with vertices labelled $W-X-Y-Z$ in counterclockwise order.

(5) If there exists a series of points $A=X_0, X_1, X_2, \cdots, X_n, X_{n+1}=B$ such that $X_kX_{k+1}=\lambda$, $0\leqslant k\leqslant n$, then we say that this is a λ - bridge between A and B.

2. THE SOLUTION TO PROBLEM 1

The main difficulty we face first is that the distance between two given points A and B may be too large for our rusty compass. So we would build a bridge to connect the two points:

RC1. *Given two points A and B, a series of points $\{X_k\}$ can be found as a 1 - bridge between A, B.* (Of course, using only a compass which is rusty. This remark will be omitted in all *RC* throughout the paper.)

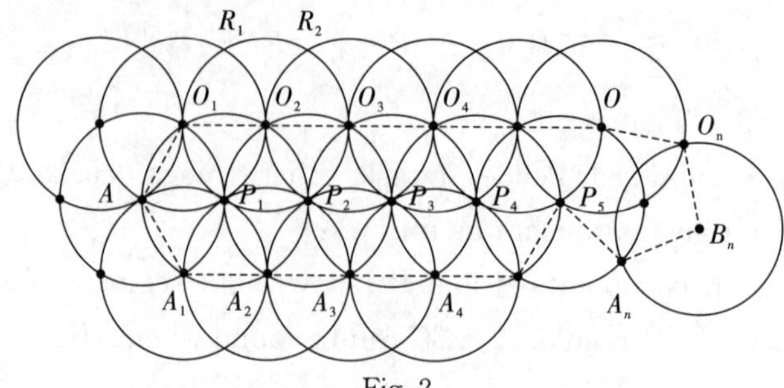

Fig. 2

Construction. Take any one point $P_1 \in \mathrm{UC}(A)$; let Q_1 be the intersection point of UC(A) and UC(P_1) such that $Q_1 = A \nabla P_1$. Then draw $Q_2 = Q_1 \nabla P_1$, $P_2 = Q_2 \nabla P_1$, $Q_3 = Q_2 \nabla P_2$, $R_1 = Q_1 \nabla Q_2$, $\cdots$ So we can obtain a point array which consists of all the vertices of these regular triangles, covering whole plane as in Figure 2. These regular triangles have unit side, so the point - array will be called a cobweb with parameter 1. It is easy to find a point Q in the cobweb such that $\mathrm{UC}(Q) \cap \mathrm{UC}(B) \neq \varnothing$. Then the bridge has been built.

RC2. *Given three points A, B and C, the point D can be found such that $D = \square(ABC)$*[11].

Construction. By RCI, we can obtain $\{P_k\}$ and $\{Q_l\}$ as 1 - bridges between B, A and B, C respectively. Then our work will consist of a series of constructions of a rhombus with unit sides like Figure 3; and induction can be used to $m+n$.

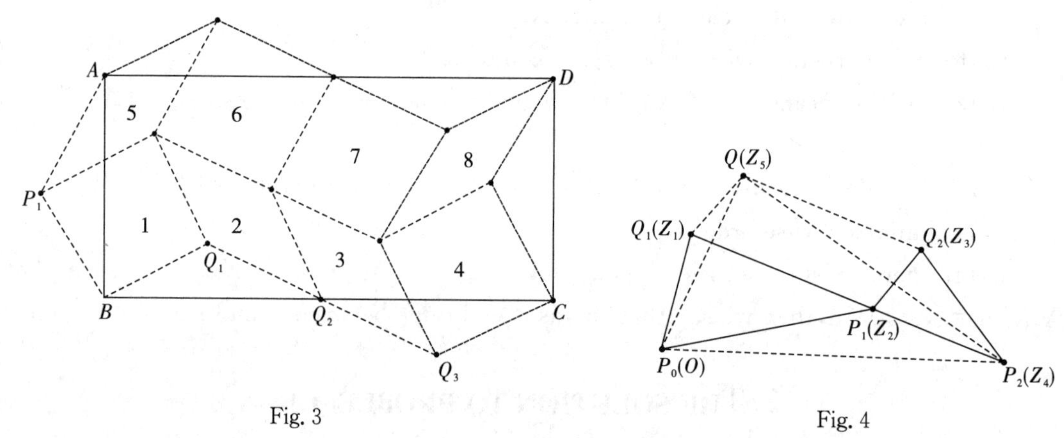

Fig. 3 Fig. 4

RC3. *Suppose there are two series of points $\{P_0, P_1, \cdots, P_n\}$ and $\{Q_1, Q_2, \cdots, Q_n\}$ such that*

1. $\triangle Q_k P_{k-1} P_k \sim \triangle Q_{k+1} P_k P_{k+1}$, $(k=1,2,\cdots,n-1)$.

2. $Q_k - P_{k-1} - P_k$ *are in counterclockwise order along the boundary of* $\triangle Q_k P_{k-1} P_k$. *Then the point Q can be found such that* $\triangle Q P_0 P_n \sim \triangle Q_1 P_0 P_1$.

Construction. First see the case of $n=2$. By RC2 we can find a point $Q = \square(Q_1 P_1 Q_2)$. Then Q is the point we want. The proof could be taken as an exercise in elementary geometry. Here we give a simple proof using complex numbers. Let P_0 be the origin of the complex plane, and Q_1, P_1, Q_2, P_2 are represented by complex numbers Z_1, Z_2, Z_3, Z_4, respectively. So conditions 1 and 2 can be represented as:

$$\frac{Z_1}{Z_2}=\frac{Z_3-Z_2}{Z_4-Z_2}=Z^*.$$

Set Q as Z_5, then the fact that $Q=\square(Q_1P_1Q_2)$ can be represented as $Z_5-Z_1=Z_3-Z_2$. So we have

$$\frac{Z_5}{Z_4}=\frac{Z_5-Z_1+Z_1}{Z_4-Z_2+Z_2}=\frac{Z_3-Z_2+Z_1}{Z_4-Z_2+Z_2}=Z^*.$$

This means that $\triangle QP_0P_2\sim\triangle Q_1P_0P_1$ and the vertices are around in the same direction respectively(Figure 4).

For the case $n\geqslant 3$, the induction can be used as in Figure 5. ①

RC4(The solution to Problem 1). *Given two points A and B, one can find $C=A\nabla B$.*

Construction. By RC1, find$\{P_k\}$as a 1-bridge between A, B. It is easy to find $Q_k=P_{k-1}\nabla P_k$. Then use RC3.

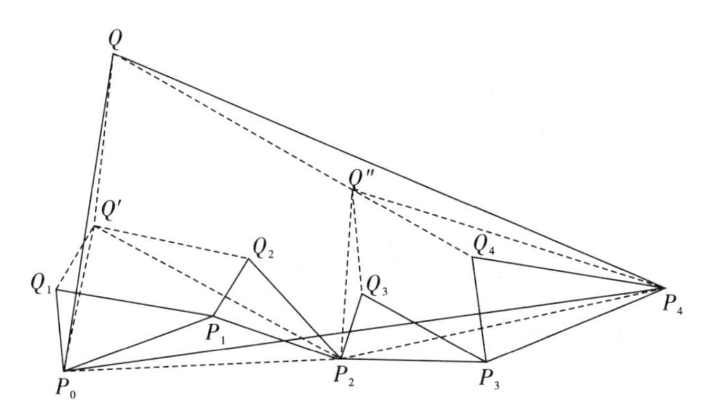

Fig. 5

In the following we shall see that RC4 is a very useful tool, and the idea in this section could be developed to solve general problems.

3. To answer the midpoint problem

RC5. *Given points X and Y, we can find Z such that*

$$XZ=2XY=2YZ$$

and the three points are collinear[12].

Construction. $Z=((X\nabla Y)\nabla Y)\nabla Y$. (Figure 6)

RC6. *Given a point X, we can find a point Y such that*

$$XY<2/\sqrt{19}.$$

Construction.

① Remark for RC3: By this proof it is obvious that RC3 will be valid when $\triangle Q_kP_{k-1}P_k$ are degenerate.

1. Take an arbitrary point $P \in UC(X)$.

2. Find a point $Q = P \nabla X \in \mathrm{UC}(X) \cap \mathrm{UC}(P)$.

3. Take an arbitrary point $A \in PQ$, distinct from P, Q.

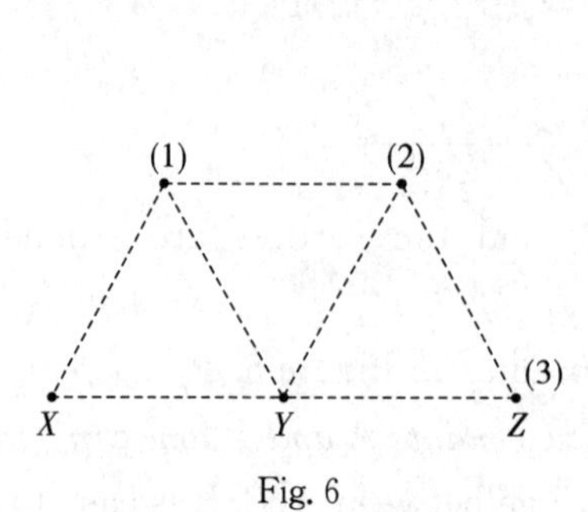

Fig. 6

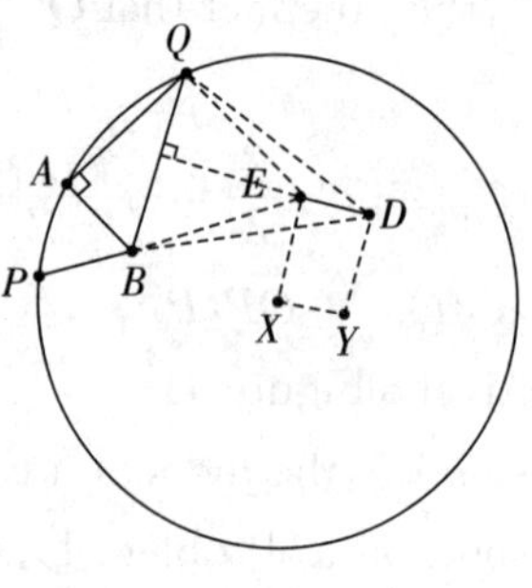

Fig. 7

4. Find $B = A \nabla P$, then $BA \perp AQ$ and $\sqrt{2}/2 < BQ < 1$. (Because:

(1) $BQ^2 = AB^2 + AQ^2 = AP^2 + AQ^2 < AP^2 + AQ^2 - 2AP \cdot AQ \cos\angle PAQ = PQ^2 = 1$

(2) $BQ^2 = AB^2 + AQ^2 \geqslant \frac{1}{2}(AB + AQ)^2 = \frac{1}{2}(AP + AQ)^2 > \frac{1}{2}$).

5. Find $E = Q \nabla B$, and take $D \in \mathrm{UC}(B) \cap \mathrm{UC}(Q)$ such that $D - Q - B$ are set in counterclockwise order(see Figure 7). Then:

$$DE < \sqrt{1 - \left(\frac{\sqrt{2}}{4}\right)^2} - \frac{\sqrt{2}}{2} \cdot \frac{\sqrt{3}}{2} = \sqrt{\frac{7}{8}} - \frac{\sqrt{6}}{4} < \frac{2}{\sqrt{19}}.$$

6. Find $Y = \square(DEX)$ by RC2.

RC7. *Given two points X and Y, we can find a point Z such that $XZ = YZ$ and $XZ = \sqrt{3k^2 + 3k + 1}XY$. Here k may be any natural number*: $k = 1, 2, 3, \cdots$

Construction. This is obvious(see Figure 8). Here we gave the cases for $k = 1, 2, 3$.

In Figure 8, the pagoda was piled up by some regular triangles with a side which equals XY. So we have:

$$\begin{aligned} XZ_1 &= \sqrt{\left(\left(1 + \frac{1}{2}\right)\sqrt{3}\right)^2 + \left(\frac{1}{2}\right)^2} XY \\ &= \sqrt{7} XY \\ XZ_2 &= \sqrt{\left(\left(2 + \frac{1}{2}\right)\sqrt{3}\right)^2 + \left(\frac{1}{2}\right)^2} XY \\ &= \sqrt{19} XY \\ XZ_3 &= \sqrt{\left(\left(3 + \frac{1}{2}\right)\sqrt{3}\right)^2 + \left(\frac{1}{2}\right)^2} XY \\ &= \sqrt{37} XY \end{aligned}$$

and, generally,

$$XZ_k = \sqrt{\left(\left(k + \frac{1}{2}\right)\sqrt{3}\right)^2 + \left(\frac{1}{2}\right)^2} XY$$

$$=\sqrt{3k^2+3k+1}XY.$$

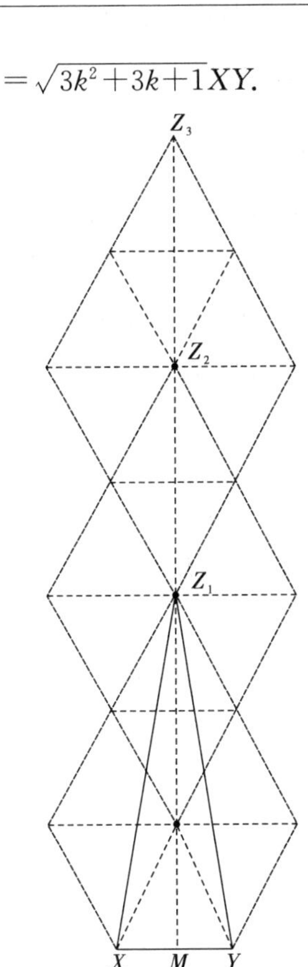

Fig. 8

RC8. *Given three points X, Y and Z such that*

$$XZ=YZ=\alpha XY\leqslant 2.$$

Then one can find a point P satisfying the condition $PX=PY=1/a$.

Construction. Suppose $X-Y-Z$ are set in counterclockwise order. Choose $A\in \mathrm{UC}(Z)\cap \mathrm{UC}(X)$ and $B\in \mathrm{UC}(Z)\cap \mathrm{UC}(Y)$ such that both $A-X-Z$ and $B-Z-Y$ are in counterclockwise order. Let $\mathrm{UC}(A)\cap \mathrm{UC}(B)\backslash Z=P$, then P is the point we want (Figure 9).

The proof is easy. It is obvious that

$$\angle XPY=2(\angle PZY+\angle PYZ)=\angle ZBY.$$

So $\triangle XPY\sim\triangle ZBY$. Then we have $PX/XY=BZ/ZY=1/ZY$, which is what we want.

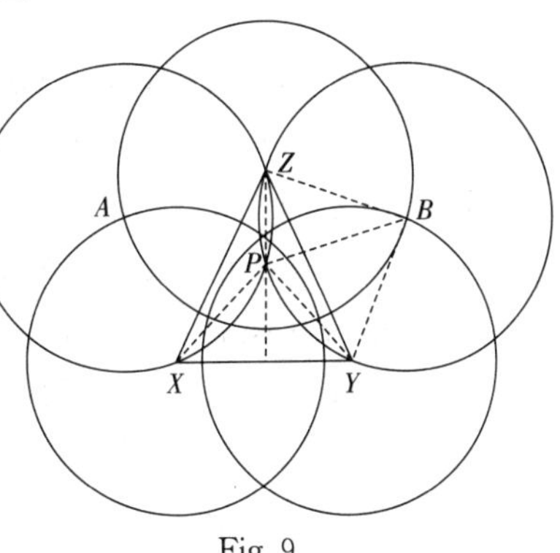

Fig. 9

RC9. *Given points X and Y with $XY=1/\sqrt{19}$, then one can find the midpoint of the segment XY.*

Construction. Find points Z and Z' such that Z', X, Y and Z are collinear and $Z'X=2XY=2YZ$ by RC5. Let $\mathrm{UC}(Z)\cap \mathrm{UC}(Z')=\{P, P'\}$, then $XP=XP'=\sqrt{15/19}$ and $YP=YP'=4/\sqrt{19}$ (Figure 10).

Find point Q such that P, Y and Q are collinear and $PY=YQ$. Take $A\in \mathrm{UC}(P)\cap \mathrm{UC}(Q)$, then $YA=\sqrt{\frac{3}{19}}$ and $AY\perp PY$. Find point B such that B, Y and A are collinear and $YA=YB$. Then take $C=B\nabla A$, C must fall on the segment YP and $YC=3/\sqrt{19}$. Find point D such that P, C, D and Y are collinear and $PD=PC+CD=3PC=3/\sqrt{19}$, so $YD=1/\sqrt{19}$. Analogously, find D' on the segment YP' and $YD'=1/\sqrt{19}$ (Figure 10).

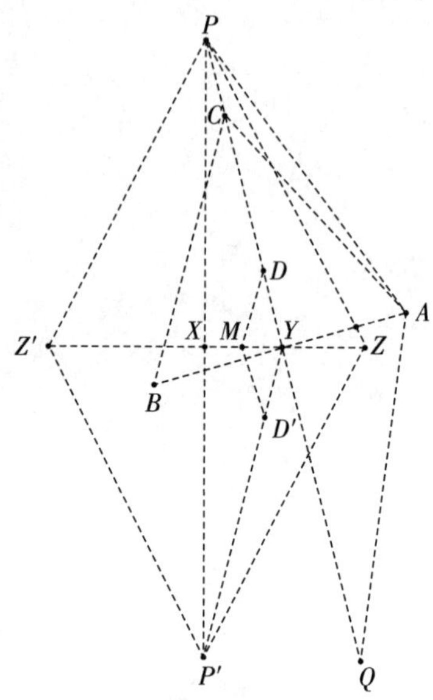

Fig. 10

Find the point $M=\square(D'YD)$, then M is the midpoint of the segment XY (Figure 10).

RC10. *Given points A and B, one can find a series of points* $\{P_k\}$ *as a* $(1/\sqrt{19})$ *bridge between A, B.*

Construction.

1. Find X_1 such that $AX_1<2/\sqrt{19}$ (by RC6).

2. Make a cobweb with parameter $d=AX_1$, starting from points A and X_1.

3. Choose points $X_0=A, X_1, X_2, \cdots, X_{l-1}$ from the cobweb such that $X_{k-1}X_k=d$ and $X_{l-1}B\leqslant d$. Let $X_l=B$.

4. Find Y_k such that $Y_kX_{k-1}=Y_kX_k=\sqrt{19}\,X_{k-1}X_k$ (*by* RC7).

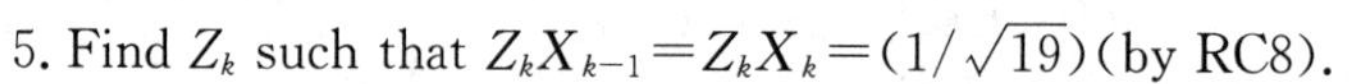

5. Find Z_k such that $Z_kX_{k-1}=Z_kX_k=(1/\sqrt{19})$ (by RC8).

Then let $P_{2m}=X_m, P_{2m+1}=Z_m$, and $\{P_k\}$ is the set we want.

RC11 (The solution to Problem 2). *Given points A and B, one can find the midpoint of the segment AB.*

Construction. By RC10, find $\{P_k\}$ as a $(1/\sqrt{19})$-bridge of A, B. Then find the midpoint Q_k of the segment $P_{k-1}P_k$ (by RC9). Finally the degenerate case of RC3 could be used.

4. THE GENERAL RESULTS

Now we represent every point X in the plane by a complex number z_x. For two given points A and B it will always be assumed that $z_A=0$ and $z_B=t$; here $t>0$ is a real number. If the point $z=z^*t$ can be found using only a rusty compass, we say $z^*\in L(t)$. If $z^*\in L(t)$ for every real number $t>0$, we say $z^*\in L$.

Our main result is the following:

THE GENERAL RC THEOREM. *Let set L be as above, then*:

(1) $\sqrt{(-1)^k}\sqrt{n}\in L$ for any natural numbers k and n;

(2) $z_1\pm z_2\in L$ if $z_1\in L$ and $z_2\in L$;

(3) $z_1 \cdot z_2 \in L$ if $z_1 \in L$ and $z_2 \in L$;

(4) if $z \in L$, then $\bar{z} \in L$;

(5) if $z \in L$ and $z \neq 0$, then $1/z \in L$;

(6) if $z \in L$, then $\sqrt{z} \in L$.

We need a series of lemmas.

LEMMA 1. *If $z_1 \in L(t)$, $z_2 \in L(t|z_1|)$, then $z_1 \cdot z_2 \in L(t)$.*

LEMMA 2. *If $z_1 \in L(t)$ and $z_2 \in L(t)$, then $z_1 \pm z_2 \in L(t)$.*

LEMMA 3. *If $z \in L(t)$, then $\bar{z} \in L(t)$.*

These lemmas are obvious: Lemma 1 is true by the definition of $L(t)$, Lemma 2 is by RC2 and Lemma 3 is by symmetry.

To obtain some non-trivial results, we introduce a new set S: If a d-bridge between any two points A and B can always be built with only a rusty compass for some fixed real number $d>0$, then we say $d \in S$. By RC3 we have

LEMMA 4. *If $d \in S$ and $z \in L(d)$, then $z \in L$.*

LEMMA 5. *For any integers m and k*

$$\left(m+\frac{1}{2}\right)+i\left(k+\frac{1}{2}\right)\sqrt{3} \in L, m \in L, ik\sqrt{3} \in L$$

where $i=\sqrt{-1}$ throughout the paper.

Proof. It is obvious that $1 \in S$ (RC1) and $e^{i\pi/3} \in L(1)$, so $e^{i\pi/3} \in L$ by Lemma 4. (This is the solution to Problem 1!) Notice that $e^{i\pi/3} = \frac{1}{2} + (\sqrt{3}/2)i$, we can obtain Lemma 5 by Lemma 2 and Lemma 3.

LEMMA 6. *Let $f(d,\lambda)=\sqrt{(1/d^2)-\lambda^2}$. If $0<\lambda \in L$, $d \in S$ and $0<\lambda d<1$, then $i \cdot f(d,\lambda) \in L$.*

Proof. If the point d is given, then the points $z_A=\lambda d$ and $z_B=-\lambda d$ can be found by $\lambda \in L$. So $z_P=\pm i\sqrt{1-\lambda^2 d^2}$ (bt $\mathrm{UC}(A) \cap \mathrm{UC}(B)$) can be found.

Then $z_p/d=\pm i\sqrt{d^{-2}-\lambda^2} \in L(d)$. By Lemma 4, $i\sqrt{d^{-2}-\lambda^2} \in L$.

Lemma 6 can be shown by Figure 11.

LEMMA 7. $\frac{1}{2} \in L$. (This is RC11.)

LEMMA 8. *If $\lambda>0$ and $i\lambda \in L$, then $\left|\frac{1}{2}+i\lambda\right|^{-1} \in S$.*

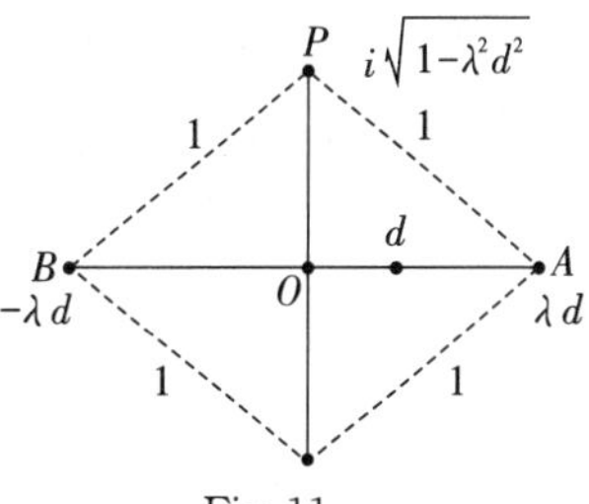

Fig. 11

Proof. Given two points A and B, there exists a series of points $A=X_0, X_1, X_2, \cdots, X_n=B$ such that $X_{k-1}X_k=(1/n)AB<2\left|\frac{1}{2}+i\lambda\right|^{-1}$. (This is by RC11.)

Since $\frac{1}{2} \in L$ and $i\lambda \in L$, we have $\frac{1}{2}+i\lambda \in L$, then we can find Y_k such that

$$Y_kX_{k-1}=Y_kX_k=\left|\frac{1}{2}+i\lambda\right|X_{k-1}X_k$$

and by RC8, we can find Z_k such that

$$Z_kX_{k-1}=Z_kX_k=\left|\frac{1}{2}+i\lambda\right|^{-1}.$$

So we obtain a$\left(\left|\frac{1}{2}+i\lambda\right|^{-1}\right)$-bridge between A and B, then $\left|\frac{1}{2}+i\lambda\right|^{-1}\in S$.

LEMMA 9. *If $d\in S$ and $i\lambda\in L$ with $\lambda>0$, then $\left|\left(\frac{1}{2}+i\lambda\right)d\right|\in S$.*

Lemma 9 is obvious.

LEMMA 10. *$(3k^2+3k+1)^{-1/2}\in S$ for every natural number k.*

By RC7, RC8 and RC11 we have Lemma 10.

LEMMA 11. *If $d\in S$ and $0<d<\frac{2}{3}$, then $((1/d^2)-2)^{-1/2}\in S$.*

Proof. By Lemma 7 and Lemma 6:

$$\pm i\sqrt{\frac{1}{d^2}-\left(1+\frac{1}{2}\right)^2}=\pm i\cdot f\left(d,1+\frac{1}{2}\right)\in L.$$

Then

$$\left(\frac{1}{d^2}-2\right)^{-1/2}=\left|\frac{1}{2}\pm i\sqrt{\frac{1}{d^2}-\left(1+\frac{1}{2}\right)^2}\right|^{-1}\in S$$

by Lemma 8.

LEMMA 12. *$(2m+1)^{-1/2}\in S$ for every natural number m.*

To prove Lemma 12, start with $d=(3k^2+3k+1)^{-1/2}\in S$ (Lemma 10) then use Lemma 11 again and again.

LEMMA 13. *$(i)^k\sqrt{n}\in L$ for any natural number k and n.*

Proof. Take natural numbers m and l such that $l<\sqrt{2m+1}$, then $(2m+1)^{-1/2}\in S$ by Lemma 12 and $l\in L$ by Lemma 5. And by $l(2m+1)^{-1/2}<1$ we can use Lemma 6, so

$$i\sqrt{2m+1-l^2}=i\cdot f((2m+1)^{-1/2},l)\in L.$$

We can choose m and l such that $2m+1-l^2=n$ for any natural number n.

And we have $i\in L$ while $m=l=2$.

LEMMA 14. *If $\lambda\in L$ and $\frac{1}{2}\leqslant\lambda<2$, then $1/\lambda\in S\cap L$.*

Proof. By $\lambda\in L$, and Lemma 8, Lemma 13, Lemma 6:

$$\sqrt{\frac{1}{2}-\lambda^2+m^2}=f\left(\frac{1}{\left|\frac{1}{2}+(i/2)\sqrt{4m^2+1}\right|},\lambda\right)\in L.$$

Here m is some natural number greater than λ. For instance, we can take $m=[\lambda]+2$. So by Lemma 6 and Lemma 8:

$$\sqrt{\lambda^2-\frac{1}{4}}=f\left(\frac{1}{\left|\frac{1}{2}+im\right|},\sqrt{\frac{1}{2}-\lambda^2+m^2}\right)\in L$$

so by Lemma 8:

$$\left|\frac{1}{2}+i\sqrt{\lambda^2-\frac{1}{4}}\right|^{-1}=\frac{1}{\lambda}\in S$$

notice $1\in S$, by Lemma 9

$$\left|\frac{1}{2}+i\sqrt{\lambda^2-\frac{1}{4}}\right|=\lambda\in S$$

so by Lemma 6:

$$\sqrt{\frac{1}{\lambda^2}-\frac{1}{4}}=f\left(\lambda,\frac{1}{2}\right)\in L.$$

So

$$\frac{1}{2}\pm i\sqrt{\frac{1}{\lambda^2}-\frac{1}{4}}\in L$$

and

$$\frac{1}{\lambda}=\lambda\cdot\frac{1}{\lambda^2}=\lambda\cdot\left(\frac{1}{2}+i\sqrt{\frac{1}{\lambda^2}-\frac{1}{4}}\right)\left(\frac{1}{2}-i\sqrt{\frac{1}{\lambda^2}-\frac{1}{4}}\right)\in L.$$

LEMMA 15. *If a real number $\lambda\in L$ and $\lambda\neq 0$, then $1/\lambda\in L$.*

Proof. We can suppose $\lambda>0$ because $-1\in L$.

Take some integer k such that $\frac{1}{2}\leqslant 2^k\lambda<2$, then $2^k\lambda\in L$ for $2\in L$ and $\frac{1}{2}\in L$. By Lemma 14 we have $1/(2^k\lambda)\in L$, then $1/\lambda=2^k\cdot 1/(2^k\lambda)\in L$.

LEMMA 16. *If $0<\lambda\in L$, then $\sqrt{\lambda}\in L$.*

Proof. Suppose $\lambda<1$ and $\lambda\neq\frac{1}{2}$. (When $\lambda=\frac{1}{2}$ we know $1/\sqrt{2}\in L$ by Lemma 13 and Lemma 15. When $\lambda>1$ we can also use Lemma 15 and replace λ by $(1/\lambda)<1$.)

Because $\lambda\pm\frac{1}{2}\in L$ and $\frac{1}{2}<\lambda+\frac{1}{2}<2$ so $1/\left(\lambda+\frac{1}{2}\right)\in S$ by Lemma 14. Then using Lemma 6 we have

$$i\sqrt{2\lambda}=i\cdot f\left(\frac{1}{\lambda+\frac{1}{2}},\lambda-\frac{1}{2}\right)\in L$$

and $\sqrt{\lambda}\in L$ because $i\in L$, $1/\sqrt{2}\in L$ and $-1\in L$.

LEMMA 17. *If $z\in L$ and $z\neq 0$, then $1/z\in L$.*

Proof. For $z\in L$ we have $z\cdot\bar{z}=|z|^2\in L$, so $|z|^{-2}\in L$ by Lemma 15. Then $1/z=\bar{z}\cdot|z|^{-2}\in L$.

LEMMA 18. *If $z\in L$, then $\sqrt{z}\in L$.*

Proof. Let $z=\lambda e^{i\theta}=\lambda(\cos\theta+i\sin\theta)$ and $\lambda>0$. So

$$\sqrt{z}=\sqrt{\lambda}\left(\cos\frac{\theta}{2}+i\sin\frac{\theta}{2}\right).$$

Since $z\in L$ we have $|z|^2=z\cdot\bar{z}\in L$ and so $\lambda=\sqrt{|z|^2}\in L$ by Lemma 16, and $1/\lambda\in L$ by Lemma 15 so $(\cos\theta+i\sin\theta)\in L$. Then $\cos\theta\in L$, $\sin\theta\in L$. And so

$$\cos\frac{\theta}{2}=\sqrt{\frac{1+\cos\theta}{2}}\in L,\ \sin\frac{\theta}{2}=\sqrt{\frac{1-\cos\theta}{2}}\in L.$$

Now it is obvious that $\sqrt{z}\in L$.

Thus, the general *RC* theorem has been proved.

ACKNOWLEDGEMENTS

The authors wish to express their thanks to the referee for valuable comments and suggestions. One of the authors, J. Zhang, would like to thank Professor Abdus Salam, the International Atomic Agency and UNESCO for hospitality at the International Centre for Theoretical Physics.

Reference

[1]Crux Math. 1983, 9(7): 217 - 218.

[2]Dan Pedoe. Queen's Quarterly, 1983, 90(2): 449 - 456.

[3]Crux Math. 1982, 8(6): 79 - 81.

[4]Crux Math. 1982, 8(8): 174.

[5]Zhang Jingzhong and Yang Lu. Essays in Math. 6, Education Publishing House of Shanghai, 1983, pp. 37 - 43(in Chinese).

[6]Xiao Renwu. Math. Commun., 1983(2): 26(in Chinese).

[7]Zhang Jingzhong. Nature, 1984, 7(12): 927 - 932(in Chinese).

[8]Zhang Jingzhong. Nature, 1986, 9(4): 288 - 296(in Chinese).

[9]Hou Xiaorong. J. China Univ. Sci. Tech. 1987, 17(1): 110 - 115(in Chinese).

[10]Kostovskii A N. Geometrical Constructions Using Compasses Only. Blaisdell, New York, 1961, pp. 69 - 70.

[11]Hjelsmlev, Johannes. Kcostruktion ved Passer med fast Indstilling, uden Burg af Lineal'. Mat. Tidsskr. A 1938, 77 - 85.

[12]Bieberbach, Ludwig. Theorie der geometrischen Konstruktionen, Basel, 1952.

Authors' addresses:

Zhang Jingzhong,

International Centre for

Theoretical Physics,

Trieste,
Italy.

Yang Lu,
Institute of Mathematical Sciences,
Chengdu Branch,
Academia Sinica,
610015 *Chengdu*,
People's Republic of China.

Hou Xiaorong.
University of Science and
Technology of China,
People's Republic of China.
(Received, February 20, 1990; revised version, June 19, 1990)

Searching Dependency between Algebraic Equations: an Algorithm Applied to Automated Reasoning

Yang Lu

International Centre for Theoretical Physics, Trieste, Italy

Zhang Jing-zhong

Institute of Mathematical Sciences, Academia Sinica

610015 Chengdu, Sichuan, People's Republic of China

ABSTRACT

An efficient computer algorithm is given to decide how many branches of the solution to a system of algebraic also solve another equation. As one of the applications, this can be used in practice to verify a conjecture with hypotheses and conclusion expressed by algebraic equations, despite the variety reducible or irreducible.

1 Introduction

Given a system of algebraic equations

$$\begin{cases} F_1(u_1, u_2, \cdots, u_n, x_1, x_2, \cdots, x_s)=0, \\ F_2(u_1, u_2, \cdots, u_n, x_1, x_2, \cdots, x_s)=0, \\ \cdots\cdots \\ F_s(u_1, u_2, \cdots, u_n, x_1, x_2, \cdots, x_s)=0, \end{cases} \tag{1}$$

and another one

$$G(u_1, u_2, \cdots, u_n, x_1, x_2, \cdots, x_s)=0, \tag{2}$$

where u_i and x_j we regard as parameters and indeterminates for $i=1,\cdots,n$ and $j=1,\cdots,s$, respectively, the problem which this paper concerns is whether the solution to $\{F_1=0,\cdots, F_s=0\}$ also solves $G=0$ or not? In other words, regarding $\{F_1=0,\cdots,F_s=0\}$ as the

---

本文刊于《International Centre for Theoretical Physics》,1991 年.

hypothesis of a theorem to be proved, and $G=0$ the desired conclusion, the task is to verify the theorem is true or not. It is easy to see that an efficient algorithm for such a problem is of significance both in theory and in practice.

In general, the solution to $\{F_1=0,\cdots,F_s=0\}$ depends upon n parameters as following,

$$\begin{cases} x_1=x_1(u_1,u_2,\cdots,u_n)=0 \\ x_2=x_2(u_1,u_2,\cdots,u_n)=0 \\ \cdots\cdots \\ x_s=x_s(u_1,u_2,\cdots,u_n)=0 \end{cases} \tag{3}$$

which consists of a number of branches. It is possible in some cases that $G=0$ holds for several branches of the solution but doesn't hold for branches else.

The algorithms have been given in the case that variety $\{F_1,\cdots,F_s\}$ is irreducible. [1—3][6—9] Otherwise, one used to do decomposition to transform a reducible variety into irreducible subvarieties then examine all of them one by one. [1] As we know, there was no algorithm for reducible case without doing decomposition until recently the authors (with Hou Xiaorong) [10] gave a decision procedure by which we can find directly how many branches of the solution to $\{F_1=0,\cdots F_s=0\}$ also solve $G=0$, despite the variety reducible or irreducible. In this paper, it is demonstrated how the theoretical result is implemented on computer and how efficient the algorithm is.

2 Algorithm

To deal with this problem, we suppose $\{F_1=0,\cdots,F_s=0\}$ has been transformed into the "triangular form":

$$\begin{cases} f_1(u_1,u_2,\cdots,u_n,x_1), \\ f_2(u_1,u_2,\cdots,u_n,x_1,x_2), \\ \cdots\cdots \\ f_s(u_1,u_2,\cdots,u_n,x_1,x_2,\cdots,x_s), \end{cases} \tag{4}$$

that is, for every equation $f_j=0$, only one new indeterminate x_j is introduced. We call $\{f_1,\cdots,f_s\}$ an "ascending chain". This assumption known as "well ordering principle" works without loss of generality because an algorithm was given some years ago [4—6][1] to transform a system of polynomials into an ascending chain which possesses the same set of zeros in any extended field.

Definition 1. By lcoeff(f_j,x_j) or l_j denote the leading coefficient of f_j with respect to x_j, for $j=1,\cdots,s$. An ascending chain $\{f_1,\cdots,f_s\}$ is called *a proper ascending chain* if any branch of the solution to $\{f_1=0,\cdots,f_s=0\}$ solves none of $\{l_1=0,\cdots,l_s=0\}$.

Definition 2. By resultant(p,q,x) denote the resultant of any two polynomials, p and q, with respect to indeterminate x, which is a determinant in terms of the coefficients of $p(x)$ and $q(x)$ to verify if the two polynomials have a common solution. Given an ascending chain $\{f_1,\cdots f_s\}$ and another polynomial G, putting

$$r_{s-1}:=\text{resultant}(G,f_s,x_s),$$
$$r_{s-2}:=\text{resultant}(r_{s-1},f_{s-1},x_{s-1}),$$
$$r_{s-3}:=\text{resultant}(r_{s-2},f_{s-2},x_{s-2}),$$
$$\cdots\cdots$$
$$r_1:=\text{resultant}(r_2,f_2,x_2),$$
$$r_0:=\text{resultant}(r_1,f_1,x_1),$$

we call r_0 *the resultant of ascending chain* $\{f_1,\cdots,f_s\}$ *with respect to polynomial* G and denote it simply by

$$\text{res}(f_1,\cdots,f_s,G).$$

Referring to MAPLE, by degree(p, T) and ldegree(p, T) denote the highest and lowest degrees of a polynomial p with respect to variable T, respectively. We have the following criterion:

Theorem 1. Given a proper ascending chain AS: $\{f_1,\cdots,f_s\}$, a polynomial g and a new variable T, put

$$p(T):=\text{res}(f_1,\cdots,f_s,g+T).$$

All the branches of solution to $\{f_1=0,\cdots,f_s=0\}$ also solve $g=0$ if and only if

$$\text{degree}(p(T),T)=\text{ldegree}(p(T),T). \tag{5}$$

The next one works for general cases:

Theorem 2. Given a proper ascending chain AS: $\{f_1,\cdots,f_s\}$, a polynomial g and a new variable T, put

$$k:=\text{ldegree}(\text{res}(f_1,\cdots,f_s,g+T),T), \tag{6}$$

then, there are exactly k branches of the solution to $\{f_1=0,\cdots,f_s=0\}$ also solve $g=0$.

Applying Theorem 1 to automated theorem proving, we have

Corollary 1. Assume $\{f_1,\cdots,f_s\}$ is a proper ascending chain. Let $\{f_1=0,\cdots,f_s=0\}$ and $g=0$ be the hypothesis and conclusion of a conjecture to be verified, respectively. The conjecture is true in general if and only if

$$\text{degree}(p(T),T)=\text{ldegree}(p(T),T),$$

where $p(T)$ means res$(f_1,\cdots,f_s,g+T)$.

Corollary 2. Assume $\{f_1,\cdots,f_s\}$ is a proper ascending chain. Let $\{f_1=0,\cdots,f_s=0\}$ and $g=0$ be the hypothesis and conclusion of a conjecture to be verified, respectively. The conjecture is true for exact k of the branches of the algebraic variety which expresses the hypothesis, if and only if

$$k=\text{ldegree}(\text{res}(f_1,\cdots,f_s,g+T),T). \tag{7}$$

Corollary 3. Given a proper ascending chain $\{f_1,\cdots,f_s\}$ and a polynomial g, none of the branches of solution to $\{f_1=0,\cdots,f_s=0\}$ solves $g=0$ if and only if

$$\text{res}(f_1,\cdots,f_s,g)\neq 0. \tag{8}$$

The last one gives following algorithm to check an ascending chain to be proper or not:

Corollary 4. An ascending chain $\{f_1,\cdots,f_s\}$ is proper if and only if $l_1\neq 0$ and for $j=2,3,\cdots,s$,

$$\text{res}(f_1,\cdots,f_{j-1},l_j)\neq 0, \tag{9}$$

where l_j denote the leading coefficient of f_j with respect to x_j.

3 Proof

The key is to prove Theorem 2 which other ones are inferred from.

Lemma 1. Let $f(x)$ and $g(x)$ be polynomials with degrees μ and ν, roots $x^{(1)},x^{(2)},\cdots,x^{(\mu)}$ and $\underline{x}^{(1)},\underline{x}^{(2)},\cdots\cdots,\underline{x}^{(\nu)}$, and leading coefficients l and $\underline{l}$, respectively. We have

$$\text{resultant}(f,g,x)=l^{\nu}\,\underline{l}^{\mu}\prod_{i=1}^{\mu}\prod_{j=1}^{\nu}(x^{(i)}-\underline{x}^{(j)}). \tag{10}$$

This well-known fact can be found in algebra textbooks.

Lemma 2. Given notations as above, then

$$\text{resultant}(f,g,x)=l^{\nu}\prod_{i=1}^{\mu}g(x^{(i)}). \tag{11}$$

It is easy to prove by using Lemma 1.

Now we need more notations. Put $\mu_j:=\text{degree}(f_j,x_j)$ for $j=1,2,\cdots,s$. Since the system of equations (4) is in a triangular form, theoretically speaking, all indeterminates $x_1,x_2,\cdots,x_s$ can be solved one by one, with several branches each. Thus, we can denote the full solution to (4) by

$$\begin{cases} x_1=x_1^{(i_1)}(u_1,\cdots,u_n)\\ x_2=x_2^{(i_1i_2)}(u_1,\cdots,u_n)\\ x_3=x_3^{(i_1i_2i_3)}(u_1,\cdots,u_n)\\ \cdots\cdots\\ x_s=x_s^{(i_1i_2\cdots i_s)}(u_1,\cdots,u_n)\end{cases} \tag{12}$$

where i_j ranges from 1 to μ_j, for $j=1,2,\cdots,s$, if the ascending chain is proper.

For simplicity, we put

$$\vec{u} := (u_1, \cdots, u_n);$$

$$\vec{x} := (x_1, \cdots, x_s);$$

and

$$\vec{x}^{(i_1 i_2 \cdots i_s)} := (x_1^{(i_1)}(\vec{u}), x_2^{(i_1 i_2)}(\vec{u}), \cdots, x_s^{(i_1 i_2 \cdots i_s)}(\vec{u})). \tag{13}$$

Lemma 3. Given notations as above, put $\nu_s := \text{degree}(G, x_s)$,

$$\nu_j := \text{degree}(\text{res}(f_j+1, f_j+2, \cdots, f_s, G), x_j) \tag{14}$$

for $j=1,2,\cdots,s-1$, and

$$L := l_1(\vec{u})^{\nu_1} \prod_{j=2}^{s} \prod_{i_1=1}^{\mu_1} \prod_{i_2=1}^{\mu_2} \cdots \prod_{i_{j-1}=1}^{\mu_{j-1}} (l_j(\vec{u}, \vec{x}^{(i_1 i_2 \cdots i_s)}))^{\nu_j} \tag{15}$$

where every $l_j(\vec{u}, \vec{x}^{(i_1 \cdots i_s)})$ is an algebraic function of $\vec{u}$ obtained by substituting a branch of the solution, $\vec{x}^{(i_1 \cdots i_s)}$, for the $\vec{x}$ in $I_j(\vec{u}, \vec{x})$, the initial of f_j. Then, we have

$$\text{res}(f_1, \cdots, f_s, G) = L \prod_{i_1=1}^{\mu_1} \prod_{i_2=1}^{\mu_2} \cdots \prod_{i_s=1}^{\mu_s} G(\vec{u}, \vec{x}^{(i_1 i_2 \cdots i_s)}). \tag{16}$$

It is not difficult to prove Lemma 3 by using Lemma 2 successively.

Now let us go to prove Theorem 2. Replacing G in (16) with $g+T$, the right hand side becomes a polynomial $p(T)$ with roots

$$-g(\vec{u}, \vec{x}^{(i_1 i_2 \cdots i_s)})$$

where i_j ranges from 1 to μ_j for $j=1,2,\cdots,s$. The lowest degree of a polynomial is nothing but the number of the vanishing roots, so that among the $\mu_1 \mu_2 \cdots \mu_s$ roots, $-g(\vec{u}, \vec{x}^{(i_1 \cdots i_s)})$, exactly k of them are vanishing. This completes the proof.

4 Examples

The resultant computation is supported by current softwares of computer algebra such as MAPLE, MACSYMA, REDUCE and MATHEMATICA. Somebody thought of it can do nothing because the computation would be too big, but in practice the complexity is acceptable for quite a number of nontrivial examples. All the ascending chain appeared in examples of this section have been checked to be proper by Corollary 4.

Example 1. A difficult geometry theorem was conjectured by Thebault in 1938 and proved by Taylor in 1983 with a proof of 26 pages! Then it was verified [1] on a SYMBOLICS 3600 computer by using *successive pseudo division* to all the irreducible subvarieties obtained from decomposition to a reducible variety which expresses hypothesis of the theorem; the program took 44 hours CPU time, while more than 500 theorems else took only a few seconds each. The wonderful theorem is stated as follows:

Given a triangle ABC with incenter w and circumcircle Γ, take any point D on edge

BC. Let w_1 be the center of the circle which contacts *DB*, *DA* and Γ, and w_2 the center of the circle which contacts *DC*, *DA* and Γ. To prove w_1, w_2 and w are collinear.

We took one algebraic interpretation somewhat different from that taken in [1] for the Thebault-Taylor theorem and ran a program in MAPLE (version 4. 3) on a SUN386*i* station with CPU time 1041. 58 seconds, and then on a CONVEX C210 with CPU time 268. 35 seconds only. The answer the screen showed was: *The conjecture is true for 2 of the 8 branches*. Our program follows below:

```
tt:=proc()
f1:=4*x1^2*u1^2*u2^4+6*u1^2*u2^4-2*u2^2-2*u2^6-2*u2^2*u1^4-2*u2
   ^6*u1^4+u1^2+u1^2*u2^8+4*u2^4*u3-4*u2^4*u1^4*u3-4*u2^4*u3^2*
   u1^2;
f2:=(4*u2^4-8*u2^4*u3*u1^2-4*u1^4*u2^4+8*x1*u1^2*u2^4)*x2^2+(-4
   *u2^2*u1^4+8*u2^4+4*u1^2-4*u2^6*u1^4-4*u2^6+8*u1^4*u2^4+4*u1
   ^2*u2^8+8*u1^2*u2^4-8*u1^2*u2^2-4*u2^2-8*u2^6*u1^2)*x2+(-2*u1
   ^2+4*u1^2*u2^4-2*u1^2*u2^8)*x1-1-8*u2^4*u3+8*u2^2*u3*u1^2+u1
   ^4+2*u2^4+4*u2^6*u1^4*u3+4*u1^4*u3*u2^2-12*u2^4*u3*u1^2-2*u1
   ^4*u2^4-2*u2^8*u3*u1^2-8*u2^4*u1^4*u3-2*u3*u1^2+u1^4*u2^8+4*
   u3*u2^2-u2^8+8*u2^6*u3*u1^2+4*u2^6*u3;
f3:=(4*u2^4-4*u1^4*u2^4-8*u2^4*u3*u1^2-8*x1*u1^2*u2^4)*x3^2+(8*
   u1^4*u2^4+4*u1^2-4*u2^2*u1^4-8*u1^2*u2^2-4*u2^6*u1^4+8*u1^2*
   u2^4-8*u2^6*u1^2+4*u1^2*u2^8-4*u2^2+8*u2^4-4*u2^6)*x3+(2*u1^2
   -4*u1^2*u2^4+2*u1^2*u2^8)*x1-1+8*u2^2*u3*u1^2+8*u2^6*u3*u1^2+
   u1^4+2*u2^4-u2^8+4*u3*u2^2-12*u2^4*u3*u1^2-2*u1^4*u2^4-8*u2^4
   *u3-8*u2^4*u1^4*u3-2*u3*u1^2+u1^4*u2^8+4*u1^4*u3*u2^2+4*u2^6*
   u3+4*u2^6*u1^4*u3-2*u2^8*u3*u1^2;
f4:=x4*u1*u2+1-u1^2*u2^2;
f5:=x5*u1*u2-u1^2+u2^2;
f6:=(2*u1^2-2*u1^2*u2^4+2*x4*u1^2*u2^2+2*x5*u1^2*u2^2)*x6+(-u1
    ^2-u1^2*u2^4)*x5+(-u1^2-u1^2*u2^4)*x4-u1^4+u2^4-1+u1^4*u2^4;
f7:=(-u2^2+u1^2*u2^4-u2^2*u1^4+u1^2)*x7+(u2^2-u2^2*u1^4-u1^2+u1^2*u2
   ^4-2*x5*u1^2*u2^2)*x6+(u1^2*u2^4+u1^2)*x5-u2^4+u1^4;
g:=(4*x2*u2^4+4*x3*u2^4)*x7+(-4*u2^2-4*u2^6+8*u2^4)*x6-4*u2^2-
  4*u2^6+1+6*u2^4-4*x2*x3*u2^4+u2^8;
r6:=resultant(g+T,f7,x7);
r5:=resultant(r6,f6,x6);
```

```
r4:=resultant(r5,f5,x5);
r3:=resultant(r4,f4,x4);
r2:=resultant(r3,f3,x3);
r1:=resultant(r2,f2,x2);
r0:=resultant(r1,f1,x1);
lprint('The conjecture is true for',ldegree(r0,T),'of the',degree(r0,T),'branches')
end;
```

Example 2. The famous Feuerbach Theorem: A circle which contacts 3 lines has to contact the nine-point-circle of the triangle formed by these lines. For this we ran a program in MAPLE(version 4. 3) on a SUN386*i* station with CPU time 6. 63 seconds, and then on a CONVEX C210 with 1. 72 seconds. The answer the screen showed was: *The conjecture is true for* 4 *of the* 4 *branches*, that means the theorem is true in general. In fact, for any triangle, one inscribed circle and three escribed circles, each of the four contacts the nine-point-circle. The program follows below:

```
fb:=proc()
f1:=-u1^3+4*u1^2*x1+4*u2^2*x1^2*u1-4*u2*x1^2*u1-4*u1*x1^2-8*u2
   ^2*x1^3+8*u2*x1^3+4*x1^2*u1^3-8*x1^3*u1^2+4*u1*x1^4;
f2:=-2*u2*x1-2*x1*x2+2*x1+2*u1*x2-u1;
f3:=4*u1*x3-u1^2-u2+u2^2;
f4:=4*x4-1-2*u2;
f5:=u1^4-2*u1^2*u2+2*u1^2*u2^2+u2^2-2*u2^3+u2^4+u1^2-16*u1^2*x5;
g:=normal(((x1-x3)^2+(x2-x4)^2)^2+x5^2+x1^4-2*((x1-x3)^2+(x2-
   x4)^2)*(x5+x1^2)-2*x5*x1^2);
r4:=resultant(g+T,f5,x5);
r3:=resultant(r4,f4,x4);
r2:=resultant(r3,f3,x3);
r1:=resultant(r2,f2,x2);
r0:=resultant(r1,f1,x1);
lprint('The conjecture is true for',ldegree(r0,T),'of the',degree(r0,T),'branches')
end;
```

Example 3. On three sides of a triangle ABC, three equilateral triangles BCA_1, CAB_1 and ABC_1 are drawn. To prove lines AA_1, BB_1 and CC_1 are concurrent. To this theorem we use a modified algorithm by replacing g in resultant computation with the *final pseudo remainder* obtained from Wu's algorithm, successive pseudo division. The answer the screen showed is *The conjecture is true for* 2 *of the* 8 *branches*. In fact, there are 8 instances as some of the equilateral triangles may be inward to triangle *ABC* and some else may be outward. The conjecture is true only for two instances, that is, three equilateral triangles all inward or all outward. The program in MAPLE follows below, with CPU time (including that for well-ordering and successive division) 26.37 seconds and 5.47 seconds on a SUN386*i* station and a CONVEX C210, respectively.

```
et:=proc()
F1:=2*x1-1;
F2:=4*x2^2-3;
F3:=x3^2+x4^2-u1^2-u2^2;
F4:=x3^2+x4^2-(x3-u1)^2-(x4-u2)^2;
F5:=(x5-1)^2+x6^2-(u1-1)^2-u2^2;
F6:=(x5-1)^2+x6^2-(x5-u1)^2-(x6-u2)^2;
f1:=F1;
f2:=F2;
f3:=resultant(F3,F4,x4);
f4:=normal(F4);
f5:=resultant(F5,F6,x6);
f6:=normal(F6);
A:=array(1..3,1..3,[[x2-u2,u1-x1,(x2-u2)*u1-(x1-u1)*u2],[x4,1-
   x3,x4],[x6,-x5,0]]);
with(linalg,det):
g:=normal(det(A));
p5:=prem(g,f6,x6);
p4:=prem(p5,f5,x5);
p3:=prem(p4,f4,x4);
p2:=prem(p3,f3,x3);
p1:=prem(p2,f2,x2);
p0:=prem(p1,f1,x1);
if p0=0 then print('The conjecture is true in general') else
r5:=resultant(p0+T,f6,x6);
r4:=resultant(r5,f5,x5);
r3:=resultant(r4,f4,x4);
r2:=resultant(r3,f3,x3);
r1:=resultant(r2,f2,x2);
r0:=resultant(r1,f1,x1);
```

```
lprint('The conjecture is true for',ldegree(r0,T),'of the',degree(r0,T),'branches');
fi
end;
```

5 Conclusion: Algorithm in Large

Among the existing algorithms for searching dependency between algebraic equations, what is the essential difference of that suggested in this paper from others? In a word, it is a *global algorithm*. To see this more clearly, we consider a problem about transversality, a global property, between algebraic submanifolds. Let

$$\begin{cases} F_1(x_1,x_2,x_3)=0 \\ F_2(x_1,x_2,x_3)=0 \end{cases}$$

be an algebraic curve and

$$F_3(x_1,x_2,x_3)=0$$

an algebraic surface. We want to know whether the curve transversally intersects the surface or not. Since the tangent vector of the curve is $\nabla F_1 \times \nabla F_2$ and the normal vector of the surface is ∇F_3, the transversal condition means

$$(\nabla F_1 \times \nabla F_2) \cdot \nabla F_3 \neq 0$$

for all the intersections.

Assume $\{F_1, F_2, F_3\}$ is transformed into a proper ascending chain, $\{f_1, f_2, f_3\}$, and put

$$g:=(\nabla F_1 \times \nabla F_2) \cdot \nabla F_3.$$

Then, by Corollary 3, the curve intersects the surface transversally if and only if

$$\mathrm{res}(f_1,f_2,f_3,g) \neq 0.$$

Otherwise, with Theorem 2, the number of the tangent points is given by

$$\mathrm{ldegree}(\mathrm{res}(f_1,f_2,f_3,g+T),T).$$

ACKNOWLEDGEMENTS

One of the authors (Yang L.) would like to thank Professor Abdus Salam, the International Atomic Energy Agency and UNESCO for hospitality at the International Centre for Theoretical Physics, Trieste.

References

[1] Chou S C. Mechanical Geometry Theorem Proving. Amsterdam D. Reidel Publishing

Company, 1988.

[2]Kapur D. Using Gröbner bases to reason about geometry problems. J. Symbolic Computation, 1986(2):399 - 408.

[3]Kutzler B, Stifter S. On the application of Buchberger's algorithm to automated geometry theorem proving. J. Symbolic Computation, 1986(2):389 - 397.

[4]Ritt J F. Differential Equations from the Algebraic Standpoint. Amer. Math. Soc., 1932.

[5]Ritt J F. Differential Algebra. Amer. Math. Soc., 1950.

[6]Wu Wen-tsun. On the decision problem and the mechanization of theorem proving in elementary geometry. Scientia Sinica, 1978(21):157 - 179.

[7] Yang Lu. A new method of automated theorem proving in The Math-ematical Revolution Inspired by Computing, J. H. Johnson and M. J. Loomes (eds). Oxford University Press, 1991:115 - 126.

[8]Zhang Jingzhong, Yang Lu. Principles of parallel numerical method and single-instance method of mechanical theorem proving. Math. in Practice and Theory (in Chinese), 1989(1):34 - 43.

[9]Zhang Jingzhong, Yang Lu, Deng Mike. The parallel numerical method of mechanical theorem proving. Theoretical Computer Science, 1990(74):253 - 271.

[10]Zhang Jingzhong, Yang Lu, Hou Xiaorong. A criterion for dependency of algebraic equations, with applications to automated theorem proving, (submitted).

A Prover for Parallel Numerical Verification of a Class of Constructive Geometry Theorems

Yang Lu, Zhang Jing-zhong and Li Chuan-zhong

International Centre for Theoretical Physics, Trieste, Italy

1 Introduction

The following issue is of considerable significance. Can a mathematical theorem be proved by carrying out a series of experiments as people used to do for physical laws? By "mathematical experiment" we mean the numerical verification of one certain instance. And the word "prove" we use in a deterministic sense, of course, not in probablistic sense[3]. This becomes possible in practice since a new algorithm for automated theorem proving due to the first two authors[4][5][6][7]. This is the *Parallel Numerical Algorithm* which replaces most of the symbolic algebra used in existing methods by numerical computation. In this way, for example, we can prove a theorem concerning general triangles by verifying it numerically for a finite number of triangles with given side-lengths. A lot of numerical, sometimes with high complexity, is often needed for our purpose. This is why such a method had not been suggested before the computer age. On the other hand, however, if one restricts the theorems within a special class, the corresponding algorithm may be of lower complexity so that it can be implemented on lower-level microcomputers without symbolic algebra softwares. As an example, here we deal with a class of theorems whose hypotheses can be expressed by the following "triangular form" system.

$$\begin{cases} f_1(u_1, u_2, \cdots, u_d, x_1) = 0, \\ f_2(u_1, u_2, \cdots, u_d, x_1, x_2) = 0, \\ \cdots\cdots \\ f_s(u_1, u_2, \cdots, u_d, x_1, x_2, \cdots, x_s) = 0, \end{cases} \tag{1}$$

where $f_1 := a_1(u_1, \cdots, u_d)x_1 - b_1(u_1, \cdots, u_d)$,

$$f_j := a_j(u_1, \cdots, u_d, x_1, \cdots, x_{j-1})x_j - b_j(u_1, \cdots, u_d, x_1, \cdots, x_{j-1}), \tag{2}$$

本文刊于《国际数学机械化研讨会》,1992 年.

for $j=2,\cdots,s$, and the desired conclusion is expressed by

$$g(u_1,u_2,\cdots,u_d,x_1,x_2,\cdots,x_s)=0, \tag{3}$$

where a_j,b_j and g are polynomials with integral coefficients. For convenience, let $\vec{u}:=(u_1,\cdots,u_d)$ and name it the *parameter* of AS. A value of $\vec{u}$ is called a *parameter point*.

Define a series of polynomials in $\vec{u}$, namely, $p_1, q_1, \cdots, p_s, q_s$, as follows. Put $p_1:=a_1(\vec{u}), q_1:=b_1(\vec{u})$. If $p_1, q_1, \cdots, p_{j-1}, q_{j-1}$ are defined, then we define p_j, q_j such that

$$p_j x+q_j=\text{numerator}\left(a_j\left(\vec{u},\frac{q_1}{p_1},\cdots,\frac{q_{j-1}}{p_{j-1}}\right)x+b_j\left(\vec{u},\frac{q_1}{p_1},\cdots,\frac{q_{j-1}}{p_{j-1}}\right)\right), \tag{4}$$

i. e. p_j, q_j are coefficients of the linear function of x on the right hand side, where function 'numerator (•)' means taking the numerator of a rational fraction. So p_j, q_j are polynomials in $\vec{u}$ with integral coefficients, for $j=1,\cdots,s$.

Define a polynomial Q in elements $\vec{u}, p_1, q_1, \cdots, p_s, q_s$ as follows,

$$Q:=\text{numerator}\left(g\left(\vec{u},\frac{q_1}{p_1},\cdots,\frac{q_s}{p_s}\right)\right). \tag{5}$$

By substituting (4) into (5), Q becomes a polynomial in $\vec{u}$ only, with integral coefficients. We denote it by $\Phi(\vec{u})$, that is,

$$\Phi(u_1,\cdots,u_d)=\Phi(\vec{u}):=Q(\vec{u},p_1(\vec{u}),q_1(\vec{u}),\cdots,p_s(\vec{u}),q_s(\vec{u})) \tag{6}$$

where $p_j(\vec{u}), q_j(\vec{u})$ for $j=2,\cdots,s$ are determined by (4).

Put $x_j:=q_j/p_j$ for $j=1,\cdots,s$. Then, the verified theorem is true for such zeros of system (1) that conditions

$$a_j(\vec{u},x_1,\cdots,x_{j-1})\neq 0 \tag{7}$$

hold at them, for $j=1,..,s$, if and only if $\Phi(\vec{u})$ is identical to 0. So the problem is reduced to verification of algebraic identities and one can uses a parallel numerical algorithm for this purpose.

2 Basic Principles of Numerical Verification

Our algorithm theoretically based upon two lemmas following below.

Definition 1. Let $S_1, S_2, \cdots, S_d$ be d finite subsets of a set K and t_i denote the cardinal number of S_i, for $i=1,\cdots,d$. We call the Cartesian product of the above d subsets

$$S=S_1\times S_2\times\cdots S_d$$

a *d-dimentional lattice array on* K, with *size*$(t_1,t_2,\cdots,t_d)$.

Lemma 1. Let $\Phi(u_1,u_2,\cdots,u_d)$ be a polynomial over a field K of characteristic 0. Assume the degree of Φ in u_i is not greater than n_i, for $i=1,\cdots,d$. If there exists a d-dimensional lattice array S on K, with size$(n_1+1,n_2+1,\cdots,n_d+1)$, such that Φ vanishes over S, i. e. for any $(\vec{u},\cdots,\vec{u}_d)\in S$,

$$\Phi(\vec{u}_1,\cdots,\vec{u}_d)=0$$

then Φ is identical to zero. This is easy to prove to d by induction.

Now let us to estimate the upper bound of $\deg(\Phi, u_i)$ for $i=1, \cdots, d$. Let $m_{ji} := \max\{\deg(p_j, u_i), \deg(q_j, u_i)\}$, $c_{ji} := \deg(f_j, u_i)$, and $d_{ji} := \deg(f_j, x_i)$.
We have

Lemma 2. It holds for m_{ji} the recursion inequalities

$$m_{ji} \leqslant c_{ji} + \sum_{k=1}^{j-1} d_{jk} \cdot m_{ki} \tag{8}$$

where $d_{jk} := \deg(f_j, x_k)$ as defined above. And

$$\deg(\Phi, u_i) \leqslant \deg(g, u_i) + \sum_{j=1}^{s} m_{ji} \deg(g, x_j) \tag{9}$$

where Φ is well defined in (6). This lemma is an obvious corollary of the recursion formula (4).

3 Description for the Algorithm

Step 1 Given hypotheses as (1) (2) and the desired conclusion as (3), estimate the upper bounds of $\deg(\Phi, u_i)$ for $i=1, \cdots, s$. Of course we need not get the detailed expression for $\Phi(\vec{u})$. Lemma 2 gives a simple algorithm for this step, without symbolic computations. It is very easy to write a program for this, and sometimes, the procedure is so simple that it can be done by hand.

Step 2 Established $\deg(\Phi(\vec{u}), u_i) \leqslant n_i$ for $i=1, \cdots, d$, let S be a lattice array on integral ring $\mathbf{Z}$, with size$(n_1+1, \cdots, n_d+1)$. The elements of S can be chosen arbitrarily, only the size is important. For every $\vec{u}_0 \in S$, do what follows: Substitute $\vec{u}_0$ for $\vec{u}$ in a_1, b_1, i, e. p_1, q_1, and then compute

$$p_2(\vec{u}_0), q_2(\vec{u}_0), \cdots, p_s(\vec{u}_0), q_s(\vec{u}_0) \tag{9}$$

by recursion formula (4), successively. Substitute the values (9) for $p_1, q_1, \cdots, p_s, q_s$ in Q (which is well defined in (5)) to check whether

$$\Phi(\vec{u}_0) = Q(\vec{u}_0, p_1(\vec{u}_0), q_1(\vec{u}_0), \cdots, p_s(\vec{u}_0), q_s(\vec{u}_0)) = 0 \tag{10}$$

or not.

Step 3 In this way, examine all the points of the lattice array S one by one. If for every point we examine, the equality $\Phi=0$ is always true, then the theorem is generically proved. Otherwise, the theorem disproved.

The algorithm described above involves integer operation only, because both the input and output of every step are integers. Our prover employs a sub-program to ensure *exact arithmetic* for large integers, the results obtained are error-free. This algorithm has the advantage that the program is simpler than the usual ones which use symbolic algebra only and suitable for parallel computing. If one uses the algorithm on a parallel computer, then every processor deals with a few sample points of the lattice array and the running time would be reduced accordingly. Also the numerical computations occupy fewer memories than symbolic algebra so it can be used to prove non-trivial theorems on a microcomputer or even on a pocket computer. In practice, we implemented the algorithm efficiently on a PB 700 pocket computer and not only proved a lot of well-known geometry theorems, but discovered new ones as well, see[4][6][7].

4 A Class of Constructive Geometry Theorems

The algorithm introduced in § 3 is applicable for a special class of geometry theorems we call it "class L" whose definition follows later.

Definition 2. Let Λ be a collection consisting of a finite number of points, lines and circles. A point P is said to be *constructed directly from* Λ if it is given in one of the following manners:

- Let P be an arbitrary point.
- Let P be an arbitrary point on a line in Λ.
- Let P be an arbitrary point on a single circle specially designated in Λ.
- Let P be the intersection of two lines in Λ.
- Let P be one intersection of a line and a circle both in Λ, while the other intersection is in Λ.
- Let P be one intersection of two circles in Λ, while the other intersection is in Λ.
- Let P be a dividing point of a segment whose endpoints A, B are both in Λ, with rational ratio, i. e. $PA:PB=m:n$ where m, n are integers.

A line l is said to be constructed directly from Λ if it is given in one of the following manners:

- Let l be the line joining two points in Λ.
- Let l be the line passing through a point in Λ and parallel to a line in Λ.
- Let l be the line passing through a point in Λ and perpendicular to a line in Λ.
- Let l be the perpendicular bisector of two points in Λ.
- Let l be the radical axis of two circles in Λ.

A circle c is said to be constructed directly from Λ if its center is in Λ and its radius equals the length of a segment whose endpoints are both in Λ.

Definition 3. A *directly constructive series* means a finite series of points, lines and circles, $e_1, e_2, \cdots, e_r$, where e_{j+1} is constructed directly from the collection $\Lambda_j := \{e_1, \cdots, e_j\}$, for $j=1, \cdots, r-1$.

A subseries of a directly constructive series is called a *constructive series*.

Definition 4. A couple (H, C) is called a proposition of class L, where H is a constructive series and C is either an equation, $P=0$, where P is a polynomial whose elements are the Cartesian coordinates of some points in H, or an assertion equivalent to such an algebraic equation.

Furthermore, H and C are called the hypothesis and conclusion of this proposition, respectively.

Obviously, any proposition of class L can be translated into an algebraic version with hypothesis and conclusion described as (1)~(3), so one can verifies it numerically, using the algorithm given in last section.

5 Examples

Our prover translates geometric statements into algebraic ones automatically. The class L in which every proposition can be verified simply by integer operation is narrower than a class of geometry propositions of constructive type discussed in [1][2]where symbolic algebra was employed. In practice, however, many well-known and non-trivial theorems can be proven by our prover. Amongest the 512 theorems proven by Chou's prover[1], most of them can be stated as theorems in class L.

Example 1. *Simson Theorem*. Let D, A, B, C be points on a circle and E, F, G the perpendicular foots of D on lines BC, CA, AB, respectively. Show that E, F, G are collinear.

Input:

circularpoint(D)
circularpoint(A)
circularpoint(B)
circularpoint(C)
foot(D,B,C,E)
foot(D,C,A,F)
foot(D,A,B,G)
collinear(E,F,G)

Output: "The theorem is generically true".

The running time is 3.019 seconds, on an AST 386SX/20 microcomputer with DOS 2.10.

Example 2. *Butterfly Theorem*. Let A, D be points on circle O and M the midpoint of AD. Draw two secant lines of circle O, passing through M and CE intersect AD at P and Q. Show that M is the midpoint of segment PQ.

Input:

circle(O)
circularpoint(A)
circularpoint(D)

midpoint(A,M,D)
circularpoint(B)
circularpoint(C)
secant(O,B,E,M)
secant(O,C,F,M)
intersection(B,F,A,D,P)
intersection(C,E,A,D,Q)
midpoint(P,M,Q)

Output: "The theorem is generically true".

The running time is 43.106 seconds, on AST 386SX/20 with DOS 2.10.

Example 3. *Pascal Theorem*. Let A,B,C,D,E,F be points on a circle and $P := AB \cap DE$, $Q := BC \cap EF$, $R := CD \cap FA$. Show that P,Q,R are collinear.

Input:

circularpoint(A)
circularpoint(B)
circularpoint(C)
circularpoint(D)
circularpoint(E)
circularpoint(F)
intersection(A,B,D,E,P)
intersection(B,C,E,F,Q)
intersection(C,D,F,A,R)
collinear(P,Q,R)

Output: "The theorem is generically true".

The running time is 101.117 seconds, on AST 386SX/20 with DOS 2.10.

Example 4. *Nine-point circle*. Let D, E, F be midpoints of the segments BC, CA, AB, respectively, and L the perpendicular foot of A on BC. Show that D,E,F,L are concyclic.

Input:

points(A,B,C)
midpoint(B,D,C)
midpoint(C,E,A)
midpoint(A,F,B)
foot(A,B,C,L)
concyclic(D,E,F,L)

Output: "The theorem is generically true".

The running time is 2.359 seconds, on AST 386SX/20 with DOS 2.10.

Example 5. *Euler line*. Let O, G, H be the circumcenter, centroid and orthocenter, respectively. Show that O,G,H are collinear.

Input:

points(A,B,C)
circumcenter(A,B,C,O)
midpoint(B,D,C)
midpoint(C,E,A)
intersection(A,D,B,E,G)
foot(A,B,C,L)
foot(B,C,A,J)
intersection(A,L,B,J,H)
collinear(O,G,H)

Output: "The theorem is generically true".

The running time is 1.156 seconds, on AST 386SX/20 with DOS 2.10.

Example 6. Let A,B,C,D be points on a circle and l_1,l_2,l_3,l_4 be Simson lines of A, B,C,D with respect to triangles BCD,CDA,DAB,ABC, respectively. Show that l_1,l_2,l_3, l_4 are concurrent.

Input:

circularpoint(A)
circularpoint(B)
circularpoint(C)
circularpoint(D)
foot(B,A,C,E)
foot(B,A,D,F)
foot(D,A,C,G)
foot(D,A,B,H)
foot(C,A,B,J)
foot(C,A,D,K)
intersection(E,F,G,H,I)
collinear(I,J,K)

Output: "The theorem is generically true".

The running time is 28.390 seconds, on AST 386SX/20 with DOS 2.10.

For simplicity, we often use an optional statement for taking intersections of various lines—

• option(A, B, C, 1, D, E, F, 1, G) defines G as the intersection of two lines, one passing through A and parallel to BC, the other passing through D and parallel to EF.

• option(A, B, C, 1, D, E, F, 2, G) defines G as the intersection of two lines, one passing through A and parallel to BC, the other passing through D and perpendicular to EF.

• option$(A, B, C, 2, D, E, F, 1, G)$ defines G as the intersection of two lines, one passing through A and perpendicular to BC, the other passing through D and parallel to EF.

• option$(A, B, C, 2, D, E, F, 2, G)$ defines G as the intersection of two lines, one passing through A and perpendicular to BC, the other passing through D and perpendicular to EF.

Example 7. *Steiner line.* The orthocenters of the four triangles of a complete quadrilateral are collinear. (*Remark*: This line is called Steiner line.)

Input:

points(A,B,C)
collinear(A,B,D)
collinear(B,C,E)
option$(A,B,C,2,B,C,A,2,H)$
option$(A,D,E,2,D,C,A,2,H_1)$
option$(E,A,B,2,D,B,C,2,H_2)$
collinear(H,H_1,H_2)

Output: "The theorem is generically true".

The running time is 7.847 seconds, on AST 386SX/20 with DOS 2.10.

Example 8. Show that the Steiner line of a complete quadrilateral is perpendicular to its Gauss line. (*Remark*: the midpoints of 3 diagonals of a complete quadrilateral are collinear and that line is called Gauss line.)

Input:

points(A,B,C)
collinear(A,B,D)
collinear(B,C,E)
option$(A,B,C,2,B,C,A,2,H)$
option$(A,D,E,2,D,C,A,2,H_1)$
midpoint(A,M,E)
midpoint(C,N,D)
perpendicular(M,N,H,H_1)

Output: "The theorem is generically true".

The running time is 5.550 seconds, on AST 386SX/20 with DOS 2.10.

Example 9. Let ABC be a triangle and P a point. Draw lines AP, BP, CP to intersect BC, CA, AB at D, E, F, respectively. Take another point Q and draw lines DQ, EQ, FQ to intersect EF, FD, DE at L, M, N, respectively. Show that lines AL, BM, CN are concurrent. (*Remark*: Mr Hou contributed this as a new theorem, but I am not sure.)

Input:

points(A,B,C)
point(P)
intersection(A,P,B,C,D)
intersection(B,P,C,A,E)
intersection(C,P,A,B,F)
point(Q)
intersection(D,Q,E,F,L)
intersection(E,Q,F,D,M)
intersection(F,Q,D,E,N)
intersection(A,L,B,M,R)
collinear(C,N,R)

Output: "The theorem is generically true".

The running time is 21.312 seconds, on AST 386SX/20 with DOS 2.10.

Example 10. *Butterfly Theorem for quadrilaterals*. Let $ABCD$ be a quadrilateral whose diagonal BD bisects segment AC at M. Draw two lines passing through M, one intersects AB, CD at E, G, the other intersects BC, AD at F, H, respectively. Assume EH and FG intersect AC at P and Q, respectively. Show that M also bisects PQ.

Input:

points(A,B,C)
midpoint(A,M,C)
collinear(A,B,E)
collinear(B,C,F)
collinear(B,M,D)
intersection(E,M,C,D,G)
intersection(F,M,A,D,H)
intersection(E,H,A,C,P)
intersection(F,G,A,C,Q)
midpoint(P,M,Q)

Output: "The theorem is generically true".

The running time is 0.820 seconds, on AST 386SX/20 with DOS 2.10.

Example 11. Given a triangle ABC, take point M on AB, point N on BC and point P on CA. Draw three lines, the first passing through M, parallel to BC and meeting CA at M_1, the second passing through N, parallel to CA and meeting AB at N_1, the third passing through P, parallel to AB and meeting BC at P_1. Show that the triangles MNP and $M_1N_1P_1$ have the same area.

Input:

points(A, B, C)
collinear(A, B, M)
collinear(B, C, N)
collinear(C, A, P)
option(M, B, C, 1, C, C, A, 1, M_1)
option(N, C, A, 1, A, A, B, 1, N_1)
option(P, A, B, 1, B, B, C, 1, P_1)
$\Rightarrow$
area(M, N, P)-area(M_1, N_1, P_1)

Output: "The theorem is generically true".

The running time is 0.328 seconds, on AST 386SX/20 with DOS 2.10.

Example 12. Let H and r be the orthocenter and circumradius of a triangle ABC. Show that

$$12r^2=(BC)^2+(CA)^2+(AB)^2+(AH)^2+(BH)^2+(CH)^2.$$

Input:

points(A, B, C)
option(A, B, C, 2, B, C, A, 2, H)
circumcenter(A, B, C, O)
$\Rightarrow$
$12(OA)^2-(BC)^2-(CA)^2-(AB)^2-(AH)^2-(BH)^2-(CH)^2$

Output: "The theorem is generically true".

The running time is 6.820 seconds, on AST 386SX/20 with DOS 2.10.

References

[1] Chou S C. Mechanical Geometry Theorem Proving. D Reidel Publishing Company, Amsterdam, 1988.

[2] Chou S C, Gao X S. A class of geometry statements of constructive type and geometry theorem proving. TR-89-37. Department of Computer Sciences, University of Texas at Austin, 1989.

[3] Schwartz J T. Fast probablistic algorithms for verification of polynomial identities. J. ACM., 1980(27): 701-717.

[4] Yang Lu. A new method of automated theorem proving. in The Mathematical Revolution Inspired by Computing, J. H. Johnson and M. J. Loomes (eds). Oxford University Press, 1991: 115-126.

[5] Yang Lu, Zhang Jingzhong, Hou Xiaorong. A criterion of dependency between algebraic equations and its applications. preprint for this conference.

[6] Zhang Jingzhong, Yang Lu. Principles of parallel numerical method and single-instance

method of mechanical theorem proving. Math . in Practice and Theory(in Chinese), 1989(1):34 - 43.

[7]Zhang Jingzhong, Yang Lu, Deng Mike. The parallel numerical method of mechanical theorem proving. Theoretical Computer Science, 1990(74):253 - 271.

最初几个 Heilbronn 数的猜想和计算

杨　路　张景中　曾振柄
（中国科学院成都分院数理科学研究室）

摘　要

本文首先结合保面积仿射变换和小扰动方法，对任意凸区域 K 及点数 n 较小的 Heilbronn 问题进行了较一般的讨论；然后分别利用进一步的计算，得出了 K 为正方形，$n=5,6$ 的 Heilbronn 数，并给出相应的最优图形；最后提出正方形中 7 点的 Heilbronn 数的一个猜测，文中的方法和主要引理可用于其它形状凸区域 K 的类似问题.

1. 引　言

在一个正方形里放上 6 个点，这 6 个点所组成的三角形中，至少有一个三角形的面积不超过该正方形面积的 1/8，而且这个界是可以达到的. 以上是 M. Goldberg 在 1972 年发表的若干未加证明的断言之一. 中国科技大学的一个学生在 1978 年又独立地提出了这个猜想. 随即发现对这一猜想的验证竟是出乎意料地困难[11]！这一类型的问题通常叫做 Heilbronn 问题.

一般说来，设 K 是一个平面凸体（即平面上一个具有非空内部的凸集），用 $|K|$ 表示它的面积；对于任意一个三角形 $r_1r_2r_3$，用 $(r_1r_2r_3)$ 表示其面积；再令

$$(r_1r_2\cdots r_n)=\min\{(r_ir_jr_k)\mid i,j,k=1,\cdots,n\};$$

$$H_n(K)=\frac{1}{|K|}\max\{(r_1r_2\cdots\cdots r_n)\mid r_i\in K;i=1,\cdots,n\}.$$

这样定义的 $H_n(K)$，$n=3,4,\cdots$，叫做凸体 K 的 Heilbronn 数. 设 $\{r_1,r_2,\cdots,r_n\}$ 是 K 的某个确定的子集，如果正好有

$$\frac{1}{|K|}(r_1r_2\cdots r_n)=H_n(K),$$

我们就说 $\{r_1,r_2,\cdots,r_n\}$ 或 $r_1r_2\cdots r_n$ 是关于 K 的一个 Heilbronn 分布，简称为关于 K 的 H 分布.

本文中约定，当 K 是一个正方形或平行四边形时，略去 $H_n(K)$ 中的 K 而将其 Heilbronn 数简记为 H_n. 关于 H_n 的阶的估计已经有大量的工作[2~10].

本文刊于《数学年刊 A 辑（中文版）》，第 4 期，1992 年.

1950 年 Heilbronn 猜想存在常数 c 使得

$$H_n < c/n^2,$$

但在 1982 年 J. Komlos 等的联名文章证明了存在常数 c 使得

$$c(\log n)/n^2 < H_n,$$

这就推翻了 Holibronn 原先的猜想. 另一方面，Komlos 等在 1981 年证明了

$$H_n < \frac{c}{n^\mu}$$

这里 $\mu = \frac{8}{7} - \varepsilon = 1.1428\cdots - \varepsilon, \varepsilon > 0$.

要想对一般的 n 找到 H_n 的一个解析表达式看来至少在本世纪是没有希望的事. 即使对一个确定的并不大的 n，准确地计算出这个特定的 H_n 也有想当大的困难.

M. Goldberg[1] 考虑了正方形的最初几个 Heilbronn 数的准确值. 除 $H_3 = H_4 = \frac{1}{2}$ 是平凡的结果外，他断言当 $n < 8$ 时，H_n 可以在某一个仿射正 n 角形上实现，即在正方形内存在一个仿射正 n 角形 $r_1 r_2 \cdots r_n$，使得

$$\frac{1}{|K|}(r_1 r_2 \cdots r_n) = H_n (n < 8).$$

并且他给出了这几个 Heilbronn 数的值：

$$H_5 = \frac{3-\sqrt{5}}{4} = 0.1909\cdots,$$

$$H_6 = \frac{1}{8} = 0.125,$$

$$H_7 = 0.0794\cdots.$$

不过所有这些断语他都没有给出证明.

本文检验了 Goldberg 的上述断言，证明了他的三个结论中只有一个(关于 H_6)是正确的；而其余两个关于 H_5 和 H_7 的结论都是错误的. 事实上任何仿射正 5 角形和仿射正 7 角形不可能实现 H_5 和 H_7.

$$H_5 = \frac{\sqrt{3}}{9} = 0.19245\cdots,$$

$$H_6 = \frac{1}{8} = 0.125.$$

前者否定了 M. Goldberg 关于 H_5 的猜想，后者证实了他关于 H_6 的猜想. 本文还举例证明了

$$H_7 \geqslant \frac{1}{12} = 0.0833\cdots > 0.0794\cdots,$$

从而又否定了 Goldberg 关于 H_7 的猜想. 这样就结束了仿射正多角形关于正方形是否 Heilbronn 分布的课题. 这些也是关于 Heilbronn 数准确值的第一批获得严格证明的结果.

2. $H_5(K)$

设 K 是任一平面凸区域，首先讨论含于 K 的一个凸 n 角形 $r_1 r_2 \cdots r_n$，特别当 $n=5$，是其

Heilbronn 分布所必须满足的性质. 以下不妨假设$|K|=1$.

引理 1 凸 n 角形 $r_1r_2\cdots r_n\subset K$ 是 K 的 H 分布,则满足关系$(r_ir_{i+1}r_{i+2})=H_n(K)$的三角形 $r_ir_{i+1}r_{i+2}$至少有三个.

证 为表达方便,将下标相连的顶点构成的三角形称为 $r_1r_2\cdots r_n$ 的外围△,易知使$(r_ir_jr_k)=(r_1r_2\cdots r_n)$成立的三角形必是其外围△.

如果 $r_1r_2\cdots r_n\subset K$ 是 H 分布但只有一个外围△,不妨设 $r_1r_2r_3$,使$(r_1r_2r_3)=H_n(K)$,则如图 1(其中 $n=5$),作 $r'_1=r_1+\varepsilon(r_4-r_1)$,$\varepsilon>0$,使$(r'_1r_2r_3)>\min\{(r_1r_2r_3),(r_4r_2r_3)\}=H_n(K)$;并保持$(r'_1r_2r_n)$,$(r'_1r_{n-1}r_n)>H_n(K)$. 于是$(r'_1r_2\cdots r_n)>(r_1r_2\cdots r_n)$. 与 $r_1r_2\cdots r_n\subset K$ 是 H 分布矛盾.

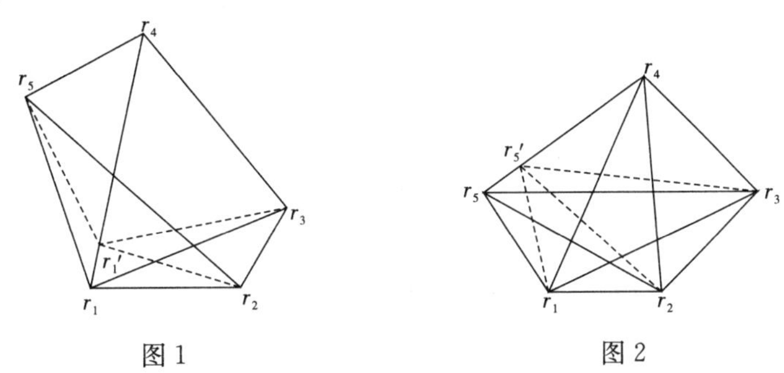

图 1　　　　图 2

如果 $r_1r_2\cdots r_n\subset K$ 是 H 分布但只有两个外围△的面积取到 $H_n(K)$,则这两个三角形或者有一公共边,或者有一公共顶点,不然同前面一样导出矛盾. 当两个仅有的面积最小的三角形有一公共边,不妨就设

$$(r_1r_2r_3)=(r_1r_2r_5)=H_n(K),$$

如图 2(其中 $n=5$),作

$$r'_n=r_n+\varepsilon(r'_{n-1}-r_n),\varepsilon>0,$$

使
$$(r'_nr_1r_2)>\min\{(r_nr_1r_2),(r_3r_1r_2)\}=H_n(K),$$

并保持
$$(r'_nr_1r_{n-1}),(r'_nr_{n-2}r_{n-1})>H_n(K).$$

这样
$$(r_1r_2\cdots r_{n-1}r'_n)=(r_1r_2\cdots r_n).$$

但 $r_1r_2\cdots r_{n-1}r'_n$只有一个外围△的面积取到 $H_n(K)$,矛盾. 当两个仅有的面积最小的三角形有一公共顶点,不妨设

$$(r_2r_3r_4)=(r_4r_5r_1)=H_n(K),$$

如图 3(其中 $n=5$),作

$$r'_1=r_1+\varepsilon(r_2-r_1),\varepsilon>0$$

使
$$(r'_1r_4r_5)>\min\{(r_1r_4r_5),(r_2r_4r_5)\}=H_n(K),$$

并保持
$$(r'_1r_2r_5),(r'_1r_2r_3)>H_n(K).$$

这样
$$(r'_1r_2\cdots r_n)=(r_1r_2\cdots r_n)$$

但 $r'_1r_2\cdots r_n$ 只有一个外围△的面积取到 $H_n(K)$,矛盾. 引理 1 获证.

引理 2 设 K 是任意凸区域,凸 5 角形 $r_1r_2\cdots r_5\subset K$ 是其 H 分布,则下列之一成立.

1. $r_1r_2\cdots r_5$ 至少有四个外围△取到 $H_5(K)$,

2. $r_1r_2\cdots r_5$ 只有三个外围△取到 $H_5(K)$,另外两个共一顶点.

证　引理 1 保证 $r_1r_2\cdots r_5$ 至少有三个外围△取到 $H_5(K)$. 假若仅有三个取到 $H_5(K)$ 而剩下的两个外围△不是共一顶点，则它们必共一边，如图 4 所示，设这两个△是 $r_1r_2r_3$，

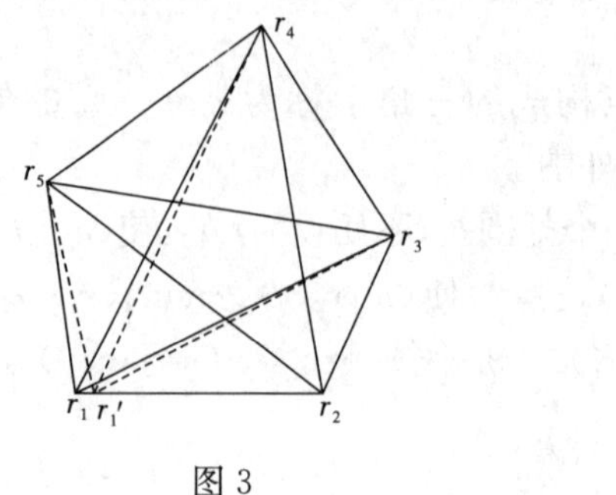

图 3

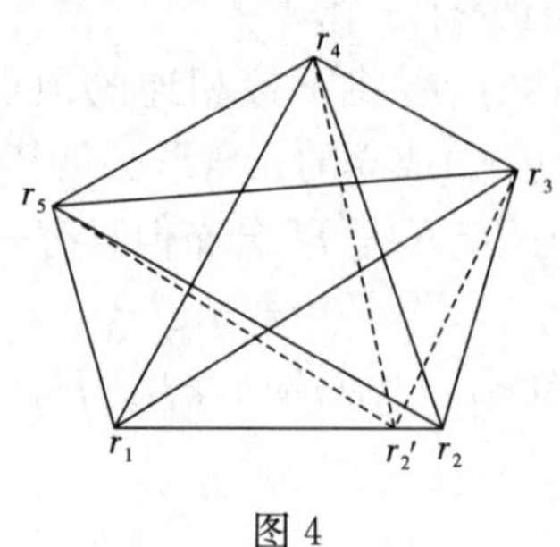

图 4

$r_1r_2r_5$. 作

$$r_2'=r_2+\varepsilon(r_1-r_2),\varepsilon>0,$$

使

$$(r_2'r_3r_4)>(r_2r_3r_4)=H_5(K),$$

并保持

$$(r_1r_2'r_3),(r_1r_2'r_5)>H_5(K),$$

则

$$(r_1r_2'r_3r_4r_5)=(r_1r_2\cdots r_5),$$

但 $r_1r_2'r_3r_4r_5$ 只有两个外围△取到 $H_5(K)$，由引理 1 它不能是 H 分布，于是 $r_1r_2\cdots r_5$ 也不是，矛盾. 引理 2 证毕.

引理 3　设 K 是任意凸区域，凸 5 角形 $r_1r_2\cdots r_5\subset K$ 是其 H 分布，则下列两项都成立.

1. $r_1r_2\cdots r_3$ 的任意两个相邻内角之和大于 π（简称 $r_1r_2\cdots r_5$ 是无大底的）；

2. $r_1,r_2,\cdots,r_5\in\partial K$.

证　1. 若一个凸 5 角形 $r_1r_2\cdots r_5$ 有两个相邻内角之和 $\leqslant\pi$，易知 $r_1r_2\cdots r_5$ 有两个外围△其面积 $>(r_1r_2\cdots r_5)$ 且它们共有一边. 由引理 2 这时 $r_1r_2\cdots r_5$ 不可能是 H 分布（见图 5）.

2. 不然，例如设 $r_1\notin\partial K$，则因 r_2r_3 和 r_5r_4 的交点 r 不可能和 r_1 在 r_3r_4 的同侧，作

$$r_1'=r_1+\varepsilon(r_1-r),\varepsilon>0,$$

使 $r_1'\in K$，则

$$(r_1'r_2r_5)>(r_1r_2r_5)\geqslant H_5(K),$$

$$(r_1'r_2r_3)>(r_1r_2r_3)\geqslant H_5(K),(r_1'r_4r_5)>(r_1r_4r_5)\geqslant H_5(K),$$

$$\min\{(r_2r_3r_4),(r_3r_4r_5)\}=H_5(K)\text{（由引理 2）}$$

所以 $(r_1'r_2\cdots r_5)=(r_1r_2\cdots r_5)$ 但 $r_1'r_2\cdots r_5$ 至多两个外围△取到 $H_5(K)$，由引理 1 它不能是 H 分布，从而 $r_1r_2\cdots r_5$ 也不是，矛盾（见图 6）. 引理 3 证毕.

引理 4　设 K 是任意凸区域，$r_1r_2\cdots r_5\subset K$ 是其 H 分布，则下列二项之一成立.

1. $r_1r_2\cdots r_5$ 至少有四个外围△取到 $H_5(K)$.

2. $r_1r_2\cdots r_5$ 只有三个外围△取到 $H_5(K)$，其余两个的公共顶点在 ∂K 上，而且如果 K 在这一点的支承线是唯一的，则它必平行于 $r_1r_2\cdots r_5$ 的一底和一对角线.

证　为方便起见，将第 2 项里两个较大外围△的公共顶点称为大顶点. 引理 2 保证它是唯一的，引理 3 保证它必然在 ∂K 上.

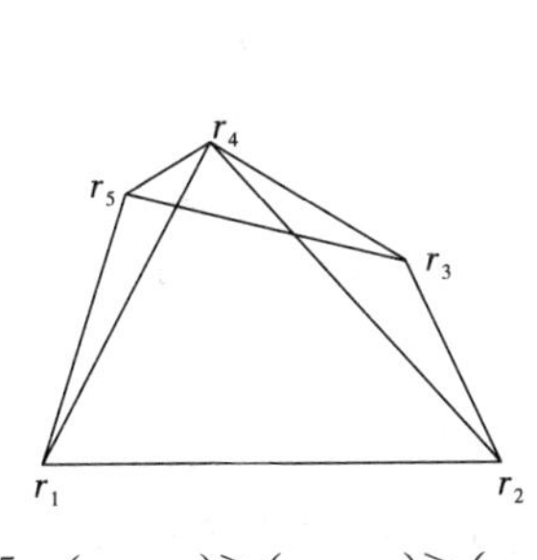

图 5　$(r_1r_2r_5)>(r_1r_5r_3)>(r_1r_5r_4)$
$(r_2r_3r_1)>(r_2r_3r_5)>(r_2r_3r_4)$

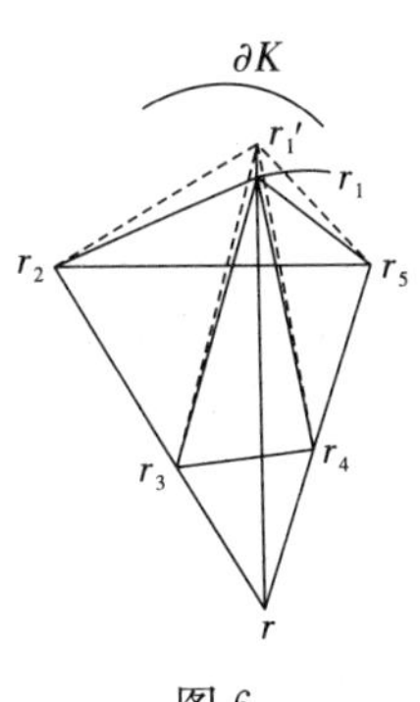

图 6

如图 7,设 r_1 是大顶点,p 是∂K 在 r_1 的支承线,$l/\!/r_2r_5$,假若 l 与 p 不重合,则在 l 的上侧与 p 的下侧交成的区域中含有 ∂K 的一段弧,因而可取 $r_1'\in K$ 使 $r_1'\in K$ 使 r_1'在 l 的上侧,此小扰动可保持不等式

$$(r_1'r_4r_5)>H_5(K),(r_1'r_2r_8)>H_5(K),$$

并使

$$(r_1'r_2r_5)>(r_1r_2r_5)=H_5(K),$$

于是

$$(r_1'r_2\cdots r_5)=(r_1r_2\cdots r_5)$$

但 $r_1'r_2\cdots r_5$ 只有两个外围△取到 $H_5(K)$,所以 $r_1'r_2\cdots r_5$ 不是 H 分布,$r_1r_2\cdots r_5$ 也不是.矛盾,所以

$$l=p/\!/r_2r_5/\!/r_3r_4.$$

引理 4 证毕.

引理 5　设 $K=\square$或△,凸 5 角形 $r_1r_2\cdots r_5\subset K$ 是其 H 分布,若 $r_1r_2\cdots r_5$ 含有 K 的顶点,则它至少有四个外围△取到 $H_5(K)$.

证　由于 $r_1r_2\cdots r_5$ 无大底,故至多含有 K 的一个顶点,我们证明 $r_1,r_2,\cdots,r_5$ 都不可能是大顶点,从而由引理 4 即推出结论.

Ⅰ. $K=\square s_1s_2s_3s_4$.不妨设 $r_5=s_1$.易知当 $r_1r_2\cdots r_5\subset\square s_1s_2s_3s_4$ 是 H 分布时其余四点分别位于 $s_1s_2s_3s_4$ 四条边的内部:$r_i\in s_is_{i+1}(1\leqslant i\leqslant 4)$,如图 8 所示.而且 r_2 不比 r_4 偏高;r_3 不比 r_1 偏右(不然,如图 9,r_2 高于 r_4,过 r_2,r_4 作平行线得$\square s_1's_2's_3's_4'\supset r_1r_2\cdots r_5$ 使$|s_1's_2's_3's_4'|<|s_1s_2s_3s_4|$),于是

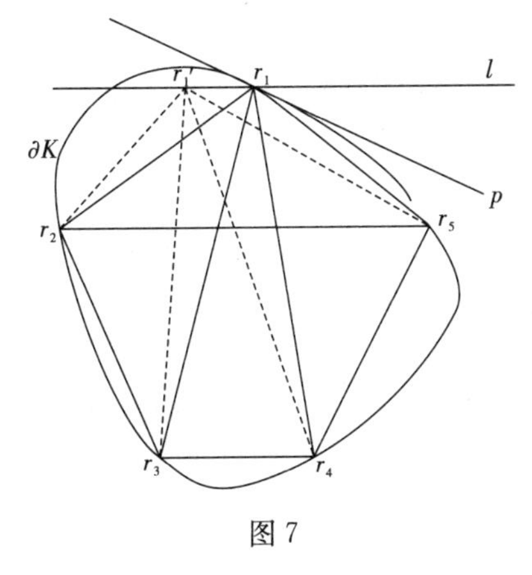

图 7

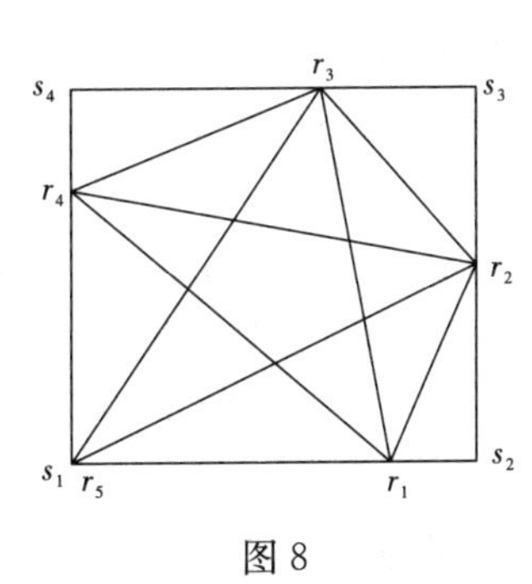

图 8

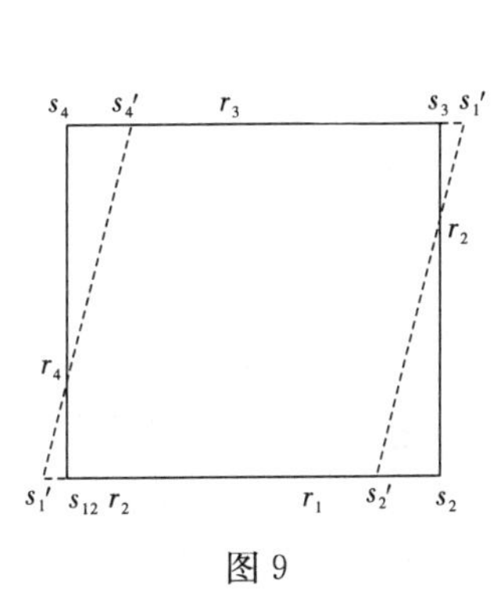

图 9

$$(r_1r_5r_4)\geqslant(r_1r_5r_2),(r_1r_5r_4)\geqslant(r_1r_3r_4).$$

这说明 r_5 不可能是大顶点；同理 r_2，r_3 也不是大顶点；又显然 $K=\square$ 在 r_1 的支承线不平行于 r_5r_2，知 r_1 不会是大顶点，同理 r_4 也不是，所以 $K=\square$ 时引理得证.

Ⅱ. $K=\triangle s_1s_2s_3$. 不妨设 $r_5=s_1$. 易知当 $r_1r_2\cdots r_5\subset\triangle s_1s_2s_3$ 是 H 分布时，$\triangle s_1s_2s_3$ 的每边各含 $r_1r_2\cdots r_5$ 的两个点，如图 10，而且该边中点位于这两点之间（不然，如图 11，将可作 $\triangle s_1's_2's_3'\supset r_1r_2\cdots r_5$ 但 $|s_1's_2's_3'|<|s_1s_2s_3|$）. 于是

$$(r_1r_5r_4)\geqslant(r_1r_5r_2),(r_1r_5r_4)\geqslant(r_1r_3r_4).$$

由此，r_5，r_2，r_8 均不可能是大顶点. 又显然 r_1，r_4 也不可能是大顶点. $K=\triangle$ 时引理成立.

为方便起见，称Ⅰ、Ⅱ中的外围 $\triangle r_1r_5r_4$ 为顶点△.

推论 1　在引理 5 的条件下，唯一可能不取 $H_5(K)$ 的外围△是顶角△.

证　只要证明当 $r_1r_2\cdots r_5\subset K=\square$ 或△是 H 分布时

$$(r_2r_3r_1)=(r_2r_3r_4)=H_5(K).$$

详略（参见图 12，13）.

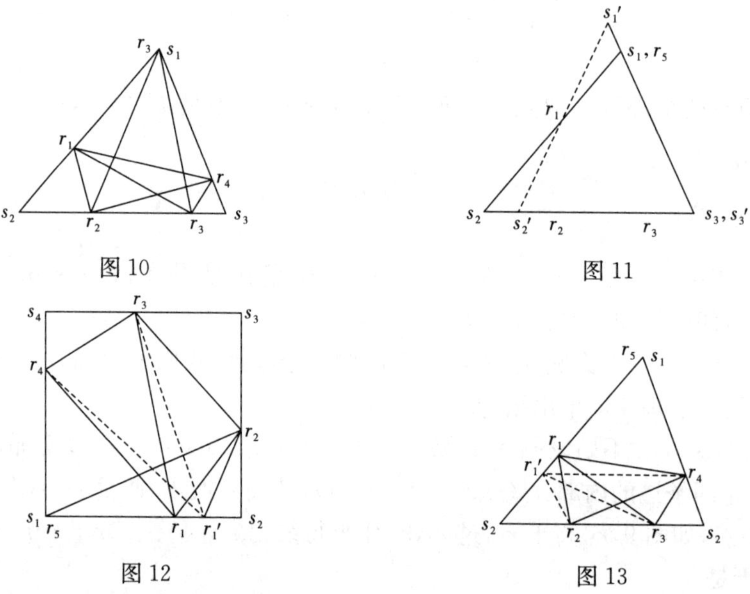

图 10　　图 11

图 12　　图 13

为计算 $H_5=H_5(\square)$，需要证明

引理 6　若凸 5 角形 $r_1r_2\cdots r_5\subset\square r_1r_2r_3r_4$ 是其 H 分布，则存在 $\square s_1's_2's_3's_4'$ 使 $r_1r_2\cdots r_5\subset\square s_1's_2's_3's_4'$ 是其中也是 H 分布，而且 $r_1r_2\cdots r_5$ 含有 $\square s_1's_2's_3's_4'$ 的一个顶点.

证　若 $r_1r_2\cdots r_5$ 不含 $s_1s_2s_3s_4$ 的顶点，则不妨设 $r_i\in s_is_{i+1}(1\leqslant i\leqslant 4)$，$r_5\in s_1s_2$，如图 14 所示，而且有 $r_2r_4/\!/r_1r_5$（不然，可作 $\square s_1''s_2''s_3''s_4''\supset r_1r_2\cdots r_5$ 但 $|s_1''s_2''s_3''s_4''|<|s_1s_2s_3s_4|$，见图 15）. 因 $\angle r_2r_3r_4+\angle r_3r_4r_5>\pi$，可作 $s_1's_2's_3's_4'\supset r_1r_2\cdots r_5$，使 $r_5=s_1'$，$r_2\in s_2's_3'$ 并且 $|s_1's_2's_3's_4'|=|s_1s_2s_3s_4|$，所以 $r_1r_2r\cdots r_5$ 在 $s_1's_2's_3's_4'$ 中也是 H 分布. 引理 6 证毕.

推论 2　若一凸 5 角形 $r_1r_2\cdots r_5$ 不是仿射正 5 角形并且它不含 $\square s_1s_2s_3s_4$ 的顶点，则 $r_1r_2\cdots r_5$ 不是 $s_1s_2s_3s_4$ 的 H 分布.

证　用反证法，由推论 1 和引理 6. 详略.

现在,计算 $H_5(\square)$并找出其 H 分布已是水到渠成.

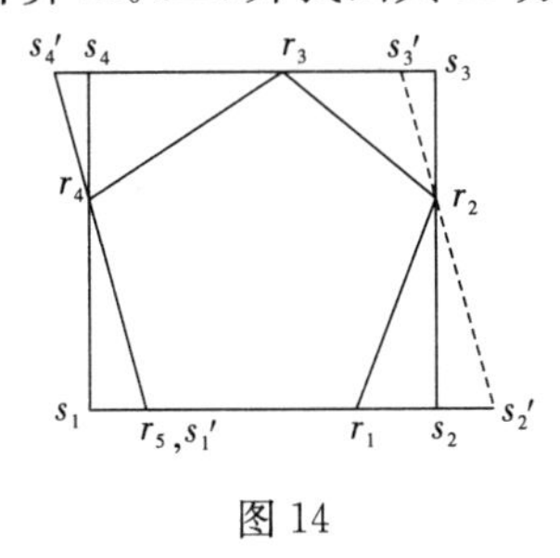

图 14

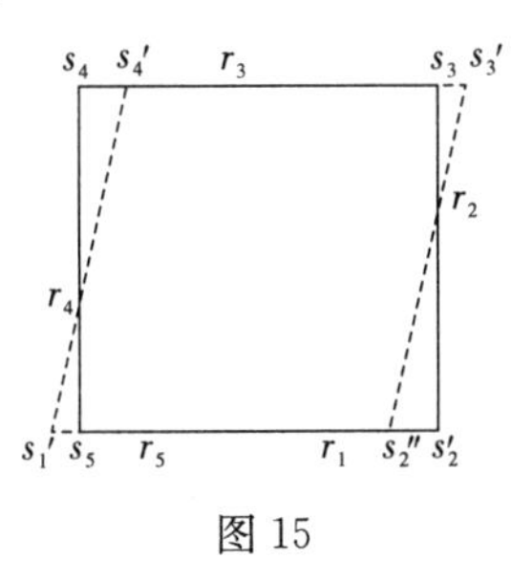

图 15

定理 1　$H_5=\frac{\sqrt{3}}{9}$.

证　$\forall r_1r_2\cdots r_5\subset\square s_1s_2s_3s_4$,若 $r_1r_2\cdots r_5$ 的凸包不是 5 角形,易知$(r_1r_2\cdots r_5)\leqslant\frac{1}{6}$;另一方面,Goldberg[1]给出 $s_1s_2s_3s_4$ 内仿射正 5 角形可以取到$(r_1r_2\cdots r_5)=\frac{3-\sqrt{5}}{4}>\frac{1}{6}$. 所以当 $r_1r_2\cdots r_5\subset\square$是 H 分布时必然是凸 5 角形. 为求 $H_5=H_5(\square)$的值,根据引理 6 不妨就取含有 $s_1s_2s_3s_4$ 的顶点的凸 5 角形 $r_1r_2\cdots r_5$ 来计算. 取 $s_i(1\leqslant i\leqslant 4)$,$r_i(1\leqslant i\leqslant 5)$,各点坐标如图 16 所示. 根据引理 5 及推论 1,$r_2r_3/\!/r_1r_4$,$r_3r_4/\!/r_2r_5$,$r_1r_2/\!/r_5r_3$,即

$$\frac{1-b}{a-1}=-\frac{y}{x},\frac{1-y}{a}=b,\frac{b}{1-x}=\frac{1}{a},$$

可化简为 $x=y,a=b,x=1-a^2$.

现在 $H_5(\square)$的计算化为求单重最优化问题

$$\max(r_1r_2\cdots r_5)=\frac{1}{2}a(1-a^2)$$

$$s.t.\ (r_1r_5r_4)\geqslant(r_1r_5r_2),\text{即 } a\leqslant\frac{1}{2}(1-a^2),$$

它的解是

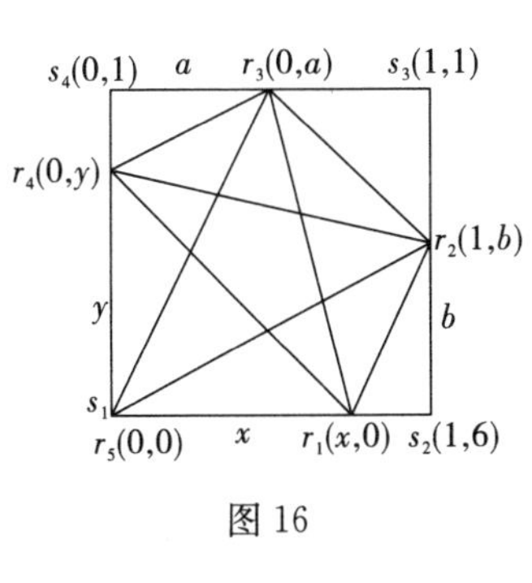

图 16

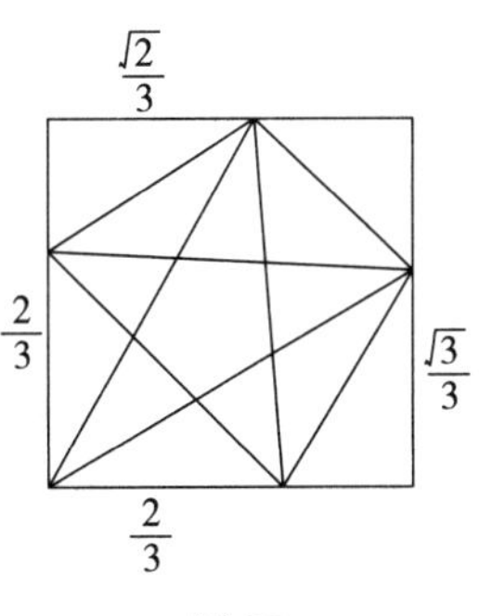

图 17

$$H_5=\max\{(r_1r_2\cdots r_5),r_1,r_2,\cdots,r_5\in\square s_1s_2s_3s_4\}=\frac{\sqrt{3}}{9},$$

最大值唯当 $a=\frac{\sqrt{3}}{3}$,$x=\frac{2}{3}$时取得,这时 $r_1r_2\cdots r_5$ 的位置如图 17. 定理 1 证毕.

定理 2　仿射不变的意义下,5 个点在正方形内只有一种 Heilbronn 分布.

证　由推论 2 即得(参见图 18).

定理 1,2 解决了 $n=5$ 时正方形的 Heilbronn 问题. 它否定了 Goldberg 的猜测.

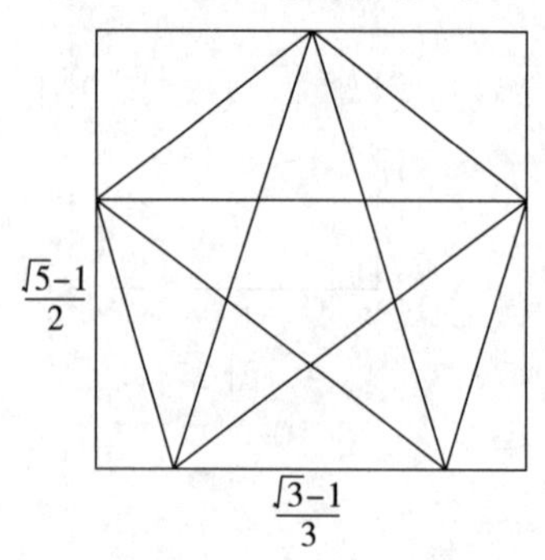

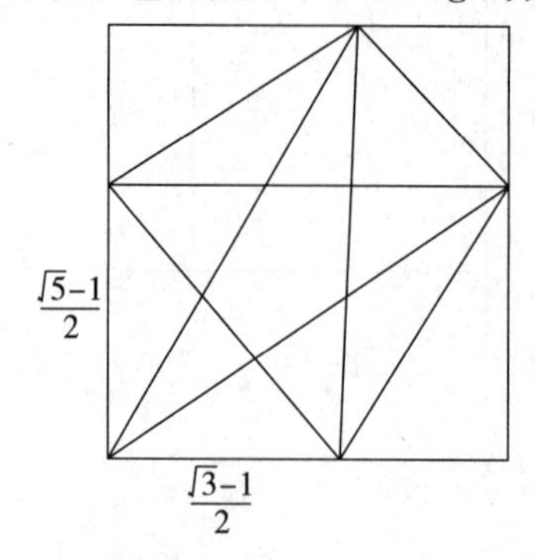

图 18　Goldberg 猜测的五点在正方形内的 Heilbronn 分布,仿射正五角形. 比较其中后一图和下图 19

3. $H_6(K)$

六点的 Heilbronn 问题,比之于五点的情况,其困难主要在于不能用简单的方法断定这时 Heilbronn 分布的凸包是什么图形. 即使对于很规则的 K 也是如此. 因此只能就凸包的各种可能形状逐一分析.

引理 7　若 $r_1r_2\cdots r_6\subset K$ 且凸包是凸四角形,则

$$(r_1r_2\cdots r_6)\leqslant\frac{1}{8}|K|.$$

证　设 $r_1r_2\cdots r_6$ 有凸包是 $r_1r_2r_3r_4$,不妨设 r_1,r_2,r_3,r_4 的仿射坐标分别是$(0,0),(1,0),(a,b),(0,1)$;并且 $a+b>1,0<a\leqslant b\leqslant 1$. 如图 20 所示.

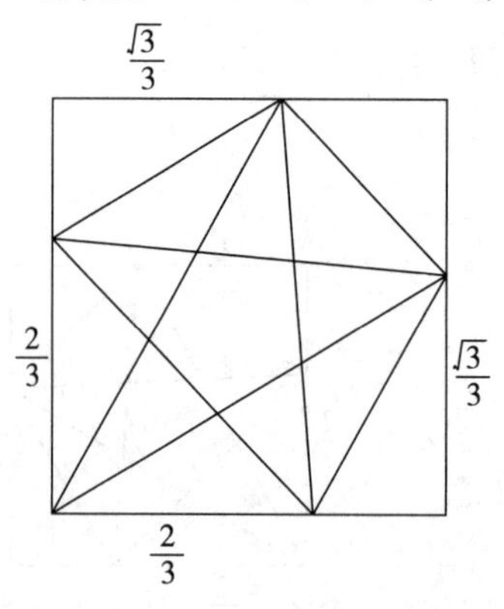

图 19　正方形内五点的 Heilbronn 分布

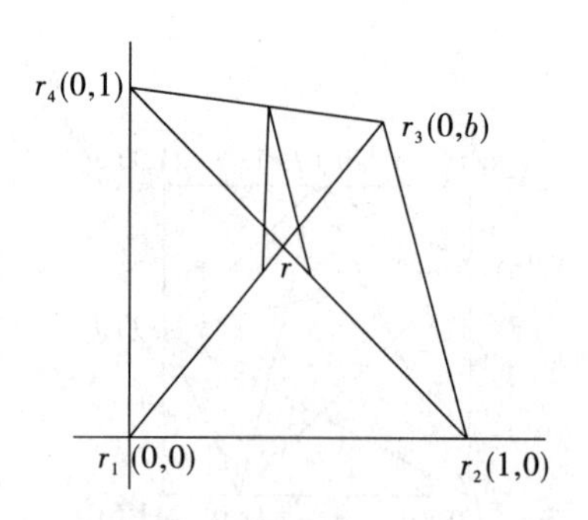

图 20

假如 $r_5,r_6\in$ 某$\triangle r_ir_jr_k$,$1\leqslant i<j\leqslant k\leqslant 4$,则易知本引理由严格不等号成立. 设 $r=r_1r_3\cap r_2r_4$,则由 a,b 的假设范围知 r_1r_3,r_2r_4 的中点均在 r 的下侧. 于是若有 $r_i\in\triangle rr_3r_4(i=5,6)$,则 r_i 在$\triangle r_2r_3r_4$ 或$\triangle r_1r_3r_4$ 中位线上方,便有

$$\begin{aligned}(r_1r_2r_3r_4r_i)&\leqslant\min\{(r_2r_3r_4r_i),(r_1r_3r_4r_i)\}\\&\leqslant\min\left\{\frac{1}{4}(r_2r_3r_4),\frac{1}{4}(r_1r_3r_4)\right\}\leqslant\frac{1}{8}|r_1r_2r_3r_4|\leqslant\frac{1}{8}|K|.\end{aligned}$$

故以下假定 $r_5\in\triangle rr_1r_4$，$r_6\in\triangle rr_2r_3$，如图 21，作直线 l_1,l_2,l_3,l_4 使 l_1 上的点和 r_1,r_3 组成的$\triangle l_1r_1r_3$ 以及$\triangle l_2r_2r_3$，$\triangle l_3r_1r_3$，$\triangle l_4r_1r_4$ 的面积都等于

$$\frac{1}{8}|r_1r_2r_3r_4|=\frac{1}{16}(a+b);t=l_1\cap l_2,s=l_3\cap l_4;$$

t',t'',s',s''是 l_1,l_2,l_3,l_4 和 r_2r_4 的交点，于是若 r_5,r_6 使

$$(r_1r_2\cdots r_6)\geqslant\frac{1}{8}|K|\geqslant\frac{1}{8}|r_1r_2r_3r_4|,$$

必须 $r_5\in\triangle ss's''$，$r_6\in tt't''$，故有

$$|sr_2r_4|\geqslant\frac{1}{8}|r_1r_2r_3r_4|,|tr_2r_4|\geqslant\frac{1}{8}|r_1r_2r_3r_4|.$$

可写出 t,s 的坐标是

$$t=\left(a-\frac{a(a+b)}{4b}+\frac{a+b}{8b},b-\frac{a+b}{4}\right),$$

$$s=\left(\frac{a+b}{8},\frac{b(a+b)}{8a}+\frac{a+b}{8a}\right)$$

代入前面两个不等式，即得

$$\frac{1}{2}\left\{a-\frac{a(a+b)}{4b}+\frac{a+b}{8b}+b-\frac{a+b}{4}-1\right\}\geqslant\frac{1}{8}\cdot\frac{1}{2}(a+b),$$

$$\frac{1}{2}\left\{1-\frac{a+b}{8}-\frac{b(a+b)}{8b}-\frac{a+b}{8a}\right\}\geqslant\frac{1}{8}\cdot\frac{1}{2}(a+b).$$

化简可得

$$-2a^2+3ab+5b^2+a-7b\geqslant0,$$

$$-2a^2-3ab-b^2+7a-b\geqslant0,$$

将此二式相加得$-a^2+b^2+2(a-b)\geqslant0$，即

$$(a-b)(a+b-2)\leqslant0.$$

但因 $0<a\leqslant b\leqslant1$，$a+b\leqslant2$，最后一式成立必须 $a=b=1$. 这时 $r_1r_2r_3r_4$ 是正方形，$r_5=\left(\frac{1}{4},\frac{1}{2}\right)$，$r_6=\left(\frac{3}{4},\frac{1}{2}\right)$. 引理中不等式成立. 引理 7 证毕.

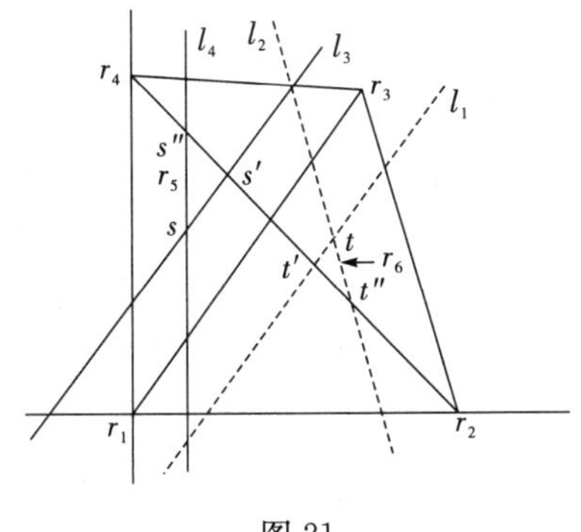

图 21

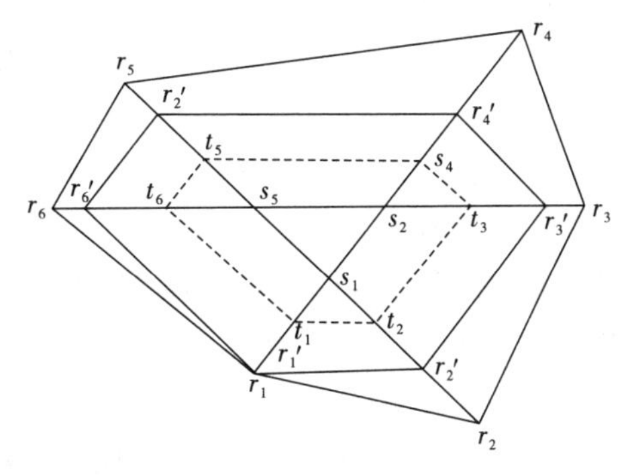

图 22

引理 8　任给凸 6 角形 $r_1r_2\cdots r_6$，有凸 6 角形 $r_1'r_2\cdots r_6'\subset r_1r_2\cdots r_6$ 使

1. $(r_1'r_2'\cdots r_6')\geqslant(r_1r_2\cdots r_6)$，

2. $r_1'r_2'\cdots r_6'$的六个外围$\triangle$面积都等于$(r_1'r_2'\cdots r_6')$.

证 图 22 给出 $r_1'r_2'\cdots r_6'$ 的构造方法，为方便起见，将称满足 2 的凸 6 角形为等围 6 角形.

引理 9 $\forall$ 凸 4 角形 $r_1r_2r_3r_4\subset\triangle s_1s_2s_3$，有

$$(r_1r_2r_3r_4)\leqslant\frac{1}{4}|s_1s_2s_3|.$$

证 对平行四边形 $r_1r_2r_3r_4$，这一结论是熟知的. 对于一般凸 4 角形 $r_1r_2r_3r_4$，可构造 $\square r_1'r_2'r_3'r_4'\subset r_1r_2r_3r_4$ 使 $(r_1'r_2'r_3'r_4')\geqslant(r_1r_2r_3r_4)$，如图 23 所示，其中假设

$$|r_1r_2r_4|=(r_1r_2r_3r_4),r_1'=r_1,r_3'=r_3.$$

引理 9 证毕.

引理 10 若 $K=s_1s_2s_3s_4$ 是凸 4 角形满足 $|s_1s_2s_4|=|s_2s_3s_4|$，则 $\forall$ 凸 6 角形 $r_1r_2\cdots r_6\subset K$，有

$$(r_1r_2\cdots r_6)\leqslant\frac{1}{8}|K|.$$

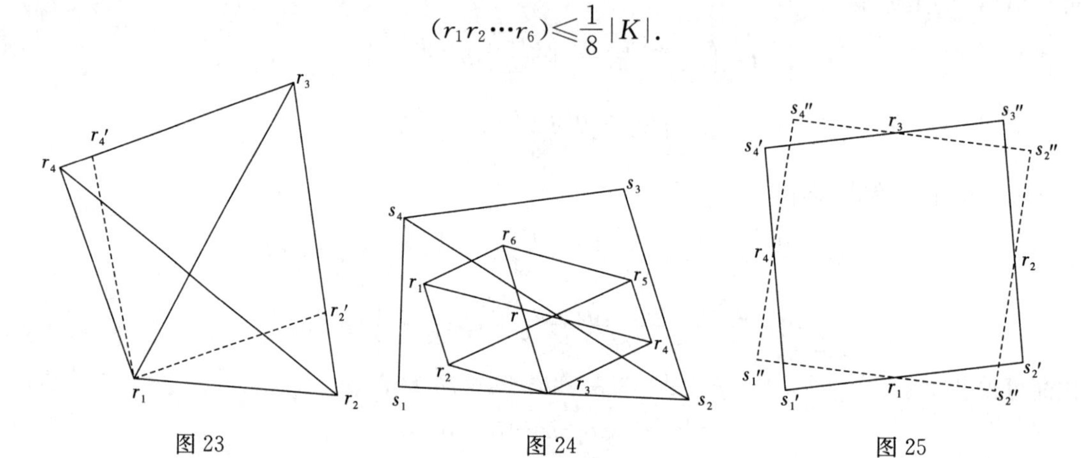

图 23　　图 24　　图 25

证 不妨设 $r_1r_2\cdots r_6$ 是等围 6 角形，则不论 $\triangle s_1s_2s_4$，$\triangle s_2s_3s_4$ 各含 $r_1,r_2,\cdots,r_0$ 的三个点（如图 24）还是其中之一含 $r_1,r_2,\cdots,r_6$ 的四个或更多的点，使用引理 9 都可证得

$$(r_1r_2\cdots r_6)\leqslant\frac{1}{8}|K|.$$

引理 11 若凸 5 角形 $r_1r_2\cdots r_5\subset K=\square s_1s_2s_3s_4$；则存在 $K'=\square s_1's_2's_3's_4'$，使 $|K'|\leqslant|K|$，$r_1r_2\cdots r_5\subset K'$，而且有 $i,j(1\leqslant i\leqslant5,1\leqslant j\leqslant4)$ 满足 $r_i=s_j'$.

证 设 $K'=\square s_1's_2's_3's_4'$ 是使 $r_1r_2\cdots r_5\subset K'$ 的面积最小的$\square$之一，但 $\forall i,j(1\leqslant i\leqslant5,1\leqslant j\leqslant4)$，$r_i\neq s_j'$，则 $s_1's_2's_3's_4'$ 的每一边内部至少含有 $r_1,r_2,\cdots,r_5$ 中的一点. 假如 $s_1's_2's_3's_4'$ 每一边内部恰含 $r_1,r_2,\cdots,r_5$ 中的一点，设 $r_i\in s_i's_{i+1}'(1\leqslant i\leqslant4)$，则不论 $r_1r_3/\!/s_1's_4'$，$r_2r_4/\!/s_1's_2'$（如图 25）还是相反（参见图 9,15）都可作面积更小的$\square s_1''s_2''s_3''s_4''$含有 $r_1r_2\cdots r_5$，所以 $s_1's_2's_3's_4'$ 某一边内部含有 $r_1,r_2,\cdots,r_5$ 中的两点. 平行于引理 6 的证明，即可构造 $K''=\square s_1''s_2''s_3''s_4''$，使 $|K''|=|K'|$，$r_1r_2\cdots r_5\subset K''$ 而且有 $i,j(1\leqslant i\leqslant5,1\leqslant j\leqslant4)$ 满足 $r_i=s''_j$. 引理 11 证毕.

引理 12 若 $r_1r_2\cdots r_6\subset\square s_1s_2s_3s_4$ 而且 $r_1r_2\cdots r_6$ 的凸包是凸 5 角形，则有严格不等式

$$(r_1r_2\cdots r_6)<\frac{1}{8}|s_1s_2s_3s_4|.$$

证 设 $r_1r_2\cdots r_6$ 的凸包是 $r_1r_2\cdots r_5$，由引理 11 存在$\square s_1's_2's_3's_4'$使 $r_1r_2\cdots r_5\subset\square s_1's_2's_3's_4'$，$|s_1's_2's_3's_4'|\leqslant|s_1s_2s_3s_4|$，而且 $r_1r_2\cdots r_5$ 在其中的位置如图 26. a)，26. b)，26. b')所示.

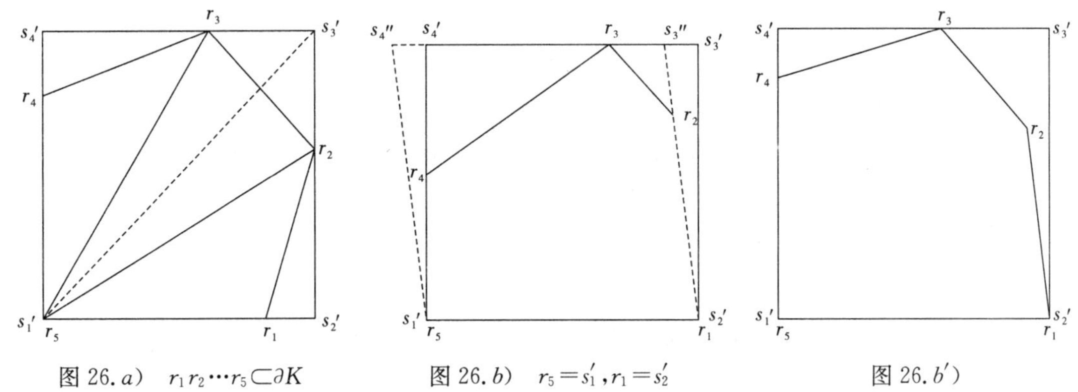

图 26.*a*)　$r_1r_2\cdots r_5\subset\partial K$　　图 26.*b*)　$r_5=s_1'$,$r_1=s_2'$　　图 26.*b*′)

情形 a. 如图 26. a). 若 $r_6\in\triangle r_5r_2r_3$,则不论 r_6 在 $s_1's_3'$ 的哪一侧,均有

$$(r_1r_2\cdots r_6)\leqslant\min\{(r_1r_2r_5r_6),(r_3r_4r_5r_6)\}$$

$$\leqslant\frac{1}{4}\cdot\frac{1}{2}|s_1's_2's_3's_4'|\leqslant\frac{1}{8}|s_1s_2\cdots s_4|;$$

若 $r_6\in\triangle r_3r_4r_5\cup\triangle r_1r_2r_5$,则不妨设 $r_6\in\triangle r_3r_4r_5$,这时

$$|r_3r_4r_5|+|r_1r_2r_3|\geqslant 3(r_1r_2\cdots r_6)+(r_1r_2\cdots r_6),$$

故

$$|r_ss_d's_1'|+|r_3s_3's_2'|\geqslant 4(r_1r_2\cdots r_6),$$

即

$$(r_1r_2\cdots r_6)\leqslant\frac{1}{4}\cdot\frac{1}{2}|s_1's_2's_3's_4'|\leqslant\frac{1}{8}|s_1s_2s_3s_4|.$$

情形 b. 如图 26. b),不妨设 r_2 不比 r_4 偏高,平行于情形 *a* 的证明;若 r_3 比 r_4 偏高,如图 26. *b*′),可作$\square s''_1s''_2s''_3s''_4$,即化为图 26. *b*)的情况.

不难利用引理 9 中等式成立的条件证明这里只有严格不等式.

综合引理 7,10,12,得到

定理 3　$H_6(\square)=\dfrac{1}{8}$.

定理 4　正方形内 6 个点的 Heilbronn 分布在仿射不变意义下只有两种.

证　用引理 7,10 中等号成立的条件.

图 27 是单位正方形内 6 点的 Heilbronn 分布图形. 定理 3、4 证实了 Goldberg 猜测 $n=6$ 的情形. 这一问题的解决首推[11],其中提出的方法沿用至今.

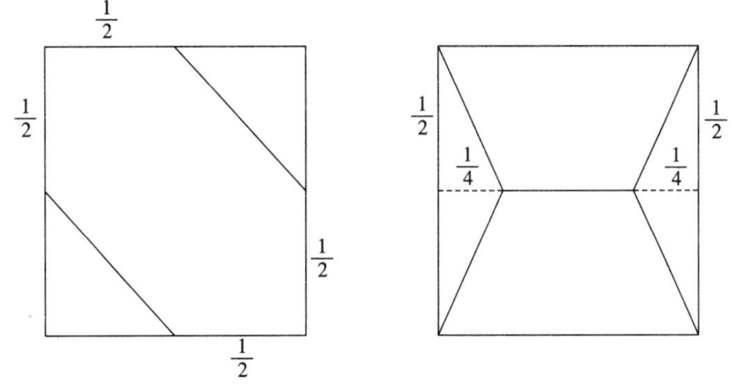

图 27　单位正方形内六点的 Heilbronn 分布的两种图形

§ 4. $H_7 \geqslant \frac{1}{12}$

按照 Goldberg 的猜测，正方形内 7 个点的 Heilbronn 分布是一仿射正 7 角形，即有

$$H_7=\frac{\left(\sin\frac{2\pi}{7}\right)^2\cdot\tan\frac{\pi}{7}}{2\left(1+\cos\frac{\pi}{7}\right)\sin\left(\frac{3\pi}{7}\right)}=0.0794\cdots$$

(见[1]). 下面的图形否定了他的推测，如图 29，其最小三角形面积是

$$\frac{1}{12}=0.0833\cdots>0.0794\cdots,$$

即

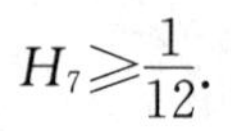

$$H_7\geqslant\frac{1}{12}.$$

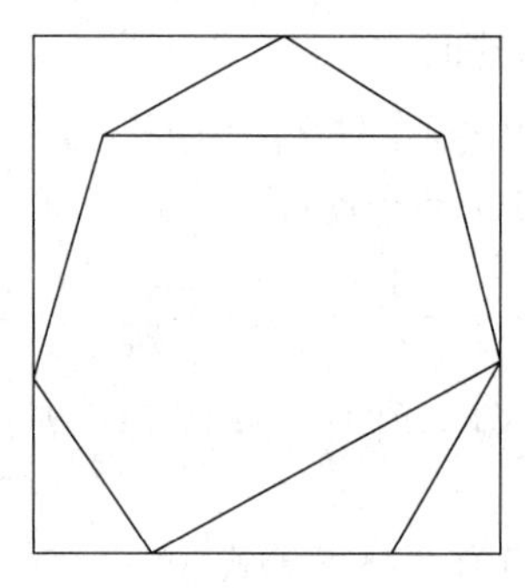

图 28　Goldberg 猜测的 Heilbronn 分布，仿射正七角形

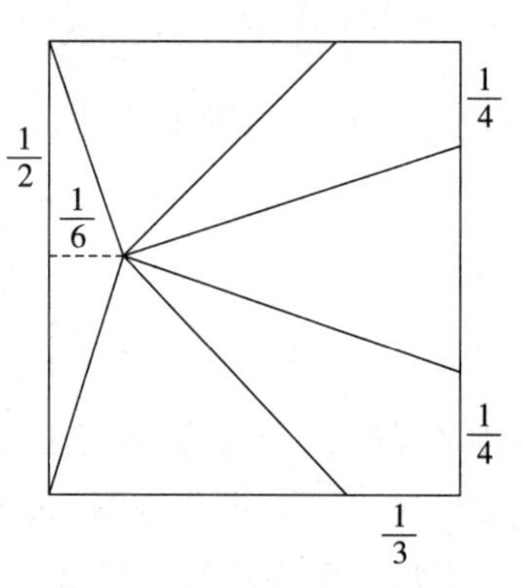

图 29　最小三角形的面积是平行四边形面积的$\frac{1}{12}$

参考文献

[1]Goldberg M. Maximizing the smallest triangle made by points in a square. Math. May., 1972(45):135 - 144.

[2]Komlos J, et al. On Heilbronn's problem. J. London Math, soc, 1981, 24(2):385 - 396.

[3]Komlos J, et al. A lower bound for Heilbronn's problem. J. London Math. Soc., 1982, 25 (2):13 - 24.

[4]Moser W. 100 Research Problems in Discrete Geometry, 1986.

[5]Rotb K F. On a problem of Heilbronn. J. London Math. Soc., 1951(26):198 - 204.

[6]Rotb K F. On a problem of Heilbronn Ⅱ. Proc. London Math. Soc., 1972(25):193 - 212.

[7]Rotb K F. On a problem of Heilbronn Ⅲ. Proc. London Math. Soc., 1972(25):543 - 549.

[8]Rotb K F. Estimation of the area of the smallest triangle obtained by selecting three out of n points in a disc of unit area. Amer. Math. Soc, Proc. Symp. Pure Math., 1973(24):

251 - 262.
[9]Roth K F. Developments in Heilbronn's triangle problem. Advances in Math. ,1976(22):364 - 385.
[10]Schmidt W M. On a problem of Heilbronn. J. London Math. Soc. ,1971(4):545 - 550.
[11]杨路,张景中. 正方形内六点问题. 数学讲座(下),四川人民出版社,1980:159 - 175.

A Criterion for Dependency of Algebraic Equations with Applications to Automated Theorem Proving

Zhang Jing-zhong, Yang Lu and Hou Xiao-rong

(*Centre for Mathematical Sciences, CICA, Academia Sinica, Chengdu* 610015, PRC)

Abstract: In this paper, a characteristic-set-based criterion is given to verify whether a polynomial is vanishing over an algebraic variety. The algorithm is feasible in practice, and does not depend on the decomposition of the variety.

Keywords: well-ordering, ascending chain, resultant.

1 Introduction

Given a system of polynomials $\{F_1(x_1, x_2, \cdots, x_n), i=1, \cdots, r\}$ and a polynomial $G(x_1, x_2, \cdots, x_n)$ over a field K, a fundamental issue is to determine whether

$$C: G(x_1, x_2, \cdots, x_n) = 0 \tag{1.1}$$

holds under the hypotheses

$$H: F_i(x_l, x_2, \cdots, x_n)=0, i=1, \cdots, r. \tag{1.2}$$

It is of importance not only for the various applications such as Automated Theorem Proving[1], Qualitative Theory of Differential Equations[2], but also for the essential role in the theory for systems of algebraic equations with higher degrees.

Let $\overline{K}$ be the algebraically closed field of K and let x_j take values in $\overline{K}$. Then, (1.1) and (1.2) express submanifolds Γ_C and Γ_H of $\overline{K}^n$, respectively. In a sense of algebraic geometry, the above problem is just to ask whether $\Gamma_H \subset \Gamma_C$.

A substantial advance is due to Wu Wen-tsun[1,3]. Using Wu's well-ordering, one can transform (1.2) into some systems whereof each one is an ascending chain; that is, a system like the following

$$H^*: f_i(u, x_1, x_2, \cdots, x_i)=0, i=1, \cdots, s, \tag{1.3}$$

where $u=(u_1, \cdots, u_d)$. Accordingly (1.1) should be written as

本文刊于《中国科学(A辑)》,第37卷第5期,1994年.

$$C^* : g(u, x_1, x_2, \cdots, x_s) = 0. \tag{1.4}$$

Then, using the remainder formula, Wu[1] gave a sufficient condition to verify H⇒C. This condition is used to prove geometry theorems automatically and has achieved a great success. The condition given in Ref. [1] is also necessary for H⇒C if the ascending chain $\{f_i\}$ is irreducible. If not, that is, $\{f_i\}$ is reducible, some geometrical discussion[4] or a pre-decomposition of the ascending chain would be required. Recently we suggest some methods to overcome the reducibility difficulty, such as WE algorithm[6] and WR algorithm[7] where no pre-decomposition is required. In this paper, a sufficient and necessary condition is established to determine whether or not H⇒C without decompositions of ascending chains in any sense.

Let $H(f_i)$ denote the degree of f_i in x_i, and let I_i denote the i-th initial; that is, the coefficient of the term of f_i with the highest degree in x_i, for $i = 1, \cdots, s$. Generally speaking, I_i is a polynomial in elements $u_1, \cdots, u_d$ and $x_i \cdots, x_{i-1}$. For simplicity, we may write m_k instead of $H(f_k)$ for $k=1, \cdots, s$.

Given a constant value of u in $K^d : \tilde{u} = (\tilde{u}_1, \cdots, \tilde{u}_l)$, if $I_1(\tilde{u}) \neq 0$, then the number of the roots of equation $f_1(\tilde{u}, x_1) = 0$ in $\overline{K}$ is m_1. Of course, these roots need not be distinct. Denote them by $x_1^{(1)}, x_2^{(2)}, \cdots, x_1^{(m_1)}$. If $I_1(\tilde{u}, x_1^{(l_1)}) \neq 0$ for $i_l = 1, \cdots, m_1$, solve equations $f_2(\tilde{u}, x_1^{(i_1)}, x_2) = 0$ and denote the solutions by $x_2^{(i_1, i_2)}$ where $i_1 = 1, \cdots, m_1$ and $i_2 = 1, \cdots, m_2$. Inductively, if for $j < l$ the solution of system $\{f_i(\tilde{u}, x_1, \cdots, x_i) = 0, i = 1, 1, \cdots j\}$ have been well-defined as

$$x^{(i_1, i_2, \cdots, i_j)} = \{x_1^{(i_1)} x^{(i_1, i_2)}, \cdots, x_j^{(i_1, i_2, \cdots, i_j)}\}, \tag{1.5}$$

where $i_k = 1, \cdots, m_k$ and $k = 1, \cdots, j$, and if

$$I_{j+1}(\tilde{u}, x^{(i_1, i_2, \cdots, i_j}) = 0 \tag{1.6}$$

for $j = 1, \cdots, l-1$, then, solving

$$f_1(\tilde{u}, x^{(i_1, i_2, \cdots, i_{l-1}}, x_l) = 0, \tag{1.7}$$

one can obtain the solutions to system $\{f_i(\tilde{u}, x_1, \cdots, x_i) = 0, i = 1, \cdots, l\}$, and denote them as $x^{(i_1, i_2, \cdots, i_l)}$, where $i_k = 1, \cdots, m_k$ and $k = 1, \cdots, l$.

If the inductive procedure can be done up to $l = s$, we have all the $m = m_1 m_2 \cdots m_s$ solutions

$$f_i(\tilde{u}, x_1, \cdots, x_l) = 0, i = 1, \cdots, s$$

and denote them as

$$x^{(i_1, i_2, \cdots, i_s)} = \{x_1^{(i_1)}, x_2^{(i_1, i_2)}, \cdots, x_s^{(i_1, i_2, \cdots, i_s)}\}, \tag{1.8}$$

where $i_k = 1, \cdots, m_k$ and $k = 1, \cdots, s$. In this case, it holds that

$$I_j(\tilde{u}, x^{(i_1, i_2, \cdots, i_j)}) \neq 0 \tag{1.9}$$

for $j = 1, \cdots, s$, and we call $\tilde{u}$ a normal parameter point of the ascending chain $\{f_i\}$.

Assume K is a number field and $\overline{K}$ the complex field from now on. Given a normal parameter point $\tilde{u}$, there is a neighbourhood of $\tilde{u}$ in $\overline{K}^d$ consisting of normal parameter points. An ascending chain is called a normal ascending chain if it has a normal parameter point. A solution $(\tilde{u}, \tilde{x})$ of system (1.3) is called a normal solution if it makes $I_j(\tilde{u}, \tilde{x}) \neq 0$

for $j=1,\cdots,s$. Of course, $\tilde{u}$ has to be a normal parameter point in this case. If every normal solution of (1.3) makes $g(\tilde{u},\tilde{x})=0$, we say that (1.4) generically holds under (1.3). Given in Sec. 3 is a sufficient and necessary condition for (1.4) under (1.3) generically.

2 Lemmas on Resultant

Let f and g be polynomials in v over a commutative ring A,

$$\begin{cases} f=a_nv^n+\cdots+a_1v+a_0, & a_n\neq 0, \\ g=b_kv^k+\cdots+b_1v+b_0, & a_i,b_j\in A. \end{cases} \tag{2.1}$$

The determinant

$$\operatorname{Res}(f,g,v)=\begin{bmatrix} a_n\, a_{n-1} & \cdots & \cdots & a_0 & & & \\ & a_n\, a_{n-1} & \cdots & \cdots & a_0 & & \\ & & & \cdots & \cdots & & \\ & & & a_n\, a_{n-1} & \cdots & \cdots & a_0 \\ b_k\, b_{k-1} \cdots & \cdots\, b_0 & & & & & \\ & b_k\, b_{k-1} \cdots & \cdots\, b_0 & & & & \\ & & \cdots\ \cdots & & & & \\ & & b_k\, b_{k-1} \cdots & \cdots\, b_0 & & & \end{bmatrix} \tag{2.2}$$

is called the resultant of g to f in element v. We call k the formal degree of g in v. The resultant $\operatorname{Res}(f,g,v)$ can be defined even if $b_k=0$.

Lemma 2.1 *Given notations as in* (2.1), *let* $\alpha_1,\cdots,\alpha_n$ *be all the roots of* $f=0$. *Then*

$$\operatorname{Res}(f,g,v)=a_n^k\prod_{i=1}^{n}g(\alpha_i). \tag{2.3}$$

It is well known, so the proof is omitted.

Lemma 2.2 *Given notations as in* (2.1), *if* $f=f_1\cdot f_2$ *where* f_1 *and* f_2 *are also polynomials over* A, *then*

$$\operatorname{Res}(f,g,v)=\operatorname{Res}(f_1,g,v)\cdot\operatorname{Res}(f_2,g,v), \tag{2.4}$$

where g has the same formal degree in the three resultants.

This follows from Lemma 2.1 immediately.

For the ascending chain $\{f_1,f_2,\cdots,f_s\}$ of (1.3), we define the resultant of g to $\{f_1,f_2,\cdots,f_s\}$ as follows:

$$\begin{cases} R_{s-1}=\operatorname{Res}(g,f_s,x_s), \\ R_{s-2}=\operatorname{Res}(R_{s-1},f_{s-1},x_{s-1}) \\ R_{s-3}=\operatorname{Res}(R_{s-2},f_{s-2},x_{s-2}), \\ \cdots \\ R_1=\operatorname{Res}(R_2,f_2,x_2), \\ R_0=\operatorname{Res}(R_1,f_1,x_1). \end{cases} \tag{2.5}$$

We call R_0 the resultant of g to the ascending chain $\{f_1, \cdots, f_s\}$ and denote it simply by

$$\mathrm{Res}(f_1, \cdots, f_s, g).$$

That is a polynomial in u and uniquely defined if the formal degrees of $R_{s-1}, \cdots, R_1$ and g are well-defined.

Lemma 2.3 *Let $\tilde{u}$ be a normal parameter point of ascending chain $\{f_1, \cdots, f_s'\}$, let m_i be the degree of f_i in x_i for $i=1, \cdots, s$, and let $x^{(i_1, i_2, \cdots, i_s)}$ be the solutions of (1.3) when $u=\tilde{u}$, as mentioned in (1.8). Then, we have*

$$\mathrm{Res}(f_1, \cdots, f_s, g)(\tilde{u}) = p(\tilde{u}) \prod_{i_1=1}^{m_1} \prod_{i_2=1}^{m_2} \cdots \prod_{i_s=1}^{m_s} g(\tilde{u}, x^{(i_1, i_2, \cdots, i_s)}), \tag{2.6}$$

where $P(\tilde{u})$ is a non-vanishing number obtained by substituting $\tilde{u}$ into some polynomial $P(u)$ for u.

Proof Do induction to s. For $s=1$, it is nothing but Lemma 2.1. Assume the conclusion holds for $s-1$ and consider the resultant of g to ascending chain $\{f_2, \cdots, f_s\}$. Regarding both u and x_1 as parameters, we have

$$\mathrm{Res}(f_2, \cdots, f_s, g)(\tilde{u}, x_1^{(i_1)}) = P^*(\tilde{u}, x_1^{(i_1)}) \prod_{i_2=1}^{m_2} \cdots \prod_{i_s=1}^{m_s} g(\tilde{u}, x^{(i_1, i_2, \cdots, i_s)}), \tag{2.7}$$

where $P^*(\tilde{u}, x_1^{(i_1)})$ is a non-vanishing number obtained by substituting $\tilde{u}$ and $x_1^{(i_1)}$ into polynomial $P^*(u, x_1)$ for u and x_1, respectively.

Putting $i_l=1,2,\cdots,m_1$ in (2.7) successively and taking the continued multiplication of the products, and applying Lemma 2.1, we have

$$\begin{aligned}
\mathrm{Res}(f_1, \cdots, f_s, g)(\tilde{u}) &= \mathrm{Res}(R_1, f_1, x_1)(\tilde{u}) \\
&= (I_1(\tilde{u}))^N \prod_{i_1=1}^{m_1} (P^*(\tilde{u}, x_1^{(i_1)}) \prod_{i_2=1}^{m_2} \cdots \prod_{i_s=1}^{m_s} g(\tilde{u}, x^{(i_1, t_2, \cdots, i_s)})) \\
&= (I_1(\tilde{u}))^N \left[\prod_{t_1=1}^{m_1} P^*(\tilde{u}, x_1^{(l_1)}) \right] \prod_{t_1=1}^{m_1} \cdots \prod_{i_s=1}^{m_s} g(\tilde{u}, x^{(i_1, i_2, \cdots, i_s)})
\end{aligned} \tag{2.8}$$

where N is a positive integer, $(I_1(\tilde{u}))^N (\prod_{t_1=1}^{m_1} P^*(\tilde{u}, x_1^{(t_1)}))$ is a polynomial in $\tilde{u}$ because $x_1^{(t_1)}$, where i_1 ranges from 1 to m_1, are all roots of $f_1(\tilde{u}, x_1)$.

It can be seen from (2.6) that, for a normal parameter point $\tilde{u}$, a sufficient and necessary condition for Res $(f_1, \cdots, f_s, g)(\tilde{u})$ to be a null function of $\tilde{u}$ is that, among the factors of the right hand side of (2.6), $g(\tilde{u}, x^{(i_1, i_2, \cdots, i_s)})$, at least one of them is a null function of $\tilde{u}$. It is not known how to determine if all the factors vanish from (2.6). This remains to be answered in the next section.

3 Parametric Resultant

Introduce a new element λ, and consider the resultant of the polynomial $g(u, x)+\lambda$ to the ascending chain $f=\{f_1,\cdots,f_s\}$,

$$\mathrm{Res}(f,g+\lambda)=\mathrm{Res}(f_1,\cdots,f_s,g+\lambda). \tag{3.1}$$

By Lemma 2.3, for a normal parameter point $\tilde{u}$, we have

$$\mathrm{Res}(f,g+\lambda)(\tilde{u}) = P(\tilde{u})\prod_{i_1=1}^{m_1}\prod_{i_2=1}^{m_2}\cdots\prod_{i_s=1}^{m_s} g(\tilde{u},x^{(i_1,i_2,\cdots,i_s)}+\lambda). \tag{3.2}$$

Since the two sides of (3.2) are both polynomials in $\tilde{u}$ and the set of normal parameter points is a non-empty open set, it follows immediately that

Theorem 3.1 *Assume $f=\{f_1,\cdots,f_s\}$ is a normal ascending chain. Then* (1.4) *generically holds under* (1.3) *if and only if*

$$\mathrm{Res}(f,g+\lambda)(u,\lambda)=P(u)\lambda^{m_1m_2\cdots m_s}. \tag{3.3}$$

Furthermore, one can obtain more information from $\mathrm{Res}(f,g+\lambda)$. For a normal parameter point u, the m solutions of (1.3), $x^{(i_1,i_2,\cdots,i_s)}$, can be regarded as m algebraic functions of u and we call them m branches of (1.3). Of course, the m branches need not be distinct from one another. How many branches of (1.3) satisfy (1.4)? It is inferred from (3.2) that

Theorem 3.2 *Assume $f=\{f_1\cdots f_s\}$ is a normal ascending chain. Let k be the greatest integer such that $Res(f,g+\lambda)$ is the product of λ^k and a polynomial $Q(u,\lambda)$:*

$$\mathrm{Res}(f,g+\lambda)=\lambda^k\cdot Q(u,\lambda). \tag{3.4}$$

Then there are exactly k branches of (1.3) *satisfying* (1.4).

Thus, it can be seen from the expression of $\mathrm{Res}(f,g+\lambda)$ to what degree $H^*\Rightarrow C^*$ holds.

By the way, sometimes one may use some of the equalities of (1.3) to simplify the computing of $\mathrm{Res}(f,g+\lambda)$ without changing the result.

The algorithm proposed here is feasible and efficient in practice. And the resultant computation is supported by current software of computer algebra. Using MAPLE programs, we have implemented it successfully in verifying complicated propositions on SUN386 work stations[7,8].

4 Parallel Numerical Method

Based on (3.4), we can explain more clearly and briefly the principles of the parallel numerical method[9] for automated theorem proving.

Step 1. Estimate the upper bounds of the degrees of the polynomial $\mathrm{Res}(f,g+\lambda)$ in elements u_i for $i=1,\cdots,d$, respectively. These upper bounds, named n_i for u_l, can be obtained from the definition of the resultant without computing the exact expression.

Step 2. Take d subsets $S_1,\cdots,S_d$ of the base field K that the cardinal numbers of S_i are n_i+1 for $i=1,\cdots,d$, respectively. Let $S=S_1\times\cdots\times S_d$ which we call a lattice array. For every $(u_1,\cdots,u_d)\in S$, solve (1.3) to get all the solutions.

Step 3. If every solution of (1.3) obtained from Step 2 makes $g=0$, and if every point of S is a normal parameter point of (1.3), then (1.4) generically holds under (1.3).

Otherwise, $g=0$ does not generically hold. Referring to (3.2)—(3.4), we have

$$Q(u,\lambda) = P(u)\prod_{\alpha} g(u,x^{(\alpha)}+\lambda), \tag{4.1}$$

where $\{x^{(\alpha)}\}$ stand for all the solutions of (1.3) whereof each makes $g\neq 0$. Put $\lambda=0$.

$$Q(u,0) = P(u)\prod_{\alpha} g(u,x^{(\alpha)}) \tag{4.2}$$

should be a polynomial of degrees in u_i f not greater than n_i for $i=1,\cdots,d$, but it vanishes over S, contradiction! Thus, (1.4) generically holds under (1.3).

We required here every point of S to be a normal parameter point of (1.3). This requirement can be removed and we will discuss it in another article.

By (3.2) we have got a method for floating-point computations to check if g exactly vanishes or it is only with a small absolute value. A so-called "Gap Theorem" is established. See Ref. [7] for details.

A demonstrator of geometry theorem proving[10] is made based on the parallel numerical method. It runs on common PC computers efficiently.

5 On Reducibility of Ascending Chains

By Lemma 2.2 we have

Lemma 5.1 *Given an ascending chain* $\{f_1,\cdots,f_s\}$ *whereof some* f_i *can be decomposed under conditions* $f_1=0,\cdots,f_{i-1}=0$ *as the product of two polynomials in* x_i,

$$f_i=f_i^{(1)}\cdot f_i^{(2)}, \tag{5.1}$$

it holds accordingly that

$$\mathrm{Res}(f_1,\cdots,f_s,g)=\mathrm{Res}(f_1,\cdots,f_i^{(1)},\cdots,f_s,g)\cdot \mathrm{Res}(f_1,\cdots,f_i^{(2)},\cdots,f_s,g), \tag{5.2}$$

where the three resultants should have the same formal degree in x_i.

It should be pointed out that the "decomposing" here means the one in a strict sense. For example, given $\{f_1=x_1^2-1, f_2=x_1x_2^2+1\}$ over a rational field, we have $f_1=0$ decomposed into $x_1+1=0$ and $x_1-1=0$; clearly f_2 is decomposable when $x_1+1=0$ but non-decomposable when $x_1-1=0$, so that this lemma can be applied to f_1, but not to f_2.

It follows immediately that

Theorem 5.1 *If there is a polynomial g such that* $Res(f,g+\lambda)$, *the resultant of* $g+\lambda$ *to ascending chain* $f=\{f_1,\cdots,f_s\}$ *is irreducible in* λ, *then* f *is an irreducible ascending chain.*

The reason is that if f is reducible, there exists a smallest i such that f_i is decomposable under conditions $f_1=0,\cdots,f_{i-1}=0$, so $\mathrm{Res}(f,g+\lambda)$ is reducible in λ.

Conversely, we have

Theorem 5.2 *If* $Res(f,g+\lambda)$, *the resultant of* $g+\lambda$ *to ascending chain* $f=\{f_1,\cdots,f_s\}$, *can be expressed as* (3.4) *but not as* (3.3), *then* f *is a reducible ascending chain.*

Furthermore, by Theorem 5.1 we have: If $Res(f,g+\lambda)$ is decomposed into

$$Res(f,g+\lambda)=\varphi_1(u,\lambda)\cdot\varphi_1(u,\lambda), \tag{5.3}$$

and there is a rational fraction $\lambda(u)$ such that $\varphi_1(u,\lambda)\equiv 0$ but $\varphi_1(u,\lambda)\not\equiv 0$, then f is a reducible ascending chain.

It is interesting and helpful to find some g whereby to determine whether the given ascending chain is reducible or not. Such a g may be called a "key" for that ascending chain.

6 Examples

Example 1. $f_1=x_1-u, f_2=x_2^2-2x_1x_2+u^2, g=x_2-x_1$.

Solution. Here all values of u are normal parameter points.

$$\text{Res}\,(f_2,g+\lambda,x_2)=\begin{vmatrix}1 & 1 & 0\\ -2x_1 & \lambda-x_1 & 1\\ u^2 & 0 & \lambda-x_1\end{vmatrix}=-x_1^2+\lambda^2+u^2,$$

$$\text{Res}(f_1,x_1-x^2+\lambda^2+u^2,x_1)=\begin{vmatrix}1 & 0 & -1\\ -u & 1 & 0\\ 0 & -u & \lambda^2+u^2\end{vmatrix}=\lambda^2,$$

so $g=0$ holds under $\{f_1=0, f_2=0\}$.

Example 2. $f_1=x_1^2+ux_1, f_2=x_2^2+(x_1^2-u)x_2+u^2x_1, g=x_2^2+x_1(1-u)x_2-ux_1^2$.

Solution. Using Wu's algorithm we find the final remainder of g to $\{f_1,f_2\}$.

$$\text{prem}(f_1,f_2,g)=(x_1+u)x_2\neq 0.$$

Using $(x_1+u)x_2$ instead of g to compute the parametric resultant,

$$R_1=\text{Res}((x_1+u)x_2+\lambda,f_2,x_2)=\lambda ux_1+\lambda(\lambda+u^2),$$

where we do some reduction using condition $f_1=0$, and

$$\text{Res}(R_1,f_1,x_1)=\lambda^3(\lambda+u^2);$$

that is, the ascending chain $\{f_1,f_2\}$ is reducible and $g=0$ is generically false for one of the two branches.

For much more complicated examples, which we treat with computer algebra MAPLE, see Ref. [8].

References

[1]Wu Wen-tsun. Scientia Sinica. 1978(21):157.

[2]Liu Li, Wang Dongming, Lu Zhengyi. Science in China(in Chinese), 1990, A20(8):799.

[3] Wu Wen-tsun. Basic principles of Mechanical Theorem Proving in Geometries (in

Chinese), Beijing, 1984.

[4]Wu Wen-tsun. Quart. J. Math. ,1987,2(2):1.

[5]Chou S C. Mechanical Geometry Theorem Proving. D. Reidel Publishing Company, Amsterdam, 1988.

[6] Zhang Jingzhong, Yang Lu. A method to overcome the redulibility difficulty in mechanical theorem proving. I. C. T. P. preprint IC/89/263.

[7]Yang Lu, Zhang Jingzhong, Hou Xiaorong. in Proceedings of the 1992 International Workshop on Mathematics Mechanization/International Academic Publishers, 1992: 110 -134.

[8] Yang Lu, Zhang Jingzhong. Searching dependency between algebraic equations: An algorithm applied to automated reasoning. ICTP preprint IC/91/6, 1991.

[9]Zhang Jingzhong, Yang Lu, Deng Mike. Theoretical Computer Science. 1990(74):253.

[10] Yang Lu, Zhang Jingzhong, Li Chuanzhong. in Proceedings of the 1992 International Workshop on Mathematics Mechanization. International Academic Publishers, 1992: 244 - 250.

The Realization of Elementary Configurations in Euclidean Space

Zhang Jing-zhong Yang Lu and Yang Xiao-chun

(*Centre for Mathematical Sciences, CICA. Academia Sinica, Chengdu* 640041, PRC)

Abstract: By "an elementary conflguration" we mean a set of a finite number of points, oriented hyperplanes and oriented hyperspheres. In this paper, a complete solution of the following problems is given. Does there exist in Euclidean space a certain elementary configuration with a prescribed metric for each pair of its elements? If so, how can one find the coordinate representations of the elements of such a configuration.

Keywords: elementary configuration, sub-eigenvalue, pseudo-Euclidean space.

1 Introduction

Can a geometric figure be realized in a certain space with some given geometric conditions? This kind of problems always attracts extensive attention of mathematicians. For example, the realization of complexes in Euclidean space was solved perfectly in [1]: the realization of a point set with prescribed distance for each pair of points in a certain constant curvature space was systematically studied in [2]; a sufficient and necessary condition for realization of a simplex with prescribed dihedral angles in Euclidean space is given in [3]; etc. The word embedding is often used here instead of realization.

Until recent years, the studies on embedding with metrical conditions were confined to those involving only one kind of geometrical quantities, for example, either distances or angles. The mixed conditions for metrical embedding were considered by Yang Lu and Zhang Jing-zhong[4,5] in order to develop the algorithms for automated geometric reasoning using invariant method. In this paper, we establish a sufficient and necessary condition for embedding an elementary configuration into a Euclidean space with distinct kinds of metrics prescribed.

本文刊于《中国科学(A辑)》,第37卷第1期,1994年.

A so-called "elementary configuration" is a set of a finite number of elementary elements. e_1, e_2, $\cdots$, e_N, where an elementary element e_1 may be a point, an oriented hyperplane or an oriented hypersphere in a Euclidean space. Throughout this paper. the hyperplanes or hyperspheres mentioned always refer to the oriented ones. Let us give the geometrical expression of elementary elements as follows. If e_i is a hypersphere or a point, it is represented as

$$e_i=(O_i, r_i), \tag{1.1}$$

where $O_i \in E^n$, the centre of e_i, and $r_i \in \mathbf{R}$, the signed radius of e_i, whose sign is determined by its orientation, $r_i=0$ in case e_i is a point. If e_1 is a hyperplane. it is represented as

$$e_i=(P_i, u_i), \tag{1.2}$$

where $P_i \in E^n$ may be any point in hyperplane e_i, and u_i the unit normal vector of e_i.

Having notations as above, for each pair of elements $\{e_i, e_j\}$, we define a "metric", which is denoted by $g_{ij}=g(e_i, e_j)=g(e_j, e_i)=g_{ji}$, as follows.

$$g_{ij}:=\begin{cases}\cos\angle(e_i, e_j)=\boldsymbol{u}_i \cdot \boldsymbol{u}_j, & \text{if both } e_i \text{ and } e_j \text{ are hyperplanes,}\\ \operatorname{dist}(O_i, e_j)=\boldsymbol{u}_j \cdot \boldsymbol{P_jO_i}, & \text{if only } e_j \text{ is a hyperplane,}\\ \frac{1}{2}(r_i-r_j)^2-\boldsymbol{O_iO_j}, & \text{if neither } e_i \text{ nor } e_j \text{ is a hyperplane,}\end{cases} \tag{1.3}$$

where $\angle(e_i, e_j)$ stands for the angle formed by hyperplanes e_i and e_j: $\operatorname{dist}(O_i, e_j)$ the signed distance from point O_i, to hyperplane e_j, positive only if O_i is in the positive half-space of e_j. If both e_i and e_j are hyperspheres, g_{ij} equals half the squared length of the common tangent. In case that one of them degenerates into a point (null sphere), the common tangent becomes a tangent of a hypersphere from a point. And in case that both e_i and e_j degenerate into points, the common tangent becomes the segment with them as end-points. Whenever g_{ij} and the signed radii of e_i, e_j are given, the relative position of e_i to e_j is determined accordingly.

To state our main result, we need the concept of sub-eigenvalues of a matrix, as introduced in [6].

Definition 1 Let A be an $(n+1)\times(n+1)$ complex matrix. The roots of the equation in λ

$$\det(a-\lambda\widetilde{E})=0, \tag{1.4}$$

where

$$\widetilde{E}:=\begin{bmatrix}0 & & & & & & \\ & 1 & & & & & \\ & & 1 & & & & \\ & & & \cdot & & & \\ & & & & \cdot & & \\ & & & & & \cdot & \\ & & & & & & 1\end{bmatrix}$$

are called sub-eigenvalues of A.

Our main result is stated as follows.

Theorem 1 *Given $\frac{1}{2}N(N-1)$ real mumber α_{ij}, $(1\leqslant i<j\leqslant n)$, and non-negative integers l, m, k where $l+m+k=N$ and $l\geqslant 1$, assume $\alpha_{ij}\neq 0$ for $j=l+1, l+2, \cdots, l+m$. Then, there are points e_1, e_2, $\cdots$, e_l, oriented hyperplanes e_{l+1}, e_{l+2}, $\cdots$, e_{l+m} and hyperspheres*[①] *$e_{l+m+1}, \cdots, e_{l+m+k}$ in E^n with $g_{ij}=a_{ij}$ for $1\leqslant i\leqslant j\leqslant N$, if and only if both the following (i) and (ii) hold.*

(*i*) *Among the sub-eigenvalues of the $(N+1)\times(N+1)$ matrix*

$$M:=\begin{bmatrix} 0 & 1 & \cdots & 1 \\ 1 & & & \\ \cdot & & & \\ \cdot & & \beta_{ij} & \\ \cdot & & & \\ 1 & & & \end{bmatrix}, \qquad (i,j=1,\cdots,N), \tag{1.5}$$

there are at most n of them to be positive, at most one negative, and the others to be zero;

(ii) *Among the sub-eigenvalues of the $(1+m+1)\times(l+m+1)$ matrix*

$$R:=\begin{bmatrix} 0 & 1 & \cdots & 1 \\ 1 & & & \\ \cdot & & & \\ \cdot & & \beta_{ij} & \\ \cdot & & & \\ 1 & & & \end{bmatrix}, \qquad (i,j=1,\cdots,l+m), \tag{1.6}$$

there are at most n of them to be positive, and the others to be zero; where (in (1.5) and (1.6)) β_{ij} are defined by setting $\alpha_{ji}:=\alpha_{ij}$ for $1\leqslant i<j\leqslant N$ and

$$\beta_{ij}:=\begin{cases} \alpha_{ij}, \text{if } i\leqslant l \text{ or } i>l+m, j\leqslant l \text{ or } j\leqslant 1+m, i\neq j, \\ -\frac{1}{2}(\alpha_{1i}^2+\alpha_{1j}^2-2\alpha_{1l}\alpha_{1l}\alpha_{lj}), \text{if } l+1\leqslant i,j\leqslant l+m, i\neq j, \\ \alpha_{1l}+\frac{1}{2}\alpha_{1l}^2-\alpha_{1l}\alpha_{ij}, \text{if } l+1\leqslant j\leqslant l+m, i\leqslant l \text{ or } i>l+m, \\ 0, \text{if } i=j. \end{cases} \tag{1.7}$$

We call M and R the discrimination matrices of $\{e_1, \cdots, e_N\}$.

It will be shown in the following sections that the required configuration $\{e_1, e_2 \cdots\cdots, e_N\}$ can be established in a constructive way, if both (i) and (ii) hold. And it is not difficult to verify whether the two conditions hold or not.

---

① Throughout this paper, a hypersphere may be a point (null sphere).

2 Concepts and Lemmas

We first review some concepts in order to apply the results in [6].

Definition 2 A pseudo-Euclidean space $E_{n+k,k}$, with index k, is the set of the complex vectors $x=(x_1,\cdots,x_n,x_{n+1},\cdots,x_{n+k})$ whose first n components are real numbers and the other k components are pure imaginary.

Sometimes a pseudo-Euclidean space is also referred to as a linear space over real field. On $E_{n+h,h}$, we can define a pseudo-Euclidean inner product

$$\boldsymbol{x}\cdot\boldsymbol{y}:=\sum_{j=1}^{n+k}x_jy_j. \tag{2.1}$$

where

$$\boldsymbol{x}=(x_1,\cdots,x_{n+k})\in E_{n+k,k},y=(y_1,\cdots,y_{n+k})\in E_{n+k,k}.$$

Accordingly, we can further define the orthogonality of two vectors and the pseudo norm $\|\boldsymbol{x}\|$ of a vector $\boldsymbol{x}$,

$$\|\boldsymbol{x}\|:=\sqrt{\boldsymbol{x}\cdot\boldsymbol{x}}.$$

Here $\|\boldsymbol{x}\|$ is a positive number, zero, or the product of a positive number and $\sqrt{-1}$. The vectors with pseudo-norm 1 and $\sqrt{-1}$ are called real unit vectors and imaginary unit vectors, respectively; they are generally called unit vectors. A basis formed by orthogonal unit vectors is called an orthonormal basis. The essential properties of pseudo-Euclidean spaces used here are well known(see [7]), for example

Lemma 1 *Given a set of orthogonal unit vectors $\boldsymbol{v}_1,\cdots,\boldsymbol{v}_t$ in space $E_{n,k}$, where $t<m$, we can extend it to an orthonormal basis in a constructive way.*

Definition 3 A set $S=\{p_1,p_2,\cdots.\}$ with a "distance" $d_{ij}=d(p_i,p_j)$ satisfying

(i) $d_{ij}^2\in\mathbf{R}$, i. e. d_{ij} is a real number or pure imaginary;

(ii) $d_{ii}=0$;

(iii) $d_{ij}=d_{ji}$,

denoted by $\{S,d\}$, is called a pseudo-Euclidean point set. If there is a mapping $F:S\rightarrow E_{m,k}$ such that

$$d^2(p,q)=(\boldsymbol{x}-\boldsymbol{y})\cdot(\boldsymbol{x}-\boldsymbol{y}), \tag{2.2}$$

where $\boldsymbol{x}=F(p)$, $\boldsymbol{y}=F(q)$, we say that $\{S,d\}$ can be congruently embedded in $E_{m,k}$, and call F a congruent embedding of S to $E_{m,k}$.

Given a pseudo-Euclidean point set $S=\{p_1\cdots,p_N\}$, the $(N+1)\times(N+1)$ matrix

$$P_s:=\begin{bmatrix}0 & 1 & \cdots & 1\\ 1 & & & \\ \cdot & & & \\ \cdot & & -\frac{1}{2}d_{ij}^2 & \\ \cdot & & & \\ 1 & & & \end{bmatrix}\quad(i,j=1,\cdots,N), \tag{2.3}$$

is called the CM matrix (Cayley-Menger matrix) of $\{S,d\}$. The sub-eigenvalues of P_s are real since it is a real symmetric matrix. A sufficient and necessary condition for congruently embedding $\{S,d\}$ in E_{nl} is given in [6]as follows.

Lemma 2 *Given a pseudo-Euclidean point set $\{S,d\}$, if among the sub-eigenvalues of its CM matrix. m of them are non-vanishing and k of them are negative, then $\{S,d\}$ can be congruently embedded in $E_{n,l}$ if and only if $n-l\geqslant m-k$ and $l\geqslant k$.*

It is well known that a hypersphere in E^n can be regarded as a point in $E_{n+1,1}$, by regarding the coordinate components of the centre as the first n components of a point in $E_{n+1,1}$, and regarding the product of the radius and $\sqrt{-1}$ as the last component of that point. Thus, the length of a common tangent of two hyperspheres is just the pseudo-Euclidean distance between the corresponding points in $E_{n+1,1}$, and may be a pure imaginary. Consequently, the realization of an elementary configuration without hyperplanes can be easily transformed into the congruent embedding of a pseudo-Euclidean point set to a pseudo-Euclidean space. As for the cases with hyperplanes, we can reduce an elementary configuration to the one without hyperplanes by the following lemma.

Lemma 3 *Given an elementary configuration $\{e_1,\cdots,e_N\}$, where e_1 and e_i are distinct points. e_l and e_k hyperplanes which do not contain e_1, e_l a hyperlane containing e_1 and e_m a hypersphere, by $\tilde{e}_j$ and $\tilde{e}_k$ denoting the feet of perpendiculars of e_1 to e_j and e_k, respectively, we have*

(i) $g(e_1,\tilde{e}_j)=-\frac{1}{2}g^2(e_1,e_j)$,

(ii) $g(e_i,\tilde{e}_j)=g(e_1,e_j)+\frac{1}{2}g^2(e_1,e_j)-g(e_1,e_j)g(e_i,e_j)$.

(iii) $g(\tilde{e}_j,\tilde{e}_k)=-\frac{1}{2}[g^2(e_1,e_j)+g^2(e_1,e_k)-2g(e_1,e_j)g(e_1,e_k)g(e_j,e_k)]$,

(iv) $g(\tilde{e}_j,e_m)=g(e_1,e_m)+\frac{1}{2}g^2(e_1,e_j)-g(e_1,e_j)g(e_m,e_j)$,

(v) $g(\tilde{e}_j,e_l)=-g(e_1,e_l)g(e_j,e_l)$.

Proof Both (i) and (v) hold clearly by definition of the metric, (1. 3). And (iii) follows from the cosine law. We need only to prove (iv) because (ii) is a particular case of (iv). Let O and r be the centre and radius of e_m, $\tilde{P}:=e_j$. We have

$$g(\tilde{e}_j,e_m)=-\frac{1}{2}[PO^2-r^2]. \tag{2.4}$$

Let Q be the foot of the perpendicular of O to e_j, $E:=e_1$. It follows that

$$\begin{cases}\boldsymbol{PO}^2=\boldsymbol{PQ}^2+\boldsymbol{OQ}^2=\boldsymbol{PQ}^2+g^2(e_m,e_j),\\ \boldsymbol{PQ}^2=\boldsymbol{EO}^2-(g(e_1,e_j)-g(e_m,e_j))^2,\\ \boldsymbol{EO}^2=r^2-2g(e_1,e_m).\end{cases} \tag{2.5}$$

Substituting (2. 5) into (2. 4) , we have (iv). This completes the proof.

Lemma 4 *Given* $\frac{1}{2}N(N-1)$ *real numbers* α_{ij} $(1\leqslant i<j\leqslant N)$, *and non-negative integers* l,k *where* $l+k=N$, *put* $\alpha_{ll}:=0$ *for* $i=1,\cdots,N$ *and* $\alpha_{ji}:=\alpha_{ij}$ *for* $1\leqslant i<j\leqslant N$. *Then there are points* $e_1,e_2,\cdots,e_l$ *and oriented hyperspheres* $e_{l+1},e_{l+2},\cdots,e_{l+k}$ *in* E^n *with* $g_{ij}=\alpha_{ij}$ *for* $i,j=1,\cdots,N$, *if and only if both the following* (i) *and* (ii) *hold*:

(i) *Among the sub-eigenvalues of the* $(N+1)\times(N+1)$ *matrix*

$$\begin{bmatrix} 0 & 1 & \cdots & 1 \\ 1 & & & \\ \cdot & & & \\ \cdot & & \alpha_{ij} & \\ 1 & & & \end{bmatrix}, \qquad (i,j=1,\cdots,N), \tag{2.6}$$

there are at most n *of them to be positive, at most one negative, and the others to be zero*;

(ii) *Among the sub-eigenvalues of the* $(l+1)\times(l+1)$ *matrix*

$$\begin{bmatrix} 0 & 1 & \cdots & 1 \\ 1 & & & \\ \cdot & & & \\ \cdot & & \alpha_{ij} & \\ \cdot & & & \\ 1 & & & \end{bmatrix}, \qquad (i,j=1,\cdots,l), \tag{2.7}$$

there are at most n *of them to be positive, and the others to be zero.*

Proof. By Lemma 2 we know that condition (i) is sufficient and necessary for the realization of $e_1,\cdots,e_N$ as hyperspheres in E^n, and condition (ii) is sufficient and necessary for the realization of $e_1\cdots,e_l$ as points in E^n. Thus, (i) and (ii) are both necessary for the realization of $\{e_1,\cdots,e_N\}$ as an elementary configuration in E^n, in which $e_1,\cdots,e_l$ are specified as points and the others specified as hyperspheres.

To show the conditions to be sufficient, we need to prove that there is an embedding of $\{e_1,\cdots,e_N\}$ in $E_{n+1,1}$ not only with the prescribed metric, but also making $e_1,\cdots,e_l$ laid in E^n, the real subspace of $E_{n+1,1}$. Now let us construct such and embedding as follows.

We regard $\{e_1,\cdots,e_N\}$ as a pseudo-Euclidean point set with distances $d_{ij}^2=-2\alpha_{ij}$ for i, $j=1,\cdots,N$. By [6], an embedding F can be constructed such that $\alpha_{ij}=-\frac{1}{2}\left(\boldsymbol{x}_i-\boldsymbol{x}_j\right)^2$, where $x_i=F(e_i)\in E_{n+1,1}$ for $i=1,\cdots,N$, since (i) holds. Also an embedding G can be constructed such that $\alpha_{ij}=-\frac{1}{2}\left(\boldsymbol{y}_i-\boldsymbol{y}_j\right)^2$, where $\boldsymbol{y}_i=\boldsymbol{G}(\boldsymbol{e}_i)\in E^n\subset E_{n+1,1}$ for $i=1,\cdots,l$, since (ii) holds.

Consider vectors $\boldsymbol{y}_i:=\boldsymbol{y}_{i+1}-\boldsymbol{y}_1$ for $i=1,\cdots,l-1$. Since $\{\boldsymbol{y}_i\}$ is a set of real vectors in E^n, by Schmidt's orthogonalization, we obtain a set of orthogonal unit vectors $\boldsymbol{u}_1,\cdots,\boldsymbol{u}_t$, where $t\leqslant l-1$, and there is a set of real numbers $\{r_{ij}\}$ such that

$$u_i = \sum_{j=1}^{l-1} r_{ij} v_j \qquad (j = 1, \cdots, t), \tag{2.8}$$

and there is also a set of real numbers $\{S_{ij}\}$ such that

$$v_j = \sum_{i=1}^{t} s_{ij} u_i \qquad (j = 1, \cdots, l-1). \tag{2.9}$$

Using Lemma 1, we extend $\{u_1, \cdots, u_t\}$ to an orthonormal basis $U := \{u_1, \cdots, u_{n+1}\}$ in $E_{n+1,1}$.

Correspondingly, consider vectors $v_i^* := x_{i+1} - x_1$ for $i = 1, \cdots, l-1$. $\{v_i^*\}$ is a set of vectors in $E_{n+1,i}$. Let

$$u_i^* = \sum_{j=1}^{l-1} r_{ij} v_j^* \qquad (i = 1, \cdots, t). \tag{2.10}$$

since $(x_i - x_j)^2 = (y_i - y_j)^2 = -2\alpha_{ij}$, we have

$$\begin{aligned} v_i^* \cdot v_j^* &= (x_{i+1} - x_1) \cdot (x_{j+1} - x_1) \\ &= -\frac{1}{2}((x_{i+1} - x_{j+1})^2 - (x_{i+1} - x_1)^2 - (x_{j+1} - x_1)^2) \\ &= \alpha_{i+1,j+1} - \alpha_{1,i+i} - \alpha_{1,j+1} \\ &= -\frac{1}{2}((y_{i+1} - y_{j+1})^2 - (y_{i+1} - y_1)^2 - (y_{j+1} - y_1)^2) \\ &= (y_{i+1} - y_1) \cdot (y_{j+1} - y_1) = v_i \cdot v_j. \end{aligned} \tag{2.11}$$

Consequently,

$$u_i^* \cdot u_j^* = u_i \cdot u_j \qquad (i, j = 1, \cdots, t), \tag{2.12}$$

so that $\{u_i^*\}$ is also a set of real orthogonal unit vectors and it holds obviously that

$$v_j^* = \sum_{i=1}^{t} s_{ij} u_i^* \qquad (j = 1, \cdots, l-1). \tag{2.13}$$

Using Lemma 1, we extend $\{u_1^*, \cdots, u_t^*\}$ to an orthonormal basis $U^* = \{u_1^*, \cdots, u_{n+1}^*\}$ in $E_{n+1,1}$.

It is a well-known fact about pseudo-Euclidean space that any orthonormal basis of $E_{n+1.1}$ exactly contains one imaginary unit vector. Without loss of generality, let u_{n+1}^* and u_{n+1} be the imaginary unit vectors in U^* and U, respectively. Put $v_i^* := x_{i+1} - x_1$ for $i = 1, \cdots, N-1$ and expand every v_i^* on basis U^*,

$$v_i^* = \sum_{j=1}^{n+1} t_{ij} u_j^* \qquad (i = 1, \cdots, N-1). \tag{2.14}$$

Let $t_i := (t_{1\,i-1}, t_{2\,i-1}, \cdots, t_{n\,i-1}, \sqrt{-1}\, t_{n+1\,i-1})$ for $i = 1, \cdots, N$. Then.

$$t_i \cdot t_j = v_{i-1}^* \cdot v_{j-1}^* = (x_i - x_1) \cdot (x_j - x_1) \quad (i, j = 1, \cdots, N),$$

hence $(t_i - t_j)^2 = (x_i - x_j)^2$. Thus, $F^* : e_i \mapsto t_i$ is also a congruent embedding of $\{e_1, \cdots, e_{v_s^t}\}$ in $E_{n+1,1}$. By the uniqueness of expanding a vector on a basis, we have

$$t_{n+1,l1} = 0 \qquad (i = 1, \cdots, l), \tag{2.15}$$

that is, the embedding F^* lies $\{e_1, \cdots, e_l\}$ in E_n the real subspace of $E_{n+1,1}$. This is what we need to prove.

3 Proof of Main Theorem

Now let us prove Theorem 1. If, $m=0$. i. e. no hyperplanes are in the desired configuration, the matrices M and R are just A and B in Lemma 4, respectively; accordingly the conclusion holds in this case.

Consider the case that $m>0$. Since $a_{l_j}\neq 0$ for $l+1\leqslant j\leqslant l+m$, there should exist a point e_1 which does not belong to any hyperplane e_j. Replacing every e_j by $\tilde{e}_j$, the foot of the perpendicular of e_1 to e_j, we obtain an elementary configuration formed by points and hyperspheres and call it a simplified configuration of $\{e_1,\cdots,e_N\}$. From Lemma 3 we can infer that the discrimination matrices of $\{e_1,\cdots,e_N\}$ and those of its simplified configuration are the same, so the conditions (i) and (ii) in Theorem 1 are sufficient and necessary for realizing the simplified configuration. It is obvious that $\{e_1,\cdots,e_n\}$ can be realized in E_n if and only if the simplified configuration can be realized, because hyperplane e_j is uniquely determined by $\{e_1,\tilde{e}_j\}$ and $g(e_1,e_j)$. The orientation of e_1 is determined by $g(e_1,e_j)$. Thus, Theorem 1 can be inferred from Lemma 4 which is proved in the last section.

In Theorem 1, the realized configuration $\{e_1,\cdots,e_N)$ is confined to having a point which does not belong to any hyperplanes of it. This is without loss of generality.

(1) If all $e_1,\cdots,e_N$ are hyperplanes, extend them by adding a point e_0, the origin. The realization of the extended configuration $\{e_0,e_1,\cdots,e_N\}$ is equivalent to that of the original one.

(2) If $e_1,\cdots,e_N$ are hyperplanes and hyperspheres, and have no points, there is no harm in replacing some hypersphere by its centre. The reason is as follows. Adding the same real number to the radii of hyperspheres and keeping hyperplanes fixed, we obtain a new configuration $\{e_1',\cdots,e_N'\}$. It holds clearly that $g(e_i',e_j')=g(e_i,e_j)$ for $i,j=1,\cdots,N$, so that the realizations of both the configurations are equivalent (Conversely, any elementary configuration can be replaced by some configuration formed by hyperplanes and non-degenerate hyperspheres, keeping the metric entire).

Thus, we can assume $\{e_1,\cdots,e_N\}$ to have a point anyway.

(3) How can we do if there is no point(of $\{e_1,\cdots,e_N\}$) which does not belong to any hyperplane(of $\{e_1,\cdots,e_N\}$)? For a point e_1 on some hyperplane e_j, we translate e_j towards its negative side with unit distance, which results in e_j^*, a new hyperplane. Clearly, $g(e_j^*,e_i)=1+g(e_j,e_i)$ whenever e_i is a point or hypersphere; and $g(e_j^*,e_i)=g(e_j,e_i)$ when e_i is a hyperplane. Do translations in this way for all the hyperplanes which contain e_1 and replace $\{e_1,\cdots,e_N\}$ by the new configuration where in the distances of all the hyperplanes from e_1 are non-vanishing. The realization of $\{e_1,\cdots,e_N\}$, therefore, is equivalent to that of the new

configuration.

We have reduced the realization of a general configuration to the case to which Theorem 1 is applicable.

4 Further Problems

The studies cause a series of problems not only theoretically interesting but also concerning varied fields such as computer graphics, molecular conformation, etc.

(1) Non-degenerate embedding. What are the conditions that guarantee the hyperspheres to be non-degenerate, i. e. with non-vanishing radii?

(2) Embedding with extra conditions. Is it consistent if extra requirements such as radii of some hyperspheres, distances between centers of some hyperspheres, angles between hyperplanes, and hyperspheres, are posed?

(3) Embedding with incomplete information. Given a subset of $\{\alpha_{ij}\}$, how can one define the rest and realize the configuration? What can one say about the realization if the rest cannot be defined exactly?

(4) Embedding with approximate data. Given real number sets $\{\alpha_{ij}\}$ and $\{b_{ij}\}$, is there a configuration with metric g such that $a_{ij} \leqslant g(e_i, e_j) \leqslant b_{ij}$ for every pair of elements. This is a very difficult open problem even if the embedding involves points only. It is referred to as "the fundamental problem of distance geometry"[8].

These problems are closely related to solving multivariate algebraic equations with parameters. Here Wu's method[9] for automated reasoning would be helpful.

Appendix

An Example of Embedding an Elementary Configuration in E^2

Given six real numbers $\alpha_{12}=-4, \alpha_{13}=-8, \alpha_{23}=-4, \alpha_{14}=0(i=1,2,3)$, check whether there exist points e_1, e_2, e_3, and oriented circle e_4 in E^2, such that $g(e_i, e_j)=\alpha_{ij}$, where metric g is defined by (1. 3). If so, give a coordinate representation.

Step 1. Compute the sub-eigenvalues of the discrimination matrices

$$M=\begin{bmatrix} 0 & 1 & 1 & 1 & 1 \\ 1 & 0 & -4 & -8 & 0 \\ 1 & -4 & 0 & -4 & 0 \\ 1 & -8 & -4 & 0 & 0 \\ 1 & 0 & 0 & 0 & 0 \end{bmatrix}, \quad R=\begin{bmatrix} 0 & 1 & 1 & 1 \\ 1 & 0 & -4 & -8 \\ 1 & -4 & 0 & -4 \\ 1 & -8 & -4 & 0 \end{bmatrix}.$$

We obtain the sub-eigenvalues of M,

$$\lambda_1=8, \lambda_2=2\sqrt{2}, \lambda_3=-2\sqrt{2},$$

and those of R.

$$\mu_1=8, \mu_2=\frac{8}{3}.$$

By Theorem 1, the required configuration exists.

Step 2. Regarding $\{e_1, e_2, e_3, e_4\}$ as a pseudo-Euclidean point set and embedding it in $E_{3.1}$, by the algorithm suggested in [6], we get the unit sub-eigenvectors of M

$$\begin{cases} \boldsymbol{p}_1 = \left(\frac{1}{\sqrt{2}}, 0, \frac{-1}{\sqrt{2}}, 0\right), \\ \boldsymbol{p}_2 = \left(\frac{1}{\sqrt{8}}, \frac{-(\sqrt{2}+1)}{\sqrt{8}}, \frac{1}{\sqrt{8}}, \frac{\sqrt{2}-1}{\sqrt{8}}\right), \\ \boldsymbol{p}_3 = \left(\frac{1}{\sqrt{8}}, \frac{\sqrt{2}-1}{\sqrt{8}}, \frac{1}{\sqrt{8}}, \frac{-(\sqrt{2}+1)}{\sqrt{8}}\right), \end{cases}$$

corresponding to $\lambda_1, \lambda_2, \lambda_3$, respectively; we also get the unit sub-eigenvectors of R

$$\begin{cases} \boldsymbol{q}_1 = \left(\frac{1}{\sqrt{2}}, 0, \frac{-1}{\sqrt{2}}\right), \\ \boldsymbol{q}_2 = \left(\frac{1}{\sqrt{6}}, \frac{-2}{\sqrt{6}}, \frac{1}{\sqrt{6}}\right), \end{cases}$$

corresponding to μ_1, μ_2, respectively. Let

$$\begin{cases} \boldsymbol{x}_1 = \left(2, \frac{1}{\sqrt[4]{8}}, \frac{1}{\sqrt[4]{8}} I\right), \\ \boldsymbol{x}_2 = \left(0, \frac{-(\sqrt{2}+1)}{\sqrt[4]{8}}, \frac{\sqrt{2}-1}{\sqrt[4]{8}} I\right), \\ \boldsymbol{x}_3 = \left(-2, \frac{1}{\sqrt[4]{8}}, \frac{1}{\sqrt[4]{8}} I\right) \\ \boldsymbol{x}_4 = \left(0, \frac{\sqrt{2}-1}{\sqrt[4]{8}}, \frac{-(\sqrt{2}+1)}{\sqrt[4]{8}} I\right). \end{cases} \qquad \begin{cases} \boldsymbol{y}_1 = \left(2, \frac{2}{3}\right), \\ \boldsymbol{y}_2 = \left(0, \frac{-4}{3}\right), \\ \boldsymbol{y}_3 = \left(-2, \frac{2}{3}\right), \end{cases}$$

where $I = \sqrt{-1}$. Then the mapping $e_i \mapsto \boldsymbol{x}_i$ makes $x_{ij} = -\frac{1}{2}\left(\boldsymbol{x}_i - \boldsymbol{x}_j\right)^2$ for $1 \leqslant i < j \leqslant 4$.

Step 3. Translate e_1, e_2, e_3 into the real plane. At first, let

$$\begin{cases} \boldsymbol{v}_1^* = \boldsymbol{x}_2 - \boldsymbol{x}_1 = \left(-2, \frac{-(\sqrt{2}+2)}{\sqrt[4]{8}}, \frac{\sqrt{2}}{\sqrt[4]{8}} I\right), \\ \boldsymbol{v}_2^* = \boldsymbol{x}_3 - \boldsymbol{x}_1 = (-4, 0.0), \\ \boldsymbol{v}_3^* = \boldsymbol{x}_4 - \boldsymbol{x}_1 = \left(-2, \frac{\sqrt{2}-2}{\sqrt[4]{8}}, \frac{-(\sqrt{2}+2)}{\sqrt[4]{8}} I\right) \end{cases} \qquad \begin{cases} \boldsymbol{v}_1 = \boldsymbol{y}_2 - \boldsymbol{y}_1 = (-2, -2), \\ \boldsymbol{v}_2 = \boldsymbol{y}_3 - \boldsymbol{y}_1 = (-4, 0). \end{cases}$$

Subsequently, apply Schmidt's orthogonalization to $\{\boldsymbol{v}_1, \boldsymbol{v}_2\}$ and obtain an orthonormal basis of E^2

$$\{\boldsymbol{u}_1 = (0, 1), \boldsymbol{u}_2 = (1, 0)\}, \text{namely}, \boldsymbol{u}_1 = -\frac{1}{2}\boldsymbol{v}_1 + \frac{1}{4}\boldsymbol{v}_2, \boldsymbol{u}_2 = -\frac{1}{4}\boldsymbol{v}_2.$$

Let

$$\boldsymbol{u}_1^* = -\frac{1}{2}\boldsymbol{v}_1^* + \frac{1}{4}\boldsymbol{v}_2^*, \boldsymbol{u}_2^* = -\frac{1}{4}\boldsymbol{v}_2^*.$$

Two orthogonal unit vectors in $E_{3,1}$ are obtained

$$u_1^* = \left(0, \frac{2+\sqrt{2}}{2\sqrt[4]{8}}, \frac{2-\sqrt{2}}{2\sqrt[4]{8}} I\right), u_2^* = (1,0,0).$$

Then extend $\{u_1^*, u_2^*\}$ to getting an orthonormal basis of $E_{3,1}$, $U^* = \{u_1^*, u_2^*, u_3^*\}$, where

$$u_3^* = \left(0, \frac{3-2\sqrt{2}}{2\sqrt{3\sqrt{2}-4}}, \frac{1}{2\sqrt{3\sqrt{2}-4}} I\right).$$

Expand $v_i^* = x_{i+1} - x_i (i=0,1,2,3)$ on basis U^*,

$$v_0^* = 0, v_1^* = -2u_1^* - 2u_2^*,$$

$$v_2^* = -4u_2^*, v_3^* = -2u_2^* - 2u_3^*;$$

accordingly t_i(i=1,2,3,4) are defined

$$t_1 = (0,0,0), t_2 = (-2,-2,0),$$

$$t_3 = (0,-4,0), t_4 = (0,-2,-2I).$$

Thus, the mapping $e_i \mapsto t_i$ makes $\{e_1, e_2, e_3, e_4\}$ congruently embedded in $E_{3.1}$ and the first three elements laid on the real plane.

Now, we have had a realization of $\{e_1, e_2, e_3, e_4\}$ with prescribed metric, that is, e_1 is the origin (0,0), e_2 and e_3 the points (−2, −2) and (0, −4); and e_4 the circle with centre (0, −2) and radius −2.

References

[1]Wu Wen-tsun. Acta Math Sinica(in Chinese), 1955(5): 505, 1957(7): 79, 1958(8): 79.

[2] Blumenthal L M. Theory and Applications of Distance Geometry. 2nd ed. New York, 1970.

[3]Yang Lu, Zhang Jingzhong. Acta Math Sinica(in Chinese), 1983(26): 250.

[4]Yang Lu, Zhang Jingzhong. J. Chin, Univ. Sci. Tech(in Chinese), 1980, 10(4): 52.

[5] Yang Lu, Zhang Jingzhong. Metric Equations in Geometry and Their, Applications. I C T P preprint JC/89/281, Trieste, 1989.

[6]Zhang Jingzhong, Yang Lu. Acta Math. Sinica(in Chinese), 1981(24): 481.

[7]Rashevskii P K. Riemannian Geometry and Tensor Analysis(translated from Russian), Beijing, 1953.

[8]Dress A W M, Havel T F. Computer-Aided Geometric Reasoning(Ed. Crapo. H). 1 N. R I A.. Roequencourt, 1987.

[9] Wu Wen-tsun. Basic Principles of Mechanical Theorem Proving in Geometries (in Chinese), Beijing, 1984.

关于三角形区域的 Heilbronn 数

杨　路　张景中　曾振柄

(中国科学院成都计算机应用研究所,成都　610041)

摘　要:设 K 是一个具有面积 $|K|$ 的平面凸体. 用 $(r_1r_2r_3)$ 表示三角形 $r_1r_2r_3$ 的面积,并令

$$(r_1r_2\cdots r_n)=\min\{(r_ir_jr_k)\mid 1\leqslant i<j<k\leqslant n\}.$$

然后将 K 中 n 个点的 Heilbronn 数定义为

$$H_n(K)=\frac{1}{|K|}\max\{(r_1r_2\cdots r_n)\mid r_i\in K;i=1,\cdots,n\}.$$

对于三角形区域 Δ,我们证明了

$$H_5(\Delta)=3-2\sqrt{2},H_6(\Delta)=\frac{1}{8}.$$

关键词: Heilbronn 数、Heilbronn 分布、凸包

一、引　　言

设 K 是一个平面凸体(即平面上一个有非空内部的凸紧集),用 $|K|$ 表示其面积;对于任一个三角形 $r_1r_2r_3$,用 $(r_1r_2r_3)$ 表示其面积;再令

$$(r_1r_2\cdots r_n)=\min\{(r_ir_jr_k)\mid 1\leqslant i<j<k\leqslant n\},$$

$$H_n(K)=\frac{1}{|K|}\max\{(r_1r_2\cdots r_n)\mid r_i\in K;i=1,\cdots,n\}.$$

这样定义的 $H_n(K)$,$n=3,4,\cdots$,叫做凸体 K 的 Heilbronn 数. 设 $\{r_1,\cdots,r_n\}$ 是 K 的某个确定的子集,如果正好有

$$\frac{1}{|K|}(r_1\cdots r_n)=H_n(K),$$

我们就说 $\{r_1,\cdots,r_n\}$ 或 $r_1\cdots r_n$ 实现 $H_n(K)$ 或者说它是关于 K 的一个 Heilbronn 分布,简称为关于 K 的一个 H 分布. 按以上定义,仿射等价的平面凸体显然具有相同的 Heilbronn 数.

固定 K 而使 n 变动时,关于 $H_n(K)$ 的界的估计已经有大量的工作[2-10]. 设 K 是个圆盘或正方形区域,1950 年 Heilbronn 猜想有常数 c 使

$$H_n(K)<c/n^2$$

但 1982 年 J. Komlos 等的文章证明有常数 c 使得

本文刊于《数学学报》,第 37 卷 5 期,1994 年.

$$c(\log n)/n^2 < H_n(K),$$

这就推翻了 Heilbronn 猜想. 另一方面, Komlos 等在 1981 年证明了

$$H_n(K) < c/n^{\mu},$$

这里 $\mu=\frac{8}{7}-\varepsilon, \varepsilon>0$.

这是迄今所知上界和下界两方面最好的结果. 即使对不大的 n, 计算 $H_n(K)$ 的准确值目前也没有成熟的方法, 且是相当困难的.

M. Goldberg[1] 考虑了正方形区域的最初几个 Heilbronn 数的准确值. 除 $H_3=H_4=\frac{1}{2}$ 是平凡的结果外, 他未加严格证明地断言

$$H_5=0.191\cdots, \qquad H_6=0.125, \qquad H_7=0.079\cdots.$$

作者在[11－12]中检验了 Goldberg 的上述断言, 严格证明了其中之一, $H_6=0.125$, 而否定了其余两个结论. 通过细致的分析, 在[12]中我们给出了 H_5 的准确值,

$$H_5=\frac{\sqrt{3}}{9}=0.19245\cdots,$$

并且证明了

$$H_7 \geqslant \frac{1}{12}=0.083>0.079\cdots,$$

这是关于 Heilbronn 数准确值的第一批获得严格证明的结果.

本文考虑了三角形区域 Δ 的 Heilbronn 数的准确值, 证明了

$$H5(\Delta)=3-2\sqrt{2}, \qquad H_6(\Delta)=\frac{1}{8}=0.125.$$

二、$H_5(\Delta)=3-2\sqrt{2}$

引理 1　若 $r_4, r_5 \in \Delta r_1 r_2 r_3$, 则

$$(r_1\cdots r_5) \leqslant \frac{1}{4+2\sqrt{3}}(r_1 r_2 r_3).$$

证明　记 $a=\sup\{(r_1\cdots r_5)/(r_1 r_2 r_3) \mid r_4, r_5 \in \Delta r_1 r_2 r_3\}$. 若图形 $r_1\cdots r_5$ 使上界 a 达到, 则诸三角形 $r_i r_j r_k$ 中面积可能取到 $a(r_1 r_2 r_3)$ 的仅有 $\Delta r_1 r_2 r_5$, $\Delta r_1 r_5 r_4$, $\Delta r_1 r_4 r_3$, $\Delta r_4 r_5 r_2$, $\Delta r_4 r_5 r_3$. 如图 1.

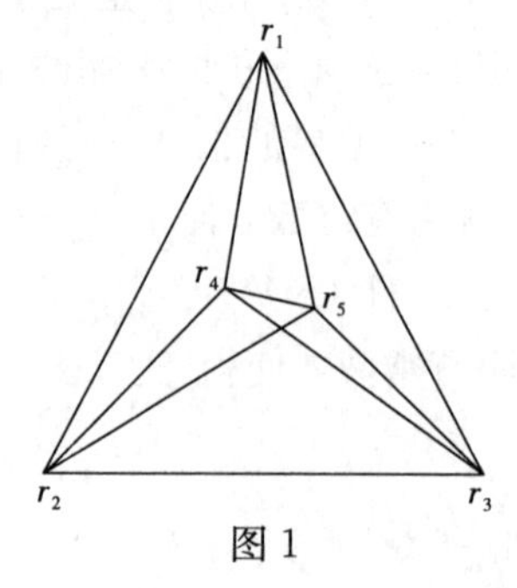

图 1

(1)若$(r_1r_2r_5)>(r_1\cdots r_5)$,则令$r_2'=r_2+\varepsilon(r_3-r_2)$,$\varepsilon>0$,则可保持$\Delta r_1r_5r_4$,$\Delta r_1r_4r_3$,$\Delta r_4r_5r_3$的面积不变而$(r_1r_2'r_5)>(r_1\cdots r_5)$;

$$(r_4r_5r_2')\geqslant\min\{(r_2r_4r_5),(r_3r_4r_5)\}\geqslant(r_1\cdots r_5).$$

这样,$(r_1r_2'r_3r_4r_5)/(r_1r_2'r_3)>a$,与假设矛盾.故$\Delta r_1r_2r_5$,$\Delta r_1r_4r_3$面积为$(r_1\cdots r_5)$.

(2)若$(r_4r_5r_2)>(r_1\cdots r_5,)$令$r_5'=r_5+\varepsilon(r_5-r_1)$,$\varepsilon>0$,可使$\Delta r_1r_4r_3$,$\Delta r_4r_5r_3$的面积不变或增大而且$(r_4r_5'r_2)>(r_1\cdots r_5)$,$(r_1r_4r_5')>(r_1\cdots r_5)$,$(r_1r_2r_5')>(r_1\cdots r_5)$,这样图形$r_1r_2r_3r_4r_5'$也实现上界$a$.但由(1)知这时必须有$(r_1r_2r_5')=(r_1r_2r_3r_4r_5')=(r_1\cdots r_5)$,矛盾!

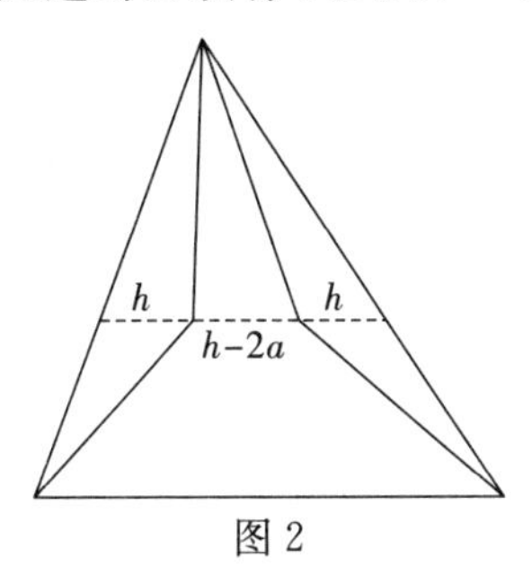

图 2

于是,$r_1\cdots r_5$若实现a则必须有

$$\frac{(r_1r_2r_5)}{(r_1r_2r_3)}=\frac{(r_1r_3r_4)}{(r_1r_2r_3)}=\frac{(r_4r_5r_2)}{(r_1r_2r_3)}=\frac{(r_4r_5r_3)}{(r_1r_2r_3)}=a.$$

现在求a的具体值.设$(r_1r_2r_3)=1$,$r_2r_3=1$,又设直线r_4r_5在$\Delta r_1r_2r_3$内部的部分长度为h,如图 2 所示,关系$(r_4r_5r_2)/(r_1r_2r_3)=\alpha$可以写为$(1-h)(h-2\alpha)=\alpha$,而条件$(r_1r_5r_4)/(r_1r_2r_3)\geqslant\alpha$可写为$h(h-2\alpha)\geqslant\alpha$.

易解得

$$\alpha=\frac{h-h^2}{3-2h},a=\sup_{h\in[0,1]}\alpha(h)=\frac{1}{4+2\sqrt{3}},\text{当 } h=\frac{1}{2}\left(3-\sqrt{3}\right).$$

引理 1 证毕.

引理 2 若r_5在凸四角形$r_1r_2r_3r_4$中,则

$$(r_1\cdots r_5)\leqslant\frac{1}{2+2\sqrt{2}}|r_1r_2r_3r_4|.$$

证明见[13].

引理 3 若$r_1\cdots r_5\subset\Delta s_1s_2s_3$,且$r_1\cdots r_5$的凸包是四角形或五角形,则有

$$(r_1\cdots r_5)\leqslant\frac{1}{3+2\sqrt{2}}(s_1s_2s_3).$$

证明 不妨设$\Delta s_1s_2s_3$三边都含有$\{r_1,\cdots,r_5\}$的点;而且若某边只含有其中的一点,则此点必是该边的中点(否则可用面积较小的$\Delta s_1's_2's_3'$包含$r_1\cdots r_5$).首先考虑$r_1\cdots r_5$的凸包是凸四角形的情况,这又可分为两种子情况:

A_1:三角形$s_1s_2s_3$的某边只含$\{r_1,\cdots,r_5\}$的一个点:不妨设r_2为s_1s_2的中点,易见$(r_1r_2r_3)<(r_1s_1r_2)+(r_2s_2r_3)$,故$(s_1s_2s_3)-|r_1\cdots r_5|>(r_1\cdots r_5)$,再由引理 2 可得(见图 3)

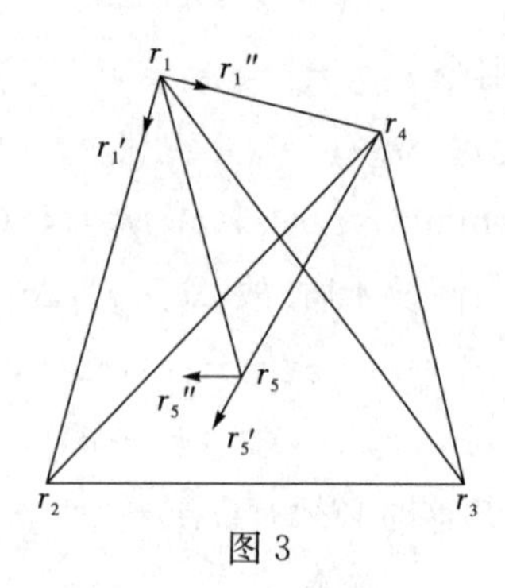

图 3

$$(s_1s_2s_3)>(2+2\sqrt{2})(r_1\cdots r_5)+(r_1\cdots r_5)=(3+2\sqrt{2})(r_1\cdots r_5).$$

A_2:三角形 $s_1s_2s_3$ 每边都含有$\{r_1,\cdots,r_5\}$的两点.这里 $r_1\cdots r_5$ 中必有两点为 $\Delta s_1s_2s_3$ 某一边的两个端点,如图 4.

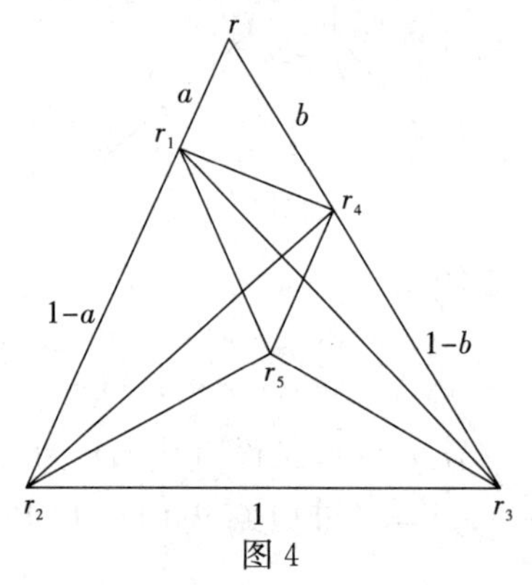

图 4

可设 r_1,r_4 分别在 s_1s_3,s_2s_3 中点或上侧(详见[12]引理 5 之证明Ⅱ).于是有 $r\in\Delta s_1s_2s_3$ 使得 $rr_4s_3r_1$ 是平行四边形.如果 $r_5\in\Delta r_1r_2r_4\cup\Delta r_1r_3r_4$,则$(r_1\cdots r_5)\leqslant\frac{1}{6}(s_1s_2s_3)$;若 $r_5\in\Delta r_3r_1r\cup\Delta r_4r_2r$,则$(r_1r_4s_3)>(r_1\cdots r_5)$,即$(s_1s_2s_3)-|r_1\cdots r_5|>(r_1\cdots r_5)$,这时有$(s_1s_2s_3)>(3+2\sqrt{2})(r_1\cdots r_5)$;若 $r_5\in r_2r_3r$ 但$(s_1s_2s_3)-|r_1\cdots r_5|\leqslant(r_1\cdots r_5)$,即$(r_1r_4s_3)\leqslant(r_2r_3r_5)\leqslant(r_2r_3r)$,这相当于

$$\frac{1}{2}\Big(1-a-b\Big)\geqslant\frac{1}{2}ab,\qquad(a,b\text{ 含义如图 4 所示.})$$

从而 $ab\leqslant3-2\sqrt{2}$,于是$(r_1\cdots r_5)\leqslant(3-2\sqrt{2})(s_1s_2s_3)$.注意这里等号成立的条件是 $a=b=\sqrt{2}-1$ 而且 $r_4r_5/\!/r_1r_2,r_1r_5/\!/r_4r_3$.

现在考虑 $r_1\cdots r_5$ 的凸包是凸五角形的情况,这也可分为两种子情况:

B_1:三角形 $s_1s_2s_3$ 的某边只含$\{r_1\cdots r_5\}$的一个点.设 r_2 为 s_1s_2 的中点,如图 3.易见$|r_1s_1s_2r_3|>2(r_1r_2r_3)\geqslant2(r_1\cdots r_5)$,因而$(r_1r_3s_3)<(s_1s_2s_3)-2(r_1\cdots r_5)$.注意到 $r_1r_3r_4r_5$ 是包含在三角形 $r_1r_3s_3$ 中的凸四角形,引用[12]之引理 9:"对于任何含在 $\Delta s_1s_2s_3$ 中凸四角形 $r_1r_2r_3r_4$,有$(r_1r_2r_3r_4)\leqslant\frac{1}{4}(s_1s_2s_3)$".得到

$$(r_1\cdots r_5)\leqslant(r_1r_3r_4r_5)\leqslant\frac{1}{4}(r_1r_3s_3)<\frac{1}{4}((s_1s_2s_3)-2(r_1\cdots r_5)),$$

即$(r_1\cdots r_5)<\frac{1}{6}(s_1s_2s_3)$.

B_2:三角形 $s_1s_2s_3$ 每边均含有$\{r_1,\cdots,r_5\}$的两点,这时 $r_1,\cdots,r_5$ 中将有一点是 $\Delta s_1s_2s_3$ 的顶点. 引用[12]之引理 5 可知:"设凸五角形 $r_1\cdots r_5$ 被包含在三角形 Δ 中且与 Δ 有公共顶点,则当 $r_1\cdots r_5/|\Delta|$ 取得最大值时,该五角形至少有四个外围三角形面积等于$(r_1\cdots r_5)$;例外者只能是顶角三角形(推论 1)",因而在图 5 中,

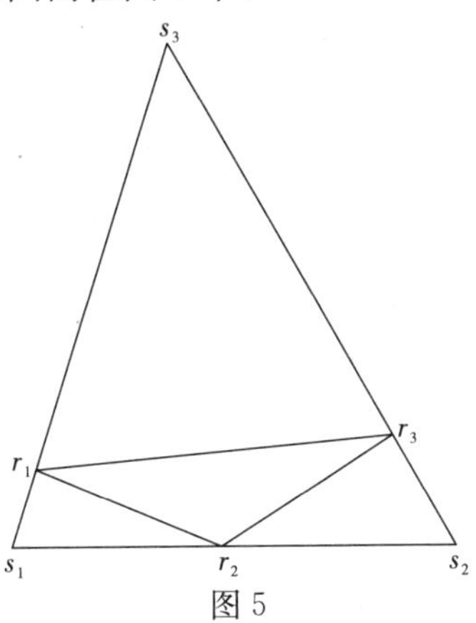

图 5

$r_1r_4 /\!/ r_2r_3, r_1r_2 /\!/ r_5r_3, r_4r_3 /\!/ r_5r_2$,即

$$\frac{1-a}{b}=\frac{a}{1-2b},\text{也即 } a+2b-ab=1,$$

由此可推出 $ab\leqslant 3-2\sqrt{2}$,于是$(1-a)(1-2b)\leqslant 3-2\sqrt{2}$. 从而$(r_1\cdots r_5)\leqslant(3-2\sqrt{2})(s_1s_2s_3)$.(这里等号成立的条件是 $a=2b=2-\sqrt{2}$.),至此引理 3 证毕.

综合引理 1 和引理 3,我们得到三角形中五个点的 Heilbronn 数和 Heilbronn 分布图形.

定理 1 $H_5(\Delta)=3-2\sqrt{2}$;五个点在三角形中的 Heilbronn 分布共有两种,如图 6,图 7.

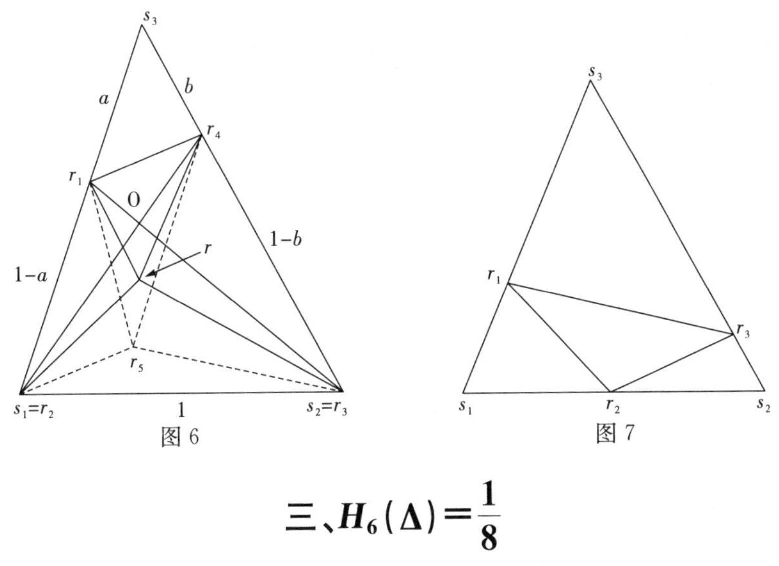

图 6　　图 7

三、$H_6(\Delta)=\frac{1}{8}$

引理 4 凸六角形 $r_1\cdots r_6\subset\Delta s_1s_2s_3$,则

$$(r_1\cdots r_6)\leqslant\frac{1}{8}(s_1s_2s_3).$$

证明　记 $c=\sup\{(r_1\cdots r_6)/(s_1s_2s_3)\mid r_1\cdots r_6$ 是凸六角形而且 $r_1\cdots r_6\subset s_1s_2s_3\}$. 则由[12]引理8可知:"任给凸六角形 $r_1\cdots r_6$,有等围六角形(即六个外围三角形皆相等的凸六角形)$r_1'\cdots r_6'\subset r_1\cdots r_6$ 使$(r_1'\cdots r_6')\geqslant(r_1\cdots r_6)$";此外 $r_1\cdots r_6$ 若能使上界 c 实现,其必为等围六角形(详见其证明). 而且在仿射等价意义下可以认为等围六角形的中心三角形是正三角形,如图8的 $s_1's_2's_3'$.

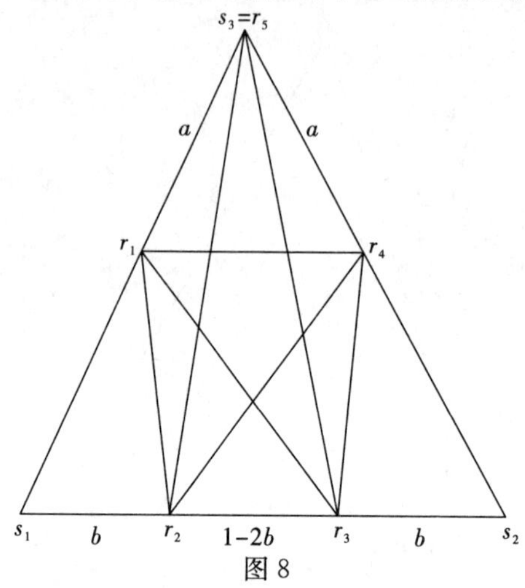

图8

并不妨设 s_1 在$\angle r_1s_1'r_6$ 内部. 直线 s_1s_1'将 $\Delta s_1s_2s_3$ 分布两个小三角形 $s_1s_1''s_2$, $s_1s_1''s_3$,它们分别含有平行四边形 $r_1r_2r_3s_1'$和 $r_4r_5r_6s_1'$. 注意到

$$\min\{(s_1s_1''s_2),(s_1s_1''s_3)\}\leqslant\frac{1}{2}(s_1s_2s_3),$$

则如引理3证明中那样,援引[12]引理9可得

$$(r_1\cdots r_6)=(r_1r_2r_3s_1')=(r_4r_5r_6s_1')\leqslant\frac{1}{2}(s_1s_2s_3).$$

等号成立的必要条件是,$r_1,\cdots,r_6$ 都在 $\Delta s_1s_2s_3$ 的边界上,即 $s_1=r_2r_1\cap r_5r_6$, $s_2=r_1r_2\cap r_4r_3$, $s_3=r_3r_4\cap r_6r_5$. 由简单的计算可知,这时候 $r_1,\cdots,r_6$ 均是 $\Delta s_1s_2s_3$ 各边的四分点,如图9. 引理4证毕.

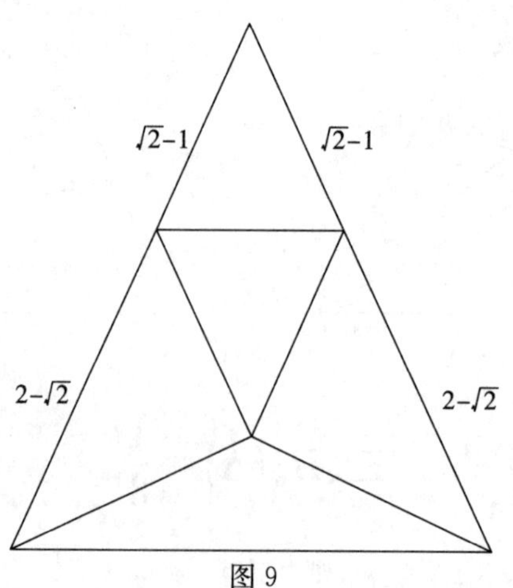

图9

引理 5 设 $r_1\cdots r_6$ 的凸包是凸五角形且含在 $\Delta s_1s_2s_3$ 中，则$(r_1\cdots r_6)\leqslant\frac{1}{8}(s_1s_2s_3)$.

证明 我们只考虑使$(r_1\cdots r_6)/(s_1s_2s_3)$取得上确界的图形，故不妨设 $\Delta s_1s_2s_3$ 每边都含有 $r_1\cdots r_6$ 的点；而且如果只含一点则此点就是该边的中点，如果含有两点则该边中点位于这两点之间. 于是可分以下三种情形来考虑：(1)$r_1,\cdots,r_6$ 中没有点是 $\Delta s_1s_2s_3$，某边的中点；(2)$r_1,\cdots,r_6$ 中有一点是 $\Delta s_1s_2s_3$ 某边的中点；(3)$r_1,\cdots,r_6$ 中有两点或三点是 $\Delta s_1s_2s_3$ 某些边的中点. 对这三种情形中的任何一种，不妨设 $s_1s_2=s_2s_3=s_3s_1=1$.

(1)这时 $r_1\cdots r_6$ 的凸包的顶点 $r_1,\cdots,r_5$ 都在 $\Delta s_1s_2s_3$ 的边界上而且两者有公共顶点，比方如 $r_5=s_1$，如图 10.

令 $c=(r_1\cdots r_6)/(s_1s_2s_3)$. 若 $r_6\in\Delta r_1r_2r_5\cup\Delta r_3r_4r_5$，则$(r_1r_2r_5)+(r_3r_4r_5)\geqslant 4c(s_1s_2s_3)$，因而$(r_2r_3r_5)\leqslant(1-4c)(s_1s_2s_3)$，$r_2r_3\leqslant 1-4c$. 又 $\min\{(r_1r_2r_3),(r_2r_3r_4)\}\geqslant c(s_1s_2s_3)$，故有 $\min\{r_1s_2,r_4s_3\}\geqslant\frac{c}{1-4c}$，从而 $\max\{r_1r_5,r_4r_5\}\leqslant 1-\frac{c}{1-4c}$；再将此式代入 $\max\{(r_1r_2r_5),(r_3r_4r_5)\}\geqslant 3c$，得 $\max\{r_2s_2,r_3s_3\}\geqslant 3c\Big/\left(1-\frac{c}{1-4c}\right)$. 注意到 $\max\{r_2s_2,r_3s_3\}<\frac{1}{2}$，得不等式

$$3c<\frac{1}{2}\left(1-\frac{c}{1-4c}\right)\text{，从而 }c<\frac{1}{8}.$$

若 $r_6\in\Delta r_2r_3r_5$，设 r 为 s_2s_3 的中点，则不论 r_6 在中线 r_5r 的哪一侧均有

$$(r_1\cdots r_6)\leqslant\min\{(r_1r_2r_6r_5),(r_3r_4r_5r_6)\}\leqslant\frac{1}{8}(s_1s_2s_3).$$

这里等号成立的条件是，r_6 是 r_5r 的中点，而且 $r_1,\cdots,r_4$ 是 $\Delta s_1s_2s_3$ 各边的四等分点，如图 10.

(2)$r_1,\cdots,r_6$ 中有一点是 $\Delta s_1s_2s_3$ 的中点，各点的位置如图 11.

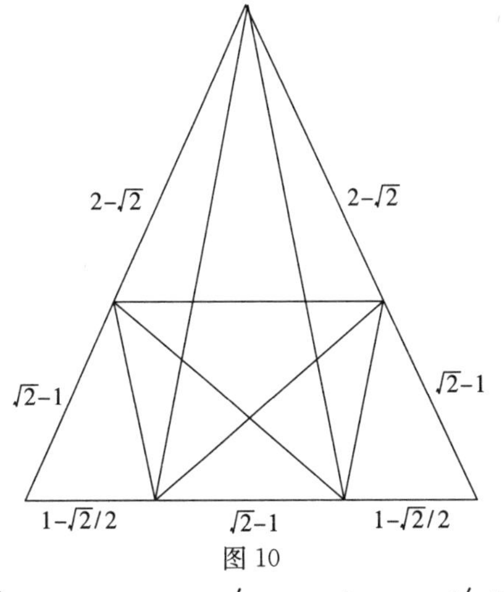

图 10

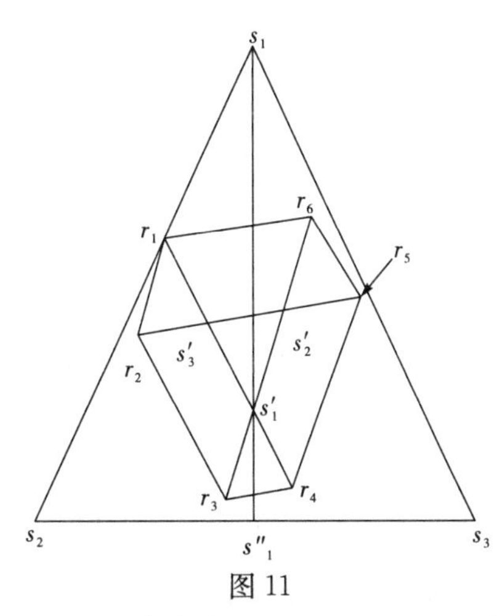

图 11

令 $s_1r_1=a,s_2r_2=a',s_1r_5=b,s_3r_4=b'$. 又设 s_1s_3,s_1s_2 的中点为 t_1,t_2，及 t_1r_3,t_2r_3 的中点为 u_1,u_2. 注意到凸包 $r_1\cdots r_5$ 被下列 9 个三角形所覆盖：$\Delta u_1r_3r_4,\Delta u_1r_3r_5,\Delta u_2r_3r_2,\Delta u_2r_3r_1,\Delta u_1r_1r_4,\Delta u_2r_5r_2;\Delta r_1r_3r_5,\Delta r_1r_2r_5,\Delta r_1r_4r_5$. 易见前 4 个三角形的面积都不超过$\frac{1}{8}(s_1s_2s_3)$，

故当 r_6 属于前 4 个三角形区域时有 $(r_1\cdots r_6)\leqslant\frac{1}{8}(s_1s_2s_3)$. 而当 r_6 属于第 5,第 6,或第 7 个三角形区域时,仅由[12]之引理 9 可知 $(r_1\cdots r_6)\leqslant\frac{1}{8}\left(s_1s_2s_3\right)$. 最后考虑 r_6 属于最后两个三角形区域的情况. 由于

$$\frac{(r_1r_2r_5)+(r_1r_4r_5)}{(s_1s_2s_3)}=b(1-a-a')+a(1-b-b')<a+b-2ab=-2\left(\frac{1}{2}-a\right)\left(\frac{1}{2}-b\right)+\frac{1}{2}\leqslant\frac{1}{2},$$

故当 $r_6\in\Delta r_1r_2r_5\cup\Delta r_1r_4r_5$ 时成立着

$$3(r_1\cdots r_6)+(r_1\cdots r_6)\leqslant(r_1r_2r_5)+(r_1r_4r_5)<\frac{1}{2}(s_1s_2s_3),$$

即 $(r_1\cdots r_6)<\frac{1}{8}(s_1s_2s_3)$. 对情况(2)等号不可能成立.

(3)设 r_1,r_4 分别为 s_1s_2,s_1s_3 的中点,其余各点位置如图 12 所示.

若 $r_6\in\Delta r_1r_4r_5\cup\Delta r_3r_4r_1$,则 $(r_1\cdots r_6)\leqslant\frac{1}{3}\max\{(r_1r_4r_5),(r_3r_4r_1)\}\leqslant\frac{1}{12}(s_1s_2s_3)$. 若 $r_6\in\Delta r_1r_2r_3$,则 $(s_1r_2r_3)=2(r_1r_2r_3)\geqslant6(r_1\cdots r_6)$. 令

$$p=\frac{r_5r_2'}{r_2'r_2}=\frac{r_5r_3'}{r_3'r_3}=4\,\frac{(r_1r_4r_5)}{(s_1s_2s_3)}.$$

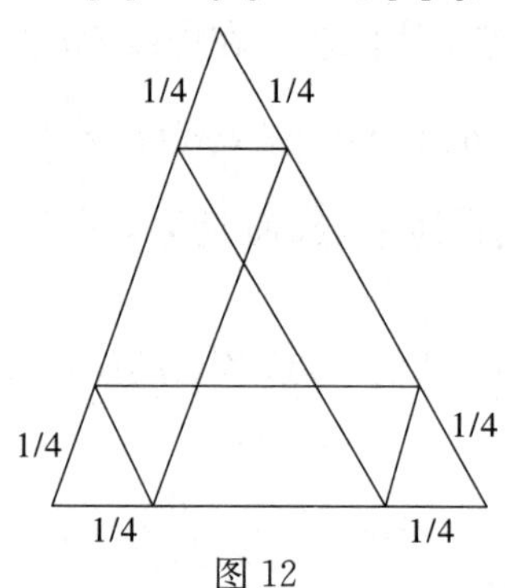

图 12

如果 $(r_1\cdots r_6)\geqslant\frac{1}{8}(s_1s_2s_3)$,则 $r_2'r_3'\geqslant\frac{3}{4}\cdot\frac{p}{1+p}$,从而 $r_1r_2'+r_3'r_4=\frac{1}{2}-r_2'r_3'\leqslant\frac{2-p}{4(1+p)}$,同时易见 $\min\{r_1r_2',r_3'r_4\}\geqslant\frac{1}{4(1+p)}$,矛盾! 至于 $\Delta s_1s_2s_3$ 的三边中点都是 $r_1\cdots r_5$ 顶点的情形是平凡的,在该情形显然有 $(r_1\cdots r_6)\leqslant\frac{1}{12}(s_1s_2s_3)$. 至此引理 5 证毕.

引理 6　设 $r_1\cdots r_6$ 的凸包是四角形或三角形且含在 $\Delta s_1s_2s_3$ 中,则 $(r_1\cdots r_6)\leqslant\frac{1}{8}(s_1s_2s_3)$.

证明　对于凸包为四角形的情况,只须引用[12]的引理 7:"设 $r_1\cdots r_6$ 的凸包是四角形且含在平面凸体 K 中,则有 $(r_1\cdots r_6)\leqslant\frac{1}{8}|K|$".

对于凸包为三角形的情况,不妨设其顶点为 r_1,r_2,r_3 且与 s_1,s_2,s_3 相重合,当 $r_1r_2r_4r_5r_6$ 的凸包为三角形或五角形时,援用引理 1 可得

$$(r_1r_2r_3)\geqslant2(r_1\cdots r_6)+(4+2\sqrt{3})(r_1\cdots r_6)<8(r_1\cdots r_6).$$

当 $r_1r_2r_4r_5r_6$ 的凸包为凸四角形 $r_1r_2r_4r_5$ 时，引进仿射坐标系令 $r_3=(0,0)$，$r_1=(1,0)$，$r_2=(0,1)$，$r_4=(u,y)$，$r_5=(x,v)$，$r=r_1r_5\cap r_2r_4=(a,b)$，如图 13 所示.

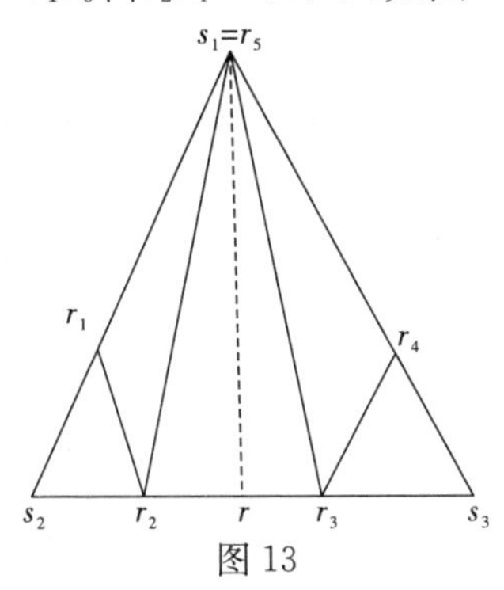

图 13

如果设 $(r_1\cdots r_6)\geqslant\frac{1}{8}(s_1s_2s_3)$，则有

$$\min\{x,y,u,v,xy-uv\}\geqslant\frac{1}{8},$$

而且 $a+b>u+v\geqslant\frac{1}{4}$. 又由引理 3 可知

$$(r_1r_2r)\geqslant(3+2\sqrt{2})(r_1r_2r_4r_5r_6)\geqslant(3+2\sqrt{2})(r_1\cdots r_6)\geqslant\frac{3+2\sqrt{2}}{8}(s_1s_2s_3),$$

即 $1-a-b\geqslant\frac{1}{8}\left(3+2\sqrt{2}\right)$. 于是 $\frac{1}{4}<a+b\leqslant\frac{1}{8}(5-2\sqrt{2})$. 另一方面，由于 $xy\geqslant uv+\frac{1}{8}\geqslant\frac{9}{64}$，并有

$$x=1-\frac{1-a}{b}v\leqslant1-\frac{1-a}{8b},y=1-\frac{1-b}{a}u\leqslant1-\frac{1-b}{8a},$$

从而 $\left(1-\frac{1-a}{8b}\right)\left(1-\frac{1-b}{8a}\right)\geqslant\frac{9}{64}$. 由最后一式可推出 $a+b\leqslant\frac{1}{6}$ 或 $a+b\geqslant\frac{1}{3}$，与前面矛盾. 故必须有 $(r_1\cdots r_6)<\frac{1}{8}(s_1s_2s_3)$. 至此引理 6 证毕.

综合引理 4,5,6，我们完全清楚了六个点在三角形中的 Heilbronn 分布.

定理 2 六个点在三角形中的 Heilbronn 分布共有两种，如图 9,10，$H_6(\Delta)=\frac{1}{8}$.

参考文献

[1]Goldberg M. Maximizing the smallest triangle made by points in a square. Math. Mag., 1972(45):135－144.

[2]Komlos J, et al. On Heilbronn's problem. J. London Math. Soc., 1981, 24(2):385－396.

[3]Komlos J, et al. A lower bound for Heilbronn's problem. J. London Math. Soc., 1982 (25):13－24.

[4]Moser W. 100 Research Problems in Discrete Geometry, 1986.

[5]Roth K F. On a problem of Heilbronn. J. London Math. Soc., 1951(26):198－204.

[6]Roth K F. On a problem of Heilbronn Ⅱ. J. London Math. Soc. ,1972(25):193-212.
[7]Roth K F. On a problem of Heilbronn Ⅲ. J. London Math. Soc. ,1972(25):543-549.
[8]Roth K F. Estimation of the area of the smallest triangle obtained by selecting three out of n points in a disc of unit area. Amer. Math. Soc. Proc. Symp. Pure Math. ,1973(24):251-262.
[9]Roth K F. Developments in Heilbronn's triangle problem. Advance in Math. ,1976(22):365-384.
[10]Schmidt W M. On a Problem of Heilbronn. J. London Math. Soc. ,1971(24):545-550.
[11]杨路,张景中. 正方形内六点问题. 数学讲座(下),四川人民出版社,1980:159-175.
[12]杨路,张景中,曾振柄. 关于最初几个 Heilbronn 数的猜想和计算. 数学年刊(A),1992,13(4):503-515.
[13]马援,王振. 关于多边形的 H_5 的最佳估计和计算. 蛙鸣数学杂志,1988(34):11-29.

几何定理可读证明的自动生成

张景中　杨　路
（中国科学院成都计算机应用研究所　成都　610041）
高小山
（中国科学院系统科学研究所　北京　100080）
周咸青
（美国 Wichita 州立大学计算机系）

摘　要：用计算机能生成几何定理的易为人们理解的证明吗？这个几十年来进展很小的难题，自 1992 年以来有了突破性进展．对于一大类欧氏几何命题——构造性几何命题，已有了相当有效的算法．基于此算法所编制的程序，已证明了 500 多条非平凡的几何命题．对其中大多数命题，机器自动生成的证明是简明而易于理解的．本文是对这一领域近三年来取得的进展的综述，包含了在非欧几何可读证明方面的最新成果．

关键词：可读机器证明，面积法，消点法，勾股差，幅．

AUTOMATED GENERATION OF READABLE PROOFS IN GEOMETRY

Zhang Jingzhong and Yang Lu
(*Chengdu Institute of Computer Applications. The Chinese Academy of Sciences, Chengdu* 610041)
Gao Xiaoshan
(*Institute of System Science, The Chinese Academy of Sciences, Beijing* 100080)
Chou Shangching
(*Department of Computer Science. The Wichita State University, Wichita, Kansas* 67208, USA)

Abstract Can the geometric theorem proofs which are easily understood by people be automatically produced by a computer? This problem did not take any remarkable advances for several decades until 1992 when a break-through took place. For a large class of Euclidean geometry propositions, there is an efficient decision algorithm and then accordingly a computer program by which more than 500 non-trivial geometry theorems

---

本文刊于《计算机学报》，第 18 卷第 5 期，1995 年．

were proven. Amongest the proofs produced automatically by computers, most are brief, easily understood, some even nice. This paper is a survey of advances in automated production of "readable" proofs during the past three years. Also included is the similar achievement in non-Euclidean geometry at the most recent.

Keywords Readable machine-proof, area method, point-eliminating method, Pythagorean difference, argument.

1 背 景

由于吴文俊院士的突破性工作[1,2]，几何定理机器证明的研究近十几年有很大发展. 继吴法之后，又出现了 Gröbner 基法、数值并行法等有效算法[3~8]. 这些方法，都属于代数方法.

代数方法在几何定理机器证明中的成功，自然地使人们对机器证明提出了更高的要求. 在近年来多次有关机器证明的国际学术会议上，都有人提出这样的问题：能不能由计算机给出易于理解、易于检验的证明，即所谓可读的证明？

希望用计算机自动生成几何定理的可读证明，这一方向的研究工作早在 1960 年已开始[9]. 30 多年来，这方面已发表不少著作[10~15]，并写出了一些试验性程序或用于计算机辅助教学程序[16,17]. 所用的方法主要是模拟人在证明几何定理时的思考过程：从已知条件和预知的几何引理出发，用逻辑推理规则作各种可能的探索，以期得到所要的结论. 由于遇到搜索空间过于庞大和添加辅助线的困难，这种方法至今未能形成能处理成类的几何命题的有效算法. 从已发表的 20 多个例题来看，多属于中学课程内容，且仅涉及直线形. 对这些例题而言，可读证明生成常需要预先输入某些必要的辅助线，而不是完全自动的.

在证明中使用辅助线，是传统几何方法的特色. 这一特色使传统几何证题方法显得格外美妙隽永，但却给几何证明的机械化增添了很大的困难. 在文献[18]中指出，辅助线在解题中的必要性，是由于欧氏几何传统方法以全等三角形和相似三角形为基本工具. 多数几何题中，常常没有现成的全等三角形和相似三角形，因而要添加辅助线为使用基本工具创造条件. 想要实现可读证明的自动生成，道路不外两条：一是找寻加辅助线的规律，另一条是改造传统的几何方法使适应于机械化. 前一条路，目前还难以走通；后一条路，却大有希望. 事实上，代数方法的几何基础——坐标法，就是对传统几何方法的一种改造方案. 但坐标法的代数味道太浓，所产生的证明缺少直观性，虽有利于机械化但可读性较差. 几何定理可读证明机械化的实现，是由面积方法的应用开始突破的.

面积方法本是一种古老的几何解题技巧，被称为几何学的基石的勾股定理，它的最早的证法，在中国和古希腊都用的是面积法. 但长期以来，面积方法仅被看成是一种特殊技巧. 近十几年来，在文献[18]，[19]等中，提出了作为平面几何的一般工具的系统的面积方法. 这种系统的面积方法与消点法相结合，使几何定理可读证明的计算机生成的研究得到突破.

1992 年,文献[20]中使用了面积方法中的少量基本命题,给出了适用于 Hilbert 交点型的平面几何命题类的机器证明算法. 这一算法用 Lisp 语言实现,能有效地生成可读证明. 随后,文献[21]把这一工作推广到三维空间,用体积代替面积取得了类似的成功.

从向量运算的观点看,面积相当于向量的外积. 用面积法能做的,向量法当然也能做,不过在依样画葫芦的基础上略加调整就是了. 但是,只用向量外积难于表达和解决有关角度和长度的几何问题,也就是说,只用面积这个几何量还不足以处理平面几何中涉及角度大小的许多问题. 为了克服这一困难,文献[19]中曾引入另一个几何量——勾股差. 勾股差相当于向量的内积. 用上勾股差,可以扩展文献[20]中的算法,使能处理一切可构造几何命题(见文献[22]).

如果进一步引入全角,并加上数据库技术,能够大大加强算法的力量,改善所产生的证明的可读性. 这一进一步的研究方向. 在有关专著[23]中已可见端倪. 我们相信,关于机器产生几何定理可读证明的研究,今后几年将趋于活跃,并出现新的进展.

2 思　　路

在文献[20]中提出了用计算机生成几何定理可读证明的基本思想:消点方法. 下面的例子虽十分简单,但已足以说明这种基本思想的轮廓.

例 1　求证:平行四边形的对角线相互平分.

为了使计算机能处理这个问题,应当把它写成某种规范形式. 由于命题涉及平行四边形的四个顶点和对角线交点共五个点,输入于计算机的规范形式中应说明这五个点之间的关系. 我们采用如下顺序引进的方式:

C_1:(A,B,C);(任取三点 A,B,C)

C_2:$(D, DA/\!/BC, DC/\!/AB)$;

(取 D 使 $ABCD$ 为平行四边形)

C_3:$(O, O\in AC, O\in BD)$;(AC,BD 交于 O)

这样,命题的假设部分就交代清楚了. 至于要证的结论,则可表示为

$$G: \frac{\overline{AO}}{\overline{OC}}=1$$

于是,对机器的输入为

$$(C_1, C_2, C_3, G) \tag{2.1}$$

当程序收到这一输入时,它如何操作呢?

首先,它检查 G 的右端出现的点中,哪一个是最后引进的. 在此例中,程序发现点 O 是最后引进的. 于是,程序确定了下一步的任务,要从 G 的右端消去点 O.

用什么办法消去点 O 呢? 这决定于两个因素:O 是如何被引进的,以及 O 出现在什么样的几何量之中. 程序检查 O 的引入方法 G,知道 O 是 AC,BD 的交点;检查 O 在 G 的右端所处的环境,知道 O 出现在共线线段比 $\frac{\overline{AO}}{\overline{OC}}$ 之中. 根据这两条信息,程序在事先编好的消点公式

库中找出一个适用的公式①

$$\frac{\overline{AO}}{\overline{OC}}=\frac{S_{ABD}}{S_{BCD}} \tag{2.2}$$

并且用公式(2.2)的右端代替 G 的左端,得到新的结论形式

$$G_1:\frac{S_{ABD}}{S_{BCD}}=1 \tag{2.3}$$

这里,S_{ABD},S_{BCD} 分别表示$\triangle ABD$ 和$\triangle BCD$ 的带号面积. 由于点 O 已不在 G_1 中出现. 其引入方式 C_3 已不必考虑,故程序将输入的命题改写为

$$(C_1,C_2,G_1) \tag{2.4}$$

重新执行上述步骤.

这一次,G_1 中最后被引进的是点 D. 程序查出 D 应满足的条件是 $DA\,/\!/\,BC$ 和 $DC\,/\!/\,AB$. 而包含 D 的几何量则为 S_{ABD} 和 S_{BCD},程序在消点公式库中找到

$$S_{ABD}=S_{ABC}\qquad(当\ DC/\!/AB) \tag{2.5}$$

$$S_{BCD}=S_{ABC}\qquad(当\ DA/\!/BC)$$

及

便把 G_1,改写成

$$G_2:\frac{S_{ABC}}{S_{ABC}}=1 \tag{2.6}$$

这时,程序经检查发现 G_2 是恒等式,便输出命题成立的信号. 程序把记录下的 G,$G1$,$G2$ 连在一起,生成一个证明:

$$\frac{\overline{AO}}{\overline{OC}}=\frac{S_{ABD}}{S_{BCD}}=\frac{S_{ABC}}{S_{ABC}}=1 \tag{2.7}$$

如果用户要查询(2.7)中推理的依据,程序可以提供出所用过的消点公式,选择这些公式的理由以及这些公式的来历.

有了上述思路,便会认识到:要具体设计出一个能自动生成几何定理证明的算法需要而且只需要构作出三个集合:

C:作图法之集(即引入点的方法之集).

Q:允许使用的几何量之集.

E:消点公式之集.

其中,C 用于表述命题的假设,Q 用于表述命题的结论,E 用于消点过程以生成证明.

为了处理 Hilbert 交点型几何命题,文献[20]中引入的作图法之集实质上包含了下列作图:

H_1:在平面上任取一点 Y.

① 这个公式就是面积法的基本工具之一:共边定理.

H_2:已知两点 A,B,在直线 AB 上取一点 Y,使$\overline{AY}=\lambda\overline{AB}$. 这里 λ 可以是独立参数,有理数(数值)或已知点构成的几何量的有理表达式.

H_3:已知四点 A,B,P,Q,取直线 AB 与直线 PQ 的交点 Y.

H_4:已知五点 A,B,P,Q,W,取直线 AB 上一点 Y 使 $WY/\!/PQ$.

H_5:已知六点 A,B,U 和 P,Q,V,取点 Y 使 $UY/\!/AB$ 且 $VY/\!/PQ$.

至于允许的几何量,只取两类:一是三角形的带号面积,一是平行或共线的有向线段之比. 当然,有了三角形带号面积,便也有了四边形带号面积.

所用的消点公式,则基于如下基本命题. 这些基本命题正是文献[18],[19]中所发展的系统的面积方法的主要工具:

B_1:若直线 AB 与 PQ 交于 M,则

$$\frac{S_{PAB}}{S_{QAB}}=\frac{\overline{PM}}{\overline{QM}},\frac{S_{PAB}}{S_{PAQB}}=\frac{\overline{PM}}{\overline{PQ}}$$

B_2:若 Y 为直线 PQ 上一点,$\overline{PY}=\lambda\overline{PQ}$,则对任两点 A,B,有

$$S_{ABY}=\lambda S_{ABQ}+(1-\lambda)S_{ABP}$$

B_3:直线 $AB/\!/PQ$ 的充要条件是

$$S_{PAB}=S_{QAB}$$

在文献[20]中证明了:对应于上述作图法 $H_1\sim H_5$ 和两种几何量——带号面积与平行(共线)线段比,由 B_1,B_2,B_3 出发,确实能建立所需要的消点公式,下面以 Pappus 定理为例,说明如何由消点法生成证明.

例 2 已知 A,B,C 三点在一直线上,X,Y,Z 三点也在一直线上,直线 BX,AY 交于 P,BZ,CY 交于 Q,AZ,CX 交于 R. 求证:P,Q,R 共直线.

为证明 P,Q,R 共直线,可用下述表达方式:设 CX 与 PQ 交于 U,AZ 与 PQ 交于 V,只要证明两点 U,V 重合,即证明

$$\frac{\overline{PU}}{\overline{QU}}=\frac{\overline{PV}}{\overline{QV}}$$

即可,这种命题转换方式,我们已写入程序.

于是,命题的输入为:

(1)取任意点 A,B,X,Y;

(2)在直线 AB 上取 C;

(3)在直线 XY 上取 Z;

(4)AY 与 BX 交于 P;

(5)BZ 与 CY 交于 Q;

(6)PQ 与 XC 交于 U;

(7)PQ 与 AZ 交于 V;

结论:$\frac{\overline{PU}}{\overline{QU}}\cdot\frac{\overline{QV}}{\overline{PV}}=1$.

计算机执行消点法而产生的证明为:

$$\frac{\overline{PU}}{\overline{QU}}=\frac{\overline{QV}}{\overline{PV}}=\frac{S_{PXC}\cdot S_{QAZ}}{S_{QXC}\cdot S_{PAZ}} \qquad (\text{消去}U,V)$$

$$=\frac{S_{BCZY}}{S_{BCZ}\cdot S_{YXC}}\cdot\frac{S_{YZC}\cdot S_{ABZ}}{S_{BCZY}}\cdot\frac{S_{PXC}}{S_{PAZ}} \qquad (\text{消去}Q)$$

$$=\frac{S_{YZC}\cdot S_{ABZ}}{S_{BCZ}\cdot S_{YXC}}\cdot\frac{S_{AYX}\cdot S_{BXC}}{S_{ABYX}}\cdot\frac{S_{ABYZ}}{S_{ABX}\cdot S_{AZY}} \qquad (\text{消去}P)$$

$$=\frac{S_{AYX}}{S_{AZY}}\cdot\frac{S_{BXC}}{S_{ABX}}\cdot\frac{S_{YZC}}{S_{YXC}}\cdot\frac{S_{ABZ}}{S_{BCZ}} \qquad (\text{整理})$$

$$=\frac{\overline{YX}}{\overline{ZY}}\cdot\frac{\overline{CB}}{\overline{AB}}\cdot\frac{\overline{YZ}}{\overline{YX}}\cdot\frac{\overline{AB}}{\overline{BC}}=1 \qquad (\text{化简})$$

这里,最后两步的整理与化简自由辅助程序完成. 关于辅助程序见第 5 节.

上述证明中,消去 P,Q 的过程需要加说明. 以由 S_{PXC} 中消去 P 为例:由于 P 是 AY 与 BX 的交点,可得:

$$S_{PXC}=\frac{S_{PXC}}{S_{BXC}}\cdot S_{BXC}=\frac{\overline{PX}}{\overline{BX}}\cdot S_{BXC}=\frac{S_{AYZ}}{S_{ABYX}}\cdot S_{BXC}$$

类似地,可处理 S_{QXC},S_{QAZ},S_{PAZ}.

3 扩充——勾股差的引入

上一节提出的思想,奠定了几何定理可读证明自动生成算法的基础. 但要真正实现这一算法,要做的工作仍是庞大的. 其中首要的是扩大作图方法之集和允许使用的几何量之集. 如果只使用上节所述的作图 $H_1\sim H_5$ 和两种几何量,很多常见的几何命题都难以处理.

在文献[21]中引入了空间的某些作图,如取直线与平面的交点;又引入了几何量体积,这就使文献[20]的结果平行地推广到了三维空间. 当然,也可以无困难地推广到更高维. 但是,文献[20]中结果的实质性的推广是文献[22]中起用了几何量勾股差.

作为一个几何量,勾股差是由文献[19]引入的. 对任三点 A,B,C,量 $\overline{AB}^2+\overline{BC}^2-\overline{AC}^2$ 叫做三点 A,B,C 的勾股差,记作

$$P_{ABC}=\overline{AB}^2+\overline{BC}^2-\overline{AC}^2.$$

在文献[22]中把 P_{ABC} 叫做 Pythagoras Differences. 显见 P_{ABC} 有下列性质:

P_1:$P_{AAB}=0$,$P_{ABC}=P_{CBA}$.

P_2:$P_{ABA}=2\,\overline{AB}^2$,$P_{ABC}+P_{ACB}=2\,\overline{BC}^2$.

P_3:当 A,B,C 共直线时,$P_{ABC}=2\,\overline{BA}\cdot\overline{BC}$.

P_4:当$\angle ABC=90°$时,$P_{ABC}=0$.

容易注意到,P_{ABC} 与 S_{ABC} 有一个明显的不同. 当 A,B,C 顺序改变时,S_{ABC} 的绝对值保持不变,而 P_{ABC} 的绝对值一般会变化. 勾股差的这个性质颇不便于使用. 但当引入四点勾股差时,情形大为改观. 类似于四边形面积

$$S_{ABCD}=S_{ABC}+S_{ACD}=S_{ABD}-S_{CBD}$$

在文献[24]中引入了四点 A,B,C,D 的勾股差：

$$P_{ABCD}=P_{ABD}-P_{CBD}=\overline{AB}^2+\overline{CD}^2-\overline{BC}^2-\overline{DA}^2$$

很快发现，P_{ABCD} 与 S_{ABCD} 有不少类似之处. 它的基本性质有：

P_5：$P_{ABCD}=-P_{ADCB}=P_{BADC}=-P_{BCDA}=P_{CDAB}$

$=-P_{CBAD}=P_{DCBA}=-P_{DABC}$；

P_6：$P_{ABCD}=0$ 的充要条件是：$AC\perp BD$，或 A 与 C 重合，或 B 与 D 重合.

P_7：若 P,Q 分别为 B,D 在直线 AC 上的正投影，则 $P_{ABCD}=P_{APCQ}$.

但是，当 P,Q 在直线 AC 上时，易知：

$$\begin{aligned}P_{APCQ}&=\overline{AP}^2+\overline{CQ}^2-\overline{PC}^2-\overline{QA}^2\\&=(\overline{AP}+\overline{PC})(\overline{AP}-\overline{PC})+(\overline{CQ}+\overline{QA})(\overline{CQ}-\overline{QA})\\&=\overline{AC}(\overline{AP}-\overline{PC}-\overline{CQ}+\overline{QA})\\&=-2\,\overline{AC}\cdot\overline{PQ}=2\,\overline{AC}\cdot\overline{QP}.\end{aligned}$$

这表明，勾股差的本质是内积，即一般有：

$$P_{ABCD}=2\,\overline{AC}\cdot\overline{DB}.$$

由此又得到勾股差的类似于面积的可分性：

P_8：$P_{ABCX}+P_{AXCD}=P_{ABCD}$.

以及垂线上点的可换性：

P_9：若 $BE\perp AC$，$DF\perp AC$，则

$$P_{ABCD}=P_{AECF}\text{（这是 }P_7\text{ 的推广）}.$$

有了勾股差作工具，便可以相应地扩大作图法之集. 在文献[22]中引入的新的作图法，实质上是增加了下列作图：

H_6：已知三点 P,A,B，自 P 作 AB 的垂线，得垂足 Y.

有了 H_6 容易构造出更多的作图，如

H_7：已知两点 A,B，作 Y 使 $AY\perp AB$，且 $\dfrac{2S_{ABY}}{\overline{AB}^2}=r$，这里 r 可以是独立参数，某个有理系数多项式的根，或已知几何量的有理式.

这里，参数 r 的绝对值恰为 $\operatorname{tg}\angle ABY$，而其符号刻画了点 Y 与有向直线 AB 的位置关系（正侧或负侧）.

H_8：已知直线 AB 和过点 B 的⊙O，求取⊙O 与 AB 的另一交点 Y.

H_9：已知交于 P 的两圆⊙O_1 和⊙O_2，求取两圆的另一交点 Y.

还可以引入更多的作图. 作图法越多，写程度的工作量越大，但有可能得到更简捷优美的证明. 理论上，只要有 H_1,H_2,H_3,H_6，便能构造出 $H_2\sim H_9$ 中的其它作图. 例如，H_4 可经过下列步骤完成：

(1)取 WQ 中点 M（作图 H_2）；

(2)在 PM 上取点 N 使 $\overline{PN}=2\,\overline{PM}$（作图 H_2）；

(3)取 WN 与 AB 交点 Y.

更进一步，H_3,H_6 均可由 H_1,H_2 构成. 其中，H_3 可表为："在 AB 上取一点 Y，使 $\dfrac{\overline{AY}}{\overline{AB}}=$

λ，而 λ 取值 S_{APQ}/S_{APBQ}”. 而 H_6 则可表为：“在 AB 上取一点 Y，使$\frac{\overline{AY}}{\overline{AB}}=\lambda=\frac{P_{PAB}}{2\,\overline{AB}^2}$”. 这样，从根本上说，只用 H_1，H_2 两种作图就够了. 但如果这样做，命题的叙述将变得复杂而不自然，所产生的证明也会相应地变得冗繁起来. 因此，我们还是采取更多的作图法.

与几何量勾股差及作图法 $H_6 \sim H_9$ 相应，文献[22]中引入更多的基本命题作为消点公式的基础. 最基本的命题是

B_4：若 Y 在 AB 上且 $PY \perp AB$，则：

$$\frac{\overline{AY}}{\overline{YB}}=\frac{P_{PAB}}{P_{PBA}}.$$

由此命题提供了消去垂足的方法.

在文献[22]中证明了算法对构造性几何命题类的完全性，作了复杂度上界分析，并且指出了非退化条件的生成方法. 不过，从可读证明生成研究的传统看来，更重要的是实际生成的证明的简捷、易理解以至优美的程度. 这里主要不是理论问题，而是实践效果问题. 以下我们着重讨论如何提高算法所产生的证明的质量这个关键问题作为本节的结束. 用下面这个简单的例子表明，勾股差的使用带来很大好处.

例 3 已知$\triangle ABC$ 中，高 AF，BE 交于 H，则有 $CH \perp AB$.

证明 $\frac{P_{ACH}}{P_{BCH}}=\frac{P_{ACB}}{P_{ACB}}=1$，证毕.

（这里，结论 $CH \perp AB$，即 $P_{ACBH}=0$，但 $P_{ACBH}=P_{ACH}-P_{BCH}$，故只要证 $P_{ACH}=P_{BCH}$. 由于 $BH \perp AC$，得 $P_{ACH}=P_{ACB}$，类似地由 $AH \perp BC$ 可得 $P_{BCH}=P_{BCA}=P_{ACB}$. 上述证明是计算机自动生成的. 见文献[23].）

4 全角与“几何信息库”

使用角的概念推证定理，是传统几何方法中最精彩的部分. 如果在发展机器证明算法时不使用角度概念，所产生的证明在很多情形下难于简捷优美，但传统的角的概念由于涉及序关系，对机械化带来较大困难. 用“全角”代替传统的角，是一条较易的路.

吴文俊院士在专著[2]中，为发展无序几何的机器证明系统而引入了全角，但在证明几何定理时用的全角的正切而不是全角本身. 在文献[23]中，进一步发展了文献[2]中的这一精辟思想，提出了直接在机器证明使用全角的设想和方法，得到初步的成功.

在文献[2]中把两相交或重合的直线构成的有序对称为一个全角. 在文献[23]中，则引入更一般的定义：由直线 l，m 构成的有序组叫做一个全角，记为$\angle[l,m]$. 两个全角$\angle[l,m]$与$\angle[u,v]$相等，意为存在一旋转 K，使 $K(l) /\!/ u$ 并且 $K(m) /\!/ v$.

当 $l \perp m$ 时，称$\angle[l,m]=\angle[1]$；而当 $l /\!/ m$ 或 l 与 m 重合时，称$\angle[l,m]=\angle[0]$.

在文献[23]中引入了全角的加法，并导出了有关的运算律，如

$\angle[1]+\angle[1]=\angle[0]$，

$\angle[u,v]+\angle[0]=\angle[u,v]$，

$\angle[u,x]+\angle[x,v]=\angle[u,v]$

等,对无序几何,全角比传统角方便.例如,关于共圆四点,有如下基本定理:

全角的圆周角定理 四点 A,B,C,D 共圆(或共直线)的充要条件是$\angle[AB,BC]=\angle[AD,DC]$.这就不必分成相等与互补两种情形来考虑了.

下面的例子说明全角如何用于几何定理:

例 3 两圆$\odot O$与$\odot Q$交于两点A,B.一直线过A与$\odot O$,$\odot Q$分别交于C,E;另一直线过B与$\odot O$,$\odot Q$分别交于D,F,则$CD/\!/EF$.

证明
$$\begin{aligned}\angle[CD,EF]&=\angle[CD,DB]+\angle[DB,FE]\\&=\angle[AC,AB]+\angle[FB,FE] &&(1)\\&=\angle[AE,AB]+\angle[AB,AE] &&(2)\\&=\angle[AE,AE]=\angle[0]\end{aligned}$$
证毕.

在推证中,(1)是由于A,B,C,D共圆,故$\angle[CD,DB]=\angle[AC,AB]$,且$D,B,F$共线,故$\angle[FB,FE]=\angle[DB,FE]$.(2)是由于$A,B,E,F$共圆,故$\angle[FB,FE]=\angle[AB,AE]$,且$A,C,E$共线,故$\angle[AC,AB]=\angle[AE,AB]$.

上述证明是机器生成的,它不依赖于图形.如用传统方法来证,要根据不同情形画四个图加以讨论.如用解析几何方法来证,证明要繁得多.

为了应用全角证题,计算机在执行程序生成证明时应当知道尽可能多的关于平行、垂直、共线、共圆的信息.为解决这一问题,文献[23]中建议采用“几何信息库”方法.按照这一建议,我们的算法中增加了一个生成几何信息库的子程序.当命题的假设条件输入之后,这一子程序即自动运行,从这些假设条件中收集各种几何信息生成几何信息库,并从已知几何信息中(应用某些预先设置的基本引理)推出新的信息以扩大几何信息库.我们把生成几何信息库的过程叫做向前推理过程,而在第2节中所述的消点过程叫做向后推理过程.当向前推理过程运行到一定程度时,程序将检查一下结论是否已被推出,若否,则开始向后推理.在后向推理过程中,随时查询已生成的几何信息库,以便使用尽可能简捷的消点公式.在前推过程中,主要使用几何引理(如圆周角定理);在后推过程中,主要使用代数运算(如整理化简)和等式代换(如共边定理).这样前后推理结合,几何代数并用,显著地改进了算法的能力.在最近的一个研究报告中①,公布了作者用SB-Prolog所写的程序在NexT工作站上对106个非平凡几何命题运行的结果,包括了机器生成的证明.其效果令人鼓舞.有些命题,用代数方法要十几个小时或甚至溢出,这里却能在几秒或几十秒内给出简短的证明.

这一方向的研究仅仅是开始.看来是一个有希望的方向.

5 辅助技巧

为了产生更为简短、可读性更强的证明,文献[20]和[22]中建议使用一些辅助技巧.其基本思想是对一些常见的几何图形,规定一些典型的辅助算法.程序运行中遇到符合条件的图形,即自动引用辅助算法中指明的处理方法,迄今用过的辅助技巧有:

① 感兴趣的同行可与本文作者联系索取此报告的副本.

1. 多点共圆时的处理方法，见文献[22].

2. 所有的点都在两条直线上时的处理方法，见文献[20].

3. 所有的点都是自由点（无几何约束）时的处理方法，见文献[20]和[22].

4. 对三角形巧合点的处理方法.

但是，对辅助技巧的运用，目前尚无系统的探讨，需要进行进一步研究.

6　经验数据及展望

在文献[23]中公布了作者根据第2节和第3节中所介绍的方法用Common Lisp（AKCL）语言写成的程序运行的情形. 书中列出了478条非平凡平面几何定理及其输入形式，内有283条定理也给出了计算机自动生成的证明. 所用的机器是Next Turbo工作站（25MIPS）. 这一程序尚未使用全角，故这478条定理不包括第4节中提到的106条，也不含立体几何定理. 有兴趣的同行可用ftp调取这一程序，地址为

emcity. cs. twsu. edu: pub/geometry

我们用三个指标来描述程序运行的效果：时间 t，多项式长度 m 和推理步数 l. 这里，多项式长度 m 指的是证明过程中出现的项数最多的多项式的项数；推理步数 l 指的是证明过程中使用引理（即消点公式）的次数①. 表1是关于478条定理运行情形的统计. 从这个统计表可见，程序效率很高：在478个定理中有85.5%运行时间不超过1秒. 这个表在一定程度上反映了几何定理可读证明自动生成的研究目前所达到的水平.

表1

时间 t（秒）		多项式长度 m		推理步数 l	
$t \leqslant 0.1$	45.3%	$m=1$	16.9%	$l \leqslant 3$	7.1%
$t \leqslant 0.5$	68.8%	$m \leqslant 2$	33.0%	$l \leqslant 5$	16.7%
$t \leqslant 1$	85.5%	$m \leqslant 5$	66.9%	$l \leqslant 10$	42.6%
$t \leqslant 5$	97.45%	$m \leqslant 10$	81.7%	$l \leqslant 20$	73.2%
$t \leqslant 10$	98.9%	$m \leqslant 100$	98.7%	$l \leqslant 50$	95.1%
$t < 1087$	100%	$m \leqslant 3125$	100%	$l \leqslant 137$	100%

由于面积、勾股差与向量的外积、内积实质上是一样的，由面积方法的成功会想到用向量法来实现可读机器证明. 由向量与复数的共同点又可想到用复数为工具实现可读证明的机械化. 在文献[23]和[24]中，作者做了这方面的研究. 此外，文献[25]中把Grobner基方法用于向量空间，文献[26]中把吴法与Cliffórd代数表示（实质上也是向量方法）相结合，对某些几何定理也能生成可读证明. 值得一提的是，文献[26]中还研究了微分几何的可读证明. 这些方法的实际效果如何，目前尚未见到有关运行结果的报道.

从人工智能的观点来看，关于几何定理可读证明自动生成的研究，目前尚处于起步阶

① 同一引理使用 n 次作 n 次计算.

段.这是由于,机器生成的证明与两千多年来人们所创造的巧妙证明相比,仍有很大的差距.这主要是由于,我们还不知道如何让机器实现

(1)添加辅助线,

(2)用反证法和合同法,

(3)自动总结出有用的引理,

(4)多种几何量之间的灵活变换

等行之有效的证题方法.至于几何不等式的证明,即使不要求可读性,也仍然是定理机器证明研究中的一大难题.这些问题的存在,表明这一研究方向的前景广阔.

关于几何定理可读证明自动生成的研究的意义,则应从两方面看:一方面,它在数学教育和科学研究中有潜在的实用价值;另一方面,它是机器证明的一般原理研究中最具体、最为丰富多彩的模型.可读证明的研究需要动用的不仅是代数方法,而是数学中和计算机科学中使用过的一切可能的方法,因而使机器证明真正成为数学与计算机科学的边缘学科,对人工智能的研究有更重要的意义.

7 最新进展:非欧几何

关于几何定理机器证明的各种有效算法及通用程序出现之后,人们自然要问:用这些新的算法和程序能否发现新定理,而勿须绞尽脑汁?特别是,是否已经用这些手段发现了有意思的新定理?或者只能限于成批地验证已知的几何定理,对原有的命题库进行不同的处理?

回答是肯定的,80年代以来,借助于基于各种不同算法的"证明器",一些欧氏几何的新定理被发现了.其中有的看上去颇为精彩;有些定理如果不用机器而用人工证明是相当繁难的.有的新定理,如果完全模拟前面介绍的可读证明算法,人工也可写出简短优美的证明.

遗憾的是:欧氏几何有几千年的丰富积累,历代人们发现的不计其数的定理没有一部完备详尽的大辞书供查询.因此,当人们宣布他发现了一条欧氏几何新定理,往往在不久之后就查到它并非新的.

于是人们自然想到向非欧几何领域试一试自动推理(机器证明)算法和程序的威力.非欧几何只有一个半世纪的历史,命题积累不多.历年来人们着重研究非欧几何的公理基础及其物理和数学模型,对其中有趣的定理知之甚少.不能像欧氏几何那样借助于直观,这也是各种非欧几何著作中其定理寥寥可数的原因之一.

近年来开始了用吴法研究非欧几何[27]及借助数值并行法发现非欧几何新定理[28]的研究.但上述工作没有形成通用程序.事实上,非欧几何处理的是非零曲率的空间,与之对应的机器证明算法从而具有更高的复杂度.

直到最近,在欧氏几何定理可读证明自动生成取得突破的基础上,继续发展消点技术,出现了世界上第一个非欧几何定理证明自动生成的通用程序和算法[29,30],该算法对于相当大一类构造性非欧几何定理是完备的和有效的.该程序目前已经有工作站上用的由Common Lisp写成的以及微机上用的由Maple语言写成的两种不同的版本.用这个程序已经产生了一百多个非欧几何定理的可读证明.其中一半以上是从未见过或未被证明过的,它

们很可能是新定理. 这也许可以说,是用计算机自动推理程序来系统地发展一门数学学科分支的一次成功的尝试. 在这里,新的非平凡的定理不是个别的,而是成批地被"产生"出来.

非欧几何可读证明自动生成需要解决的关键问题是找到一些基本的几何不变量,它们将像欧氏几何中的面积和勾股差那样,在消点过程中起决定性的作用.

不幸的是,非欧几何中的面积不再具有我们所需要的那种良好的线性运算性质. 例如:设 A,B,P,Q 为非欧平面上任意四点,Y 为线段 PQ 上一点. 一般说来,在三个非欧三角形 ABY,ABP,ABQ 的面积之间,并不存在如第 2 节中 B2 那样的简单的线性关系.

在文献[29]中找到了这样一个线性不变量并被叫做"幅"(argument). 在非欧双曲平面上一个四角形的幅定义作

$$S_{ABCD}=\sinh(AC)\cdot\sinh(BD)\cdot\sin\angle(AC,BD),$$

而一个三角形的幅定义作

$$S_{ABC}=\sinh(AB)\cdot\sinh(BC)\cdot\sin\angle(AB,BC),$$

注意这里的 S 表示幅,不再表示面积. 双曲平面上 A,B,C 三点共直线的充分必要条件是其幅为零,即 $S_{ABC}=0$.

在上面那些记号和约定下,成立如下的简单而重要的线性关系:

$$S_{ABY}=\frac{\sinh(PY)}{\sinh(PQ)}S_{ABQ}+\frac{\sinh(YQ)}{\sinh(PQ)}S_{ABP}.$$

找到的第二个基本的线性不变量是所谓的 Pythagoras 差,其地位和作用相当于欧氏几何中的勾股差. 双曲平面上四角形与三角形的 Pythagoras 差分别定义为

$$P_{ABCD}=\cosh(AD)\cdot\cosh(BC)-\cosh(AB)\cdot\cosh(CD)$$

$$P_{ABC}=\cosh(AC)-\cosh(AB)\cdot\cosh(BC)$$

两条直线 AC 和 BD 相互垂直的充分必要条件是 $P_{ABCD}=0$.

非欧几何的基于不变量计算的研究方向一直受到重视,参见文献[31],[32],但此前未能形成自动推理的完备算法和通用程序. 正是由于幅和 Pythagoras 差这样经过精心提炼的不变量的有效运用导致这样一个算法和程序的产生.

为了发现非欧几何新定理,可以从成千条已知的欧氏几何的"构造性"的定理中任取一条,如果该定理的叙述在非欧几何中也有意义的话. 然后用我们的计算机程序对它的"非欧版本"进行检验,无论其成立与否,我们都会知道新的东西. 在文献[30]中收集的非欧几何新定理,大部分都是以这种方式获取的.

例如,欧氏几何中有这样一条非常著名的定理:任何一个三角形的外接圆心、重心和垂心三点共线. 这条直线被叫做 Euler 线. 经我们的程序检验,该命题在非欧情形一般是不成立的. 而且获知,命题(非欧)成立的充分必要条件是,该三角形至少是等腰的.

对于欧氏几何中另外两个著名的定理,蝴蝶(Butterfly)定理和 Simson 定理,经程序检验,前者的非欧版本也是真的,并由机器给出了可读证明;后者的非欧版本一般是不成立的.

再举一个不太熟知的例:设 A,B,C,D 四点共圆,其圆心为 O. 又设直线 AB 与 CD 交于 E;AD 与 BC 交于 F;AC 与 BD 交于 G. 求证 O 点是三角形 EFG 的垂心. 作为一个欧氏几何定理其证明也是比较困难的,但机器肯定了它的非欧版本并给出了可读证明.

有些新发现的非欧几何定理在欧氏几何中找不到有意义的对应物. 如下面一例：非欧平面上四点 A,B,C,D 共圆(或超圆，或极限圆)的充分必要条件是

$$S_{ABC}-S_{BCD}+S_{CDA}-S_{DAB}=0,$$

这里的 S 表示三角形的幅.

从非欧氏几何可读证明自动生成的通用程序的完成到文献[30]中的 90 个定理被一一验证通过，这中间只经历很短的周期. 仅就数量而言，似乎在一夜之间，计算机告诉我们的非欧几何新定理比过去许多年的积累还要多. 当然，这只是一个比喻.

参考文献

[1]Wu Wen-tsun. On the decision problem and the mechanization of theorem in elementary geometry. Scientia Sinica，1979，21(2)：159 - 172.

[2]吴文俊. 几何定理机器证明的基本原理. 北京：科学出版社. 1984.

[3]Chou S C. Mechanical Geometry Theorem Proving. Dordrecht：D. Reidel Publishing Company，1988.

[4]Chou SC. Automated proving in geometries using the characteristic set and Gröbner basis method. In：Proc ISSAC-90. 1990：255 - 260.

[5]Kapur D. Using Gröbner bases to reason about geometry problems. J of Symbolic Computation，1985，2(3)：399 - 408.

[6] Zhang J Z，Yang L，Deng M K. The parallel numerical method of mechanical theorem proving. Theoretical Computer Science. 1990，74(2)：253 - 271.

[7]Yang L，Zhang J Z，Li C Z. A prover for parallel numerical verification of a class of constructive geometry theorems. In：Proc 1992 International Workshop on Mathematica Mechanization. Beijing，1992：244 - 250.

[8]张景中，杨路. 数值并行法与单点例证法原理概述. 数学的实践与认识，1989(1)，34 - 47.

[9] Gelernter H，et al. Empirical explorations of the geometry theorem proving machine. In：Proc West Joint Computer Conf. 1960：143 - 147.

[10]Gelernter H. Realization of a geometry-theorem proving machine. In：Computers and Thought. Feigenbaum E A & Feldman J(eds)，Mcgraw Hill. 134 - 152.

[11] Goldstein H. Elementary geometry theorem proving. MIT. Tech Rep：AI—LAB—memo—280，1973.

[12]Gilmore P C. An examination of the geometry theorem proving machine. Artificial Intelligence，1970，1(2)：171 - 187.

[13] Nevins A J. Plane geometry theorem proving using forward chaining. Artificial Intelligence，1975，6(1)：1 - 23.

[14]Anderson J R. Tuning of search of the problem space for geometry. In：Proc Int Joint Conf Artificial Intelligence，Vancouver，1981：165 - 170.

[15]Coelho H，Pereira L M. Automated reasoning in geometry with Prolog. J of Automated

Reasoning,2(3):329 - 390.
[16]Hardzikadic M,et al. An application of knowledge based technology in education: a geometry theorem prover. In:Proc SYMSAC'86. ACM Press,141 - 147.
[17]Anderson J R,et al. The geometry tutor. In:Proc Int Joint Conf Artificial Intelligence, Los Angeles:1985,1 - 7.
[18]张景中. 面积关系帮你解题. 上海:上海教育出版社,1982.
[19]张景中. 平面几何新路. 成都:四川教育出版社. 1992.
[20] Zhang J Z. Chou S C, Gao X S. Automated production of traditional proofs for theorems in Euclidean geometry, I. The Hilbert intersection point theorems. Dept of Computer Science,WSU,Tech Rep:TR - 92 - 3,1992.
[21] Chou S C, Gao X S, Zhang J Z. Automated production of traditional proofs for theorems in Euclidean geometry, II. The volume method. Dept of Computer Science, WSU,Tech Rep:IR - 92 - 5,1992.
[22] Chou S C, Gao X S, Zhang J Z. Automated production of traditional proofs for constructive geometry theorems. In: Proc of 8th IEEE Symp on Logic in Computer Science,1993,48 - 56.
[23] Chou S C, Gao X S, Zhang J Z. Machine Proofs in Geometry. Singapore: World Scientific,1994.
[24]Chou S C,Gao X S,Zhang J Z. Mechanical theorem proving by vector calculation. In: Proc ISSAC—93. Keiv. 1993:284 - 291.
[25]Stifter S. Geometry theorem proving in vector spaces by means of Gröbner bases. In: Proc ISSAC—93,Keiv,1993:301 - 310.
[26]李洪波. 几何定理机器证明的新探讨[博士论文]. 北京大学数学系,北京,1994.
[27] Gao X S. Transcendental functions and mechanical theorem proving in elementary geometries. *J of Automated Reasoning*,1990,6(3):403 - 417.
[28]Yang L. Computer-aided proving for new theorems of non-Euclidean geometry. Mathematics Research Section,Australian National University,Tech Rep:No. 4—1989,1989.
[29]Chou S C,Gao X S,Yang L,Zhang J Z. Automated production of readable of proofs for theorems in non-Euclidean geometries. Dept of Computer Science, Wichita State University,Kansas,Tech Rep:WSUCS—94—9,1994.
[30] Chou S C, Gao X S, Yang L, Zhang J Z. A collection of 90 mechanically solved geometry problems from non-Euclidean geometries. Dept of Computer Science,Wichita State University,Kansas,Tech Rep:WSUCS—94—10,1994.
[31]杨路,张景中. 双曲型空间紧致集的覆盖半径. 中国科学,1982,25(8):683 - 692.
[32]杨路,张景中. 非欧双曲几何的若干度量问题. 中国科技大学学报,1983,13(数学专集): 123 - 134.

Some Advances on Functional Equations

Zhang Jing-zhong　Yang Lu　Zhang Wei-nian

(*Centre for Mathematical*, *CICA*, *Academia Sinica*, *Chengdu*, *Sichuan*, 610041)

Abstract In this paper some new advances on functional equations in a single real variable are introduced and surveyed, including iterative roots, Schröder's equation, and iterated equations of polynomial type. The basic contents are listed as follows. Ⅰ. introduction: iteration and related problems; Ⅱ. iterative roots: existence; Ⅲ. iterative roots: uniqueness, differentiability & bifurcation; Ⅳ. iterated equations of polynomial type.

Key words iterative equation; iterative root; embedding flow; discrete dynamical system

1 Introduction: Iteration and Related Problems

What is iteration? Usually for a set X and a mapping $f: X \to X$ we denote $f^n(x) = f \cdot f^{n-1}(x)$ and $f^0(x) = x$. We refer to the composited mapping $f^n(x)$ as an n-order iteration of $f(x)$.

Iteration is a common phenomenon in natural sciences and even in human life. In experimental sciences we often analyze the regularity of a system from the observed initial state to the present state. Sometimes, the whole process of the transition from state to state is a result repeated actions of according to a certain rule. As long as the rule is discovered, we can determine and predict the evolution of this system. For examples, the population model

$$x_{n+1} = x_n(a - bx_n), \qquad 0 \leqslant x_n \leqslant a/b, \tag{1.1}$$

of insects, studied in ecology, is just an iteration of the function

$$f(x) = x(a - bx). \tag{1.2}$$

The penetration of X-ray can be regarded as the iteration of decay rate. We can also list a lot of other examples of iteration such as infiltration of liquid, transimition of heat, and other dynamical models. In mathematics and computer sciences, we more than often rewrite differential equations into iterations by numerical methods so as to design programmes for their computations, discuss the periodic orbits and even chaos of dynamical systems by analyzing the iteration of their Poincaré mappings in phase spaces, and treat cyclic

本文刊于《数学进展》,第 24 卷第 5 期,1995 年.

computations with iteration of the same variable. In daily life we often need to calculate the amount of loans with a rate of interest, namely a principal P placed at a rate of interest r of n years accumulates to an amount

$$A_n = P(1+nr), \qquad \text{if the loan with a simple interest,} \tag{1.3a}$$

or

$$A_n = P(1+r)^n, \qquad \text{if with interest compounded annually.} \tag{1.3b}$$

Obviously A_n is an n-order iteration of $a(x)=x+rP$ or $a(x)=x(1+r)$ respectively.

Consider an iteration of a function $f(x)$, defined on X. It is easy to see that

(1) $f^0 = I$, the identity operator;

(2) $f^m \cdot f^n = f^{m+n}$.

Obviously, $\{f^n : n \in \mathbf{Z}_+\}$ is a semi-group, and is called a discrete (semi-) dynamical system as usual. There are three basic problems there into be worthily discussing:

(1) how to compute or estimate the n-order iteration $f^n(x)$, and what about the long-term behaviours of its orbit $\mathrm{Orb}_f(x) = \{f^n(x) : n=0,1,2,\cdots\}$, for example, $\omega_f(x)$;

(2) how to determine the n-order iterative roots $f(x)$ of a given function $F(x)$, i. e.

$$f^n(x) = F(x), \ \forall x \in X; \tag{1.4}$$

(3) how to embed this discrete dynamical system into a flow, and even find a vector field $V(x)$ on X such that $f(t,x)$ satisfies

$$(df/dt)(t,x) = V(x), t \in \mathbf{R}, x \in X, \tag{1.5}$$

and $f(n,x) = f^n(x)$ for $n \in \mathbf{Z}_+$.

These problems were researched early in 1871 by E. Schröder[1] and N. H. Abel[2]. It has been said in [3] that C. Babbage's work involved iteration of functions earlier. It has been pointed out that the above mentioned three basic problem can be discussed with solving the so-called Schröder's functional equation

$$h(f(x)) = ch(x), \tag{1.6}$$

when $f(x)$ has a fixed point, or Abel's functional equation

$$h(f(x)) = h(x) + c, \tag{1.7}$$

where c is a constant. Hence functional equation, including the problems of iterative roots, plays an important role in the theory of iteration.

In 1968 M. Kuczma, a Polish mathematician, published a nice work[4] on functional equations in a single variable. Around that time there were also many relevant papers published such as [5, 10]. Recently, some new advances has been made on existence, uniqueness, stability and differentiability of iterative roots as well as solutions of iterated functional equations. These new advances concerning iterated roots are to be developed and surveyed in our next sections 2 and 3. Recent research works on the iterated equation of polynomial type will be reviewed in the final section 4.

2 Iterative Roots: Existence

2.1 Classical Results: Strictly Increasing Case

Let $I=[a,b]$ and

$$CI(I,I)=\{f(x)\in C^0(I,I): f(a)=a, f(b)=b, \text{and } f(x) \text{ is strictly increasing}\}. \tag{2.1}$$

Theorem 2.1 Suppose $F(x)$ and $f(x)$ belong to $CI(I,I)$, and n is a positive integer. If

$$f^n(x)=F(x), \qquad \forall x\in I, \tag{2.2}$$

then there is a function $h(x)$ in $CI(I,I)$ such that

$$h(f(x))=F(h(x)), \qquad \forall x\in I, \tag{2.3}$$

i. e. an iterative root in $CI(I,I)$ must be conjugate to its originally given function.

Proof If there is a fixed point c of $f(x)$ in (a,b), we can discuss separately in $[a,c]$ and $[c,b]$ since $f(x)$ is increasing. So we suppose, without loss of generality, that there is no fixed point of $f(x)$ in (a,b), that is, either $f(x)>x$ or $f(x)<x$ in (a,b). Explicitly, it suffices to prove the theorem under the assumption that $f(x)>x$. The opposite can be discussed in the same way with $f_1(x)=(b+a)-f(b+a-x)$ in place of $f(x)$.

First of all, we choose arbitrarily a point x_0 in I and let

$$x_i=f^i(x_0), \qquad i=0,\pm1,\pm2,\cdots \tag{2.4a}$$

$$y_0=f(x_0), y_i=f^{ni}(y_0), \qquad i=0,\pm1,\pm2,\cdots \tag{2.4b}$$

Since $f(x)>x$ in (a,b) and $f(x)$ is increasing, we see that both $\{x_i\}$ and $\{y_i\}$ are increasing and approach a as $i\to-\infty$ or approach b as $i\to+\infty$. Define

$$h_0(x)=y_0+\{(y_1-y_0)/(x_1-x_0)\}(x-x_0), \qquad \forall x\in[x_0,x_1]. \tag{2.5}$$

Clearly $h_0(x)$ is a continuous and increasing function such that

$$h_0(x_0)=y_0 \qquad \text{and} \qquad h_0(x_1)=y_1. \tag{2.6}$$

Naturally, $h_0(f(x_0))=h_0(x_1)=y_1=f^n(y_0)=f^n(h_0(x_0))=F(h_0(x_0))$. Next, we define successively for $i=1,2,\cdots$ that

$$h_i(x)=f^n[h_{i-1}(f^{-1}(x))], \qquad x\in[x_i,x_{i+1}], \tag{2.7a}$$

$$h_{-i}(x)=(f^n)^{-1}[h_{-i+1}(f(x))], \qquad x\in[x_{-i},x_{-i+1}]. \tag{2.7b}$$

By induction we can prove that both $h_i(x)$ and $h_{-i}(x)$ are continuous and increasing functions such that

$$h_i(x_i)=y_i, \qquad \text{and} \qquad h_i(x_{i+1})=y_{i+1}; \tag{2.8a}$$

$$h_{-i}(x_{-i})=y_{-i}, \qquad \text{and} \qquad h_{-i}(x_{-i+1})=y_{-i+1}. \tag{2.8b}$$

Finally, we define

$$h(x)=\begin{cases} x, \text{for } x=a \text{ or } b, & (2.9a) \\ h_i(x), \text{for } x \text{ in } [x_i,x_{i+1}], \quad i=0,\pm1,\pm2,\cdots & (2.9b) \end{cases}$$

It is easy to verify that $h(x)$ belongs to $CI(I,I)$ satisfying

$$h(f(x))=F(h(x)), \qquad i.e. \qquad f(x)=h^{-1}[F(h(x))], \tag{2.10}$$

for x in I. The proof of the theorem is now complete.

The so-called **successively defining method**, as shown in the above proof, is very useful to construct iterative roots.

Theorem 2.2(Hardy-Böedewadt) For any $F(x)\in CI(I,I)$ and any integer $n>0$, there is an n-order interative root $f(x)$ of $F(x)$ in $CI(I,I)$.

Proof By theorem 2.1 there is a function $h(x)$ in $CI(I,I)$ such that

$$F(x)=h^{-1}[F^n(h(x))], \qquad \forall x \text{ in } I. \tag{2.11}$$

We let $f(x)=h^{-1}[F(h(x))]$. Obviously $f(a)=a, f(b)=b$, and $f(x)$ is also a continous and increasing function on I such that

$$f^n(x)=h^{-1}[F^n(h(x))]=F(x)(\text{from}(2.11)), \tag{2.12}$$

that is, $f(x)$ satisfies (2.2). The proof is complete.

The above proof, found in[10—13], can be also given directly by defining successively $f^i(x)$ on subinterval $[x,x_{i+1}]$, $i=0,\pm1,\pm2,\cdots$ with $\bigcup_{i=-\infty}^{\infty}[x_i,x_{i+1}]=I$. It is found that the reason why a continuous iterative root misses its uniqueness is just that there are many choices of initial function of iteration.

Theorem 2.3 $F(x)\in C^0(I,I)$ with only one fixed point x_0. If in a certain small neighbourhood $(x_0-\delta,x_0+\delta)$

$$F(x)>x_0 \qquad \text{for } x \text{ in } (x_0-\delta,x_0) \tag{2.13a}$$

and

$$F(x)\leqslant x_0 \qquad \text{for } x \text{ in } (x_0,x_0+\delta) \tag{2.13b}$$

then for any even positive integer $2m$ there is no $2m$-order iterative root of $F(x)$ in $C^0(I,I)$. Furthermore, any strictly decreasing function has no continuous $2m$-order iterative root.

Theorem 2.4 Suppose a continuous function $F(x)$ maps I onto I with an extreme point c in I such that $F(x)$ is strictly monotone on both $[a,c]$ and $[c,b]$. Then for any integer $n\geqslant2$ there is no n-order continuous iterative root of $F(x)$.

Theorems 2.3 and 2.4, whose proofs can be found in[12], indicate that the problems of iterative roots become complicated without the strictly increasing property.

2.2 Strictly Piecewise Monotone Case

Definition 2.1 $x_0\in(a,b)$ is referred to be a monotone point of $F(x): I\to I$, if $F(x)$ is strictly monotone in a neighbourhood of x_0. Otherwise, x_0 is referred to be a **non-monotone point.**

Definition 2.2 $F(x)\in C^0(I,I)$ is referred to be a **strictly piecewise monotone continuous function**, of **S-function** simply, if $F(x)$ has only finite extreme points in I and is strictly monotone between every two consecutive extreme points. Let $S(I,I)$ consist of such functions.

Apparently $F(x)\in S(I,I)$ if and only if $F(x)\in C^0(I,I)$ and has only finite non-monotone points; $F^n(x)\in S(I,I)$ if and only if $F(x)\in S(I,I)$.

Let $N(F)$ denote the number of extreme points of S-function F in I. Obviously

$$0=N(F^0)\leqslant N(F^1)\leqslant N(F^2)\leqslant\cdots\leqslant N(F^m)\leqslant\cdots \tag{2.14}$$

Especially $N(F)=0$ when $F(x)$ is strictly monotone on I. Let $H(F)$ denote the minimum of positive integer m such that $N(F^m)=N(F^{m+1})$. Note that $H(F)=\infty$ means the sequence $\{N(F^m)\}$ is strictly increasing, and when $H(F)=m<\infty$ we have $N(F^m)=N(F^{m+k})$ for any natural number k. Moreover, for any natural number n

$$H(F^n)=[m/n]+\operatorname{sgn}(m/n-[m/n]). \tag{2.15}$$

Theorem 2.5 (Zhang & Yang[13]) If $F(x)\in S(I,I)$, and $H(F)>1$, then for any $n>N(f)$, $F(x)$ has no continuous n-order iterative root.

Proof Assume $f(x)\in C^0(I,I)$ is an n-order iterative root of $F(x)$. Naturally $f(x)\in S(I,I)$. Since $H(F)>1$, $N(F^2)>N(f)$, i. e. $N(f^{2n})>N(f^n)$. Thus $H(f)>n$ and $0=N(f^0)<N(f)<N(f^2)<\cdots<N(f^n)$. This implies $N(f^n)=N(F)\geqslant n$, which conflicts with the assumption at the beginning.

It is easy to see that theorem 2. 4 is a corollary of theorem 2. 5. In fact, the function $F(x)$, satisfying the conditions in theorem 2. 4, is called a **unimodal mapping** on I, i. e. $N(F)=1$. Since $F(x)$ maps I onto I we assume, without loss of generality, that $F(c)=b$, the maximum, and $F(a)=a$, the minimum. By the continuity of $F(x)$, $c<F(x_0)<b$ for $\forall x_0\in B(c,\delta)$, a small neighbourhood of c. Since $F(x)$ must be strictly decreasing on $[c,d]$ we see easily that $F^2(x_0)=F(F(x_0))>F(b)=F^2(c)$, i. e. $F^2(c)$ is a minimal value. However, $F^2(a)=a$ is still the minimum, so there must be another extreme point c' in (a,c). This implies $N(F^2)\geqslant 2>N(F)$, i. e. $H(F)>1$. By theorem 2. 5 we obtain easily the conclusion of theorem 2. 4.

Definition 2.3 Suppose $F(x)\in S(I,I)$ and $H(F)\leqslant 1$. Let $A=\min\{F(x):x\in I\}$ and $B=\max\{F(x):x\in I\}$. We refer to $[a',b']$ as a **characteristic interval** of $F(x)$, if a' and b' are two consecutive extreme points of $F(x)$ in I and $[A,B]\subset[a',b']\subset[a,b]$.

Obviously, $H(F)\leqslant 1$, i. e. $N(F)=N(F(F(x)))$, if and only if $F(x)$ is strictly monotone on $[A,B]$. In fact, if $F(x)$ has an extreme point $c\in[A,B]$, by the continuity there must be a monotone point $x_0\in(a,b)$ such that $F(x_0)=c$. Thus x_0 is another extreme point of $F^2(x)$ and $N(F(F(x)))>N(F)$. This makes a contradiction. Therefore, the above definition is reasonable. Naturally $F(x)$ is also strictly monotone on its characteristic interval $[a',b']$. In the case that $H(F)\leqslant 1$, a series of results, stated as follows, has been proved by Zhang J. and Yang L. in[13].

Theorem 2.6 Given $F(x)\in S(I,I)$, and $H(F)\leqslant 1$. Suppose

(a) $F(x)$ is increasing on its characteristic interval $[a',b']$, and

(b) $F(x)$ cannot reach a' and b' on I if $F(a')\neq a'$ and $F(b')\neq b'$.

Then for any integer $n\geqslant 2$, $F(x)$ has a continuous n-order iterative root. Moreover, these conditions are necessary for $n>N(F)+1$.

Theorem 2.7 Suppose $F(x)\in C^0(I,I)$ is strictly increasing with a fixed point $x_0\in(a,b)$, and either $F(a)=a$, $F(b)=b$ or $a<F(x)<b$ on I. Let E' consist of fixed points of $F(x)$ in $[a,x_0]$ and E'' consist of fixed points of $F(x)$ in $[x_0,b]$. If

(a) there exists a reversing one-to-one correspondence $D: E'\to E''$, i. e. $D(e)=D(e')$ such that $e_1'>e_2'$, as $e_1<e_2$, and

(b) $F(x)-x$ is negative (or positive) on $(e_2'>e_1')$ when it is positive (or negative) on (e_1,e_2), then for any even number $2m\geqslant 2$, $F(x)$ has a decreasing $2m$-order iterative root in $C^0(I,I)$.

Theorem 2.8 Suppose $F(x)\in C^0(I,I)$ is strictly decreasing, and either $F(a)=b$, $F(b)=a$ or $a<F(x)<b$ on I. Then for any odd number $2m+1\geqslant 3$, $F(x)$ has a decreasing $2m+1$-order iterative root in $C^0(I,I)$.

Theorem 2.9 $F(x)\in S(I,I)$ and $H(F)\leqslant 1$. If

(a) $F(x)$ is decreasing on its characteristic interval $[a',b']$, and

(b) either $F(a')=b'$, $F(b')=a'$, or $a'<F(x)<b'$ on I,

then for any odd number $n>0$, $F(x)$ has an n-order iterative root in $C^0(I,I)$.

Similar results for mappings on S^1 we also given by He L. and Niu D. [17].

Theorem 2.10 Assume $F: S^1\to S^1$ continuous and of deg $(F)=0$. Then for any integer $n\geqslant 2$, $F(x)$ has a continuous n-order iterative root, if and only if F has a characteristic interval $[a',b']\subset S^1$ such that F is strictly increasing on $[a',b']$.

Theorem 2.11 Assume $F: S^1\to S^1$ continuous and of deg $(F)=0$. If F has a characteristic interval $[a',b']\subset S^1$ on which F is strictly decreasing, then

(a) F has not an n-order iterative root for even n;

(b) F has an n-order iterative root for odd n, provided that $F(a')=b'$, $F(b')=a'$ or that $F(a')<b'$ and $F(b')>a'$.

Some open questions:

1. Does $F(x)$ have n-order iterative roots when $N(F)<N(F^2)$ and $n\leqslant N(F)$?

2. Does $F(x)$ have n-order iterative roots when $N(F)=N(F^2)$, $n\leqslant N(F)+1$ and $F(x)$ can reach a', b' on I but not on $[a',b']$?

2.3 Method of Embedding Flow

The method of embedding flow is an effective way to study iterative roots. If the given mapping $F(x)$ is proved to be embedded into a flow (or semiflow) $\{F^t(x)\}$, its iterative roots can be defined easily by $f(x)=F^{\lambda/n}(x)$ where λ is the index of embedding of F. Recently there were some results about embedding flow in [15−16, 18−21]. It is worthy to mention that an equivalence is given in [17]by theorem 2.10 and some results in [18].

Theorem 2.12 Assume $F: S^1 \to S^1$ continuous and of deg $(F)=0$. F has an n-order iterative root for any integer $n \geqslant 2$ if and only if F can be embedded into a quasi-semiflow.

In [20-21]Zhang M. and Li W. discussed TDF, the abbreviation of **Time Difference Function** which is an important invariant of smooth conjugacy, to obtain some new results about iterative roots. Suppose f, an orientation preserving C^r diffeomorphism, has no fixed points on $I=[a,b]$ except the endpoints and has some regularity at its fixed points. Let $I^\alpha=[a,b)$, $I^\omega=(a,b]$, and $I^0=(a,b)$. In[20]the authors proved that f has a unique continuous embedding flow $\{f_\sigma^t\}(t \in \mathbf{R})$ on I such that $\{f_\sigma^t\}$ is C^r on I^σ, $\sigma=\alpha,\omega$. Let V^σ be the vector field of $\{f_\alpha^t\}$, i.e.,

$$V^\sigma(x)=\frac{\partial}{\partial t} f_\sigma^t(x)|_{t=0}, \qquad x \in I, \sigma=\alpha,\omega. \tag{2.16}$$

For any $c \in I^0$ we define

$$T_c^\sigma(x)=\int_c^x \frac{ds}{V^\sigma(s)}, \qquad x \in I^0, \qquad \sigma=\alpha,\omega. \tag{2.17}$$

Then the embedding flows is given by

$$f_\sigma^t(x)=(T_c^\sigma)^{-1}(T_c^\sigma(x)+t), \qquad x \in I^0, t \in \mathbf{R}, \sigma=\alpha,\omega. \tag{2.18}$$

Let $T_c(t)=T_c^\alpha((T_c^w)^{-1}(t))$, $\hat{T}_c(t)=T_c^w((T_c^\alpha)^{-1}(t))$, and

$$\psi_c(t)=T_c(t)-t-\int_0^1 (T_c(t)-t)dt, \qquad t \in \mathbf{R}, \tag{2.19}$$

$$\hat{\psi}_c(t)=\hat{T}_c(t)-t-\int_0^1 (\hat{T}_c(t)-t)dt, \qquad t \in \mathbf{R}. \tag{2.20}$$

We call $\psi^f:=\psi_{c_0}$ and $\hat{\psi}^f:=\hat{\psi}_{c_0}$, where $c_0=(a+b)/2$, as the first and second time difference function of f respectively. Using TDF we can obtain the structure of smooth centralizer $Z(f)=\{h \in \mathrm{Diff}^r(M): hf=fh\}$ of C^r Morse-Smale diffeomorphism f on one-dimensional compact manifold M.

Theorem 2.13 (Zhang M.[21]) Let $f \in \mathrm{Diff}^r(S^1)$ $(2 \leqslant r \leqslant \omega)$ be a Morse-Smale diffeomorphism. Then the centralizer $Z(f)$ is a solvable subgroup of $\mathrm{Diff}^r(S^1)$. In fact the commutator $[Z(f),Z(f)]=\{h\bar{h}h^{-1}\bar{h}^{-1}: h,\bar{h} \in Z(f)\}$ is always a commutative subgroup of $\mathrm{Diff}^r(S^1)$.

Furthermore, in discussing the rotation number $\rho(f)=M/N$, where $0 \leqslant M < N$ and $(M,N)=1$, the rotation degree d_f of f and $D_f=\rho(f)d_f$, in[21]the author also gave the expressions of isomorphic groups for $Z(f)$, by which iterative roots can be investigated for they should be in $Z(f)$.

Theorem 2.14 Let f be a Morse-Smale orientation preserving C^r diffeomorphism on S^1, $r \geqslant 2$. Then f has m-order iterative roots which are orientation preserving C^r diffeomorpic on S^1 if and only if $m|m_f$ and $(m,d_f)|D_f$, where m_f is the prime degree of f.

The sufficient and necessary condition for orientation reversing iterative roots is also given in[21].

Theorem 2.15 Let f be a Morse-Smale orientation reversing C^r diffeomorphism on S^1, $r\geqslant 2$, and $m\neq 0$ be an integer. Then f has no orientation preserving m-order iterative roots. Moreover f has (an unique) orientation reversing m-order iterative roots if and only if m is odd and $m|m_{f*}$.

3 Iterative Roots: Uniqueness, Differentiability and Bifurcation

Theorem 3.1(Böedewadt) Let $I=[a,b]$. Suppose $F(x)\in C^\infty(I,I)$, $F'(x)>0$, and $F(x)>x$. Then for any integer $n\geqslant 2$, $F(x)$ has an n-order iterative root in $C^\infty(I,I)$.

Recently C^k-iterative roots and their uniqueness were discussed further by Zhang J. et al[14], Jiang X.[22], and Zhang W.[23,24].

3.1 Differentiability at One Fixed Point

Definition 3.1 $G(x,t)$ is said to be a **normal iterative family** (abbreviated by **NIF**) on $I=[a,b]$, if for $t\in\mathbf{R}_+$,

(1) $G(x,t)$ is continuous in x and t, such that $G(x,0)=x$, $G(G(x,t_1),t_2)=G(x,t_1+t_2)$,

(2) $G(x_1,t)<G(x_2,t)$ as $x_1<x_2$, and

(3) $a<x<G(x,t)<b$ for $x\in(a,b)$.

For example, $G_0(x,t)=x+t$, $x\in\mathbf{R}$; $G_1(x,t)=a^t x$, $x\in\mathbf{R}_-$, where $0<a<1$; $G_2(x,t)=x^{2t}$, $1<x<+\infty$; and $G_3(x,t)=x/\sqrt{1+tx^2}$, $-1<x<0$. It is easy to prove that for an NIF there is a continuous function $t=T(x,y)$, increasing in y and decreasing in x such that $y=G(x,T(x,y))$.

Definition 3.2 A continuous function $F(x):(a,b)\to(a,b)$ is said to be G-**metrable** for an NIF $G(x,t)$, if

(1) $F(x)\geqslant x$ in a neighbourhood of b, and

(2) $T(x,F(x))$ converges as $x\to b$ to a positive number α_F which is called a G-**index** of $F(x)$.

Furthermore, $F(x)$ is **below G-metrable** if $F_1(x)=(b+a)-F(b+a-x)$ is G-metrable; $F(x)$ is **reversing G-metrable** (or reversing **below G-metrable**) if $F^{-1}(x)$ is G-metrable (or below G-metrable).

With these definitions Zhang J. and Yang L. proved following results.

Theorem 3.2 Let $F(x):(a,b)\to(a,b)$ be continuous and increasing, $F(x)>x$ and G-metrable for an NIF $G(x,t)$. Then the following propositions are equivalent to one another:

(a) There is a sequence $\{m_i\}\subset\mathbf{Z}_+$, $m_1<m_2<\cdots<m_i<\cdots$, such that for any natural number i, $F(x)$ has a continuous m_i-order iterative root $f_i(x)$, which is G-metrable and tends to x as $i \to +\infty$.

(b) For any natural number m, $F(x)$ has a unique continuous G-metrable m-order iterative root $f(x)$.

(c) $P_n(x,y) := T(F^n(x), F^n(y))$ converges to a continuous and increasing function $P(x,y)$ as $n \to +\infty$, in the region $\{(x,y): a<x<b, x \leqslant y<b\}$.

(d) If $g(x):(a,b) \to (a,b)$ is G-metrable, then $H_n(x) := F^{-n} \circ g \circ F^n(x)$ approaches a G-metrable $H(x)$ and $\alpha_H = \alpha_g$, satisfying $P(x, H(x)) = \alpha_H$.

(e) For any constant $c \in (0, +\infty)$ there is a unique function $g(x,c)$, monotone in $x \in (a,b)$ and G-metrable with $\alpha_{g(x,c)} = c$, such that $g(F(x),c) = F(g(x,c))$.

Similarly, the corresponding results can be also given for the below G-metrable, the reversing G-metrable, or the reversing below G-metrable functions.

Corollary 3.1 J is an interval in **R**. Suppose

(1) $F \in C^1(J,J)$ and $F'(x)>0$, $\forall x \in J$;

(2) $F(x)$ has a unique fixed point x_0 in J such that $F'(x_0) \neq 1$;

(3) $F''(x_0)$ is defined.

Then for any integer $k>1$, $F(x)$ has a unique strictly increasing k-order iterative root $f(x)$ in $C^1(J,J)$. Moreover, $F(x)$ can be embedded uniquely into a C^1 semi-flow $f(t,x)$ on J, or even C^1 flow $f(t,x)$ on J only if $F(J)=J$.

The proof of corollary 3.1 was given in [12]. Note that $F'(x_0) \neq 1$ means actually the hyperbolicity of the fixed point x_0, and is equivalent to that $F(x)$ is G-metrable for NIF $G(x,t) = (F'(0))^t x$.

Remark 3.1 Suppose $I=[0,1]$and $F(x) \in C^1(I,I)$ be such that

(1′) $F'(t)>0$, $\forall x \in I$;

(2′) $F(x)$ has only two fixed points 0 and 1, and $F'(0) \neq 1$ and $F' \neq 1$;

(3′) $F''(0)$ and $F''(1)$ are defined.

Then $F(x)$ also satisfies conditions stated in corollary 3.1 on $J=[0,1)$ (or $J=(0,1]$). Thus for any integer $k>1$ there is unique k-order iterative root of $F(x)$ in $C^1[0,1)$ or $C^1(0,1]$. Particularly, on a small neighbourhood $[0,\delta)$ (or $(1-\delta,1]$) there is a unique continuously defferentiable k-order iterative root of $F(x)$. In fact,

(a) if $F'(0<1)$, there is a small neighbourhood $[0,\delta)$ such that F maps $[0,\delta)$ into itself, and so $F \in C^1(J,J)$, where $J=[0,\delta)$ satisfies the above mentioned conditions;

(b) if $F'(0)>1$, then

$$(F^{-1})'(0) = 1/F'(0) < 1, \tag{3.1}$$

and

$$(F^{-1})''(0) = -F''(0)/F'(0)^3 \tag{3.2}$$

is defined, so F^{-1} maps a small neighbourhood $[0,\delta)$ into itself, and satisfies the above mentioned conditions. Thus F^{-1} has also a unique continuously differentiable k-order iterative root $g(x)$ on $J=[0,\delta)$. Let $f=g^{-1}$. Obviously the strict monotonicity of $g(x)$ means that $f\in C^1(J,J)$ is strictly increasing and

$$(f^k)^{-1}=(\underbrace{g^{-1}\cdot g^{-1}\cdot\cdots\cdot g^{-1}}_{k})^{-1}=\underbrace{g\cdot g\cdot\cdots\cdot g}_{k}=F^{-1}, \tag{3.3}$$

i. e. $f^k(x)=F(x)$.

Another corollary is devoted to the case that the unique fixed point is not hyperbolic.

Corollary 3.2 Suppose $F(x):[-1,0]\to[-1,0]$ is continuous and increasing, and $F(x)\geqslant x$, where the equality sign is valid if and only if $x=0$. If $F'(0)=1$ and at $x=0_-$, $F(x)$ has an expansion

$$F(x)=x-cx^{k+1}+o(x^{k+1}),\qquad k\in\mathbf{Z}_+,\qquad (-1)^k c>0, \tag{3.4}$$

then the following statements are equivalent to one another:

(a) there is a sequence $\{m_i\}\subset\mathbf{Z}_+$, $m_1<m_2<\cdots<m_i<\cdots$, such that for any natural number i, $F(x)$ has a continuous m_i-order iterative root $f_i(x):[-1,0]\to[-1,0]$, satisfying that $f_i'(0)=1$, $f_i(x)$ has up to $k+1$-order generalized derivatives at $x=0_-$, and $f_i(x)$ tends to x as $i\to+\infty$;

(b) for any natural number m, $F(x)$ has a unique continuous m-order iterative root $f(x)$, which has up to $k+1$-order generalized derivatives;

(c) $P_n(x,y):=(F^n(y))^{-k}-(F^n(x))^{-k}$ converges to a continuous and monotone function $P(x,y)$ as $n\to+\infty$, in the region $[(x,y):-1<x<0, x\leqslant y<0]$.

Considering high order differentiability we have the following theorem.

Theorem 3.3[22] Let $I=[a,b]$. Suppose $F\in C^r(I,I)$ has a unique fixed point $x_0\in I$, $F'(x_0)=c$, and $F'(x)\neq 0$ on I, where $r\geqslant 2$ and $0<c<1$. Then for any integer $n>0$, $F(x)$ has a unique increasing n-order iterative root $f(x)$ in $C^r(I,I)$.

To prove this theorem, by the idea stated in section I it suffices to discuss the proof of the C^r solutions of the corresponding Schröder's equation.

Lemma 3.1 Suppose $r\geqslant 2$. $F(x)\in C^r(I,I)$ has a unique fixed point $0\in I$. $F'(0)=c$ where $0<|c|<1$. Then Schröder's equations

$$h\circ F(x)=ch(x) \tag{3.5}$$

has a solution $h(x)$ in $C^r(I,I)$ with $h'(0)\neq 0$.

Sketch of the Proof Let $S(x)=F(x)-cx$. We claim

$$x+\sum_{n=0}^{\infty}(1/c^{n+1})S(F^n(x))\to h(x)\ \text{, in } C^r(I,I), \tag{3.6}$$

uniformly with respect to $x\in I$, as $n\to+\infty$. Obviously it is convergent for $x=0$. To prove its series of termwise differentiation convergent uniformly, we let $x_n=F^n(x)$. By induction we have

$$(S(F^n(x)))'=S'(x_n)F'(x_{n-1})F'(x_{n-2})\cdots F'(x), \tag{3.7a}$$

and

$$(S(F^n(x)))^{(k)} = \sum_t S^{(i_0)}(x_n)F^{(i_1)}(x_{j_1})F^{(i_2)}(x_{j_2})\cdots F^{(i_{s_t})}(x_{j_{s_t}}) \tag{3.7b}$$

where

(a) the sum formula consists of at most $k!\ (n+1)^k$ terms;

(b) any term consists of at most $k(n+1)$ factors but at least $i_0 \cdot n+1$ factors, and

(c) in any term, the number of included factors $F^{(i)}(x_j)$ with $i \geqslant 2$ is not greater than k, and the number of those with $j<n$ is not greater than $k(n-1)$.

Since F and S belong to $C^r(I,I)$, $F(0)=S(0)=0$, $0<|F'(0)|=|c|<1$ and $S'(0)=0$ there is a small constant $d>0$ such that

$$|F'(x)| \leqslant |c|^{2/3}, \qquad |F^m(x)| \leqslant |c|^{2m/3}|x|,$$

and

$$|F(x)| \leqslant |c|^{2/3}|x|, \qquad |S'(x)| \leqslant M|x|, \text{for } x \text{ in } B(0,d). \tag{3.8}$$

Clearly $x=0$ is a unique and stable fixed point, so

$$F^n(x) \to 0, \text{as } n \to +\infty \text{ uniformly to } x \in I. \tag{3.9}$$

Thus there is an integer $N>0$ such that for any $x \in I$ and $n \geqslant N$

$$|F'(F^n(x))| \leqslant |c|^{2/3}, \qquad |F^n(x)| \leqslant |c|^{2(n-N)/3}d,$$

and

$$|S'(F^n(x))| \leqslant M|F^n(x)| \leqslant M|c|^{2(n-N)/3}d. \tag{3.10}$$

By (3.10) we can estimate (3.7) and verify the convergence of series (3.6).

So far in the above, only results on iterative roots, which is defferentiable at one fixed point, are given. However, the existence of iterative roots, which is defferentiable at two fixed points, is just a key step for us to explore the global differentiability of iterative roots. In [24] an interesting generic property that there is no differentiable iterative root on the closed interval $I=[0,1]$ was given for a kind of strictly increasing C^1-smooth functions with two hyperbolic fixed points 0 and 1.

3.2 Differentiability at Two Fixed Points

Let $I=[0,1]$. Consider a subspace of $C^1(I,I)$,

$$A^1(I)=\{F(x)\in C^1(I,I): F'(x)>0 \text{ on } I, F(x)\neq x \text{ on } (0,1), \text{and}$$
$$F(y)=y, F'(y)\neq 1, F''(y) \text{ exists for } y=0 \text{ or } 1\}, \tag{3.11}$$

with the topology induced by the C^1-norm $\|\cdot\|_1$, and a closed subspace of $C^1(I,I)$

$$E^1(I;m,M)=\{f(x)\in C^1(I,I): f(0)=0, f(1)=1, \text{and}$$
$$m \leqslant f'(x) \leqslant M, |f'(x_1)-f'(x_2)| \leqslant M \quad \text{for } x,x_1,x_2,\in I\}, \tag{3.12}$$

with the topology induced by the C^1-norm $\|\cdot\|_1$, where m and M are certain positive constants.

Theorem 3.4[24] Given integer $k \geqslant 2$. It is generic for C^1-smooth function in $A^1(I)$ that there is none of its differentiable k-order iterative root in $E^1(I;m,M)$.

Remark 3.2 A proposition P is said to be generic in a space X if the proposition P is valid in a countable intersection of dense open sets in X.

Sketch of the Proof Let A^1 and E^1 denote simply $A^1(I)$ and $E^1(I;m,M)$ respectively, and let

$$W=\{F(x)\in A^1: F(x) \text{ has none of } k\text{-order iterative root in } E^1\}.$$

It suffices to show that W is a dense open subset in A^1.

Step 1 Prove that W is an open subset, i. e. for any given $F(x)$ in W there is a sufficiently small neighbourhood $V(F(x))\subset W$. By reduction to absurdity, we assume that in each neighbourhood $B=B(F(x),d/2^i)$, $i=1,2,\cdots,d>0$, there exists a function $G_i(x)$ belonging to $A^1\backslash W$. This means that $G_i(x)$ has a k-order iterative root $g_i(x)$ in E^1, $i=1,2,\cdots$ Clearly $G_i(x)\to F(x)$ in $\|\cdot\|_1$ as $i\to+\infty$. On the other hand, it is not difficult to see that the sequence $\{g_i\}$ is uniformly bounded and equicontinuous. By Ascoli-Arzela's principle we see that there is a subsequence of $\{g_i\}$, denoted simply by itself, convergent in $\|\cdot\|_1$ to $f(x)\in E^1$, a closed space, as $i\to+\infty$. It is also evident by induction for iterations that

$$\lim_{i\to+\infty} g_i^k(x)=f^k(x),\qquad \forall x\in I, \tag{3.13}$$

where k is an iterative order. Therefore

$$f^k(x)=\lim_{i\to+\infty} g_i^k(x)=\lim_{i\to+\infty} G_i(x)=F(x),\qquad \forall x\in I, \tag{3.14}$$

that is, $F(x)\notin W$. This contradiction implies that W is an open subset.

Step 2 Prove that W is dense in A^1, i. e. for each $F(x)$ in A^1 and any given $d>0$ there must be function G belonging to W in the neighbourhood $B(F(x),d)$. Without loss of generality, we let $F(x)\in A^1\backslash W$, that is, $F(x)$ has a k-order iterative root $f(x)$ in E^1.

We assume $F(x)>x$ on $(0,1)$, and the opposite can be discussed in the same way with $F_1(x)=1-F(1-x)$ in place of $F(x)$. In this case the iterative root $f(x)$ of $F(x)$ in E^1 must satisfy $f(x)>x$ on $(0,1)$ since $f(x)$ is strictly increasing. Selecting a point $c\in(0,1)$ arbitrarily, we construct a small perturbation of $F(x)$

$$G(x)\begin{cases}>F(x), & x\in B(c,\delta),\\ =F(x), & x\notin B(c,\delta),\end{cases} \tag{3.15}$$

where $\delta>0$ sufficiently small, such that

$$\delta<\min\{(f(c)-c)/2,(c-f^{-1}(c))/2\} \tag{3.16}$$

and $G\in A^1$,

$$\|G-F\|_1<d. \tag{3.17}$$

In order to prove $G\in W$, by reduction to absurdity we assume $G\in A^1\backslash W$, i. e. G has also a k-order iterative root $g(x)$ in E^1. Since $G(x)=F(x)$ in small neighbourhoods of both 0 and 1, we can affirm by the uniqueness in remark 3. 1 that

$$f(x)=g(x) \text{ in small neighbourhoods of both 0 and 1.} \tag{3.18}$$

On the other hand, we can construct g uniquely on I as follows. Consider a sufficiently small neighbourhood $B(0,\varepsilon)$ and $x_0\in B(0,\varepsilon)$ such that $F(x_0)\in B(0,\varepsilon)$. Let

$$x_i = f^i(x_0), \qquad i=0,1,\cdots,k; \tag{3.19a}$$

$$x_{k+i} = F(x_i), \qquad x_{-i} = F^{-1}(x_{k-i}), \qquad i=1,2,\cdots \tag{3.19b}$$

Clearly $x_0 < x_1 < \cdots < x_{k-1} < x_k = f^k(x_0) = F(x_0)$, since f is strictly increasing. Define successively

$$g_i(x) = \begin{cases} f(x), & x \in [x_i, x_{i+1}), i=0,1,\cdots,k-2; \\ G \circ g_{i-k+1}^{-1} \circ \cdots \circ g_{i-2}^{-1} \circ g_{i-1}^{-1}(x), & x \in [x_i, x_{i+1}), i=k-1,k,\cdots; \\ g_{i+1}^{-1} \circ g_{i+2}^{-1} \circ \cdots \circ g_{i+k-1}^{-1} \circ G(x), & x \in [x_i, x_{i+1}), i=-1,-2,\cdots \end{cases} \tag{3.20}$$

and define

$$g(x) = g_i(x), \qquad x \in [x_i, x_{i+1}), \qquad i=0,\pm 1,\pm 2,\cdots \tag{3.21}$$

Obviously $g(x)$ is strictly increasing, and

$$g(x) > x, \qquad g^k(x) = G(x), \qquad \forall x \text{ in } I. \tag{3.22}$$

Furthermore, (3.16) implies $f^{-1}(c) \notin B(c,\delta)$ and $f^{-1}(c) < c$. By (3.20) $g^{-1}(c) = f^{-1}(c)$. Thus

$$f^{k-1}(c) = F \circ f^{-1}(c) = G \circ g^{-1}(c) = g^{k-1}(c) := c_0, \tag{3.23a}$$

where ":=" means "denote". Clearly $c_0 > c$. Also (3.16) implies $f(c) \notin B(c,\delta)$. Since $c_0 = f^{k-1}(c) \geqslant f(c)$ we see easily that $c_0 \notin B(c,\delta)$. It follows that for any integer $i > 0$

$$f^{ki}(c_0) = F^i(c_0) = G^i(c_0) = g^{ki}(c_0) := c_i. \tag{3.23b}$$

Naturally

$$c < c_0 < c_1 < \cdots < c_i < c_{i+1} < \cdots \to 1, \text{as } i \to +\infty, \tag{3.24}$$

because $F(x) > x$ and the end point 1 becomes a stable fixed point. However, $G(c) = g^k(c) > g^{k-1}(c) = c_0$, i. e. $G(c) \notin B(c,\delta)$, so we have

$$\begin{aligned} f(c_i) &= f \circ f^{ki}(c_0) = f \circ f^{ki} \circ f^{k-1}(c) = F^{i+1}(c) = F^i \circ F(c) \\ &< (3.15) < F^i \circ G(c) = G^i \circ G(c) = G^{i+1}(c) = g(c_i). \end{aligned} \tag{3.25}$$

Summarizing (3.24) and (3.25), we see that there is a strictly increasing sequence $\{c_i\}$ in I, approaching 1 as $i \to +\infty$, such that $f(c_i) < g(c_i)$. Obviously it is impossible that $f(x) = g(x)$ in a small neighbourhood of fixed point 1. This contradicts (3.18). Therefore the small perturbation G of F, defined in (3.15), satisfies $G \in B(F(x), d)$ and $G \in W$. Thus we have proved that W is dense in A^1.

From step 1 and step 2 we can see the generic property as stated in this theorem. This proof is now complete.

Remark 3.3 If $F(x)$ has more than two fixed points, we can still obtain a similar result. In fact, it suffices to consider any two consecutive fixed points y', y'' with $y' < y''$ and discuss the restriction of $F(x)$ on $[y', y'']$.

Remark 3.4 The idea in this proof was once given partly by the author in [23].

3.3 A Note on Local Uniqueness and Bifurcation

Consider a function $F(x): I \to I, I = [a,b]$. Let $E(F)$ be a set of some fixed points of $F(x)$.

Definition 3.3 For a positive integer m, the function $F(x)$ is said to possess m-**order local uniqueness in** $E(F)$ **on a set** H **of functions**, or simply to possess m-**order** H-**local uniqueness in** $E(F)$, if $F(x)$ has a unique m-order iterative root which is fixed at each point of $E(F)$ and satisfies the property of H in a neighbourhood of each point of $E(F)$.

For example, for any given integer $m \geqslant 2$, under conditions stated in corollary 3.1 the function $F(x)$ possesses m-order $C^1(J,J)$-local uniqueness at its unique fixed point, and under conditions stated in theorem 3.3 the same $F(x)$ also possesses m-order $C^r(I,I)$-local uniqueness at its unique fixed point.

Theorem 3.5[23] Suppose $F(x): I \to I$, strictly increasing, has a set $E(F)$ of fixed points with more than one elements, and possesses m-order H-local uniqueness in $E(F)$ for a given integer $m \geqslant 2$. Then $F(x)$ is structurally unstable, i.e. its small perturbation $F_\delta(x)$, given arbitrarily, does not possess such a property. Obviously, bifurcation will arise from its perturbations.

The idea of its proof is the same as in step 2 in the proof of theorem 3.4. In fact, from the proof given above we can summarize some interesting results.

Corollary 3.3 Suppose both $F(x)$ and $f(x)$ belong to $CI(I,I)$ with no fixed points in I except the two end points, and $f^m(x)=F(x)$. Then for $m \geqslant 2$,

(a) any m-order iterative root $\tilde{f}(x) \in CI(I,I)$ of perturbation $F_\delta(x)$, where $F_\delta(x)$ is defined simply as $F_\delta(x) >$ (or $<$) $F(x)$ for $x \in B(c,\delta)$, but $=F(x)$ for $x \notin B(c,\delta)$, herein δ is sufficiently small and $c \in \operatorname{int} I$ and is chosen arbitrarily, does not identically equal $f(x)$ in neighbourhoods of the two end fixed points;

(b) if $\tilde{f}(x)=f(x)$ in a neighbourhood of one of the two end fixed points, the sign of $g(x):=\tilde{f}(x)-f(x)$ must be alternating near the other end point, i.e. there are sequences $\{a_i\}$ and $\{b_i\}$ such that either $a_i<b_i<a_{i+1}<b_{i+1}$ or $a_i>b_i>a_{i+1}>b_{i+1}$ for integer $i>0$ and $g(a_i)g(a_j)>0, g(b_i)g(b_j)>0, g(a_i)g(b_j)<0$ for integer $i,j>0$.

Now we give out a useful notation $G(H;E)$ as follows. Let E be a subset of I, G and H be sets of functions from I into I, we denote

$$G(H;E)=\{F \in G: F(x) \text{ possesses } m\text{-order } H\text{-local uniqueness in } E \text{ for an integer } m \geqslant 2\}. \tag{3.26}$$

Consider the following sets:

$$G_0=\{F(x) \in C^2[0,1]: F'(x)>0 \text{ on } [0,1], F(x) \neq x \text{ on } (0,1), \text{ and } F(y)=y, F'(y) \neq 1, F''(y) \text{ exists for } y=0 \text{ or } 1\} \tag{3.27}$$

$$C^1=C^1[0,1], \tag{3.28}$$

$G_1=\{F(x) \in C^0[0,1]$: is increasing, where $F(0)=0, F(1)=1, F'(0)=F'(1)=1$, and satisfies that

(1) $F(x)>x, \forall x \in (0,1)$;

(2) F^{-1} has an expansion $F^{-1}(x)=x-cx^{k+1}+o(x^{k+1})$ at $x=0_+$ where $k \in \mathbf{Z}_+$ and $c>$

0, and $(F^{-n}(y))^{-k}-(F^{-n}(x))^{-k}$ converges in $\{0<x<1, x\leqslant y<1\}$ to a C^0 increasing function; and

(3) $\widetilde{F}(x)=F(x+1)-1$ has an expansion $\widetilde{F}(x)=x-cx^{j+1}+o(x^{j+1})$ at $x=0_-$, where $j\in\mathbf{Z}_+$ and $(-1)^j\cdot c>0$, and $(\widetilde{F}^n(y))^{-j}-(\widetilde{F}^n(x))^{-j}$ converges in $\{-1<x<0, x\leqslant y<0\}$ to a C^0 increasing function$\}$. (3.29)

$G_2=\{F(x)\in C^0[0,1]$: is increasing, $F(0)=0$, $F(1)=1$, $F'(0)=F'(1)=1$, and satisfies that

(1) $F(x)<x$, $\forall x\in(0,1)$;

(2) F has an expansion $F(x)=x-cx^{k+1}+o(x^{k+1})$ at $x=0_+$, $k\in\mathbf{Z}_+$ and $c>0$, and $(F^n(y))^{-k}-(F^n(x))^{-k}$ converges in $\{0<x<1, 0<y\leqslant x\}$ to a C^0 increasing function; and

(3) $\widetilde{F}(x)=F^{-1}(x+1)-1$ has an expansion $\widetilde{F}(x)=x-cx^{j+1}+o(x^{j+1})$ at $x=0_-$, $j\in\mathbf{Z}_+$ and $(-1)^j\cdot c>0$, and $(\widetilde{F}^n(y))^{-j}-(\widetilde{F}^n(x))^{-j}$ converges in $\{-1<x<0, x\leqslant y<0\}$ to a c^0 increasing function$\}$. (3.30)

$G_3=\{F(x)\in C^0[0,1]$: $F(0)=0$, $F(1)=1$, $F'(0)=1$, $F'(1)=0$, and satisfies that

(1) $F(x)>x$, $\forall x\in(0,1)$;

(2) F^{-1} has an expansion $F^{-1}(x)=x-cx^{k+1}+o(x^{k+1})$ at $x=0_+$, where $k\in\mathbf{Z}_+$ and $c>0$, and $(F^{-n}(y))^{-k}-(F^{-n}(x))^{-k}$ converges in $\{0<x<1, x\leqslant y<1\}$ to a C^0 increasing function; and

(3) $\widetilde{F}(x)=F(x+1)-1$ has an expansion $\widetilde{F}(x)=(-A)^{1-j}x^{j+o(1)}$ at $x=0_-$, where $j\in\mathbf{Z}_+$ and $A>0$, and $[\text{In}(-\widetilde{F}^n(y))-\text{In }A]/[\text{In}(-\widetilde{F}^n(x))-\text{In }A]$ converges in $\{-1<x<0, x\leqslant y<0\}$ to a C^0 increasing function. $\}$ (3.31)

$G_4=\{F(x)\in C^0[0,1]$: $F(0)=0$, $F(1)=1$, $F'(0)=0$, $F'(1)=1$, and satisfies that

(1) $F(x)<x$, $\forall x\in(0,1)$;

(2) F has an expansion $F(x)=Ax^{k+0(1)}$ at $x=0_+$, where $k\in\mathbf{Z}_+$ and $A>0$, and $[\text{In}(F^n(y))-\text{In }A]/[\text{In}(F^n(x))-\text{In }A]$ converges in $\{0<x<1, 0<y\leqslant x\}$ to a C^0 increasing function; and

(3) $\widetilde{F}(x)=F^{-1}(x+1)-1$ has an expansion $\widetilde{F}(x)=x-cx^{j+1}+o(x^{j+1})$ at $x=0_-$, where $j\in\mathbf{Z}_+$ and $(-1)^j\cdot c>0$, and $(\widetilde{F}n(y))^{-j}-(\widetilde{F}^n(x))^{-j}$ converges in $\{-1<x<0, x\leqslant y<0\}$ to a C^0 increasing function. $\}$ (3.23)

$C_*^p=\{F\in C^0[0,1]$: $F(x)$ has up to p-order generalized derivatives at 0 and $1\}$, where $p=\min\{j,k\}$.

It was proved in [14] and [23] that the classes of functions $G_0(C^1;0)$, $G_1(C_*^p;0)$, $G_2(C_*^p;0)$, $G_3(C_*^p;0)$, $G_4(C_*^p;0)$, $G_0(C^1;1)$, $G_1(C_*^p;1)$, $G_2(C_*^p;1)$, $G_3(C_*^p;1)$, and, $G_4(C_*^p;1)$ are nonempty. By theorem 3.5, $G_0(C^1;0,1)$, $G_1(G_*^p;0,1)$, $G_2(G_*^p;0,1)$, $G_3(G_*^p;0,1)$, *and*, $G_4(G_*^p;0,1)$ are structurally unstable.

Deep research on bifurcations of functional equations, worthily mentioning here, is very significant and also encouraging.

4 Iterated Equations of Polynomial Type

As a natural generalization of the problem of iterative roots, a class of iterated functional equations of polynomial type, say

$$c_1 f(x)+c_2 f^2(x)+\cdots+c_n f^n(x)=F(x), \qquad x\in I=[a,b], \tag{4.1}$$

where $f^0(x)=x$, and $f^n(x)=f\circ f^{n-1}(x)$, always fascinates many mathematicians. In 1977 J. G. Dhombres [31] studied the equation

$$f^2(x)=af(x)+(1-a)x. \tag{4.2}$$

In 1983 Zhao Liren[32] gave conclusions for the equation

$$c_1 f(x)+c_2 f^2(x)=F(x), \tag{4.3}$$

and in the same year A. Mukherjea and J. S. Ratti[33] discussed (4.1) for some special function $F(x)$. Despite of their nice constructive proofs, the classical methods prevent them from obtaining more fruitful conclusions of existence, uniqueness, stability, and differentiability of solution for Eq. (4.1).

In 1986, Zhang Weinian[25] constructed an interesting operator, say "structural operator" $L: f\mapsto L_f(x)$, for Eq. (4.1), and used the fixed point theory in Banach spaces to overcome the difficulties encountered by the formers. By means of this theory and operator method, Zhang Weinian and Si Jianguo made a series of works [26—30] and [34—37] concerning these qualitative problems.

4.1 Results

Consider Eq. (4.1), where $c_1>0, c_i\geqslant 0, i=2,3,\cdots,n$, and $\sum_{i=1}^{n} c_i=1$. Let

$$X(I;\delta,M)=\{f(x)\in C^0(I,I): f(a)=a, f(b)=b \text{ and } \forall x_1>x_2 \\ \delta(x_1-x_2)\leqslant f(x_1)-f(x_2)\leqslant M(x_1-x_2)\}, \tag{4.4}$$

$$X'(I;\delta,M,M')=\{f(x)\in C^1(I,I): f(a)=a, f(b)=b, \text{and } \forall x,x_1,x_2 \\ \delta\leqslant f'(x)\leqslant M, |f'(x_1)-f'(x_2)|\leqslant M'|x_1-x_2|\}, \tag{4.5}$$

where δ, M and M' are positive constants.

Theorem 4.1 $M>0$. For $F(x)\in X(I;0,M)$, there exists a solution $f(x)$ of Eq. (4.1) in $X(I;0,M/c_1)$.

Theorem 4.2 $\delta>0, M>0$. For $F(x)\in X(I;\delta,c_1M)$, there is a solution $f(x)$ of Eq. (4.1) in $X(I;0,M)$. Furthermore, if

$$c_1\geqslant 1-\delta/(\sum_{i=1}^{n-1}M^i), \tag{4.6}$$

then the solution is unique and stable, i. e. continuously dependent on $F(x)$.

Proofs were given in [25—27].

Theorem 4.3[28] Let $\delta>0, M>0, M'>0$, and $K_0=\sum_{j=1}^{n-1}c_{j+1}(\sum_{i=j-1}^{2j-2}M^i)$. If $c_1>$

K_0M^2, then for $F(x)\in X'(I;\delta,c_1M,M')$, there is a solution $f(x)$ of Eq. (4.1) in $X'(I;0,M,M'/(c_1-K_0M^2))$. Furthermore, the solution $f(x)$ is unique and stable, provided by

$$E_1\cdot E_2\cdot E_3<1, \tag{4.7}$$

where

$$K_1=\sum_{i=1}^{n}c_iM^{i-1},$$
$$E_1=\max\{M+K_1MM'/\delta(c_1-K_0M^2);M^2\},$$
$$E_2=1/\delta+K_1M'/\delta^3, E_3=\sum_{j=1}^{n-1}c_{j+1}E_3^{(j)},$$

and $E_3^{(j)}=\max\{(\sum_{i=1}^{j}M^{i-1}+Q(j)(M'/(c_1-K_0M^2))(\sum_{i=1}^{j-1}(j-i)M^{j+i-2});jM^{j-1}\}$. Here $Q(s)=0$ as $s=1$ and $Q(s)=1$ as $s=2,3,\cdots$

Results on C_1 solutions were also given in [29]but with a slight difference.

For example, we consider the equation

$$cf(x)+(1-c)f^2(x)=F(x),\qquad 0<c<1, x\in I, \tag{4.8}$$

where $F(x)=a(\exp(x)-1)$, $0<a<1$, $I=[0,r]$ and $r=a(\exp(r)-1)>0$. It is easy to verify that $F(x)\in X'(I;a,a\exp(r)/c,a\exp(r))$. By theorem 4.3, Eq. (4.8) has a unique, stable and continuously differentiable solution if $1-\delta<c<1$ and $\delta>0$ small enough.

Although the C^2-smooth solutions of Eq. (4.1) have been discussed by Si[35], so far we know only a little about smoothness of its solutions, which remains to be studied further.

4.2 Method of Structural Operator

To show the method of structural operator, we give a comprehensive sketch of proof of theorem 4.1.

First of all, we define the structural operator $L:f\to L_f(x)$ for Eq. (4.1) on $X(I;0,M/c_1)$ such that

$$L_f(x)=c_1x+c_2f(x)+c_3f^2(x)+\cdots+c_nf^{n-1}(x). \tag{4.9}$$

Evidently $L_f(a)=a$, $L_f(b)=b$, and

$$0<L_f^{-1}(x_1)-L_f^{-1}(x_2)<(1/c_1)(x_1-x_2),\qquad \forall x_1>x_2. \tag{4.10}$$

Then we define the mapping T on $X(I;0,M/c_1)$ such that

$$T(f)=L_f^{-1}\circ F,\qquad \text{for } f \text{ in } X(I;0,M/c_1). \tag{4.11}$$

It suffices to prove the existence of fixed points of T.

Secondly, we prove T maps $X(I;0,M/c_1)$ into itself and prove its continuity. In fact we have

$$\begin{aligned}0&\leqslant T(f)(x_1)-T(f)(x_2)=L_f^{-1}\circ F(x_1)-L_f^{-1}\circ F(x_2)\\&\leqslant(1/c_1)(F(x_1)-F(x_2))\leqslant(M/c_1)(x_1-x_2)\end{aligned} \tag{4.12a}$$

for any $x_1>x_2$ in I, and

$$\|T(f_1)-T(f_2)\|\leqslant(1/c_1)(\sum_{i=1}^{n-1}c_{i+1}\sum_{j=1}^{i}(M/c_1)^{j-1})\|f_1-f_2\| \tag{4.12b}$$

for any f_1, f_2 in $X(I;0,M/c_1)$.

Thirdly, using Ascoli-Arzela's principle we show that $X(I;0,M/c_1)$ is a compact convex set in $C^0(I,I)$.

Finally, by Schauder's fixed point theorem, the mapping T has a fixed point $f(x)$ in $X(I;0,M/c_1)$, which is just a solution of Eq. (4.1).

4.3 Some Generalizations

Using the method of structural operator, we can give some similar results for more general functional equation

$$G(f(x), f^{n_1}(x), \cdots, f^{n_k}(x)) = F(x), x \in I = [a,b], \tag{4.13}$$

where integer $k > 0$, and $n_1, n_2, \cdots, n_k \geqslant 2$.

Let $Y(I^{k+1}; c_i, C_i)$ consist of functions $G(y_0, y_1, \cdots, y_k) \in C^0(I^{k+1}, I)$, $I^{k+1} = \underbrace{I \times I \times I \times \cdots \times I}_{k+1}$, satisfying

(H1) $G(a,a,\cdots,a) = a, G(b,b,\cdots,b) = b$;

(H2) there are constants $c_0 > 0, c_i \geqslant 0 (i=1,2,\cdots,k)$, and $C_i > 0$ such that for any $y_i \geqslant y'_i, i=0,1,\cdots,k$,

$$\sum_{i=0}^{k} c_i (y_i - y'_i) \leqslant G(y_0, y_1, \cdots, y_k) - G(y'_0, y'_1, \cdots, y'_k) \leqslant \sum_{i=0}^{k} C_i (y_i - y'_i).$$

Let $Y'(I^{k+1}; c_i, C_i, N_{ji})$ consist of functions $G(y_0, y_1, \cdots, y_k) \in C^1(I^{k+1}, I)$, $I^{k+1} = \underbrace{I \times I \times \cdots \times I}_{k+1}$, satisfying (H1) and

(H3) there are constants $c_0 > 0$ and $C_j > 0 (j=0,1,\cdots,k)$ such that

$$c_0 \leqslant G'(y_0, y_1, \cdots, y_k) \leqslant C_0, 0 \leqslant G'(y_0, y_1, \cdots, y_k) \leqslant C_j, j=1,2,\cdots,k;$$

(H4) there are constants $N_{ji} \geqslant 0 (j, i=0,1,\cdots,k)$ such that

$$| G'_{y_j}(y_0, y_1, \cdots, y_k) - G'_{y_j}(y'_0, y'_1, \cdots, y'_k) | \leqslant \sum_{i=0}^{k} N_{ji} | y_i - y'_i |, j = 0,1,\cdots,k.$$

Theorem 4.4[36] Let $M > 0$. Suppose $F(x) \in X(I;0,c_0 M)$ and $G(x) \in Y(I^{k+1}; c_i, C_i)$. Then Eq. (4.13) has a solution $f(x)$ in $X(I;0,M)$. Furthermore, this solution $f(x)$ is unique and continuously dependent on F and G if

$$\sum_{i=1}^{k} C_i [Q(n_i - 1) \sum_{j=1}^{n_i - 2} M^j + 1] < c_0. \tag{4.14}$$

Theorem 4.5[37] Let $M > 0$ and $M' > 0$. Suppose $F(x) \in X'(I;0,c_0 M, M')$, $G(x) \in Y'(I^{k+1}; c_i, C_i, N_{ji})$, and

$$c_0 > \sum_{i=1}^{k} (C_i \sum_{j=n_i-2}^{2n_i-4} M^{j+2}). \tag{4.15}$$

Then Eq. (4.13) has a solution $f(x)$ in $X'(I;0,M,M')$, where

$$M' \geqslant [\sum_{i=0}^{k} N_{\alpha i} M^{n_i+1} + \sum_{j=1}^{k}\sum_{i=0}^{k} N_{ji} M^{n_j+n_i} + M'] / [c_0 - \sum_{i=1}^{k}(C_i \sum_{j=n_i-2}^{2n_i-4} M^{j+2})].$$

Furthermore, this solution $f(x)$ is unique and continuously dependent on F and G if

$$E := \max\{1/c_0 + MK_3/c_0^2\} \sum_{i=1}^{k}(C_i \sum_{j=1}^{n_i-1} M^{j-1}) + MK_4/c_0,$$

$$(M/c_0)\sum_{i=1}^{k}(C_i(n_i-1)M^{n_i-2}) < 1, \tag{4.16}$$

where

$$K_3 = \sum_{i=0}^{k} N_{\alpha i} M^{n_i-1} + \sum_{j=1}^{k}(M^{n_j-1}\sum_{i=0}^{k} N_{ji} M^{n_i-1} + C_j M'' \sum_{i=n_j-2}^{2n_j-4} M^i),$$

and

$$K_4 = \sum_{j=0}^{k}(M^{n_j-1}\sum_{i=0}^{k} N_{ji}\sum_{l=1}^{n_i-1} M^{l-1}) + M'\sum_{i=1}^{k}[C_i Q(n_i-1)\sum_{j=1}^{n_i-2}(n_i-j-1)M^{n_i+j-3}].$$

For example, the equation

$$cf(x) + (1-c)(f^2(x))^2 = F(x), \qquad x \in [0,1], \tag{4.17}$$

where $F(x) = \text{In}(x+1) - x\ln(2/e)$ and $9/10 < c < 1$, has a unique continuous solution. Particularly, Eq. (4.17) has a unique differentiable solution if $1-\delta < c < 1$ and $\delta > 0$ small enough.

4.4 Discussion on a Critical Restriction

In spite of some improvements in the above works, the necessity of the assumption $c_1 > 0$ in section 4.1 or $c_0 > 0$ in section 4.3, still appears questionable and critical. These assumptions seem irrational because the first order iterated term $f(x)$ should not be necessary for an iterated equation of n-order polynomial type. Nevertheless, it has become an open problem to eliminate these assumptions because great difficulties would arise and involve painstaking effort to overcome, if there is no such assumption.

Recently, the iterated functional equation

$$c_1 g^k(x) + c_2 g^{2k}(x) + \cdots + c_n g^{nk}(x) = F(x), x \in I, \tag{4.18}$$

where $F(x)$ is a given continuous function on I and integer $k \geqslant 2, c_1 > 0, c_i \geqslant 0, i = 2, \cdots, n, \sum_{i=1}^{n} c_i = 1$, was successfully discussed by Zhang W. in [30]. Obviously (4.18) does not include the first order iterative term $g(x)$. Here the difficulties resulted from the missing of the first order iterative term are overcome by utilization of Hardy-Böedewadt's theorem 2.2.

Theorem 4.6 For given continuous function $F(x)$ on the interval $I = [a,b]$ satisfying $F(a) = a, F(b) = b$ and for $x_1 > x_2$ in I

$$0 < F(x_1) - F(x_2) \leqslant M(x_1 - x_2), \tag{4.19}$$

where M is a positive constant, there is a solution $g(x)$ of Eq. (4.18) in $CI(I,I)$, defined in (2.1).

Proof Let $f(x)=g^k(x)$. Eq. (4.18) can be rewritten into Eq. (4.1). Evidently $F(x)$ satisfies the conditions in theorem 4.1, so Eq. (4.1) has a solution $f(x)$ in $X(I;0,M/c_1)$. Particularly (4.19) implies $F(x)$ is strictly increasing. It follows from (4.12a) in the proof of theorem 4.1 that $f(x)$ is also strictly increasing, that is, $f(x)\in CI(I,I)$. By theorem 2.2 we see that there is a function $g(x)\in CI(I,I)$ such that $g^k(x)=f(x)$. Hence $g(x)$ is a solution of Eq. (4.18) in $CI(I,I)$. Thus the proof is complete.

4.5 Several Open Problems

The following problems are difficult but very interesting:

1. When does Eq. (4.1) with $c_1>0, c_i\geqslant 0$ and $\sum_{i=1}^{n} c_i = 1$ have $C^r (r\geqslant 3)$ solutions?

2. Is there a decreasing solution of Eq. (4.1) under above assumptions?

3. As our ultimate aim, we need discuss Eq. (4.1) only with $c_n>0$. More generally we need discuss the equation.

$$G(f^{n_0}(x), f^{n_1}(x), \cdots, f^{n_k}(x))=F(x), \qquad x\in I=[a,b], \tag{4.20}$$

where integer $k>0$ and $n_0, n_1, \cdots, n_k\geqslant 2$.

References

[1]Schröder E. Ueber iterate funktionen. Math. Ann., 1871(3):296 - 322.

[2]Abel N H. Oeuvres Complétes. Vol. Ⅱ, Christiana, 1881, 36 - 39.

[3]Dubbey J M. The Mathematical Work of Charles Babbage. Cambridge Univ. Press, 1978.

[4]Kuczma M. Functional Equations in a Single Variables. Monografie Matematyczne. Tom 46, Warszwa, 1968.

[5]Kuczma M. Fractional iteration of differentiable functions. Ann. Polon. Math., 1969/1970(22):217 - 227.

[6]Riggert G. n-te iterative Wurzeln von beliebigen Abbildungen (abstract). Aequationes Math., 1976(14):208.

[7]Rice R E. An upper bound for the order of an iterative root of function. Notices Amer. Math. Soc., 1976, 23(A-609, Abstract):738 - 810.

[8]Rice R E, Schweizer B, Sklar A. When is $f(f(z))=az^2+bz+c$? Amer. Math. Monthly, 1980(87):252 - 263.

[9]Isaacs R. Iterates of fractional order. Canad. J. Math., 1950(2):409 - 416.

[10]Böedewadt U T. Zur iteration realler funktionen. Zeittsc. Math., 1944, 49(3): 497 - 516.

[11]Hardy G H. Order of Infinity, Cambridge. 1924:31.

[12]Zhang Jingzhong, Xiong Jincheng. Iteration of Functions and One-Dimensional Dynamical Systems (In Chinese). Sichuan Educational Press, 1992.

[13]Zhang Jingzhong, Yang Lu. Discussion on iterative roots of piecewise monotone functions. In

Chinese, Acta Math. Sinica, 1983, 26(4): 398 - 412.

[14] Zhang Jingzhong, Yang Lu. On criterion of existence and uniqueness of real iterative groups in a single variable. In Chinese, Acta Sci. Nat. Univ. pekin., 1982(6): 23 - 45.

[15] Zhang Jingzhong, Yang Lu. Problems on homeomorphic embedding flows and asymptotic embedding flows. (In Chinese). Sci. Sinica (A), 1985(1): 32 - 43.

[16] Yang Lu, Zhang Jingzhong. Sufficient and necessary condition of embedding semiflows for C^0 self-mappings on intervals. (In Chinese). Acta Math. Sinica, 1986, 29(2): 180 - 183.

[17] He Lianfa, Niu Dongxiao. The iterative roots for a class of self-mappings on S^1. J. of Math. Research & Exposition, 1991, 11(2): 305 - 310.

[18] Chen Zaoping, He Lianfa. Sufficient and necessary condition of embedding semiflows for C^0 self-mappings on S^1. (In Chinese). Acta Math. Sinica, 1987, 30(6): 729 - 732.

[19] He Lianfa. Sufficient and necessary condition of embedding quasi-semiflows for C^0 self-mappings on intervals. (In Chinese). Acta Math. Sinica, 1988, 31(2): 258 - 261.

[20] Zhang Meirong, Li Weigu. One dimensional dynamics: adopting embedding flow method. Advances Math. (China), 1992(21): 245 - 246.

[21] Zhang Meirong. Centralizers and iterate radicals of Morse-Smale diffeomorphisms of the circle. Acta Math. Sinica (N. S.), accepted to appear.

[22] Jiang Xingyao. Iterative roots of multi-order differentiable functions and Böedewadt's conjecture. (In Chinese). J. of Math., 1986, 6(4): 433 - 438.

[23] Zhang Weinian. On the problem of potentiality of the existence of iterative roots which are differentiable at both end points of an interval. (In Chinese). Chinese Quarterly J. of Math., 1989, 4(2): 31 - 38.

[24] Zhang Weinian. A generic property of globally smooth iterative roots. Sci. sinica(A), 1995, 38(3): 267 - 272.

[25] Zhang Weinian. Discussion on the existence of solutions of the iterated equation $\sum_{i=1}^{n}\lambda_i f^i(x) = F(x)$. (In Chinese). Chin. Sci. Bul., 1986, 31(17): 1290 - 1295.

[26] Zhang Weinian. Discussion on the iterated equation $\sum_{i=1}^{n}\lambda_i f^i(x) = F(x)$. Chin. Sci. Bul., 1987, 32(21): 1444 - 1451.

[27] Zhang Weinian. Stability of the solution of the iterated equation $\sum_{i=1}^{n}\lambda_i f^i(x) = F(x)$. Acta Math. Sci., 1988, 8(4): 421 - 424.

[28] Zhang Weinian. Discussion on the differentiable solutions of iterated equation $\sum_{i=1}^{n}\lambda_i f^i(x) = F(x)$. Nonlin. Analysis TMA., 1990, 15(4): 387 - 398.

[29] Zhang Weinain. Differentiability on the solutions of the iterated equation $\sum_{i=1}^{n}\lambda_i f^i(x) = F(x)$. (In Chinese). Acta Math. Sinica, 1989, 32(1): 98 - 109.

[30] Zhang Weinian. An application of Hrady-Böedewadt's theorem to iterated functional

equations. Acta Math. Sci. ,1995,15(3),in press.

[31]Dhombres J G. Itération d'order deux. Publ. Math. Debrecen. ,1977,24(3-4):277-287.

[32] Zhao Liren. A theorem concerning the existence and uniqueness of solutions of functional equation $\lambda_1 f(x)+\lambda_2 f^2(x)=F(x)$. (In Chinese). J Univ. Sci. Tech. China, Special Issue of Math. ,1983:21-27.

[33]Mukherjea A,Ratti J S. On a functional equation involving iterates of a bijection on the unit interval. Nonlin. Analysis TMA. ,1983(7):899-908.

[34]Si Jianguo. Existence and uniqueness of solutions of an iterative system. In Chinese. Pure & Applied Math. ,1990,6(2):38-42.

[35]Si Jianguo. C^2 solutions of the iterated equation $\sum_{i=1}^{n}\lambda_i f^i(x)=F(x)$. In Chinese. Acta Math. Sinica,1993,36(3):348-357.

[36]Si Jianguo. Discussion on the C^0 solutions of the iterated equation $G(f^{n_0}(x),f^{n_1}(x),\cdots,f^{n_k}(x))=F(x)$. Preprint,1993,in Chinese.

[37]Si Jianguo. Discussion on the C^1 solutions of a class of iterated equations. Preprint, 1993,in Chinese.

An Efficient Decomposition Algorithm for Geometry Theorem Proving without Factorization

Yang Lu Zhang Jing-zhong and Hou Xiao-rong
Laboratory for Automated Reasoning & Programming Chengdu Institute of Computer Applications, Academia Sinica 610041 Chengdu, People's Republic of China

Sometimes the hypothesis of a geometry theorem to be verified is represented by a reducible ascending chain. To decompose it into components such that on each component the theorem is either generically true or generically false, an efficient computer algorithm is given that employs Wu's division and sub-resultant computations only. No factorization is needed. This algorithm is successfully implemented on microcomputers with very complicated examples.

1 Introduction

On proving a geometry theorem using characteristic-set-based method, the hypothesis is usually represented by a system of algebraic equations in "triangular form":

$$\begin{cases} f_1(u_1,u_2,\cdots,u_d,x_1)=0, \\ f_2(u_1,u_2,\cdots,u_d,x_1,x_2)=0, \\ \cdots\cdots \\ f_s(u_1,u_2,\cdots,u_d,x_1,x_2,\cdots,x_s)=0, \end{cases} \tag{1}$$

where $u_1,\cdots,u_d$ are regarded as parameters; x_j is called the *principal element of* f_j and the coefficient of the highest term of f_j with respect to it's principal element x_j the *initial* of f_j, for $j=1,2,\cdots,s$. And the set of polynomials $\{f_1,f_2,\cdots,f_s\}$ is called an *ascending chain*. The conclusion is represented by another equation,

$$g(u_1,\cdots,u_d,x_1,\cdots,x_s)=0. \tag{2}$$

Definition 1.1 By resultant (p,q,x) denote the resultant of any two polynomials, p

---

本文刊于《亚洲计算机数学研讨会》,第 37 卷第 5 期,1994 年.

and q, with respect to indeterminate x, which is a determinant in terms of the coefficients of $p(x)$ and $q(x)$ to verify if the two polynomials have a common solution.

Given an ascending chain $\{f_1, \cdots, f_s\}$ and another polynomial G, putting

$$r_{s-1} := \text{resultant }(G, f_s, x_s),$$
$$r_{s-2} := \text{resultant }(r_{s-1}, f_{s-1}, x_{s-1}),$$
$$r_{s-3} := \text{resultant }(r_{s-2}, f_{s-2}, x_{s-2}),$$
$$\cdots\cdots$$
$$r_1 := \text{resultant }(r_2, f_2, x_2),$$
$$r_0 := \text{resultant }(r_1, f_1, x_1),$$

we call r_0 *the resultant of ascending chain* $\{f_1, \cdots, f_s\}$ *with respect to polynomial* G and denote it simply by

$$\text{res}(f_1, \cdots, f_s, G).$$

Definition 1.2 An ascending chain $\{f_1, \cdots, f_s\}$ is called a *normal ascending chain* if $I_1 \neq 0$ and for every $j = 2, 3, \cdots, s$,

$$\text{res}(f_1, \cdots, f_{j-1}, I_j) \neq 0, \tag{3}$$

where I_j denote the initial of f_j, i. e. the leading coefficient of f_j in x_j, for $j = 1, \cdots, s$.

While an ascending chain is mentioned hereafter, we always mean a normal one, throughout this paper.

It is well-known that the theorem represented by (1)—(2) is generically true if the *pseudo remainder of* $\{f_1, \cdots, f_s\}$ *with respect to* g, obtained by doing Wu's[2]successive pseudo division, is equal to zero, i. e.

$$\text{prem}(f_1, \cdots, f_s, g) = 0. \tag{4}$$

And if the ascending chain $\{f_1, \cdots, f_s\}$ is *irreducible*, then (4) is a sufficient and necessary condition making the theorem generically true.

On the other hand, it has been pointed out [4][5]that the theorem represented by (1)—(2) is generically false if the resultant of $\{f_1, \cdots, f_s\}$ with respect to g is not equal to zero, i. e.

$$\text{res}(f_1, \cdots, f_s, g) \neq 0. \tag{5}$$

Definition 1.3 A normal ascending chain $\{f_1, \cdots, f_s\}$ is said to be *simplicial with respect to* g if the theorem represented by (1)—(2) is either generically true or generically false.

Consequently, an ascending chain $\{f_1, \cdots, f_s\}$ is simplicial with respect to g, provided either

$$\text{prem}(f_1, \cdots, f_s, g) = 0$$

or

$$\text{res}(f_1, \cdots, f_s, g) \neq 0.$$

Generally speaking, however, the converse does not hold. In fact, a criterion was established by us for checking an ascending chain simplicial or not. That is

Theorem 1.1 Assume $\{f_1,\cdots,f_s\}$ is a normal ascending chain. Let $\{f_1=0,\cdots,f_s=0\}$ and $g=0$ be the hypothesis and conclusion of a conjecture to be verified, respectively. The conjecture is generically true if and only if

$$\mathrm{res}\{f_1,\cdots,f_s,g+T\}=L(u_1,\cdots,u_d)T^k \tag{6}$$

where T is another variable, k some positive integer, and $L(u_1,\cdots,u_d)$ is something independent of T. (see Corollary 1 in [4] or Corollary 2.1.1 in [5], or Theorem 3.1 in [9].)

And the conjecture is generically false if and only if

$$\mathrm{res}\{f_1,\cdots,f_s,g\}\neq 0.$$

(see Corollary 3 in [4] or Corollary 2.1.3 in [5].)

An irreducible ascending chain is always simplicial with respect to any polynomial. In general case, however, it is often required to decompose a reducible ascending chain into components simplicial with respect to some polynomial. The concept of simplicial component is of importance in various fields. On proving a theorem, if the ascending chain which represents the hypothesis has been decomposed into simplicial components, with respect to the conclusion polynomial g, then, the theorem is either true or false over all generic points of a simplicial component.

For example, consider a well-known geometry theorem stated ambiguously as follows: *On three sides of a triangle ABC, three equilateral triangles BCA_1, CAB_1 and ABC_1 are erected. Show lines AA_1, BB_1 and CC_1 are concurrent.* This statement has not specified which equilateral triangles are outward against triangle ABC and which inward. As a result, provided $A=(0,0)$, $B=(1,0)$, $C=(u_1,u_2)$, $C_1=(x_1,x_2)$, $B_1=(x_3,x_4)$, $A_1=(x_5,x_6)$, the hypothesis is represented by a reducible ascending chain:

```
f1=-1+2*x1;
f2=-3+4*x2^2;
f3=u1^4-2*u1^2*u2^2-3*u2^4-4*u1^3*x3-4*u1*u2^2*x3+4*u1^2*x3^2+
    4*u2^2*x3^2;
f4=-u1^2-u2^2+2*u1*x3+2*u2*x4;
f5=1-2*u1^2+u1^4-2*u2^2+8*u1*u2^2-2*u1^2*u2^2-3*u2^4-4*x5+4*u1
    *x5+4*u1^2*x5-4*u1^3*x5-4*u2^2*x5-4*u1*u2^2*x5+4*x5^2-8*
    u1*x5^2+4*u1^2*x5^2+4*u2^2*x5^2;
f6=1-u1^2-u2^2-2*x5+2*u1*x5+2*u2*x6;
```

while the conclusion polynomial is

```
g=-u2*x4*x5+u2*x1*x4*x5+x2*x4*x5-u1*x2*x4*x5+u2*x1*x6-u1*
   x2*x6-u2*x1*x3*x6+u1*x2*x3*x6+u1*x4*x6-x1*x4*x6;
```

It is easy to check that ascending chain $\{f_1,f_2,f_3,f_4,f_5,f_6\}$ is not simplicial with respect to g. In order to produce a perfect proof, we should decompose the former into simplicial components, with respect to g.

An efficient decomposition algorithm that employs Wu's division and sub-resultant computations is given in Section 2. No factorization is required. And it has been implemented successfully on an IBM PC microcomputer with the programs written in MATHEMATICA, see Section 3.

2 New Decomposition Algorithm

An elimination introduced by Wu Wensün[2]in 1977 has been demonstrated to be an efficient tool for nonlinear problems in various fields. One of the important applications is the Wu's Algorithm for automated geometry theorem proving. This method was expounded in Chou's monograph [1]where 512 non-trivial geometry theorems were proven by Wu's characteristic-set-based method. Using Wu's method to deal with theorem-proving or other issues, however, one may meet with the so-called "reducibility difficulty" [3][6][7]if involved ascending chains are reducible. In that case one usually need decompose the ascending chain into irreducible components. Some existing algorithms depend upon the factorization of polynomials in several elements over an algebraic number field.

In this section, an algorithm is given for a normal ascending chain AS and a polynomial g to decompose AS into *simplicial components*, i. e. the subvarieties each of them is an ascending chain and their union is AS, such that every component is simplicial with respect to g. Using this algorithm, we need not factorize AS beforehand. In fact, we need neither do factorization over an algebraic number field, nor do that in any sense.

Firstly we introduce notations. Let D be a domain, Q the quotient field of D. Put

$$f(x):=a_nx^n+a_{n-1}x^{n-1}+\cdots+a_1x+a_0\in D[x],$$
$$g(x):=b_mx^m+b_{m-1}x^{m-1}+\cdots+b_1x+b_0\in D[x],$$

and $m_i(g,f,x)$ denote the following $(m+n-2i)\times(m+n-i)$ matrix,

$$\begin{bmatrix} a_n & a_{n-1} & \cdots & \cdots & a_0 & & & \\ & a_n & a_{n-1} & \cdots & \cdots & a_0 & & \\ & & & & \cdots & \cdots & & \\ & & & a_n & a_{n-1} & \cdots & \cdots a_0 \\ b_m & b_{m-1} & \cdots & \cdots & b_0 & & & \\ & b_m & b_{m-1} & \cdots & \cdots & b_0 & & \\ & & & & \cdots & \cdots & & \\ & & & b_m & b_{m-1} & \cdots & \cdots b_0 \end{bmatrix}$$

where the first $m-i$ rows in terms of the coefficients of $f(x)$ and the rest $n-i$ rows in terms of those of $g(x)$, for $i=0,1,\cdots,\min(m,n)-1$. Then, by $M_j^i(g,f,x)$ denote the determinant of the sub-matrix which consists of the first $m+n-2i-1$ columns and the $(m+n-i-j)$-th column of $m_i(g,f,x)$ for $j=0,1,\cdots,i$.

The *sub-resultant polynomial series* of $f(x),g(x)$ means a polynomial series

$$\{P_i(g,f,x)\}_{i=0,1,\cdots,\min(m,n)-1}$$

where

$$P_i(g,f,x):=\sum_{j=0}^{i} M_j^i(g,f,x)x^j \tag{7}$$

for $i=0,1,\cdots,\min(m,n)-1$. For simplicity, put $s_i(g,f,x):=M_i^i(g,f,x)$ and call it the i-th *principal sub-resultant* of $f(x),g(x)$. We have:

Theorem 2.1 Defined notations as above, If at least one of a_n,b_m is non-vanishing and it holds that

$$s_0(g,f,x)=\cdots=s_{k-1}(g,f,x)=0,\qquad s_k(g,f,x)\neq 0, \tag{8}$$

then $P_k(g,f,x)$ is the greatest common divisor of $f(x),g(x)$ in $\mathbf{Q}[x]$.
The theorem is not new, for example, a proof can be found in [8].

Let $\{f_1(\vec{u},x_1),\cdots,f_s(\vec{u},x_1,\cdots,x_j)\}$ be a normal ascending chain consisting of polynomials over a domain D and

$$f(x):=a_ny^n+a_{n-1}y^{n-1}+\cdots+a_1y+a_0,$$
$$g(x):=b_my^m+b_{m-1}y^{m-1}+\cdots+b_1y+b_0,$$

two polynomials over $D(\vec{u})[x_1,\cdots,x_j]$. And put

$$R_i:=\mathrm{prem}(f_1,\cdots,f_j,s_i(g,f,y))$$

where $s_i(g,f,y)$, the i-th principal sub-resultant of $f(y)$ and $g(y)$, is a polynomial in $D(\vec{u})[x_1,\cdots,x_j]$. Our algorithm employs the following

Theorem 2.2 If at least one of

$$\mathrm{res}(f_1,\cdots,f_j,a_n)\qquad \text{and}\qquad \mathrm{res}(f_1,\cdots,f_j,b_m)$$

doesn't equal 0 and

$$R_0=\cdots=R_{k-1}=0,\qquad \mathrm{res}(f_1,\cdots,f_j,R_k)\neq 0,$$

then, $P_k(g,f,y)$ is the greatest common divisor of $f(y),g(y)$ over the quotient ring

$$D(\vec{u})[x_1,\cdots,x_j]/(f_1,\cdots,f_j).$$

Proof: Let $\{f_1^*,\cdots,f_j^*\}$ be an irreducible component of $\{f_1,\cdots,f_j\}$. Equalities $R_i=0$ for $i=0,\cdots,k-1$ mean that

$$s_i(g,f,y)=0\qquad \mathrm{mod}(f_1,\cdots,f_j)$$

for $i=0,\cdots,k-1$, hence

$$s_i(g,f,y)=0\qquad \mathrm{mod}(f_1^*,\cdots,f_j^*)$$

for $i=0,\cdots,k-1$. On the other hand, $\mathrm{res}(f_1,\cdots,f_j,R_k)\neq 0$ means

$$s_k(g,f,y)\neq 0\qquad \mathrm{mod}(f_1^*,\cdots,f_j^*).$$

By assumption, at least one of $\mathrm{res}(f_1,\cdots,f_j,a_n)$ and $\mathrm{res}(f_1,\cdots,f_j,b_m)$ doesn't equal 0, so that at least one of a_n and b_m doesn't equal 0 in the field

$$D(\vec{u})[x_1,\cdots,x_j]/(f_1^*,\cdots,f_j^*).$$

Applying Theorem 2.1, we know that then, $P_k(g,f,y)$ is the greatest common divisor of $f(y),g(y)$ over the quotient ring

$$D(\vec{u})[x_1,\cdots,x_j]/(f_1^*,\cdots,f_j^*).$$

Since the choice for $\{f_1^*,\cdots,f_j^*\}$ is arbitrarily, it is also the g. c. d. of $f(y),g(y)$ over

$$D(\vec{u})[x_1,\cdots,x_j]/(f_1,\cdots,f_j).$$

Now, let us describe the steps of our algorithm in detail as follows.

For an ascending chain of single polynomial, $\{f_1\}$, it is simplicial with respect to g if either prem $(g,f_1,x_1)=0$ or res $(g,f_1,x_1)\neq 0$. Otherwise, $\{f_1\}$ can be decomposed into $\{f_1'\}$ and $\{f_2'\}$ where f_1' is the greatest common divisor of f_1 and g; and f_2' is the pseudo quotient of f_1 by f_1'. In general, we have

Step 1 Compute prem$(f_1,\cdots,f_s,g)$, the pseudo remainder of AS with respect to g. If it equals 0, stop. Otherwise, let either $g^*:=g$ or $g^*:=$ prem$(f_1,\cdots,f_s,g)$. If res$(f_1,\cdots,f_s,g^*)\neq 0$, stop.

Step 2 Assume res$(f_1,\cdots,f_s,g^*)=0$. Let k be the smallest integer such that

$$\text{prem}(f_1,\cdots,f_{s-1},s_k(g^*,f_s,x_s))\neq 0.$$

If

$$\text{res}(f_1,\cdots,f_{s-1},s_k(g^*,f_s,x_s))\neq 0,$$

then by Theorem 2.2, over the quotient ring,

$$D(u_1,\cdots,u_d)[x_1,\cdots,x_{s-1}]/(f_1,\cdots,f_{s-1}),$$

we can find the g. c. d. of f_s and g^*, namely, $P_k(g^*,f_s,x_s)$, which we denote by f_s'. Let f_s'' be the pseudo quotient of f_s by f_s'. Then, return two ascending chains

$$\{f_1,\cdots,f_{s-1},f_s'\},\qquad \{f_1,\cdots,f_{s-1},f_s''\}$$

to be decomposed.

Step 3 Let k be the smallest number such that

$$\text{prem}(f_1,\cdots,f_{s-1},s_k(g^*,f_s,x_s))\neq 0.$$

If

$$\text{res}(f_1,\cdots,f_{s-1},s_k(g^*,f_s,x_s))=0.$$

Replace AS and g by $\{f_1,\cdots,f_{s-1}\}$ and $s_k(g^*,f_s,x_s)$, i. e. to decompose $\{f_1,\cdots,f_{s-1}\}$ with respect to $s_k\{g^*,f_s,x_s\}$. Then, if $\{f_1,\cdots,f_{s-1}\}$ is well decomposed with respect to $s_k\{g^*,f_s,x_s\}$ and $\{\overline{f}_1,\cdots,\overline{f}_{s-1}\}$ is a component, we return the ascending chain $\{\overline{f}_1,\cdots,\overline{f}_{s-1},f_s\}$ to be decomposed.

Since every return of the steps reduces the degrees of a polynomial in the ascending chain we deal with unless it's simplicial already, the algorithm ends in finite returns and gives a decomposition consisting of ascending chains every of which is simplicial with respect to g.

Example 1. To decompose the following ascending chain

$$f_1 := 4x_1^2 - 3$$
$$f_2 := 2x_2 - 1$$
$$f_3 := x_3 - 1$$
$$f_4 := x_4^2 - 3$$
$$f_5 := 4x_5^2 - 8x_5 + 1$$
$$f_6 := 2x_6 - 4x_5 + 3$$
$$f_7 := ((4 - 2x_1)x_4 + 2x_1 - 3)x_7 + 2 - 2x_1$$
$$f_8 := 2(2 - x_1)x_8 + x_7 - 2$$

into simplicial components with respect to $g := x_5x_8 - x_6x_7$.

First, by computing,

$$g^* := \mathrm{prem}(f_1, \cdots, f_8, g) = (36 - 44x_1 + (16 - 8x_1)x_4)x_5 + 36x_1 - 33 \neq 0,$$

while $\mathrm{res}(f_1, \cdots, f_8, g^*) = 0$. Using Step 3 thrice, we lay f_8, f_7, f_6 aside and decompose $\{f_1, \cdots, f_5\}$ with respect to g^*. Compute the principal sub-resultant,

$$s_0(g^*, f_5, x_5) = 4(16x_1^2x_4^2 - 400x_1^2x_4 - 1388x_1^2 - 64x_1x_4^2 + 1184x_1x_4 + 2328x_1 + 64x_4^2 - 768x_4 - 963),$$

which is non-vanishing. Then, since

$$g_1^* := \mathrm{prem}(f_1, \cdots, f_4, s_0(g^*, f_5, x_5)) = 16((296x_1 - 267)x_4 + 534x_1 - 444) \neq 0$$

while $\mathrm{res}(f_1, \cdots, f_4, g_1^*) = 0$, using Step 3 once again, we lay f_5 aside and decompose $\{f_1, \cdots, f_4\}$ with respect to g_1^*. Compute the principal sub-resultant,

$$s_0(g_1^*, f_4, x_4) = 89232(4x_1^2 - 3),$$

that leads to $\mathrm{prem}(f_1, f_2, f_3, s_0(g_1^*, f_4, x_4)) = 0$. Return to the next principal sub-resultant,

$$s_1(g_1^*, f_4, x_4) = 16(296x_1 - 267),$$

so $\mathrm{prem}(f_1, f_2, f_3, s_1(g_1^*, f_4, x_4)) \neq 0$, $\mathrm{res}(f_1, f_2, f_3, s_1(g_1^*, f_4, x_4)) \neq 0$. By Step 2, f_4 and g_1^* have a g. c. d., namely, g_1^* itself. Let

$$f_4' := (296x_1 - 267)x_4 + 534x_1 - 444$$

and f_4'' be the pseudo quotient of f_4 by f_4', i. e.

$$f_4'' := (296x_1 - 267)x_4 - 534x_1 + 444,$$

and then decompose $\{f_1, f_2, f_3, f_4', f_5\}$ and $\{f_1, f_2, f_3, f_4'', f_5\}$ with respect to g^*. By computing,

$$\mathrm{prem}(f_1, f_2, f_3, f_4', s_0(g^*, f_5, x_5)) = 0.$$

Return to the next principal sub-resultant,

$$s_1(g^*, f_5, x_5) = 36 - 44x_1 + (16 - 8x_1)x_4,$$

so $\mathrm{prem}(f_1, f_2, f_3, f_4', s_1(g^*, f_5, x_5)) \neq 0$, $\mathrm{res}(f_1, f_2, f_3, f_4', s_1(g^*, f_5, x_5)) \neq 0$. By Step 2, f_5 and g^* have a g. c. d., namely, g^* itself. Let $f_5' := g^*$ and f_5'' be the pseudo quotient of f_5 by f_5', i. e.

$f_5'':=(36-44x_1+(16-8x_1)x_4)x_5-(32-16x_1)x_4+52x_1-39.$

Now, verified by Step 1, the ascending chain $\{f_1,\cdots,f_8\}$ is well decomposed into three simplicial components, (with respect to g), which are

$$C_1: f_1, f_2, f_3, f_4', f_5', f_6, f_7, f_8$$
$$C_2: f_1, f_2, f_3, f_4', f_5'', f_6, f_7, f_8$$
$$C_3: f_1, f_2, f_3, f_4'', f_5, f_6, f_7, f_8$$

where

$$f_4':=(296x_1-267)x_4+534x_1-444$$
$$f_4'':=(296x_1-267)x_4-534x_1+444$$
$$f_5':=(36-44x_1+(16-8x_1)x_4)x_5+36x_1-33$$
$$f_5'':=(36-44x_1+(16-8x_1)x_4)x_5-(32-16x_1)x_4+52x_1-39.$$

In detail, the C_1 is generically dependent on g, while C_2 and C_3 are independent of g.

Example 2. Consider the geometry theorem involving three equilateral triangles stated in Section 1. It has been represented as follows:

```
f1=-1+2*x1;
f2=-3+4*x2^2;
f3=u1^4-2*u1^2*u2^2-3*u2^4-4*u1^3*x3-4*u1*u2^2*x3+4*u1^2*x3^2+
    4*u2^2*x3^2;
f4=-u1^2-u2^2+2*u1*x3+2*u2*x4;
f5=1-2*u1^2+u1^4-2*u2^2+8*u1*u2^2-2*u1^2*u2^2-3*u2^4-4*x5+4*u1
    *x5+4*u1^2*x5-4*u1^3*x5-4*u2^2*x5-4*u1*u2^2*x5+4*x5^2-8*
    u1*x5^2+4*u1^2*x5^2+4*u2^2*x5^2;
f6=1-u1^2-u2^2-2*x5+2*u1*x5+2*u2*x6;
g=-u2*x4*x5+u2*x1*x4*x5+x2*x4*x5-u1*x2*x4*x5+u2*x1*x6-u1*
    x2*x6-u2*x1*x3*x6+u1*x2*x3*x6+u1*x4*x6-x1*x4*x6;
```

Applying our decomposition algorithm to $\{f_1,\cdots,f_6\}$ with respect to g, in an interactive manner, we have a program in MATHEMATICA 2.0:

```
prem[a_,b_,c_]:=
    Numerator[Together[PolynomialRemainder[a,b,c]]];
i1=prem[prem[prem[prem[prem[prem[g,f6,x6],f5,x5],f4,x4],f3,x3],f2,x2],f1,x1];
i2=Resultant[i1,f5,x5];
i3=Resultant[Resultant[i2,f3,x3],f2,x2];
i4=prem[prem[prem[i2,f3,x3],f2,x2],f1,x1];
i5=Resultant[i4,f3,x3];
i6=prem[prem[i5,f2,x2],f1,x1];
i7=Coefficient[i4,x3,1];
i8=prem[prem[i7,f2,x2],f1,x1];
```

```
i9=Resultant[Resultant[i7,f2,x2],f1,x1];
i10=Numerator[Together[PolynomialQuotient[f3,i4,x3]]];
i11=prem[prem[prem[prem[prem[prem[g,f6,x6],f5,x5],f4,x4],i4,x3],f2,x2],f1,x1];
i12=Resultant[i11,f5,x5];
i13=Resultant[i12,f2,x2];
i14=prem[i12,f2,x2];
i15=Coefficient[i11,x5,1];
i16=prem[i15,f2,x2];
i17=Resultant[i15,f2,x2];
i18=Numerator[Together[PolynomialQuotient[f5,i11,x5]]];
i19=prem[prem[prem[prem[prem[prem[g,f6,x6],i11,x5],f4,x4],i4,x3],f2,x2],f1,x1];
i20=prem[prem[prem[prem[prem[prem[g,f6,x6],i18,x5],f4,x4],i4,x3],f2,x2],f1,x1];
i21=Resultant[i20,f2,x2];
i22=prem[prem[prem[prem[prem[prem[g,f6,x6],i11,x5],f4,x4],i10,x3],f2,x2],f1,x1];
i23=Resultant[i22,f2,x2];
i24=prem[prem[prem[prem[prem[prem[g,f6,x6],i18,x5],f4,x4],i10,x3],f2,x2],f1,x1];
i25=Resultant[i24,f2,x2];
```

Let us explain this interactive program in detail. The first line gives a definition of pseudo remainder that has not been well-defined in MATHEMATICA. We found the chain non-simplicial since $i_1 \neq 0$ and $i_3=0$. Secondly, $i_4 \neq 0$ and $i_3=0$ mean

$$\operatorname{prem}(f_1,\cdots,f_4,s_0(i_1,f_5,x_5)) \neq 0,$$
$$\operatorname{res}(f_1,\cdots,f_4,s_0(i_1,f_5,x_5))=0,$$

so, by Step 3, we just decompose $\{f_1,f_2,f_3\}$ with respect to i_2 or i_4 instead. Subsequently, $i_6=0$ means

$$\operatorname{prem}(f_1,f_2,s_0(i_4,f_3,x_3))=0,$$

so we consider the next principal sub-resultant, $s_1(i_4,f_3,x_3)$, that is, the leading coefficient of i_4 in x_3, we denote it by i_7. And $i_8 \neq 0, i_9 \neq 0$ mean

$$\operatorname{prem}(f_1,f_2,s_1(i_4,f_3,x_3)) \neq 0,$$
$$\operatorname{res}(f_1,f_2,s_1(i_4,f_3,x_3)) \neq 0,$$

so, by Step 2, f_3 and i_4 have a g. c. d. which should be i_4 itself since the degree of i_4 in x_3 is 1. Taking the pseudo quotient of f_3 divided by i_4 in x_3 as the other factor, namely, i_{10}, we have f_3 decomposed into

```
i4=16*(-1+2*u1)*u2^3*(-2*u1+u1^2+u2^2)*(1-2*u1+u1^2+u2^2)*(-3
   *u1*u2+9*u1^2*u2-6*u1^3*u2+3*u2^3-6*u1*u2^3-3*u1^2*x2+6*u1
   ^3*x2-3*u1^4*x2+3*u2^2*x2+2*u1*u2^2*x2-6*u1^2*u2^2*x2-3*u2^
   4*x2-3*u2*x3+3*u1^2*u2*x3+3*u2^3*x3+6*u1*x2*x3-12*u1^2*x2
   *x3+6*u1^3*x2*x3-4*u2^2*x2*x3+6*u1*u2^2*x2*x3);
```

i10=(u1^2+u2^2) * (6 * u1 * u2-9 * u1^2 * u2+3 * u1^3 * u2-3 * u2^3+3 * u1 * u2^3-3 * u1^2 * x2+6 * u1^3 * x2-3 * u1^4 * x2-3 * u2^2 * x2+2 * u1 * u2^2 * x2+3 * u2^4 * x2-3 * u2 * x3+3 * u1^2 * u2 * x3+3 * u2^3 * x3+6 * u1 * x2 * x3-12 * u1^2 * x2 * x3+6 * u1^3 * x2 * x3-4 * u2^2 * x2 * x3+6 * u1 * u2^2 * x2 * x3);

Then, decompose $\{f_1, f_2, i_4, f_4, f_5, f_6\}$ with respect to g, analogously. The chain is non-simplicial since $i_{11} \neq 0$ and $i_{13} = 0$. Subsequently, $i_{14} = 0$ means

$$\text{prem}(f_1, f_2, i_4, f_4, s_0(i_{11}, f_5, x_5)) = 0,$$

so we turn to the next principal sub-resultant, $s_1(i_{11}, f_5, x_5)$, that is, the leading coefficient of i_{11} in x_5, we denote it by i_{15}. And $i_{16} \neq 0$, $i_{17} \neq 0$ mean

$$\text{prem}(f_1, f_2, i_4, f_4, s_1(i_{11}, f_5, x_5)) \neq 0,$$
$$\text{res}(f_1, f_2, i_4, f_4, s_1(i_{11}, f_5, x_5)) \neq 0,$$

so, by Step 2, f_5 and i_{11} have a g. c. d. which should be i_{11} itself since the degree of i_{11} in x_5 is 1. Taking the pseudo quotient of f_5 divided by i_{11} in x_5 as the other factor, namely, i_{18}, we have f_5 decomposed into

i11=u2 * (-1+u1^2+u2^2) * (-9 * u1^2+9 * u1^3+9 * u1^4-9 * u1^5-3 * u2^2-15 * u1 * u2^2+18 * u1^2 * u2^2-6 * u1^3 * u2^2+9 * u2^4+3 * u1 * u2^4+12 * u1 * u2 * x2+18 * u1^2 * u2 * x2-48 * u1^3 * u2 * x2+18 * u1^4 * u2 * x2+2 * u2^3 * x2-16 * u1 * u2^3 * x2+12 * u1^2 * u2^3 * x2-6 * u2^5 * x2+18 * u1^2 * x5-36 * u1^3 * x5+18 * u1^4 * x5+6 * u2^2 * x5-12 * u1 * u2^2 * x5+12 * u1^2 * u2^2 * x5-6 * u2^4 * x5-24 * u1 * u2 * x2 * x5+24 * u1^2 * u2 * x2 * x5+8 * u2^3 * x2 * x5);

i18=(1-2 * u1+u1^2+u2^2) * (-9 * u1^2+9 * u1^3+9 * u1^4-9 * u1^5-3 * u2^2+21 * u1 * u2^2-18 * u1^2 * u2^2-6 * u1^3 * u2^2-3 * u2^4+3 * u1 * u2^4+12 * u1 * u2 * x2-18 * u1^2 * u2 * x2+24 * u1^3 * u2 * x2-18 * u1^4 * u2 * x2-10 * u2^3 * x2+8 * u1 * u2^3 * x2-12 * u1^2 * u2^3 * x2+6 * u2^5 * x2+18 * u1^2 * x5-36 * u1^3 * x5+18 * u1^4 * x5+6 * u2^2 * x5-12 * u1 * u2^2 * x5+12 * u1^2 * u2^2 * x5-6 * u2^4 * x5-24 * u1 * u2 * x2 * x5+24 * u1^2 * u2 * x2 * x5+8 * u2^3 * x2 * x5);

Now, we have 4 components to be checked. $\{f_1, f_2, i_4, f_4, i_{11}, f_6\}$, $\{f_1, f_2, i_4, f_4, i_{18}, f_6\}$, $\{f_1, f_2, i_{10}, f_4, i_{11}, f_6\}$ and $\{f_1, f_2, i_{10}, f_4, i_{18}, f_6\}$. Obviously, by Step 1, all the 4 components are simplicial with respect to g because $i_{19} = 0$, $i_{21} \neq 0$, $i_{23} \neq 0$ and $i_{25} \neq 0$. The program ends.

We ran it on an AITCAN 486/33 computer with CPU time 213. 88 seconds. For simplicity, cancelled all the factors independent of x's, we have $f_{31}, f_{32}, f_{51}, f_{52}$ instead of $i_4, i_{10}, i_{11}, i_{18}$.

f31=-3 * u1 * u2+9 * u1^2 * u2-6 * u1^3 * u2+3 * u2^3-6 * u1 * u2^3-3 * u1^2 * x2+6 * u1^3 * x2-3 * u1^4 * x2+3 * u2^2 * x2+2 * u1 * u2^2 * x2-6 * u1^2 * u2^2 * x2-3 * u2^4 * x2-3 * u2 * x3+3 * u1^2 * u2 * x3+3 * u2^3 * x3+6 * u1 * x2 * x3-12 * u1 ^2 * x2 * x3+6 * u1^3 * x2 * x3-4 * u2^2 * x2 * x3+6 * u1 * u2^2 * x2 * x3;

```
f32=6*u1*u2-9*u1^2*u2+3*u1^3*u2-3*u2^3+3*u1*u2^3-3*u1^2*x2+6*
    u1^3*x2-3*u1^4*x2-3*u2^2*x2+2*u1*u2^2*x2+3*u2^4*x2-3*u2*
    x3+3*u1^2*u2*x3+3*u2^3*x3+6*u1*x2*x3-12*u1^2*x2*x3+6*u1^3
    *x2*x3-4*u2^2*x2*x3+6*u1*u2^2*x2*x3;
f51=-9*u1^2+9*u1^3+9*u1^4-9*u1^5-3*u2^2-15*u1*u2^2+18*u1^2*u2^2
    -6*u1^3*u2^2+9*u2^4+3*u1*u2^4+12*u1*u2*x2+18*u1^2*u2*x2-48
    *u1^3*u2*x2+18*u1^4*u2*x2+2*u2^3*x2-16*u1*u2^3*x2+12*u1^2*
    u2^3*x2-6*u2^5*x2+18*u1^2*x5-36*u1^3*x5+18*u1^4*x5+6*u2^2*
    x5-12*u1*u2^2*x5+12*u1^2*u2^2*x5-6*u2^4*x5-24*u1*u2*x2*x5
    +24*u1^2*u2*x2*x5+8*u2^3*x2*x5;
f52=-9*u1^2+9*u1^3+9*u1^4-9*u1^5-3*u2^2+21*u1*u2^2-18*u1^2*u2^2
    -6*u1^3*u2^2-3*u2^4+3*u1*u2^4+12*u1*u2*x2-18*u1^2*u2*x2+24
    *u1^3*u2*x2-18*u1^4*u2*x2-10*u2^3*x2+8*u1*u2^3*x2-12*u1^2*
    u2^3*x2+6*u2^5*x2+18*u1^2*x5-36*u1^3*x5+18*u1^4*x5+6*u2^2*
    x5-12*u1*u2^2*x5+12*u1^2*u2^2*x5-6*u2^4*x5-24*u1*u2*x2*x5
    +24*u1^2*u2*x2*x5+8*u2^3*x2*x5;
```

The factorization is still not required. For example, f_{31} can be obtained from i_4 by the following procedure,

```
CoefficientList[i4,{x3,x2,x1}];
Flatten[%];
Apply[PolynomialGCD,%];
f31=Cancel[i4/%]
```

The last lines, i_{19} to i_{25}, indicate that theorem is generically true for one out of the 4 components, $\{f_1, f_2, f_{31}, f_4, f_{51}, f_6\}$, and generically false for others.

Example 3. A difficult geometry theorem was conjectured by Thébault in 1938 and proved by Taylor in 1983 with a proof of 26 pages! Then it was verified [1] on a SYMBOLICS 3600 computer by using *Ritt's decomposition algorithm* to a reducible variety which represents the hypothesis of the theorem; the program took 44 hours CPU time, while more than 500 theorems else took only a few seconds each. The wonderful theorem is stated as follows:

Given a triangle ABC with incenter w and circumcircle Γ, take any point D on edge BC. Let w_1 be the center of the circle which contacts DB, DA and Γ, and w_2 the center of the circle which contacts DC, DA and Γ. Show that w_1, w_2 and w are collinear.

Taking one algebraic interpretation somewhat different from that taken in [1] for the Thébault-Taylor Theorem, we represent the hypothesis and conclusion as follows:

```
f1=4*x1^2*u1^2*u2^4+6*u1^2*u2^4-2*u2^2-2*u2^6-2*u2^2*u1^4-2*u2^6*
    u1^4+u1^2+u1^2*u2^8+4*u2^4*u3-4*u2^4*u1^4*u3-4*u2^4*u3^2*u1^2;
f2=(4*u2^4-8*u2^4*u3*u1^2-4*u1^4*u2^4+8*x1*u1^2*u2^4)*x2^2+(-4*
```

u2^2*u1^4+8*u2^4+4*u1^2-4*u2^6*u1^4-4*u2^6+8*u1^4*u2^4+4*u1^2*
u2^8+8*u1^2*u2^4-8*u1^2*u2^2-4*u2^2-8*u2^6*u1^2)*x2+(-2*u1^2+
4*u1^2*u2^4-2*u1^2*u2^8)*x1-1-8*u2^4*u3+8*u2^2*u3*u1^2+u1^4+2
*u2^4+4*u2^6*u1^4*u3+4*u1^4*u3*u2^2-12*u2^4*u3*u1^2-2*u1^4*u2
^4-2*u2^8*u3*u1^2-8*u2^4*u1^4*u3-2*u3*u1^2+u1^4*u2^8+4*u3*u2^
2-u2^8+8*u2^6*u3*u1^2+4*u2^6*u3;

f3=(4*u2^4-4*u1^4*u2^4-8*u2^4*u3*u1^2-8*x1*u1^2*u2^4)*x3^2+(8*u1^
4*u2^4+4*u1^2-4*u2^2*u1^4-8*u1^2*u2^2-4*u2^6*u1^4+8*u1^2*u2^4-
8*u2^6*u1^2+4*u1^2*u2^8-4*u2^2+8*u2^4-4*u2^6)*x3+(2*u1^2-4*u1
^2*u2^4+2*u1^2*u2^8)*x1-1+8*u2^2*u3*u1^2+8*u2^6*u3*u1^2+u1^4+
2*u2^4-u2^8+4*u3*u2^2-12*u2^4*u3*u1^2-2*u1^4*u2^4-8*u2^4*u3-8
*u2^4*u1^4*u3-2*u3*u1^2+u1^4*u2^8+4*u1^4*u3*u2^2+4*u2^6*u3+4
*u2^6*u1^4*u3-2*u2^8*u3*u1^2;

f4=x4*u1*u2+1-u1^2*u2^2;

f5=x5*u1*u2-u1^2+u2^2;

f6=(2*u1^2-2*u1^2*u2^4+2*x4*u1^2*u2^2+2*x5*u1^2*u2^2)*x6+(-u1^2-
u1^2*u 2^4)*x5+(-u1^2-u1^2*u2^4)*x4-u1^4+u2^4-1+u1^4*u2^4;

f7=(-u2^2+u1^2*u2^4-u2^2*u1^4+u1^2)*x7+(u2^2-u2^2*u1^4-u1^2+u1^2*u2^
4-2*x5*u1^2*u2^2)*x6+(u1^2*u2^4+u1^2)*x5-u2^4+u1^4;

g=(4*x2*u2^4+4*x3*u2^4)*x7+(-4*u2^2-4*u2^6+8*u2^4)*x6-4*u2^2-4
*u2^6+1+6*u2^4-4*x2*x3*u2^4+u2^8;

To decompose the ascending chain $\{f_1,\cdots,f_7\}$ by our algorithm, in an interactive manner, we ran a program in MATHEMATICA 2.0 on an AITCAN 486/33 computer with CPU time 923.3 seconds and have our program follows below:

prem[a_,b_,c_]:=
Numerator[Together[PolynomialRemainder[a,b,c]]];

i1=prem[prem[prem[prem[prem[prem[prem[g,f7,x7],f6,x6],f5,x5],f4,x4],f3,x3],
f2,x2],f1,x1];

i2=CoefficientList[i1,{x3,x2,x1}];

i3=Flatten[i2];

i4=Apply[PolynomialGCD,i3];

i5=Cancel[i1/i4];

i6=Resultant[i5,f3,x3];

i7=CoefficientList[i6,{x2,x1}];

i8=Flatten[i7];

i9=Apply[PolynomialGCD,i8];

i10=Cancel[i6/i9];

```
i11=prem[prem[i10,f2,x2],f1,x1];
i12=Resultant[i10,f2,x2];
i13=CoefficientList[i12,x1]:
i14=Apply[PolynomialGCD,i13];
i15=Cancel[i12/i14];
i16=Resultant[i15,f1,x1];
i17=CoefficientList[i11,{x2,x1}];
i18=Flatten[i17];
i19=Apply[PolynomialGCD,i18];
i20=Cancel[i11/i19];
i21=Resultant[i20,f2,x2];
i22=CoefficientList[i21,x1]:
i23=Apply[PolynomialGCD,i22];
i24=Cancel[i21/i23];
i25=prem[i24,f1,x1];
i26=prem[Coefficient[i20,x2,1],f1,x1];
i27=Resultant[Coefficient[i20,x2,1],f1,x1];
i28=Numerator[Together[PolynomialQuotient[f2,i20,x2]]];
Print["f21=",i20];
Print["f22=",i28];
i29=prem[prem[prem[prem[prem[prem[prem[g,f7,x7],f6,x6],f5,x5],f4,x4],f3,x3],
    i20,x2],f1,x1];
i30=Resultant[g,f7,x7];
i31=Resultant[i30,f6,x6];
i32=CoefficientList[i31,{x5,x4,x3,x2,x1}];
i33=Flatten[i32];
i34=Apply[PolynomialGCD,i33];
i35=Cancel[i31/i34];
i36=Resultant[i35,f5,x5];
i37=CoefficientList[i36,{x4,x3,x2,x1}];
i38=Flatten[i37];
i39=Apply[PolynomialGCD,i38];
i40=Cancel[i36/i39];
i41=Resultant[i40,f4,x4];
i42=CoefficientList[i41,{x3,x2,x1}];
i43=Flatten[i42];
i44=Apply[PolynomialGCD,i43];
```

```
i45=Cancel[i41/i44];
i46=Resultant[i45,f3,x3];
i47=CoefficientList[i46,{x2,x1}];
i48=Flatten[i47];
i49=Apply[PolynomialGCD,i48];
i50=Cancel[i46/i49];
i51=Resultant[i50,i20,x2];
i52=CoefficientList[i51,x1];
i53=Apply[PolynomialGCD,i52];
i54=Cancel[i51/i53];
i55=Resultant[i54,f1,x1];
i56=CoefficientList[i29,{x3,x2,x1}];
i57=Flatten[i56];
i58=Apply[PolynomialGCD,i57];
i59=Cancel[i29/i58];
i60=Resultant[i59,f3,x3];
i61=CoefficientList[i60,{x2,x1}];
i62=Flatten[i61];
i63=Apply[PolynomialGCD,i62];
i64=Cancel[i60/i63];
i65=prem[prem[i64,i20,x2],f1,x1];
i66=prem[prem[Coefficient[i59,x3,1],i20,x2],f1,x1];
i67=Resultant[Resultant[Coefficient[i59,x3,1],i20,x2],f1,x1];
i68=Numerator[Together[PolynomialQuotient[f3,i59,x3]]];
Print["f31=",i59];Print["f32=",i68];
```

Now, we have $\{f_1,\cdots,f_7\}$ decomposed into 4 components:

$$C_1:=\{f_1,f_{21},f_{31},f_4,f_5,f_6,f_7\}$$
$$C_2:=\{f_1,f_{21},f_{32},f_4,f_5,f_6,f_7\}$$
$$C_3:=\{f_1,f_{22},f_{31},f_4,f_5,f_6,f_7\}$$
$$C_4:=\{f_1,f_{22},f_{32},f_4,f_5,f_6,f_7\}$$

where

```
f21=u1^2+u1*u2-u1^3*u2-u2^2-2*u1^2*u2^2-u1^4*u2^2+u1*u2^3-u1^3*u2^
    3+2*u2^4+2*u1^2*u2^4+2*u1^4*u2^4-u1*u2^5+u1^3*u2^5-u2^6-2*u1^2
    *u2^6-u1^4*u2^6-u1*u2^7+u1^3*u2^7+u1^2*u2^8-2*u1*u2^3*u3-2*u1^
    3*u2^3*u3+2*u1*u2^5*u3+2*u1^3*u2^5*u3+2*u1*u2^3*x1+2*u1^3*
    u2^3*x1-2*u1*u2^5*x1-2*u1^3*u2^5*x1+2*u2^4*x2-2*u1^4*u2^4*
    x24*u1^2*u2^4*u3*x2+4*u1^2*u2^4*x1*x2;
```

f22=-u1^2+u1 * u2-u1^3 * u2+u2^2+2 * u1^2 * u2^2+u1^4 * u2^2+u1 * u2^3-u1^3 *
u2^3-2 * u2^4-2 * u1^2 * u2^4-2 * u1^4 * u2^4-u1 * u2^5+u1^3 * u2^5+u2^6+2 *
u1^2 * u2^6+u1^4 * u2^6-u1 * u2^7+u1^3 * u2^7-u1^2 * u2^8-2 * u1 * u2^3 * u3-2
* u1^3 * u2^3 * u3+2 * u1 * u2^5 * u3+2 * u1^3 * u2^5 * u3+2 * u1 * u2^3 * x1+2 *
u1^3 * u2^3 * x1-2 * u1 * u2^5 * x1-2 * u1^3 * u2^5 * x1-2 * u2^4 * x2+2 * u1^4 *
u2^4 * x2+4 * u1^2 * u2^4 * u3 * x2-4 * u1^2 * u2^4 * x1 * x2;

f31=1+u1^2-2 * u2^2-2 * u1^2 * u2^2+u2^4+u1^2 * u2^42 * u1 * u2 * u3+2 * u1 * u2^3
* u3+2 * u1 * u2 * x1-2 * u1 * u2^3 * x1+2 * u1 * u2 * x3-2 * u2^2 * x3+2 * u1^2
* u2^2 * x3-2 * u1 * u2^3 * x3;

f32=2 * u1^3-2 * u1^2 * u2+2 * u1^4 * u2-2 * u1 * u2^2-6 * u1^3 * u2^2-2 * u1^5 * u2^2
+u2^3+u1^2 * u2^3-u1^4 * u2^3-u1^6 * u2^3+6 * u1 * u2^4+8 * u1^3 * u2^4+6 * u1
^5 * u2^4-2 * u2^5+2 * u1^2 * u2^5-2 * u1^4 * u2^5+2 * u1^6 * u2^5-6 * u1 * u2^6-
8 * u1^3 * u2^6-6 * u1^5 * u2^6+u2^7+u1^2 * u2^7-u1^4 * u2^7-u1^6 * u2^7+2 * u1
* u2^8+6 * u1^3 * u2^8+2 * u1^5 * u2^8-2 * u1^2 * u2^9+2 * u1^4 * u2^9-2 * u1^3 *
u2^10+2 * u1^2 * u2^3 * u3+2 * u1^4 * u2^3 * u3+2 * u1 * u2^4 * u3-2 * u1^5 * u2^4
* u3-4 * u1^2 * u2^5 * u3-4 * u1^4 * u2^5 * u3-2 * u1 * u2^6 * u3+2 * u1^5 * u2^6
* u3+2 * u1^2 * u2^7 * u3+2 * u1^4 * u2^7 * u3-4 * u1^3 * u2^4 * u3^2+4 * u1^3 * u2
^6 * u3^2+2 * u1^2 * u2^3 * x1+2 * u1^4 * u2^3 * x1-2 * u1 * u2^4 * x1+2 * u1^5 * u2
^4 * x1-4 * u1^2 * u2^5 * x1-4 * u1^4 * u2^5 * x1+2 * u1 * u2^6 * x1-2 * u1^5 * u2^6
* x1+2 * u1^2 * u2^7 * x1+2 * u1^4 * u2^7 * x1+4 * u1^3 * u2^4 * x1^2-4 * u1^3 * u2
^6 * x1^2+2 * u1 * u2^4 * x3-2 * u1^5 * u2^4 * x3-2 * u2^5 * x3+2 * u1^2 * u2^5 * x3
+2 * u1^4 * u2^5 * x3-2 * u1^6 * u2^5 * x3-2 * u1 * u2^6 * x3+2 * u1^5 * u2^6 * x3
-4 * u1^3 * u2^4 * u3 * x3+4 * u1^2 * u2^5 * u3 * x3-4 * u1^4 * u2^5 * u3 * x3+4 *
u1^3 * u2^6 * u3 * x3-4 * u1^3 * u2^4 * x1 * x3+4 * u1^2 * u2^5 * x1 * x3-4 * u1^4 *
u2^5 * x1 * x3+4 * u1^3 * u2^6 * x1 * x3;

It is easy to check that Thébault-Taylor Theorem is generically true for C_1, one out of the 4 components, and generically false for others.

Recently, a generic program for this algorithm written in MAPLE is available from authors, which runs much faster than those written in MATHEMATICA does. The CPU time of running the MAPLE program for Example 3 on the same 486/33 PC computer was about 30 seconds.

References

[1] Chou S C. Mechanical Geometry Theorem Proving. D. Reidel Publishing Company, (Amsterdam) 1988.

[2] Wu Wen-tsun. On the decision problem and the mechanization of theorem proving in elementary geometry. Scientia Sinica, 1978(21): 157 - 179.

[3]Wu Wen-tsun. On reducibility problem in mechanical theorem proving of elementary geometries. Chinese Quarterly J. Math. 1987,2(2):1-19.

[4]Yang Lu, Zhang Jingzhong. Searching dependency between algebraic equations: An algorithm applied to automated reasoning. in Articial Intelligence in Mathematics, IMA Conference Proceedings, Oxford University Press, 1994:147-156.

[5]Yang Lu, Zhang Jingzhong, Hou Xiaorong. A criterion of dependency between algebraic equations and its applications. Proceedings of the 1992 International Workshop on Mathematics Mechanization, Wu and Cheng (eds), International Academic Publishers, 1992:110-134.

[6]Zhang Jingzhong, Yang Lu. A method to overcome the reducibility difficulty in mechanical theorem proving. I. C. T. P. preprint IC/89/263.

[7]Zhang Jingzhong, Yang Lu, Hou Xiaorong. W. E. complete method for automated theorem proving in geometry. J. Sys. Sci. and Math. Sci. (Chinese ed.), 1995, 15(3), to appear.

[8]Zhang Jingzhong, Yang Lu, Hou Xiaorong. The sub-resultant method for automated theorem proving. J. Sys. Sci. and Math. Sci. (Chinese ed.), 1995, 15(2):10-15.

[9]Zhang Jingzhong, Yang Lu, Hou Xiaorong. A criterion for dependency of algebraic equations with applications to automated theorem proving. Science in China A 1994, 37 (5):547-554.

几何定理机器证明的 WE 完全方法

张景中　杨　路　侯晓荣

（中国科学院成都计算机应用研究所）

摘　要：在几何定理机器证明的各种方法中，吴氏方法获得了显著的成功. 如预先把有关代数簇分解为不可约簇，则吴氏方法可成为完全方法. 本文在吴法的基础上，以辗转伪除法为辅助工具，发展出一种不必预先分解代数簇的完全方法，并给出一些手算实例.

关键词：机器证明，完全方法，辗转伪除法.

1 引　　言

一般而言，要检验的一个几何命题，其假设部分对应于某个代数簇，这个代数簇如为可约，有时它可以分解为两个或更多的非退化子簇. 所谓完全方法是：用它可以判明命题的结论究竟在命题的假设所对应代数簇的哪些非退化子簇上成立，在哪些子簇上不成立，而不是简单地回答结论是否一般成立.

吴法对不可约升列是充分必要的，所以，如果预先把有关代数簇分解为若干不可约簇，再分别一一用吴法检验，就是一种完全方法. 在吴文俊教授的专著中[2]，已经阐明确实有分解代数簇为不可约簇的机械方法，但工作量较大. 周咸青成功地发展了这一方法，并用它解决了一些需要预先分解代数簇的繁难的几何问题. 在他的专著文[13]中，称这种预先把有关代数簇分解为不可约簇，然后用吴法检验的一套方法为"吴氏完全方法"

但是，这种分解工作基本上是"盲目"的，很可能做一些不必要的"虚功"，即对一些实际上已不必分解的簇再作分解；同时在初等几何范围之外，代数簇的预先分解有时要相当大的工作量. 所以，从理论、效率以及吴法在更广大范围的运用上来看，预先分解代数簇的方法并非理想的方法.

本文用辗转伪除法补充吴法，发展出一种不必预先将代数簇分解为不可约簇的完全方法. 即：WE 法. 这里 WE 意指"用欧几里得除法补充了的吴氏方法". 使用 WE 法，一般而言，对有关代数簇仅作必要的分解，而且分解步骤也简单得多.

本文刊于《系统科学与数学》，第 15 卷 3 期，1995 年.

2 升列与机器证明

设 D 为交换整环,K 为 D 的比域. 给定升列 AS

$$\begin{cases} f_1=f_1(u,x_1), \\ f_2=f_2(u,x_1,x_2) \\ \quad\cdots \\ f_s=f_s(u,x_1,x_2,\cdots,x_s). \end{cases}$$

这里 $f_i\in D[u,x_1,\cdots,x_i]$,用 I_i 记 f_i 关于 x_i 最高次项的系数. 如果每个 f_i 在多项式环 $K(u)[x_1,\cdots,x_i]/(f_1,\cdots,f_{i-1})$中不可约,则称升列 AS 是不可约的. 此时序列

$$F_0=K(u),F_1=F_0[x_1]/(f_1),\cdots,F_s=F_{s-1}[x_s]/(f_s)=F_0[x]/(f_1,\cdots,f_s)$$

是域的扩张塔.

一个升列 AS:$f_1,f_2,\cdots,f_s$,如果对每个 $j=2,\cdots,s$,I_j 不是环 $K(u)[x_1,\cdots,x_j]/(f_1,\cdots,f_{j-1})$中的零因子,且 I_1 不是零,则称其为一个真升列,记为 p-AS.

在定理机器证明的吴氏方法中,升列的概念实际上应被真升列来取代,这是因为,吴氏非退化条件有可能与其他的假设相矛盾. 在文[13]中指出了这种矛盾情形出现的可能性,同时也说明,对于大量的实际上机判断的几何命题而言,出现这种矛盾现象是极罕见的. 但在理论上这仍是一个缺口. 利用后文将给出的 WE 分解算法,可以把一个升列分解为一些真升列,在这个意义下,吴氏定理证明方法总是导致如下问题:给定真升列 p-AS:$f_1,f_2,\cdots,f_s$,以及多项式 g,在 $f_1=0,f_2=0,\cdots,f_s=0$ 的条件下,是否有 $g=0$? 根据吴氏余式公式. 若 $\mathrm{prem}(g,f_s,\cdots,f_2,f_1)=0$,则 $g=0$ 一般成立;若 $\mathrm{prem}(g,f_s,\cdots,f_2,f_1)\neq 0$,且 p-AS 不可约,则 $g=0$ 一般不成立;而若 $\mathrm{prem}(g,f_s,\cdots,f_2,f_1)\neq 0$,但 p-AS 可约,就要另作考虑,这就是所谓的可约性困难.

3 辗转伪除与最大公因式

设 D,K 如前,$f(x)$,$g(x)\in D[x]$.

若 $f\neq 0$,令 $r_1=\mathrm{prem}(g,f,x)$. 若 $r_1\neq 0$,令 $r_2=\mathrm{prem}(f,r_1,x)$. 若 $r_2\neq 0$,令 $r_3=\mathrm{prem}(r_1,r_2,x)$. 如此继续下去,因

$$\deg(f,x)>\deg(r_1,x)>\deg(r_2,x)>\deg(r_3,x)>\cdots,$$

并且 $\deg(f,x)$为一非负整数,所以这样作了有限次后,有

$$r_1=p_1g+q_1f,\qquad r_2=p_2f+q_2r_1,\qquad r_3=p_3r_1+q_3r_2,$$
$$\cdots,$$
$$r_{s-1}=p_{s-1}r_{s-3}+q_{s-1}r_{s-2},\qquad r_s=p_sr_{s-2}+q_sr_{s-1}.$$

且 $r_s\in D$,$r_{s-1}\notin D$.

易知,有 $p(x)$,$q(x)$,$u(x)$,$v(x)$,使得

$$r_{s-1}=p(x)g(x)+q(x)f(x),\qquad r_s=u(x)g(x)+v(x)f(x).$$

上述由 f,g 得到 r_s 的过程称为辗转伪除法.

定理 1 记号如上. 如果 $r_s=0$,则 $f(x),g(x)$在 $K[x]$中的最大公因子是 r_{s-1}. 否则, $f(x),g(x)$没有关于 x 的公因子.

给定真升列 $p-AS: f_1,f_2,\cdots,f_s$ 和多项式 $f,g\in D[x,x_1,\cdots,x_s]$,令

$$r_0=r_0(g,f,x):=\text{prem}(f,f_s,\cdots,f_1).$$
$$r_1=r_1(g,f,x):=\text{prem}(\text{prem}(g,r_0,x),f_s,\cdots,f_1),$$
$$r_2=r_2(g,f,x):=\text{prem}(\text{prem}(r_0,r_1,x),f_s,\cdots,f_1),$$
$$r_3=r_3(g,f,x):=\text{prem}(\text{prem}(r_1,r_2,x),f_s,\cdots,f_1),$$
$$\cdots$$
$$r_{t-1}=r_{t-1}(g,f,x):=\text{prem}(\text{prem}(r_{t-3},r_{t-2},x),f_s,\cdots,f_1),$$
$$r_t=r_t(g,f,x):=\text{prem}(\text{prem}(r_{t-2},r_{t-1},x),f_s,\cdots,f_1),$$

其中 $r_t\in D,r_{t-1}\notin D$.

特别记

$$r_{t-1}=\gcd(g,f,x),\qquad r_t=\text{Eprem}(g,f,x),$$

这里 $\text{Eprem}(g,f,x)$称为 g,f(模$(f_1,\cdots,f_s)$)关于变元 x 的辗转伪余式. 又令

$$e_s:=g,\qquad e_{s-1}:=\text{Eprem}(e_s,f_s,x_s),$$
$$e_{s-2}:=\text{Eprem}(e_{s-1},f_{s-1},x_{s-1}),$$
$$\cdots$$
$$e_1:=\text{Eprem}(e_2,f_2,x_2),\qquad e_0:=\text{Eprem}(e_1,f_1,x_1).$$

e_0 称为 g 关于 p-AS 的辗转伪余式,记为 $\text{Eprem}(g,f_s,\cdots,f_2,f_1)$. 易知

定理 2 记号如上. 如果 $\text{Eprem}(g,f_s,\cdots,f_2,f_1)\neq 0$,则在条件 $f_1=0,f_2=0,\cdots,f_s=0$ 之下,g 一般不为零. 即在$\{f_1=0,f_2=0,\cdots,f_s=0\}$的所有非退化分支中,没有一个使得 $g=0$. 设

$$f_1(u,x_1),f_2(u,x_1,x_2),\cdots,f_s(u,x_1,x_2,\cdots,x_s)$$

是一真升列,$f_i\in D[u,x_1,\cdots,x_i]$,且 $f(y),g(y)\in D[u,x_1,\cdots,x_s][y]$,又设 J_i 为 $r_i(g,f,y)$的最高次项系数. $R_i:=\text{Eprem}(r_i(g,f,y),f_s,\cdots,f_2,f_1)$.

定理 3 如果对所有的 i,$\text{Eprem}(J_i,f_{s-1},\cdots,f_2,f_1)\neq 0$ 且 $R_k\neq 0,k<j,r_j(g,f,y)=0$,则 $r_{j-1}(g,f,y)$是 $f(y),g(y)$在环

$$D(u)[x_1,\cdots,x_s]/(f_1,\cdots,f_s)$$

上的最大公因子.

证 设$\{f_1^*,\cdots,f_s^*\}$是$\{f_1,\cdots,f_s\}$的一不可约子升列,由条件可知(利用定理 2)

$$r_j(g,f,y)=0\quad \text{mod}(f_1^*,\cdots,f_s^*),\qquad r_k(g,f,y)\neq 0,\quad \text{mod}(f_1^*,\cdots,f_s^*),k<j$$

及 $J_i\neq 0 \text{ mod}(f_1^*,\cdots,f_s^*)$. 所以,由定理 1 可知 $r_{j-1}(g,f,y)$是 $f(y),g(y)$在域

$$D(u)[x_1,\cdots,x_s]/(f_1^*,\cdots,f_s^*)$$

上的最大公因子. 又根据$\{f_1^*,\cdots,f_s^*\}$选择的任意性,$r_{j-1}(g,f,y)$也是 $f(y),g(y)$在环

$$D(u)[x_1,\cdots,x_s]/(f_1,\cdots,f_s)$$

上的最大公因子.

4 WE 分解算法

给定真升列 p-AS:$f_1,\cdots,f_s$ 和多项式 g. 如果 $\text{prem}(g,f_s,\cdots,f_1)=0$ 或 $\text{Eprem}(g,f_s,\cdots,f_1)\neq 0$ 两者必居其一,就称 p-AS 关于 g 是 WE 单纯的.

若真升列 p-AS 被分解为 $C_1,C_2,\cdots,C_m(m\geqslant 1)$,使得每个 $C_i(1\leqslant i\leqslant m)$关于 g 是 WE 单纯的,则称$\{C_1,C_2,\cdots,G_m\}$是 p-AS 关于 g 的一个 WE 分解. 特别地,p-AS 本身可以是 p-AS 关于 g 的一个 WE 分解.

下面给出 p-AS 关于 g 的一个 WE 分解算法.

如果 p-AS 只有一个多项式$\{f_1\}$,则如果 $\text{prem}(g,f_1,x_1)=0$ 或 $\text{Eprem}(g,f_1,x_1)\neq 0$,那么$\{f_1\}$本身即为$\{f_1\}$关于 g 的一个 WE 分解;否则 $\text{prem}(g,f_1,x_1)\neq 0$,$\text{Eprem}(g,f_1,x_1)=0$. 由定理 1 可知 $\gcd(g,f_1,x_1)$是 $g(u,x_1)$,$f_1(u,x_1)$在 $K(u)$上的最大公因子,因而$\{f_1\}$可被分解为$\{f'_1\}$,$\{f''_1\}$,其中 $f'_1=\gcd(g,f_1,x_1)$,f''_1是 f'_1除 f_1 的伪商. $\{f'_1\}$,$\{f''_1\}$即为$\{f_1\}$的一个 WE 分解. 一般地,有

步骤 1 计算 $\text{prem}(g,f_s,\cdots,f_1)$,若其为零,结束. 否则,令 $g^*:=\text{prem}(g,f_s,\cdots,f_1)$,如果 $\text{Eprem}(g^*,f_s,\cdots,f_1)\neq 0$,结束.

步骤 2 如果 $\text{Eprem}(g^*,f_s,\cdots,f_1)=0$,且设 k 是使得

$$\text{Eprem}(r_k(g^*,f_s,x_s),f_{s-1},\cdots,f_1)\neq 0$$

的最大整数. 若

$$r_{k+1}(g^*,f_s,x_s)=0,$$

而 $\text{Eprem}(J_i,f_{s-1},\cdots,f_1)\neq 0$,$i\leqslant k$(其中 J_i 为 $r_i(g^*,f_s,x_s)$)关于 x_s 最高次项的系数),那么,由定理 3,在环 $D(u)[x_1,x_2,\cdots,x_{s-1},x_s]/(f_1,\cdots,f_{s-1})$上 g^* 与 f_s 有最大公因子 $r_k(g^*,f_s,x_s)$. 记 $f'_s=r_k(g^*,f_s,x_s)$,f''_s为 f'_s除 f_s 的伪商,从而得到两个升列

$$\{f_1,\cdots,f_{s-1},f'_s\},\qquad \{f_1,\cdots,f_{s-1},f''_s\},$$

返回步骤 2 继续分解.

步骤 3 如果 $\text{Eprem}(r_k(g^*,f_s,x_s),f_{s-1},\cdots,f_1)\neq 0$,$r_{k+1}=0$,但有 $i\leqslant k$,使得 $\text{Eprem}(J_i,f_{i-1},\cdots,f_1)=0$,则任选这样的一个 J_i,并用 J_i 代替 g,用$\{f_1,\cdots,f_{s-1}\}$代替$\{f_1,\cdots,f_s\}$,返回步骤 1 继续分解. 设$\{\overline{f}_1,\cdots,\overline{f}_{s-1}\}$为其一个子升列,则返回升列$\{\overline{f}_1,\cdots,\overline{f}_{s-1},f_s\}$到步骤 2,继续分解.

步骤 4 如果 $\text{Eprem}(r_k(g^*,f_s,x_s),f_{s-1},\cdots,f_1)\neq 0$,$\text{Eprem}(r_{k+1}(g^*,f_s,x_s),f_{s-1},\cdots,f_1)=0$,$r_{k+1}(g^*,f_s,x_s)\neq 0$,则取 $r_{k+1}(g^*,f_s,x_s)$关于 x_s 任一幂次的系数,并用它代替 g,用$\{f_1,\cdots,f_{s-1}\}$代替$\{f_1,\cdots,f_s\}$返回步骤 1,对$\{f_1,\cdots,f_{s-1}\}$进行分解,设$\{\overline{f}_1,\cdots,\overline{f}_{s-1}\}$为其一子升列,则返回升列$\{\overline{f}_1,\cdots,\overline{f}_{s-1},f_s\}$到步骤 2,继续分解.

由于上述每一步返回或者使 s 减少或者使升列中多项式的次数降低,故有限步后必然终止.

为了把一个升列 AS:$f_1,\cdots,f_s$ 分解为关于 g 是单纯的一些子升列,在步骤 1 之前,需要首先作$\{f_1,\cdots,f_j\}$关于 I_{j+1}(即 f_{j+1}关于 x_{i+1}的最高次项系数)的 WE 分解. 分解应按 $j=1,2,\cdots,s-1$ 次序逐个进行.

目前,WE 方法尚未作上机试验. 附录提供几个手算实例以说明其应用步骤.

附录一　几个例子

例 1　设

$$H:\begin{cases}f_1=x_1-u=0,\\ f_2=x_2^2-2x_1x_2+u^2=0.\end{cases}$$

检验等式 $g=x_2-x_1=0$.

解　因 $\mathrm{prem}(g,f_1,f_2)=x_2-u\neq0$,不能作断言.继续用 WE 法.将 $g^*=x_2-u$ 与 f_2 辗转相除,得余式

$$R_1=\mathrm{Eprem}(g^*,f_2,x_2)=ux_1-u^2.$$

对 R_1 作零因子检验,发现 $\mathrm{prem}(R_1,f_1,x_1)=0$,从而得到 f_2 之分解式(在 $f_1=0$ 条件下) $f_2=(x_2-u)(x_2-2x_1+u)$.于是得到两个升列

$$C_1=\{x_1-u,x_2-u\},\qquad C_2=\{x_1-u,x_2-2x_1+u\},$$

而且

$$\mathrm{prem}(g,x_1-u,x_2-u)=0,\qquad \mathrm{prem}(g,x_1-u,x_2-2x_1+u)=0.$$

这表明 $g=0$ 在 $V(C_1)$与 $V(C_2)$上皆真.从而在假设 H 下有 $g=0$.

此题也可不计算 $\mathrm{prem}(g,f_1,f_2)$,直接将 g 与 f_2 辗转相除,得 $R_1=\mathrm{Eprem}(g,f_2,x_2)=u^2-x_1u$,然后由 $\mathrm{prem}(R_1,f_1,x_1)=0$ 得到 $f_2=(x_2-x_1)^2$,从而把升列$\{f_1,f_2\}$分解为两个相同的升例$\{f_1,x_2-x_1\}$,更为简捷;但一般情形下,仍应先计算 $\mathrm{prem}(g,f_1,\cdots,f_s)$.

例 2

$$H:\begin{cases}f_1=x_1^2-ux_1=0,\\ f_2=x_2^2+x_1x_2-u^2=0.\end{cases}$$

$$G:g=x_1+x_2+1=0.$$

解　$\mathrm{Eprem}(g,f_1,f_2)=(u^2-1)(u^2-u-1)\neq0$,故结论 G 一般不真.

例 3

$$H:\begin{cases}f_1=x_1^2+ux_1,\\ f_2=x_2^2+(x_1^2-u)x_2+u^2x_1=0.\end{cases}$$

$$G:g=x_2^2+x_1(1-u)x_2-ux_1^2=0.$$

解　$\mathrm{prem}(g,f_1,f_2)=(x_1+u)x_2\neq0$,不能作结论.把 $g^*=(x_1+u)x_2$ 与 f_2 辗转相除

$$r_{-1}=f_2,\qquad r_0=g^*,\qquad r_1=-u(x_1+u)x_1^2.$$

对 $lc(r_0)=x_1+u$ 作零因子检验:将 x_1+u 与 f_1 辗转相除,余式为 0,乃得 f_1 的分解 $f_1=x_1(x_1+u)$.于是得到两个升列

$$C_1=\{x_1,f_2\},\qquad C_2=\{x_1+u,f_2\}.$$

把 $V(G_2)$上有 $\mathrm{prem}(g,x_1+u,f_2)=0$,结论 G 为真.在 $V(C_1)$上 $\mathrm{prem}(g,x_1,f_2)=ux_2\neq0$,把 ux_2 与 f_2 辗转相除

$$r_{-1}=f_2,r_0=ux_2,r_1=u^3x_1.$$

对 r_1 作零因子检验,发现在 $V(C_1)$上有 $r_1=0$,于是得到 f_2 的分解

$$f_2=x_2(x_2+(x_1^2-u)).$$

从而把 C_1 又分解为

$$C_3=\{x_1,x_2\},C_4=\{x_1,x_2+(x_1^2-u)\}.$$

由于 $V(C_3)$上有 $\mathrm{prem}(g,x_1,x_2)=0$,在 $V(C_4)$上有 $\mathrm{Eprem}(g,x_1,x_2+(x_1^2-u))=u^2\neq0$,故知结论 G 在 $V(C_3)$上真,在 $V(G_4)$上假.

最后结论:$\{f_1,f_2\}$关于 g 有 WE 分解

$$C_2=\{x_1+u,f_2\},C_3=\{x_1,x_2\},C_4=\{x_1,x_2+(x_1^2-u)\}.$$

而结论 $G:g=0$ 在 $V(G_2)$,$V(G_3)$ 上真,在 $V(G_4)$ 上假.

例 4 在△ABC 两边 AC,BC 上分别作正方形 $ACDE$ 和 $BCFG$,设 M 是 AB 中点. 问是否有等式 $DF=2CM$.

解 按文[13],令 $A=(u_1,0)$,$B=(u_2,u_3)$,$C=(0,0)$,$D=(0,u_1)$,$F=(x_1,x_2)$,$M=(x_3,x_4)$,则诸假设条件为

$$h_1=x_2^2+x_1^2-u_2^2-u_3^2=0(CF=BC),$$

$$h_2=u_3x_2+u_2x_1=0(CF\perp BC),$$

$$\left.\begin{aligned}h_3=2x_3-u_2-u_1=0,\\ h_4=2x_4-u_3=0,\end{aligned}\right\}M\text{ 是 }AB\text{ 中点}.$$

结论为

$$g=4x_4^2+4x_3^2-x_2^2+2u_1x_2-x_1^2-u_1^2=0(DF=2CM).$$

从 $\{h_1,\cdots,h_4\}$ 得到升列 $\{f_1,\cdots,f_4\}$,使假设成为

$$f_1=(u_3^2+u_2^2)x_1^2-u_3^4-u_2^2u_3^2=0,f_2=u_3x_2+u_2x_1=0,$$

$$f_3=2x_3-u_2-u_1=0,f_4=2x_4-u_3=0,$$

求出 $\text{prem}(g,f_1,\cdots,f_4)=2u_1u_2u_3(u_3-x_1)\neq 0$,故尚不能作定论. 在非退化条件 $u_1u_2u_3\neq 0$ 之下,取 $g^*=x_1-u_3$. 因 g^* 中不含 x_2,x_3,x_4,故直接将 g^* 与 f_1 辗转相除

$$r_{-1}=f_1,r_0=x_1-u_3,r_1=0.$$

从而将 f_1 分解:$f_1=(u_3^2+u_2^2)(x_1+u_3)(x_1-u_3)$. 于是升列 $\{f_1,f_2\}$ 分解为

$$C_1=\{(u_3^2+u_2^2)x_1+u_3,f_2,f_3,f_4\},\qquad C_2=\{(x_1-u_3),f_2,f_3,f_4\},$$

在 $V(C_1)$ 上有 $\text{Eprem}(g^*,x_1+u_3,f_2,f_3,f_4)\neq 0$,在 C_2 上有 $\text{prem}\{g^*,x_1-u_3,f_2,f_3,f_4\}=0$. 故结论在 $V(C_2)$ 上真,在 $V(C_1)$ 上假.

这一结论的几何意义是:两正方形应同时向外或同时向内,才能保证结论成立.

例 5(蝴蝶定理) 设 A,B,C,D 是⊙O 上的 4 个点,E 是 AC,BD 的交点,过 E 作 OE 之垂线交 AD 于 F,交 BC 于 G. 检验是否有 $EG=FE$.

解 按文[13],令 $E=(0,0)$,$O=(u_1,0)$,$A=(u_2,u_3)$,$B=(x_1,u_4)$,$C=(x_3,x_2)$,$D=(x_5,x_4)$,$F=(0,x_6)$,$G=(0,x_7)$,则假设条件为

$$h_1=-x_1^2+2u_1x_1-u_4^2+u_3^2+u_2^2-2u_1u_2=0(OA=OB),$$

$$h_2=-x_3^2+2u_1x_3-x_2^2+u_3^2+u_2^2-2u_1u_2=0(OA=OC),$$

$$h_3=-u_3x_3+u_2x_2=0(C,A,E\text{ 共线}),$$

$$h_4=-x_5^2+2u_1x_5-x_4^2+u_3^2+u_2^2-2u_1u_2=0(OA=OD),$$

$$h_5=-u_4x_5+x_1x_4=0(D,B,E\text{ 共线}),$$

$$h_6=(-x_5+u_2)x_6+u_3x_5-u_2x_4=0(F,A,D,\text{共线}),$$

$$h_7=(-x_3+x_1)x_7+u_4x_3-x_1x_2=0(G,B,C,\text{共线}).$$

结论为 $g=x_7+x_6=0$. 从 $h_1,\cdots,h_7$ 得到升列 $\{f_1,\cdots,f_7\}$,使得假设条件成为

$$f_1=-x_1^2+2u_1x_1-u_4^2+u_3^2+u_2^2-2u_1u_2=0,$$
$$f_2=-(u_3^2+u_2^2)x_2^2+2u_1u_2u_3x_2+u_3^4+(u_2^2-2u_1u_2)u_3^2=0,$$
$$f_3=-u_3x_3+u_2x_2=0,$$
$$f_4=-(2u_1x_1+u_3^2+u_2^2-2u_1u_2)u_4^2+2u_1u_4x_1x_4+(u_3^2+u_2^2-2u_1u_2)u_4^2=0,$$
$$f_5=-u_4x_5+x_1x_4=0,$$
$$f_6=(-x_1x_4+u_2u_4)x_6+u_4u_3x_5-u_2u_4x_4=0,$$
$$f_7=(-u_2x_2+u_3x_1)x_7+u_4u_3x_3-u_3x_1x_2=0,$$

求得

$$\text{prem}(g,f_1,\cdots,f_7)=(u_3x_1-u_2u_4)[((x_1-u_2)x_2+u_3x_1)x_4-u_2u_4x_2]\neq 0,$$

尚不能作定论. 令 $g^*=\text{prem}(g,f_1,\cdots,f_7)$,在 g^* 中不含 x_5,x_6,x_7. 故直接将 g^* 与 f_4 辗转相除. 由于 g^* 中已分出因子 $(u_3x_1-u_2u_4)$,可先对它作零因子检验:求出 $\text{Eprem}(u_3x_1-u_2u_4,f_1,x_1)\neq 0$,知其不是零因子,令 $g^{**}=g^*/(u_3x_1-u_2u_4)$,将 g^{**} 与 f_4 辗转相除得

$$r_{-1}=f_4,r_0=g^{**}=((x_1-u_2)x_2+u_3x_1)x_4-u_2u_4x_2,$$
$$r_1=2u_3u_4^2(x_1-u_2)x_1[(u_2^2+u_3^2)x_2+u_3(u_3^2+u_3^2-2u_1u_2)].$$

对 r_1 作零因子检验,发现 (x_2-u_2),x_1 均不是零因子. 于是取 $r_1^*=r_1/(2u_3u_4^2(x_1-u_2)x_1)$. 对 r_1^* 的检验发现 $\text{Eprem}(r_1^*,f_2,x_2)=0$,从而得到 f_2 的分解 $f_2=(-x_2+u_3)r_1^*$,同时把 $\{f_1,\cdots,f_7\}$ 分解为两个升列

$$C_1=\{f_1,-x_2+u_3,f_3,\cdots,f_7\},C_2=\{f_1,r_1^*,f_3,\cdots,f_7\},$$

在 $V(C_1)$ 上有 $\text{Eprem}(g^*,f_1,-x_2+u_3,f_3,\cdots,f_7)\neq 0$. 在 $V(C_2)$ 上有 $r_1=0$,从而可以将 f_4 分解. 这里,因 r_1^* 关于 x_2 为 1 次. 故 $r_0=g^{**}$ 的系数可约化,改写为

$$\tilde{r}_0=(2u_1x_1+u_2^2+u_3^2-2u_1u_2)x_4+u_4(u_2^2+u_3^2-2u_1u_2).$$

于是 f_4 的分解有形式 $f_4=(-x_4+u_4)\tilde{r}_0$. 这就把升列 C_2 又分解为

$$C_3=\{f_1,r_1^*,f_3,-x_4+u_4,f_5,f_6,f_7\},\qquad C_4=\{f_1,r_1^*,f_3,\tilde{r}_0,f_5,f_6,f_7\}.$$

显然有 $\text{prem}\{g^{**},f_1,r_1^*,f_3,\tilde{r}_0\}=0$,而 $\text{Eprem}\{g^{**},f_1,r_1^*,f_3,-x_4+u_4\}\neq 0$,从而得出结论:升列 $\{f_1,\cdots,f_7\}$ 关于 g 有 WE 分解:

$$C_1=\{f_1,-x_2+u_3,f_3,\cdots,f_7\},$$
$$C_3=\{f_1,(u_2^2+u_3^2)x_2+u_3(u_2^2+u_3^2-2u_1u_2),f_3,-x_4+u_4,f_5,f_6,f_7\},$$
$$C_4=\{f_1,(u_2^2+u_3^2)x_2+u_3(u_2^2+u_3^2-2u_1u_2),f_3,(2u_1x_1+u_2^2+u_3^2-2u_1u_2)x_4+u_4(u_2^2+u_3^2-2u_1u_2),f_5,f_6,f_7\}.$$

而结论 $g=0$ 仅在 $V(C_4)$ 上为真.

在文[13]与[20]中均曾指出:几何定理机器证明产生上述例 4 和例 5 的情形,原因常常是人们把几何条件转述为代数形式时的疏忽. 本文表明,用 WE 方法可以机械地发现并纠正这种疏忽,把一些不合理的情形清除.

此外,WE 分解比起预先进行的不可约分解来说,既易于进行,也易于检验.

参考文献

[1]Wu Wen-tsun. On the decision problem and the mechanization of theorem proving in elementary geometry. Scientia Sinica,1978(21):157－179.

[2]吴文俊. 几何定理机器证明的基本原理. 北京:科学出版社,1984.

[3]Hu Sen and Wang Dongming. Fast factorization of polynomials over rational number field or its extension fields. Kexue Tongbao,1986(31):150－156.

[4]高小山. 三角恒等式与初等几何定理的机械化证明. 系统科学与数学,1987,7(3):264－272.

[5]Wu Wen-tsun. On zeros of algebraic equation: an application of Ritt principle. Kexue Tongbao,1986(31):1－5.

[6]Li Ziming. On the triangulation for any finite polynomials set(Ⅰ), M-M research preprints. 1987(2):48－54.

[7]Liu Zhuojun. Algorithm of decomposing high degree polynomials, M-M research preprints. 1987(2):62－67.

[8]洪加威. 能用例证法来证明几何定理吗？中国科学,1986(3):234－242.

[9]洪加威. 近似计算有效数字的增长不超过几何级数. 中国科学,1986(3):225－233.

[10]张景中,杨路. 定理机械化证明的数值并行及单点例证法原理概述. 数学的实践与认识,1989(1):34－43.

[11]邓米克. 证明构造性几何定理的数值并行法. 科学通报,1988(24):1851－1854.

[12]Chou S C and Schelter W F. Proving geometry theorems with rewrite rules. Journal of Automated Reasoning,1986,2(4):253－273.

[13]Chou S C. Mechanical Geometry Theorem Proving. D. Reidel Publishing Company,1987.

[14]Chou S C,Gao X S. Mechanical Formula Derivation in Elementary Geometries. TR-89-21(Preprint),1989.

[15]Chou S C, Gao X S. Ritt-Wu's Decompostion Algorithm and Geometry Theorem Proving. TR-89-09(preprint),1980.

[16]Chou S C, Gao X S. A Collection of 120 Computer Solved Geometry Problems in Mechanical Formula Derivation, TR-89-22(preprint),1989.

[17]Chou S C,Gao X S. Mechanical Theorem Proving in Differential Geometry. TR-89-08 (preprint),1989.

[18]Chou S C,Gao X S. Automated Reasoning in Mechanics Using Ritt-Wu's Method. TR-89-11(preprint),1989.

[19]Chou S C,Gao X S. A Class of Geometry Statements and Geometry Theorem Proving. TR-89-37(preprint),1989.

[20]Wu Wen-tsun. On reducibility problem in mechanical theorem proving of elementary geometries. 数学年刊,1987,2(2):1－19.

几何定理机器证明的结式矩阵法

张景中　杨　路　侯晓荣

（中国科学院成都计算机应用研究所）

摘　要：本文提出了一种不必预先分解升列为不可约子列而克服所谓“可约性困难”的方法. 由于使用了吴除法及子结式计算，我们也称这种方法为 WR 分解算法.

关键词：升列，吴除法，子结式.

1　引　　言

吴文俊教授所建立的处理多元代数方程组的整序算法——吴消元法[1,2]，是解决非线性数学问题的一个重要手段. 著名的几何定理机器证明的吴法，是其重要应用之一. S. C. Chou 在其专著[3]中，详细地阐述了吴法，并用丰富的资料(例如，列举了用吴法在计算机上证明了的 512 条非平凡的欧氏几何定理)表明吴法的成功.

用吴法处理定理证明或其它问题时，若涉及的多项式组的特征列为可约升列时，有可能遇到困难. 对这一困难，吴文俊教授在文[4]中曾提出过有效的处理手段，但从理论上彻底解决这一问题，有赖于多元多项式在代数域上的因式分解.

本文给出的方法，我们也称其为 WR 法，是一种不用事先的因式分解(且不要求有分解算法)而克服可约性困难的方法. 顺便说，这里 W 是指吴氏伪除，R 是指结式.

2　子结式与公因式

设 D 是一整环，F 是其商域，K 为 F 的代数封闭域. 令

$$f=a_nx^n+a_{n-1}x^{n-1}+\cdots+a_0\in D[x],\quad g=b_mx^m+b_{m-1}x^{m-1}+\cdots+b_0\in D[x],$$

$$m_i(f,n,g,m)=\underbrace{\begin{bmatrix} a_n & a_{n-1} & \cdots & a_0 & & & \\ & a_n & a_{n-1} & \cdots & a_0 & & \\ & & & \cdots & & & \\ & & & a_n & a_{n-1} & \cdots & a_0 \\ b_m & b_{m-1} & \cdots & b_0 & & & \\ & b_m & b_{m-1} & \cdots & b_0 & & \\ & & & \cdots & & & \\ & & & b_m & b_{m-1} & \cdots & b_0 \end{bmatrix}}_{m+n-i\text{列}}\begin{matrix} \left.\vphantom{\begin{matrix}a\\a\\a\\a\end{matrix}}\right\}m-i\text{行} \\ \left.\vphantom{\begin{matrix}a\\a\\a\\a\end{matrix}}\right\}n-i\text{行}\end{matrix}$$

本文刊于《系统科学与数学》，第 15 卷第 1 期，1995 年.

矩阵 $m_i(f,n,g,m)$ 的第 $1,2,\cdots,m+n-2i-1,m+n-i-j(j\leqslant i)$ 列组成的方阵的行列式记为 $M_j^i(f,g)$.

关于 (f,n,g,m) 的子结式序列是 $D[x]$ 中的一列多项式

$$\{S_{\mathrm{res}i}(f,n,g,m)\}_{i\in\{0,1,\cdots,\inf(m,n)-1\}},$$

其中 $S_{\mathrm{res}i}(f,n,g,m)=\sum_{j=0}^{i}M_j^i(f,g)x^j=M_i^i(f,g)x^i+\cdots+M_1^i(f,g)x+M_0^i(f,g)$.

对任一 $i\in\{0,1,\cdots,\inf(m,n)-1\}$，简记

$$P_i(f,g)=S_{\mathrm{res}i}(f,n,g,m),s_i(f,g)=M_j^i(f,g).$$

关于 (f,n,g,m) 的子结式序列有如下的重要性质：

引理 如果 a_n,b_m 不同时为零，且

$$s_0(f,g)=\cdots=s_{k-1}(f,g)=0,s_k(f,g)\neq 0.$$

则 $P_k(f,g)$ 为 f,g 在 $F[x]$ 中的最大公因子.

证 考虑 $k=1$ 时的情形.

设 $f=(x+c_0)f',g=(x+c_0)g'$,

$$f'=a_nx^{n-1}+a'_{n-1}x^{n-2}+\cdots+a'_1,g'=b_mx^{m-1}+b'_{m-1}x^{m-2}+\cdots+b'_1,$$

则

$$f=a_nx^n+(a'_{n-1}+c_0a_n)x^{n-1}+(a'_{n-2}+c_0a'_{n-1})x^{n-2}+\cdots+(+a'_1c_0),$$

$$g=b_mx^m+(b'_{m-1}+c_0b_m)x^{m-1}+(b'_{m-2}+c_0b'_{m-1})x^{m-2}+\cdots+(+b'_1c_0).$$

现在已知 f,g 有公因子 $x+c_0$，$s_0(f,g)$ 即为 f,g 的结式，这等价于 $s_0(f,g)=0$. f,g 有其他公因子的充要条件是 $s_0(f',g')=0$，而

$$s_0(f',g')=\begin{vmatrix} a_n & a'_{n-1} & \cdots & a'_1 & & & \\ & a_n & a'_{n-1} & \cdots & a'_1 & & \\ & & \cdots & & & & \\ & & & a_n & a'_{n-1} & \cdots & a'_1 \\ b_m & b_{m-1} & \cdots & b'_1 & & & \\ & b_m & b'_{m-1} & \cdots & b'_1 & & \\ & & \cdots & & & & \\ & & & b_m & b'_{m-1} & \cdots & b'_1 \end{vmatrix}\begin{matrix} \left.\vphantom{\begin{matrix}a\\a\\a\\a\end{matrix}}\right\}m-1\text{行} \\ \left.\vphantom{\begin{matrix}a\\a\\a\\a\end{matrix}}\right\}n-1\text{行}\end{matrix}$$

对该行列式施行初等列变换：倒数第 2 列乘以 c_0 加到倒数第 1 列，然后倒数第 3 列乘以 c_0 加到倒数第 2 列，如此变换下去，一直到第 1 列乘以 c_0 加到第 2 列，可得

$$s_0(f',g')=\begin{vmatrix} a_n & a'_{m-1}+c_0a_n & \cdots & c_0a'_1 & & & \\ & a_n & a'_{n-1}+c_0a_n & \cdots & c_0a'_1 & & \\ & & \cdots\cdots & & & & \\ & & & a_n & a'_{n-1}+c_0a_n & \cdots & c_0a'_1 \\ & & & & a_n & a'_{n-1}+c_0a_n\cdots a'_1+c_0a'_2 & \\ b_m & b'_{m-1}+c_0b_m & \cdots & c_0b'_1 & & & \\ & b_m & b'_{m-1}+c_0b_m & \cdots & c_0b'_1 & & \\ & & \cdots\cdots & & & & \\ & & & b_m & b'_{m-1}+c_0b_m & \cdots & c_0b'_1 \\ & & & & b_m & b'_{m-1}+c_0b_m\cdots b'_1+c_0b'_2 & \end{vmatrix}$$

$$=s_1(f,g).$$

即 f,g 有其他公因子的充要条件是 $s_1(f,g)=0$. 如果 $s_1(f,g)\neq 0$,则最大公因子为 $x+c_0$,而 $M_0^1(f,g)=c_0M_1^1(f,g)$(这可由 M_j^i 的定义,用前述列变换证得),所以此时命题成立.

当 $k>1$ 时,设 f,g 的最大公因子为

$$c_kx^k+c_{k-1}x^{k-1}+\cdots+c_0(c_k=1).$$

则用归纳法,依照前面对 $k=1$ 时的证明,可知此时 $s_0(f,g)=\cdots=s_{k-1}(f,g)=0,s_k(f,g)\neq 0$,并且 $M_i^k(f,g)=c_iM_k^k$,与 $k=1$ 时相同,此时,上述条件是充分必要的,所以原命题成立. 证毕.

子结式多项式序列及有关概念的定义,可以自然地推广到 D 不是整环的情形,在此不再赘述.

3　结式、真升列

我们用 $\mathrm{Res}(f,g,x)$ 表示 $s_0(f,g)=M_0^0(f,g)$,即 f,g 关于 x 的结式.

给定升列 AS:

$$\begin{cases} f_1=f_1(u,x_1), \\ f_2=f_2(u,x_1,x_2), \\ \quad\cdots \\ f_s=f_s(u,x_1,x_2,\cdots,x_s). \end{cases}$$

这里 $f_i\in D[u,x]$(D 为整环),用 I_i 记 f_i 关于 x_i 最高次项的系数. 如果每个 f_i 在多项式环

$$K(u)[x_1,\cdots,x_i]/(f_1,\cdots,f_{i-1})$$

中不可约,则称升列 AS 是不可约的. 此时序列 $F_0=K(u),F_1=F_0[x]/(f_1),\cdots,F_s=F_{s-1}[x_s]/(f_s)=F_0[x]/(f_1,\cdots,f_s)$是域的扩张塔.

设 $g\in D[u,x]$,令

$r_{s-1}:=\mathrm{Res}(g,f_s,x_s),r_{s-2}=\mathrm{Res}(r_{s-1},f_{s-1},x_{s-1})$,

　　…

$r_1:=\mathrm{Res}(r_2,f_2,x_2),r_0:=\mathrm{Res}(r_1,f_1,x_1)$.

称 r_0 为 g 关于升列 $AS:f_1,f_2,\cdots,f_s$ 的结式,记为

$$r_0=\mathrm{Res}(g,f_1,f_2,\cdots,f_s).$$

可以证明[5],若 $\mathrm{Res}(g,f_1,f_2,\cdots,f_s)\neq 0$,则 $g(u,x)=0$ 在条件$\{f_1=0,f_2=0,\cdots,f_s=0\}$之下一般不成立. 换言之,$g$ 不是 $K(u)[x]/(f_1,f_2,\cdots,f_s)$中的零因子或零.

一个升列 $AS:f_1,f_2,\cdots,f_s$,如果对每个 $j=2,\cdots,s$ 有

$$\mathrm{Res}(I_j,f_1,\cdots,f_{j-1})\neq 0,$$

且 I_1 不是零,则称其为一个真升列,记为 p-AS.

4　定理证明的 WR 法

根据 Ritt-吴整序原理,可以把条件多项式组化为一升列 $AS:f_1,f_2,\cdots,f_s$. 为了简单,不妨设升列 AS 为真升列 p-AS.

对真升列 p-AS，可顺序用 $f_s,f_{s-1},\cdots,f_1$ 去除结论多项式 g，可得 g 对 p-AS 的余式 $\mathrm{prem}(g,\mathrm{f}_1,f_2,\cdots,f_s)$，根据吴余式公式，若 $\mathrm{prem}(g,f_1,f_2,\cdots,f_s)=0$，则 $g=0$ 一般成立；如果 $\mathrm{prem}(g,f_1,f_2,\cdots,f_s)\neq 0$，$p$-$AS$ 不可约，则 $g=0$ 一般不成立；但是如果 $\mathrm{prem}(g,f_1,f_2,\cdots,f_s)\neq 0$，而 p-AS 可约，就需另作考虑，这就是所谓的可约性困难.

根据前面几节的讨论，我们提出如下的定理证明的方法，用它可以克服可约性困难而不必进行事先的因式分解.

步骤 1　$s=1$. 如果 $\mathrm{prem}(g,f_1)=0$，则 $g=0$ 一般成立，结束. 否则令 $g:=\mathrm{prem}(g,f_1)$，作结式 $\mathrm{Res}(g,f_1)$. 如果 $\mathrm{Res}(g,f_1)\neq 0$，$g=0$ 一般不成立，结束；否则，可由引理求得 f_1 的分解 $f_1=\overline{f}_1\cdot\overline{\overline{f}}_1$，分别令 $f_1:=\overline{f}_1$ 及 $f_1:=\overline{\overline{f}}_1$，返回步骤 1.

步骤 2　$s>1$. 求 $\mathrm{prem}(g,f_1,f_2,\cdots,f_s)$. 如果其为零，结束；否则，令 $g:=\mathrm{prem}(g,f_1,f_2,\cdots,f_s)$.

步骤 3　记 $r_i=\mathrm{prem}(s_i(g,f_s),f_1,\cdots,f_{s-1})$，$i\in\{0,1,\cdots,\inf(p,q)-1\}$，其中 $p=\deg(g,x_s)$，$q=\deg(f_s,x_s)$. 如果 $r_0=\cdots=r_{k-1}=0$，$r_k\neq 0$，且 $\mathrm{Res}(r_k,f_1,\cdots,f_{s-1})\neq 0$，则根据引理及后面将要说明的理由，可得 f_s 的两因子，其中之一关于 x_i 的次数为 k，然后将所得因子分别代替 f_s 返回步骤 2. 否则，如果 $r_0=\cdots=r_{k-1}=0$，$r_k\neq 0$，且 $\mathrm{Res}(r_k,f_1,\cdots,f_{s-1})=0$，则令 r_k 代替 g，$f_1,\cdots,f_{s-1}$ 代替 $f_1,\cdots,f_{s-1},f_s$，返回步骤 1. 最终可得升列的一种分解，然后分别用子升列代替原升列，返回步骤 2.

由于上述每一步返回或者使 s 减小，或者使升列中的多项式次数降低，故有限步后必然终止.

上面我们假定了升列 AS 是真升列 p-AS，因此在步骤 1 之前，应检查 AS 是否为 p-AS，如果 AS 不是 p-AS，则令 I_k（这里 k 是使得 $\mathrm{Res}(I_k,f_1,\cdots,f_{k-1})=0$ 的最小数）代替 g，而以 $f_1,\cdots,f_{k-1}$ 代替 $f_1,\cdots,f_s$，利用步骤 1～3 对 AS 进行分解，直至得到的子升列都是真升列.

在步骤 3 中留下来一个问题，现在加以说明. 假定 AS' 为 AS 的一不可约子升列，令 AS' 由下列多项式组成：$f'_1,f'_2,\cdots,f'_s$，由于 $s_i(g,f_s)$，$i=0,\cdots,k-1$，在 $f_1=0,\cdots,f_s=0$ 之下为零，故它们在 $f'_1=0,\cdots,f'_s=0$ 之下也为零. 又 $\mathrm{Res}(r_k,f_1,\cdots,f_{s-1})\neq 0$，即 $s_k(g,f_s)$ 在 $f_1=0,\cdots,f_s=0$ 之下一般不为零，故 $s_k(g,f_s)$ 在 $f'_1=0,\cdots,f'_s=0$ 之下一般不为零. 而 $K(u)[x]/(f'_1,\cdots,f'_s)$ 是一域，故令之为 D，则引理的条件全都满足，g,f_s 的公因子为 $P_k(g,f)$. 又由 AS' 的任意性可知，g 和 f_s 在 $K(u)[x]/(f_1,\cdots,f_s)$ 中的公因子就是 $P_k(g,f_s)$.

5　几个例子

例 1　$f_1=x_1^4-x_1^2$，$f_2=x_1x_2-u_1x_1$，$g=x_1^4$.

解　因 $\mathrm{Res}(I_2,f_1)=0$，故 f_1,f_2 不是真升列，分解并约化后得到两个真升列：$\{x_1\}$，$\{x_1^2-1,x_1x_2-u_1x_1\}$ 在前簇上 $g=0$ 真，后簇上 $g=0$ 假.

例 2　$f_1=x_1-u$，$f_2=x_2^2-2x_1x_2+u^2$，$g=x_2-x_1$.

解　f_1,f_2 显然为真升列. 令 $g:=\mathrm{prem}(g,f_1,f_2)=x_2-u$. 因为 $s_0(g,f_2)=\mathrm{Res}(g,f_2,x_2)=2u^2-2ux_1$，$s_1=1$，$r_0=\mathrm{prem}(s_0(g,f_2),f_1,x_1)=0$，故 g,f_2 有一次公因子 x_2-x_1，因

而$\{f_1,f_2\}$分解为相同的两个升列：$\{x_1-u,x_2-x_1\}$. $g=0$ 真.

例 3 $f_1=x_1^2-ux_1,f_2=x_2^2+x_1x_2-u^2,g=x_2+x_1+1$.

解 $\text{Res}(g,f_1,f_2)=(1-u^2)(1+u-u^2)\neq 0$. 故 $g=0$ 假.

例 4 $f_1=x_1^2+ux_1,f_2=x_2^2+(x_1^2-u)x_2+u^2x_1,g=x_2^2+x_1(1-u)x_2-ux_1^2$.

解 令 $g:=\text{prem}(g,f_1,f_2)=(x_1+u)x_2,s_0(g,f_2)=u^2x_1(x_1+u)^2,s_1(g,f_2)=x_1+u$, $r_0=0,r_1=x_1+u,\text{Res}(r_1,f_1)=0$. 故 r_1 与 f_1 有公因子 x_1+u，从而$\{f_1,f_2\}$分解为 $C_1:\{x_1,x_2^2-ux_2\},C_2:\{x_1+u,x_2^2+(u^2-u)x_2-u^3\}$. 在 C_1 上，$\text{Res}(r_1,f_1)\neq 0$，故 $g=ux_2$ 与 $f_2=x_2^2$ 有公因子 x_2，C_1 分解为 $C_1^*:\{x_1,x_2\},C_1^{**}:\{x_1,x_2-u\}$. 在 C_1^* 上 $g=0$，在 C_1^{**} 上 $g=u^2\neq 0$，在 C_2 上 $g:=\text{prem}(g,x_1+u)=0$.

例 5 在$\triangle ABC$ 两边 AC,BC 上分别作正方形 $ACDE$ 和 $BCFG$. 设 M 是 AB 之中点. 试证：$DF=2CM$.

证明 令 $A=(u_1,0),B=(u_2,u_3),C(0,0),D(0,u_1),F=(x_1,x_2),M=(x_1,x_4)$，则假设条件为：

$$h_1=x_2^2+x_1^2-u_2^2-u_3^2=0(CF=BC),$$
$$h_2=u_3x_2+u_2x_1=0(CF\perp BC),$$
$$\left.\begin{aligned}h_3&=2x_3-u_2-u_1=0,\\ h_4&=2x_4-u_3=0,\end{aligned}\right\}M\text{ 是 }AB\text{ 中点.}$$

结论

$$g=4x_4^2+4x_3^2-x_2^2+2u_1x_2-x_1^2-u_1^2=0(DF=2CM).$$

将$\{h_1,h_2,h_3,h_4\}$整序得

$$f_1=(u_2^2+u_3^2)x_1^2-u_3^2(u_2^2+u_3^2)=0,$$
$$f_2=u_3x_2+u_2x_1=0,$$
$$f_3=2x_3-u_2-u_1=0,$$
$$f_4=2x_4-u_3=0,$$

令

$$g:=\text{prem}(g,f_1,f_2,f_3,f_4)=2u_1u_2u_3(u_3-x_1),\text{Res}(g,f_1,f_2,f_3,f_4)=0,$$

g 与 f_1 有公因式 x_1-u_3，得两升列

$$C_1:\{x_1-u_3,f_2,f_3,f_4\},$$
$$C_2:\{(u_2^2+u_3^2)(x_1+u_3),f_2,f_3,f_4\}.$$

$g=0$ 在 C_1 上一般真，在 C_2 上一般假.

参考文献

[1]Wu Wen-tsun. On the decision problem and the mechanization of theorem proving in elementary geometry. Scientia Sinica, 1978(21): 157 - 179.

[2]吴文俊，几何定理机器证明的基本原理. 北京：科学出版社，1984.

[3] Chou Shang-Ching Mechanical Geometry Theorem Proving, D. Reidel Publishing Company, 1987.

[4] Wu Wen-tsun. On the reducibility problem in mechanical theorem proving of elementary geometries. 数学季刊,1978,2(2):1 - 19.

[5] Yang Lu, Zhang Jingzhong. Searching Dependency between Algebraic Equations: An Algorithm Applied to Automated Reasoning, preprint: IC/91/6, ICTP (1991). Also in "Artificial intelligence in mathematics", Oxford University Press, 1994.

Automated Production of Readable Proofs for Theorems in Non-Euclidean Geometries

Yang Lu[1], Gao Xiao-shan[2], Chou Shang-ching[3], and Zhang Jing-zhong[1]

[1] Chengdu Institute of Computer Applications, Academia Sinica

[2] Institute of Systems Science, Academia Sinica

[3] Department of Computer Science, The Wichita State University

Abstract, We present a complete method which can be used to produce short and human readable proofs for a class of constructive geometry statements in non-Euclidean geometry. The method is a substantial extension of the area method for Euclidean geometry. The method is an *elimination algorithm* which is similar to the *variable elimination method* of Wu used for proving geometry theorems. The difference is that instead of eliminating coordinates of points from general algebraic expressions, our method eliminates points from high level geometry invariants. As a result the proofs produced by our method are generally short and each step of the elimination has clear geometric meaning. A computer program based on this method has been used to prove more than 90 theorems from non-Euclidean geometries including many new ones. The proofs produced by the program are generally very short and readable.

1 Introduction

Two of the main approaches to automated geometry reasoning are the approach based on *synthetic deduction* or heuristic methods [9, 14, 13, 6] and the approach based on *algebraic computation* which first transforms a geometry statement into an algebraic statement via coordinates and then deals with the algebraic statement using algebraic techniques such as Wu's method[23,1,20,21,26] and the Gröbner basis method [1,15,16, 17,18]. Generally speaking, the algebraic approaches are decision procedures and are more powerful, while the synthetic approaches are not decision procedures but can be used to produce proofs in traditional style.

本文刊于《Automated Deduction in Geometry》, 1997 年.

Automated theorem proving in non-Euclidean geometries with algebraic method has been reported in a few works. A method for non-Euclidean geometries was introduced by W. T. Wu[23] as a direct consequence of his algebraic method transformation theorems between hyperbolic and elliptic geometries were studied in [3,8]. In[24], properties of trigonometric functions were used to derive new properties of non-Euclidean geometries. The polynomials involved in the proofs are generally larger than those in Euclidean geometry.

In[2], we introduced the *area method* for Euclidean geometry which is a combination of algebraic computation and synthetic deduction: it uses simple algebraic operations to achieve efficiency and uses synthetic deduction to preserve the geometric meaning of the proofs. The area method has been used to produce short and *human-readable proofs* automatically for hundreds of difficult geometry theorems from Euclidean geometry [2]. By *human-readable proofs*, we mean that the proofs are short enough for human to write down without difficulty and each step of the proofs has clear geometric meaning.

The method presented in this paper is a substantial extension of the area method for Euclidean geometry to two non-Euclidean geometries: the hyperbolic (or the Bolyai-Lobachevsky) geometry and the elliptic (or the Riemann-Cayley) geometry. The main difficulty of the extension is that the concept of area in these two non-Euclidean geometries does not have the properties that make the area method possible in Euclidean geometry. We thus need to find new types of geometry quantities. The basic quantities used by us in non-Euclidean geometries are the *argument* (roughly speaking, the sine of the area) and the *cosine of distance*. Furthermore, not all of the properties of these quantities needed in our method are in classical books of non-Euclidean geometry [7,10]. We think some of the geometry results proved us, such as Propositions 7 and 16, are new.

The method is based on an *elimination approach* which is similar to the *variable elimination method* of Wu used for proving geometry theorems [23]. The difference is that instead of eliminating point coordinates from algebraic expressions, our method eliminates points from high level geometry invariants or quantities. In other words, it is an elimination method at a higher level and as a result the proofs produced are generally human-readable. Our experiments confirm this: the proofs of the 90 theorems produced by our program are short. Actually, the average term of the maximal polynomials in these proofs is 2.2 (see Section 4), while using the coordinate approach, proofs involving polynomials with more than one hundred terms are not uncommon. The geometric meaning of each step of the proofs for the 90 theorems is also clear.

The format of proofs produced by our method is similar to that produced by rewrite rule methods. But there is a major difference. A typical rewrite method will first generate a canonical set of rules and then use these rules to reduce the conclusion. In our case, the rules are generated dynamically, i. e., they are generated only when they are needed during the process of proofs. Generating all the rules for a geometry statement is much time consuming and not needed.

We have implemented the algorithm using Common Lisp on a NeXT workstation. The prover is available via ftp at emcity. cs. twsu. edu: pub/geometry/software/euc. tar. Z. (The default method in the prover is for Euclidean geometry. To use the method for non-Euclidean geometries, you need to do (setq non-Euclidean t).) The prover has been used to prove 90 theorems totally automatically[5]. Of the 90 theorems, about 40 belong to the projective geometry and thus are true in non-Euclidean geometries. The validity of the others are generally non-trivial. Notably, some extensions of the Ceva and Menelaus theorems discovered recently[11] were extended to hyperbolic and elliptic geometries using the method developed in this paper.

The rest of the paper is organized as follows. Section 2 presents an outline of the method. Section 3 presents the detailed elimination method. Section 4 presents the experimental results and examples. Section 5 presents the conclusion remarks.

2 Outline of the Method for Hyperbolic Geometry

In this section, we give an outline of the method, leaving the detailed elimination method to Section 3. For the basic concepts of non-Euclidean geometry, the reader may consult [7].

2. 1. An Example. We first use an example to show how the method works.

Example 1(*Ceva's Theorem*). In Figure 1, ABC is a triangle; P is any point. Points D, E, F are the intersection points of lines AP, BP, CP with lines BC, CA, AB respectively. Show that

$$\frac{\sinh(AF)}{\sinh(FB)}\frac{\sinh(BD)}{\sinh(DC)}\frac{\sinh(CE)}{\sinh(EA)}=1,$$

where $\sinh(AF)$ is the hyperbolic sine of segment AF.

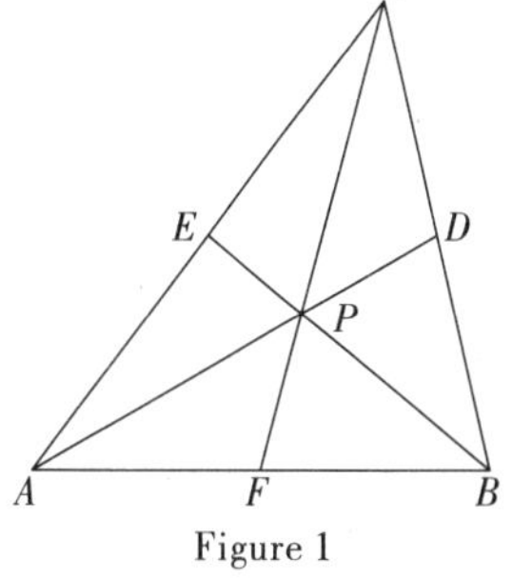

Figure 1

To prove this theorem, we use a geometry invariant or quantity: *the argument*. The absolute value of the argument of a triangle ABC is defined to be

$$|S_{ABC}| = |\sinh(AB).\sinh(BC).\sin\angle(AB,BC)|.$$

We also assume that the argument of a clockwised triangle is positive and

$$S_{ABC}=S_{CAB}=S_{BCA}=-S_{ACB}=-S_{BAC}=-S_{CBA}.$$

It is clear that $S_{ABC}=0$ iff A, B, and C are on the same line. Another quantity is the *ratio of hyperbolic sines* of directed line segments $\frac{\sinh(AB)}{\sinh(PQ)}$ where A, B, P, Q are collinear. Then it is clear that

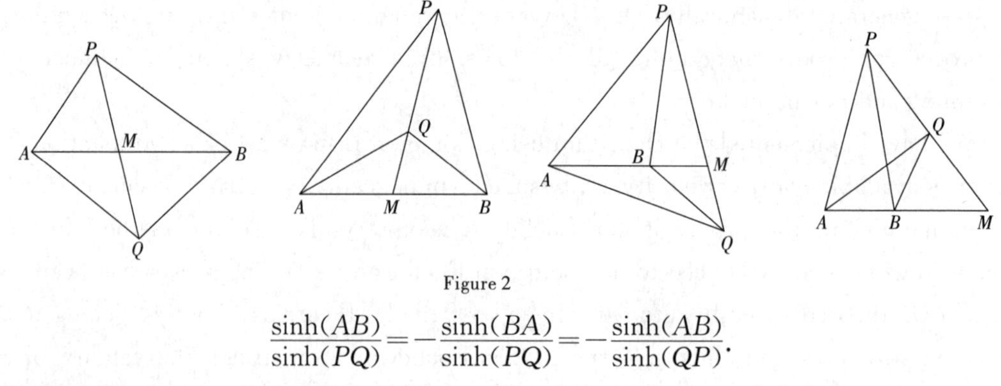

Figure 2

$$\frac{\sinh(AB)}{\sinh(PQ)}=-\frac{\sinh(BA)}{\sinh(PQ)}=-\frac{\sinh(AB)}{\sinh(QP)}.$$

the following result is a direct consequence of the definition of the argument.

Proposition 1 (The Co-side Theorem). *Let M be the intersection of two lines AB and PQ and* $Q\neq M$. *Then* $\frac{\sinh(PM)}{\sinh(QM)}=\frac{S_{PAB}}{S_{QAB}}$. *Note that the result is true in all the four cases in Figure* 2.

Proof of Example 1. In our method, we need to give an order for the points in the geometry statement such that each point can be constructed from the previous points. In this example, a construction order of points is A,B,C,P,D,E,F: first taking four free points A,B,C,P, then taking intersection points of lines AP,BP,CP with lines BC,CA, AB respectively.

Our proof method is to *eliminate points* from the conclusion. By the co-side theorem, we can eliminate points F,E and D from the conclusion respectively;

$$\frac{\sinh(AF)}{\sinh(FB)}=\frac{S_{ACP}}{S_{BPC}},\frac{\sinh(CE)}{\sinh(EA)}=\frac{S_{BCP}}{S_{BPA}},\frac{\sinh(BD)}{\sinh(DC)}=\frac{S_{ABP}}{S_{APC}}.$$

Then it is clear that

$$\frac{\sinh(AF)}{\sinh(FB)}\frac{\sinh(BD)}{\sinh(DC)}\frac{\sinh(CE)}{\sinh(EA)}=\frac{S_{ACP}S_{BCP}S_{ABP}}{S_{BPC}S_{BPA}S_{APC}}=1.$$

2. 2. Constructive Geometry Statements. We have mentioned that the proving method is for constructive geometry statements, in which points are introduced by some constructions. A *construction* is one of the following ways of introducing new points. For each construction, we also give its *non-degenerate* (ndg) conditions which guarantee the construction introduces a point properly.

C1 (POINT[S], $Y_1,\cdots,Y_l$). Take arbitrary points $Y_1,\cdots,Y_l$ in the plane.

C2 (ON, Y, (LINE, U, V)). Take a point Y on the line passing through points U and V. The ndg condition of C2 is $U\neq V$.

C3 (MRATIO, Y,U,V,r), Take a point Y on line UV such that $r=\frac{\sinh(UY)}{\sinh(YV)}$. The quantity r could be a number, an indeterminate, or an expression in geometry quantities. The ndg condition is $U\neq V$.

C4 (LRATIO, Y, U, V, r_1, r_2). Take a point Y on a line UV such that $r_1=\frac{\sinh(UY)}{\sinh(UV)}$, $r_2=\frac{\sinh(YV)}{\sinh(UV)}$. The quantities r_1 and r_2 could be numbers or indeterminates. The ndg condition is $U\neq V$.

C5 (INTER, Y, (LINE, U, V), (LINE, P, Q)). Point Y is the intersection point of line PQ and line UV. The ndg condition is that line UV and line PQ have a proper intersection point. The algebraic form is $S_{UPVQ}\neq 0$. The argument of a quadrilateral is defined in Section 3. 1.

C6 (FOOT, Y, P, U, V). Y is the foot of the perpendicular line drawn from P to line UV. The ndg condition is $U\neq V$.

C7 (INTER, Y, (LINE, U, V), (TLINE, P, Q)). Point Y is the intersection of line UV and the line passing through P and perpendicular to PQ. The ndg condition is that the two lines have a proper intersection point.

C8 (INTER, Y, (LINE, P, Q), (CIR, O, P)). Point Y is the intersection point of line PQ and circle (CIR, O, P)(i. e. , the circle with center O and passing through point P). The ndg conditions are $Y\neq P$, $O\neq P$, and $P\neq Q$.

C9 (INTER, Y, (CIR, O_1, P), (CIR, O_2, P)). Point Y is the intersection point of circle (CIR, O_1, P), and circle (CIR, O_2, P) other than point P. The ndg condition is that O_1, O_2, and P are not collinear.

Now class C, the class of *constructive geometry statements*, is defined as follows. A statement in class C is a list $S=(C_1\cdots, C_k, G)$ where $C_i, i=1,\cdots,k$, are constructions such that each C_i introduces a new point from the points introduced before; and G, which is either a geometry predicate collinear and perpendicular or an algebraic equation geometric quantities, is the conclusion of the statement.

Let $S=(C_1,\cdots,C_k,G)$ be a statement in C. The *ndg condition* of S is the set of ndg conditions of the C_i plus the condition that the denominators of the geometry quantities in G are not equal to zero.

The constructive description for Ceva's theorem(Example 1)is as follows.

((POINTS, A, B, C, P)

(INTER, D, (LINE, B, C), (LINE, P, A))

(INTER, E, (LINE, A, C), (LINE, P, B))

(INTER, F, (LINE, AB), (LINE, P, C))

$\left(\frac{\sinh(AF)}{\sinh(FB)}\frac{\sinh(BD)}{\sinh(DC)}\frac{\sinh(CE)}{\sinh(EA)}, =, 1)\right)$

The ndg conditions for Ceva's theorem are: AP intersects BC properly; BP intersects AC properly; CP intersects AB properly; $F\neq B$; $D\neq C$; and $E\neq A$.

It is obvious that each construction can be reduced to one or two *geometry predicates*.

For example. construction C6 is equivalent to two predicates: collinear (Y, U, V) and perpendicular(Y,P,P,Q). For a geometry statement S in class C, let Pr be the conjunction of the predicates derived from the constructions and Nd be the set of the non-degenerate conditions. Then the *predicate form of* S is

$$\forall P_i[(P_r \wedge Nd) \Rightarrow G],$$

where P_i are the points in S and G is the conclusion of S.

2.3 The Algorithm. We have mentioned in Example 1 that the proving method is to eliminate points from the basic geometry quantities Formally, we can describe the method as follows.

INPUT: $S=(C_1, C_2, \cdots, C_k, E)$ is a statement C, where E is an equation in geometric quantities.

OUTPUT: The algorithm tells whether S is true or not, and if it is true, produces a proof for S.

S1. For $i=k, \cdots, 1$, do S2, S3, S4, and finally S5.

S2. Check whether the ndg conditions of C_i are satisfied. For instance, if the ndg condition is $A \neq B$, we will check whether $\cosh(AB)=1$ with this algorithm. In other words, we need to prove a theorem whose conclusion is the negation of the ndg. If an ndg condition of a geometry statement is not satisfied, the statement is *trivially true*. The algorithm terminates.

S3. Let $G_1, \cdots, G_s$ be the geometric quantities occurring in E. For $j=1, \cdots, s$ do S4.

S4. Let $H_j=\mathrm{ELIM}(G_j, C_i)$ where ELIM is an algorithm which eliminates the point introduced by construction C_i from G_j using methods given in Section 3 to obtain a formula H_j. Replace G_j by H_j in E to obtain a new E.

S5. Now E is an equation in hyperbolic trigonometric functions of independent variables. We can use the method described in the last paragraph of Section 3.4 to check whether E is a hyperbolic trigonometric identity. If E is an identity then S is valid. Otherwise S is false.

Proof of the correctness. Only the last step needs explanation. If E is an identity, the statement is obviously true. Note that the ndg conditions ensure that the denominators of all the expressions occurring in the proof do not vanish. If E is not an identity. We divide it into two cases. If all the elimination is for linear quantities then the geometric quantities left in E are free parameters, i.e., in the geometric configuration of S they can take arbitrary values. Since E is not an identity, we can take some integer values for these quantities such that when replacing these quantities by the corresponding values in E, we obtain a non-vanishing number. In other words, we obtain a counter example in real geometry for S. In this case S is false in the real hyperbolic geometry. If some quadratic equations are used in the elimination by a theorem in [22], the statement is false in the complex geometry. We do not know whether the statement is valid over the field of real numbers.

It is clear that the key step of the algorithm is algorithm ELIM in Step S4, i. e., *how to eliminate points from geometry quantities*. This will be presented in detail in Section 3.

3 Elimination Methods for Hyperbolic Geometry

In Section 3. 1, we give the geometric facts which will serve as the deduction basis or axioms of our method. Some of these properties are known or easy to prove, but some of them such as Propositions 7 and 16 are new. In the rest of this section, we present the elimination algorithm ELIM. We will try to give the proofs for some of the frequently used elimination steps. Proofs of others can be found in our full report [4]. Readers who are not concerned with the technical details may skip this section.

3. 1. Geometry Preliminaries. We use capital English letters to denote points in the hyperbolic plane. The absolute value of *argument of a quadrilateral* $ABCD$ is defined to be

$$|S_{ABCD}| = |\sinh(AC) \cdot \sinh(BD) \cdot \sin\angle(AC, BD)|.$$

We also assume that $S_{ABCD} = -S_{CBAD} = -S_{BADC}$. A more general form of the co-side theorem is as follows.

Proposition 2 (The Co-side Theorem). *Let* M *be the intersection of two lines* AB *and* PQ *and* $Q \neq M$. *Then*

$$\frac{\sinh(PM)}{\sinh(QM)} = \frac{S_{PAB}}{S_{QAB}};\ \frac{\sinh(PM)}{\sinh(PQ)} = \frac{S_{PAB}}{S_{PAQB}};\ \frac{\sinh(QM)}{\sinh(PQ)} = \frac{S_{QAB}}{S_{PAQB}}.$$

Proposition 3. *Let* R *be a point on line* PQ. *Then for two points* A *and* B,

$$S_{RAB} = \frac{\sinh(PR)}{\sinh(PQ)} S_{QAB} + \frac{\sinh(RQ)}{\sinh(PQ)} S_{PAB}.$$

Another basic geometry quantity is the hyperbolic cosine of a line segment.

Proposition 4. *Let* R *be a point on line* PQ. *Then for any point* A, *we have*:

$$\cosh(RA) = \frac{\sinh(PR)}{\sinh(PQ)} \cosh(QA) + \frac{\sinh(PQ)}{\sinh(PQ)} \cosh(PA).$$

Similar to the area method in Euclidean geometry, we introduce another geometry quantity: the *Pythagorean difference* for quadrilateral $ABCD$ is defined to be

$$P_{ABCD} = \cosh(AD) \cdot \cosh(BC) - \cosh(AB) \cdot \cosh(CD).$$

Then the *Pythagorean difference* for a triangle ABC is

$$P_{ABC} = P_{ABBC} = \cosh(AC) - \cosh(AB)\cosh(CB).$$

Proposition 5. $AB \perp PQ$ *iff* $P_{APBQ} = 0$.

Proposition 6. *Let* Y *be the foot of the perpendicular from point* P *to* UV. *Then*

$$\frac{\sinh(UY)}{\sinh(YV)} = \frac{P_{PUV}}{P_{PVU}},\ \frac{\sinh(UY)}{\sinh(UV)} = \frac{P_{PUV}}{f_1(P,U,V)},\ \frac{\sinh(YV)}{\sinh(UV)} = \frac{P_{PVU}}{f_1(P,U,V)},$$

where $f_1(P,U,V)^2 = \sinh(UV)^2 (2\cosh(PU)\cosh(PV)\cosh(UV) - \cosh(PU)^2$

$-\cosh(PV)^2)$.

Proposition 7. *Let R be the intersection of line UV and the line passing through point P and perpendicular to PQ. Then*

$$\frac{\sinh(UR)}{\sinh(VR)}=\frac{P_{UPQ}}{P_{VPQ}},\frac{\sinh(UR)}{\sinh(UV)}=\frac{P_{UPQ}}{f_2(P,Q,U,V)},\frac{\sinh(VR)}{\sinh(UV)}=\frac{P_{VPQ}}{f_2(P,Q,U,V)},$$

where $f_2(P,Q,U,V)^2=P_{UPQ}^2+P_{VPQ}^2-2P_{UPQ}P_{VPQ}\cosh(UV)$.

The following proposition gives the relations among the parameters r, r_1, and r_2 in constructions C3 and C4.

Proposition 8. *Let R be a point on line UV*, $r=\frac{\sinh(UR)}{\sinh(RV)}$; $r_1=\frac{\sinh(UR)}{\sinh(UV)}$, $r_2=\frac{\sinh(RV)}{\sinh(UV)}$. *Then*

1. $r_1^2+2r_1r_2\cosh(UV)+r_2^2=1$.

2. $r_1=rr_2, r_2^2=\frac{1}{1+2r\cosh(UV)+r^2}$.

We have defined four *geometry quantities*: the argument, the Pythagorean difference, the hyperbolic cosine of line segments, and the ratio of hyperbolic sines of directed line segments.

In the rest of this section, we will present algorithm ELIM. We need to show how to eliminate points introduced by constructions C1 − C8 from the three basic geometry quantities: ratios, arguments, and hyperbolic cosines. We first show that the nine constructions are not independent. For instance, C2 can be reduced to C3: taking an arbitrary point on a line UV is equivalent to taking a point on UV such that $\frac{\sinh(UY)}{\sinh(UV)}$ is an indeterminate. In fact, constructions C3, C8 and C9 can also be reduced to other constructions [4]. Therefore, we need only consider five constructions C1, C4, C5, C6, and C7. Algorithm ELIM will be presented as eight elimination lemmas.

3. 2. Eliminating Points from Ratios. Now we consider how to eliminate points from the ratios.

Lemma 9. *ELIM*(f,c): $f=\frac{\sinh(AY)}{\sinh(BC)}$ *and c is construction* $(LRATIO, Y, U, V, r_1, r_2)$, *where A, B, C are points on line UV. We have*:

$$f=\frac{\sinh(AU)}{\sinh(BC)}(r_1\cosh(UV)+r_2)+r_1\frac{\sinh(UV)}{\sinh(BC)}\cosh(AU).$$

Proof. By proposition 4,

$$\frac{\sinh(AY)}{\sinh(BC)}=\frac{\sinh(AU+UY)}{\sinh(BC)}$$

$$=\frac{\sinh(AU)}{\sinh(BC)}\cosh(UY)+\frac{\sinh(UY)}{\sinh(BC)}\cosh(AU)$$

$$=\frac{\sinh(AU)}{\sinh(BC)}(r_1\cosh(UV)+r_2)+r_1\frac{\sinh(UV)}{\sinh(BC)}\cosh(AU).$$

Lemma 10. *$ELIM(f,c)$: $f=\frac{\sinh(AY)}{\sinh(CD)}$ and c is $(INTER,Y,(LINE,U,V,),(LINE,P,Q))$, where points A and Y are on line CD. We have:*

$$f=\begin{cases}\frac{S_{AUV}}{S_{CUDV}} & if\ A\ is\ not\ on\ UV;\\ \frac{S_{APQ}}{S_{CPDQ}} & otherwise.\end{cases}$$

Proof. This proposition can be similarly proved as the co-side theorem.

Lemma 11. *$ELIM(f,c)$: $f=\frac{\sinh(AY)}{\sinh(CD)}$ and c is $(FOOT,Y,P,U,V,)$, where points A and Y are on CD. We have:*

$$f=\begin{cases}\frac{\sinh(AV)}{\sinh(CD)}\frac{P_{PAV}}{f_1(P,A,V)} & if\ A\in UV\ and\ A\neq V;\\ \frac{S_{AUV}}{S_{CUDV}} & if\ A\notin UV.\end{cases}$$

Proof. The first and second cases can be proved similarly as Proposition 6 and the co-side theorem respectively.

Lemma 12. *$ELIM(f,c)$: $f=\frac{\sinh(AY)}{\sinh(CD)}$ and c is $(INTER,Y,(LINE,U,V),(TLINE,P,Q))$, where points A and Y are on line CD. We have:*

$$\frac{\sinh(AY)}{\sinh(CD)}=\begin{cases}\frac{\sinh(AV)}{\sinh(CD)}\frac{P_{APQ}}{f_2(P,Q,A,V)} & if\ A\in UV\ and\ A\neq V;\\ \frac{S_{AUV}}{S_{CUDV}} & if\ A\notin UV.\end{cases}$$

Proof. The first and second cases can be similarly proved as Proposition 7 and the co-side theorem respectively.

3.3. Eliminating Points from Linear Quantities. Let $G(Y)$ be S_{ABY}, $\cosh(AY)$, P_{ABY}, or P_{ABCY} for different points A,B,C, and Y. Then for three collinear points Y,U, and V, by Propositions 3 and 4 we have:

$$(\mathrm{I})\qquad G(Y)=\frac{\sinh(UY)}{\sinh(UV)}G(V)+\frac{\sinh(YV)}{\sinh(UV)}G(U).$$

We call $G(Y)$ a *linear geometry quantity* of point Y. Elimination procedures for linear geometry quantities are similar for constructions C4—C6.

Lemma 13. *$ELIM(G(Y),c)$: Let $G(Y)$ be a linear geometry quantity. Then $G(Y)$ equals to*

$$\begin{cases} r_1G(V)+r_2G(U) & if\ c=(LRATIO,Y,U,V,r_1,r_2); \\ \dfrac{S_{UPQ}G(V)-S_{VPQ}G(U)}{S_{UPVQ}} & if\ c=(INTER,Y,(LINE,U,V),(LINE,P,Q)); \\ \dfrac{P_{PUV}G(V)+P_{PVU}G(U)}{f_1(PUV)} & if\ c=(FOOT,Y,P,U,V); \\ \dfrac{P_{UPQ}G(V)-P_{VPQ}G(U)}{f_2(P,Q,U,V)} & if\ c=(INTER,Y,(LINE,U,V),(TLINE,P,Q)). \end{cases}$$

Proof. Since point Y is on line UV in each case, equation (Ⅰ) is true. The first case is just (Ⅰ). For the second case, by the co-side theorem

$$\frac{\sinh(UY)}{\sinh(UV)}=\frac{S_{UPQ}}{S_{UPVQ}},\frac{\sinh(YV)}{\sinh(UV)}=-\frac{S_{VPQ}}{S_{UPVQ}}.$$

Substituting these into (Ⅰ), we prove the result. The third and fourth cases are consequences of (Ⅰ) and Propositions 6 and 7. Note the non-degenerate conditions of each construction guarantee that the denominator of the new expression does not vanish.

3.4. Eliminating Points from Quadratic Quantities. The functions f_1 and f_2 in Lemmas 13, 11, and 12 satisfy quadratic equations. We call such quantities quadratic quantities. Now, we present a mechanical method of eliminating points from them. First, another quadratic quantity is S_{ABCD}.

Proposition 14. We *have*:

$$S_{ABCD}^2=S_{ABD}S_{BCA}\cosh(CD)+S_{BCA}S_{CDB},$$
$$\cosh(AD)+S_{CDB}S_{DAC}\cosh(AB)+S_{ABD}S_{DAC}\cosh(BC).$$

To eliminate geometric quantities satisfying quadratic equations completely, we need more algebraic tools which can be found in our full report [4]. In what follows, we only describe briefly how this method works by showing how to prove hyperbolic identities.

Let P be a polynomial in hyperbolic trigonometric functions of variables $\pm x_1,\cdots,\pm x_n,\pm\frac{1}{2}x_1,\cdots,\pm\frac{1}{2}x_n,2x_1,\cdots,2x_n,\cdots$. We want to know whether $P=0$ is an identity. We set $yi=\frac{1}{m}x_i,i=1,\cdots,n$ such that each hyperbolic trigonometric function in P has the form $\sinh(\sum k_iy_i)$ or $\cosh(\sum s_iy_i)$ for integers k_i and s_i. First, we can easily represent P as a polynomial R in $z_i=\sinh(y_i)$ and $w_i=\cosh(y_i),i=1,\cdots,n$. Then $R=0$ is an identity iff R will become zero when replacing w_i^2 by $1+z_i^2$ (for a proof see [8]).

3.5. Eliminating Free Points. We now show to eliminate the free points (points introduced by construction C1). To do that, we need the concept of *argument coordinates*. Let A,O,U, and V be four points such that O,U, and V are not collinear. The argument coordinates of A with respect to OUV are

$$x_A=\frac{S_{AUV}}{S_{OUV}},y_A=\frac{S_{OAV}}{S_{OUV}},z_A=\frac{S_{OUA}}{S_{OUV}}.$$

It is clear that the points in the plane are in a one to one correspondence with their argument coordinates. The following lemma reduces a linear quantity of free points to their argument coordinates.

Lemma 15. *ELIM(G(Y), c): G(Y) is a linear geometry quantity G(Y); c is one of the constructions C4—C7. Let O, U, and V be three non-collinear points. Then we have:*

$$G(Y)=\frac{S_{YUV}}{S_{OUV}}G(O)+\frac{S_{OYV}}{S_{OUV}}G(U)+\frac{S_{OUY}}{S_{OUV}}G(V).$$

Proof. Without loss of generality, let OY intersect UV at T. If OY does not intersect UV, we may consider the intersection of UY and OV or that of VY and OU since one of them must exist. By (Ⅰ),

$$G(Y)=\frac{\sinh(OY)}{\sinh(OT)}G(T)+\frac{\sinh(YT)}{\sinh(OT)}G(O)$$
$$=\frac{\sinh(OY)}{\sinh(OT)}\left(\frac{\sinh(UT)}{\sinh(UV)}G(V)+\frac{\sinh(TV)}{\sinh(UV)}G(U)\right)+\frac{\sinh(YT)}{\sinh(OT)}G(O).$$

By the co-side theorem,

$$\frac{\sinh(YT)}{\sinh(OT)}=\frac{S_{YUV}}{S_{OUV}};\ \frac{\sinh(OY)}{\sinh(OT)}=\frac{S_{OUYV}}{S_{OUV}};$$
$$\frac{\sinh(UT)}{\sinh(UV)}=\frac{S_{OUY}}{S_{OUYV}};\ \frac{\sinh(TV)}{\sinh(UV)}=\frac{S_{OYV}}{S_{OUYV}}.$$

Substituting these into the above formula, we obtain the desired result.

Using Lemma 15, any expression in geometric quantities can be written as an expression in $\cosh(OU)$, $\cosh(OV)$, $\cosh(UV)$, S_{OUV}, and the argument coordinates of the free points. These quantities are still not independent. First, it is known that [7].

$$S_{OUV}^2=1-\cosh(OU)^2-\cosh(OV)^2-\cosh(UV)^2$$
$$+2\cosh(OU)\cosh(OV)\cosh(UV).$$

Second, the three argument coordinates of point satisfy the following property.

Proposition 16. *Let x_A, y_A, z_A be the argument coordinates of A with respect to OUV. Then*

$$x_A^2+y_A^2+z_A^2+2y_Az_A\cosh(UV)+2x_Az_A\cosh(OV)+2x_Ay_A\cosh(OU)=1.$$

Proof. We will prove this result with the aid of our program. We introduce five points with the following constructions:

(POINTS, O, U, V)

(LRATIO, D, U, V, r_1, r_2)

(LRATIO, A, O, D, s_1, s_2).

Using Lemma 13, we can eliminate points D and A from

$$f=x_A^2+y_A^2+z_A^2+2y_Az_A\cosh UV+2x_Az_A\cosh OV+2x_Ay_A\cosh OU-1$$

where $x_4=\frac{S_{AUV}}{S_{OUV}}$, $y_A=\frac{S_{OAV}}{S_{OUV}}$, $z_A=\frac{S_{OUA}}{S_{OUV}}$. Denote the output by g. Let

$$h_1=r_1^2+r_2^2+2r_1r_2\cosh UV-1,$$
$$h_2=s_1^2+s_2^2+2s_1s_2\cosh OD-1,$$
$$h_3=r_1\cosh OV+r_2\cosh OU-\cosh OD.$$

According to Propositions 4 and 8, $h_1=0, h_3=0$. Doing successive pseudo-division of g with h_3, h_2, h_1, we know that g is a linear combination of h_1, h_2, h_3. This proves the result.

Since the argument coordinates satisfy quadratic equations of triangular form, we can use the techniques presented in Section 3 to eliminate one of the three coordinates for each point.

3.6. Eliminating Cyclic Points. To deal with theorems involving cyclic points efficiently, we introduce a new construction.

C9 (CIRCLE, $A_1, \cdots, A_s$)($s\geqslant 3$), points $A_1 \cdots A_s$ are on the same circle. There is no ndg condition for this construction.

Let $A_1, \cdots, A_s$ be points on circle with center O. We choose a point, say A_1, as the reference point. Then each point A_i is uniquely determined by the oriented angle $\frac{\angle A_1OA_i}{2}$ (we assume that all angles have values from $-\pi$ to π).

Lemma 17. *Let A, B, C, D be points on a circle with center O and radius δ, and A the reference point. we denote $\frac{\angle AOB}{2}$ by $\angle B$. Then*

1. $S_{BCD}=\dfrac{4\sinh\left(\angle\frac{BOC}{2}\right)\cdot\sinh\left(\angle\frac{CD}{2}\right)\cdot\sinh\left(\angle\frac{BD}{2}\right)}{\tanh(\delta)}$.

2. $\sinh\left(\angle\frac{BC}{2}\right)=\sinh(\delta)\ \sin\ (\angle C-\angle B)$.

Using Lemma 17, an expression in arguments and Pythagorean differences of points on a circle can be reduced to an expression in the radius of the circle and hyperbolic trigonometric functions of independent angles. By Example 3, we can check whether such an expression is an identity. We thus have a complete proving method for this construction. Note that this construction can be used only as the first construction in a geometry statement.

4 Experimental Results and Extensions

4.1. Experimental Results. We have incorporated the method into the Common Lisp prover for the area method [2]. [5] is a collection of 90 geometry theorems proved by our prover. Of the 90 theorems, about 40 belong to the projective geometry and thus are true in non-Euclidean geometries. The validity of the other 50 theorems are not trivial. Many of the theorems such as Example 3 are new.

The following table contains the timing and proof length statistics for the 90 geometry problems solved by our computer program. Maxterm means the number of terms of the maximal polynomial occurring in a proof. Lemmano is the number of elimination lemmas used to eliminate points from geometry quantities. In other words, lemmano is the number of deduction steps in the proof.

Proving Time		Proof Length		Deduction Step	
Time(secs)	% of Thms	Maxterm	% of Thms	Lemmano	% of Thms
$t\leqslant 0.05$	46%	$m=1$	62%	$l\leqslant 5$	14%
$t\leqslant 0.1$	62%	$m\leqslant 2$	82%	$l\leqslant 10$	46%
$t\leqslant 1$	94%	$m\leqslant 5$	91%	$l\leqslant 20$	77%
$t\leqslant 5$	97%	$m\leqslant 10$	96%	$l\leqslant 30$	90%
$t\leqslant 6$	100%	$m\leqslant 15$	100%	$l\leqslant 69$	100%
0.27	average	2.2	average	15.12	average

Table 1. Statistics for the 90 geometry problems

From Table 1, we can see that our program is very fast. More importantly, the average length of the maximal polynomials in the proofs for the 90 theorems is only 2.2. Considering the fact that the 90 theorems are quite non-trivial, for instance, Examples 2 and 4 are among the moderately difficult ones in the 90 theorems, we can say that the proofs produced by our prover are much shorter than the coordinate based methods.

4.2. Two Examples. In [11], B. Grünbaum and G. C. Shephard discovered many new results in Euclidean geometry about polygons using numerical search. We will prove the corresponding results in non-Euclidean geometries using the method developed in this paper. Furthermore, for any polygon with concrete number of sides, our prover may be used to prove the corresponding theorems automatically.

Example 2. *Let $ABCDE$ be a pentagon, $P=AD\cap BE$, $Q=AC\cap BE$, $R=BD\cap AC$, $S=CE\cap BD$, and $T=AD\cap CE$. Then*

$$\frac{\sinh(AP)}{\sinh(TD)}\frac{\sinh(DS)}{\sinh(RB)}\frac{\sinh(BQ)}{\sinh(PE)}\frac{\sinh(ET)}{\sinh(SC)}\frac{\sinh(CR)}{\sinh(QA)}=1,$$

$$\frac{\sinh(AT)}{\sinh(PD)}\frac{\sinh(DR)}{\sinh(SB)}\frac{\sinh(BP)}{\sinh(QE)}\frac{\sinh(ES)}{\sinh(TC)}\frac{\sinh(CQ)}{\sinh(RA)}=1.$$

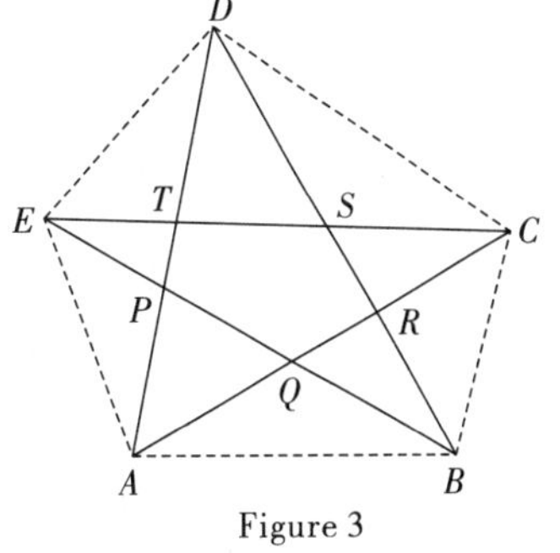

Figure 3

In general case, let $V_1,\cdots,V_m$ be a polygon and $1\leqslant d\leqslant \frac{m}{2}$, $1\leqslant j\leqslant\frac{m}{2}$ integers. We denote by $P_{a,j,i}$ the intersection of lines V_iV_{i+d} and lines $V_{i+j}V_{i+j+d}$, $i=1,\cdots,m$. Then $P_{d,j,i-j}$ is the intersection of line $V_{i-j}V_{i-j+d}$ and line V_iV_{i+d}. Let

$$T(m,d,j)=\prod_{i=1}^{m}\frac{\sinh(V_iP_{d,j,i})}{\sinh(P_{d,j,i-j}V_{i+d})};S(m,d,j)=\prod_{i=1}^{m}\frac{\sinh(V_iP_{d,j,i-j})}{\sinh(P_{d,j,i}V_{i+d})}.$$

Then Example 2 is a special case of the following theorem.

Example 3. (1) $T(m,d,j)=1$ *iff* $d+2j=m$ *or* $2d+j=m$. (2) $S(m,d,j)=1$ *iff* $d+2j=m$, $2d=j$, *or* $2j=d$.

All specific cases of Example 3 can be proved with our prover. For instance, the proof for the first case of Example 2 is as follows.

The machine proof:

$$\frac{-\frac{\sinh(ET)}{\sinh(CE)}\cdot\frac{\sinh(DS)}{\sinh(BD)}\cdot\frac{\sinh(CR)}{\sinh(AC)}\cdot\frac{\sinh(BQ)}{\sinh(BE)}\cdot\frac{\sinh(AP)}{\sinh(AD)}}{\frac{\sinh(EP)}{\sinh(BE)}\cdot\frac{\sinh(DT)}{\sinh(AD)}\cdot\frac{\sinh(CS)}{\sinh(CE)}\cdot\frac{\sinh(BR)}{\sinh(BD)}\cdot\frac{\sinh(AQ)}{\sinh(AC)}}$$

$$\overset{T}{=}\frac{S_{ADE}\cdot S_{ACDE}\cdot\left(-\frac{\sinh(DS)}{\sinh(BD)}\right)\cdot\frac{\sinh(CR)}{\sinh(AC)}\cdot\frac{\sinh(BQ)}{\sinh(BE)}\cdot\frac{\sinh(AP)}{\sinh(AD)}}{\left(-\frac{\sinh(EP)}{\sinh(BE)}\cdot\frac{\sinh(CS)}{\sinh(CE)}\cdot\frac{\sinh(BR)}{\sinh(BD)}\cdot\frac{\sinh(AQ)}{\sinh(AC)}\cdot S_{CDE}\right)\cdot(-S_{ACDE})}$$

$$\overset{simplify}{=}\frac{-S_{ADE}\cdot\frac{\sinh(AP)}{\sinh(AD)}\cdot\frac{\sinh(BQ)}{\sinh(BE)}\cdot\frac{\sinh(CR)}{\sinh(AC)}\cdot\frac{\sinh(DS)}{\sinh(BD)}}{S_{CDE}\cdot\frac{\sinh(AQ)}{\sinh(AC)}\cdot\frac{\sinh(BR)}{\sinh(BD)}\cdot\frac{\sinh(CS)}{\sinh(CE)}\cdot\frac{\sinh(EP)}{\sinh(BE)}}$$

$$\overset{S}{=}\frac{-S_{ADE}\cdot(-S_{CDE})\cdot(-S_{BCDE})}{S_{CDE}\cdot\frac{\sinh(AQ)}{\sinh(AC)}\cdot\frac{\sinh(BR)}{\sinh(BD)}\cdot(-S_{BCD})\cdot\frac{\sinh(EP)}{\sinh(BE)}\cdot S_{BCDE}}\cdot\frac{\sinh(AP)}{\sinh(AD)}\cdot\frac{\sinh(BQ)}{\sinh(BE)}\cdot\frac{\sinh(CR)}{\sinh(AC)}$$

$$\overset{simplify}{=}\frac{S_{ADE}}{\frac{\sinh(BP)}{\sinh(BE)}\cdot S_{BCD}\cdot\frac{\sinh(BR)}{\sinh(BD)}\cdot\frac{\sinh(AQ)}{\sinh(AC)}}\cdot\frac{\sinh(CR)}{\sinh(AC)}\cdot\frac{\sinh(BQ)}{\sinh(BE)}\cdot\frac{\sinh(AP)}{\sinh(AD)}$$

$$\overset{R}{=}\frac{(-S_{BCD})\cdot S_{ADE}\cdot(-S_{ABCD})}{\frac{\sinh(EP)}{\sinh(BE)}\cdot S_{BCD}\cdot(-S_{ABC})\cdot\frac{\sinh(AQ)}{\sinh(AC)}\cdot S_{ABCD}}\cdot\frac{\sinh(RQ)}{\sinh(BE)}\cdot\frac{\sinh(AP)}{\sinh(AD)}$$

$$\overset{simplify}{=}\frac{-S_{ADE}}{\frac{\sinh(AQ)}{\sinh(AC)}\cdot S_{ABC}\cdot\frac{\sinh(EP)}{\sinh(BE)}}\cdot\frac{\sinh(AP)}{\sinh(AD)}\cdot\frac{\sinh(BQ)}{\sinh(BE)}$$

$$\overset{Q}{=}\frac{-S_{ADE}\cdot(-S_{ABC})\cdot S_{ABCE}}{S_{ABE}\cdot S_{ABC}\cdot\frac{\sinh(EP)}{\sinh(BE)}\cdot(-S_{ABCE})}\cdot\frac{\sinh(AP)}{\sinh(AD)}$$

$$\overset{simplify}{=}\frac{-S_{ADE}}{\frac{\sinh(EP)}{\sinh(BE)}\cdot S_{ABE}}\cdot\frac{\sinh(AP)}{\sinh(AD)}$$

$$\overset{P}{=}\frac{-S_{ABE}\cdot S_{ADE}\cdot(-S_{ABDE})}{S_{ADE}\cdot S_{ABE}\cdot S_{ABDE}}\overset{simplify}{=}$$

The elimination lemmas used in the proof:

$$\frac{\sinh(DT)}{\sinh(AD)}\overset{T}{=}\frac{-S_{CDE}}{S_{ACDE}},\frac{\sinh(ET)}{\sinh(CE)}\overset{T}{=}\frac{S_{ADE}}{-S_{ACDE}},$$

$$\frac{\sinh(CS)}{\sinh(CE)} \overset{S}{=} \frac{S_{BCD}}{S_{BCDE}}, \frac{\sinh(DS)}{\sinh(BD)} \overset{S}{=} \frac{-S_{CDE}}{S_{BCDE}},$$

$$\frac{\sinh(BR)}{\sinh(BD)} \overset{R}{=} \frac{S_{ABC}}{S_{ABCD}}, \frac{\sinh(CR)}{\sinh(AC)} \overset{R}{=} \frac{-S_{BCD}}{S_{ABCD}},$$

$$\frac{\sinh(AQ)}{\sinh(AC)} \overset{Q}{=} \frac{S_{ABE}}{S_{ABCE}}, \frac{\sinh(BQ)}{\sinh(BE)} \overset{Q}{=} \frac{S_{ABC}}{S_{ABCE}},$$

$$\frac{\sinh(EP)}{\sinh(BE)} \overset{P}{=} \frac{S_{ADE}}{-S_{ABDE}}, \frac{\sinh(AP)}{\sinh(AD)} \overset{P}{=} \frac{S_{ABE}}{S_{ABDE}},$$

In the above *machine produced proof*, $a \overset{T}{=} b$ means that b is the result obtained by eliminating point T from a; $a \overset{simplify}{=} b$ means that b is obtained by canceling some common factors from the denominator and numerator of a; "eliminants" are the results obtained by eliminating points from separate geometry quantities.

Example 4 (*Pascal's Theorem*). *Let* A, B, C, D, E, *and* F *be six points on a circle. Let* $P = AB \cap DF$, $Q = BC \cap EF$, *and* $S = CD \cap EA$. *Show that* P, Q, *and* S *are collinear.*

Constructive description:

((CIRCLE, A, B, C, A_1, B_1, C_1)

(INTER, P, (LINE, A_1, B)(LINE, A, B_1))

(INTER, Q, (LINE, A, C_1)(LINE, A_1, C))

(INTER, S, (LINE, B_1, C)(LINE, B, C_1))

(INTER, Z_S, (LINE, Q, P)(LINE, B_1, C))

$\left(\frac{\sinh(B_1S)}{\sinh(CS)} = \frac{\sinh(B_1Z_S)}{\sinh(CZ_S)}\right)$

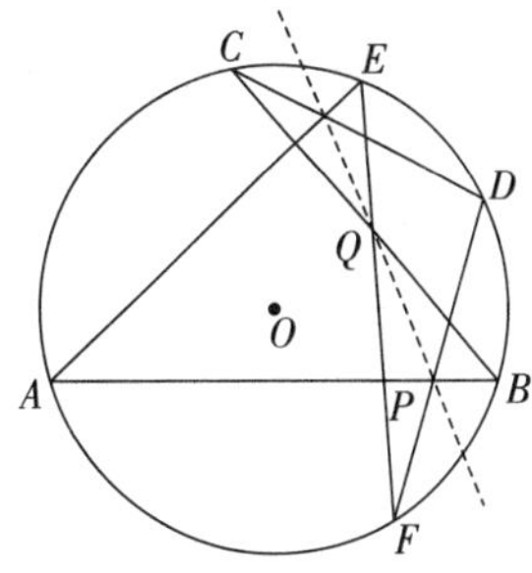

Figure 4

In the above description, by introducing a new Z_S, the fact that P, Q, and S are collinear is equivalent to the fact $S = Z_S$ or $\frac{\sinh(B_1S)}{\sinh(CS)} = \frac{\sinh(B_1Z_S)}{\sinh(CZ_S)}$. We always use this trick if the conclusion of a geometry statement is collinear. The machine proof:

$$\left(\frac{\sinh(B_1S)}{\sinh(CS)}\right) \Big/ \left(\frac{\sinh(B_1Z_S)}{\sinh(CZ_S)}\right)$$

$$\overset{Z_S}{=} \frac{-S_{CPQ}}{-S_{B_1PQ}} \cdot \frac{\sinh(B_1S)}{\sinh(CS)}$$

$$\overset{S}{=} \frac{(-S_{BB_1C_1}) \cdot S_{CPQ}}{S_{B_1PQ} \cdot (-S_{BCC_1})}$$

$$\overset{Q}{=} \frac{S_{CA_1P} \cdot S_{ACC_1} \cdot S_{BB_1C_1} \cdot S_{ACC_1A_1}}{(-S_{B_1C_1P} \cdot S_{ACA_1}) \cdot S_{BCC_1} \cdot (-S_{ACC_1A_1})}$$

$$\overset{simplify}{=} \frac{S_{BB_1C_1} \cdot S_{ACC_1} \cdot S_{CA_1P}}{S_{BCC_1} \cdot S_{ACA_1} \cdot S_{B_1C_1P}}$$

$$P = \frac{S_{BB_1C_1}\cdot S_{ACC_1}\cdot S_{BCA_1}\cdot S_{AA_1B_1}\cdot S_{ABB_1A_1}}{S_{BCC_1}\cdot S_{ACA_1}\cdot(-S_{BA_1B_1}\cdot S_{AB_1C_1})\cdot(-S_{ABB_1A_1})}$$

$$\stackrel{simplify}{=} \frac{S_{AA_1B_1}\cdot S_{BCA_1}}{S_{AB_1C_1}\cdot S_{BA_1B}}\frac{S_{ACC_1}\cdot S_{BB_1C_1}}{S_{ACA_1}\cdot S_{BCC_1}}$$

$$= \frac{\left(\mathrm{sh}\left(\frac{A_1B_1}{2}\right)\cdot \mathrm{sh}\left(\frac{AB_1}{2}\right)\cdot \mathrm{sh}\left(\frac{AA_1}{2}\right)\right)\cdot\left(\mathrm{sh}\left(\frac{CA_1}{2}\right)\cdot \mathrm{sh}\left(\frac{BA_1}{2}\right)\cdot \mathrm{sh}\left(\frac{BC}{2}\right)\right)}{\left(\mathrm{sh}(\frac{B_1C_1}{2}\cdot \mathrm{sh}\left(\frac{AC_1}{2}\right)\cdot \mathrm{sh}\left(\frac{AB_1}{2}\right)\right)\cdot\left(\mathrm{sh}\left(\frac{A_1B_1}{2}\right)\cdot \mathrm{sh}\left(\frac{BB_1}{2}\right)\cdot \mathrm{sh}\left(\frac{BA_1}{2}\right)\right)}$$

$$\frac{\left(\mathrm{sh}\left(\frac{CC_1}{2}\right)\cdot \mathrm{sh}\left(\frac{AC_1}{2}\right)\cdot \mathrm{sh}\left(\frac{AC}{2}\right)\right)\cdot\left(\mathrm{sh}\left(\frac{B_1C_1}{2}\right)\cdot \mathrm{sh}\left(\frac{BC_1}{2}\right)\cdot \mathrm{sh}\left(\frac{BB_1}{2}\right)\right)\cdot(\tan(\delta))^4}{\left(\mathrm{sh}\left(\frac{CA_1}{2}\right)\cdot \mathrm{sh}\left(\frac{AA_1}{2}\right)\cdot \mathrm{sh}\left(\frac{AC}{2}\right)\cdot\left(\mathrm{sh}\left(\frac{CC_1}{2}\right)\cdot \mathrm{sh}\left(\frac{BC_1}{2}\right)\cdot \mathrm{sh}\left(\frac{BC}{2}\right)\right)\cdot(\tan(\delta)\right)^4}$$

$$\stackrel{simplify}{=} 1$$

The elimination lemmas used in the proof.

$$\frac{\sinh(B_1Z_S)}{\sinh(CZ_S)} \stackrel{Z_S}{=} \frac{S_{B_1PQ}}{S_{CPQ}}, \frac{\sinh(B_1S)}{\sinh(CS)} \stackrel{S}{=} \frac{S_{BB_1C_1}}{S_{BCC_1}}$$

$$S_{B_1PQ} \stackrel{Q}{=} \frac{-S_{B_1C_1P}\cdot S_{ACA_1}}{S_{ACC_1A_1}}, S_{CPQ} \stackrel{Q}{=} \frac{S_{CA_1P}\cdot S_{ACC_1}}{-S_{ACC_1A_1}}$$

$$S_{B_1C_1P} \stackrel{P}{=} \frac{-S_{BA_1B_1}\cdot S_{AB_1C_1}}{S_{ABB_1A_1}}, S_{CA_1P} \stackrel{P}{=} \frac{S_{BCA_1}\cdot S_{AA_1B_1}}{-S_{ABB_1A_1}}$$

$$S_{BCC_1} = \frac{4\cdot \mathrm{sh}\left(\frac{BC_1}{2}\right)\cdot \mathrm{sh}\left(\frac{CC_1}{2}\right)\cdot \mathrm{sh}\left(\frac{BC}{2}\right)}{-\tan(\delta)}, S_{ACA_1} = \frac{4\cdot \mathrm{sh}\left(\frac{AA_1}{2}\right)\cdot \mathrm{sh}\left(\frac{CA_1}{2}\right)\cdot \mathrm{sh}\left(\frac{AC}{2}\right)}{-\tan(\delta)}$$

$$S_{BA_1B_1} = \frac{4\cdot \mathrm{sh}\left(\frac{BB_1}{2}\right)\cdot \mathrm{sh}\left(\frac{A_1B_1}{2}\right)\cdot \mathrm{sh}\left(\frac{BA_1}{2}\right)}{-\tan(\delta)},$$

$$S_{AB_1C_1} = \frac{4\cdot \mathrm{sh}\left(\frac{AC_1}{2}\right)\cdot \mathrm{sh}\left(\frac{B_1C_1}{2}\right)\cdot \mathrm{sh}\left(\frac{AB_1}{2}\right)}{-\tan(\delta)}$$

$$S_{BB1C1} = \frac{4\cdot \mathrm{sh}\left(\frac{BC_1}{2}\right)\cdot \mathrm{sh}\left(\frac{B_1C_1}{2}\right)\cdot \mathrm{sh}\left(\frac{BB_1}{2}\right)}{-\tan(\delta)},$$

$$S_{ACC_1} = \frac{4\cdot \mathrm{sh}\left(\frac{AC_1}{2}\right)\cdot \mathrm{sh}\left(\frac{CC_1}{2}\right)\cdot \mathrm{sh}\left(\frac{AC}{2}\right)}{-\tan(\delta)}$$

$$S_{BCA_1} = \frac{4\cdot \mathrm{sh}\left(\frac{BA_1}{2}\right)\cdot \mathrm{sh}\left(\frac{CA_1}{2}\right)\cdot \mathrm{sh}\left(\frac{BC}{2}\right)}{-\tan(\delta)},$$

$$S_{AA_1B_1} = \frac{4\cdot \mathrm{sh}\left(\frac{AB_1}{2}\right)\cdot \mathrm{sh}\left(\frac{A_1B_1}{2}\right)\cdot \mathrm{sh}\left(\frac{AA_1}{2}\right)}{-\tan(\delta)}$$

In the last step of the above, Lemma 17 is used. Note that we use sh to represent sinh in order to save printing space.

4. 3. Proving Theorems in Elliptic Geometry. The proving method pre-sented by now is for hyperbolic geometry. The proving method for constructive geometry statements in elliptic geometry is quite similar to that of the hyperbolic geometry. First, we have the following result[8].

Theorem 18. *Let points $A, B, P, R\cdots$ satisfy the same geometry conditions in hyperbolic and elliptic geometries. Then*

$$P(\sin(AB), \cos(AB), \sin(\angle PQR), \cos(\angle PQR), \cdots)=0$$

is a valid algebraic equation in elliptic geometry if and only if

$$P(\sqrt{-1}\sinh(AB), \cosh(AB), \sin(\angle PQR), \cos(\angle PQR), \cdots)=0$$

is a valid algebraic equation in hyperbolic geometry.

By the above theorem, all the basic propositions and elimination lemmas proved in this paper can be transformed to similar results in elliptic geometry. We thus have a similar algorithm of theorem proving in elliptic geometry.

Another consequence of Theorem 18 is that a constructive geometry statement with conclusion $P(\sin(AB), \cosh(AB), \cdots)=0$ is true in elliptic geometry iff the statement with same hypotheses and with conclusion $P(\sqrt{-1}\sinh(AB), \cosh(AB), \cdots)=0$ is true in hyperbolic geometry. For instance, Ceva' theorem(Example 1 with conclusion $\frac{\sinh(AF)}{\sinh(FB)} \cdot \frac{\sinh(BD)}{\sinh(DC)} \cdot \frac{\sinh(CE)}{\sinh(EA)}=1$) is true in elliptic geometry.

5 Conclusion Remarks

Conventionally, studies on non-Euclidean geometries are mainly focused on the axiom systems, the representations of these geometries in Euclidean geometry, etc. The topic of proving interesting and difficult geometry problems is generally overlooked. Despite of this, we believe that an efficient mechanical proving method for non-Euclidean geometries like the one given in this paper is of importance: besides the fact that the method/program is used to find many new theorems, it is also important for educational purpose. One of the authors was once asked by a professor during the 1993 International Mathematical Olympiad about how to prove the centroid theorem in hyperbolic geometry, which is not available in common textbooks of non-Euclidean geometries. We take this as one indication of lack of a general method of solving problems in non-Euclidean geometries. The method developed in this paper may provide such a tool, because the proofs produced by the method are generally short and in a shape that a student of mathematics could easily learn to design with pencil and paper.

There are still many problems not solved or solved unsatisfactorily for this approach. Though a large portion of the geometry theorems in textbooks can be proved by our prover, there are still ones which are not in class C, e. g. , theorems which cannot be described constructively. A moderate goal is to extend the elimination method to more constructions, such as the ones considered in [2]. The main difficulty: for more complicated constructions, the elimination formulas needed may become large and obscure. In this case, the goal of producing short and readable proofs may be dampened. Also the current method is limited to theorems of equational type: geometry inequalities cannot be handled by the current method.

Acknowledgement. The work reported here was supported in part by the NSF Grant CCR-9420857 and the Chinese National Science Foundation.

References

[1] S C Chou. Mechanical Geometry Theorem Proving, D. Reidel Publishing Company, Dordrecht, Netherlands, 1988.

[2] S C Chou, X S Gao, J Z Zhang. Machine Proofs in Geometry. World Scientific Pub. , Singapore, 1994.

[3] S C Chou, X S Gao, Mechanical Theorem Proving in Riemann Geometry. Computer Mathematics, W. T. Wu ed. , World Scientific Pub. , Singapore, pp. 136 – 157, 1993.

[4] S C Chou, X S Gao, L Yang, J Z Zhang. Automated Production of Readable Proofs for Theorems in non-Euclidean Geometries. TR-WSU-94 – 9, 1994.

[5] S C Chou, X S Gao, L Yang, J Z Zhang. A Collection of 90 Automatically Proved Geometry Theorems in non-Euclidean Geometries. TR-WSU-94 – 10, 1994.

[6] H Coelho, L M Pereira. Automated Reasoning in Geometry with Prolog. J. of Automated Reasoning, 1987(2): 329 – 390.

[7] H S M Coxeter. Non-Euclidean Geometry, Univ of Toronto Press, 1968.

[8] X S Gao. Transcendental Functions and Mechanical Theorem Proving in Elemen-tary Geometries. J. of Automated Reasoning, 1990(6): 403 – 417.

[9] H Gelernter, J R Hanson, D W Loveland. Empirical Explorations of the Geometry-theorem Proving Machine. Proc West Joint Computer Conf. , 1960: 143 – 147.

[10] M J Greenberg. Euclidean Geometry and non-Euclidean Geometry, Development and History, Freeman, 1980.

[11] B Grünbaum, G C Shephard. From Mennelaus to Computer Assisted Proofs in Geometry, Preprint, Washington State University, 1993.

[12] T Havel. Some Examples of the Use of Distances as Coordinates for Euclidean Geometry. J. of Symbolic Computation, 1991(11): 579 – 594.

[13] K R Koedinger, J R Anderson. Abstract Planning and Perceptual Chunks: Elements of

Expertise in Geometry. Cognitive Science, 1990(14): 511 - 550.

[14] A J Nevins. Plane Geometry Theorem Proving Using Forward Chaining. Artificial Intelligence, 1976: 1 - 23.

[15] D Kapur. Geometry Theorem Proving Using Hilbert's Nullstellensatz. Proc. of SYMSAC'86, Waterloo, 1986: 202 - 208.

[16] B Kutzler, S Stifter. A Geometry Theorem Prover Based on Buchberger's Al - gorithmProc CADE-7, Oxford, ed. J. H. Siekmann, LNCS, no 230, 1987: 693 - 694.

[17] S Stifter. Geometry Theorem Proving in Vector Spaces by Means of Gröbner Basees, Proc of ISSAC-93, Kiev, ACM Press, 1993: 301 - 310.

[18] T E Stokes. On the Algebraic and Algorithmic Properties of Some Generalized Algebras, Ph. D thesis, University of Tasmania, 1990.

[19] A Tarski. A Decision Method for Elementary Algebra and Geometry, Univ of California Press, Berkeley, 1951.

[20] D M Wang. On Wu's Method for Proving Constructive Geometry Theorems. Proc. IJCAI' 89, Detroit, 1989: 419 - 424.

[21] D M Wang. Elimination Procedures for Mechanical Theorem Proving in Geometry. Ann. of Math. AI, 1995(13): 1 - 24.

[22] W T Wu. On the Decision Problem and the Mechanization of Theorem Proving in Elementary Geometry, Scientia Sinica 21, 1978: 159 - 172; Also in Automated Theorem Proving: After 25 years, A. M. S. , Contemporary Mathematics, 29, 1984: 213 - 234.

[23] W T Wu. Basic Principles of Mechanical Theorem Proving in Geometries, Volume I: Part of Elementary Geometries, Science Press, Beijing (in Chinese), 1984. English Edition, Springer-Verlag, 1994.

[24] L Yang. Computer-Aided Proving for New Theorems of non-Euclidean Geometry, Research Report, No. 4, Mathematics Research Section, Australian National University, 1989.

[25] J Z Zhang, L Yang, M Deng. The Parallel Numerical Method of Mechanical Theorem Proving, Theoretical Computer Science, 1990(74): 253 - 271.

[26] J Z Zhang, L Yang, R X Hou. A Criterion for Depenency of Algedbraic Equations, with Applications to Automated Theorem Proving, Science in China Ser. A, 1994(37): 547 - 554.

A Set of Geometric Invariants for Kinematic Analysis of 6R Manipulators

Fu Hong-guang Yang Lu Zhang Jing-zhong
Chengdu Institute of Computer Applications
Academia Sinica, Chengdu, 610041, China

Abstract

This paper proposes four geometric invariants to be used to eliminate joint variables in closure equations for a 6R manipulator instead of the Gaussian elimination used in other studies. The geometric invariants are determined by the structure of any three consecutive joints in space. They have specific geometric meanings such as angle, length, area, volume. *For a general 6R manipulator, its four basic closure equations containing only three angular variables can be directly constructed in a few minutes from the geometric invariants using Maple on a general personal computer. These resulting equations have no extraneous roots and are algebraically independent. Because the basic closure equations are obtained from the geometric invariants, they have the most simple forms and provide a very good chance to solve the input-output equation for the inverse kinematics problem. As a result, we use the set of basic closure equations to derive symbolic 16th degree input-output equation and compute the input-output equation for a special case with 16 different real solutions. All position and orientation coordinates of the end-effector may be symbolic parameters. The definition of the geometric invariants is independent of joint types and can be applied to manipulators with any serial geometry.*

KEY WORDS—inverse kinematics, geometric invariants, basic closure equations, 6R manipulator

1. Introduction

The inverse kinematics problem of a serial manipulator has been a very significant and

---

本文刊于《International Journal of Robotics Research》,第 19 卷第 8 期,2000 年.

important problem for computer-controlled robots. It corresponds to solving a system of polynomial equations (called closure equations). The most difficult case has been the serial manipulator with six revolute joints (i. e. , 6R). As we know, a very important result is that Lee and Liang (1988) and Raghavan and Roth (1989) developed different methods to reduce the closure equations of 6R to a univariate polynomial of 16th degree (called the input-output equation). To get a set of basic closure equations with only three joint variables in Lee and Liang (1988) and Raghavan and Roth (1989), they had to eliminate two joint variables from 14 equations by Gaussian elimination. In fact, for a general 6R manipulator, when the position and orientation coordinates of the end-effector and all its structural parameters are not specific real numbers, the complexity of the above Gaussian elimination is so high that the computation is far beyond the capacity of existing computers. Besides, the process may bring about extraneous roots and become degenerate for special cases (Mavroidis and Roth 1994). Therefore, we want to know if there is a direct method to get a set of new basic closure equations and overcome the problem of extraneous roots and degeneration.

Our answer is yes. By extending our geometric invariant approach for 3 joint placeable manipulators in Fu, Yang, and Zhou (1998), we found four geometric invariants determined by the geometric structure of any three consecutive joints in space, which can be used to directly construct four basic closure equations. For a general 6R manipulator, we used a PC 686 to get the basic closure equations using Maple software (Heck 1993) in a few minutes and obtained the 16th degree symbolic polynomial in the tan-half-angle of the output displacement from the set of basic closure equations. The input-output equation of a general 6R manipulator can be expressed as a ratio $\frac{det\ (M)}{(1+t_3^2)^4 D}$. The numerator $det\ (M)$ cannot be expanded by a personal computer; it is expressed as a 16×16 determinate whose every nonzero entry is a symbolic polynomial of degree 2; the denominator $(1+t_3^2)^4 D$ is just a factor of the numerator, and D is a positive symbolic polynomial of 8th degree. At the end, we will give an example with 16 different real solutions as in Wampler and Morgan (1991) and derive its input-output equation. It is unique that we assume the position and orientation parameters of the end-effector are all symbolic; this kind of assumption is necessary for obtaining the solutions in symbolic form. As far as we know, no previous methods can solve such a complicated problem.

2. Problem Formulation

The links of the 6R manipulator are numbered from 1 to 7, the fixed or base link being 1 and the outermost link or end-effector being 7. A coordinate system attached to the ith link is numbered i. Serial manipulators are modeled using the Denavit-Hartenberg formalism. In particular, there is a 4×4 transformation matrix A_i relating the coordinate systems $i+1$ and i. The matrix is

$$A_i=\begin{pmatrix} c_i & -s_i\lambda_i & s_i\mu_i & a_ic_i \\ s_i & c_i\lambda_i & -c_i\mu_i & a_is_i \\ 0 & \mu_i & \lambda_i & d_i \\ 0 & 0 & 0 & 1 \end{pmatrix}$$

where

$s_i=\sin\theta_i, c_i=\cos\theta_i$,

$\lambda_i=\cos\alpha_i, \mu_i=\sin\alpha_i$,

a_i is the length of link $i+1$,

α_i is the twist angle between the axes of joints i and $i+1$,

d_i is the offset distance at joint i,

θ_i is the joint rotation angle at joint i.

For a given 6R manipulator, each A_i contains θ_i as its unknown variable; the others are parameters. The position and orientation of the end-effector is also represented as a 4×4 transformation matrix H with respect to the coordination of the base. As a result, the problem of inverse kinematics corresponds to computing the unknowns, which satisfy the following matrix equation (closure equation):

$$A_1A_2A_3A_4A_5A_6=H.$$

Although the matrix equation includes 12 nontrivial scalar equations, only 6 of them are independent. In fact, substituting cos and sin for the tangent of the half angle, the problem of inverse kinematics is reduced to solving 12 polynomial equations in 6 unknowns.

Without loss of generality, we can assume $a_6=0, d_1=d_6=0$, and $\alpha_6=0$ by properly selecting coordinate systems above link 1 and link 7 (Fu, Gozalves, and Lee 1987). All structural parameters of a general 6R manipulator can be described by the following table:

$$\begin{bmatrix} a_1 & 0 & \tan\left(\frac{\alpha_1}{2}\right) \\ a_2 & d_2 & \tan\left(\frac{\alpha_2}{2}\right) \\ a_3 & d_3 & \tan\left(\frac{\alpha_3}{2}\right) \\ a_4 & d_4 & \tan\left(\frac{\alpha_4}{2}\right) \\ a_5 & d_5 & \tan\left(\frac{\alpha_5}{2}\right) \\ 0 & 0 & 0 \end{bmatrix}$$

To simplify the symbolic computation, we use Cayley's formula (Craig 1989) for an orthonormal matrix to express H. In this paper, $[x, y, z]$ and $[u_1, u_2, u_3]$ stand for the position and orientation of hand, respectively, then we have,

$$H=\begin{pmatrix} & R & & x \\ & & & y \\ & & & z \\ 0 & 0 & 0 & 1 \end{pmatrix}$$

where R is a proper orthonormal matrix and expressed as

$$R=(I_3-S)^{-1}(I_3+S);$$

here, I_3 is a 3×3 unit matrix and S is a skew-symmetric matrix, which is specified by three orientation parameters $[u_1, u_2, u_3]$ as

$$S=\begin{pmatrix} 0 & -u_3 & u_2 \\ u_3 & 0 & -u_1 \\ -u_2 & u_1 & 0 \end{pmatrix}$$

The position and orientation parameters of the end-effector are $[x, y, z, u_1, u_2, u_3]$. Therefore the closure equations contain 20 parameters and 6 joint variables. It is obvious that the complexity of this algebraic system is very high.

3. Geometric Invariants

Consider any three consecutive revolute joints in space. We can assume that $A_{i-1}\cdot A_iA_{i+1}=A_h$, where V_{i-1} and P_{i-1} stand for the first three elements of the 3rd and 4th columns of A_{i-1}. So do V_h and P_h stand for A_h. Then, their geometric relations can be demonstrated by Figure 1.

In Fu, Yang, and Zhou (1998), we proposed an area ratio invariant for 3R manipulators. Based on the same idea, we will introduce more geometric variables involving *angle*, *projective length*, *square distance*, *area*, *volume*.

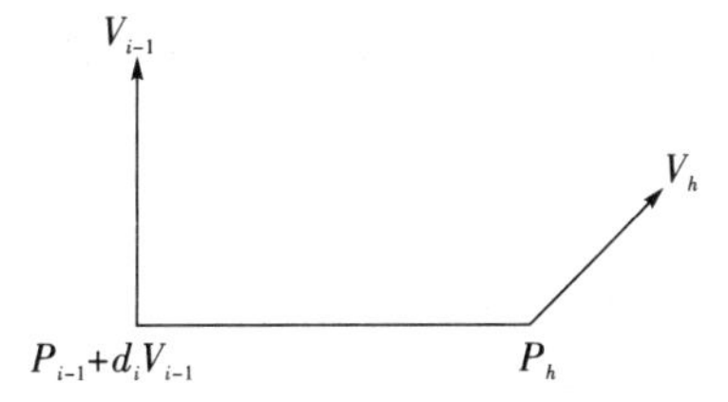

Fig. 1. The geometric relation between A_{i-1} and A_h

DEFINITION 1.

$v_{i-1}\triangleq\langle V_{i-1}, V_h\rangle$

$h_{i-1}\triangleq\langle P_h-P_{i-1}-d_iV_{i-1}, V_{i-1}\rangle$

$h'_{i-1}\triangleq\langle P_h-P_{i-1}-d_iV_{i-1}, V_h\rangle$

$r_{i-1}^2 \triangleq <P_h - P_{i-1} - d_i V_{i-1}, P_h - P_{i-1} - d_i V_{i-1}>$

$W_{i-1} \triangleq <P_h - P_{i-1} - d_i V_{i-1}[V_{i-1}, V_h]>$

$S_{i-1} \triangleq < r_{i-1}^2 v_{i-1} - 2h_{i-1} h'_{i-1}.$

Refer to Figure 1: their geometric meanings are very obvious. $P_{i-1}+d_iV_{i-1}$ and P_h are the position coordinates of joint $i-1$ and $i+1$. V_{i-1} and V_h are unit vectors of the axes of revolute joints. v_{i-1} stands for the cosine of the angle between V_{i-1} and V_h. h_{i-1} and h'_{i-1} stand for the projected length of vector $P_h - P_{i-1} - d_iV_{i-1}$ on V_{i-1} and V_h. r_{i-1}^2 stands for the square of the distance between $P_{i-1}+d_iV_{i-1}$ and P_h. W_{i-1} stands for the volume of the tetrahedron formed by the three vectors v_{i-1}, v_h, and $P_h - P_{i-1} - d_iV_{i-1}$. S_{i-1} stands for the area difference between $r_{i-1}^2 v_{i-1}$ and $2h_{i-1}h'_{i-1}$.

First, we calculate all the geometric variables in Definition 1 using a computer. It is fortunate that the resulting trigonometric expressions are all linear polynomials of $\sin\theta_i$ and $\cos\theta_i$. Then we can discover that some combinations of these geometric variables satisfy the following invariant relations by directly eliminating $\sin\theta_i$ and $\cos\theta_i$.

THEOREM 1. Let $k_i = \frac{\sin\alpha_i}{l_i}$, $k_{i+1} = \frac{\sin\alpha_{i+1}}{l_{i+1}}$, then for any three consecutive joints, the following equations hold.

$$\frac{h_{i-1} - d_{i+1}\cos\alpha_i}{h'_{i-1} - d_{i+1}\cos\alpha_{i+1}} = \frac{k_i}{k_{i+1}} \tag{1}$$

$$\frac{2(v_{i-1} - \cos\alpha_i\cos\alpha_{i+1})}{l_i^2 + l_{i+1}^2 + d_{i+1}^2 - r_{i-1}^2} = k_i k_{i+1} \tag{2}$$

$$W_{i-1} + l_i\sin\alpha_i\cos\alpha_{i+1} + l_{i+1}\sin\alpha_{i+1}\cos\alpha_i = k_{i+1}d_{i+1}(h_{i-1} - d_{i+1}\cos a_i) + \left(\frac{\cos\alpha_i}{k_i} + \frac{\cos\alpha_{i+1}}{k_{i+1}}\right)(v_{i-1} - \cos\alpha_i\cos\alpha_{i+1}) \tag{3}$$

$$\begin{aligned} &S_{i-1} + (d_{i+1}^2 - l_{i+1}^2 - l_i^2)\cos\alpha_i\cos\alpha_{i+1} + 2l_il_{i+1}\sin a_i\sin\alpha_{i+1} \\ &= \left(l_i^2 + l_{i+1}^2 + d_{i+1}^2 - 2\frac{\cos\alpha_i\cos\alpha_{i+1}}{k_ik_{i+1}}\right)(v_{i-1} - \cos\alpha_i\cos\alpha_{i+1}) \\ &\quad - 2\left(d_{i+1}k_{i+1}\frac{\cos a_i}{k_i} + d_{i+1}\cos\alpha_{i+1}\right)(h_{i-1} - d_{i+1}\cos\alpha_i). \end{aligned} \tag{4}$$

Proof. Because the geometric relations of any three consecutive joints are independent of any matrix transformation, without loss of generality we can assume $i = 2$. Then, we have

$$A_1 = \begin{bmatrix} \cos\theta_1 & -\sin\theta_1\cos\alpha_1 & \sin\theta_1\sin\alpha_1 & l_1\cos\theta_1 \\ \sin\theta_1 & \cos\theta_1\cos\alpha_1 & -\cos\theta_1\sin\alpha_1 & l_1\sin\theta_1 \\ 0 & \sin\alpha_1 & \cos\alpha_1 & 0 \\ 0 & 0 & 0 & 1 \end{bmatrix}$$

$$A_2 = \begin{bmatrix} \cos\theta_2 & -\sin\theta_2\cos\alpha_2 & \sin\theta_2\sin\alpha_2 & l_2\cos\theta_2 \\ \sin\theta_2 & \cos\theta_2\cos\alpha_2 & -\cos\theta_2\sin\alpha_2 & l_2\sin\theta_2 \\ 0 & \sin\alpha_2 & \cos\alpha_2 & d_2 \\ 0 & 0 & 0 & 1 \end{bmatrix}$$

and

$$A_3=\begin{bmatrix}\cos\theta_3 & -\sin\theta_3\cos\alpha_3 & \sin\theta_3\sin\alpha_3 & l_3\cos\theta_3\\ \sin\theta_3 & \cos\theta_3\cos\alpha_3 & -\cos\theta_3\sin\alpha_3 & l_3\sin\theta_3\\ 0 & \sin\alpha_3 & \cos\alpha_3 & d_3\\ 0 & 0 & 0 & 1\end{bmatrix}$$

According to A_1, A_2, and A_3, we can get the positions and orientations of the two ends of the joints, i. e., V_1, P_1 and V_h, P_h:

$$V_1=[\sin\theta_1\sin\alpha_1, -\cos\theta_1\sin\alpha_1, \cos\alpha_1]$$

$$P_1=[l_1\cos\theta_1, l_1\sin\theta_1, 0]$$

$$\begin{aligned}V_h=[&(\cos\theta_1\cos\theta_2-\sin\theta_1\cos\alpha_1\sin\theta_2)\sin\theta_3\sin\alpha_3\\&-(-\cos\theta_1\sin\theta_2\cos\alpha_2-\sin\theta_1\cos\alpha_1\cos\theta_2\cos\alpha_2\\&+\sin\theta_1\sin\alpha_1\sin\alpha_2)\cos\theta_3\sin\alpha_3\\&+(\cos\theta_1\sin\theta_2\sin\alpha_2+\sin\theta_1\cos\alpha_1\cos\theta_2\sin\alpha_2\\&+\sin\theta_1\sin\alpha_1\cos\alpha_2)\cos\alpha_3, \sin\theta_1\cos\theta_2\\&+\cos\theta_1\cos\alpha_1\sin\theta_2\sin\theta_3\sin\alpha_3\\&-(-\sin\theta_1\sin\theta_2\cos\alpha_2+\cos\theta_1\cos\alpha_1\cos\theta_2\cos\alpha_2\\&-\cos\theta_1\sin\alpha_1\sin\alpha_2)\cos\theta_3\sin\alpha_3\\&+(\sin\theta_1\sin\theta_2\sin\alpha_2-\cos\theta_1\cos\alpha_1\cos\theta_2\sin\alpha_2\\&-\cos\theta_1\sin\alpha_1\cos\alpha_2)\cos\alpha_3, \sin\alpha_1\sin\theta_2\sin\theta_3\sin\alpha_3\\&-(\sin\alpha_1\cos\theta_2\cos\alpha_2+\cos\alpha_1\sin\alpha_2\cos\theta_3)\sin\alpha_3\\&+(-\sin\alpha_1\cos\theta_2\sin\alpha_2+\cos\alpha_1\cos\alpha_2)\cos\alpha_3]\end{aligned}$$

$$\begin{aligned}P_h=[&(\cos\theta_1\cos\theta_2-\sin\theta_1\cos\alpha_1\sin\theta_2)l_3\cos\theta_3\\&+(-\cos\theta_1\sin\theta_2\cos\alpha_2-\sin\theta_1\cos\alpha_1\cos\theta_2\cos\alpha_2\\&+\sin\theta_1\sin\alpha_1\sin\alpha_2)l_3\sin\theta_3(\cos\theta_1\sin\theta_2\sin\alpha_2\\&+\sin\theta_1\cos\alpha_1\cos\theta_2\sin\alpha_2+\sin\theta_1\sin\alpha_1\cos\alpha_2 d_3\\&+\cos\theta_1 l_2\cos\theta_2-\sin\theta_1\cos\alpha_1 l_2\sin\theta_2\\&+\sin\theta_1\sin\alpha_1 d_2+l_1\cos\theta_1, (\sin\theta_1\cos\theta_2\\&+\cos\theta_1\cos\alpha_1\sin\theta_2)l_3\cos\theta_3\\&+(-\sin\theta_1\sin\theta_2\cos\alpha_2+\cos\theta_1\cos\alpha_1\cos\theta_2\cos\alpha_2\\&-\cos\theta_1\sin\alpha_1\sin\alpha_2)l_3\sin\theta_3+(\sin\theta_1\sin\theta_2\sin\alpha_2\\&-\cos\theta_1\cos\alpha_1\cos\theta_2\sin\alpha_2-\cos\theta_1\sin\alpha_1\cos\alpha_2)d_3\\&+\sin\theta_1 l_2\cos\theta_2+\cos\theta_1\cos\alpha_1 l_2\sin\theta_2\\&-\cos\theta_1\sin\alpha_1 d_2+l_1\sin\theta_1, \sin\alpha_1\sin\theta_2 l_3\cos\theta_3\\&+(\sin\alpha_1\cos\theta_2\cos\alpha_2+\cos\alpha_1\sin\alpha_2)l_3\sin\theta_3\\&+(-\sin\alpha_1\cos\theta_2\sin\alpha_2+\cos\alpha_1\cos\alpha_2)d_3\\&+\sin\alpha_1 l_2\sin\theta_2+\cos\alpha_1 d_2].\end{aligned}$$

Let

$$P_1'=P_1+d_2V_1=[\sin\theta_1\sin\alpha_1 d_2+l_1\cos(\theta_1),-\cos\theta_1\sin\alpha_1 d_2+l_1\sin\theta_1,\cos\alpha_1 d_2]$$

and

$$\begin{aligned}P_h-P_1'=[&l_3\cos\theta_3\cos\theta_1\cos\theta_2\\&-l_3\cos\theta_3\sin\theta_1\cos\alpha_1\sin\theta_2\\&-l_3\sin\theta_3\cos\theta_1\sin\theta_2\cos\alpha_2\\&-l_3\sin\theta_3\sin\theta_1\cos\alpha_1\cos\theta_2\cos\alpha_2\\&+l_3\sin\theta_3\sin\theta_1\sin\alpha_1\sin\alpha_2\\&+d_3\cos\theta_1\sin\theta_2\sin\alpha_2\\&+d_3\sin\theta_1\cos\alpha_1\cos\theta_2\sin\alpha_2\\&+d_3\sin\theta_1\sin\alpha_1\cos\alpha_2+\cos\theta_1 l_2\cos\theta_2\\&-\sin\theta_1\cos\alpha_1 l_2\sin\theta_2,l_3\cos\theta_3\sin\theta_1\cos\theta_2\\&+l_3\cos\theta_3\cos\theta_1\cos\alpha_1\sin\theta_2\\&-l_3\sin\theta_3\sin\theta_1\sin\theta_2\cos\alpha_2\\&+l_3\sin\theta_3\cos\theta_1\cos\alpha_1\cos\theta_2\cos\alpha_2\\&-l_3\sin\theta_3\cos\theta_1\sin\alpha_1\sin\alpha_2\\&+d_3\sin\theta_1\sin\theta_2\sin\alpha_2\\&-d_3\cos\theta_1\cos\alpha_1\cos\theta_2\sin\alpha_2\\&-d_3\cos\theta_1\sin\alpha_1\cos\alpha_2+\sin\theta_1 l_2\cos\theta_2\\&+\cos\theta_1\cos\alpha_1 l_2\sin\theta_2,\sin\alpha_1\sin\theta_2 l_3\cos\theta_3\\&+l_3\sin\theta_3\sin\alpha_1\cos\theta_2\cos\alpha_2\\&+\sin\alpha_1 l_2\sin\theta_2+l_3\sin\theta_3\cos\alpha_1\sin\alpha_2\\&-d_3\sin\alpha_1\cos\theta_2\sin\alpha_2+d_3\cos\alpha_1\cos\alpha_2],\end{aligned}$$

then compute the geometric variables according to Definition 1 with Maple on a general PC 686, the results are

$$v_1=\langle V_1,V_h\rangle=\cos\alpha_2\cos\alpha_3-\cos\theta_3\sin\alpha_2\sin\alpha_3$$
$$h_1=\langle P_h-P_1',V_1\rangle=d_3\cos\alpha_2+l_3\sin\alpha_2\sin\theta_3$$
$$h_1'=\langle P_h-P_1,V_h\rangle=d_3\cos\alpha_3+l_2\sin\alpha_3\sin\theta_3$$
$$r_1^2=\langle P_h-P_1',P_h-P_1'\rangle=d_3^2+l_2^2+l_3^2+2l_2l_3\cos\theta_3$$

$$\begin{aligned}W_1&=\langle P_h-P_1',[V_{i-1},V_h]\rangle\\&=l_2\cos\alpha_3\sin\alpha_2+l_3\cos\alpha_2\sin\alpha_3\\&\quad+l_3\cos\alpha_3\sin\alpha_2\cos\theta_3+l_2\sin\alpha_3\cos\alpha_2\cos\theta_3\\&\quad-d_3\sin\alpha_2\sin\alpha_3\sin\theta_3\end{aligned}$$

$$\begin{aligned}S_1&=r_1^2v_1-2h_1h_1'\\&=2l_2l_3\cos\theta_3\cos\alpha_3\cos\alpha_2-l_2^2\cos\theta_3\sin\alpha_3\sin\alpha_2\\&\quad-d_3^2\cos\theta_3\sin\alpha_3\sin\alpha_2+l_3^2\cos\alpha_3\cos\alpha_2\\&\quad-2l_3\sin\theta_3 d_3\cos\alpha_3\sin\alpha_2-2l_2l_3\sin\alpha_2\sin\alpha_3\\&\quad-l_3^2\sin\alpha_2\cos\theta_3\sin\alpha_3-2l_2d_3\cos\alpha_2\sin\theta_3\sin\alpha_3\end{aligned}$$

$+l_2^2 \cos\alpha_3 \cos\alpha_2 - d_3^2 \cos\alpha_3 \cos\alpha_2$.

By substituting $v_1, h_1, h_1', r_1, W_1, S_1$ in eqs. (1), (2), (3), and (4), we can verify these equations hold regardless of the joint angles. So eqs. (1), (2), (3), and (4) are a set of invariants determined by the geometric structure of any three consecutive joints.

In fact, without the aid of a computer, the proof would be very difficulty because the symbolic computations are too large.

4. Basic Closure Equations of a General 6R Manipulator

To make use of the geometric invariants to eliminate the joint variables, first we rewrite the closure equation of a general 6R manipulator

$$A_1A_2A_3A_4A_5A_6 = H$$

as

$$A_{123}A_4A_5 = A_h;$$

here, $A_{123} = A_1A_2A_3$, $A_h = HA_6^{-1}$.

It is obvious that the 3rd column and the 4th column of A_h do not include any joint variables (Raghavan and Roth 1989). Hence, we pay attention to the the 3rd column and the 4th column of the closure equations. Because A_{123} can be considered as the matrix for an equivalent joint, we can use Theorem 1 to construct four basic closure equations:

$$\begin{cases} f_1(c_1, s_1, c_2, s_2, c_3, s_3, u) = 0 \\ f_2(c_1, s_1, c_2, s_2, c_3, s_3, u) = 0 \\ f_3(c_1, s_1, c_2, s_2, c_3, s_3, u) = 0 \\ f_4(c_1, s_1, c_2, s_2, c_3, s_3, u) = 0; \end{cases} \tag{5}$$

here, $u = [l_1, l_2, l_3, l_4, l_5, d_2, d_3, d_4, d_5, \alpha_1, \alpha_2, \alpha_3, \alpha_4, \alpha_5, x, y, z, u_1, u_2, u_3]$.

Using a PC 686, we spent about 400 s to produce these equations. f_1 is listed in the appendix; f_2, f_3, and f_4 have 100, 198, and 727 terms, respectively, but they are omitted from the appendix.

It is obvious that only three variables $\theta_1, \theta_2, \theta_3$ are included in f_1, f_2, f_3, f_4 and they are linear in every $c_1, s_1, c_2, s_2, c_3, s_3$. Let

$$\begin{cases} t_1 = \tan\dfrac{\theta_1}{2} \\ t_2 = \tan\dfrac{\theta_2}{2} \\ t_3 = \tan\dfrac{\theta_3}{2} \end{cases}$$

and substitute sin and cos for tan of half angle in (5) to get the following polynomial equations:

$$\begin{cases} p_1(t_1,t_2,t_3,u)=0 \\ p_2(t_1,t_2,t_3,u)=0 \\ p_3(t_1,t_2,t_3,u)=0 \\ p_4(t_1,t_2,t_3,u)=0, \end{cases} \tag{6}$$

where $degree(p_i,t_j)=2$ and $degree(p_i,\{t_1,t_2,t_3\})=6$, $i=1,2,3,4;j=1,2,3$.

By the dialytic method, we multiply (6) by t_1,t_2, and t_1t_2, and obtain the following 16 equations:

$$\begin{cases} p_1(t_1,t_2,t_3,u)=0 \\ p_2(t_1,t_2,t_3,u)=0 \\ p_3(t_1,t_2,t_3,u)=0 \\ p_4(t_1,t_2,t_3,u)=0 \\ p_1(t_1,t_2,t_3,u)*t_1=0 \\ p_2(t_1,t_2,t_3,u)*t_1=0 \\ p_3(t_1,t_2,t_3,u)*t_1=0 \\ p_4(t_1,t_2,t_3,u)*t_1=0 \\ p_1(t_1,t_2,t_3,u)*t_2=0 \\ p_2(t_1,t_2,t_3,u)*t_2=0 \\ p_3(t_1,t_2,t_3,u)*t_2=0 \\ p_4(t_1,t_2,t_3,u)*t_2=0 \\ p_1(t_1,t_2,t_3,u)*t_1*t_2=0 \\ p_2(t_1,t_2,t_3,u)*t_1*t_2=0 \\ p_3(t_1,t_2,t_3,u)*t_1*t_2=0 \\ p_4(t_1,t_2,t_3,u)*t_1*t_2=0 \end{cases} \tag{7}$$

For the sequence of variables

$$X=[t_1^3t_1^3,t_1^3t_2^2,t_1^3t_2,t_1^3,t_1^2t_2^3,t_1^2t_2^2,t_1^2t_2,t_1^2,t_1t_2^3,t_1t_2^2,t_1t_2,t_1,t_2^3,t_2^2,t_2,1],$$

we can rewrite (7) as a matrix equation:

$$MX=0; \tag{8}$$

here, M is a 16×16 matrix.

It is necessary that *det* $(M)=0$. Because every nonzero entry of M is a polynomial of degree 2, *det* (M) is a polynomial of 32 degrees in t_3. We have tried to expand the *det* (M), but the symbolic expansion is too large for the capability of our PC 686.

As is known, the input-output equation of a general 6R manipulator is a 16th polynomial equation. Therefore, we must solve the following two problems:

- The basic closure equations must generally be algebraically independent.
- A method is needed to remove the extraneous factors in *det* (M).

The first problem is not difficult. By substituting the parameters for a set of rational numbers in *det* (M) and expanding *det* (M), we will see it is generally not identical to zero.

For the second problem, we found the extraneous factors by solving the basic closure eqs. (6) directly. To remove these extraneous complex roots, let $t_1=i$ and $t_1=-i$ in eq. (6), then solve the two sets of equations directly. We can determine a factor D by Maple, which is a positive polynomial of 8th degree in t_3. D is listed in the appendix.

In the same way, let $t_3=i$ and $t_3=-i$ in eq. (6), then solve the two sets of equations directly. We can determine four sets of solutions for every set of equations. Hence, $(t3-i)^4(t_3+i)^4=(t_3^2+1)^4$ is another factor of $det\ (M)$.

THEOREM 2. The input-output equation of a general 6R manipulator is

$$\frac{det\ (M)}{(1+t_3^2)^4 D}=0, \tag{9}$$

which is a 16th degree polynomial equation. Here, D is a positive polynomial of 8th degree in t_3, and $(1+t_3^2)^4 D$ is a factor of $det\ (M)$.

It is noticeable that the input-output eq. (9) is a symbolic polynomial, which is different from the numerical input-output equation obtained by numerical calculation for a special position and orientation of the end-effector. As soon as it has been obtained, the symbolic equation can be repeatedly used to derive the input-output equation for a special manipulator by substituting the parameters for specific numbers.

5. A Case with 16 Different Real Solutions

We use problem 6 in Wampler and Morgan (1991), which has 16 different real solutions, to illustrate how to obtain the input-output equation using the geometric invariants. We use a slight variation in the twist angles and assume position and orientation parameters of the end-effector are $[x, y, z]$ and $[r_1, r_2, r_3]$. The symbolic parameters would make the problem become very difficult.

The link parameters are

$$\begin{bmatrix} 3 & 0 & 1 \\ 10 & 0 & 2 \\ 0 & 2 & 1 \\ 15 & 0 & 2 \\ 0 & 0 & 1 \\ 0 & 0 & 0 \end{bmatrix}$$

The position coordinates of the end-effector are $[x, y, z]$, and the orientation parameters of the Cayley formula in Section 2 are r_1, r_2, and r_3.

Step 1: We rewrite the closure equation in the following form:

$$(A_1 A_2) A_3 A_4 = H A_6^{-1} A_5^{-1}.$$

According to (1),(2),and (3) in Theorem 1,we get three equations,

$$p_1=560+100s_2z-300c_2-5x^2+30c_1x+30s_1y-5z^2+100c_1c_2x+100s_1c_2y-5y^2-48s_2-16zc_2+12yc_1-12xs_1+16xc_1s_2+16ys_1s_2;$$

$$p_2=-50u_1c_1c_2-15c_1u_1+5u_1x-8u_1c_1s_2+6u_1s_1-50u_2s_1c_2-15s_1u_2+5u_2y-8u_2s_1s_2-6u_2c_1-50u_3s_2+5u_3z+8u_3c_2;$$

$$p_3=-295u_2c_1+60u_3c_2+295u_1s_1-36u_3-60u_2s_1s_2+12yu_3s_1+16xu_3s_1s_2-12zu_2s_1+12xu_3c_1-16yu_3c_1s_2-16zu_1s_1s_2-48c_1c_2u_2-60u_1c_1s_2+48s_1c_2u_1-16yc_2u_1-12zu_1c_1+16xc_2u_2+16zu_2c_1s_2;$$

here,$[u_1,u_2,u_3,0]^T$ is the 3rd column of $HA_6^{-1}A_5^{-1}$,and

$$\begin{cases} u_1=\dfrac{-\sin\theta_6-\sin\theta_6r_1^2+\sin\theta_6r_3^2+\sin\theta_6r_2^2+2\cos\theta_6r_3-2\cos\theta_6r_2r_1}{1+r_1^2+r_3^2+r_2^2} \\ u_2=\dfrac{-(-2\sin\theta_6r_3-2\sin\theta_6r_2r_1+\cos\theta_6r_3^2-\cos\theta_6-\cos\theta_6r_2^2+\cos\theta_6r_1^2}{1+r_1^2+r_3^2+r_2^2} \\ u_3=\dfrac{2(\sin\theta_6r_3r_1-\sin\theta_6r_2+\cos\theta_6r_1+\cos\theta_6r_3r_2)}{1+r_1^2+r_3^2+r_2^2} \end{cases}$$

By eliminating $\cos\theta_6$ and $\sin\theta_6$ in the above equations,we get an additional equation:

$$p_4=(2r_2+2r_3r_1)u_1+(2r_2r_3-2r_1)u_2+(1-r_1^2-r_2^2+r_3^2)u_3.$$

Step 2: It is obvious that p_2,p_3,p_4 are homogeneous polynomials of u_1,u_2,and u_3. Let $u_1'=\dfrac{u_1}{u_3}$, $u_2'=\dfrac{u_2}{u_3}$,then rewrite p_2,p_3,and p_4 as

$$p_2'=-50u_1'c_1c_2-15c_1u_1'+5u_1'x-8u_1'c_1s_2+6u_1's_1-50u_2's_1c_2-15s_1u_2'+5u_2'y-8u_2's_1s_2-6u_2'c_1-50s_2+5z+8c_2$$

$$p_3'=-295u_2'c_1+60c_2+295u_1's_1-36-60u_2's_1s_2+12ys_1+16xs_1s_2-12zu_2's_1+12xc_1-16yc_1s_2-16zu_1's_1s_2-48c_1c_2u_2'-60u_1'c_1s_2+48s_1c_2u_1'-16yc_2u_1'-12zu_1'c_1+16xc_2u_2'+16zu_2'c_1s_2$$

$$p_4'=(2r_2+2r_3r_1)u_1'+(-2r_1+2r_2r_3)u_2'+(1-r_1^2-r_2^2+r_3^2).$$

Step 3: Solving $p_4'=0$,we can get $u_2'=ku_1'+b$;here,$k=\dfrac{2r_2+2r_3r_1}{r_1-r_2r_3}$,$b=\dfrac{1-r_1^2-r_2^2+r_3^2}{r_1-r_2r_3}$. After substituting u_2' for $ku_1'+b$ in p_1' and p_3',we compute the resultant of p_1' and p_3' in u_1' to derive a new polynomial p_5 without u_1' and u_2'.

$$p_5=14750s_1s_2-180yk+216c_1k+540s_1k-180x-1475zs_1-3000s_1c_2^2k-72zs_1^2b+2000c_1^2bs_2$$

$$+15470c_1^2bc_2-300zc_1s_2-216s_1+2400s_2s_1c_2$$
$$+288s_1s_2k+900s_1c_2k+96c_2zc_1+80y^2bc_2$$
$$+2400s_1^2c_2^2b+15470s_1^2bc_2+80z^2s_1s_2$$
$$+60z^2s_1k-800zs_1s_2^2-3000s_1s_2^2k$$
$$-72c_1^2bz+384c_2^2c_1k-128c_2^2xk+60s_1yx$$
$$+4425s_1^2b-180s_1yc_1+300c_2yk-1475c_1bx$$
$$+2400c_2^2c_1^2b+128yc_1^2s_2^2+240yc_1^2s_2$$
$$-72c_1^2xk-96c_1^2xs_2-600c_1^2xc_2$$
$$+2000s_1^2s_2b+60s_1y^2k-180s_1^2yk+288c_1s_2$$
$$+80c_2x^2b+96xs_1^2s_2+80x^2s_1s_2-384s_1c_2^2$$
$$+128c_2^2y+540c_1-3000c_1s_2^2-1475ybs_1+80zc_2y$$
$$+60z^2c_1+1475zc_1k-240zs_1c_2-800s_2c_2y$$
$$-14750s_2c_1k+900c_2c_1-180c_1^2x+60c_1x^2$$
$$2000s_1c_2-800c_2^2xbc_1+72s_1^2y-600s_1^2yc_2k$$
$$-192s_1yc_1s_2-128c_2xbc_1s_2-192xs_1s_2c_1k$$
$$-600s_1yc_2c_1-800xs_1^2s_2c_2k-240xs_1^2s_2k$$
$$-128xs_1^2s_2^2k-128xs_1s_2^2c_1-800xs_1s_2c_2c_1$$
$$-3000c_2^2c_1+4425c_1^2b+96c_2xbs_1-480c_2xbc_1$$
$$-240xs_1s_2c_1-60zs_1bx-128zc_1^2s_2^2b$$
$$-240zc_1^2s_2b-800zc_1^2s_2bc_2+800yc_1s_2s_1c_2k$$
$$-80y^2c_1s_2k+96yc_1^2s_2k+128yc_1s_2^2s_1k$$
$$+240yc_1s_2s_1k-180c_1xs_1k+800yc_1^2s_2c_2+60c_1xyk$$
$$-600c_1xs_1c_2k-96s_1^2ys_2k-72s_1yc_1k-480s_1bc_2y$$
$$-240s_1^2bxs_2-80z^2c_1s_2k+800zc_1s_2^2k$$
$$-96c_1bc_2y+384c_2c_1^2bs_2+80zc_1s_2bx-300s_1s_2bx$$
$$+72c_1xs_1-128s_1s_2bc_2y+384s_1^2s_2bc_2+60ybzc_1$$
$$-128s_1^2s_2^2bz-800s_1^2c_2bzs_2-800s_1c_2^2by$$
$$-2400s_2c_2c_1k+240zc_2c_1k-80zc_2xk-300s_2zs_1k$$
$$+800s_2c_2xk-128c_2zc_1s_2k+80ybzs_1s_2$$
$$+128c_2zs_1s_2+300ybc_1s_2+96c_2zs_1k+80s_2s_1yxk$$
$$-80s_2c_1xy+300c_2x+2000c_2c_1k$$

Step 4: It is obvious that p_1 does not include any u_1', u_2' either. Substituting $c_2=\dfrac{1-t_2^2}{1+t_2^2}$, $s_2=\dfrac{2t_2}{1+t_2^2}$ in p_1 and p_5, we get two equations f and g including only variables s_1, c_1, and t_2. To reduce the complexity of computation, we can assume

$f=m_2t_2^2+m_1t_2^2+m_0$; here,

$m_2=860-70s_1y-12xs_1+16z-5z^2-5x^2-5y^2+12yc_1-70c_1x$

$m_1=-96+200z+32s_1y+32c_1x$

$m_0=-16z+260-5x^2-5z^2-5y^2+130c_1x+130s_1y+12yc_1-12xs_1$

Step 5: Computing the resultant of f and g in t_2, we derive a polynomial equation:

$$F(s_1,c_1,x,y,z,k,b,m_2,m_1,m_0)=0.$$

It is an 8-degree polynomial in s_1,c_1 and has 8672 terms. The first 5 terms are

$$692160m_0^4s_1^3ky-20480m_0^4y^3b-624000m_2^4s_1x+16384m_2^4x^2k^2+80470324m_0^3m_2c_1^4b^2.$$

Step 6: To verify that the case has exactly 16 real roots, we substitute $[x,y,z,r_1,r_2,r_3]$ for the specific numbers:

$$[x,y,z]=[-10,1,2]$$

$$[r_1,r_2,r_3]=[1,2,3]$$

in F and get

$$4102400(53t_1{}^4+28t_1{}^3-170t_1{}^2-52t_1+173)(610482474025\,t_1{}^8+16564766109400\,t_1{}^7-4101066327804\,t_1{}^6-127766501350008\,t_1{}^5+23678385069478\,t_1{}^4+250730315526984\,t_1{}^3-44842533943068\,t_1{}^2-79825500168040\,t_1-13498947998375)(1+t_1{}^2)^2=0.$$

There are only 8 real roots of t_1. As soon as t_1 has been determined, t_2 and t_3 can be easily derived from the above basic equations. We can see that t_3 has 16 different real roots. This fact illustrates that the ordering of variables is very important for cases with special geometry.

t_1 =[−27.09557054, −2.299681297, −1.637272369, −.3135454638, −.2535029725, .8511608737, 1.557179609, 2.057339192]

t_2 =[−1.901483673, −1.286137813, −1.224274836, −.1627925866, .4766135817, 1.260217169, 2.791439153, 3.878055099]

t_3 =[−3.783973126, −3.517486509, −3.101713223, −2.456210216, −.4742738138, −.3192256649, −.2856396727, −.1706670042, .2642724899, .2842939120, .3224024686, .4071312763, 2.108486640, 3.132580209, 3.500914248, 5.859363413].

6. Conclusion

Although we have known that the closure equations can be reduced to a univariate polynomial by the dialytic method, for a special case the ordering of joint variables plays a very important role in solving closure equations. Because the input-output equation produced by the dialytic method is just a necessary condition, the equation may contain extraneous roots when the resulting equation has more factors. By trying some examples aided by a computer, we have found it is a very efficient and simple approach using symbolic algebra methods such as Grobner basis and Wu's characteristic sets in Maple to solve the closure eqs. (5) directly instead of using the dialytic approach for special cases.

Besides, according to the definition of the geometric invariants, they are independent of joint type and can be applied to any serial manipulator. We believe the set of geometric invariants will be helpful in other branches of robotics also.

Appendix

f_1 in the basic closure equations (5):

$$
\begin{aligned}
f_1 = {} & l_4 b_5 d_4 + l_4 b_5 z c_3 b_3 a_1 b_2 - l_4 b_5 z a_3 a_1 a_2 \\
& + l_4 b_5 z c_3 b_3 b_1 c_2 a_2 + l_5 b_4 u_2 l_3 s_3 s_1 s_2 a_2 \\
& - l_5 b_4 u_3 d_3 a_1 a_2 + l_4 b_5 y a_3 c_1 b_1 a_2 + l_4 b_5 d_5 a_4 \\
& + l_5 b_4 u_3 d_4 a_3 b_1 c_2 b_2 + l_5 b_4 u_2 d_4 a_3 c_1 b_1 a_2 \\
& + l_5 b_4 u_1 x - l_5 b_4 u_1 d_4 c_3 b_3 c_1 s_2 a_2 \\
& + l_5 b_4 u_1 d_4 c_3 b_s b_1 b_2 + l_5 b_4 u_3 z \\
& + l_5 b_4 u_2 d_4 c_3 b_3 c_1 a_1 c_2 a_2 - l_5 b_4 u_3 d_4 a_3 a_1 a_2 \\
& + l_5 b_4 u_1 l_3 c_3 s_1 a_1 s_2 + l_5 b_4 u_1 l_3 s_3 c_1 s_2 a_2 \\
& + l_5 b_4 u_3 d_4 c_3 b_3 a_1 b_2 - l_5 b_4 u_2 d_4 c_3 b_3 c_1 b_1 b_2 \\
& - l_5 b_4 u_2 d_4 s_3 b_3 s_1 c_2 + l_5 b_4 u_2 l_3 s_3 c_1 b_1 b_2 \\
& - l_5 b_4 u_1 d_4 a_3 s_1 a_1 c_2 b_2 - l_5 b_4 u_3 b_1 l_2 s_2 \\
& + l_5 b_4 u_1 l_3 s_3 s_1 a_1 c_2 a_2 + l_5 b_4 u_2 d_4 a_3 c_1 a_1 c_2 b_2 \\
& + l_5 b_4 u_2 d_3 c_1 b_1 a_2 + l_5 b_4 u_2 c_1 b_1 d_2 \\
& + l_5 b_4 u_3 d_4 c_3 b_3 b_1 c_2 a_2 + l_5 b_4 u_3 d_3 b_1 c_2 b_2 \\
& - l_5 b_4 u_3 b_1 s_2 l_3 c_3 - l_5 b_4 u_3 l_3 s_3 b_1 c_2 a_2 - l_5 b_4 u_2 l_1 s_1 \\
& - l_5 b_4 u_1 d_4 c_3 b_3 s_1 a_1 c_2 a_2 - l_5 b_4 u_2 d_4 a_3 s_1 s_2 b_2 \\
& - l_5 b_4 u_2 c_1 a_1 l_2 s_2 + l_5 b_4 u_2 d_3 c_1 a_1 c_2 b_2 \\
& - l_5 b_4 u_1 s_1 b_1 d_2 - l_5 b_4 u_2 s_1 l_2 c_2 + l_4 b_5 d_3 a_3 + l_5 b_4 u_2 y \\
& - l_5 b_4 u_1 l_1 c_1 - l_5 b_4 u_3 a_1 d_2 - l_5 b_4 u_1 d_3 s_1 b_1 a_2 \\
& - l_5 b_4 u_1 d_3 s_1 a_1 c_2 b_2 - l_5 b_4 u_1 l_3 c_3 c_1 c_2 \\
& - l_5 b_4 u_1 l_3 s_3 s_1 b_1 b_2 - l_5 b_4 u_3 l_3 s_3 a_1 b_2 \\
& + l_4 b_5 d_2 a_3 a_2 - l_5 b_4 u_1 d_4 s_3 b_3 c_1 c_2 \\
& - l_5 b_4 u_1 d_4 a_3 s_1 b_1 a_2 - l_5 b_4 u_1 d_3 c_1 s_2 b_2 \\
& - l_5 b_4 u_2 l_3 c_3 s_1 c_2 - l_5 b_4 u_2 d_4 c_3 b_3 s_1 s_2 a_2 \\
& - l_5 b_4 u_1 c_1 l_2 c_2 - l_5 b_4 u_1 d_4 a_3 c_1 s_2 b_2
\end{aligned}
$$

$-l_4 b_5 y a_3 s_1 s_2 b_2 - l_5 b_4 u_2 d_4 s_3 b_3 c_1 a_1 s_2$

$-l_4 b_5 y s_3 b_3 c_1 a_1 s_2 - l_5 b_4 u_2 l_3 c_3 c_1 a_1 s_2$

$-l_5 b_4 d_5 a_5 - l_4 b_5 y c_3 b_3 c_1 b_1 b_2 - l_4 b_5 y s_3 b_3 s_1 c_2$

$+l_4 b_5 l_2 s_3 b_3 + l_4 b_5 x c_3 b_3 s_1 b_1 b_2 + l_4 b_5 l_1 c_3 b_3 s_2 a_2$

$+l_4 b_5 l_1 a_3 s_2 b_2 + l_4 b_5 y a_3 c_1 a_1 c_2 b_2$

$+l_4 b_5 z a_3 b_1 c_2 b_2 + l_4 b_5 l_1 s_3 b_3 c_2$

$+l_4 b_5 y c_3 b_3 c_1 a_1 c_2 a_2 + l_5 b_4 u_1 s_1 a_1 l_2 s_2$

$-l_5 b_4 u_3 d_4 b_1 s_2 s_3 b_3 - l_4 b_5 y c_3 b_3 s_1 s_2 a_2$

$-l_4 b_5 d_2 c_3 b_3 b_2 - l_4 b_5 x a_3 s_1 a_1 c_2 b_2$

$-l_4 b_5 x a_3 c_1 s_2 b_2 - l_4 b_5 x a_3 s_1 b_1 a_2$

$+l_5 b_4 u_1 d_4 s_3 b_3 s_1 a_1 s_2 - l_4 b_5 z b_1 s_2 s_3 b_3$

$+l_4 b_5 x s_3 b_3 s_1 a_1 s_2 - l_4 b_5 x c_3 b_3 s_1 a_1 c_2 a_2$

$-l_4 b_5 x s_3 b_3 c_1 c_2 - l_4 b_5 x c_3 b_3 c_1 s_2 a_2$

$-l_5 b_4 u_2 l_3 s_3 c_1 a_1 c_2 a_2 - l_5 b_4 u_2 d_3 s_1 s_2 b_2.$

D in Theorem 2:

$D = 4l_1^2 b_1^2(-2b_3 t_3 a_2 a_3 l_3 - 2b_3 t_3^3 a_2 a_3 l_2$

$+2b_2 a_3^2 a_2^2 t_3^3 l_3 + 2b_2 a_3^2 a_2^2 t_3 l_3 - 2b_3 t_3 a_2 a_3 l_2$

$-2t_3^2 b_2^2 b_3^2 a_2 d_3 - t_3^4 b_2^3 b_3 a_3 d_3 + b_2 a_3 d_3 a_2^2 b_3$

$+2b_2 a_3 a_2 b_3 d_2 + 2b_3 t_3^3 a_2 a_3 l_3 + t_3^4 a_2 b_3^2 b_2^2 d_3$

$+2b_3 a_3 a_2^3 t_3 l_3 - 2b_3 a_3 a_2^3 t_3^3 l_3 + a_2 b_3^2 b_2^2 d_3$

$-2t_3^2 b_3^2 a_2^3 d_3 + t_3^4 a_2^2 b_3^2 d_2 - 2t_3^3 b_2 b_3^2 l_2$

$+2t_3 b_2 b_3^2 l_2 + a_3 b_3 b_2^3 d_3 + 4b_3^2 t_3^2 d_2 + 2t_3^2 b_2^2 a_3^2 d_2$

$+t_3^4 b_2^2 a_3^2 d_2 + a_2^2 b_3^2 d_2 + d_3 a_2^3 b_3^2 + t_3^4 d_3 a_2^3 b_3^2$

$+2t_3 b_2 b_3^2 l_3 + 2t_3^3 l_3 b_2^3 a_3^2 + 2b_3^2 b_2 t_3^3 l_3 + 2t_3 l_3 b_2^3 a_3^2$

$+4b_3^2 t_3^2 a_2 d_3 - 2t_3^2 b_3^2 a_2^2 d_2 + b_2^2 a_3^2 d_2$

$+2t_3 l_3 b_2^2 a_3 a_2 b_3 - 2t_3^4 b_3 a_2 b_2 d_2 - t_3^4 b_3 a_2^2 a_3 b_2 d_3$

$-2t_3^3 l_3 b_2^2 a_3 a_2 b_3)^2 + (-2d_2^2 b_1^2 a_2^2 t_3^2 b_3^2$

$-2d_2 d_3 b_2^3 b_3 b_1^2 a_3 t_3^4 - 2d_2 d_3 a_3 b_2 b_1^2 b_3 a_2^2 t_3^4$

$-4d_2 d_3 b_2^2 b_3^2 b_1^2 a_2 t_3^2 - 4d_2 d_3 b_3^2 a_2^3 b_1^2 t_3^2$

$-2d_2^2 t_3^4 a_3 b_2 b_1^2 b_3 a_2 - 2l_2 b_3^2 b_1^2 b_2^2 l_3 t_3^4$

$+2l_3^2 b_3 b_1^2 b_2 a_3 a_2 t_3^4 + 2b_3^2 a_2^2 l_2^2 a_1^2 t_3^2$

$-2b_3^2 a_2^2 l_2^2 t_3^2 + 2a_1^2 l_2 a_3^2 b_2^2 l_3 t_3^4$

$+2l_3^2 a_1^2 a_3^2 b_2^2 t_3^2 - 2b_3 a_2 l_3^2 a_3 b_2 t_3^4$

$-2b_3^2 a_2^2 l_3 l_2 t_3^4 + 2b_3^2 a_2^2 l_1^2 t_3^2 - 2a_3^2 b_2^2 l_3^2 t_3^2$

$+2b_3 a_2 l_3^2 a_1^2 a_3 b_2 t_3^4 + 2b_3^2 a_2^2 l_3 a_1^2 l_2 t_3^4$

$+2b_3 a_2 l_2^2 a_1^2 a_3 b_2 t_3^4 - 2l_3^2 a_3^2 b_1^2 a_2^2 t_3^2$

$-2l_2^2 b_3^2 b_1^2 b_2^2 t_3^2 + 2b_3 a_2 l_1^2 a_3 b_2 t_3^4$

$$-2b_3a_2l_2{}^2a_3b_2t_3{}^4+2l_2^2a_3b_1{}^2a_2b_3b_2t_3{}^4$$
$$-2a_3{}^2b_2{}^2l_2l_3t_3{}^4-2l_2a_3{}^2b_1{}^2a_2{}^2l_3t_3{}^4+b_3{}^2a_2{}^2l_3{}^2$$
$$-b_3{}^2a_2{}^2l_3{}^2+l_3{}^2b_3{}^2b_1{}^2b_2{}^2$$
$$+4d_2l_3t_3a_2{}^2a_3{}^2b_1{}^2-4d_2b_1{}^2t_3b_3l_2a_3a_2$$
$$+4d_2a_3{}^2b_2{}^3b_1{}^2l_3t_3-2d_3^2b_3{}^2a_2{}^4b_1{}^2t_3{}^2$$
$$-2d_3^2b_2{}^4b_3{}^2b_1{}^2t_3{}^2+4a_3{}^2+b_2{}^4l_3{}^2t_3{}^2b_1{}^2$$
$$+4d_3^2b_2{}^2t_3{}^2b_3{}^2+4d_2^2b_1{}^2t_3{}^2b_3{}^2$$
$$+8d_2d_3b_1{}^2a_2t_3{}^2b_3{}^2+l_3{}^2t_3{}^4a_2{}^2b_3{}^2$$
$$-4l_3t_3{}^3a_2b_3{}^3d_3b_2+4l_3t_3{}^3a_2b_3{}^2a_1{}^2d_3b_2$$
$$+4d_2l_3t_3{}^3b_1{}^2b_2b_3{}^2+4d_3^2b_1{}^2a_2{}^2t_3{}^2b_3{}^2$$
$$-4l_1{}^2t_3{}^2b_3{}^2-4d_3b_2{}^2t_3b_3a_3l_2-4d_3b_1{}^2a_2{}^2t_3b_3l_2a_3$$
$$+8l_3{}^2t_3{}^2a_2{}^2a_3{}^2b_1{}^2b_2{}^2+4l_3{}^2t_3{}^2t_3{}^2a_2{}^4a_3{}^2b_1{}^2$$
$$+4d_3b_2{}^2t_3b_3a_1{}^2l_2a_3-4d_3^2b_2{}^2t_3{}^2b_3{}^2a_1{}^2$$
$$+l_3{}^2t_3{}^4b_1{}^2b_2{}^2b_3{}^2-l_3{}^2t_3{}^4a_2{}^2b_3{}^2a_1{}^2$$
$$+4l_3t_3{}^3b_1{}^2b_2b_3{}^2d_3a_2-2a_3{}^2b_2{}^2l_1{}^2t_3{}^2$$
$$-a_3{}^2b_2{}^2l_1{}^2t_3{}^4+2l_3{}^2b_3{}^2b_1{}^2b_2{}^2t_3{}^2$$
$$+2l_3{}^2t_3{}^2a_2{}^2b_3{}^2-2b_3{}^2a_2{}^2l_3{}^2a_1{}^2t_3{}^2$$
$$+4l_3t_3a_2{}^4a_3b_1{}^2b_3d_3-8l_3t_3{}^3a_2{}^2a_3b_1{}^2d_3b_2b_3$$
$$+8l_3t_3a_2{}^2a_3b_1{}^2d_3b_2{}^2b_3-4l_3t_3{}^3a_2{}^4a_3b_1{}^2b_3d_3$$
$$-4d_3b_1{}^2a_2t_3{}^3b_3{}^2l_2b_2+4d_3b_1{}^2a_2t_3b_3{}^2l_2b_2$$
$$-4d_3{}^2b_2{}^2b_3{}^2b_1{}^2a_2{}^2t_3{}^2-2b_3{}^2a_2{}^2l_3a_1{}^2l_2$$
$$-2b_3a_2l_3{}^2a_1{}^2a_3b_2+4d_3b_2t_3b_3{}^2l_3a_1{}^2a_2$$
$$-4l_3t_3{}^4a_2b_3a_1{}^2l_2a_3b_2+a_3{}^2b_2{}^2l_3{}^2t_3{}^4+a_3{}^2b_2{}^2l_3{}^2$$
$$-4d_3b_2t_3b_3{}^2a_2l_3+4d_3b_2{}^2t_3{}^3b_3a_3l_3$$
$$-4d_3b_2{}^2t_3b_3a_3l_3+4d_3b_2t_3{}^3b_3{}^2a_2l_2$$
$$-4d_3b_2t_3b_3{}^2a_2l_2-4d_3b_2{}^2t_3{}^3b_3a_3l_2$$
$$+2d_2b_3{}^2a_2{}^3b_1{}^2d_3-4d_3b_2{}^2t_3{}^3b_3l_3a_1{}^2a_3$$
$$+4d_3b_2{}^2t_3b_3l_3a_1{}^2a_3-b_3{}^2a_2{}^2l_1{}^2t_3{}^4$$
$$-b_3{}^2a_2{}^2l_1{}^2+2b_3{}^2a_2{}^2l_3l_2-l_3{}^2a_1{}^2a_3{}^2b_2{}^2t_3{}^4$$
$$-l_3{}^2a_1{}^2a_3{}^2b_2{}^2+2b_3a_2l_3{}^2a_3b_2-2a_1{}^2l_2a_3{}^2b_2{}^2l_3$$
$$+b_3{}^2a_2{}^2l_2^2t_3{}^4+b_3{}^2a_2{}^2l_2^2+2d_2a_3b_2b_1{}^2b_3a_2{}^2d_3$$
$$+2d_2^2a_3b_2b_1{}^2b_3a_2+d_2^2b_3{}^2a_2{}^2b_1{}^2t_3{}^4$$
$$+d_2^2b_3{}^2a_2{}^2b_1{}^2+2d_2b_3{}^2a_2{}^3b_1{}^2d_3t_3{}^4$$
$$-4d_3l_3t_3{}^3a_2{}^3a_3b_1{}^2b_3+4d_2l_3t_3a_2{}^3a_3b_1{}^2b_3$$
$$+2d_2d_3b_2{}^3b_3b_1{}^2a_3+2d_2d_3b_2{}^2b_3{}^2b_1{}^2a_2t_3{}^4$$
$$+2d_2d_3b_2{}^2b_3{}^2b_1{}^2a_2-4d_2a_3b_2{}^2l_3t_3{}^3b_1{}^2b_3a_2$$

$+4d_2a_3b_2{}^2l_3t_3b_1{}^2b_3a_2-4d_2b_1{}^2t_3{}^3b_3l_2a_3a_2$
$+4d_2l_3t_3{}^3a_2{}^2a_3{}^2b_1{}^2b_2+l_2^2a_3{}^2b_1{}^2a_2{}^2t_3{}^4$
$+2l_2^2a_3{}^2b_1{}^2a_2{}^2t_3{}^2+4d_2b_1{}^2t_3{}^3b_3l_3a_3a_2$
$-d_2b_1{}^2t_3b_3l_3a_3a_2+d_2^2a_3{}^2b_2{}^2b_1{}^2t_3{}^4$
$+2d_2^2a_3{}^2b_2{}^2b_1{}^2t_3{}^2+4d_2a_3{}^2b_2{}^3b_1{}^2l_3t_3{}^3$
$-4a_1{}^2l_2^2a_3{}^2b_2{}^2t_3{}^4-2a_1{}^2l_2^2a_3{}^2b_2{}^2t_3{}^2$
$+2a_3{}^2b_2{}^2l_2^2t_3{}^2+a_3{}^2b_2{}^2l_2^2t_3{}^4+4b_3a_2l_3a_3b_2l_2$
$-2l_3{}^2b_3b_1{}^2b_2a_3a_2+4b_3a_2l_3a_3b_2l_2t_3{}^4$
$+d_2^2a_3{}^2b_2{}^2b_1{}^2-b_3{}^2a_2{}^2l_2^2a_1{}^2-4d_3b_1{}^2a_2{}^2t_3{}^3b_3l_2a_3$
$-4b_3a_2l_3a_1{}^2l_2a_3b_2+2l_2b_3{}^2b_1{}^2b_2{}^2l_3$
$+d_3^2b_2{}^4b_3{}^2b_1{}^2t_3{}^4+d_3^2b_2{}^4b_3{}^2b_1{}^2$
$-b_3{}^2a_2{}^2l_2^2a_1{}^2t_3{}^4+2d_3^2b_2{}^2b_3{}^2b_1{}^2a_2{}^2t_3{}^4$
$+2d_3^2b_2{}^2b_3{}^2b_1{}^2a_2{}^2+4d_3b_1{}^2a_2t_3b_3{}^2l_3b_2$
$+4d_3b_2{}^2t_3{}^3b_3a_1{}^2l_2a_3+4d_2b_1{}^2t_3b_3{}^2l_2b_2$
$-4l_3t_3{}^4b_1{}^2b_2b_3l_2a_3a_2-a_1{}^2l_2^2a_3{}^2b_2{}^2$
$+l_2^2a_3{}^2b_1{}^2a_2{}^2-4d_3b_1{}^2a_2{}^2t_3b_3l_3a_3$
$+4d_3b_1{}^2a_2{}^2t_3{}^3b_3l_3a_3+a_3{}^2b_2{}^2l_2^2-a_3{}^2b_2{}^2l_1{}^2$
$-4d_2b_1{}^2t_3{}^3b_3{}^2l_2b_2+4d_2b_1{}^2t_3b_3{}^2l_3b_2$
$+2l_2a_3{}^2b_1{}^2a_2{}^2l_3+4d_3b_2t_3b_3{}^2a_1{}^2l_2a_2$
$-4d_3b_2t_3{}^3b_3{}^2a_1{}^2l_2a_2+2a_3{}^2b_2{}^2l_2l_3$
$+4d_3b_2{}^4b_3b_1{}^2a_3l_3t_3-4d_3b_2{}^4b_3b_1{}^2a_3l_3t_3{}^3$
$-2l_2^2a_3b_1{}^2a_2b_3b_2+2b_3a_2l_2^2a_3b_2$
$+l_2^2b_3{}^2b_1{}^2b_2{}^2+l_2^2b_3{}^2b_1{}^2b_2{}^2t_3{}^4$
$-4l_2a_3b_1{}^2a_2l_3b_3b_2+l_3{}^2a_3{}^2b_1{}^2a_2{}^2$
$+l_3{}^2a_3{}^2b_1{}^2a_2{}^2t_3{}^4+b_3{}^2a_2{}^4d_3{}^2b_1{}^2$
$+b_3{}^2a_2{}^4d_3^2b_1{}^2t_3{}^4-2b_3a_2l_1^2a_3b_2$
$-2b_3a_2l_2^2a_1{}^2a_3b_2)^2.$.

Acknowledgements

The authors wish to express gratitude to Professor Zhou Chaochen in IIST/UNU and Dr. Zeng Zhenbing for their valuable suggestions. Research supported in part by National '973' Project of China and by '95' Key Project on Fundamental Research of Academia Sinica.

References

[1]Craig J J. Introduction to Robotics. Reading, MA: Addison-Wesley, 1989.
[2]Fu H G, Yang L, Zhou, C C. A geometric approach to inverse kinematics. Journal of Robotic Systems, 1998, 15(3): 131 - 143.
[3]Fu K S, Gozalves R C, Lee C S G. Robotics: Control, Sensing, Vision and Intelligence. New York: McGraw-Hill, 1987.
[4]Heck A. Introduction to Maple. New York: Springer-Verlag. Lee, H. Y. , and Liang, C. G. 1988. Displacement analysis of the general spatial 7-link 7R mechanism. Mechanisms and Machine Theory, 1993, 23(3): 219 - 226.
[5]Mavroidis C, Roth B. Structural parameters which reduce the number of manipulator configurations. J. of Mechanical Design, Trans. ASME 1994(116): 3 - 10.
[6] Raghavan M, Roth B. Kimematic analysis of the 6R manipulator of general geometry. International Symposium on Robotics Research, Tokyo , 1989: 314 - 320.
[7]Wampler C, Morgan A P. Solving the 6R inverse position problem using a generic-case solution methodology. Mechanisms and Machine Theory, 1991, 26(1): 91 - 106.

On Number of Circles Intersected by a Line

Yang Lu Zhang Jing-zhong

Laboratory of Automatic Reasoning, CICA, Academia Sinica Chengdu, Sichuan 610041, *People's Republic of China*

Zhang Wei-nian

Department of Mathematics, Sichuan University Chengdu, Sichuan 610064, *People's Republic of China*

Consider a set U of circles in the plane such that any line intersects at least one of those circles. For a given natural number m, is there a line in the plane intersecting at least m circles in U? In this paper this problem is solved. Our result is also generalized to compact convex subsets and to higher dimensional cases.

Key Words: combinatorial geometry, circle, distribution, series, divergence

1. INTRODUCTION

In combinatorial geometry the Sylvester-type problems are very important and attractive. A century ago, Sylvester [6] posed a question: For a finite set of points in the plane such that the line through any two of them passes through a third point of the set, must all the points lie on one line? Later it resurfaced as a conjecture by Paul Erdös [2]: If a finite set of points in the plane is not on one line then there is a line through exactly two of the points. Since then there has appeared a substantial literature (seen in [1]) on the problem and its generalizations. For example, in [5] Motzkin considered n points in the plane, not all on a line and not all on a circle, and showed that there is either a circle or a line containing exactly three of the points. Herzog and Kelly [4] also proved that for given n pairwise disjoint compact sets in $\mathbf{R}^d$, which are not all contained in a line and at least one of which contains infinitely many points, there is a line intersecting exactly two of them.

本文刊于《Journal of Combinatorial Theory》,第 19 卷第 2 期,2002 年.

Related to the above mentioned, another problem says the following: *For a set U of some unit circles in the plane $\mathbf{R}^2$ such that any line in $\mathbf{R}^2$ intersects at least one of those circles, given natural number m, is there a line in $\mathbf{R}^2$ which intersects at least m circles in U?* Although it has been proposed for a long time and known extensively, no published answer is found yet. It is too hard to search where this was stated originally but one version states that it was once raised in a personal letter of P. Erdös to Y. Q. Yin. A closer result is Proposition 93 in [3], which tells that if a collection of mutually congruent convex bodies is not "extremely sparsely distributed", then for any natural number m there is a line which intersects more than m bodies of the family. However, it does not give a full answer to the problem because it requires a different assumption that the collection of circles are not extremely sparsely distributed. No matter where it comes from, this problem is interesting and has been puzzling us since we heard of it.

In this paper this problem is solved by reducing to divergent series, the same idea as used for Proposition 93 in [3]. We state the main result in next section. By lemmas given in Section 3, we prove the result in Section 4. The basic result is generalized in Section 5 to compact convex subsets and to higher dimensional cases. For convenience, let $\mathbf{R}^2$ and $\mathbf{N}$ denote the plane (2-dimensional Euclidean space) and the set of natural numbers, respectively. Let $C(Q,r)$ denote the circle of radius r centered at Q. Let $|PQ|$ represent the distance between P and Q in Euclidean spaces if P,Q stand for points and $|B|$ represents the area of B if B stands for a set.

2. MAIN RESULT

Theorem 1. *Let U be a set of circles in the plane $\mathbf{R}^2$ such that any line in $\mathbf{R}^2$ intersects at least a circle in U. Then for any $m \in \mathbf{N}$ and any point $P \in \mathbf{R}^2$ there exists a line in $\mathbf{R}^2$ through P intersecting at least m circles in U.*

This result is stronger than the original problem hoped since we do not require the circles of U to be unitary. Moreover, different from [3] we do not require the collection of circles to be congruent. Consider $U=\{C(O,2^k) : k\in \mathrm{N}\}$. As in [3], let $\mathcal{N}(R)$ denote the number of circles in U which lie entirely inside the disk of radius R about the reference point O. Then $\liminf_{R\to\infty}\mathcal{N}(R)/R=0$. Therefore U is extremely sparsely distributed and as in the case of Proposition 93 in [3] it cannot give an answer to our problem. However our theorem works in this case; actually, each line through a point $P\in\mathbf{R}^2$ intersects infinitely many circles in this U.

For another example, consider U to be a set U consisting of a circle centered at the origin and some small circles along the hyperbolas $x^2-y^2=\pm 1$ such that along the same branch of the hyperbolas each circle is tangent to its consecutive two circles; two circles respectively along different branches do not intersect, and no circle along a branch intersects the circle centered at the origin. Such U satisfies the condition of Theorem 1. Notice that no line in the plane intersects infinitely many circles in such a U.

Solving this problem would be easier if the set U contains more circles. We mainly prove Theorem 1 when U is countable. The case of uncountable U is simple and its proof is a standard argument. In the following, we suppose that U is countable; i. e. , $U=\{C(A_k, r_k) : A_k \in \mathbf{R}^2, r_k > 0, k=1,2,\cdots\}$. Let $d_k = |A_k O|$, where O is the origin of $\mathbf{R}^2$. We only need to consider the case that

$$r_k/d_k < 1, \qquad \forall k \geqslant k_0, \tag{2.1}$$

for some $k_0 > 0$; otherwise there is a subsequence k_i such that $r_{ki}/d_{ki} \geqslant 1$ $(i=1,2,\cdots)$ and thus Theorem 1 holds naturally because every line through O intersects all circles $C(Q_{ki}, r_{ki}), i=1,2,\cdots$

LEMMA 1. *Suppose* (2. 1) *holds. If* $\sum_{k=k_0}^{+\infty} r_k/d_k$ *diverges, then for any* $m \in \mathrm{N}$ *there is a line through* O *intersecting at least* m *circles in* U.

Proof. When $r_k/d_k < 1$, the origin O is outside the circle $C(A_k, r_k)$. Let $\emptyset_k$ denote the scope-angle of O to $C(A_k, r_k)$, namely, the angle between the two tangents from O to $C(A_k, r_k)$. Clearly, $\emptyset_k = 2\arcsin(r_k/d_k) > 2r_k/d_k$, so $\sum_{k=k_0}^{\infty} \emptyset_k$ also diverges. Thus for any natural number m there exists a natural number N such that $\sum_{k=k_0}^{N} \emptyset_k > 2m\pi$. By the drawer principle, there is at least a line through O intersecting at least m circles in U.

Lemma 1 gives a way to reduce our problem to divergence of a series.

3. SOME LEMMAS

For a line l in $\mathbf{R}^2$, let F_l denote the intersection point of l with its normal through O. Obviously, $F_l=0$ if l is through O. F_l is unique and $F_l \neq F_{l'}$ if neither l nor l' is through O and $l \neq l'$. Let

$$\Omega_k := \{F_l : l \cap C(A_k, r_k) \neq \emptyset\}. \tag{3.2}$$

Under the condition in Theorem 1,

$$\mathbf{R}^2 = \bigcup_{k=1}^{\infty} \Omega_k. \tag{3.3}$$

In fact, for each $P \in \mathbf{R}^2$ there is a line l such that $F_l = P$. It is assumed in Theorem 1 that l intersects a circle in U, so $F_l \in \Omega_k$ for some $k \in \mathbf{N}$.

We still need more geometric properties of Ω_k. Let S_k be the circle of diameter $|OA_k|$ through 0 and A_k and let Q be the center of S_k. Then $S_k \subset \Omega_k$ since $S_k = \{F_l : A_k \in l\}$. For $A' \in S_k$, let

$$\theta(A') := \angle A' A_k O \tag{3.4}$$

and let $C(A')$ denote the circle of radius $r_k \sin\theta(A')$ centered at A'.

Lemma 2. $\Omega_k \subset \bigcup_{A' \in S_k} C(A')$.

Proof. Let the line l satisfy that $l \cap C(A_k, r_k) \neq \emptyset$ and let A^* be one of two intersection points of l and $C(A_k, r_k)$ arbitrarily fixed. Take $A \in S_k$ such that the rectangular triangle $\triangle OAA_k$ is similar to $\triangle OF_lA^*$ with the same orientation, as in Fig 1. Clearly A exists uniquely. Thus $\angle A^* OF_l = \angle A_kOA$ and $|OA|/|OF_l| = |OA_k|/|OA^*|$. It follows that $\angle AOF_l = \angle A^* OF_l + \angle AOA^* = \angle A_kOA + \angle AOA^* = \angle A_kOA^*$ and $|OA|/|OA_k| = |OF_l|/|OA^*|$. Hence $\triangle OAF_l \sim \triangle OA_kA^*$ and thus

$$|AF_l| = |A_kA^*| \cdot |OA|/|OA_k| = r_k \sin\theta(A), \tag{3.5}$$

where $\theta(A) = \angle AA_kO$. Therefore, $F_l \in C(A)$. Since $A \in S_k$, we get $F_l \in \bigcup_{A \in S_k} C(A')$.

Let $|\Omega_k|$ represent the area (or out measure) of Ω_k, which clearly has its area. By Lemma 2, each Ω_k is covered by an annular region between a circle of radius $d_k/2 + r_k$ and a circle of radius $d_k/2 - r_k$, which are both centered at Q. Thus

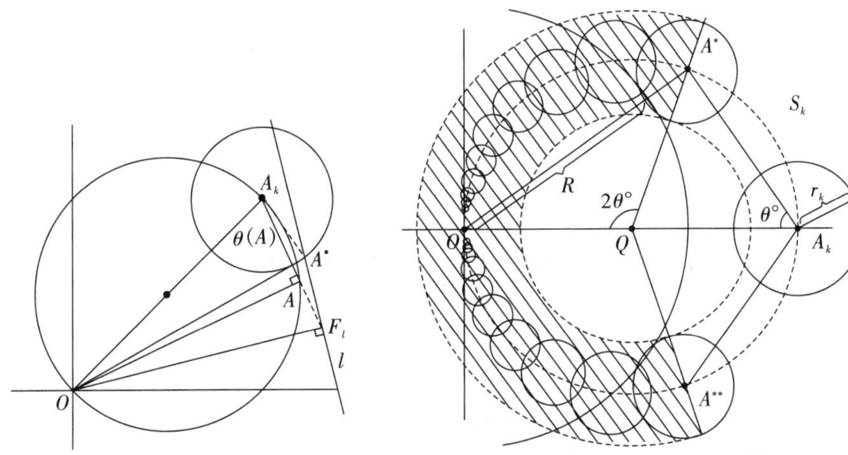

FIG. 1. $\triangle OAF_l \sim \triangle OA_kA*$ FIG. 2. What is $\Omega_{k,R}$ covered by?

$$|\Omega_k| \leqslant \pi\left(\frac{d_k}{2} + r_k\right)^2 - \pi\left(\frac{d_k}{2} - r_k\right)^2 = 2\pi r_k d_k, \tag{3.6}$$

if $d_k \geqslant 2r_k$.

For a given $R > 0$, let $B(O,R)$ be the open disk of radius R centered at O and let

$$\Omega_{k,R} := \Omega_k \cap \overline{B}(O,R), k = 1, 2, \cdots, \tag{3.7}$$

where $\overline{B}(O,R)$ is the closure of $B(O,R)$ (see Fig. 2).

Lemma 3. *If* $d_k \geqslant \max\{R + r_k, 2r_k\}$, *then* $|\Omega_{k,R}| \leqslant 8\pi R^2 r_k / d_k$.

Proof. For $A \in S_k$, $C(A) \cap B(O,R) = \emptyset$ if and only if

$$|OA| > R + r_k \sin\theta(A). \tag{3.8}$$

Note that $|OA|$ varies continuously from 0 to d_k when A goes from O to A_k. Thus we can take $A^* \in S_k$ appropriately such that

$$|OA^*| = R + r_k \sin\theta(A^*); \tag{3.9}$$

i. e., $C(A^*)$ is externally tangent to $B(O,R)$. Substituting $|OA| = d_k \sin\theta(A)$ in (3.8) and (3.9) separately we obtain

$$\sin\theta(A) > \frac{R}{d_k - r_k} = \sin\theta(A^*). \tag{3.10}$$

Hence $C(A)\cap\Omega_{k,R}=\emptyset$ when $\sin\theta(A) > \sin\theta(A^*)$. Furthermore, let A^{**} be a symmetric point of A^* to the line OA_k. Then in the angular region centered at Q, the middle of OA_k, and faced by the arc $A^*\hat{A}_kA^{**}$, there is no point of $\Omega_{k,R}$ because the distances between Q and points on $C(A)$ are greater than $d_k/2-r_k\sin\theta(A^*)$ when $\sin\theta(A) < \sin\theta(A^*)$. Therefore, $\Omega_{k,R}$ is covered by both the closed annulus of width $2r_k\sin\theta(A^*)$ along S_k and the angular region of angle $4\theta(A^*)$ centered at Q and faced by the arc $A^*\hat{O}A^{**}$. Notice that Q is the center of S_k. Applying the known inequalities $d_k-r_k\geqslant d_k/2$ and $|\theta|\leqslant|\frac{\pi}{2}\sin\theta|$ for $|\theta|\leqslant\frac{\pi}{2}$ and applying (3.10), we obtain

$$\begin{aligned}|\Omega_{k,R}| &\leqslant 4\theta(A^*)\cdot\frac{d_k}{2}\cdot 2r_k\sin\theta(A^*)\\ &\leqslant\left(4\arcsin\frac{R}{d_k-r_k}\right)\cdot\frac{d_k}{2}\cdot 2r_k\cdot\frac{R}{d_k-r_k}\\ &\leqslant 4\cdot\frac{\pi}{2}\cdot\frac{R}{d_k-r_k}\cdot\frac{Rd_kr_k}{d_k-r_k}\leqslant\frac{8\pi R^2r_k}{d_k}.\end{aligned} \tag{3.11}$$

4. PROOF OF THEOREM 1

If there is a subsequence $\{k_i,k_{i+1},\cdots\}$ such that $r_j/d_j > 1/2$, $j=k_i,k_{i+1},\cdots$, then

$$\sum_{j=1}^{\infty}\frac{r_j}{d_j} > \sum_{i=0}^{\infty}\frac{r_{ki}}{d_{ki}} > \frac{1}{2}+\frac{1}{2}\cdots=\infty, \tag{4.12}$$

implying the result of Theorem 1 by Lemma 1. Thus in what follows we suppose that there exists a natural number K such that

$$\frac{r_k}{d_k}\leqslant\frac{1}{2},\ \forall k>K. \tag{4.13}$$

Take $R>0$ large enough such that

$$R>2\max_{0\leqslant k\leqslant K}\{d_k+r_k\}. \tag{4.14}$$

By (3.3) we have

$$\left(\bigcup_{k=1}^{\infty}\Omega_k\right)\cap\mathscr{A}=\mathscr{A}, \tag{4.15}$$

where $\mathscr{A}$ is the annulus between two circles both centered at O and respectively of radii R and $R/2$. The definition of $\mathscr{A}$ implies that $(\bigcup_{d_k+r_k<R/2}\Omega_k)\cap\mathscr{A}=\emptyset$. Hence $(\bigcup_{d_k+r_k\geqslant R/2}\Omega_k)\cap\mathscr{A}=((\bigcup_{d_k+r_k\geqslant R/2}\Omega_k)\cup(\bigcup_{d_k+r_k<R/2}\Omega_k))\cap\mathscr{A}=\mathscr{A}$; that is,

$$\begin{aligned}&\left(\bigcup_{R/2-r_k\leqslant d_k\leqslant R+r_k}\Omega_k\right)\cup\left\{\left(\bigcup_{d_k>R+r_k}\Omega_k\right)\cap\mathscr{A}\right\}\\ &\supset\left\{\left(\bigcup_{R/2-r_k\leqslant d_k\leqslant R+r_k}\Omega_k\right)\cap\mathscr{A}\right\}\cup\left\{\left(\bigcup_{d_k>R+r_k}\Omega_k\right)\cap\mathscr{A}\right\}\end{aligned}$$

$$= \Big(\bigcup_{R/2-r_k\leqslant d_k} \Omega_k\Big) \cap \mathscr{A} = \mathscr{A}. \tag{4.16}$$

Estimating areas for both sides of (4.16) we get

$$\sum_{R/2-r_k\leqslant d_k\leqslant R+r_k} |\Omega_k| + \sum_{d_k>R+r_k} |\Omega_{k,R}| > \frac{3}{4}\pi R^2. \tag{4.17}$$

It follows from (3.6) and Lemma 3 that

$$\frac{2}{R^2}\sum_{R/2-r_k\leqslant d_k\leqslant R+r_k} r_k d_k + 8\sum_{d_k>R+r_k}\frac{r_k}{d_k} > \frac{3}{4}. \tag{4.18}$$

Now we claim that (4.18) implies the divergence of $\sum_{k=1}^{\infty} r_k/d_k$.

Assume this series converges. Then the second sum in (4.18) is arbitrarily small as R is large enough; i.e., there is a constant $M>0$ such that for all $R>M$,

$$\frac{2}{R^2}\sum_{R/2-r_k\leqslant d_k\leqslant R+r_k} r_k d_k > \frac{1}{2}. \tag{4.19}$$

For sufficiently large R such that

$$R>M^* := \max\{M, 2\max_{0,\leqslant k\leqslant k}\{dk+rk\}\}, \tag{4.20}$$

by (4.14), all those k satisfying $d_k+r_k\geqslant R/2$ must be greater than K. From (4.13) we see that $d_k\geqslant 2r_k$ holds for all k such that $d_k+r_k\geqslant R/2$. Thus the condition $R/2-r_k\leqslant d_k\leqslant R+r_k$ of summation in (4.19) implies that

$$\frac{d_k}{2}\leqslant d_k - r_k \leqslant R. \tag{4.21}$$

It follows that $d_k/R\leqslant 2$ and $r_k d_k/R^2\leqslant 4r_k/d_k$. Therefore, from (4.19) we get

$$8\sum_{R/2-r_k\leqslant d_k\leqslant R+r_k}\frac{r_k}{d_k} \geqslant \frac{2}{R^2}\sum_{R/2-r_k\leqslant d_k\leqslant R+r_k} r_k d_k > \frac{1}{2}, \forall R>M^*. \tag{4.22}$$

Clearly, for arbitrarily large K in (4.13) we can take a correspondingly large R. Thus

$$\sum_{k=k+1}^{\infty}\frac{r_k}{d_k} \geqslant \sum_{R/2-r_k\leqslant d_k\leqslant R+r_k}\frac{r_k}{d_k} > \frac{1}{16}. \tag{4.23}$$

This contradicts to the assumption of convergence of $\sum_{k=1}^{\infty} r_k/d_k$. Lemma 1, we obtain the conclusion of Theorem 1.

5. GENERALIZATION

Corollary 1. *Let U and U^* be sets of circles in $\mathbf{R}^2$ and let there be a mapping $f: U^* \to U$ defined by $C(A_\alpha^*, r_\alpha^*) \mapsto C(A_\alpha, r_\alpha)$ such that*

$$\frac{r_\alpha + |A_\alpha^* A_\alpha|}{r_\alpha^*} < L, \tag{5.24}$$

where L is a positive constant independent of α. If any line in $\mathbf{R}^2$ intersects at least one circle in U, then for any $m\in\mathbf{N}$ and any point $P\in\mathbf{R}^2$ there exists a line in $\mathbf{R}^2$ through P intersecting at least m circles in U^*.

The proof is simple. In fact, under the assumption (5.24) the divergence of $\sum_{k=1}^{\infty} r_k/d_k$ implies the divergence of $\sum_{k=1}^{\infty} r_k^*/d_k^*$, where $d_k^* = |OA_k^*|$. Applying Corollary 1, we can generalize Theorem 1 further to compact convex subsets in the plane. We refer to the ratio between the diameter of the minimum circle containing a compact convex subset V and the diameter of the maximum circle contained in the V as the rectangular ratio of the V.

Corollary 2. *Let M consist of some compact convex subsets in $\mathbf{R}^2$ such that their rectangular ratios have a uniform upper bound $L>0$. If any line in $\mathbf{R}^2$ intersects at least a compact convex subset in M, then for any $m\in\mathbf{N}$ and any point $P\in\mathbf{R}^2$ there exists a line in $\mathbf{R}^2$ through P intersecting at least m compact convex subsets in M.*

Proof. Let U^* consist of all those circles, each of which is the maximum circle contained in a compact convex subset in M. Similarly, let U consist of all those circles, each of which is the minimum circle containing a compact convex subset in M. Define a mapping $f: U^* \to U$ such that the image of a circle C^* in U^* is that one in U which corresponds to the same compact convex subset as C^* does. Because of the uniform boundedness of rectangular ratios, f satisfies (5.24). Thus our result can be deduced directly from Corollary 1.

The uniform boundedness of rectangular ratios in Corollary 2 is indispensible. Consider a hyperbola C in R^2 and take a sequence of different points $\{P_k: k=0,\pm1,\pm2,\cdots\}$ on the same branch of C such that $|P_kP_{k-1}|=1$. Similarly take another sequence $\{A_k: k=0,\pm1,\pm2,\cdots\}$ on the other branch of C such that $|A_kA_{k-1}|=1$. Let $B(P_kP_{k-1})$ denote the closed region surrounded by the chord $\overline{P_kP_{k-1}}$ and the arc $\widehat{P_kP_{k-1}}$. Moreover, let B_0 be the closed unit disk centered at the center of C. Let $M := \{B_0, B(P_kP_{k-1}), B(A_kA_{k-1}) : k=0,\pm1,\pm2,\cdots\}$, which clearly consists of compact convex subsets but does not possess uniform boundedness of rectangular ratios. Obviously, every line in $\mathbf{R}^2$ intersects at least one but at most five in M.

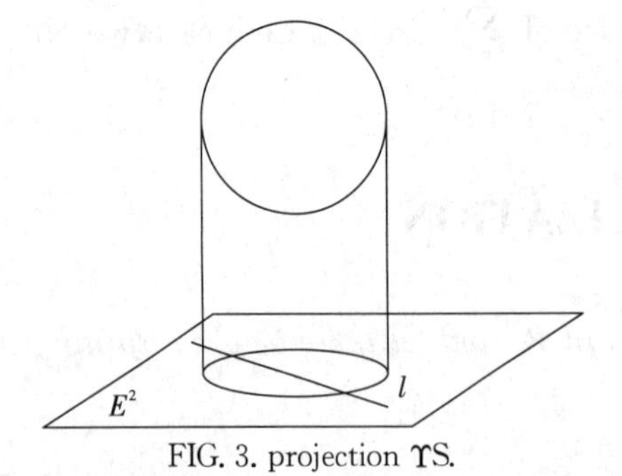

FIG. 3. projection ΥS.

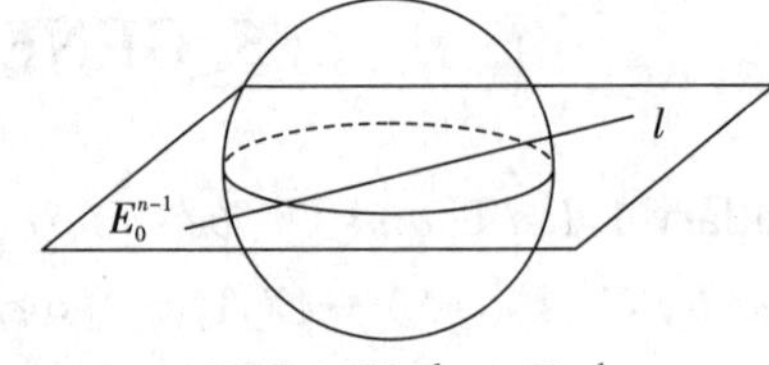

FIG. 4. U^{n-1} on E_0^{n-1}

We can also generalize our result to $\mathbf{R}^n$.

Remark 1. Let $n \geqslant 3$ and U consist of $(n-1)$-dimensional superspheres in $\mathbf{R}^n$ such that any $(n-1)$-dimensional superplane in $\mathbf{R}^n$ intersects at least one supersphere in U. Then for any given $m\in\mathbf{N}$, any plane E^2, and any $P\in E^2$, there exists a superplane E^{n-1} in $\mathbf{R}^n$ which is through P, orthogonal to E^2, and intersects at least m superspheres in U. This can be shown easily with the orthogonal projection $\Upsilon: \mathbf{R}^n \to E^2$ as in Fig. 3 since $\Upsilon U := \{\Upsilon S : S \in U\}$ is a set of some circles on the plane E^2 and Theorem 1 can be applied.

Remark 2. Let $n \geqslant 3$ and U consist of $(n-1)$-dimensional superspheres in $\mathbf{R}^n$. For given $k \in \mathbf{N}$ with $1 \leqslant k < n$, if any k-dimensional superplane in $\mathbf{R}^n$ intersects at least one supersphere in U, then for any $m \in \mathbf{N}$ and any $P \in \mathbf{R}^n$ there exists a k-dimensional superplane E^k in $\mathbf{R}^n$ through P intersecting at least m superspheres in U (Fig. 4). In fact, by Remark 1, it suffices to discuss the case of $k \leqslant n-2$. For $k = n-2$, take a $(n-1)$-dimensional superplane E_0^{n-1} in R^n through O and P. Let $\mathscr{E}_{n-1}^{n-2}$ be the set of all $(n-2)$-dimensional superplanes in E_0^{n-1} and let $U^{n-1} := \{B_n \cap E_0^{n-1} : B_n \in U\}$. For any $l \in \mathscr{E}_{n-1}^{n-2}$ there is a supersphere $B_n \in U$ such that $l \cap B_n \neq \emptyset$. Thus $B_n \cap E_0^{n-1} \neq \emptyset$ and $l \cap (B_\cap n E_0^{n-1}) \neq \emptyset$ since $l \subset E_0^{n-1}$. This means that every $(n-2)$-dimensional superplane in E_0^{n-1} intersects at least one $(n-2)$-dimensional supersphere of U^{n-1}. By Remark 1 there exists a $(n-2)$ dimensional superplane l^* in E_0^{n-1} through P intersecting m $(n-2)$-dimensional superspheres $B_n^1 \cap E_0^{n-1}, B_n^2 \cap E_0^{n-1}, \cdots, B_n^m \cap E_0^{n-1}$ in U^{n-1}. Of course, l^* intersects all $B_n^1, B_n^2, \cdots, B_n^m$.

ACKNOWLEDGEMENTS

The authors thank the referees for their helpful comments and suggestions.

Reference

[1]P Borwein, W O J Moser. A survey of Sylvester's problem and its generalizations, Aequationes Math. , 1990(40): 111 - 35.

[2]P Erdös. Problem 4065, Amer. Math. Monthly 1943(50): 65.

[3]H Hadwiger, H Debrunner. Combinatorial Geometry in the Plane. Holt, Rinehart, & Winston, New York, 1964.

[4]F Herzog, L M Kelly. A generalization of the theorem of Sylvester, Proc. Amer. Math. Soc. , 1960(11): 327 - 31.

[5]T Motzkin. The lines and planes connecting the points of a finite set. Trans. Amer. Math. Soc. , 1951(70): 451 - 64.

[6]J J Sylvester. Mathematical question 11851. Educational Times 59, 1893: 98.

Decomposing Polynomial Systems into Strong Regular Sets

Li Yong-bin

[1] Institute of Mathematics, Shantou University,

Shantou, Guangdong, 515063, P. R. China

[2] Mathematical College, Sichuan University

E-mail: yuengpen@netease.com

Zhang Jing-zhong Yang Lu

Chengdu Institute of Computer Applications,

Chinese Academy of Sciences, Chengdu, 610041, P. R. China

E-mail: zjz20001@sina.com; luyang@guangztc.edu.cn

A polynomial system is a pair of polynomial sets in $\mathbf{K}[x]$. This paper presents a method of zero decomposition for polynomial system into strong regular sets in $\mathbf{K}[x,t]$, where t is a new variable.

1 Introduction

Let $\mathbf{K}$ be a computable field of characteristic 0 and $\mathbf{K}[x_1,\cdots,x_n]$ (or $\mathbf{K}[x]$ for short) the ring of polynomials in the variables $(x_1,\cdots,x_n)$ with coefficients in $\mathbf{K}$. A *polynomial set* is a finite set $\mathbf{P}$ of nonzero polynomials in $\mathbf{K}[x]$. Any nonzero polynomial $P\notin\mathbf{K}$, the biggest index p such that $\deg(P,x_p)>0$ is called the *class*, denoted by $\mathrm{cls}(P)$, and x_p is called the *leading variable*, denoted by $\mathrm{lv}(P)$. A finite nonempty ordered set $\mathbf{T}=\{f_1,\cdots,f_s\}$ of polynomials in $\mathbf{K}[x]\backslash\mathbf{K}$ is called a *triangular set* if $\mathrm{cls}(f_1)\prec\cdots\prec\mathrm{cls}(f_s)$. Let a triangular set $\mathbf{T}$ be written as the following form

$$\mathbf{T}=\{f_1(u_1,u_2,\cdots,u_r,y_1),\cdots,f_s(u_1,u_2,\cdots,u_r,y_1,y_2,\cdots,y_s)\},$$

where $(u_1,u_2,\cdots,u_r,y_1,y_2,\cdots,y_s)$ is a permutation of $(x_1,x_2,\cdots,x_n)$. $I_j=\mathrm{ini}(f_j)$ denotes the leading coefficient of f_j in y_j for each j, and $\mathrm{ini}(\mathbf{T})$ denotes the set of all I_j. Let $\bar{\mathbf{K}}$ be an arbitrary extension field of $\mathbf{K}$. $\bar{\mathbf{K}}$ is the field of complex numbers $\mathbf{C}$ in this paper. While speaking about a *polynomial systems* [10], we refer to a pair $[\mathbf{P},\mathbf{Q}]$ of polynomial sets. Given a nonempty polynomial set $\mathbf{P}$. The famous zero decomposition was developed

本文刊于《Mathemtical Software》,2002 年.

by Wu[1]:

Theorem 1.1. Let **P** be a polynomial set. There is an algorithm, one may arrive after a finite number of steps at a zero decomposition of the form Zero(**P**) $= \bigcup_i$ Zero($\mathbf{T}_i \neq$ini($\mathbf{T}_i$)) in which $\mathbf{T}_i$ is an ascending set **P**.

Let $\boldsymbol{x}_i$ stand for $(x_1, \cdots, x_i)$ or $x_1, \cdots, x_i$, and similarly for $\bar{\boldsymbol{x}}_i$. A polynomial system [**T**, **U**] is called a *triangular* system if **T** is a triangular set and $I(\bar{\boldsymbol{x}}_i) \neq 0$ for any $I \in$ ini(**T**) of class i and $\bar{\boldsymbol{x}}_i \in$ Zero($\mathbf{T}^{(i)}/\mathbf{U}$) where $\mathbf{T}^{(i)} = \mathbf{T} \cap \mathbf{K}[x_1, \cdots, x_i]$. A triangular system [**T**, **U**] is said to be *fine* if prem(**U**, **T**) $\neq \{0\}$. The following theorem was presented by Wang[10,11,12].

Theorem 1.2. Let [**P**, **Q**] be a polynomial system. There are two methods, by which one can decompose [**P**, **Q**] into e fine triangular systems [$\mathbf{T}_i$, $\mathbf{U}_i$] such that Zero(**P**/**Q**) $= \bigcup_i^e$ Zero($\mathbf{T}_i/\mathbf{U}_i$).

Definition 1.1. Let $\{f_1, \cdots, f_s\}$ be a triangular set as above and g a polynomial in **K**[**x**]. The polynomial resultant($\cdots$resultant(g, f_s, y_s), $\cdots$, f_1, y_1) is called the *resultant* of triangular set $\{f_1, \cdots, f_s\}$ with respect to g, denoted by res($g, f_s, \cdots, f_1$) or res($g, \{f_1, \cdots, f_s\}$).

The concept of *regular sets* was introduced by Yang and Zhang[2] under the name of *proper ascending chains* in 1991. Independently, it was also introduced by Kalkbrener[5] under the name of *regular chains* in 1993.

Definition 1.2. [27] A triangular set $\mathbf{T} = \{f_1, \cdots, f_s\}$ is called a regular set, if $I_1 \neq 0$ and res($I_j, f_{j-1}, \cdots, f_1$) $\neq 0$, for $j = 2, 3; \cdots, s$.

Definition 1.3. [15] Let **T** be a triangular set and g a polynomial. **T** is called to be *strong independent* of g if Zero($\{g\}$) $\cap$ Zero(**T**) $= \emptyset$.

It is obvious that **T** is strong independent of g if res(g, **T**) $\in \mathbf{K} \setminus \{0\}$.

Kalkbrener[5] and Yang and Zhang Hou[7] are respectively presented two different algorithms for decomposing **P** into a finite number of regular sets $\mathbf{T}_1, \cdots, \mathbf{T}_e$ such that Zero(**P**) $= \bigcup_i^e$ Zero(sat($\mathbf{T}_i$)); where sat($\mathbf{T}_i$) is the saturation of $\mathbf{T}_i$. The set $\{\mathbf{T}_1, \cdots, \mathbf{T}_e\}$ is also called a *regular series of* **P**.

One of algorithms mentioned in Theorem 1.2 is RegSer, by which one can decompose any polynomial system into finitely many *regular systems*.

Definition 1.4. [12] A triangular system [**T**, **U**] is called a regular system if cls(T) $\neq$ cls(U) for any $T \in \mathbf{T}$ and $U \in \mathbf{U}$, and $I(\bar{x}_i) \neq 0$ for any $I \in$ ini(**U**) of class i and $\bar{x}_i \in$ Zero($\mathbf{T}^{(i)} \neq \mathbf{U}^{(i)}$) where $\mathbf{U}^{(i)} = \mathbf{U} \cap \mathbf{K}[x_1, \cdots, x_i]$. A triangular set **T** is said to be *regular* if there exists a polynomial set **U** such that [**T**, **U**] is a regular system. A finite set Ψ of regular systems is called a *regular series* of a polynomial system [**P**, **Q**] if

$$\mathrm{Zero}(\mathbf{P}/\mathbf{Q}) = \bigcup_{[\mathbf{T},\mathbf{U}] \in \Psi} \mathrm{Zero}(\mathbf{T}/\mathbf{U}).$$

Theorem 1.3[12,13] Let $[\mathbf{T}_1,\mathbf{U}_1],\cdots,[\mathbf{T}_e,\mathbf{U}_e]$be a a regular series of polynomial system $[\mathbf{P},\mathbf{Q}]$computed by RegSer($\mathbf{P},\mathbf{Q}$). Then

1) there exists an integer $d>0$ such that $\mathrm{prem}(P^d,\mathbf{T}_i)=0$ for all $P\in\mathbf{P}$ and $1\leqslant i\leqslant e$;

2) for any integers $m>0, 1\leqslant i\leqslant e$ and polynomial $Q\in\mathbf{Q}$, $\mathrm{prem}(Q^m,\mathbf{T}_i)\neq 0$;

3) $\mathrm{Zero}(\mathbf{P}/\mathbf{Q})=\bigcup_{i=1}^{e}\mathrm{Zero}(\mathbf{T}_i/\mathrm{ini}(\mathbf{T}_i)\cup\mathbf{Q})$.

Remark 1.1. The condition of Definition 1.4 is slightly strong in verifying $I\in\mathrm{ini}(\mathbf{U})$ of each class i. It maybe cause some trivial results such as

$$\begin{aligned}\mathrm{RegSer}(\{x_1\},\{x_2x_3x_5+x_4\})=\{&[\{x_1,x_2x_3\},\{x_2,x_4\}],[\{x_1,x_2\},\{x_4\}],\\&[\{x_1\},\{x_2x_3x_5+x_4,x_2x_3,x_2\}]\}.\end{aligned}$$

Moreover, it is likely to be true that $\mathrm{Zero}(\mathbf{T}_i/\mathbf{U}_i)\cap\mathrm{Zero}(\{Q\})\neq\emptyset$ for some $Q\in\mathbf{Q}$ and $1\leqslant i\leqslant e$ in Theorem 1.3. RegSer is very efficient in computing regular sets of nonempty polynomial set $\mathbf{P}$ by RegSer($\mathbf{P}$, $\emptyset$). In this paper, Reg($\mathbf{P}$) denotes a regular series of nonempty polynomial set $\mathbf{P}$ computed by one of above algorithms. In order to decompose any polynomial system $[\mathbf{P},\mathbf{Q}]$, we present an approach of splitting every regular set in Reg ($\mathbf{P}$) with respect to every polynomial in $\mathbf{Q}$. Because a good helpful theory in next section is used, the approach has some few advantages.

Remark 1.2. If one computes Reg($\mathbf{P}$) by RegSer($\mathbf{P}$, $\emptyset$), then all $\mathbf{U}_i$ are abandoned. It seems to cause a waste, but the $\mathbf{U}_i$ have less important in our approach as $\mathrm{Zero}(\mathbf{T}_i/\mathbf{U}_i)\subset\mathrm{Zero}(\mathbf{T}_i/\mathrm{ini}(\mathbf{T}_i))$: In fact we may obtain something similar to the $\mathbf{U}_i$ which maybe be more simpler in most cases.

2 Strong regular sets

It is well-known that there exists the problem of excluding some redundant zeros in most of zero decompositions. The theory of the *weakly nondegenerate condition*[3,4] may deduce a useful result to help forward solving the problem.

Definition 2.1.[3,4] Let $\mathbf{T}=\{f_1,\cdots,f_s\}$ be a regular set, where $f_j=c_{j0}y_j^{n_j}+c_{j1}y_j^{n_j-1}+\cdots+c_{jn_j}$, for $j=1,\cdots,s$. A zero $\mathbf{z}_0\in\mathrm{Zero}(\mathbf{T})$ is called a normal *zero* if $c_{j_0}(\mathbf{z}_0)\neq 0$ for $j=1,\cdots,s$, also said to be satisfying *nondegenerate condition*; the zero is called a *quasi-normal zero* if there exists some $k_0\in\{0,1,\cdots,n_j\}$ such that $c_{jk_0}(\mathbf{z}_0)\neq 0$ for $j=1,\cdots,s$, also said to be satisfying weakly nondegenerate condition.

Theorem 2.1. Let $\mathbf{T}=\{f_1,\cdots,f_s\}$ be a regular set and g a polynomial. If there exists an integer $d>0$ such that $\mathrm{prem}(g^d,\mathbf{T})=0$, then every quasi-normal zero of $\mathbf{T}$ is also a zero of g.

Definition 2.2. Let $\mathbf{T}$ be a regular set. $\mathbf{T}$ is called *a strong regular set* if every zero of $\mathbf{T}$ is also a quasi-normal zero.

Remark 2.1. There exists a great difference between a strong regular set and a regular system. One is a polynomial set, the other is a polynomial system. A regular system $[\mathbf{T},\emptyset]$

implies a strong regular set **T**. But a strong regular set **T** does not imply that $[T,\emptyset]$is a regular system, indeed a triangular system. For example, $\{x_2^2-x_1, x_2x_3x_4^2+x_4-x_1\}$ is a strong regular set, but $[\{x_2^2-x_1, x_2x_3x_4^2+x_4-x_1\},\emptyset]$ is not a triangular system at all. One must construct a triangular set $\mathbf{U}=\{x_2, x_2x_3\}$ such that $[\{x_2^2-x_1, x_2x_3x_4^2-x_1\},\mathbf{U}]$ is a regular system.

Proposition 2. 1. [15] Let $\mathbf{T}=\{f_1,\cdots,f_s\}$ be a regular set with the above notation. If $\mathrm{Zero}(\{c_{10},\cdots,c_{1n_1}\})=\emptyset$ and $\mathrm{Zero}(\{c_{j0},\cdots,c_jn_j\})=\emptyset$ or $\mathrm{Zero}(\{RI_{j0},\cdots,RI_{jn_j}\})=\emptyset$; for $j=2,\cdots,s$, where $\mathrm{RI}_{jk}=\mathrm{res}(c_{jk}, f_{j-1}, f_{j-2},\cdots,f_1)$ for $\mathrm{k}=0,\cdots,n_j$, then **T** is a strong regular set.

Theorem 2. 2. [15] Let **T** be a strong regular set and g a polynomial. If there exists an integer $d>0$ such that $\mathrm{prem}(g^d,\mathbf{T})=0$, then every zero of **T** is also a zero of g. Moreover, $\mathrm{Zero}(\mathrm{sat}(\mathbf{T}))=\mathrm{Zero}(\mathbf{T})$.

Example 2. 1. Let $\mathbf{P}=\{P_1,P_2,P_3\}$ with $P_1=z(x^2+y^2-c)+1, P_2=y(x^2+z^2-c)+1, P_3=x(y^2+z^2-c)+1$. This set of polynomials has been considered by Gao and Chou[9] and Wang[11,13]. Under the variable ordering $c\prec z\prec y\prec x$, **P** can be decomposed into 7 fine triangular systems $[\mathbf{T}_1,\mathbf{U}_1],\cdots,[\mathbf{T}_7,\mathbf{U}_7]$such that $\mathrm{Zero}(\mathbf{P})=\bigcup_{i=1}^{7}\mathrm{Zero}(\mathbf{T}_i/\mathbf{U}_i)$, where

$\mathbf{T}_1=\{2cz^4-2z^3-c^2z^2-2cz-1;(cz+1)y+cz^2-z,2z^2x+cz+1\}$,

$\mathbf{T}_2=\{2z^4-3cz^2+z+c^2, zy-z^2+c, x-z\}$,

$\mathbf{T}_3=\{z^3-cz-1,(z^2-c)y^2+y-cz^2+z+c^2, yx-z^2+c\}$,

$\mathbf{T}_4=\{2z^4-3cz^2+z+c^2,(2z^3-2cz+2)y-cz^2-z+c^2, P_3\}$,

$\mathbf{T}_5=\{2z^3-cz+1, y-z, 2z^2x-cx+1\}$,

$\mathbf{T}_6=\{c, 2z^3+1, y-z, 2z^2x+1\}$,

$\mathbf{T}_7=\{4c^3-27, 9z+2c^2, 6cy^2-9y-4c^2, 3yx+2c\}$;

$\mathbf{U}_1=\{c, z, cz+1\}, \mathbf{U}_2=\{z, z^2-c, 2z^2-c\}, \mathbf{U}_3=\{z^2-c, y\}$,

$\mathbf{U}_4=\{z^2-c, z^3-cz+1, z^3-cz-1\}$,

$\mathbf{U}_5=\{z, 2z^2-c\}, \mathbf{U}_6=\{z\}, \mathbf{U}_7=\{c, y\}$.

It is easy to verify that $\mathbf{T}_i$ is a strong regular set for each $i\neq 2$. Thus, we have

$$\mathrm{Zero}(\mathbf{P})=(\bigcup_{i=1,i\neq 2}^{7}\mathrm{Zero}(\mathbf{T}_i))\bigcup\mathrm{Zero}(\mathbf{T}_2/\mathbf{U}_2)$$

by Theorem 2. 2. In fact $\mathbf{T}_7$ can be removed according to the result used by Chou and Gao[8].

3 Decompose polynomial systems into strong regular sets

In order to decompose any polynomial systems into strong regular sets, we state first the following algorithm RegToStr by which one can decompose one regular set into finitely many strong regular sets without excluding any zero.

3.1 Algorithm RegToStr

Definition 3.1.1. For any polynomial set **P** in $\mathbf{K}[x_1,\cdots,x_n,t]$ (or $\mathbf{K}[x,t]$ for short) with an ordering for variables $x_1 \prec \cdots \prec x_n \prec t$, *the projection* of Zero(**P**) onto $x=(x_1,x_2,\cdots,x_n)$ is defined to be

$$\mathrm{Proj}_x \mathrm{Zero}(\mathbf{P}) = \{(\bar{x}_1,\bar{x}_2,\cdots,\bar{x}_n)\in \bar{\mathbf{K}}^n : \exists \bar{t}\ \bar{\mathbf{K}} \text{ such that } (\bar{x}_1,\bar{x}_2,\cdots,\bar{x}_n\bar{t})\in \mathrm{Zero}(\mathbf{P}).$$

Hereinafter, **T** is said a triangular set in $\mathbf{K}[x,t]$ if $\mathbf{T}=\{f_1,\cdots,f_s\}$ or $\mathbf{T}=\{f_1,\cdots,f_s,\mu_0 t-1\}$ where each f_i and $\mu_2\in\mathbf{K}[x]$.

Theorem 3.1.1. Let **T** be a regular set, but not a strong regular set in $\mathbf{K}[x,t]$ with the notation above. There exist a strong regular set $\mathbf{T}^*$ and a finite nonzero set $\{\mu_1,\cdots,\mu_{m_T}\}\subset\mathbf{K}[u_1,\cdots,u_r]$ such that

$$\mathrm{Zero}(\mathbf{T}) = \mathrm{Zero}(\mathrm{T}^*)\bigcup_{i=1}^{m_T}\mathrm{Zero}(\mathbf{T}\cup\{\mu_i\}).$$

Proof: Consider the case in which $\mathbf{T}=\{f_1,\cdots,f_s\}$. As **T** is not a strong regular set, there exists a nonempty index set η such that $\mathrm{Zero}(\{c_{10},\cdots,c_{1n_1}\}\neq\emptyset$ if $1\in\eta$ or $\mathrm{Zero}(\{c_{j0},\cdots,c_{jn_j},f_1,\cdots,f_{j-1}\})\neq\emptyset$ if $1<j\in\eta$. We pick a nonzero element from $\{c_{j0},\cdots,c_{jn_j}\}\cap\mathbf{K}[u_1,\cdots,u_r]$ or $\{RI_{j0},\cdots,RI_{jn_j}\}$, and write as μ_j^* for any $j\in\eta$. Collect all μ_j^* and denote by $\{\mu_1,\cdots,\mu_{m_T}\}$. Then, put $\mathbf{T}^*=\{f_1,\cdots,f_s,(\prod_{i=1}^{m_T}\mu_i)t-1\}$, where t is a new variable. $\mathbf{T}^*$ is a strong regular set in $\mathbf{K}[x,t]$.

The case in which $\mathbf{T}=\{f_1,\cdots,f_s,\mu_0 t-1\}$ is proved analogously, just replacing $\mathbf{T}^*$ by $\{f_1,\cdots,f_s,(\prod_{i=0}^{m_T}\mu_i)t-1\}$ in the above case.

Remark 3.1. Continue to Remark 1.2, let $[\mathbf{T},\mathbf{U}]$ be a regular system computed by RegSer($\mathbf{P},\emptyset$), Theorem 3.1.1 holds also true if one pick μ_j^* from **U**. Since some elements of **U** are likely to be superabundant and not simple, we have no choice but to abandon **U**. Refer to the example in Remark 2.1, the next zero decomposition is not necessary $\mathrm{Zero}(\{x_2^2-x_1,x_2x_3x_4^2+x_4-x_1\})=\mathrm{Zero}(\{x_2^2-x_1,x_2x_3x_4^2+x_4-x_1,x_2^2x_3t-1\})\cup\mathrm{Zero}(\{x_2^2-x_1,x_2x_3x_4^2+x_4-x_1,x_2\})\cup\mathrm{Zero}(\{x_2^2-x_1,x_2x_3x_4^2+x_4-x_1,x_2x_3\})$ if one does not abandon U.

Algorithm RegToStr: $\Psi\leftarrow$RegToStr(**T**). Given a regular set **T**, but not a strong regular set in K[x, t], this algorithm computes a finite set Ψ of strong regular sets in $\mathbf{K}[x,t]$ such that

$$\mathrm{Zero}(\mathrm{T})\subseteq\bigcup_{\mathrm{T}^*\in\Psi}\mathrm{Zero}(\mathbf{T}^*).$$

RTS1. Set $\Psi\leftarrow\emptyset$, and $\Phi\leftarrow\{\mathbf{T}\}$.

RTS2. While $\Phi\neq\emptyset$, do:

RTS2.1. Let $\mathbf{T}^*$ be an element of Φ set $\Phi\leftarrow\Phi\backslash\{\mathbf{T}^*\}$. If $\mathbf{T}^*$ is a strong regular set, then $\psi\leftarrow\psi\cup\{\mathbf{T}^*\}$ and go to RTS2.

RTS2.2. If $\mathbf{T}^*$ is not a strong regular set, then one can compute a strong regular set

$\{\mathbf{T}^{**}\}$ and a polynomial set $\{\mu_1^*, \cdots, \mu_{m_{T^*}}^*\}$ by Theorem 3.1.1. Set $\psi \leftarrow \psi \cup \{\mathbf{T}^{**}\}$ and

$$\Phi \leftarrow \Phi \cup (\bigcup_{i=1}^{m_{T^*}} \mathrm{Reg}(\mathbf{T}^* \cup \{\mu_i^*\})).$$

Example 3.1.1. Continue to Example 2.1. $\mathbf{T}_2$ is not a strong regular set in $\mathbf{Q}[c, z, y, x, t]$. According to Theorem 3.1.1, we can get $\eta = \{2\}$, and then pick $\mu_2^* = RI_{20} = c^2$, thereby obtain a strong regular set $\mathbf{T}_{2,0} = \{2z^4 - 3cz^2 + z + c^2, zy - z^2 + c, x - z, c^2t - 1\}$ and a polynomial set $\{c^2\}$ in $\mathbf{Q}[c, z, y, x, t]$.

Compute $\mathrm{Reg}(\mathbf{T}_2 \cup \{c^2\}) = \{\mathbf{T}_{2,1}, \mathbf{T}_{2,2}\}$, where $\mathbf{T}_{2,1} = \{c, z, x\}$ and $\mathbf{T}_{2,2} = \{c, 2z^3 + 1, zy - z^2, x - z\}$. Thus $\mathrm{RegToStr}(\mathbf{T}_2) = f\{\mathbf{T}_{2,0}, \mathbf{T}_{2,1}, \mathbf{T}_{2,2}\}$ and $\mathrm{Zero}(\mathbf{T}_2) \subseteq \bigcup_{i=0}^{2} \mathrm{Zero}(\mathbf{T}_{2,i})$ in $\mathbf{Q}[c, z, y, x, t]$.

3.2 Algorithm RSplit

Given an arbitrary strong regular set $\mathbf{T}$ in $\mathbf{K}[\mathbf{x}, t]$ and polynomial g in $\mathbf{K}[\mathbf{x}]$.

By the following algorithm RSplit, one can decompose $\mathbf{T}$ into two sets Φ_1 and Φ_2 of strong regular sets in $\mathbf{K}[\mathbf{x}, t]$ with respect to g.

Definition 3.2.1. Let $\mathbf{T}$ be a triangular set in $\mathbf{K}[\mathbf{x}, t]$ and g a polynomial in $\mathbf{K}[\mathbf{x}]$. One can form a sequence of polynomials $g_{-1}, g_0, g_1, \cdots, g_{m-1}, g_m$, where $g_{-1} = g$ and $g_0 = \mathrm{prem}(g_{-1}, \mathbf{T})$ by the rule: for any $i \geqslant 0$, if $g_i = 0$ and $g_{i-1} \neq 0$ then put $g_m := g_{i-1}$; else if $g_i \in \mathbf{K}[u_1, \cdots, u_r]\{0\}$ then put $g_m := g_i$; otherwise put $g_i := \mathrm{prem}(\mathrm{prem}(f_{\mathrm{cls}(g_{i-1})}, g_{i-1}, \mathrm{lv}(g_{i-1})), \mathbf{T})$. The gm is called *the expanding pseudo-remainder*[14] of g with respect to $\mathbf{T}$, denoted simply by $\mathrm{Eprem}(g, \mathbf{T})$.

Proposition 3.2.1. Let $\mathbf{T}$ be a triangular set in $\mathbf{K}[x, t]$ and g a polynomial in $\mathbf{K}[\mathbf{x}]$. If $\mathrm{Eprem}(g, \mathbf{T}) \in \mathbf{K} \backslash \{0\}$, then $\mathrm{Proj}_x \mathrm{Zero}(\mathbf{T}) \cap \mathrm{Zero}(\{g\}) = \emptyset$.

Theorem 3.2.1. Let $\mathbf{T}$ be a strong regular set in $\mathbf{K}[\mathbf{x}, t]$ and g a polynomial in $\mathbf{K}[\mathbf{x}]$ such that $\mathrm{prem}(g, \mathbf{T}) \neq 0$. Assume $\mathrm{lv}(g^*) = y_k$ $(1 \leqslant k \leqslant s)$, where $g^* = \mathrm{Eprem}(g, \mathbf{T})$. If $\mathrm{res}(\mathrm{ini}(g^*), \mathbf{T}) \in \mathbf{K} \backslash \{0\}$, then $\mathbf{T}$ can be split up into two strong regular sets $\mathbf{T}_1$ and $\mathbf{T}_2$ in $\mathbf{K}[\mathbf{x}, t]$ with respect to g such that $\mathrm{Zero}(\mathbf{T}) = \mathrm{Zero}(\mathbf{T}_1) \cup \mathrm{Zero}(\mathbf{T}_2)$.

Proof: Consider the case in which $\mathbf{T} = \{f_1, \cdots, f_s\}$. From the definition of the expanding pseudo-remainder, the form $g^* = \mathrm{Eprem}(g, \mathbf{T})$ implies that $\mathrm{prem}(\mathrm{prem}(f_k, g^*, y_k), \mathbf{T}) = 0$. It follows that

$$I_0^{c_0} f_k = f_{k,1} f_{k,2} + R, \mathrm{prem}(R, \{f_1, \cdots, f_{k-1}\}) = 0,$$

where $f_{k,1} = g^*$ $I_0 = \mathrm{ini}(f_{k,1})$, $f_{k,2} \in \mathbf{K}[\mathbf{x}]$ and c_0 is some nonnegative integer. Obtain two triangular sets $\mathbf{T}_1$ and $\mathbf{T}_2$ by substituting f_k in $\mathbf{T}$ for $f_{k,1}$ and $f_{k,2}$ respectively.

Next we prove that $\mathrm{Zero}(\mathbf{T}) = \mathrm{Zero}(\mathbf{T}_1) \cup \mathrm{Zero}(\mathbf{T}_2)$. For any $\mathbf{z}_0 \in \mathrm{Zero}(\mathbf{T})$. It follows from Theorem 2.2 that $f_{k,1}(\mathbf{z}_0) f_{k,2}(\mathbf{z}_0) = 0$. Thus we have $\mathrm{Zero}(\mathbf{T}) \subset \mathrm{Zero}(\mathbf{T}_1) \cup \mathrm{Zero}(\mathbf{T}_2)$. In other direction, consider any $\mathbf{z}_0 \in \mathrm{Zero}(\mathbf{T}_1) \cup \mathrm{Zero}(\mathbf{T}_2)$. The fact $\mathrm{res}(I_0, \mathbf{T}) \in \mathbf{K} \backslash \{0\}$ implies $f_k(\mathbf{z}_0) = 0$, thus $\mathrm{Zero}(\mathbf{T}) \supset \mathrm{Zero}(\mathbf{T}_1) \cup \mathrm{Zero}(\mathbf{T}_2)$. Therefore, $\mathrm{Zero}(\mathbf{T}) = \mathrm{Zero}(\mathbf{T}_1) \cup \mathrm{Zero}(\mathbf{T}_2)$. The proof of the assertion that $\mathbf{T}_1$ and $\mathbf{T}_2$ are strong regular sets is somewhat simple and trivial so we omit the details.

The case in which $\mathbf{T}=\{f_1,\cdots,f_s,\mu_0 t-1\}$ is proved similarly.

Theorem 3.2.2. Let $\mathbf{T}$ be a strong regular set in $\mathbf{K}[\mathbf{x},t]$ and g a polynomial in $\mathbf{K}[\mathbf{x}]$ such that $\mathrm{prem}(g,\mathbf{T})\neq 0$. If g^* or $\mathrm{res}(\mathrm{ini}(\mathbf{g}^*),\mathbf{T})\in\mathbf{K}[u_1,\cdots,u_r]\backslash\mathbf{K}$ with $g^*=\mathrm{Eprem}(g,\mathrm{T})$, then there exist a strong regular set $\mathbf{T}_1$ and a triangular set $\mathbf{T}_2$ with respect to g such that $\mathrm{Zero}(\mathbf{T})=\mathrm{Zero}(\mathbf{T}_1)\cup\mathrm{Zero}(\mathbf{T}_2)$:

Proof: Consider the case in which $\mathbf{T}=\{f_1,\cdots,f_s\}$. Set $c_0=g^*$ if $g^*\in\mathbf{K}[u_1,\cdots,u_r]\backslash\mathbf{K}$, $c_0=\mathrm{res}(\mathrm{ini}(g^*),\mathbf{T})$ otherwise. Put $\mathbf{T}_1=\{f_1,\cdots,f_s,c_0t-1\}$ and $\mathbf{T}_2=\mathbf{T}\cup\{c_0\}$. $\mathbf{T}_1$ is a strong regular set in $\mathbf{K}[\mathbf{x},t]$.

The case in which $\mathbf{T}=\{f_1,\cdots,f_s,\mu_0^t-1\}$ is proved analogously, just replacing $\mathbf{T}_1$ by $\{f_1,\cdots,f_s,c_0\mu_0 t-1\}$ in the above case.

Theorem 3.2.3. Let $\mathbf{T}$ be a strong regular set in $\mathbf{K}[\mathbf{x},t]$ and g a polynomial in $\mathbf{K}[\mathbf{x}]$ such that $\mathrm{prem}(g,\mathbf{T})\neq 0$. If $\mathrm{lv}(g^*)=y_k, 1\leqslant k\leqslant s$ and $\mathrm{res}(\mathrm{ini}(g^*),\mathbf{T})=0$ with $g^*=\mathrm{Eprem}(g,\mathbf{T})$, then there exist two triangular sets $\mathbf{T}_1$ and $\mathbf{T}_2$ with respect to g such that $\mathrm{Zero}(\mathbf{T})=\mathrm{Zero}(\mathbf{T}_1)\cup\mathrm{Zero}(\mathbf{T}_2)$.

Proof: Consider the case in which $\mathbf{T}=\{f_1,\cdots,f_s\}$. Put $\mathbf{T}_1=\{f_1,\cdots,f_s,c_0t-1\}$ and $\mathbf{T}_2=\mathbf{T}\cup\{c_0\}$, where $c_0=\mathrm{res}(\mathrm{ini}(g^*),\mathbf{T})$.

The case in which $\mathbf{T}=\{f_1,\cdots,f_s,\mu_0 t-1\}$ is proved analogously, just replacing $\mathbf{T}_1$ by $\{f_1,\cdots,f_s,c_0\mu_0 t-1\}$ in the above case.

Algorithm RSplit: $[\Phi_1,\Phi_2]\leftarrow\mathrm{RSplit}(\mathbf{T}_0,P)$. Given a strong regular set $\mathbf{T}_0$ in $\mathbf{K}[\mathbf{x},t]$ and a polynomial P in $\mathbf{K}[\mathbf{x}]$, this algorithm computes two sets Φ_1 and Φ_2 of strong regular sets in $\mathbf{K}[\mathbf{x},t]$ such that

$$\mathrm{Zero}(\mathbf{T}_0)\subseteq\bigcup_{\mathrm{T}^*\in\Phi_1\cup\Phi_2}\mathrm{Zero}(\mathbf{T}^*),$$

and there exists an integer $d>0$ such that $\mathrm{prem}(P^d,\mathbf{T}^*)=0$ for any $\mathbf{T}^*\in\Phi_1$; $\mathrm{Proj}_{\mathbf{x}}\mathrm{Zero}(\mathbf{T}^*)\cap\mathrm{Zero}(\{P\})=\emptyset$ for any $\mathbf{T}^*\in\Phi_2$.

SP1. Set $\Phi_1\leftarrow\emptyset$, $\Phi_2\leftarrow\emptyset$ and $\Psi\leftarrow\{\mathbf{T}_0\}$:

SP2. While Compute $\Psi=\emptyset$, do:

SP2.1. Let $\mathbf{T}$ be an element of Ψ set $\Psi\leftarrow\Psi\backslash\{\mathbf{T}\}$. If there exists an integer $d_0>0$ such that $\mathrm{prem}(P^{d_0},\mathbf{T})=0$, then set $\Phi_1\leftarrow\Phi_1\cup\{\mathbf{T}\}$, and go to SP2.

SP2.2. If $\mathrm{Eprem}(P,\mathbf{T})\in\mathbf{K}\backslash\{0\}$, then set $\Phi_2\leftarrow\Phi_2\cup\{\mathbf{T}\}$ and go to SP2.

SP2.3. The condition of Theorem 3.2.1 holds, then one can obtain two strong regular sets $\mathbf{T}_1$ and $\mathbf{T}_2$ by Theorem 3.2.1. Set $\Psi\leftarrow\Psi\cup\{\mathbf{T}_1,\mathbf{T}_2\}$ and go to SP2.

SP2.4. The condition of Theorem 3.2.2 holds, then one can obtain a strong regular sets $\mathbf{T}_1$ and a triangular set $\mathbf{T}_2$ by Theorem 3.2.2. Compute $\mathrm{Reg}(\mathbf{T}_2)=\Phi^*$, then set

$$\Psi\leftarrow\Psi\cup\Big(\bigcup_{\mathrm{T}^*\in\Phi^*}\mathrm{RegToStr}(\mathrm{T}^*)\cup\{\mathbf{T}_1\}.$$

SP2.5. The condition of Theorem 3.2.2 holds, then one can obtain two triangular sets

$\mathbf{T}_1$ and $\mathbf{T}_2$ by Theorem 3.2.3. Compute $\mathrm{Reg}(\mathbf{T}_1)=\Phi_1^*$ and $\mathrm{Reg}(\mathbf{T}_2)=\Phi_2^*$, then set

$$\Psi \leftarrow \Psi \cup \left(\bigcup_{T^* \in \Phi_1^* \cup \Phi_2^*} \mathrm{RegToStr}(T^*) \right)$$

Example 3.2.1. Continue to Example 3.1.1, $\mathrm{RSplit}(\mathbf{T}_{2,0}, P)=[\{\mathbf{T}_2, 0\}, \emptyset]$; $\mathrm{RSplit}(\mathbf{T}_{2,1}, P)=[\emptyset, \{\mathrm{T}_{2,1}\}]$; $\mathrm{RSplit}(\mathbf{T}_{2,2}, P)=[\{\mathbf{T}_{2,2}\}, \emptyset]$, for any $P \in \mathbf{P}$.

3.3 Algorithm Dec

Algorithm Dec: $\Phi \leftarrow \mathrm{Dec}(\mathbf{P}, \mathbf{Q})$. Given a polynomial system $[\mathbf{P}, \mathbf{Q}]$ in $\mathbf{K}[\mathrm{x}]$, this algorithm computes a set Φ of strong regular sets in $\mathbf{K}[\mathbf{x}, t]$ such that

$$\mathrm{Zero}(\mathbf{P}/\mathbf{Q}) = \bigcup_{\mathrm{T} \in \Phi} \mathrm{Proj}_{\mathbf{x}} \mathrm{Zero}(\mathbf{T}),$$

and there exists an integer $\mathrm{d}>0$ such that $\mathrm{prem}(P^d, \mathbf{T})=0$ and $\mathrm{Proj}_{\mathbf{x}}\mathrm{Zero}(\mathbf{T}) \cap \mathrm{Zero}(\{\mathbf{Q}\})=\emptyset$ for any $\mathbf{T} \in \Phi$, $\mathrm{P} \in \mathbf{P}$ and $\mathrm{Q} \in \mathbf{Q}$.

D1. Compute $\mathrm{Reg}(\mathbf{P})=\Phi_0$ and set

$$\Phi \leftarrow \emptyset, \Psi \leftarrow \bigcup_{\mathrm{T} \in \Phi_0} \mathrm{RegToStr}(\mathbf{T}).$$

D2. While Compute $\Psi \neq \emptyset$, do:

D2.1. Let $\mathbf{T}$ be an element of Ψ set $\Psi \leftarrow \Psi \backslash \{\mathbf{T}\}$.

D2.2. For any $P \in \mathbf{P}$ and $Q \in \mathbf{Q}$ do:

D2.1.1. If $\mathrm{Eprem}(P; \mathbf{T}) \in \mathbf{K} \backslash \{0\}$, then go to D2.

D2.1.2. If there exists an integer $k>0$ such that $\mathrm{prem}(Q^k, \mathbf{T})=0$, then go to D2.

D2.1.3. If $\mathrm{prem}(P, \mathbf{T}) \neq 0$, then compute $\mathrm{RSplit}(\mathrm{T}, \mathrm{P})=[\Phi_1, \Phi_2]$. Set $\Psi \leftarrow \Psi \cup \Phi_1$, then go to D2.

D2.1.4. If $\mathrm{prem}(Q, \mathbf{T}) \neq 0$ and $\mathrm{Eprem}(Q, \mathbf{T}) \notin \mathbf{K}$, then compute $\mathrm{RSplit}(\mathbf{T}, Q)=[\Phi_1, \Phi_2]$. Set $\Psi \leftarrow \Psi \cup \Phi_2$, then go to D2.

D2.3. Set $\Phi \leftarrow \Phi \cup \{\mathbf{T}\}$.

Example 3.3.1. Continue to Example 3.2.1, we have

$$\mathrm{Zero}(\mathbf{P}) = \left(\bigcup_{\mathrm{i}=1, \mathrm{i} \neq 2}^{6} \mathrm{Proj}_{\mathbf{x}} \mathrm{Zero}(\mathbf{T}_{\mathrm{i}}) \right) \cup \mathrm{Proj}_{\mathbf{x}} \mathrm{Zero}(\mathbf{T}_{2,0}) \cup \mathrm{Proj}_{\mathbf{x}} \mathrm{Zero}(\mathrm{T}_{2,2}),$$

where $\mathbf{x}=(c, z, y, x)$.

The algorithm is different with RegSer which uses subresultant subchain and allows simultaneously splitting one polynomial system into several subsystems. $\mathrm{Dec}(\mathbf{P}, \mathbf{Q})$ performs expanding pseudo-division and splitting step-by-step each regular set in $\mathrm{Reg}(\mathbf{P})$ if there exists any common zero, without computing GCDs, into finitely many strong regular sets in $\mathbf{K}[\mathbf{x}, t]$.

Remark 3.2. If a regular series $[\mathbf{T}_1, \mathbf{U}_1], \cdots, [\mathbf{T}_e, \mathbf{U}_e]$ is computed by RegSer from a polynomial system $[\mathbf{P}, \mathbf{Q}]$ in $\mathbf{K}[\mathbf{x}]$, let v_i be the product of all the polynomials in U_i and $\mathbf{T}_i^* = \mathbf{T}_i \cup \{v_i t - 1\}$ where t is a new variable, then we similarly have

$$\mathrm{Zero}(\mathbf{P}/\mathbf{Q}) = \bigcup_{i=1}^{e} \mathrm{Zero}(\mathbf{T}_i/\mathbf{U}_i) = \bigcup_{i=1}^{e} \mathrm{Proj}_{\mathbf{x}} \mathrm{Zero}(\mathbf{T}_i^*).$$

But the result of Dec($\mathbf{P},\mathbf{Q}$) is better than that of RegSer($\mathbf{P},\mathbf{Q}$), because $\mathrm{Proj}_{\mathbf{x}}\mathrm{Zero}(\mathbf{T})\cap\mathrm{Zero}(\{Q\})=\emptyset$ for any $\mathbf{T}\in\mathrm{Dec}(\mathbf{P},\mathbf{Q})$ and $Q\in\mathbf{Q}$. On the other hand, Dec overcomes the trivial mentioned in Remark 1. 1, for example

$\mathrm{Zero}(\{x_1\}/\{x_2x_3x_5+x_4\})=[\mathrm{Proj}_{\mathbf{x}}\mathrm{Zero}(\{x_1,x_2x_3,x_2x_4t-1\})$

$\cup[\,Proj_{\mathbf{x}}\mathrm{Zero}(\{x_1,x_2x3x_5+x_4)x_2^2x_3t-1\})$

$\cup[\,\mathrm{Proj}_{\mathbf{x}}\mathrm{Zero}(\{x_1,x_2,x_4t-1\})$

by RegSer($\{x_1\},\{x_2x_3x_5+x_4\}$), but

$$\mathrm{Zero}(\{x_1\}/\{x_2x_2x_5+x_4\})=\mathrm{Proj}_{\mathbf{x}}\mathrm{Zero}(\{x_1,(x_2x_3x_5+x_4)t-1\})$$

by Dec($\{x_1\},\{x_2x_3x_5+x_4\}$).

Remark 3. 3. The Dec can been implemented by mathematical software, the efficiency heavily depends upon of algorithm of Reg. We recommend one to use the algorithm RegSer. All of above examples in this paper have been computed by Maple system in interactive way. We will further experiment and compare our algorithm with some of the other efficient algorithm implemented with Maple system when the package of RegSer is available.

Acknowledgements

The authors thank specially Dr. D. M Wang for his helpful comments and suggestions. The first author is grateful to Prof. J-H. Mai for his helps.

References

[1]Wu W-T. On the decision problem and the mechanization of theorem proving in elementary geometry. Scientia Sinica. 1978(21): 159 - 172.

[2] Yang L, Zhang J-Z. Search dependency between algebraic equations: An algorithm applied to automated reasoning. Technical Report IC/91/6, International Atomic Energy, Miramare, Trieste, 1991.

[3]Zhang J-Z, Yang L, Hou X-R. A note on Wu Wen-Tsün's nondegenerate condition. Technical Report IC/91/160, International Atomic Energy, Miramare, Trieste, 1991

[4]Zhang J-Z, Yang L, Hou X-R. A note on Wu Wen-Tsün's nondegenerate condition, Chinese Science Bulletin, 1993, 38(1): 86 - 87.

[5]Kalkbrener M. A generalized Euclidean algorithm for computing triangular representations of algebraic varieties. J. Symb. Comput, 1993(15): 143 - 167.

[6]Yang L, Zhang J-Z, Hou X-R. An efficient decomposition algorithm for geometry theorem proving without factorization. in MM-Preprints, 1993(9): 115 - 131.

[7]Yang L, Zhang J-Z, Hou X-R. Non-Linear equation systems and automated theorem proving. Shanghai: Shanghai Sci. Tech. Education Publ, 1996(in Chinese).

[8]Chou S-C, Gao X-S. Ritt-Wu's decomposition algorithm and geometry theorem proving. In: Stickel, M. E. (ed.): 10th International Conference on Automated Deduction. Springer, Berlin

Heidelberg New York Tokyo, 1990: 207-220 (Lecture notes in computer science, vol. 449)
[9]Gao X-S, Chou S-C. Solving parametric algebraic systems. In: Proceedings ISSAC'92, Berkeley, July 27 - 29, 1992. Association for Computing Machinery, New York, 1992: 335 - 341.
[10]Wang D M. An elimination method for polynomial systems. J. Symb. Comput, 1993 (16): 83 - 114.
[11]Wang D M. Decomposing triangular systems into simple systems. J. Symb. Comput, 1998(25): 295 - 314.
[12]Wang D M. Decomposing triangular systems and regular systems. J. Symb. Comput, 2000(30): 221 - 236.
[13]Wang D M. Elimination methods. Wien/New York: Springer, 2001.
[14]Li Y-B. The expanding WE algorithm a mechanical geometry therorem proving. Journal of Sichuan University (Natural Science Edition), 2000 (37) 3: 331-335 (in Chinese).
[15]Li Y-B. The expanding WE algorithm and the approach of strong regular decomposition on nonlinear algebraic equation system Ph. D. thesis, Mathematical college, Sichuan University, Chengdu, China, 2001.

Linear Duration Invariants

Zhou Chao-chen,
UNU/IIST, P. O. Box 3058, Macau

Zhang Jing-zhong, Yang Lu
CICA, Academia Sinica, Chengdu, China

Li Xiao-shan
Software Institute, Academia Sinica, Beijing, China

Abstract. This paper is to present an algorithm to decide whether a real-time system satisfies a set of invariants which are constructed from linear inequalities of integrated durations of system states. Real-time systems in the paper are taken to be real-time automata which set up for each of state transitions a lower time bound and an upper time bound. The satisfaction problem can be translated into a family of *linear programming problems*. The algorithm is, according to the invariants, to reduce the infinite family of linear programming problems to an equivalent one with only finite members, and then to solve each of linear programming problems efficiently. The algorithm is so simple that no prerequisite of linear programming theory is assumed.

1 Introduction

It is the Duration Calculus (DC) [9] which uses integrated durations of states over intervals to specify and reason about real-time behaviour of software embedded systems. In DC, states are modelled as Boolean functions from reals (representing continuous *time*) to {0,1}, where 1 denotes state presence, and 0 denotes state absence. DC has been applied to a number of examples [7][8][2][6], including a gas burner. In the gas burner example, one of the safety-critical requirements of the gas burner is formulated in DC as *Req*

本文刊于《Formal Techniques in Real-Time and Fault-Tolerant Systems Volume 863 of the series Lecture Notes in Computer Science》, 2005 年.

$$\int 1 \geqslant 60\text{sec} \Rightarrow 20\int \text{Leak} \leqslant \int 1,$$

Req is an interval formula with interval variables $\int$ Leak and $\int 1$, where *Leak* is a Boolean function to represent the leak state of the gas burner, and 1 is the constant Boolean function to represent the constantly present state. For Boolean function S, interval variable $\int S$ of DC is a function from bounded and closed intervals to reals, $\int S$ stands for the integrated duration of state S over intervals. For bounded interval $[a, b]$ $(b \geqslant a)$,

$$\int S[a,b] \triangleq \int_a^b S(t)dt,$$

it follows that

$$\int 1[a,b] \triangleq \int_a^b 1dt = (b-a).$$

i. e. the length of $[a, b]$. So *Req* means that if the gas burner is observed for a time period which is not less than 60 seconds, the proportion of time when the gas burner is leak is not more than one twentieth of the whole period.

Req consists of two parts: a premise and a conclusion, $\int 1 \geqslant 60$ is the premise to specify the length of observation intervals. The conclusion is a linear inequality of the integrated durations of states: $(20\int Leak - \int 1) \leqslant 0$.

In general, we call a requirement a linear duration invariant, if the requirement consists of a premise to specify the length of observation intervals

$$T \geqslant \int 1 \geqslant t$$

and a conclusion

$$\bigwedge_{j=1}^{k} \left(\sum_{i=1}^{n} c_{ij} \int S_i \leqslant M_j\right),$$

where t, T, M_j and c_{ij} are real numbers and T may be ∞ when $T=\infty$, $(T \geqslant \int 1 \geqslant t)$ is understood as $(\int 1 \geqslant t)$ and S_i are states. ①

A system satisfies a linear duration invariant, if and only if the integrated durations of its states will always satisfy the conclusion when the system is observed for a time period which length satisfies the premise.

A designer of the gas burner may refine the above requirement into two design decisions.

---

① States include primitive states and composite states-Boolean expressions of primitive states.

—A leak should be detectable and stoppable within one second; and

—The time distance between two leaks should not be less than thirty seconds.

There have been several approaches to representing those two design decisions and verifying the correctness. They use DC [9], hybrid State Charts [5], hybrid automata [1] and real-time automata [7][10]. This paper extends the real-time automata approach in [7] [10]and develops an algorithm to decide whether a real-time automaton satisfies a linear duration invariant. This algorithm can easily conclude the correctness of the design decisions for the gas burner.

A real-time automaton is a conventional automaton with real-time constraints on its state transitions. The constraints set up for a state transition a time interval (closed-bounded or half-closed-unbounded), during which the state transition is allowed to take place. For example, a real-time automaton which represents the design decisions for the gas burner is depicted in Fig. 1.

The automaton has two states: *Leak* and *Nonleak*. *Nonleak* is the initial state, denoted with a bullet. It has two state transitions, called f (failure) and r (recovery). The failure transition can take place after the automaton has stayed at *Nonleak* for not less than 30 seconds, and the recovery transition must take place before the automaton stays at *Leak* for more than 1 second.

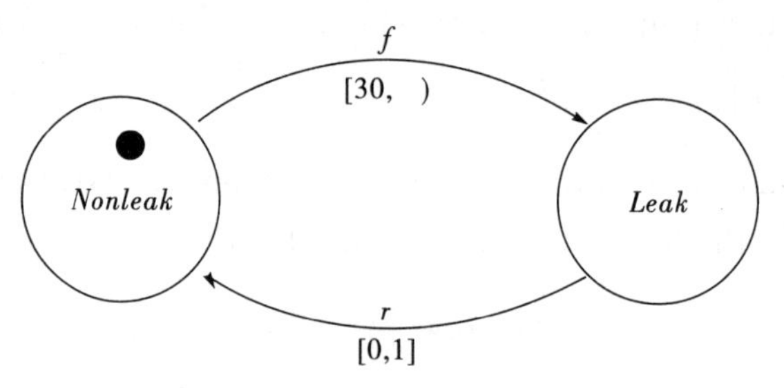

Fig 1. Gas Burner

The untimed behaviour of the automaton can be described by its transition sequences. For instance

$$f^\frown r^\frown f$$

records an untimed behaviour of the automaton: starting from *Nonleak*, transiting to *Leak* with transition f, returning to *Nonleak* with transition r and changing to *Leak* with transition f again.

Time — distance stamped transition sequences can be used to describe the timed behaviour of the automaton. For example

$$(f,31)^\frown(r,0.5)^\frown(f,50)$$

is one of possible timed behaviour of the automaton, which untimed projection is $f^\frown r^\frown f$. This sequence defines a timed evolution of the automaton: staying at *Nonleak* for 31 seconds before an f transition takes place, then staying at *Leak* for 0.5 seconds before a recovery transition r occurs, and at last having another failure transition f after 50 seconds to stay at *Nonleak*.

In general, $(f,t_1)\hat{}(r,t_2)\hat{}(f,t_3)$ is a timed behaviour of the automaton, iff

$$t_1 \geq 30, 1 \geq t_2 \geq 0 \text{ and } t_3 \geq 30.$$

For the behaviour $(f,t_1)\hat{}(r,t_2)\hat{}(f,t_3)$, the integrated duration when the automaton stays at Nonleak is (t_1+t_3). That is

$$\int Nonleak = (t_1 + t_3)$$

and similarly, for the behaviour $(f,t_1)\hat{}(r,t_2)\hat{}(f,t_3)$,

$$\int Leak = t_2.$$

As *Nonleak* and *Leak* are the only states of the automaton and they are exclusive from each other, it is sure that

$$\int 1 = (\int Nonleak + \int Leak).$$

Therefore, for the behaviour $(f,t_1)\hat{}(r,t_2)\hat{}(f,t_3)$,

$$\int 1 = (t_1 + t_2 + t_3).$$

The problem whether all the timed behaviour of the automaton which corresponds to the untimed behaviour $f\hat{}r\hat{}f$ satisfy the requirement turns out to be a typic linear programming problem: subject to the constraints

$$t_1 \geq 30, 1 \geq t_2 \geq 0, t_3 \geq 30 \text{ and} (t_1 + t_2 + t_3) \geq 60,$$

calculate the greatest value of the objective function

$$20t_2 - (t_1 + t_2 + t_3),$$

and check whether it is non-positive.

However the number of the untimed transition sequences of the automaton is infinite, and therefore the number of the corresponding linear programming problems is infinite also. In order to decide whether all the transition sequences of the automaton satisfy the requirement, this paper develops a mechanical algorithm to reduce the infinite set of the problems to a finite set, and then to solve the finite problems efficiently, the algorithm is so simple that no prerequisite of linear programming theory is assumed.

[3] also uses mathematical programming theory to study similar problems, and identifies very interesting reducible cases. In [3], the previous infinite linear programming problems are translated into a *mixed integer programming* problem, and then the mixed integer programming technique can be employed to resolve the satisfaction problem.

Section 2 is to mathematically define real-time automata, and translate satisfaction problems for linear duration invariants to families of linear programming problems. In Section 3, the concerned families of linear programming problems are re-formulated as so-called Mountaineer Problems. The reduction algorithm is presented in Section 4. Section 5 shows how to decide a single untimed behaviour to satisfy a linear duration invariant. The paper is organized in such a way that readers who are only interested in the Mountaineer Problem and the reduction algorithm can skip Section 2 and read the paper from Section 3.

2 Real-Time Automata & Linear Duration Invariants

2.1 Untimed Automata

A real-time automaton is a conventional (untimed) automaton with real-time constraints upon its state transitions. An (untimed) automaton has a finite set of states, denoted S, and a finite set of state transitions, denoted ε. Let $e\in\varepsilon$. Suppose e is a state transition from state S to state S'. We call $S(S')$ *pre-state* (*post-state*) of e, denoted $\overleftarrow{e}$ ($\overrightarrow{e}$). Some of the states are *initial* states, denoted S_0.

A behaviour of an automaton can be represented by a finite sequence of its state transitions. It records an evolution history of the automaton, starting from one of its initial states, and traveling from state to state.

Let $\in$ stand for *empty* transition sequence, $\in$ represents the behaviour of the automaton before it starts. Let $e_1 \hat{e}_2 \hat{\cdots} \hat{e}_m (m>0)$ be a non-empty transition sequence. It represents a behaviour of the automaton if and only if

1. *Initiality*:

$$\overleftarrow{e_1}\in S_0$$

i. e. the automaton starts at one of its initial states, and

2. *Consecutivity*: for $0<i<m$

$$\overrightarrow{e_i}=\overleftarrow{e_{i+1}}$$

i. e. the post-state of, a transition in the sequence is the pre-state of its successor, so that the transition sequence represents a continuous evolution of the automaton.

Example:

The automaton in Fig. 1 has

—Two states: $S=\{Nonleak, Leak\}$

—One initial state: $S_0=\{Nonleak\}$

—Two transitions: $\varepsilon=\{f, r\}$.

They mean *failure and recovery* respectively.

—Pre-states and post-states of transitions:

$$\overleftarrow{f}=Nonleak, \overrightarrow{f}=Leak$$

$$\overleftarrow{r}=Leak, \overrightarrow{r}=Nonleak$$

The following transition sequences represent some of the untimed behaviour of the automaton

$$E, f, f^\frown r, f^\frown r^\frown f, \cdots$$

The total number of transition sequences which represent the untimed behaviour of the gas burner automaton is infinite, and it becomes impossible to list them one by one. However it is well known that the set of transition sequences which characterizes the behaviour of an automaton can be expressed by a *regular expression* over state transitions [4]. A regular expression is constructed from state transitions with combinators $\oplus$, $^\frown$ and $*$, where $\oplus$ means *union*, $^\frown$ means *concatenation*, and $*$ means *repetition*. A mathematical definition of the combinators is given in Section 3.

Example (continued):

The regular expression which describes the untimed behaviour of the gas burner automaton is

$$(f^\frown r)^* \oplus (f^\frown r)^{*\frown} f.$$

2.2 Real-time Automata

A *real-time* automaton is an automaton with real-time constraint on its behaviour. The real-time constraint assigns to each state transition e ($e \in \varepsilon$) a (closed-bounded or half-closed-unbounded) time interval, denoted $[a_e, b_e]$ (when $b_e = \infty$, it is taken to be $[a_e, \infty)$). Interval $[a_e, b_e]$ constrains the automaton to reject transition e unless it has just stayed at state $\overleftarrow{e}$ for a time period t such that $b_e \geq t \geq a_e$. In Section 2.1, transition sequences are used to represent untimed behaviour of automata, which define temporal order of occurrences of transitions, but not real-time occurrences of transitions. To represent real-time behaviour of an automaton, we use sequences of *time-distance* stamped transitions. A time-distance stamped transition is a pair of transition e and nonnegative real number t. (e, t) defines an e transition which occurs at t time units after the occurrence of the last transition (or after the initiation of the automaton, if e is the first transition of a sequence).

Definition. A time-distance stamped transition sequence

$$(e_1, t_1)^\frown(e_2, t_2)^\frown\cdots^\frown(e_m, t_m) \qquad (m \geq 0)$$

represents a real-time behaviour of a real-time automaton, if and only if

1. its projection

$$e_1{}^\frown e_2{}^\frown \cdots {}^\frown e_m$$

is a (untimed) behaviour of the automaton, i. e. the projection satisfies the *Initiality* and *Consecutivity*; and

2. its time-distance stamps satisfy the real-time constraint of the automaton. That is,

$$b_{ei} \geq t_i \geq a_{ei} (1 \leq i \leq m).$$

Example (continued):

For the real-time automaton shown in Fig. 1,

$$[a_f, b_f] = [30, \infty),$$

which constrains the gas burner to stay at *Nonleak* for at least 30 seconds before it (re-)risks a flame failure.

$$[a_r, b_r] = [0, 1],$$

which constrains a flame failure to be recovered in 1 second.

The time-distance stamped transition sequence given in Section 1 describes a possible real-time behaviour of the gas burner automaton

$$(f, 31)^\frown(r, 0.5)^\frown(f, 50).$$

2.3 Linear Duration Invariant & Linear Programming

Linear duration invariants are used to specify real-time system requirements. It has been explained in Section 1 that a linear duration invariant consists of a premise and a conclusion. The premise specifies a concerned range of the evolution time of the system. Let it be

$$T \geqslant \int 1 \geqslant t$$

and the conclusion specifies a set of system invariants in terms of linear inequalities of state durations, and let it be

$$\bigwedge_{j=1}^{k} \left(\sum_{i=1}^{n} c_{ij} \int S_i \leqslant M_j \right)$$

where t, T, M_j and c_{ij} are real numbers and T may be ∞, S_i are the system states and 1 is the constant state, and $\int 1$ and $\int S_i$ are integrated durations of state 1 and S_i.

Let $(e_1, t_1)^\frown(e_2, t_2)^\frown\cdots^\frown(e_m, t_m)$ be a time-distance stamped transition sequence. It represents a timed behaviour of an automaton with real-time constraint $[a_e, b_e]$ for $e \in \varepsilon$, if $e_1{}^\frown e_2{}^\frown\cdots{}^\frown e_m$ is an untimed behaviour of the automaton, and

$$b_{eu} \geqslant t_u \geqslant a_{eu} (1 \leqslant u \leqslant m) \quad \text{(denoted } C1\text{)}.$$

Suppose $\overleftarrow{e_u} \Rightarrow S_i$. Thus, when the automaton performs the given behaviour, it must stay at S_i for t_u time units before making transition e_u. So, for the given timed behaviour, the integrated duration of state S_i, i. e. the value of $\int S_i$, can be calculated as

$$\int S_i = \sum_{u \in a_i} t_u$$

where $a_i \triangleq \{u \mid (1 \leqslant u \leqslant m) \wedge (\overleftarrow{e_u} \Rightarrow S_i)\}$. It follows that, for the given timed behaviour,

$$\int 1 = \sum_{u=1}^{m} t_u,$$

as $\{u \mid (1 \leqslant u \leqslant m) \wedge (\overleftarrow{e_u} \Rightarrow 1)\} = \{1, 2, \cdots, m\}$.

Definition. $(e_1, t_1)^\frown(e_2, t_2)^\frown\cdots^\frown(e_m, t_m)$ satisfies the linear duration invariant, if

$$\bigwedge_{j=1}^{k} \left(\sum_{i=1}^{n} c_{ij} \left(\sum_{u \in a_i} t_u \right) \leqslant M_j \right) \quad \text{(denoted } I\text{)}$$

when

$$T \geqslant \sum_{u=1}^{m} t_u \geqslant t \qquad \text{(denoted } C2\text{).}$$

Example: (continued)

An linear duration invariant for the gas burner automaton has been shown in Section 1. It has premise

$$\int 1 \geqslant 60$$

and conclusion

$$(20\int Leak - \int 1) \leqslant 0.$$

The timed behaviour of the gas burner automaton

$$(f, t_1)\hat{}(r, t_2)\hat{}(f, t_3)$$

where $t_1 \geqslant 30$, $1 \geqslant t_2 \geqslant 0$ and $t_3 \geqslant 30$, satisfies the linear duration invariant, since for this behaviour

$$\int 1 = (t_1 + t_2 + t_3) \geqslant 60 \text{ and } \int Leak = t_2 \leqslant 1.$$

Regarding $t_1, t_2, \cdots, t_m$ as real variables, $C1$, $C2$ and I constitute a linear programming problem, which has *box constraint* $C1$ and *side constraint* $C2$, and requests that subject to $C1$ and $C2$ the maximum value of the linear *objective function*

$$\sum_{i=1}^{n} c_{ij} \left(\sum_{u \in \alpha_i} t_u \right)$$

is not greater than M_j for $j = 1, 2 \cdots, k$, i. e. that, under $C1$ and $C2$, I holds. The linear programming problem is equivalent to the satisfaction problem of the linear duration invariant for the untimed behaviour

$$e_1 \hat{e}_2 \hat{} \cdots \hat{e}_m,$$

if we define:

Definition. An untimed behaviour $e_1 \hat{e}_2 \hat{} \cdots \hat{e}_m$ of a real-time automaton satisfies a linear duration invariant, if all its corresponding timed behaviour of the automaton satisfy the linear duration invariant.

Futhermore the satisfaction of a real-time automaton for a linear duration invariant can be naturally defined as

Definition. A real-time automaton satisfies a linear duration invariant, if all the untimed behaviour of the real-time automaton satisfy the linear duration invariant.

Therefore the satisfaction problem of a linear duration invariant for a real-time automaton can be translated into a family of the linear programming problems shown above. The family is determined by a regular expression which defines the untimed behaviour of the automaton. Those families of linear programming problems can be formulated in terms of geometry, and are called *Mountaineer Problems*. They are introduced in Section 3.

2.4 Note

The linear duration invariant for the gas burner intends to be satisfied by *any* observation interval which may not start from *Nonleak* state, but from *Leak* state, or may start from (or end at) half-way during a stay of the automaton at *Nonleak* or *Leak* state. However the satisfaction of the linear duration invariant by the gas burner automaton in Fig. 1 can cover only a part of the observation intervals. In order to cover all the observation intervals, we have to extend the automaton by allowing

1. *Leak* to be initial state as well as *Nonleak*, so that the observation interval can start from *Leak* state also; and

2. the first (and last) *Nonleak* state in a state transition sequence to the last for arbitrary short period. To allow so, we introduce a new state transition, denoted f'. It is a variant of f with the same pre- and post-states

$$\overleftarrow{f'} = Nonleak, \overrightarrow{f'} = Leak.$$

but the lower bound of the real-time constraint of f' is changed to 0, so that the observation interval can start (or end) at any time during the stay of the automaton in the *Nonleak* state. That is

$$a_{f'} = 0, \text{and } b_{f'} = b_f = \infty.$$

However we do not need an additional variant state of r, since its lower bound is 0 already.

The regular expression which describes the untimed behaviour of the extended automaton becomes

$$\epsilon \oplus f' \oplus f'\hat{}r\hat{}(f\hat{}r)^* \oplus f'\hat{}r\hat{}(f\hat{}r)^*\hat{}f' \oplus (r\hat{}f)^*\hat{}r\hat{}f' \oplus (r\hat{}f)^*\hat{}r.$$

Since the lower bounds of the real-time constraints for r and f' are 0, we can substantially simplify the extended regular expression. For example, $f'\hat{}r\hat{}(f\hat{}r)^*$ can be considered a special case of $f'\hat{}r\hat{}(f\hat{}r)^*\hat{}f'$, where the last f' takes zero time-distance from its previous transition. The simplified regular expression can be

$$f'\hat{}r\hat{}(f\hat{}r)^*\hat{}f'$$

3 Mountaineer Problem

In this section, we formulate the concerned families of linear programming problems in terms of geometry, and call them *Mountaineer Problems*.

Suppose a mountaineer climbs a mountain in an *n-dimensional space*. The mountaineer starts from the origin of the space, and climbs up step by step in a way directionally controlled by a *regular expression* and pacely controlled by pace constraints.

3.1 Regular Expression

Let $s_i (n \geqslant i \geqslant 1)$ denote a step in the direction of axis x_i of the space. A sequence of steps, denoted

$$s_{i_1} s_{i_2} \cdots s_{i_m} \quad (n \geqslant i_j \geqslant 1 \text{ for } j=1,\cdots,m)$$

defines a climbing path of the mountaineer: starting from the origin and climbing m steps up the mountain. The first step is in the direction of axis x_{i_1}, the second step is in the direction of axis x_{i_2}, so on so forth, and at last he/she completes the mth step in the direction of axis x_{i_m} and *stays* at the place where he/she reaches. When $m=0$, it becomes an empty sequence, denoted $\in$. $\in$ defines a void path, which means a still stay of the mountaineer at the origin.

A regular expression [4] of steps is an expression constructed from ϵ(*empty* sequence) and single step sequence $s_i (i=1,2,\cdots,n)$ with combinators $\oplus$ (union), $\hat{}$ (concatenation) and * (repetition). It defines a set of sequences of steps, which determines all permissible paths of the mountaineer. The combinators are defined as follows. Let σ, σ_1 and σ_2 be sequences of steps $s_i (i=1,2,\cdots,n)$, and R, R_1 and R_2 be sets of sequences of steps.

$$R_1 \oplus R_2 \triangleq \{\sigma \mid \sigma \in R_1 \vee \sigma \in R_2\}$$

$$R_1 \hat{} R_2 \triangleq \{\sigma_1 \hat{} \sigma_2 \mid \sigma_1 \in R_1 \wedge \sigma_2 \in R_2\}$$

$$R^* \triangleq \{\sigma_1 \hat{} \cdots \hat{} \sigma_m \mid (m \geqslant 0) \bigwedge_{i=1}^{m} (\sigma_i \in R)\}$$

where

$$\sigma_1 \hat{} \cdots \hat{} \sigma_m \triangleq \epsilon \text{ when } m=0,$$

and

$$\epsilon \hat{} \sigma \triangleq \sigma \hat{} \epsilon \triangleq \sigma.$$

3.2 Pace Constraint

A pace constraint assigns for each step a lower bound and an upper bound, formulated by $C1$:①

$$b_i \geqslant |s_i| \geqslant a_i, (i=1,2,\cdots,n)$$

where $|s_i|$ denotes the length of step s_i, and a_i is a non-negative real, b_i is a positive real or ∞ and $b_i \geqslant a_i$. The pace constraints restrict that the length of eash step in the direction of axis x_i will not be longer than b_i or shorter than a_i.

By a regular expression and a pace constraint, the reachable area of the mountaineer in the space is determined.

---

① We use the same notation of $C1$ (also $C2$ and I) as in Section 2.3, since they bear the same mathematical meaning.

3.3 Hazard Area

A *hazard* area in the mountain is defined by lower and upper bounds of climbing distance, formulated by $C2$:

$$T \geqslant \sum_{i=1}^{n} x_i \geqslant t,$$

x_i denotes the ith coordinate of a point in the space. When the mountaineer reaches the point with coordinates $(x_1, x_2, \cdots, x_n)$, coordinate x_i records the summation of all the paces of steps in the direction of the ith axis during the climbing. Therefore $\sum_{i=1}^{n} x_i$ counts his/her actually climbing distance, t is a non-negative real, and T is a positive real or ∞. If the mountaineer climbs to and *stays* at the area defined by $C2$, it would be considered a risk, and may need a rescue.

3.4 Protectable Area

However, rescue is not available everywhere in the mountain. The *protectable* area is defined by another linear inequality over the coordinates, denoted I:

$$\sum_{i=1}^{n} c_i x_i \leqslant M,$$

where c_i and M are reals.

The unprotectable, hazard area is called *dangerous* area, and otherwise called *safe* area. The mountaineer problem is asking whether the path controller-the regular expression can guarantee that the mountaineer with the pace constraint will never *slay* at the dangerous area. Let Π denote $(C1, C2, I)$, i. e. a given pace constraint $C1$, a given hazard area defined by $C2$, and a given protectable area defined by I. A regular expression R which can guarantee the safety of the mountaineer will be denoted

$$R \models \Pi$$

and will be called that R satisfies Π.

Example:

The mountaineer problem corresponding to the satisfaction problem for the linear duration invariant by the (extended) gas burner example shown in Section 2.4 has

$$R: s_3 \,\hat{}\, s_2 \,\hat{}\, (s_1 \,\hat{}\, s_2)^* \,\hat{}\, s_3$$

$$C1: \begin{cases} |s_1| \geqslant 30, \\ 1 \geqslant |s_2| \geqslant 0, \\ |s_3| \geqslant 0 \end{cases}$$

$$C2: (x_1 + x_2 + x_3) \geqslant 60$$

$$I: -x_1 + 19x_2 - x_3 \leqslant 0.$$

Given a single path, it is easy to check whether the path will lead the mountaineer to stay in the safe area. An algorithm to check the safety of a single path is presented in Section 5. However a regular expression may define infinite many paths by using combinator $*$, it is therefore impossible to solve the mountaineer problem by employing *brutal force* to check through all the paths defined by the regular expression. In the following section, we are developing a mechanical algorithm to reduce the necessary checked paths to a finite number of them only.

4 Reduction

At first we develop a reduction algorithm for the case where $T=\infty$ in $C2$, i. e.

$$C2:\left(\sum_{i=1}^{n} x_i \geqslant t\right).$$

4.1 Congruent Equivalence

A regular expression C of steps $s_1, s_2, \cdots, s_n$ and *letter* X is called a *context* of X, and denoted $C(X)$. For example,

$$C_1(X):(X^* \oplus s_1 \hat{} X).$$

Replacing X in $C(X)$ with any regular expression of the steps, say R we can obtain a regular expression, denoted $C(R)$, such as

$$C_1(s_1 \hat{} s_2):((s_1 \hat{} s_2)^* \oplus s_1 \hat{} (s_1 \hat{} s_2)).$$

Let R_1 and R_2 be regular expressions of steps $s_1, s_2, \cdots, s_n$, and Π be given $(C1, C2, I)$. We define an equivalence relation with respect to Π among regular expressions.

Definition. R_1 and R_2 are (congruently) equivalent with respect to Π, denoted as

$$R_1 \equiv_{\pi} R_2$$

if and only if for any context $C(X)$

$$C(R_1) \models \Pi \text{ iff } C(R_2) =\!\mid \Pi$$

We are listing a set of theorems for the equivalence. Those theorems justify the reduction algorithm developed in the following subsection.

Theorem 1.

$$(R_1 \hat{} R_2) \equiv_{\pi} (R_2 \hat{} R_1)$$

Proof. Trivial, since any permutation of steps in a path will not change the destination of the mountaineer.

Corollary 1.

$$(R_1 \oplus R_2)^* \equiv_{\pi} R_1^* \hat{} R_2^*$$

Proof. The right hand side is a subset of the left hand side, according to the definitions of $\oplus$ and $*$. Hence we can prove the half of the equivalence. The other half can be proved by Theorem 1, since any step sequence in the left hand side can be permuted to a step sequence belonging to the right hand side.

Corollary 2.

$$(R_1 \hat{} (R_2^*))^* \equiv_{\pi} \epsilon \oplus R_1^+ \hat{} (R_2^*),$$

where $R_1^+ \triangleq R_1^* \hat{} R_1$

Proof. The proof is similar to Corollary 1, and omitted.

The following Theorem 2 and 3 are reasoning about the equivalence among step sequences which climbing distances can degenerate to zero.

Theorem 2. Suppose $a_u=0$ in C1, and $c_u \geqslant c_v$ in I.

$$s_u^* \hat{} s_v^* \equiv_{\pi} s_u^*.$$

Proof. s_u^* is a subset of $s_u^* \hat{} s_v^*$, where the step sequence from s_v is empty. Hence we prove the half of the equivalence.

The proof of the second half can be explained below. As a_u(the lower bound of $|s_u|$) is zero, and b_u(the upper bound of $|s_u|$) is greater than zero, from any path accepted by s_u^* $\hat{} s_v^*$, we can obtain a new path accepted by s_u^* in the way that transfers the travel distance in the direction of axis x_u exactly to the travel in the direction of axis x_u by repeating step s_u with longest pace b_u, until the rest of the distance is not longer than b_u any more, and then taking one more step of s_u with the rest of the distance as its pace which belongs to $[0, b_{u_1}]$. The climbing distances of those two paths are the same. However, since c_u and c_v are coefficients of x_u and x_v in the linear function $\sum_{i=1}^{n} c_i x_i$ of I, and $c_u \geqslant c_v$ the value of the linear function for the new path must be greater than or equal to the value for the origin M path. It follows that if the new one stays in the protectable area, so does the original one. Hence the proof is completed.

Theorem 3. Suppose $a_{u_i}=0(i=1,2,\cdots,m)$ in C1.

$$(s_{u_1} \hat{} s_{u_2} \hat{}\cdots \hat{} s_{u_m})^* \equiv_{\pi} s_{u_1}^* \hat{} s_{u_2}^* \hat{}\cdots \hat{} s_{u_m}^*$$

Proof. With Theorem 1, we can prove the set defined by the regular expression in the left hand side of the equivalence is a subset of the one defined by the right hand side. Thus the first half of the proof follows. The other half can be proved as follows. Since $a_{u_i}=0$ $(i=1,2,\cdots,m)$, any path accepted by the right hand side can be extended to a new path in which all steps s_{u_i} $(i=1,2,\cdots,m)$ have same multiplicity and the paces of the extended steps are zero. Those two paths are equivalent, because they reach the same destination. However, a permutation of the new path is accepted by the left hand side. Thus using Theorem 1 again we can complete the proof.

The theorems have the following corollary. Suppose $a_{u_i}=0$ $(i=1,2,\cdots,m)$ in C1 and $c_{u_j}=\max(c_{u_1}, c_{u_2}, \cdots, c_{u_m})(m \geqslant j \geqslant 1)$ in I.

Corollary 3.

$$(s_{u_1} \hat{} s_{u_2} \hat{}\cdots \hat{} s_{u_m})^* \equiv_{\pi} s_{u_j}^*,$$

Proof. The proof can be easily derived from Theorem 3 and 2.

The following two theorems are concerned with the equivalence of step sequences with non-zero climbing distance. Suppose

$$\sigma=(s_{u_1}\hat{}s_{u_2}\hat{}\cdots\hat{}s_{u_m}).$$

We can calculate the shortest climbing distance of σ. It is denoted t_σ, and equal to the summation of the lowest bound of the step paces

$$t_\sigma=\sum_{i=1}^{m}a_{u_i}$$

The maximum value of the linear function $\sum_{i=1}^{n}c_i x_i$ for the step sequence σ can be also easily calculated. Let it be denoted $M\sigma$. Then

$$M\sigma=\sum_{i=1}^{m}c_{u_i}v_{u_i}$$

where

$$v_{u_i}=\begin{cases}b_{u_i} & \text{if } c_{u_i}>0\\ a_{u_i} & \text{otherwise}\end{cases}$$

We let $M\sigma=\infty$, if one of v_{u_i} is ∞.

Theorem 4. If $t_\sigma>0$ and $M_\sigma>0$ (inclusive of $M\sigma=\infty$), then any regular expression R which includes σ^* as sub-expression will violate Π.

Proof. Since $t\sigma$ is the shortest climbing distance of σ, each repetition of σ will increase the climbing distance with at least t_σ. By $t_\sigma>0$, after a finite number of repetitions of σ the climbing distance of the resultant path will be always greater than t, and the mountaineer will always stay at the hazard area. At the same time, for each repetition of σ, choosing v_{u_i} (if none of v_{u_i} is ∞) as the pace for step s_{u_i}, the mountaineer can increase the value of the linear function with $M\sigma$. By $M\sigma>0$, after a finite number of repetitions of σ, the value of the linear function will be always greater than M, and the mountaineer will always stay in unprotectable area. It concludes that Π is violated by the paths of R which contains a big enough number of repetitions of σ^*, so by R. The argument above can also be applied to the case where some of v_{u_i} is ∞.

Let $[x]$ be the *floor* function of real variable x. It delivers the maximum of the integers which are not greater than or equal to x. And let σ^j be the j-repetition of σ

$$\sigma^j\triangleq\underbrace{\sigma\hat{}\sigma\hat{}\cdots\hat{}\sigma}_{j}$$

where $\sigma^0\triangleq\epsilon$.

Theorem 5. If $t_\sigma>0$ and $M_\sigma\leqslant 0$, then

$$\sigma^*\equiv_\pi\bigoplus_{j=0}^{p}\sigma^j$$

where $p=([t/t_\sigma]+1)$.

Proof. The step sequences given by the expression in the right hand side of the equivalence is a subset of the sequences defined by the left hand side. It concludes the half of the proof. Below is the other half of the proof. Since t_σ is the shortest climbing distance of

σ, the climbing distance of any path which contains p occurrences of σ will be greater than t, and it will lead the mountaineer to the hazard area. So does any path with more than p occurrences of σ. However, since $M_\sigma \leqslant 0$, the value of the linear function $\sum_{i=1}^{n} c_i x_i$ for a path with more than p occurrences of σ will not be greater than the value of the linear function for the path which is obtained from the given one by cancelling some of occurrences of σ such that the number of the occurrences of σ becomes p. So the safety of the paths containing p occurrences of σ can guarantee the safety of the paths with more than p occurrences of σ.

With the theorems above, we are now ready to present the reduction algorithms.

4.2 Normal Form and Reduction Algorithms

Definition. A *normal form* N is a regular expression of the following form

$$A_1 \oplus A_2 \oplus \cdots \oplus A_q$$

where A_j is either a step sequence (such as $s_{u_1} \hat{} s_{u_2} \hat{} \cdots \hat{} s_{u_m}$), called *finite term*, or a step sequence followed by a repetition of a single step with zero as its lower pace bound (such as $s_{u_1} \hat{} s_{u_2} \hat{} \cdots \hat{} s_{u_m} \hat{} (s_v^*)$, where $a_v = 0$), called *infinite term*.

We present two reduction algorithms. Algorithm A reduces a mountaineer problem of a normal form to a finite number of mountaineer problems of finite terms. Algorithm B derives an equivalent normal form for a regular expression, if it does not discover the violation of Π by the regular expression.

Algorithm A Let N be a normal form

$$A_1 \oplus A_2 \oplus \cdots \oplus A_q.$$

The mountaineer problem of Π for N is equivalent to the mountaineer problems of Π for A_j ($j = 1, 2, \cdots, q$). That is

$$N \models \Pi \text{ iff } A_j \models \Pi (\text{for } q \geqslant j \geqslant 1)$$

where A_j is either a finite term or an infinite term.

When A_j is an infinite term

$$s_{u_1} \hat{} s_{u_2} \hat{} \cdots \hat{} s_{u_m} \hat{} (s_v^*),$$

where $a_v = 0$, and $\Pi = (C1, C2, I)$ in an n dimensional space, we construct as follows another mountaineer problem in an $(n+1)$ dimensional space, denoted $\Pi' = (C1', C2', I')$, where step sequence in s_v^* is regarded as a single step in the direction of the additional axis x_{n+l}.

The pace constraint $C1'$ is defined as

$$b_i \geqslant |s_i| \geqslant a_i \qquad i = 1, 2, \cdots, n \; |s_{n+1}| \geqslant 0,$$

i. e. $C1'$ maintains the pace constraints of $C1$ for the steps along the first n dimensions, but assigns no pace constraint for the step along the additional ($(n+1)$th) dimension. It is because by the step sequences in s_v^* the mountaineer can climb arbitrary distance between zero (since $a_v = 0$) and ∞ (since $b_v > 0$).

The hazard area of Π' t is defined by $C2'$:

$$\sum_{i=1}^{n+1} x_i \geqslant t.$$

i. e. $C2'$ uses the same real number t in $C2$ to define the hazard area.

The protectable area of Π' is defined by I':

$$\left(\sum_{i=1}^{m} c_i x_i\right) + c_v x_{n+1} \leqslant M$$

where I' uses the same number M in I to define the protectable area, and uses c_v (the coefficient for the vth dimension in I) as the coefficient for the $(n+1)$th dimension in I', since the steps in the direction of axis x_n+1 simulate the step sequences in the direction of the vth dimension. It is obvious that

Lemma A.

$$s_{u_1} \hat{} s_{u_2} \hat{} \cdots \hat{} s_{u_m} \hat{} (s_v^*) \models \Pi \text{ iff } s_{u_1} \hat{} s_{u_2} \hat{} \cdots \hat{} s_{u_m} \hat{} s_{n+1} \models \Pi'.$$

Following Lemma A we can easily design Algorithm A to reduce a mountaineer problem of a normal form to a finite number of mountaineer problems of finite terms, but we are not going into the details here.

Algorithm *B* We here only to demonstrate the essential parts of the algorithm with a series of lemmas, which explain how to proceed equivalence transformation of a regular expression from its innermost components outwards. The innermost sub-expressions of a regular expression are either ε or s_u, which are normal forms by the definition.

Lemma B. 1. ϵ and $s_u (n \geqslant u \geqslant 1)$ are normal forms.

The union operator maintains the normality.

Lemma B. 2. If N_1 and N_2 are normal forms, so is $N_1 \oplus N_2$.

The concatenation and repetition operators also maintain the normality under the congruent equivalence, although the constructive proofs are not obvious. Let $N_i (i=1,2)$ be normal forms:

$$A_{i1} \oplus A_{i2} \oplus \cdots \oplus A_i q_i \quad (i=1,2)$$

where A_{ijs} are finite or infinite terms.

Lemma B. 3. $N_1 \hat{} N_2$ is equivalent to a normal form.

Proof. The followings present an algorithm to construct the equivalent normal form for $N_1 \hat{} N_2$.

1. Distributing $\hat{}$ over $\oplus$, we can transform $N_1 \hat{} N_2$ into an equivalent regular expression

$$A_{11} \hat{} A_{21} \oplus \cdots \oplus A_1 q_1 \hat{} A_2 q_2.$$

2. If both A_{1u} and A_{2v} are finite terms, the concatenation $A_{1u} \hat{} A_{2v}$, composes also a finite term.

3. If one of A_{1u} and A_{2v} is finite term and the other is infinite, with Theorem 1, we can obtain an equivalent infinite term for $A_{1u} \hat{} A_{2v}$ by permuting (if necessary).

4. Suppose both of A_{1u} and A_{2v} are infinite term, A_{1u} is $\sigma_1 \hat{} (s_i^*)$, and A_{2v} is $\sigma_2 \hat{} (s_j^*)$,

where σ_1 and σ_2 are finite terms and $a_i = a_j = 0$. By Theorem 1, we can move s_i^* to the end of $A_{1u}\hat{}\,A_{2v}$, and obtain an equivalent expression

$$\sigma_1 \hat{}\, \sigma_2 \hat{}\, (s_j^*) \hat{}\, (s_i^*).$$

Applying Theorem 2, we can delete from the previous expression one of s_i^* and s_j^* by comparing c_i with c_j. We at last obtain an infinite term which is equivalent to $A_{1u}\hat{}\,A_{2v}$.

Let N be a normal form

$$A_1 \oplus A_2 \oplus \cdots \oplus A_q.$$

Lemma B. 4. There exists an algorithm which can either discover the violation of Π by N^*, or find out an equivalent normal form of N^*.

Proof. The algorithm can be explained as follows.

1. By Corollary 1, we can have

$$N^* \equiv_\pi (A_1^*) \hat{}\, (A_2^*) \hat{}\, \cdots \hat{}\, (A_q^*).$$

2. Transform the repetitions of infinite terms:

Suppose A_i ($q \geqslant i \geqslant 1$) be an infinite term: $\sigma\hat{}\,(s_v^*)$. Using Corollary 2, we can transform A_i^* into the equivalent expression

$$(\epsilon \oplus (\sigma^+ \hat{}\, (s_v^*))),$$

where $\sigma^+ = \sigma^* \hat{}\, \sigma$. After the transformation, we can obtain an equivalent expression without repetitions of infinite terms.

3. Transform the repetitions of finite terms:

Suppose $\sigma = (s_{u_1} \hat{}\, s_{u_2} \hat{}\, \cdots \hat{}\, s_{u_m})$, and t_σ is the shortest climbing distance of σ. If $t_\sigma = 0$, by applying Corollary 3, we can transform σ^* to an equivalent infinite term $s_{u_j}^*$ where $m \geqslant j \geqslant 1$ and $c_{u_j} = max(c_{u_1}, c_{u_2}, \cdots, c_{u_m})$. When $t_\sigma > 0$, let $M\sigma$ be the maximum value of the linear function $\sum_{i=1}^n c_i x_i$ for σ. If $M\sigma > 0$ (or equals to ∞) Theorem 4 proves that σ^* violates Π, so do N^* and the original regular expression. If $M\sigma \leqslant 0$, using Theorem 5, we can transform σ^* into an equivalent finite union of finite terms $\oplus_{j=0}^p \sigma^j$, where $p = ([t/t\sigma] + 1)$.

With the transformation, we can either discover the violation of Π by N^*, or transform N^* into an equivalent expression which is constructed from finite/infinite terms with combinators $\oplus$ and $\hat{}$.

4. Applying the algorithms given in Lemmas B. 2 and B. 3 to the resultant equivalent expression of the transformation in 3, we can at last transform N^* into an equivalent normal form.

The four lemmas sketch out Algorithm B, but we are not elaborating it here.

Example: (continued)

The mountaineer problem corresponding to the (extended) gas burner example shown in Section 2. 4 has

$$R: s_3 \hat{}\, s_2 \hat{}\, (s_1 \hat{}\, s_2)^* \hat{}\, s_3$$

$$C1:\begin{cases} |s_1| \geqslant 30, \\ 1 \geqslant |s_2| \geqslant 0, \\ |s_3| \geqslant 0 \end{cases}$$

$$C2:(x_1+x_2+x_3) \geqslant 60$$

$$I: -x_1+19x_2-x_3 \leqslant 0.$$

With Algorithm B, we can transform R to an equivalent normal form.

—Transform the repetitions of the finite term $\sigma: s_1 \hat{} s_2$.

Since

$$t_\sigma = (30+0) = 30, M_\sigma = (-30+19) = -11 \text{ and } p = \lfloor 60/30 \rfloor + 1 = 2,$$

according to Lemma B. 4, we can transform σ^* into

$$\varepsilon \oplus (s_1 \hat{} s_2) \oplus (s_1 \hat{} s_2)^2.$$

Thus we obtain

$$s_3 \oplus s_3 \hat{} s_2 \hat{} (\epsilon \oplus (s_1 \hat{} s_2) \oplus (s_1 \hat{} s_2)^2) \hat{} s_3.$$

which is equivalent to R.

—Transform the last expression into a normal form with Lemma B. 3.

Distributing ^ over $\oplus$, we can at last obtain an equivalent normal form for R. It is

$$s_3 \hat{} s_2 \hat{} s_3 \oplus s_3 \hat{} s_2 \hat{} s_1 \hat{} s_2 \hat{} s_3 \oplus s_3 \hat{} s_2 \hat{} s_1 \hat{} s_2 \hat{} s_1 \hat{} s_2 \hat{} s_3$$

4. 3. $T \neq \infty$

The previous subsections develop an algorithm to reduce a mountaineer problem with $C2$:

$$\sum_{i=1}^{n} x_i \geqslant t$$

for a regular expression to a finite number of mountaineer problems for single path when $C2$ is

$$T \geqslant \sum_{i=1}^{n} x_i \geqslant t \text{ and } T \neq \infty.$$

Algorithm A still works: but not Algorithm B, since Theorem 4 and 5 in Section 4. 1 will not hold in this case.

However the case $T \neq \infty$ is much simpler than the case $T = \infty$ and the main theorem concerning the equivalence among step sequences with non-zero climbing distance can be stated without considering the maximum values of the linear function of I for the sequences.

Let σ stand for an arbitrary given step sequence, t_σ be the shortest climbing distance of σ and $t_\sigma > 0$, and $h = \lfloor T/t\sigma \rfloor + 1$. (Read Section 4. 1 for the detailed explanation of the notations.)

Theorem.

$$\sigma^* \equiv_\pi \oplus_{j=0}^{h} \sigma^j$$

Proof. Trivial, as the mountaineer will not stay in the hazard area after repeating σ for more than h times.

Algorithm B will have a slight adaptation, which is about the algorithm to transform repetitions of finite terms shown in Lemma B. 4. 3. We will simply transform σ^* into $\bigoplus_{j=0}^{h}\sigma^j$ without a prerequisite calculation of the maximum value of the linear function for σ. Therefore the adapted algorithm will only transform a regular expression into its equivalent normal form, but not try to discover the possible violation of Π during the transformation.

5 Single Path

In this section, we present an algorithm to decide whether a single path satisfies an arbitrary given Π. By renaming the steps, we can assume a general form for singlepath mountaineer problems. That is

$$R: s_1 \hat{}\, s_2 \hat{}\cdots\hat{}\, s_n$$

$$C1: b_i \geqslant |s_i| \geqslant a_i \quad (i=1,2,\cdots,n)$$

$$C2: T \geqslant \sum_{i=1}^{n} x_i \geqslant t$$

$$I: \sum_{i=1}^{n} c_i x_i \leqslant M$$

where $b_i \geqslant a_i \geqslant 0$ and $b_i > 0$, $T \geqslant t \geqslant 0$, b_i and T may be ∞, and $c_i \neq 0$, $(i=1,2,\cdots,n)$.

Since each step s_i occurs in the path exactly once, it is not necessary to distinguish between s_i, $|s_i|$ and x_i any longer. Let us use x_i as real variable to denote the pace length of step s_i, and translate the problem into a simple linear programming problem for real variables $x_i (i=1,2,\cdots n)$, which consists of a box constraint C (a new version of $C1$) and the side constraint $C2$, and the inequality I of the linear objective function.

$$C: b_i \geqslant x_i \geqslant a_i \quad (i=1,2,\cdots,n)$$

$$C2: T \geqslant \sum_{i=1}^{n} x_i \geqslant t$$

$$I: \sum_{i=1}^{n} c_i x_i \leqslant M.$$

The algorithm is designed to find out the maximum value of the objective function $\sum_{i=1}^{n} c_i x_i$ under the constraints C and $C2$. We will present the algorithm with a list of statements including assignment statement, conditional statement/expression, and go to statement, etc. Let

$$v_i := \text{if } c_i < 0 \text{ then } a_i \text{ else } b_i$$

and

$$\text{L1}: l := \sum_{i=1}^{n} v_i;$$

$$v := \sum_{i=1}^{n} c_i v_i$$

where $l \triangleq \infty$ and $v \triangleq \infty$, if one of v_i is ∞. Then v contains the maximum value of $\sum_{i=1}^{n} c_i x_i$ under constraint C, and we can conclude

L2: If $v \leqslant M$ then *YES*

where *YES* means that the problem has a positive answer. Otherwise $v > M$, and

If $T \geqslant l \geqslant t$ then *NO*

where *NO* means that the problem has a negative answer. Otherwise $v > M$ and either $l < t$ or $l > T$. At first we consider the case: $v > M$ and $l < t$, and show how to find out the maximum value of the objective function under the constraints C and ($\sum_{i=1}^{n} x_i \geqslant t$). In order to increase l to reach t and meanwhile keep v as big as possible, we increase the value of such a v_i that is still allowed to be increased (i. e. $v_i \neq b_i$) and its coefficient c_i has the maximum value among the others. Let

$$N := \{c_i \mid v_i \neq b_i \wedge n \geqslant i \geqslant 1\}.$$

It is obvious that

If $N = \varnothing$ then *YES*

since it implies that under constraint C we can not let l reach or go beyond t. Otherwise let

$$c := max\{N\};$$

$$r := (t - l)$$

Assume $(c_j = c \wedge v_j \neq b_j \wedge r \leqslant (b_j - a_j))$ (when $b_j = \infty$, $(b_j - a_j) \triangleq \infty$). Let

$$v_j := v_j + r$$

and repeat the previous reasoning. That is

Go to L1

The statement L2 will still hold during the repetition, since the value of v at L2 keeps to be the maximum value of the objective function under the constraints C and ($\sum_{i=1}^{n} x_i \geqslant t$) .

If

$$\forall_i : n \geqslant i \geqslant 1. (c_i = c \wedge v_i \neq b_i) \Rightarrow (r > (b_i - a_i)),$$

then choose an arbitrary j such that $(c_j = c \wedge v_j \neq b_j)$, let

$$v_j := b_j$$

and repeat the reasoning from L1

Go to L1.

The procedure will at last terminate and give off an answer (positive or negative).

The decision procedure for the case where $v > M$ and $l > T$ is similar to the above one, but using

$$r' := l - T,$$

$$N' := \{c_i \mid v_i \neq a_i \wedge n > i > 1\},$$

$$c' := min\{N\},$$

$$v_j := v_j - r',$$

instead of r, N, c and $(v_j := v_j + r)$.

Example. Let

$$C: \begin{cases} x_1 \geqslant 0, \\ 1 \geqslant x_2 \geqslant 0, \\ x_3 \geqslant 30, \\ 1 \geqslant x_4 \geqslant 0, \\ x_5 \geqslant 0 \end{cases}$$

$$C2: (x_1 + x_2 + x_3 + x_4 + x_5) \geqslant 60$$

$$I: (-x_1 + 19x_2 - x_3 + 19x_4 - x_5) \leqslant 0.$$

The example is in fact the corresponding linear programming problem of the satisfaction of the path

$$s_3 \,\hat{}\, s_2 \; s_1 \,\hat{}\, s_2 \,\hat{}\, s_3$$

for the mountaineer problem shown in the example of Section 4. 2.

Applying the decision procedure to the example, we have

$$v_1 := 0;$$
$$v_2 := 1;$$
$$v_3 := 30;$$
$$v_4 = 1;$$
$$v_5 := 0;$$
$$l := 32;$$
$$v := 0 + 19 - 30 + 19 - 0 = 8.$$

Thus $(v > 0 \wedge l < 60)$. We count N, r and c, and have

$$N := \{-1\};$$
$$r := 60 - 32 = 28;$$
$$c := max\{N\} = max\{-1\} = -1.$$

Since $c_3 = c = -1, v_3 \neq b_3 (b_3 = \infty)$ and $r \leqslant (\infty - 30)$, following the procedure, we let

$$v_3 := 30 + 28 = 58$$

and repeat the procedure from L1. At last we reach

$$l := 60;$$
$$v := 0 + 19 - 58 + 19 - 0 = -20$$

and stop at L2 and obtain a positive answer: *YES*.

6　Conclusion

The algorithm provided in the paper is quite simple and uses only elementary arithmetic. However the mathematical complexity of the algorithm has not been analysed yet.

The authors share the same opinion with [3] that linear programming can help computing scientists develop model checking technique for real-time system design, and hope to see more results in this direction.

Acknowledgement

Thanks for valuable comments from Hans Rischel, Arne Stolbjerg Drud and Alexey Tikhonov.

References

[1] R Alur, C Courcoubetis, T. Henzinger, P-H Ho. Hybrid Automata: An Algorithmic Approach to the Specification and Verification of Hybrid Systems. in Hybrid Systems, LNCS 736, pp 209 - 229, R. L. Grossman, A. Nerode, A. P. Ravn and H. Rischel (eds.), Springer-Verlag, 1993.

[2] M Engel, M Kubica, J Madey, D J Parnas, A P Ravn, A J van Schouwen. A Formal Approach to Computer Systems Requirements Documentation. in Hybrid Systems, LNCS 736, pp 452 - 474, R. L. Grossman, A. Nerode, A. P. Ravn and H. Rischel (eds.), Springer-Verlag, 1993.

[3] Y Kesten, A Pnueli, J Sifakis, S Yovine. Integration Graphs: A Class of Decidable Hybrid Systems. in Hybrid Systems, LNCS 736, pp 179 - 208, R. L. Grossman, A. Nerode, A. P. Ravn and H. Rischel (eds.), Springer-Verlag, 1993.

[4] S. C. Kleene. Representation of Events in Nerve Nets and Finite Automata. in Automata Studies, pp 3 - 41, C. Shannon and J. McCarthy (eds.), Princeton Univ. Press, Princeton, N J, 1956.

[5] Z Manna, A Pnueli. Verifying Hybrid Systems. in *Hybrid Systems*, *LNCS* 736, pp 4 - 35, R. L. Grossman, A. Nerode, A. P. Ravn and H. Rischel (eds.), Springer-Verlag, 1993.

[6] A P Ravn, H Rischel. Requirements Capture for Embedded Real-Time Systems. Proc. IMA CS-MCTS'91 Symp. Modelling and Control of Technological Systems, Vol 2, pp. 147 - 152, Villeneuve d'Aseq, France, 1991.

[7] A P Ravn, H Rischel, K M Hansen. Specifying and Verifying Requirements of Real-Time Systems. IEEE Trans. Software Eng., Vol 19, No 1, pp 41 - 55, January 1993.

[8] J U Skakkebaek, A P Ravn, H Rischel, Zhou Chaochen. Specification of Embedded Real-Time Systems. Proc. 4th Euromicro Workshop on Real-Time Systems, IEEE Press, pp 116 - 121, June 1992.

[9] Zhou Chaochen, C A R Hoare, A P Ravn. A Calculus of Durations. Information Processing Letter, 40, 5, pp. 269 - 276, 1991.

[10] Zhou Chaochen, Li Xiaoshan. A Mean-Value Duration Calculus. in A Classical Mind, Essays in Honour of C. A. R. Hoare, pp 431 - 451, A. W. Roscoe (ed.), Prentice Hall International, 1994.

后　记

我的博士导师张景中、杨路两位先生是北京大学 1954 级数学力学系的同学,自大学时代起就建立了深厚友谊,也开始了他们在科学研究中的合作。青壮年时期他们历经 20 余年的磨难仍痴心不改,在当时极其恶劣的环境下仍通过书信等方式讨论数学问题。1978 年一场春风卷过全国,两位先生也由此开始了攀登科学高峰的步伐,先后在中国科学技术大学、中国科学院成都分院、广州大学等地留下他们的身影。在这长达 60 余年的岁月中,两位先生不仅在生活上互相关心、工作上互相促进,学术研究上更是密切合作,在距离几何、动力系统、机器证明、教育软件、教育数学、科普创作等领域取得了卓越成就,也铸成了他们长期并肩合作的佳话。

在两位先生 80 寿辰之际,由广州大学数学教育软件研究中心牵头将两位先生多年的合作成果整理成册,付梓《张景中杨路文集》。湖南教育出版社闻讯,欣然邀约出版该文集,在此谨向湖南教育出版社社长黄步高先生、总编辑刘新民先生、文集责任编辑钟劲松先生及其他为文集出版付出辛勤劳动的各位表示诚挚的谢意。

文集策划出版过程中得到各方面不少人士的协助。广州大学数学教育软件研究中心的饶永生助理研究员、中国科学院成都计算机应用研究所的秦小林博士提出了许多建议并提供了大力支持。中国科技大学叶向东教授为文集作序并参与原文搜集。广州大学曹忠博士在论文原文搜集、整理、校对过程中付出了许多心血。广州大学博士研究生唐松锦、陈泽桐、张红兵、黄星,中国科学院大学成都计算机应用研究所的博士研究生张国华、孙治、王会勇及硕士研究生吴定雄、王文彬、张力戈参加了文集的校对工作。特向以上各位表示衷心感谢。

黄勇

2016－9－22 于广州大学